Graduate Texts in Mathematics 96

T0202435

Graduate Texts in Mathematics

(continued after index)

John B. Conway

A Course in Functional Analysis

Second Edition

 Springer

John B. Conway
Department of Mathematics
University of Tennessee
Knoxville, Tennessee 37996
USA

With 1 Illustration.

Mathematical Subject Classification (2000): 46–01, 46L05, 47B15, 47B25

Library of Congress Cataloging-in-Publication Data
Conway, John B.
    A course in functional analysis/John B. Conway.—2nd ed.
      p.   cm. — (Graduate texts in mathematics; 96)
    Includes bibliographical references.
    ISBN 978-1-4419-3092-7
    1. Functional analysis.   I. Title.   II. Series.
QA320.C658   1990
515.7—dc20   90-9585

Printed on acid-free paper

15 14 13 12 11 10

springer.com

For Ann (of course)

Preface

Functional analysis has become a sufficiently large area of mathematics that it is possible to find two research mathematicians, both of whom call themselves functional analysts, who have great difficulty understanding the work of the other. The common thread is the existence of a linear space with a topology or two (or more). Here the paths diverge in the choice of how that topology is defined and in whether to study the geometry of the linear space, or the linear operators on the space, or both.

In this book I have tried to follow the common thread rather than any special topic. I have included some topics that a few years ago might have been thought of as specialized but which impress me as interesting and basic. Near the end of this work I gave into my natural temptation and included some operator theory that, though basic for operator theory, might be considered specialized by some functional analysts.

The word "course" in the title of this book has two meanings. The first is obvious. This book was meant as a text for a graduate course in functional analysis. The second meaning is that the book attempts to take an excursion through many of the territories that comprise functional analysis. For this purpose, a choice of several tours is offered the reader—whether he is a tourist or a student looking for a place of residence. The sections marked with an asterisk are not (strictly speaking) necessary for the rest of the book, but will offer the reader an opportunity to get more deeply involved in the subject at hand, or to see some applications to other parts of mathematics, or, perhaps, just to see some local color. Unlike many tours, it is possible to retrace your steps and cover a starred section after the chapter has been left.

There are some parts of functional analysis that are not on the tour. Most authors have to make choices due to time and space limitations, to say nothing of the financial resources of our graduate students. Two areas that are only briefly touched here, but which constitute entire areas by themselves, are topological vector spaces and ordered linear spaces. Both are beautiful theories and both have books which do them justice.

The prerequisites for this book are a thoroughly good course in measure and integration—together with some knowledge of point set topology. The appendices contain some of this material, including a discussion of nets in Appendix A. In addition, the reader should at least be taking a course in analytic function theory at the same time that he is reading this book. From the beginning, analytic functions are used to furnish some examples, but it is only in the last half of this text that analytic functions are used in the proofs of the results.

It has been traditional that a mathematics book begin with the most general set of axioms and develop the theory, with additional axioms added as the exposition progresses. To a large extent I have abandoned tradition. Thus the first two chapters are on Hilbert space, the third is on Banach spaces, and the fourth is on locally convex spaces. To be sure, this causes some repetition (though not as much as I first thought it would) and the phrase "the proof is just like the proof of ..." appears several times. But I firmly believe that this order of things develops a better intuition in the student. Historically, mathematics has gone from the particular to the general—not the reverse. There are many reasons for this, but certainly one reason is that the human mind resists abstraction unless it first sees the need to abstract.

I have tried to include as many examples as possible, even if this means introducing without explanation some other branches of mathematics (like analytic functions, Fourier series, or topological groups). There are, at the end of every section, several exercises of varying degrees of difficulty with different purposes in mind. Some exercises just remind the reader that he is to supply a proof of a result in the text; others are routine, and seek to fix some of the ideas in the reader's mind; yet others develop more examples; and some extend the theory. Examples emphasize my idea about the nature of mathematics and exercises stress my belief that doing mathematics is the way to learn mathematics.

Chapter I discusses the geometry of Hilbert spaces and Chapter II begins the theory of operators on a Hilbert space. In Sections 5–8 of Chapter II, the complete spectral theory of normal compact operators, together with a discussion of multiplicity, is worked out. This material is presented again in Chapter IX, when the Spectral Theorem for bounded normal operators is proved. The reason for this repetition is twofold. First, I wanted to design the book to be usable as a text for a one-semester course. Second, if the reader understands the Spectral Theorem for compact operators, there will be less difficulty in understanding the general case and, perhaps, this will lead to a greater appreciation of the complete theorem.

Chapter III is on Banach spaces. It has become standard to do some of this material in courses on Real Variables. In particular, the three basic principles, the Hahn-Banach Theorem, the Open Mapping Theorem, and the Principle of

Uniform Boundedness, are proved. For this reason I contemplated not proving these results here, but in the end decided that they should be proved. I did bring myself to relegate to the appendices the proofs of the representation of the dual of L^p (Appendix B) and the dual of $C_0(X)$ (Appendix C).

Chapter IV hits the bare essentials of the theory of locally convex spaces— enough to rationally discuss weak topologies. It is shown in Section 5 that the distributions are the dual of a locally convex space.

Chapter V treats the weak and weak-star topologies. This is one of my favorite topics because of the numerous uses these ideas have.

Chapter VI looks at bounded linear operators on a Banach space. Chapter VII introduces the reader to Banach algebras and spectral theory and applies this to the study of operators on a Banach space. It is in Chapter VII that the reader needs to know the elements of analytic function theory, including Liouville's Theorem and Runge's Theorem. (The latter is proved using the Hahn–Banach Theorem in Section III.8.)

When in Chapter VIII the notion of a C^*-algebra is explored, the emphasis of the book becomes the theory of operators on a Hilbert space.

Chapter IX presents the Spectral Theorem and its ramifications. This is done in the framework of a C^*-algebra. Classically, the Spectral Theorem has been thought of as a theorem about a single normal operator. This it is, but it is more. This theorem really tells us about the functional calculus for a normal operator and, hence, about the weakly closed C^*-algebra generated by the normal operator. In Section IX.8 this approach culminates in the complete description of the functional calculus for a normal operator. In Section IX.10 the multiplicity theory (a complete set of unitary invariants) for normal operators is worked out. This topic is too often ignored in books on operator theory. The ultimate goal of any branch of mathematics is to classify and characterize, and multiplicity theory achieves this goal for normal operators.

In Chapter X unbounded operators on Hilbert space are examined. The distinction between symmetric and self-adjoint operators is carefully de- lineated and the Spectral Theorem for unbounded normal operators is obtained as a consequence of the bounded case. Stone's Theorem on one parameter unitary groups is proved and the role of the Fourier transform in relating differentiation and multiplication is exhibited.

Chapter XI, which does not depend on Chapter X, proves the basic properties of the Fredholm index. Though it is possible to do this in the context of unbounded operators between two Banach spaces, this material is presented for bounded operators on a Hilbert space.

There are a few notational oddities. The empty set is denoted by $\square$. A reference number such as (8.10) means item number 10 in Section 8 of the present chapter. The reference (IX.8.10) is to (8.10) in Chapter IX. The reference (A.1.1) is to the first item in the first section of Appendix A.

There are many people who deserve my gratitude in connection with writting this book. In three separate years I gave a course based on an evolving set of notes that eventually became transfigured into this book. The students in those courses were a big help. My colleague Grahame Bennett gave me several pointers in Banach spaces. My ex-student Marc Raphael read final versions of the manuscript, pointing out mistakes and making suggestions for improvement. Two current students, Alp Eden and Paul McGuire, read the galley proofs and were extremely helpful. Elena Fraboschi typed the final manuscript.

John B. Conway

Preface to the Second Edition

The most significant difference between this edition and the first is that the last chapter, Fredholm Theory, has been completely rewritten and simplified. The major contribution to this simplification was made by Hari Bercovici who showed me the most simple and elegant development of the Fredholm index I have seen.

Other changes in this book include many additional exercises and numerous comments and bibliographical notes. Several of my friends have been helpful here. The greatest contributor, however, has been Robert B. Burckel; in addition to pointing out mistakes, he has made a number of comments that have been pertinent, scholarly, and very enlightening. Several others have made such comments and this is a good opportunity to publicly thank them: G.D. Bruechert, Stephen Dilworth, Gerald A. Edgar, Lawrence C. Ford, Fred Goodman, A.A. Jafarian, Victor Kaftall, Justin Peters, John Spraker, Joseph Stampfli, J.J. Schäffer, Waclaw Szymanski, James E. Thomson, Steve Tesser, Bruce Watson, Clifford Weil, and Pei Yuan Wu.

Bloomington, Indiana
December 7, 1989

John B. Conway

Contents

CHAPTER VII
Banach Algebras and Spectral Theory for Operators on a Banach Space

CHAPTER VIII
C*-Algebras

CHAPTER IX
Normal Operators on Hilbert Space

CHAPTER X
Unbounded Operators

Contents

CHAPTER I

Hilbert Spaces

A Hilbert space is the abstraction of the finite-dimensional Euclidean spaces of geometry. Its properties are very regular and contain few surprises, though the presence of an infinity of dimensions guarantees a certain amount of surprise. Historically, it was the properties of Hilbert spaces that guided mathematicians when they began to generalize. Some of the properties and results seen in this chapter and the next will be encountered in more general settings later in this book, or we shall see results that come close to these but fail to achieve the full power possible in the setting of Hilbert space.

§1. Elementary Properties and Examples

Throughout this book $\mathbb{F}$ will denote either the real field, $\mathbb{R}$, or the complex field, $\mathbb{C}$.

1.1. Definition. If $\mathscr{X}$ is a vector space over $\mathbb{F}$, a *semi-inner product* on $\mathscr{X}$ is a function $u\colon \mathscr{X} \times \mathscr{X} \to \mathbb{F}$ such that for all α, β in $\mathbb{F}$, and x, y, z in $\mathscr{X}$, the following are satisfied:

(a) $u(\alpha x + \beta y, z) = \alpha u(x, z) + \beta u(y, z),$
(b) $u(x, \alpha y + \beta z) = \bar{\alpha} u(x, y) + \bar{\beta} u(x, z),$
(c) $u(x, x) \geqslant 0,$
(d) $u(x, y) = \overline{u(y, x)}.$

Here, for α in $\mathbb{F}$, $\bar{\alpha} = \alpha$ if $\mathbb{F} = \mathbb{R}$ and $\bar{\alpha}$ is the complex conjugate of α if $\mathbb{F} = \mathbb{C}$. If $\alpha \in \mathbb{C}$, the statement that $\alpha \geqslant 0$ means that $\alpha \in \mathbb{R}$ and α is non-negative.

Note that if $\alpha = 0$, then property (a) implies that $u(0, y) = u(\alpha \cdot 0, y) =$

$\alpha u(0, y) = 0$ for all y in $\mathscr{X}$. This and similar reasoning shows that for a semi-inner product u,

(e) $u(x, 0) = u(0, y) = 0$ for all x, y in $\mathscr{X}$.

In particular, $u(0, 0) = 0$.

   An *inner product* on $\mathscr{X}$ is a semi-inner product that also satisfies the following:

(f) If $u(x, x) = 0$, then $x = 0$.

   An inner product in this book will be denoted by

$$\langle x, y \rangle = u(x, y).$$

There is no universally accepted notation for an inner product and the reader will often see (x, y) and $(x|y)$ used in the literature.

1.2. Example. Let $\mathscr{X}$ be the collection of all sequences $\{\alpha_n : n \geq 1\}$ of scalars α_n from $\mathbb{F}$ such that $\alpha_n = 0$ for all but a finite number of values of n. If addition and scalar multiplication are defined on $\mathscr{X}$ by

$$\{\alpha_n\} + \{\beta_n\} \equiv \{\alpha_n + \beta_n\},$$
$$\alpha\{\alpha_n\} = \{\alpha\alpha_n\},$$

then $\mathscr{X}$ is a vector space over $\mathbb{F}$.

   If $u(\{\alpha_n\}, \{\beta_n\}) \equiv \sum_{n=1}^{\infty} \alpha_{2n}\bar{\beta}_{2n}$, then u is a semi-inner product that is not an inner product. On the other hand,

$$\langle \{\alpha_n\}, \{\beta_n\} \rangle = \sum_{n=1}^{\infty} \alpha_n\bar{\beta}_n,$$

$$\langle \{\alpha_n\}, \{\beta_n\} \rangle = \sum_{n=1}^{\infty} \frac{1}{n} \alpha_n\bar{\beta}_n,$$

$$\langle \{\alpha_n\}, \{\beta_n\} \rangle = \sum_{n=1}^{\infty} n^5 \alpha_n\bar{\beta}_n,$$

all define inner products on $\mathscr{X}$.

1.3. Example. Let (X, Ω, μ) be a measure space consisting of a set X, a σ-algebra Ω of subsets of X, and a countably additive measure μ defined on Ω with values in the non-negative extended real numbers. If f and $g \in L^2(\mu) \equiv L^2(X, \Omega, \mu)$, then Hölder's inequality implies $f\bar{g} \in L^1(\mu)$. If

$$\langle f, g \rangle = \int f\bar{g} \, d\mu,$$

then this defines an inner product on $L^2(\mu)$.

   Note that Hölder's inequality also states that $|\int f\bar{g}d\mu| \leq [\int|f|^2 d\mu]^{1/2}$. $[\int|g|^2 d\mu]^{1/2}$. This is, in fact, a consequence of the following result on semi-inner products.

1.4. The Cauchy–Bunyakowsky–Schwarz Inequality. *If $\langle \cdot, \cdot \rangle$ is a semi-inner product on $\mathscr{X}$, then*

$$|\langle x, y \rangle|^2 \leqslant \langle x, x \rangle \langle y, y \rangle$$

for all x and y in $\mathscr{X}$. Moreover, equality occurs if and only if there are scalars α and β, both not 0, such that $\langle \beta x + \alpha y, \beta x + \alpha y \rangle = 0$.

PROOF. If $\alpha \in \mathbb{F}$ and x and $y \in \mathscr{X}$, then

$$0 \leqslant \langle x - \alpha y, x - \alpha y \rangle$$
$$= \langle x, x \rangle - \alpha \langle y, x \rangle - \bar{\alpha} \langle x, y \rangle + |\alpha|^2 \langle y, y \rangle.$$

Suppose $\langle y, x \rangle = be^{i\theta}$, $b \geqslant 0$, and let $\alpha = e^{-i\theta} t$, t in $\mathbb{R}$. The above inequality becomes

$$0 \leqslant \langle x, x \rangle - e^{-i\theta} t b e^{i\theta} - e^{i\theta} t b e^{-i\theta} + t^2 \langle y, y \rangle$$
$$= \langle x, x \rangle - 2bt + t^2 \langle y, y \rangle$$
$$= c - 2bt + at^2 \equiv q(t),$$

where $c = \langle x, x \rangle$ and $a = \langle y, y \rangle$. Thus $q(t)$ is a quadratic polynomial in the real variable t and $q(t) \geqslant 0$ for all t. This implies that the equation $q(t) = 0$ has at most one real solution t. From the quadratic formula we find that the discriminant is not positive; that is, $0 \geqslant 4b^2 - 4ac$. Hence

$$0 \geqslant b^2 - ac = |\langle x, y \rangle|^2 - \langle x, x \rangle \langle y, y \rangle,$$

proving the inequality.

The proof of the necessary and sufficient condition for equality is left to the reader.   ∎

The inequality in (1.4) will be referred to as the CBS inequality.

1.5. Corollary. *If $\langle \cdot, \cdot \rangle$ is a semi-inner product on $\mathscr{X}$ and $\|x\| \equiv \langle x, x \rangle^{1/2}$ for all x in $\mathscr{X}$, then*

(a) $\|x + y\| \leqslant \|x\| + \|y\|$ *for x, y in $\mathscr{X}$,*
(b) $\|\alpha x\| = |\alpha| \|x\|$ *for α in $\mathbb{F}$ and x in $\mathscr{X}$.*

If $\langle \cdot, \cdot \rangle$ is an inner product, then

(c) $\|x\| = 0$ *implies $x = 0$.*

PROOF. The proofs of (b) and (c) are left as an exercise. To see (a), note that for x and y in $\mathscr{X}$,

$$\|x + y\|^2 = \langle x + y, x + y \rangle$$
$$= \|x\|^2 + \langle y, x \rangle + \langle x, y \rangle + \|y\|^2$$
$$= \|x\|^2 + 2\,\mathrm{Re}\langle x, y \rangle + \|y\|^2.$$

By the CBS inequality, $\mathrm{Re}\langle x,y\rangle \leqslant |\langle x,y\rangle| \leqslant \|x\|\,\|y\|$. Hence,

$$\|x+y\|^2 \leqslant \|x\|^2 + 2\|x\|\,\|y\| + \|y\|^2$$
$$= (\|x\| + \|y\|)^2.$$

The inequality now follows by taking square roots. ∎

If $\langle\cdot,\cdot\rangle$ is a semi-inner product on $\mathscr{X}$ and if x, $y \in \mathscr{X}$, then as was shown in the preceding proof,

$$\|x+y\|^2 = \|x\|^2 + 2\,\mathrm{Re}\langle x,y\rangle + \|y\|^2.$$

This identity is often called the *polar identity*.

The quantity $\|x\| = \langle x,x\rangle^{1/2}$ for an inner product $\langle\cdot,\cdot\rangle$ is called the *norm* of x. If $\mathscr{X} = \mathbb{F}^d (\mathbb{R}^d$ or $\mathbb{C}^d)$ and $\langle \{\alpha_n\}, \{\beta_n\} \rangle \equiv \sum_{n=1}^d \alpha_n \bar{\beta}_n$, then the corresponding norm is $\|\{\alpha_n\}\| = [\sum_{n=1}^d |\alpha_n|^2]^{1/2}$.

The virtue of the norm on a vector space $\mathscr{X}$ is that $d(x,y) = \|x-y\|$ defines a metric on $\mathscr{X}$ [by (1.5)] so that $\mathscr{X}$ becomes a metric space. In fact, $d(x,y) = \|x-y\| = \|(x-z)+(z-y)\| \leqslant \|x-z\| + \|z-y\| = d(x,z) + d(z,y)$. The other properties of a metric follow similarly. If $\mathscr{X} = \mathbb{F}^d$ and the norm is defined as above, this distance function is the usual Euclidean metric. It is sometimes useful to note that with this metric the inner product becomes a continuous function from $\mathscr{X} \times \mathscr{X}$ into $\mathbb{R}$.

1.6. Definition. A *Hilbert space* is a vector space $\mathscr{H}$ over $\mathbb{F}$ together with an inner product $\langle\cdot,\cdot\rangle$ such that relative to the metric $d(x,y) = \|x-y\|$ induced by the norm, $\mathscr{H}$ is a complete metric space.

If $\mathscr{H} = L^2(\mu)$ and $\langle f,g\rangle = \int f\bar{g}\,d\mu$, then the associated norm is $\|f\| = [\int |f|^2 d\mu]^{1/2}$. It is a standard result of measure theory that $L^2(\mu)$ is a Hilbert space. It is also easy to see that $\mathbb{F}^d$ is a Hilbert space.

Remark. The inner products defined on $L^2(\mu)$ and $\mathbb{F}^d$ are the "usual" ones. Whenever these spaces are discussed these are the inner products referred to. The same is true of the next space.

1.7. Example. Let I be any set and let $l^2(I)$ denote the set of all functions $x: I \to \mathbb{F}$ such that $x(i) = 0$ for all but a countable number of i and $\sum_{i\in I}|x(i)|^2 < \infty$. For x and y in $l^2(I)$ define

$$\langle x,y\rangle = \sum_i x(i)\overline{y(i)}.$$

Then $l^2(I)$ is a Hilbert space (Exercise 2).

If $I = \mathbb{N}$, $l^2(I)$ is usually denoted by l^2. Note that if $\Omega = $ the set of all subsets of I and for E in Ω, $\mu(E) \equiv \infty$ if E is infinite and $\mu(E) = $ the cardinality of E if E is finite, then $l^2(I)$ and $L^2(I, \Omega, \mu)$ are equal.

Recall that an absolutely continuous function on the unit interval $[0, 1]$ has a derivative a.e. on $[0, 1]$.

1.8. Example. Let $\mathcal{H}$ = the collection of all absolutely continuous functions $f: [0, 1] \to \mathbb{F}$ such that $f(0) = 0$ and $f' \in L^2(0, 1)$. If $\langle f, g \rangle = \int_0^1 f'(t)\overline{g'(t)}dt$ for f and g in $\mathcal{H}$, then $\mathcal{H}$ is a Hilbert space (Exercise 3).

Suppose $\mathcal{X}$ is a vector space with an inner product $\langle \cdot, \cdot \rangle$ and the norm is defined by the inner product. What happens if $(\mathcal{X}, d)(d(x, y) \equiv \| x - y \|)$ is not complete?

1.9. Proposition. *If $\mathcal{X}$ is a vector space and $\langle \cdot, \cdot \rangle_{\mathcal{X}}$ is an inner product on $\mathcal{X}$ and if $\mathcal{H}$ is the completion of $\mathcal{X}$ with respect to the metric induced by the norm on $\mathcal{X}$, then there is an inner product $\langle \cdot, \cdot \rangle_{\mathcal{H}}$ on $\mathcal{H}$ such that $\langle x, y \rangle_{\mathcal{H}} = \langle x, y \rangle_{\mathcal{X}}$ for x and y in $\mathcal{X}$ and the metric on $\mathcal{H}$ is induced by this inner product. That is, the completion of $\mathcal{X}$ is a Hilbert space.*

The preceding result says that an incomplete inner product space can be completed to a Hilbert space. It is also true that a Hilbert space over $\mathbb{R}$ can be imbedded in a complex Hilbert space (see Exercise 7).

This section closes with an example of a Hilbert space from analytic function theory.

1.10. Definition. If G is an open subset of the complex plane $\mathbb{C}$, then $L_a^2(G)$ denotes the collection of all analytic functions $f: G \to \mathbb{C}$ such that

$$\iint_G |f(x + iy)|^2 dx\, dy < \infty.$$

$L_a^2(G)$ is called the *Bergman space* for G.

Several alternatives for the integral with respect to two-dimensional Lebesgue measure will be used. In addition to $\iint_G f(x + iy)dx\, dy$ we will also see

$$\iint_G f \quad \text{and} \quad \int_G f\, d\text{Area}.$$

Note that $L_a^2(G) \subseteq L^2(\mu)$, where $\mu = \text{Area}|G$, so that $L_a^2(G)$ has a natural inner product and norm from $L^2(\mu)$.

1.11. Lemma. *If f is analytic in a neighborhood of $\bar{B}(a; r)$, then*

$$f(a) = \frac{1}{\pi r^2} \iint_{B(a;r)} f.$$

[Here $B(a; r) \equiv \{z: |z - a| < r\}$ and $\bar{B}(a; r) \equiv \{z: |z - a| \leqslant r\}$.]

PROOF. By the mean value property, if $0 < t \leqslant r$, $f(a) = (1/2\pi)\int_{-\pi}^{\pi} f(a + te^{i\theta})d\theta$. Hence

$$(\pi r^2)^{-1}\iint_{B(a;r)} f = (\pi r^2)^{-1}\int_0^r t\left[\int_{-\pi}^{\pi} f(a + te^{i\theta})d\theta\right]dt$$

$$= (2/r^2)\int_0^r tf(a)dt = f(a). \qquad \blacksquare$$

1.12. Corollary. *If $f \in L_a^2(G)$, $a \in G$, and $0 < r < \text{dist}(a, \partial G)$, then*

$$|f(a)| \leqslant \frac{1}{r\sqrt{\pi}}\|f\|_2.$$

PROOF. Since $\bar{B}(a; r) \subseteq G$, the preceding lemma and the CBS inequality imply

$$|f(a)| = \frac{1}{\pi r^2}\left|\iint_{B(a;r)} f \cdot 1\right|$$

$$\leqslant \frac{1}{\pi r^2}\left[\iint_{B(a;r)} |f|^2\right]^{1/2}\left[\iint_{B(a;r)} 1^2\right]^{1/2}$$

$$\leqslant \frac{1}{\pi r^2}\|f\|_2 r\sqrt{\pi}. \qquad \blacksquare$$

1.13. Proposition. *$L_a^2(G)$ is a Hilbert space.*

PROOF. If $\mu = $ area measure on G, then $L^2(\mu)$ is a Hilbert space and $L_a^2(G) \subseteq L^2(\mu)$. So it suffices to show that $L_a^2(G)$ is closed in $L^2(\mu)$. Let $\{f_n\}$ be a sequence in $L_a^2(G)$ and let $f \in L^2(\mu)$ such that $\int |f_n - f|^2 d\mu \to 0$ as $n \to \infty$.

Suppose $\bar{B}(a; r) \subseteq G$ and let $0 < \rho < \text{dist}(B(a; r), \partial G)$. By the preceding corollary there is a constant C such that $|f_n(z) - f_m(z)| \leqslant C\|f_n - f_m\|_2$ for all n, m and for $|z - a| \leqslant \rho$. Thus $\{f_n\}$ is a uniformly Cauchy sequence on any closed disk in G. By standard results from analytic function theory (Montel's Theorem or Morera's Theorem, for example), there is an analytic function g on G such that $f_n(z) \to g(z)$ uniformly on compact subsets of G. But since $\int |f_n - f|^2 d\mu \to 0$, a result of Riesz implies there is a subsequence $\{f_{n_k}\}$ such that $f_{n_k}(z) \to f(z)$ a.e. $[\mu]$. Thus $f = g$ a.e. $[\mu]$ and so $f \in L_a^2(G)$. $\qquad \blacksquare$

EXERCISES

1. Verify the statements made in Example 1.2.

2. Verify that $l^2(I)$ (Example 1.7) is a Hilbert space.

3. Show that the space $\mathscr{H}$ in Example 1.8 is a Hilbert space.

4. Describe the Hilbert spaces obtained by completing the space $\mathscr{X}$ in Example 1.2 with respect to the norm defined by each of the inner products given there.

5. (A variation on Example 1.8) Let $n \geq 2$ and let $\mathcal{H} =$ the collection of all function $f: [0,1] \rightarrow \mathbb{F}$ such that (a) $f(0) = 0$; (b) for $1 \leq k \leq n-1$, $f^{(k)}(t)$ exists for all t in $[0,1]$ and $f^{(k)}$ is continuous on $[0,1]$; (c) $f^{(n-1)}$ is absolutely continuous and $f^{(n)} \in L^2(0,1)$. For f and g in $\mathcal{H}$, define

$$\langle f, g \rangle = \sum_{k=1}^{n} \int_0^1 f^{(k)}(t) \overline{g^{(k)}(t)} dt.$$

Show that $\mathcal{H}$ is a Hilbert space.

6. Let u be a semi-inner product on $\mathcal{X}$ and put $\mathcal{N} = \{x \in \mathcal{X}: u(x,x) = 0\}$.

(a) Show that $\mathcal{N}$ is a linear subspace of $\mathcal{X}$.
(b) Show that if

$$\langle x + \mathcal{N}, y + \mathcal{N} \rangle \equiv u(x,y)$$

for all $x + \mathcal{N}$ and $y + \mathcal{N}$ in the quotient space $\mathcal{X}/\mathcal{N}$, then $\langle \cdot, \cdot \rangle$ is a well-defined inner product on $\mathcal{X}/\mathcal{N}$.

7. Let $\mathcal{H}$ be a Hilbert space over $\mathbb{R}$ and show that there is a Hilbert space $\mathcal{X}$ over $\mathbb{C}$ and a map $U: \mathcal{H} \rightarrow \mathcal{X}$ such that (a) U is linear; (b) $\langle Uh_1, Uh_2 \rangle = \langle h_1, h_2 \rangle$ for all h_1, h_2 in $\mathcal{H}$; (c) for any k in $\mathcal{X}$ there are unique h_1, h_2 in $\mathcal{H}$ such that $k = Uh_1 + iUh_2$. ($\mathcal{X}$ is called the *complexification* of $\mathcal{H}$.)

8. If $G = \{z \in \mathbb{C}: 0 < |z| < 1\}$ show that every f in $L^2_a(G)$ has a removable singularity at $z = 0$.

9. Which functions are in $L^2_a(\mathbb{C})$?

10. Let G be an open subset of $\mathbb{C}$ and show that if $a \in G$, then $\{f \in L^2_a(G): f(a) = 0\}$ is closed in $L^2_a(G)$.

11. If $\{h_n\}$ is a sequence in a Hilbert space $\mathcal{H}$ such that $\sum_n \|h_n\| < \infty$, then show that $\sum_{n=1}^{\infty} h_n$ converges in $\mathcal{H}$.

§2. Orthogonality

The greatest advantage of a Hilbert space is its underlying concept of orthogonality.

2.1. Definition. If $\mathcal{H}$ is a Hilbert space and $f, g \in \mathcal{H}$, then f and g are *orthogonal* if $\langle f, g \rangle = 0$. In symbols, $f \perp g$. If $A, B \subseteq \mathcal{H}$, then $A \perp B$ if $f \perp g$ for every f in A and g in B.

If $\mathcal{H} = \mathbb{R}^2$, this is the correct concept. Two non-zero vectors in $\mathbb{R}^2$ are orthogonal precisely when the angle between them is $\pi/2$.

2.2. The Pythagorean Theorem. *If $f_1, f_2, \ldots, f_n$ are pairwise orthogonal vectors in $\mathcal{H}$, then*

$$\|f_1 + f_2 + \cdots + f_n\|^2 = \|f_1\|^2 + \|f_2\|^2 + \cdots + \|f_n\|^2.$$

PROOF. If $f_1 \perp f_2$, then

$$\|f_1 + f_2\|^2 = \langle f_1 + f_2, f_1 + f_2 \rangle = \|f_1\|^2 + 2 \operatorname{Re} \langle f_1, f_2 \rangle + \|f_2\|^2$$

by the polar identity. Since $f_1 \perp f_2$, this implies the result for $n = 2$. The remainder of the proof proceeds by induction and is left to the reader. ∎

Note that if $f \perp g$, then $f \perp -g$, so $\|f - g\|^2 = \|f\|^2 + \|g\|^2$. The next result is an easy consequence of the Pythagorean Theorem if f and g are orthogonal, but this assumption is not needed for its conclusion.

2.3. Parallelogram Law. *If $\mathscr{H}$ is a Hilbert space and f and $g \in \mathscr{H}$, then*

$$\|f + g\|^2 + \|f - g\|^2 = 2(\|f\|^2 + \|g\|^2).$$

PROOF. For any f and g in $\mathscr{H}$ the polar identity implies

$$\|f + g\|^2 = \|f\|^2 + 2 \operatorname{Re} \langle f, g \rangle + \|g\|^2,$$
$$\|f - g\|^2 = \|f\|^2 - 2 \operatorname{Re} \langle f, g \rangle + \|g\|^2.$$

Now add. ∎

The next property of a Hilbert space is truly pivotal. But first we need a geometric concept valid for any vector space over $\mathbb{F}$.

2.4. Definition. If $\mathscr{X}$ is any vector space over $\mathbb{F}$ and $A \subseteq \mathscr{X}$, then A is a *convex set* if for any x and y in A and $0 \leqslant t \leqslant 1$, $tx + (1 - t)y \in A$.

Note that $\{tx + (1 - t)y: 0 \leqslant t \leqslant 1\}$ is the straight-line segment joining x and y. So a convex set is a set A such that if x and $y \in A$, the entire line segment joining x and y is contained in A.

If $\mathscr{X}$ is a vector space, then any linear subspace in $\mathscr{X}$ is a convex set. A singleton set is convex. The intersection of any collection of convex sets is convex. If $\mathscr{H}$ is a Hilbert space, then every open ball $B(f; r) = \{g \in \mathscr{H}: \|f - g\| < r\}$ is convex, as is every closed ball.

2.5. Theorem. *If $\mathscr{H}$ is a Hilbert space, K is a closed convex nonempty subset of $\mathscr{H}$, and $h \in \mathscr{H}$, then there is a unique point k_0 in K such that*

$$\|h - k_0\| = \operatorname{dist}(h, K) \equiv \inf\{\|h - k\|: k \in K\}.$$

PROOF. By considering $K - h \equiv \{k - h: k \in K\}$ instead of K, it suffices to assume that $h = 0$. (Verify!) So we want to show that there is a unique vector k_0 in K such that

$$\|k_0\| = \operatorname{dist}(0, K) \equiv \inf\{\|k\|: k \in K\}.$$

Let $d = \operatorname{dist}(0, K)$. By definition, there is a sequence $\{k_n\}$ in K such that $\|k_n\| \to d$. Now the Parallelogram Law implies that

$$\left\|\frac{k_n - k_m}{2}\right\|^2 = \tfrac{1}{2}(\|k_n\|^2 + \|k_m\|^2) - \left\|\frac{k_n + k_m}{2}\right\|^2.$$

Since K is convex, $\frac{1}{2}(k_n + k_m) \in K$. Hence, $\|\frac{1}{2}(k_n + k_m)\|^2 \geq d^2$. If $\varepsilon > 0$, choose N such that for $n \geq N$, $\|k_n\|^2 < d^2 + \frac{1}{4}\varepsilon^2$. By the equation above, if $n, m \geq N$, then

$$\left\|\frac{k_n - k_m}{2}\right\|^2 < \frac{1}{2}(2d^2 + \frac{1}{2}\varepsilon^2) - d^2 = \frac{1}{4}\varepsilon^2.$$

Thus, $\|k_n - k_m\| < \varepsilon$ for $n, m \geq N$ and $\{k_n\}$ is a Cauchy sequence. Since $\mathscr{H}$ is complete and K is closed, there is a k_0 in K such that $\|k_n - k_0\| \to 0$. Also for all k_n,

$$d \leq \|k_0\| = \|k_0 - k_n + k_n\|$$
$$\leq \|k_0 - k_n\| + \|k_n\| \to d.$$

Thus $\|k_0\| = d$.

To prove that k_0 is unique, suppose $h_0 \in K$ such that $\|h_0\| = d$. By convexity, $\frac{1}{2}(k_0 + h_0) \in K$. Hence

$$d \leq \|\frac{1}{2}(h_0 + k_0)\| \leq \frac{1}{2}(\|h_0\| + \|k_0\|) = d.$$

So $\|\frac{1}{2}(h_0 + k_0)\| = d$. The Parallelogram Law implies

$$d^2 = \left\|\frac{h_0 + k_0}{2}\right\|^2 = d^2 - \left\|\frac{h_0 - k_0}{2}\right\|^2;$$

hence $h_0 = k_0$. ∎

If the convex set in the preceding theorem is in fact a closed linear subspace of $\mathscr{H}$, more can be said.

2.6. Theorem. *If $\mathscr{M}$ is a closed linear subspace of $\mathscr{H}$, $h \in \mathscr{H}$, and f_0 is the unique element of $\mathscr{M}$ such that $\|h - f_0\| = \operatorname{dist}(h, \mathscr{M})$, then $h - f_0 \perp \mathscr{M}$. Conversely, if $f_0 \in \mathscr{M}$ such that $h - f_0 \perp \mathscr{M}$, then $\|h - f_0\| = \operatorname{dist}(h, \mathscr{M})$.*

PROOF. Suppose $f_0 \in \mathscr{M}$ and $\|h - f_0\| = \operatorname{dist}(h, \mathscr{M})$. If $f \in \mathscr{M}$, then $f_0 + f \in \mathscr{M}$ and so $\|h - f_0\|^2 \leq \|h - (f_0 + f)\|^2 = \|(h - f_0) - f\|^2 = \|h - f_0\|^2 - 2\operatorname{Re}\langle h - f_0, f\rangle + \|f\|^2$. Thus

$$2\operatorname{Re}\langle h - f_0, f\rangle \leq \|f\|^2$$

for any f in $\mathscr{M}$. Fix f in $\mathscr{M}$ and substitute $te^{i\theta}f$ for f in the preceding inequality, where $\langle h - f_0, f\rangle = re^{i\theta}$, $r \geq 0$. This yields $2\operatorname{Re}\{te^{-i\theta}re^{i\theta}\} \leq t^2\|f\|^2$, or $2tr \leq t^2\|f\|^2$. Letting $t \to 0$, we see that $r = 0$; that is, $h - f_0 \perp f$.

For the converse, suppose $f_0 \in \mathscr{M}$ such that $h - f_0 \perp \mathscr{M}$. If $f \in \mathscr{M}$, then $h - f_0 \perp f_0 - f$ so that

$$\|h - f\|^2 = \|(h - f_0) + (f_0 - f)\|^2$$
$$= \|h - f_0\|^2 + \|f_0 - f\|^2$$
$$\geq \|h - f_0\|^2.$$

Thus $\|h - f_0\| = \operatorname{dist}(h, \mathscr{M})$. ∎

If $A \subseteq \mathcal{H}$, Let $A^{\perp} \equiv \{f \in \mathcal{H} : f \perp g \text{ for all } g \text{ in } A\}$. It is easy to see that $A^{\perp}$ is a closed linear subspace of $\mathcal{H}$.

Note that Theorem 2.6, together with the uniqueness statement in Theorem 2.5, shows that if $\mathcal{M}$ is a closed linear subspace of $\mathcal{H}$ and $h \in \mathcal{H}$, then there is a unique element f_0 in $\mathcal{M}$ such that $h - f_0 \in \mathcal{M}^{\perp}$. Thus a function $P: \mathcal{H} \to \mathcal{M}$ can be defined by $Ph = f_0$.

2.7. Theorem. *If $\mathcal{M}$ is a closed linear subspace of $\mathcal{H}$ and $h \in \mathcal{H}$, let Ph be the unique point in $\mathcal{M}$ such that $h - Ph \perp \mathcal{M}$. Then*

(a) *P is a linear transformation on $\mathcal{H}$,*
(b) *$\|Ph\| \leqslant \|h\|$ for every h in $\mathcal{H}$,*
(c) *$P^2 = P$ (here P^2 means the composition of P with itself),*
(d) *$\ker P = \mathcal{M}^{\perp}$ and $\operatorname{ran} P = \mathcal{M}$.*

PROOF. Keep in mind that for every h in $\mathcal{H}$, $h - Ph \in \mathcal{M}^{\perp}$ and $\|h - Ph\| = \operatorname{dist}(h, \mathcal{M})$.

(a) Let $h_1, h_2 \in \mathcal{H}$ and $\alpha_1, \alpha_2 \in \mathbb{F}$. If $f \in \mathcal{M}$, then $\langle [\alpha_1 h_1 + \alpha_2 h_2] - [\alpha_1 Ph_1 + \alpha_2 Ph_2], f \rangle = \alpha_1 \langle h_1 - Ph_1, f \rangle + \alpha_2 \langle h_2 - Ph_2, f \rangle = 0$. By the uniqueness statement of (2.6), $P(\alpha_1 h_1 + \alpha_2 h_2) = \alpha_1 Ph_1 + \alpha_2 Ph_2$.
(b) If $h \in \mathcal{H}$, then $h = (h - Ph) + Ph$, $Ph \in \mathcal{M}$, and $h - Ph \in \mathcal{M}^{\perp}$. Thus $\|h\|^2 = \|h - Ph\|^2 + \|Ph\|^2 \geqslant \|Ph\|^2$.
(c) If $f \in \mathcal{M}$, then $Pf = f$. For any h in $\mathcal{H}$, $Ph \in \mathcal{M}$; hence $P^2 h \equiv P(Ph) = Ph$. That is, $P^2 = P$.
(d) If $Ph = 0$, then $h = h - Ph \in \mathcal{M}^{\perp}$. Conversely, if $h \in M^{\perp}$, then 0 is the unique vector in $\mathcal{M}$ such that $h - 0 = h \perp \mathcal{M}$. Therefore $Ph = 0$. That $\operatorname{ran} P = \mathcal{M}$ is clear. ∎

2.8. Definition. If $\mathcal{M}$ is a closed linear subspace of $\mathcal{H}$ and P is the linear map defined in the preceding theorem, then P is called the *orthogonal projection* of $\mathcal{H}$ onto $\mathcal{M}$. If we wish to show this dependence of P on $\mathcal{M}$, we will denote the orthogonal projection of $\mathcal{H}$ onto $\mathcal{M}$ by $P_{\mathcal{M}}$.

It also seems appropriate to introduce the notation $\mathcal{M} \leqslant \mathcal{H}$ to signify that $\mathcal{M}$ is a closed linear subspace of $\mathcal{H}$. We will use the term *linear manifold* to designate a linear subspace of $\mathcal{H}$ that is not necessarily closed. A *linear subspace* of $\mathcal{H}$ will always mean a closed linear subspace.

2.9. Corollary. *If $\mathcal{M} \leqslant \mathcal{H}$, then $(\mathcal{M}^{\perp})^{\perp} = \mathcal{M}$.*

PROOF. If I is used to designate the identity operator on $\mathcal{H}$ (viz., $Ih = h$) and $P = P_{\mathcal{M}}$, then $I - P$ is the orthogonal projection of $\mathcal{H}$ onto $\mathcal{M}^{\perp}$ (Exercise 2). By part (d) of the preceding theorem, $(\mathcal{M}^{\perp})^{\perp} = \ker(I - P)$. But $0 = (I - P)h$ iff $h = Ph$. Thus $(\mathcal{M}^{\perp})^{\perp} = \ker(I - P) = \operatorname{ran} P = \mathcal{M}$. ∎

2.10. Corollary. *If $A \subseteq \mathcal{H}$, then $(A^{\perp})^{\perp}$ is the closed linear span of A in $\mathcal{H}$.*

The proof is left to the reader; see Exercise 4 for a discussion of the term "closed linear span."

2.11. Corollary. *If $\mathscr{Y}$ is a linear manifold in $\mathscr{H}$, then $\mathscr{Y}$ is dense in $\mathscr{H}$ iff $\mathscr{Y}^{\perp} = (0)$.*

PROOF. Exercise.

EXERCISES

1. Let $\mathscr{H}$ be a Hilbert space and suppose f and g are linearly independent vectors in $\mathscr{H}$ with $\|f\| = \|g\| = 1$. Show that $\|tf + (1-t)g\| < 1$ for $0 < t < 1$. What does this say about $\{h \in \mathscr{H}: \|h\| \leqslant 1\}$?

2. If $\mathscr{M} \leqslant \mathscr{H}$ and $P = P_{\mathscr{M}}$, show that $I - P$ is the orthogonal projection of $\mathscr{H}$ onto $\mathscr{M}^{\perp}$.

3. If $\mathscr{M} \leqslant \mathscr{H}$, show that $\mathscr{M} \cap \mathscr{M}^{\perp} = (0)$ and every h in $\mathscr{H}$ can be written as $h = f + g$ where $f \in \mathscr{M}$ and $g \in \mathscr{M}^{\perp}$. If $\mathscr{M} + \mathscr{M}^{\perp} \equiv \{(f,g): f \in \mathscr{M}, g \in \mathscr{M}^{\perp}\}$ and $T: \mathscr{M} + \mathscr{M}^{\perp} \to \mathscr{H}$ is defined by $T(f,g) = f + g$, show that T is a linear bijection and a homeomorphism if $\mathscr{M} + \mathscr{M}^{\perp}$ is given the product topology. (This is usually phrased by stating the $\mathscr{M}$ and $\mathscr{M}^{\perp}$ are *topologically complementary* in $\mathscr{H}$.)

4. If $A \subseteq \mathscr{H}$, let $\vee A \equiv$ the intersection of all closed linear subspaces of $\mathscr{H}$ that contain A. $\vee A$ is called the *closed linear span* of A. Prove the following:

   (a) $\vee A \leqslant \mathscr{H}$ and $\vee A$ is the smallest closed linear subspace of $\mathscr{H}$ that contains A.
   (b) $\vee A =$ the closure of $\{\sum_{k=1}^{n} \alpha_k f_k: n \geqslant 1, \alpha_k \in \mathbb{F}, f_k \in A\}$.

5. Prove Corollary 2.10.

6. Prove Corollary 2.11.

§3. The Riesz Representation Theorem

The title of this section is somewhat ambiguous as there are at least two Riesz Representation Theorems. There is one so-called theorem that represents bounded linear functionals on the space of continuous functions on a compact Hausdorff space. That theorem will be discussed later in this book. The present section deals with the representation of certain linear functionals on Hilbert space. But first we have a few preliminaries to dispose of.

3.1. Proposition. *Let $\mathscr{H}$ be a Hilbert space and $L: \mathscr{H} \to \mathbb{F}$ a linear functional. The following statements are equivalent.*

(a) *L is continuous.*
(b) *L is continuous at 0.*
(c) *L is continuous at some point.*
(d) *There is a constant $c > 0$ such that $|L(h)| \leqslant c\|h\|$ for every h in $\mathscr{H}$.*

PROOF. It is clear that (a)$\Rightarrow$(b)$\Rightarrow$(c) and (d)$\Rightarrow$(b). Let's show that (c)$\Rightarrow$(a) and (b)$\Rightarrow$(d).

(c)⇒(a): Suppose L is continuous at h_0 and h is any point in $\mathscr{H}$. If $h_n \to h$ in $\mathscr{H}$, then $h_n - h + h_0 \to h_0$. By assumption, $L(h_0) = \lim[L(h_n - h + h_0)] = \lim[L(h_n) - L(h) + L(h_0)] = \lim L(h_n) - L(h) + L(h_0)$. Hence $L(h) = \lim L(h_n)$.

(b)⇒(d): The definition of continuity at 0 implies that $L^{-1}(\{\alpha \in \mathbb{F}: |\alpha| < 1\})$ contains an open ball about 0. So there is a $\delta > 0$ such that $B(0; \delta) \subseteq L^{-1}(\{\alpha \in \mathbb{F}: |\alpha| < 1\})$. That is, $\|h\| < \delta$ implies $|L(h)| < 1$. If h is an arbitrary element of $\mathscr{H}$ and $\varepsilon > 0$, then $\|\delta(\|h\| + \varepsilon)^{-1}h\| < \delta$. Hence

$$1 > \left| L\left[\frac{\delta h}{\|h\| + \varepsilon} \right] \right| = \frac{\delta}{\|h\| + \varepsilon} |L(h)|;$$

thus

$$|L(h)| < \frac{1}{\delta}(\|h\| + \varepsilon).$$

Letting $\varepsilon \to 0$ we see that (d) holds with $c = 1/\delta$. ∎

3.2. Definition. A *bounded linear functional* L on $\mathscr{H}$ is a linear functional for which there is a constant $c > 0$ such that $|L(h)| \leqslant c\|h\|$ for all h in $\mathscr{H}$. In light of the preceding proposition, a linear functional is bounded if and only if it is continuous.

For a bounded linear functional $L: \mathscr{H} \to \mathbb{F}$, define

$$\|L\| = \sup\{|L(h)|: \|h\| \leqslant 1\}.$$

Note that by definition, $\|L\| < \infty$; $\|L\|$ is called the *norm* of L.

3.3. Proposition. *If L is a bounded linear functional, then*

$$\begin{aligned}\|L\| &= \sup\{|L(h)|: \|h\| = 1\}\\ &= \sup\{|L(h)|/\|h\|: h \in \mathscr{H}, h \neq 0\}\\ &= \inf\{c > 0: |L(h)| \leqslant c\|h\|, h \text{ in } \mathscr{H}\}.\end{aligned}$$

Also, $|L(h)| \leqslant \|L\| \|h\|$ for every h in $\mathscr{H}$.

PROOF. Let $\alpha = \inf\{c > 0: |L(h)| \leqslant c\|h\|, h \text{ in } \mathscr{H}\}$. It will be shown that $\|L\| = \alpha$; the remaining equalities are left as an exercise. If $\varepsilon > 0$, then the definition of $\|L\|$ shows that $|L((\|h\| + \varepsilon)^{-1}h)| \leqslant \|L\|$. Hence $|L(h)| \leqslant \|L\|(\|h\| + \varepsilon)$. Letting $\varepsilon \to 0$ shows that $|L(h)| \leqslant \|L\| \|h\|$ for all h. So the definition of α shows that $\alpha \leqslant \|L\|$. On the other hand, if $|L(h)| \leqslant c\|h\|$ for all h, then $\|L\| \leqslant c$. Hence $\|L\| \leqslant \alpha$. ∎

Fix an h_0 in $\mathscr{H}$ and define $L: \mathscr{H} \to \mathbb{F}$ by $L(h) = \langle h, h_0 \rangle$. It is easy to see that L is linear. Also, the CBS inequality gives that $|L(h)| = |\langle h, h_0 \rangle| \leqslant \|h\| \|h_0\|$. So L is bounded and $\|L\| \leqslant \|h_0\|$. In fact, $L(h_0/\|h_0\|) = \langle h_0/\|h_0\|, h_0 \rangle = \|h_0\|$, so that $\|L\| = \|h_0\|$. The main result of this section provides a converse to these observations.

3.4. The Riesz Representation Theorem. *If* $L: \mathcal{H} \to \mathbb{F}$ *is a bounded linear functional, then there is a unique vector* h_0 *in* $\mathcal{H}$ *such that* $L(h) = \langle h, h_0 \rangle$ *for every h in* $\mathcal{H}$. *Moreover,* $\|L\| = \|h_0\|$.

PROOF. Let $\mathcal{M} = \ker L$. Because L is continuous $\mathcal{M}$ is a closed linear subspace of $\mathcal{H}$. Since we may assume that $\mathcal{M} \neq \mathcal{H}$, $\mathcal{M}^\perp \neq (0)$. Hence there is a vector f_0 in $\mathcal{M}^\perp$ such that $L(f_0) = 1$. Now if $h \in \mathcal{H}$ and $\alpha = L(h)$, then $L(h - \alpha f_0) = L(h) - \alpha = 0$; so $h - L(h)f_0 \in \mathcal{M}$. Thus

$$0 = \langle h - L(h)f_0, f_0 \rangle$$
$$= \langle h, f_0 \rangle - L(h)\|f_0\|^2.$$

So if $h_0 = \|f_0\|^{-2} f_0$, $L(h) = \langle h, h_0 \rangle$ for all h in $\mathcal{H}$.

If $h_0' \in \mathcal{H}$ such that $\langle h, h_0 \rangle = \langle h, h_0' \rangle$ for all h, then $h_0 - h_0' \perp \mathcal{H}$. In particular, $h_0 - h_0' \perp h_0 - h_0'$ and so $h_0' = h_0$. The fact that $\|L\| = \|h_0\|$ was shown in the discussion preceding the theorem. ∎

3.5. Corollary. *If* (X, Ω, μ) *is a measure space and* $F: L^2(\mu) \to \mathbb{F}$ *is a bounded linear functional, then there is a unique* h_0 *in* $L^2(\mu)$ *such that*

$$F(h) = \int h\overline{h_0}\, d\mu$$

for every h in $L^2(\mu)$.

Of course the preceding corollary is a special case of the theorem on representing bounded linear functionals on $L^p(\mu)$, $1 \leq p < \infty$. But it is interesting to note that it is a consequence of the result for Hilbert space [and the result that $L^2(\mu)$ is a Hilbert space].

EXERCISES

1. Prove Proposition 3.3.

2. Let $\mathcal{H} = l^2(\mathbb{N})$. If $N \geq 1$ and $L: \mathcal{H} \to \mathbb{F}$ is defined by $L(\{\alpha_n\}) = \alpha_N$, find the vector h_0 in $\mathcal{H}$ such that $L(h) = \langle h, h_0 \rangle$ for every h in $\mathcal{H}$.

3. Let $\mathcal{H} = l^2(\mathbb{N} \cup \{0\})$. (a) Show that if $\{\alpha_n\} \in \mathcal{H}$, then the power series $\sum_{n=0}^\infty \alpha_n z^n$ has radius of convergence ≥ 1. (b) If $|\lambda| < 1$ and $L: \mathcal{H} \to \mathbb{F}$ is defined by $L(\{\alpha_n\}) = \sum_{n=0}^\infty \alpha_n \lambda^n$, find the vector h_0 in $\mathcal{H}$ such that $L(h) = \langle h, h_0 \rangle$ for every h in $\mathcal{H}$. (c) What is the norm of the linear functional L defined in (b)?

4. With the notation as in Exercise 3, define $L: \mathcal{H} \to \mathbb{F}$ by $L(\{\alpha_n\}) = \sum_{n=1}^\infty n\alpha_n \lambda^{n-1}$, where $|\lambda| < 1$. Find a vector h_0 in $\mathcal{H}$ such that $L(h) = \langle h, h_0 \rangle$ for every h in $\mathcal{H}$.

5. Let $\mathcal{H}$ be the Hilbert space described in Example 1.8. If $0 < t \leq 1$, define $L: \mathcal{H} \to \mathbb{F}$ by $L(h) = h(t)$. Show that L is a bounded linear functional, find $\|L\|$, and find the vector h_0 in $\mathcal{H}$ such that $L(h) = \langle h, h_0 \rangle$ for all h in $\mathcal{H}$.

6. Let $\mathcal{H} = L^2(0, 1)$ and let $C^{(1)}$ be the set of all continuous functions on $[0, 1]$ that have a continuous derivative. Let $t \in [0, 1]$ and define $L: C^{(1)} \to \mathbb{F}$ by $L(h) = h'(t)$. Show that there is no bounded linear functional on $\mathcal{H}$ that agrees with L on $C^{(1)}$.

§4. Orthonormal Sets of Vectors and Bases

It will be shown in this section that, as in Euclidean space, each Hilbert space can be coordinatized. The vehicle for introducing the coordinates is an orthonormal basis. The corresponding vectors in $\mathbb{F}^d$ are the vectors $\{e_1, e_2, \ldots, e_d\}$, where e_k is the d-tuple having a 1 in the kth place and zeros elsewhere.

4.1. Definition. An *orthonormal* subset of a Hilbert space $\mathcal{H}$ is a subset $\mathscr{E}$ having the properties: (a) for e in $\mathscr{E}$, $\|e\| = 1$; (b) if $e_1, e_2 \in \mathscr{E}$ and $e_1 \neq e_2$, then $e_1 \perp e_2$.

A *basis* for $\mathcal{H}$ is a maximal orthonormal set.

Every vector space has a Hamel basis (a maximal linearly independent set). The term "basis" for a Hilbert space is defined as above and it relates to the inner product on $\mathcal{H}$. For an infinite-dimensional Hilbert space, a basis is never a Hamel basis. This is not obvious, but the reader will be able to see this after understanding several facts about bases.

4.2. Proposition. *If $\mathscr{E}$ is an orthonormal set in $\mathcal{H}$, then there is a basis for $\mathcal{H}$ that contains $\mathscr{E}$.*

The proof of this proposition is a straightforward application of Zorn's Lemma and is left to the reader.

4.3. Example. Let $\mathcal{H} = L^2_{\mathbb{C}}[0, 2\pi]$ and for n in $\mathbb{Z}$ define e_n in $\mathcal{H}$ by $e_n(t) = (2\pi)^{-1/2} \exp(int)$. Then $\{e_n : n \in \mathbb{Z}\}$ is an orthonormal set in $\mathcal{H}$. (Here $L^2_{\mathbb{C}}[0, 2\pi]$ is the space of complex-valued square integrable functions.)

It is also true that the set in (4.3) is a basis, but this is best proved after a bit of theory.

4.4. Example. If $\mathcal{H} = \mathbb{F}^d$ and for $1 \leqslant k \leqslant d$, $e_k =$ the d-tuple with 1 in the kth place and zeros elsewhere, then $\{e_1, \ldots, e_d\}$ is a basis for $\mathcal{H}$.

4.5. Example. Let $\mathcal{H} = l^2(I)$ as in Example 1.7. For each i in I define e_i in $\mathcal{H}$ by $e_i(i) = 1$ and $e_i(j) = 0$ for $j \neq i$. Then $\{e_i : i \in I\}$ is a basis.

The proof of the next result is left as an exercise (see Exercise 5). It is very useful but the proof is not difficult.

4.6. The Gram–Schmidt Orthogonalization Process. *If $\mathcal{H}$ is a Hilbert space and $\{h_n : n \in \mathbb{N}\}$ is a linearly independent subset of $\mathcal{H}$, then there is an orthonormal set $\{e_n : n \in \mathbb{N}\}$ such that for every n, the linear space of $\{e_1, \ldots, e_n\}$ equals the linear span of $\{h_1, \ldots, h_n\}$.*

Remember that $\bigvee A$ is the closed linear span of A (Exercise 2.4).

4.7. Proposition. *Let* $\{e_1, \ldots, e_n\}$ *be an orthonormal set in* $\mathcal{H}$ *and let* $\mathcal{M} = \bigvee \{e_1, \ldots, e_n\}$. *If* P *is the orthogonal projection of* $\mathcal{H}$ *onto* $\mathcal{M}$, *then*

$$Ph = \sum_{k=1}^{n} \langle h, e_k \rangle e_k$$

for all h *in* $\mathcal{H}$.

PROOF. Let $Qh = \sum_{k=1}^{n} \langle h, e_k \rangle e_k$. If $1 \leqslant j \leqslant n$, then $\langle Qh, e_j \rangle = \sum_{k=1}^{n} \langle h, e_k \rangle \langle e_k, e_j \rangle = \langle h, e_j \rangle$ since $e_k \perp e_j$ for $k \neq j$. Thus $\langle h - Qh, e_j \rangle = 0$ for $1 \leqslant j \leqslant n$. That is, $h - Qh \perp \mathcal{M}$ for every h in $\mathcal{H}$. Since Qh is clearly a vector in $\mathcal{M}$, Qh is the unique vector h_0 in $\mathcal{M}$ such that $h - h_0 \perp \mathcal{M}$ (2.6). Hence $Qh = Ph$ for every h in $\mathcal{H}$. ∎

4.8. Bessel's Inequality. *If* $\{e_n : n \in \mathbb{N}\}$ *is an orthonormal set and* $h \in \mathcal{H}$, *then*

$$\sum_{n=1}^{\infty} |\langle h, e_n \rangle|^2 \leqslant \|h\|^2.$$

PROOF. Let $h_n = h - \sum_{k=1}^{n} \langle h, e_k \rangle e_k$. Then $h_n \perp e_k$ for $1 \leqslant k \leqslant n$ (Why?) By the Pythagorean Theorem,

$$\|h\|^2 = \|h_n\|^2 + \left\| \sum_{k=1}^{n} \langle h, e_k \rangle e_k \right\|^2$$

$$= \|h_n\|^2 + \sum_{k=1}^{n} |\langle h, e_k \rangle|^2$$

$$\geqslant \sum_{k=1}^{n} |\langle h, e_k \rangle|^2.$$

Since n was arbitrary, the result is proved. ∎

4.9. Corollary. *If* $\mathcal{E}$ *is an orthonormal set in* $\mathcal{H}$ *and* $h \in \mathcal{H}$, *then* $\langle h, e \rangle \neq 0$ *for at most a countable number of vectors* e *in* $\mathcal{E}$.

PROOF. For each $n \geqslant 1$ let $\mathcal{E}_n = \{e \in \mathcal{E} : |\langle h, e \rangle| \geqslant 1/n\}$. By Bessel's Inequality, $\mathcal{E}_n$ is finite. But $\bigcup_{n=1}^{\infty} \mathcal{E}_n = \{e \in \mathcal{E} : \langle h, e \rangle \neq 0\}$. ∎

4.10. Corollary. *If* $\mathcal{E}$ *is an orthonormal set and* $h \in \mathcal{H}$, *then*

$$\sum_{e \in \mathcal{E}} |\langle h, e \rangle|^2 \leqslant \|h\|^2.$$

This last corollary is just Bessel's Inequality together with the fact (4.9) that at most a countable number of the terms in the sum differ from zero.

Actually, the sum that appears in (4.10) can be given a better interpretation—a mathematically precise one that will be useful later. The question is, what is meant by $\sum \{h_i : i \in I\}$ if $h_i \in \mathcal{H}$ and I is an infinite, possibly uncountable, set? Let $\mathcal{F}$ be the collection of all finite subsets of I and order

$\mathscr{F}$ by inclusion, so $\mathscr{F}$ becomes a directed set. For each F in $\mathscr{F}$, define

$$h_F = \Sigma\{h_i: i \in F\}.$$

Since this is a finite sum, h_F is a well-defined element of $\mathscr{H}$. Now $\{h_F: F \in \mathscr{F}\}$ is a net in $\mathscr{H}$.

4.11. Definition. With the notation above, the sum $\Sigma\{h_i: i \in I\}$ converges if the net $\{h_F: F \in \mathscr{F}\}$ *converges*; the value of the sum is the limit of the net.

If $\mathscr{H} = \mathbb{F}$, the definition above gives meaning to an uncountable sum of scalars. Now Corollary 4.10 can be given its precise meaning; namely, $\Sigma\{|\langle h, e \rangle|^2: e \in \mathscr{E}\}$ converges and the value $\leqslant \|h\|^2$ (Exercise 9).

If the set I in Definition 4.11 is countable, then this definition of convergent sum is not the usual one. That is, if $\{h_n\}$ is a sequence in $\mathscr{H}$, then the convergence of $\Sigma\{h_n: n \in \mathbb{N}\}$ is not equivalent to the convergence of $\Sigma_{n=1}^{\infty} h_n$. The former concept of convergence is that defined in (4.11) while the latter means that the sequence $\{\Sigma_{k=1}^{n} h_k\}_{n=1}^{\infty}$ converges. Even if $\mathscr{H} = \mathbb{F}$, these concepts do not coincide (see Exercise 12). If, however, $\Sigma\{h_n: n \in \mathbb{N}\}$ converges, then $\Sigma_{n=1}^{\infty} h_n$ converges (Exercise 10). Also see Exercise 11.

4.12. Lemma. *If $\mathscr{E}$ is an orthonormal set and $h \in \mathscr{H}$, then*

$$\Sigma\{\langle h, e \rangle e: e \in \mathscr{E}\}$$

converges in $\mathscr{H}$.

PROOF. By (4.9), there are vectors $e_1, e_2, \ldots$ in $\mathscr{E}$ such that $\{e \in \mathscr{E}: \langle h, e \rangle \neq 0\} = \{e_1, e_2, \ldots\}$. We also know that $\Sigma_{n=1}^{\infty} |\langle h, e_n \rangle|^2 \leqslant \|h\|^2 < \infty$. So if $\varepsilon > 0$, there is an N such that $\Sigma_{n=N}^{\infty} |\langle h, e_n \rangle|^2 < \varepsilon^2$. Let $F_0 = \{e_1, \ldots, e_{N-1}\}$ and let $\mathscr{F} = $ all the finite subsets of $\mathscr{E}$. For F in $\mathscr{F}$ define $h_F \equiv \Sigma\{\langle h, e \rangle e: e \in F\}$. If F and $G \in \mathscr{F}$ and both contain F_0, then

$$\|h_F - h_G\|^2 = \Sigma\{|\langle h, e \rangle|^2: e \in (F \backslash G) \cup (G \backslash F)\}$$

$$\leqslant \sum_{n=N}^{\infty} |\langle h, e_n \rangle|^2$$

$$< \varepsilon^2.$$

So $\{h_F: F \in \mathscr{F}\}$ is a Cauchy net in $\mathscr{H}$. Because $\mathscr{H}$ is complete, this net converges. In fact, it converges to $\Sigma_{n=1}^{\infty} \langle h, e_n \rangle e_n$. ∎

4.13. Theorem. *If $\mathscr{E}$ is an orthonormal set in $\mathscr{H}$, then the following statements are equivalent.*

(a) *$\mathscr{E}$ is a basis for $\mathscr{H}$.*
(b) *If $h \in \mathscr{H}$ and $h \perp \mathscr{E}$, then $h = 0$.*
(c) *$\bigvee \mathscr{E} = \mathscr{H}$.*
(d) *If $h \in \mathscr{H}$, then $h = \Sigma\{\langle h, e \rangle e: e \in \mathscr{E}\}$.*

(e) *If g and h∈ℋ, then*

$$\langle g,h\rangle = \sum\{\langle g,e\rangle\langle e,h\rangle: e\in\mathscr{E}\}.$$

(f) *If h∈ℋ, then* $\|h\|^2 = \sum\{|\langle h,e\rangle|^2: e\in\mathscr{E}\}$ (*Parseval's Identity*).

PROOF. (a)⇒(b): Suppose $h\perp\mathscr{E}$ and $h\neq 0$; then $\mathscr{E}\cup\{h/\|h\|\}$ is an orthonormal set that properly contains $\mathscr{E}$, contradicting maximality.

(b)⇔(c): By Corollary 2.11, $\vee\mathscr{E} = \mathscr{H}$ if and only if $\mathscr{E}^\perp = (0)$.

(b)⇒(d): If $h\in\mathscr{H}$, then $f = h - \sum\{\langle h,e\rangle e: e\in\mathscr{E}\}$ is a well-defined vector by Lemma 4.12. If $e_1\in\mathscr{E}$, then $\langle f,e_1\rangle = \langle h,e_1\rangle - \sum\{\langle h,e\rangle\langle e,e_1\rangle: e\in\mathscr{E}\} = \langle h,e_1\rangle - \langle h,e_1\rangle = 0$. That is, $f\in\mathscr{E}^\perp$. Hence $f = 0$. (Is everything legitimate in that string of equalities? We don't want any illegitimate equalities.)

(d)⇒(e): This is left as an exercise for the reader.

(e)⇒(f): Since $\|h\|^2 = \langle h,h\rangle$, this is immediate.

(f)⇒(a): If $\mathscr{E}$ is not a basis, then there is a unit vector e_0 ($\|e_0\| = 1$) in $\mathscr{H}$ such that $e_0\perp\mathscr{E}$. Hence, $0 = \sum\{|\langle e_0,e\rangle|^2: e\in\mathscr{E}\}$, contradicting (f). ∎

Just as in finite dimensional spaces, a basis in Hilbert space can be used to define a concept of dimension. For this purpose the next result is pivotal.

4.14. Proposition. *If ℋ is a Hilbert space, any two bases have the same cardinality.*

PROOF. Let $\mathscr{E}$ and $\mathscr{F}$ be two bases for $\mathscr{H}$ and put ε = the cardinality of $\mathscr{E}$, η = the cardinality of $\mathscr{F}$. If ε or η is finite, then $\varepsilon = \eta$ (Exercise 15). Suppose both ε and η are infinite. For e in $\mathscr{E}$, let $\mathscr{F}_e \equiv \{f\in\mathscr{F}: \langle e,f\rangle \neq 0\}$; so $\mathscr{F}_e$ is countable. By (4.13b), each f in $\mathscr{F}$ belongs to at least one set $\mathscr{F}_e, e$ in $\mathscr{E}$. That is, $\mathscr{F} = \cup\{\mathscr{F}_e: e\in\mathscr{E}\}$. Hence $\eta \leq \varepsilon\cdot\aleph_0 = \varepsilon$. Similarly, $\varepsilon \leq \eta$. ∎

4.15. Definition. The *dimension* of a Hilbert space is the cardinality of a basis and is denoted by dim $\mathscr{H}$.

If (X,d) is a metric space that is separable and $\{B_i = B(x_i;\varepsilon_i): i\in I\}$ is a collection of pairwise disjoint open balls in X, then I must be countable. Indeed, if D is a countable dense subset of X, $B_i\cap D \neq \square$ for each i in I. Thus there is a point x_i in $B_i\cap D$. So $\{x_i: i\in I\}$ is a subset of D having the cardinality of I; thus I must be countable.

4.16. Proposition. *If ℋ is an infinite dimensional Hilbert space, then ℋ is separable if and only if* dim $\mathscr{H} = \aleph_0$.

PROOF. Let $\mathscr{E}$ be a basis for $\mathscr{H}$. If $e_1,e_2\in\mathscr{E}$, then $\|e_1 - e_2\|^2 = \|e_1\|^2 + \|e_2\|^2 = 2$. Hence $\{B(e; 1/\sqrt{2}): e\in\mathscr{E}\}$ is a collection of pairwise disjoint open balls in $\mathscr{H}$. From the discussion preceding this proposition, the assumption that $\mathscr{H}$ is separable implies $\mathscr{E}$ is countable. The converse is an exercise. ∎

EXERCISES

1. Verify the statements in Example 4.3.

2. Verify the statements in Example 4.4.

3. Verify the statements in Example 4.5.

4. Find an infinite orthonormal set in the Hilbert space of Example 1.8.

5. Using the notation of the Gram–Schmidt Orthogonalization Process, show that up to scalar multiple $e_1 = h_1/\|h_1\|$ and for $n \geqslant 2$, $e_n = \|h_n - f_n\|^{-1}(h_n - f_n)$, where f_n is the vector defined formally by

$$f_n = \frac{-1}{\det[\langle h_i, h_j \rangle]_{i,j=1}^{n-1}} \det \begin{bmatrix} \langle h_1, h_1 \rangle & \cdots & \langle h_{n-1}, h_1 \rangle & \langle h_n, h_1 \rangle \\ \vdots & & \vdots & \vdots \\ \langle h_1, h_{n-1} \rangle & \cdots & \langle h_{n-1}, h_{n-1} \rangle & \langle h_n, h_{n-1} \rangle \\ h_1 & \cdots & h_{n-1} & 0 \end{bmatrix}$$

In the next three exercises, the reader is asked to apply the Gram–Schmidt Orthogonalization Process to a given sequence in a Hilbert space. A reference for this material is pp. 82–96 of Courant and Hilbert [1953].

6. If the sequence $1, x, x^2, \ldots$ is orthogonalized in $L^2(-1, 1)$, the sequence $e_n(x) = [\frac{1}{2}(2n+1)]^{1/2} P_n(x)$ is obtained, where

$$P_n(x) = \frac{1}{2^n n!} \left(\frac{d}{dx} \right)^n (x^2 - 1)^n.$$

The functions $P_n(x)$ are called *Legendre polynomials*.

7. If the sequence $e^{-x^2/2}, xe^{-x^2/2}, x^2 e^{-x^2/2}, \ldots$ is orthogonalized in $L^2(-\infty, \infty)$, the sequence $e_n(x) = [2^n n! \sqrt{\pi}]^{-1/2} H_n(x) e^{-x^2/2}$ is obtained, where

$$H_n(x) = (-1)^n e^{x^2} \left(\frac{d}{dx} \right)^n e^{-x^2}.$$

The functions H_n are *Hermite polynomials* and satisfy $H'_n(x) = 2n H_{n-1}(x)$.

8. If the sequence $e^{-x/2}, xe^{-x/2}, x^2 e^{-x/2}, \ldots$ is orthogonalized in $L^2(0, \infty)$, the sequence $e_n(x) = e^{-x/2} L_n(x)/n!$ is obtained, where

$$L_n(x) = e^x \left(\frac{d}{dx} \right)^n (x^n e^{-x}).$$

The functions L_n are called *Laguerre polynomials*.

9. Prove Corollary 4.10 using Definition 4.11.

10. If $\{h_n\}$ is a sequence in Hilbert space and $\sum \{h_n : n \in \mathbb{N}\}$ converges to h (Definition 4.11), then $\lim_n \sum_{k=1}^n h_k = h$. Show that the converse is false.

11. If $\{h_n\}$ is a sequence in a Hilbert space and $\sum_{n=1}^\infty \|h_n\| < \infty$, show that $\sum \{h_n : n \in \mathbb{N}\}$ converges in the sense of Definition 4.11.

12. Let $\{\alpha_n\}$ be a sequence in $\mathbb{F}$ and prove that the following statements are equivalent: (a) $\sum \{\alpha_n : n \in \mathbb{N}\}$ converges in the sense of Definition 4.11. (b) If π is any permutation of $\mathbb{N}$, then $\sum_{n=1}^\infty \alpha_{\pi(n)}$ converges (*unconditional convergence*). (c) $\sum_{n=1}^\infty |\alpha_n| < \infty$.

13. Let $\mathscr{E}$ be an orthonormal subset of $\mathscr{H}$ and let $\mathscr{M} = \vee \mathscr{E}$. If P is the orthogonal projection of $\mathscr{H}$ onto $\mathscr{M}$, show that $Ph = \sum\{\langle h, e \rangle e : e \in \mathscr{E}\}$ for every h in $\mathscr{H}$.

14. Let $\lambda =$ Area measure on $\{z \in \mathbb{C} : |z| < 1\}$ and show that $1, z, z^2, \ldots$ are orthogonal vectors in $L^2(\lambda)$. Find $\|z^n\|$, $n \geq 0$. If $e_n = \|z^n\|^{-1} z^n$, $n \geq 0$, is $\{e_0, e_1, \ldots\}$ a basis for $L^2(\lambda)$?

15. In the proof of (4.14), show that if either ε or η is finite, then $\varepsilon = \eta$.

16. If $\mathscr{H}$ is an infinite dimensional Hilbert space, show that no orthonormal basis for $\mathscr{H}$ is a Hamel basis. Show that a Hamel basis is uncountable.

17. Let $d \geq 1$ and let μ be a regular Borel measure on $\mathbb{R}^d$. Show that $L^2(\mu)$ is separable.

18. Suppose $L^2(X, \Omega, \mu)$ is separable and $\{E_i : i \in I\}$ is a collection of pairwise disjoint subsets of X, $E_i \in \Omega$, and $0 < \mu(E_i) < \infty$ for all i.. Show that I is countable. Can you allow $\mu(E_i) = \infty$?

19. If $\{h \in \mathscr{H} : \|h\| \leq 1\}$ is compact, show that $\dim \mathscr{H} < \infty$.

20. What is the cardinality of a Hamel basis for l^2?

§5. Isomorphic Hilbert Spaces and the Fourier Transform for the Circle

Every mathematical theory has its concept of isomorphism. In topology there is homeomorphism and homotopy equivalence; algebra calls them isomorphisms. The basic idea is to define a map which preserves the basic structure of the spaces in the category.

5.1. Definition. If $\mathscr{H}$ and $\mathscr{K}$ are Hilbert spaces, an *isomorphism* between $\mathscr{H}$ and $\mathscr{K}$ is a linear surjection $U: \mathscr{H} \to \mathscr{K}$ such that

$$\langle Uh, Ug \rangle = \langle h, g \rangle$$

for all h, g in $\mathscr{H}$. In this case $\mathscr{H}$ and $\mathscr{K}$ are said to be *isomorphic*.

It is easy to see that if $U: \mathscr{H} \to \mathscr{K}$ is an isomorphism, then so is U^{-1}: $\mathscr{K} \to \mathscr{H}$. Similar such arguments show that the concept of "isomorphic" is an equivalence relation on Hilbert spaces. It is also certain that this is the correct equivalence relation since an inner product is the essential ingredient for a Hilbert space and isomorphic Hilbert spaces have the "same" inner product. One might object that completeness is another essential ingredient in the definition of a Hilbert space. So it is! However, this too is preserved by an isomorphism. An *isometry* between metric spaces is a map that preserves distance.

5.2. Proposition. *If $V: \mathscr{H} \to \mathscr{K}$ is a linear map between Hilbert spaces, then V is an isometry if and only if $\langle Vh, Vg \rangle = \langle h, g \rangle$ for all h, g in $\mathscr{H}$.*

PROOF. Assume $\langle Vh, Vg \rangle = \langle h, g \rangle$ for all h, g in $\mathcal{H}$. Then $\|Vh\|^2 = \langle Vh, Vh \rangle = \langle h, h \rangle = \|h\|^2$ and V is an isometry.

Now assume that V is an isometry. If $h, g \in \mathcal{H}$ and $\lambda \in \mathbb{F}$, then $\|h + \lambda g\|^2 = \|Vh + \lambda Vg\|^2$. Using the polar identity on both sides of this equation gives

$$\|h\|^2 + 2\operatorname{Re}\bar{\lambda}\langle h, g \rangle + |\lambda|^2 \|g\|^2 = \|Vh\|^2 + 2\operatorname{Re}\bar{\lambda}\langle Vh, Vg \rangle + |\lambda|^2 \|Vg\|^2.$$

But $\|Vh\| = \|h\|$ and $\|Vg\| = \|g\|$, so this equation becomes

$$\operatorname{Re}\bar{\lambda}\langle h, g \rangle = \operatorname{Re}\bar{\lambda}\langle Vh, Vg \rangle$$

for any λ in $\mathbb{F}$. If $\mathbb{F} = \mathbb{R}$, take $\lambda = 1$. If $\mathbb{F} = \mathbb{C}$, first take $\lambda = 1$ and then take $\lambda = i$ to find that $\langle h, g \rangle$ and $\langle Vh, Vg \rangle$ have the same real and imaginary parts. ∎

Note that an isometry between metric spaces maps Cauchy sequences into Cauchy sequences. Thus an isomorphism also preserves completeness. That is, if an inner product space is isomorphic to a Hilbert space, then it must be complete.

5.3. Example. Definite $S: l^2 \to l^2$ by $S(\alpha_1, \alpha_2, \ldots) = (0, \alpha_1, \alpha_2, \ldots)$. Then S is an isometry that is not surjective.

The preceding example shows that isometries need not be isomorphisms.

A word about terminology. Many call what we call an isomorphism a unitary operator. We shall define a *unitary operator* as a linear transformation $U: \mathcal{H} \to \mathcal{H}$ that is a surjective isometry. That is, a unitary operator is an isomorphism whose range coincides with its domain. This may seem to be a minor distinction, and in many ways it is. But experience has taught me that there is some benefit in making such a distinction, or at least in being aware of it.

5.4. Theorem. *Two Hilbert spaces are isomorphic if and only if they have the same dimension.*

PROOF. If $U: \mathcal{H} \to \mathcal{K}$ is an isomorphism and $\mathcal{E}$ is a basis for $\mathcal{H}$, then it is easy to see that $U\mathcal{E} \equiv \{Ue: e \in \mathcal{E}\}$ is a basis for $\mathcal{K}$. Hence, $\dim \mathcal{H} = \dim \mathcal{K}$.

Let $\mathcal{H}$ be a Hilbert space and let $\mathcal{E}$ be a basis for $\mathcal{H}$. Consider the Hilbert space $l^2(\mathcal{E})$. If $h \in \mathcal{H}$, define $\hat{h}: \mathcal{E} \to \mathbb{F}$ by $\hat{h}(e) = \langle h, e \rangle$. By Parseval's Identity $\hat{h} \in l^2(\mathcal{E})$ and $\|h\| = \|\hat{h}\|$. Define $U: \mathcal{H} \to l^2(\mathcal{E})$ by $Uh = \hat{h}$. Thus U is linear and an isometry. It is easy to see that $\operatorname{ran} U$ contains all the functions f in $l^2(\mathcal{E})$ such that $f(e) = 0$ for all but a finite number of e; that is, $\operatorname{ran} U$ is dense. But U, being an isometry, must have closed range. Hence $U: \mathcal{H} \to l^2(\mathcal{E})$ is an isomorphism.

If $\mathcal{H}$ is a Hilbert space with a basis $\mathcal{F}$, $\mathcal{H}$ is isomorphic to $l^2(\mathcal{F})$. If $\dim \mathcal{H} = \dim \mathcal{K}$, $\mathcal{E}$ and $\mathcal{F}$ have the same cardinality; it is easy to see that $l^2(\mathcal{E})$ and $l^2(\mathcal{F})$ must be isomorphic. Therefore $\mathcal{H}$ and $\mathcal{K}$ are isomorphic. ∎

5.5. Corollary. *All separable infinite dimensional Hilbert spaces are isomorphic.*

This section concludes with a rather important example of an isomorphism, the Fourier transform on the circle.

The proof of the next result can be found as an Exercise on p. 263 of Conway [1978]. Another proof will be given latter in this book after the Stone–Weierstrass Theorem is proved. So the reader can choose to assume this for the moment. Let $\mathbb{D} = \{z \in \mathbb{C}: |z| < 1\}$.

5.6. Theorem. *If $f: \partial \mathbb{D} \to \mathbb{C}$ is a continuous function, then there is a sequence $\{p_n(z, \bar{z})\}$ of polynomials in z and $\bar{z}$ such that $p_n(z, \bar{z}) \to f(z)$ uniformly on $\partial \mathbb{D}$.*

Note that if $z \in \partial \mathbb{D}$, $\bar{z} = z^{-1}$. Thus a polynomial in z and $\bar{z}$ on $\partial \mathbb{D}$ becomes a function of the form

$$\sum_{k=-m}^{n} \alpha_k z^k.$$

If we put $z = e^{i\theta}$, this becomes a function of the form

$$\sum_{k=-m}^{n} \alpha_k e^{ik\theta}.$$

Such functions are called *trigonometric polynomials*.

We can now show that the orthonormal set in Example 4.3 is a basis for $L^2_{\mathbb{C}}[0, 2\pi]$. This is a rather important result.

5.7. Theorem. *If for each n in $\mathbb{Z}$, $e_n(t) \equiv (2\pi)^{-1/2} \exp(int)$, then $\{e_n: n \in \mathbb{Z}\}$ is a basis for $L^2_{\mathbb{C}}[0, 2\pi]$.*

PROOF. Let $\mathcal{T} = \{\sum_{k=-n}^{n} \alpha_k e_k: \alpha_k \in \mathbb{C}, n \geq 0\}$. Then $\mathcal{T}$ is a subalgebra of $C_{\mathbb{C}}[0, 2\pi]$, the algebra of all continuous $\mathbb{C}$-valued functions on $[0, 2\pi]$. Note that if $f \in \mathcal{T}$, $f(0) = f(2\pi)$. We want to show that the uniform closure of $\mathcal{T}$ is $\mathcal{C} \equiv \{f \in C_{\mathbb{C}}[0, 2\pi]: f(0) = f(2\pi)\}$. To do this, let $f \in \mathcal{C}$ and define $F: \partial \mathbb{D} \to \mathbb{C}$ by $F(e^{it}) = f(t)$. F is continuous. (Why?) By (5.6) there is a sequence of polynomials in z and $\bar{z}$, $\{p_n(z, \bar{z})\}$, such that $p_n(z, \bar{z}) \to F(z)$ uniformly on $\partial \mathbb{D}$. Thus $p_n(e^{it}, e^{-it}) \to f(t)$ uniformly on $[0, 2\pi]$. But $p_n(e^{it}, e^{-it}) \in \mathcal{T}$.

Now the closure of $\mathcal{C}$ in $L^2_{\mathbb{C}}[0, 2\pi]$ is all of $L^2_{\mathbb{C}}[0, 2\pi]$ (Exercise 6). Hence $\bigvee \{e_n: n \in \mathbb{Z}\} = L^2_{\mathbb{C}}[0, 2\pi]$ and $\{e_n\}$ is thus a basis (4.13). ∎

Actually, it is usually preferred to normalize the measure on $[0, 2\pi]$. That is, replace dt by $(2\pi)^{-1} dt$, so that the total measure of $[0, 2\pi]$ is 1. Now define

$e_n(t) = \exp(int)$. Hence $\{e_n : n \in \mathbb{Z}\}$ is a basis for $\mathscr{H} \equiv L^2_{\mathbb{C}}([0, 2\pi], (2\pi)^{-1} dt)$. If $f \in \mathscr{H}$, then

5.8
$$\hat{f}(n) \equiv \langle f, e_n \rangle = \frac{1}{2\pi} \int_0^{2\pi} f(t)e^{-int} dt$$

is called the nth *Fourier coefficient* of f, n in $\mathbb{Z}$. By (5.7) and (4.13d),

5.9
$$f = \sum_{n=-\infty}^{\infty} \hat{f}(n)e_n,$$

where this infinite series converges to f in the metric defined by the norm of $\mathscr{H}$. This is called the *Fourier series* of f. This terminology is classical and has been adopted for a general Hilbert space.

If $\mathscr{H}$ is any Hilbert space and $\mathscr{E}$ is a basis, the scalars $\{\langle h, e \rangle; e \in \mathscr{E}\}$ are called the *Fourier coefficients* of h (relative to $\mathscr{E}$) and the series in (4.13d) is called the *Fourier expansion* of h (relative to $\mathscr{E}$).

Note that Parseval's Identity applied to (5.9) gives that $\sum_{n=-\infty}^{\infty} |\hat{f}(n)|^2 < \infty$. This proves a classical result.

5.10. The Riemann–Lebesgue Lemma. *If $f \in L^2_{\mathbb{C}}[0, 2\pi]$, then $\int_0^{2\pi} f(t)e^{-int} dt \to 0$ as $n \to \pm\infty$.*

If $f \in L^2_{\mathbb{C}}[0, 2\pi]$, then the Fourier series of f converges to f in L^2-norm. It was conjectured by Lusin that the series converges to f almost everywhere. This was proved in Carleson [1966]. Hunt [1967] showed that if $f \in L^p_{\mathbb{C}}[0, 2\pi]$, $1 < p \leqslant \infty$, then the Fourier series also converges to f a.e. Long before that, Kolmogoroff had furnished an example of a function f in $L^1_{\mathbb{C}}[0, 2\pi]$ whose Fourier series a.e. does not converge to f.

For f in $L^2_{\mathbb{C}}[0, 2\pi]$, the function $\hat{f} : \mathbb{Z} \to \mathbb{C}$ is called the *Fourier transform* of f; the map $U : L^2_{\mathbb{C}}[0, 2\pi] \to l^2(\mathbb{Z})$ defined by $Uf = \hat{f}$ is the *Fourier transform*. The results obtained so far can be applied to this situation to yield the following.

5.11. Theorem. *The Fourier transform is a linear isometry from $L^2_{\mathbb{C}}[0, 2\pi]$ onto $l^2(\mathbb{Z})$.*

PROOF. Let $U : L^2_{\mathbb{C}}[0, 2\pi] \to l^2(\mathbb{Z})$ be the Fourier transform. That U maps $L^2 \equiv L^2_{\mathbb{C}}[0, 2\pi]$ into $l^2(\mathbb{Z})$ and satisfies $\|Uf\| = \|f\|$ is a consequence of Parseval's Identity. That U is linear is an exercise. If $\{\alpha_n\} \in l^2(\mathbb{Z})$ and $\alpha_n = 0$ for all but a finite number of n, then $f = \sum_{n=-\infty}^{\infty} \alpha_n e_n \in L^2$. It is easy to check that $\hat{f}(n) = \alpha_n$ for all n, so $Uf = \{\alpha_n\}$. Thus ran U is dense in $l^2(\mathbb{Z})$. But U is an isometry, so ran U is closed; hence U is surjective. ∎

Note that functions in $L^2_{\mathbb{C}}[0, 2\pi]$ can be defined on $\partial \mathbb{D}$ by letting $f(e^{i\theta}) = f(\theta)$. The ambiguity for $\theta = 0$ and 2π (or $e^{i\theta} = 1$) might cause us to pause, but remember that elements of $L^2_{\mathbb{C}}[0, 2\pi]$ are equivalence classes of functions—not really functions. Since $\{0, 2\pi\}$ has zero measure, there is really no ambiguity. In this way $L^2_{\mathbb{C}}[0, 2\pi]$ can be identified with $L^2_{\mathbb{C}}(\partial \mathbb{D})$, where the measure on $\partial \mathbb{D}$ is normalized arc-length measure (normalized so that the total measure of $\partial \mathbb{D}$ is 1). So $L^2_{\mathbb{C}}[0, 2\pi]$ and $L^2_{\mathbb{C}}(\partial \mathbb{D})$ are (naturally) isomorphic. Thus, Theorem 5.11 is a theorem about the Fourier transform of the circle.

The importance of Theorem 5.11 is not the fact that $L^2_{\mathbb{C}}[0, 2\pi]$ and $l^2(\mathbb{Z})$ are isomorphic, but that the Fourier transform is an isomorphism. The fact that these two spaces are isomorphic follows from the abstract result that all separable infinite dimensional Hilbert spaces are isomorphic (5.5).

EXERCISES

1. Verify the statements in Example 5.3.

2. Define $V: L^2(0, \infty) \rightarrow L^2(0, \infty)$ by $(Vf)(t) = f(t-1)$ if $t > 1$ and $(Vf)(t) = 0$ if $t \leqslant 1$. Show that V is an isometry that is not surjective.

3. Define $V: L^2(\mathbb{R}) \rightarrow L^2(\mathbb{R})$ by $(Vf)(t) = f(t-1)$ and show that V is an isomorphism (a unitary operator).

4. Let $\mathscr{H}$ be the Hilbert space of Example 1.8 and define $U: \mathscr{H} \rightarrow L^2(0, 1)$ by $Uf = f'$. Show that U is an isomorphism and find a formula for U^{-1}.

5. Let (X, Ω, μ) be a σ-finite measure space and let $u: X \rightarrow \mathbb{F}$ be an Ω-measurable function such that $\sup\{|u(x)|: x \in X\} < \infty$. Show that $U: L^2(X, \Omega, \mu) \rightarrow L^2(X, \Omega, \mu)$ defined by $Uf = uf$ is an isometry if and only if $|u(x)| = 1$ a.e. $[\mu]$, in which case U is surjective.

6. Let $\mathscr{C} = \{f \in C[0, 2\pi]: f(0) = f(2\pi)\}$ and show that $\mathscr{C}$ is dense in $L^2[0, 2\pi]$.

7. Show that $\{(1/\sqrt{2\pi}), (1/\sqrt{\pi})\cos nt, (1/\sqrt{\pi})\sin nt: 1 \leqslant n < \infty\}$ is a basis for $L^2[-\pi, \pi]$.

8. Let (X, Ω) be a measurable space and let μ, ν be two σ-finite measures defined on (X, Ω). Suppose $\nu \ll \mu$ and ϕ is the Radon–Nikodym derivative of ν with respect to μ ($\phi = d\nu/d\mu$). Define $V: L^2(\nu) \rightarrow L^2(\mu)$ by $Vf = \sqrt{\phi}f$. Show that V is a well-defined linear isometry and V is an isomorphism if and only if $\mu \ll \nu$ (that is, μ and ν are mutually absolutely continuous).

9. If $\mathscr{H}$ and $\mathscr{K}$ are Hilbert spaces and $U: \mathscr{H} \rightarrow \mathscr{K}$ is a surjective function such that $\langle Uf, Ug \rangle = \langle f, g \rangle$ for all vectors f and g in $\mathscr{H}$, then U is linear.

§6. The Direct Sum of Hilbert Spaces

Suppose $\mathscr{H}$ and $\mathscr{K}$ are Hilbert spaces. We want to define $\mathscr{H} \oplus \mathscr{K}$ so that it becomes a Hilbert space. This is not a difficult assignment. For any vector spaces $\mathscr{X}$ and $\mathscr{Y}$, $\mathscr{X} \oplus \mathscr{Y}$ is defined as the Cartesian product $\mathscr{X} \times \mathscr{Y}$ where the operations are defined on $\mathscr{X} \times \mathscr{Y}$ coordinatewise. That is, if elements of

$\mathscr{X} \oplus \mathscr{Y}$ are defined as $\{x \oplus y \colon x \in \mathscr{X}, \; y \in \mathscr{Y}\}$, then $(x_1 \oplus y_1) + (x_2 \oplus y_2) \equiv (x_1 + x_2) \oplus (y_1 + y_2)$, and so on.

6.1. Definition. If $\mathscr{H}$ and $\mathscr{K}$ are Hilbert spaces, $\mathscr{H} \oplus \mathscr{K} = \{h \oplus k \colon h \in \mathscr{H}, \; k \in \mathscr{K}\}$ and

$$\langle h_1 \oplus k_1, h_2 \oplus k_2 \rangle \equiv \langle h_1, h_2 \rangle + \langle k_1, k_2 \rangle.$$

It must be shown that this defines an inner product on $\mathscr{H} \oplus \mathscr{K}$ and that $\mathscr{H} \oplus \mathscr{K}$ is complete (Exercise).

Now what happens if we want to define $\mathscr{H}_1 \oplus \mathscr{H}_2 \oplus \cdots$ for a sequence of Hilbert spaces $\mathscr{H}_1, \mathscr{H}_2, \ldots$? There is a problem about the completeness of this infinite direct sum, but this can be overcome as follows.

6.2. Proposition. *If $\mathscr{H}_1, \mathscr{H}_2, \ldots$ are Hilbert spaces, let $\mathscr{H} = \{(h_n)_{n=1}^{\infty} \colon h_n \in \mathscr{H}_n$ for all n and $\sum_{n=1}^{\infty} \|h_n\|^2 < \infty\}$. For $h = (h_n)$ and $g = (g_n)$ in $\mathscr{H}$, define*

$$6.3 \qquad\qquad \langle h, g \rangle = \sum_{n=1}^{\infty} \langle h_n, g_n \rangle.$$

Then $\langle \cdot, \cdot \rangle$ is an inner product on $\mathscr{H}$ and the norm relative to this inner product is $\|h\| = [\sum_{n=1}^{\infty} \|h_n\|^2]^{1/2}$. With this inner product $\mathscr{H}$ is a Hilbert space.

PROOF. If $h = (h_n)$ and $g = (g_n) \in \mathscr{H}$, then the CBS inequality implies $\sum |\langle h_n, g_n \rangle| \le \sum \|h_n\| \|g_n\| \le (\sum \|h_n\|^2)^{1/2} (\sum \|g_n\|^2)^{1/2} < \infty$. Hence the series in (6.3) converges absolutely. The remainder of the proof is left to the reader. ∎

6.4. Definition. If $\mathscr{H}_1, \mathscr{H}_2, \ldots$ are Hilbert spaces, the space $\mathscr{H}$ of Proposition 6.2 is called the *direct sum* of $\mathscr{H}_1, \mathscr{H}_2, \ldots$ and is denoted by $\mathscr{H} \equiv \mathscr{H}_1 \oplus \mathscr{H}_2 \oplus \cdots$.

This is part of a more general process. If $\{\mathscr{H}_i \colon i \in I\}$ is a collection of Hilbert spaces, $\mathscr{H} \equiv \oplus \{\mathscr{H}_i \colon i \in I\}$ is defined as the collection of functions $h \colon I \to \cup \{\mathscr{H}_i \colon i \in I\}$ such that $h(i) \in \mathscr{H}_i$ for all i and $\sum \{\|h(i)\|^2 \colon i \in I\} < \infty$. If $h, g \in \mathscr{H}$, $\langle h, g \rangle \equiv \sum \{\langle h(i), g(i) \rangle \colon i \in I\}$; $\mathscr{H}$ is a Hilbert space.

The main reason for considering direct sums is that they provide a way of manufacturing operators on Hilbert space. In fact, Hilbert space is a rather dull subject, except for the fact that there are numerous interesting questions about the linear operators on them that are as yet unresolved. This subject is introduced in the next chapter.

EXERCISES

1. Let $\{(X_i, \Omega_i, \mu_i) \colon i \in I\}$ be a collection of measure spaces and define X, Ω, and μ as follows. Let $X =$ the disjoint union of $\{X_i \colon i \in I\}$ and let $\Omega = \{\Delta \subseteq X \colon \Delta \cap X_i \in \Omega_i$ for all $i\}$. For Δ in Ω put $\mu(\Delta) = \sum_i \mu_i(\Delta \cap X_i)$. Show that (X, Ω, μ) is a measure space and $L^2(X, \Omega, \mu)$ is isomorphic to $\oplus \{L^2(X_i, \Omega_i, \mu_i) \colon i \in I\}$.

2. Let (X, Ω) be a measurable space, let μ_1, μ_2 be σ-finite measures defined on (X, Ω), and put $\mu = \mu_1 + \mu_2$. Show that the map $V: L^2(X, \Omega, \mu) \rightarrow L^2(X, \Omega, \mu_1) \oplus L^2(X, \Omega, \mu_2)$ defined by $Vf = f_1 \oplus f_2$, where f_j is the equivalence class of $L^2(X, \Omega, \mu_j)$ corresponding to f, is well defined, linear, and injective. Show that V is an isomorphism iff μ_1 and μ_2 are mutually singular.

CHAPTER II

Operators on Hilbert Space

A large area of current research interest is centered around the theory of operators on Hilbert space. Several other chapters in this book will be devoted to this topic.

There is a marked contrast here between Hilbert spaces and the Banach spaces that are studied in the next chapter. Essentially all of the information about the geometry of Hilbert space is contained in the preceding chapter. The geometry of Banach space lies in darkness and has attracted the attention of many talented research mathematicians. However, the theory of linear operators (linear transformations) on a Banach space has very few general results, whereas Hilbert space operators have an elegant and well-developed general theory. Indeed, the reason for this dichotomy is related to the opposite status of the geometric considerations. Questions concerning operators on Hilbert space don't necessitate or imply any geometric difficulties.

In addition to the fundamentals of operators, this chapter will also present an interesting application to differential equations in Section 6.

§1. Elementary Properties and Examples

The proof of the next proposition is similar to that of Proposition I.3.1 and is left to the reader.

1.1. Proposition. *Let $\mathscr{H}$ and $\mathscr{K}$ be Hilbert spaces and $A\colon \mathscr{H} \to \mathscr{K}$ a linear transformation. The following statements are equivalent.*

(a) *A is continuous.*

(b) *A is continuous at 0.*

(c) *A is continuous at some point.*

(d) *There is a constant $c > 0$ such that $\|Ah\| \leqslant c\|h\|$ for all h in $\mathscr{H}$.*

As in (I.3.3), if

$$\|A\| = \sup\{\|Ah\|: h \in \mathscr{H}, \|h\| \leqslant 1\},$$

then

$$\|A\| = \sup\{\|Ah\|: \|h\| = 1\}$$
$$= \sup\{\|Ah\|/\|h\|: h \neq 0\}$$
$$= \inf\{c > 0: \|Ah\| \leqslant c\|h\|, h \text{ in } \mathscr{H}\}.$$

Also, $\|Ah\| \leqslant \|A\|\|h\|$. $\|A\|$ is called the *norm* of A and a linear transformation with finite norm is called *bounded*. Let $\mathscr{B}(\mathscr{H}, \mathscr{K})$ be the set of bounded linear transformations from $\mathscr{H}$ into $\mathscr{K}$. For $\mathscr{H} = \mathscr{K}$, $\mathscr{B}(\mathscr{H}, \mathscr{H}) \equiv \mathscr{B}(\mathscr{H})$. Note that $\mathscr{B}(\mathscr{H}, \mathbb{F}) =$ all the bounded linear functionals on $\mathscr{H}$.

1.2. Proposition. (a) *If A and $B \in \mathscr{B}(\mathscr{H}, \mathscr{K})$, then $A + B \in \mathscr{B}(\mathscr{H}, \mathscr{K})$, and $\|A + B\| \leqslant \|A\| + \|B\|$.*
(b) *If $\alpha \in \mathbb{F}$ and $A \in \mathscr{B}(\mathscr{H}, \mathscr{K})$, then $\alpha A \in \mathscr{B}(\mathscr{H}, \mathscr{K})$ and $\|\alpha A\| = |\alpha|\|A\|$.*
(c) *If $A \in \mathscr{B}(\mathscr{H}, \mathscr{K})$ and $B \in \mathscr{B}(\mathscr{K}, \mathscr{L})$, then $BA \in \mathscr{B}(\mathscr{H}, \mathscr{L})$ and $\|BA\| \leqslant \|B\|\|A\|$.*

PROOF. Only (c) will be proved; the rest of the proof is left to the reader. If $k \in \mathscr{K}$, then $\|Bk\| \leqslant \|B\|\|k\|$. Hence, if $h \in \mathscr{H}$, $k = Ah \in \mathscr{K}$ and so $\|BAh\| \leqslant \|B\|\|Ah\| \leqslant \|B\|\|A\|\|h\|$. ∎

By virtue of preceding proposition, $d(A, B) = \|A - B\|$ defines a metric on $\mathscr{B}(\mathscr{H}, \mathscr{K})$. So it makes sense to consider $\mathscr{B}(\mathscr{H}, \mathscr{K})$ as a metric space. This will not be examined closely until later in the book, but later in this chapter the idea of the convergence of a sequence of operators will be used.

1.3. Example. If $\dim \mathscr{H} = n < \infty$ and $\dim \mathscr{K} = m < \infty$, let $\{e_1, \ldots, e_n\}$ be an orthonormal basis for $\mathscr{H}$ and let $\{\varepsilon_1, \ldots, \varepsilon_m\}$ be an orthonormal basis for $\mathscr{K}$. It can be shown that every linear transformation from $\mathscr{H}$ into $\mathscr{K}$ is bounded (Exercise 3). If $1 \leqslant j \leqslant n$, $1 \leqslant i \leqslant m$, let $a_{ij} = \langle Ae_j, \varepsilon_i \rangle$. Then the $m \times n$ matrix (α_{ij}) represents A and every such matrix represents an element of $\mathscr{B}(\mathscr{H}, \mathscr{K})$.

1.4. Example. Let $l^2 \equiv l^2(\mathbb{N})$ and let $e_1, e_2, \ldots$ be its usual basis. If $A \in \mathscr{B}(l^2)$, form $\alpha_{ij} = \langle Ae_j, e_i \rangle$. The infinite matrix (α_{ij}) represents A as finite matrices represent operators on finite dimensional spaces. However, this representation has limited value unless the matrix has a special form. One difficulty is that it is unknown how to find the norm of A in terms of the entries in the matrix. In fact, if $4 < n < \infty$, there is no known formula for the norm of an $n \times n$ matrix in terms of its entries. (This restriction on the values of n is due to our inability to solve polynomial equations of degree greater than 4.) See

Exercise 11. A sufficient condition for the boundedness of an infinite matrix that is useful is known. (See Exercise 9.) Another sufficient condition for a matrix to be bounded can be found on page 61 of Maddox [1980].

1.5. Theorem. *Let* (X, Ω, μ) *be a σ-finite measure space and put* $\mathscr{H} = L^2(X, \Omega, \mu) \equiv L^2(\mu)$. *If* $\phi \in L^\infty(\mu)$, *define* $M_\phi: L^2(\mu) \to L^2(\mu)$ *by* $M_\phi f = \phi f$. *Then* $M_\phi \in \mathscr{B}(L^2(\mu))$ *and* $\| M_\phi \| = \| \phi \|_\infty$.

PROOF. Here $\| \phi \|_\infty$ is the *μ-essential supremum norm.* That is,

$$\| \phi \|_\infty \equiv \inf \{ \sup \{ | \phi(x) |: x \notin N \}: N \in \Omega, \mu(N) = 0 \}$$
$$= \inf \{ c > 0: \mu(\{ x \in X: | \phi(x) | > c \}) = 0 \}.$$

Thus $\| \phi \|_\infty$ is the infimum of all $c > 0$ such that $| \phi(x) | \leqslant c$ a.e. $[\mu]$ and, moreover, $| \phi(x) | \leqslant \| \phi \|_\infty$ a.e. $[\mu]$. Thus we can, and do, assume that ϕ is a bounded measurable function and $| \phi(x) | \leqslant \| \phi \|_\infty$ for all x. So if $f \in L^2(\mu)$, then $\int | \phi f |^2 d\mu \leqslant \| \phi \|_\infty^2 \int | f |^2 d\mu$. That is, $M_\phi \in \mathscr{B}(L^2(\mu))$ and $\| M_\phi \| \leqslant \| \phi \|_\infty$. If $\varepsilon > 0$, the σ-finiteness of the measure space implies that there is a set Δ in $\Omega, 0 < \mu(\Delta) < \infty$, such that $| \phi(x) | \geqslant \| \phi \|_\infty - \varepsilon$ on Δ. (Why?) If $f = (\mu(\Delta))^{-1/2} \chi_\Delta$, then $f \in L^2(\mu)$ and $\| f \|_2 = 1$. So $\| M_\phi \|^2 \geqslant \| \phi f \|_2^2 = (\mu(\Delta))^{-1} \int_\Delta | \phi |^2 d\mu \geqslant (\| \phi \|_\infty - \varepsilon)^2$. Letting $\varepsilon \to 0$, we get that $\| M_\phi \| \geqslant \| \phi \|_\infty$. ∎

The operator M_ϕ is called a *multiplication operator.* The function ϕ is its *symbol.*

If the measure space (X, Ω, μ) is not σ-finite, then the conclusion of Theorem 1.5 is not necessarily valid. Indeed, let $\Omega =$ the Borel subsets of $[0, 1]$ and define μ on Ω by $\mu(\Delta) =$ the Lebesgue measure of Δ if $0 \notin \Delta$ and $\mu(\Delta) = \infty$ if $0 \in \Delta$. This measure has an infinite atom at 0 and, therefore, is not σ-finite. Let $\phi = \chi_{\{0\}}$. Then $\phi \in L^\infty(\mu)$ and $\| \phi \|_\infty = 1$. If $f \in L^2(\mu)$, then $\infty > \int | f |^2 d\mu \geqslant | f(0) |^2 \mu(\{0\})$. Hence every function in $L^2(\mu)$ vanishes at 0. Therefore $M_\phi = 0$ and $\| M_\phi \| < \| \phi \|_\infty$.

There are more general measure spaces for which (1.5) is valid—the decomposable measure spaces (see Kelley [1966]).

1.6. Theorem. *Let* (X, Ω, μ) *be a measure space and suppose* $k: X \times X \to \mathbb{F}$ *is an $\Omega \times \Omega$-measurable function for which there are constants c_1 and c_2 such that*

$$\int_X | k(x, y) | d\mu(y) \leqslant c_1 \quad \text{a.e.} [\mu],$$

$$\int_X | k(x, y) | d\mu(x) \leqslant c_2 \quad \text{a.e.} [\mu].$$

If $K: L^2(\mu) \to L^2(\mu)$ *is defined by*

$$(Kf)(x) = \int k(x, y) f(y) d\mu(y),$$

then K is a bounded linear operator and $\| K \| \leqslant (c_1 c_2)^{1/2}$.

PROOF. Actually it must be shown that $Kf \in L^2(\mu)$, but this will follow from the argument that demonstrates the boundedness of K. If $f \in L^2(\mu)$,

$$|Kf(x)| \leq \int |k(x, y)| \, |f(y)| \, d\mu(y)$$

$$= \int |k(x, y)|^{1/2} |k(x, y)|^{1/2} |f(y)| \, d\mu(y)$$

$$\leq \left[\int |k(x, y)| \, d\mu(y) \right]^{1/2} \left[\int |k(x, y)| \, |f(y)|^2 \, d\mu(y) \right]^{1/2}$$

$$\leq c_1^{1/2} \left[\int |k(x, y)| \, |f(y)|^2 \, d\mu(y) \right]^{1/2}.$$

Hence

$$\int |Kf(x)|^2 \, d\mu(x) \leq c_1 \int \int |k(x, y)| \, |f(y)|^2 \, d\mu(y) d\mu(x)$$

$$= c_1 \int |f(y)|^2 \int |k(x, y)| \, d\mu(x) d\mu(y)$$

$$\leq c_1 c_2 \|f\|^2.$$

Now this shows that the formula used to define Kf is finite a.e. $[\mu]$, $Kf \in L^2(\mu)$, and $\|Kf\|^2 \leq c_1 c_2 \|f\|^2$. ∎

The operator described above is called an *integral operator* and the function k is called its *kernel*. There are conditions on the kernel other than the one in (1.6) that will imply that K is bounded.

A particular example of an integral operator is the *Volterra operator* defined below.

1.7. Example. Let k: $[0, 1] \times [0, 1] \to \mathbb{R}$ be the characteristic function of $\{(x, y): y < x\}$. The corresponding operator V: $L^2(0, 1) \to L^2(0, 1)$ defined by $Vf(x) = \int_0^1 k(x, y) f(y) dy$ is called the *Volterra operator*. Note that

$$Vf(x) = \int_0^x f(y) dy.$$

Another example of an operator was defined in Example 1.5.3. The nonsurjective isometry defined there is called the *unilateral shift*. It will be studied in more detail later in this book. Note that any isometry is a bounded operator with norm 1.

EXERCISES

1. Prove Proposition 1.1.

2. Prove Proposition 1.2.

3. Suppose $\{e_1, e_2, \ldots,\}$ is an orthonormal basis for $\mathscr{H}$ and for each n there is a vector Ae_n in $\mathscr{H}$ such that $\sum \|Ae_n\| < \infty$. Show that A has an unique extension to a bounded operator on $\mathscr{H}$.

4. Proposition 1.2 says that $d(A, B) = \|A - B\|$ is a metric on $\mathscr{B}(\mathscr{H}, \mathscr{K})$. Show that $\mathscr{B}(\mathscr{H}, \mathscr{K})$ is complete relative to this metric.

5. Show that a multiplication operator M_ϕ (1.5) satisfies $M_\phi^2 = M_\phi$ if and only if ϕ is a characteristic function.

6. Let (X, Ω, μ) be a σ-finite measure space and let k_1, k_2 be two kernels satisfying the hypothesis of (1.6). Define

$$k: X \times X \to \mathbb{F} \quad \text{by} \quad k(x, y) = \int k_1(x, z) k_2(z, y) \, d\mu(z).$$

(a) Show that k also satisfies the hypothesis of (1.6). (b) If K, K_1, K_2 are the integral operators with kernels k, k_1, k_2, show that $K = K_1 K_2$. What does this remind you of? Is more going on than an analogy?

7. If (X, Ω, μ) is a measure space and $k \in L^2(\mu \times \mu)$, show that k defines a bounded integral operator.

8. Let $\{e_n\}$ be the usual basis for l^2 and let $\{\alpha_n\}$ be a sequence of scalars. Show that there is a bounded operator A on l^2 such that $Ae_n = \alpha_n e_n$ for all n if and only if $\{\alpha_n\}$ is uniformly bounded, in which case $\|A\| = \sup\{|\alpha_n|: n \geqslant 1\}$. This type of operator is called a *diagonal operator* or is said to be *diagonalizable*.

9. (Schur test) Let $\{\alpha_{ij}\}_{i,j=1}^\infty$ be an infinite matrix such that $\alpha_{ij} \geqslant 0$ for all i, j and such that there are scalars $p_i > 0$ and $\beta, \gamma > 0$ with

$$\sum_{i=1}^\infty \alpha_{ij} p_i \leqslant \beta p_j,$$

$$\sum_{j=1}^\infty \alpha_{ij} p_j \leqslant \gamma p_i$$

for all $i, j \geqslant 1$. Show that there is an operator A on $l^2(\mathbb{N})$ with $\langle Ae_j, e_i \rangle = \alpha_{ij}$ and $\|A\|^2 \leqslant \beta\gamma$.

10. (Hilbert matrix) Show that $\langle Ae_j, e_i \rangle = (i + j + 1)^{-1}$ for $0 \leqslant i, j < \infty$ defines a bounded operator on $l^2(\mathbb{N} \cup \{0\})$ with $\|A\| \leqslant \pi$. (See also Choi [1983] and Redheffer and Volkmann [1983].)

11. If $A = \begin{bmatrix} a & b \\ c & d \end{bmatrix}$, put $\alpha = [|a|^2 + |b|^2 + |c|^2 + |d|^2]^{1/2}$ and show that $\|A\| = \frac{1}{2}(\alpha^2 + \sqrt{\alpha^4 - 4\delta^2})$, where $\delta^2 = \det A^*A$.

12. (Direct sum of operators) Let $\{\mathscr{H}_i\}$ be a collection of Hilbert spaces and let $\mathscr{H} = \oplus_i \mathscr{H}_i$. Suppose $A_i \in \mathscr{B}(\mathscr{H}_i)$ for all i. Show that there is a bounded operator A on $\mathscr{H}$ such that $A|\mathscr{H}_i = A_i$ for all i if and only if $\sup_i \|A_i\| < \infty$. In this case, $\|A\| = \sup_i \|A_i\|$. The operator A is called the *direct sum* of the operators $\{A_i\}$ and is denoted by $A = \oplus_i A_i$.

13. (Bari [1951]) Call a sequence of vectors $\{f_n\}$ in a Hilbert space $\mathcal{H}$ a *Bessel sequence* if $\sum |\langle f, f_n \rangle|^2 < \infty$ for every f in $\mathcal{H}$. Show that a sequence $\{f_n\}$ of vectors in $\mathcal{H}$ is a Bessel sequence if and only if the infinite matrix $(\langle f_m, f_n \rangle)$ defines a bounded operator on l^2. (Also see Shapiro and Shields [1961], p. 524.)

§2. The Adjoint of an Operator

2.1. Definition. If $\mathcal{H}$ and $\mathcal{K}$ are Hilbert spaces, a function $u: \mathcal{H} \times \mathcal{K} \to \mathbb{F}$ is a *sesquilinear form* if for h, g in $\mathcal{H}$, k, f in $\mathcal{K}$, and α, β in $\mathbb{F}$,

(a) $u(\alpha h + \beta g, k) = \alpha u(h, k) + \beta u(g, k)$;
(b) $u(h, \alpha k + \beta f) = \bar{\alpha} u(h, k) + \bar{\beta} u(h, f)$.

The prefix "sesqui" is used because the function is linear in one variable but (for $\mathbb{F} = \mathbb{C}$) only conjugate linear in the other. ("Sesqui" means "one-and-a-half.")

A sesquilinear form is *bounded* if there is a constant M such that $|u(h, k)| \leqslant M \|h\| \|k\|$ for all h in $\mathcal{H}$ and k in $\mathcal{K}$. The constant M is called a bound for u.

Sesquilinear forms are used to study operators. If $A \in \mathcal{B}(\mathcal{H}, \mathcal{K})$, then $u(h, k) \equiv \langle Ah, k \rangle$ is a bounded sesquilinear form. Also, if $B \in \mathcal{B}(\mathcal{K}, \mathcal{H})$, $u(h, k) \equiv \langle h, Bk \rangle$ is a bounded sesquilinear form. Are there any more? Are these two forms related?

2.2. Theorem. *If $u: \mathcal{H} \times \mathcal{K} \to \mathbb{F}$ is a bounded sesquilinear form with bound M, then there are unique operators A in $\mathcal{B}(\mathcal{H}, \mathcal{K})$ and B in $\mathcal{B}(\mathcal{K}, \mathcal{H})$ such that*

2.3 $$u(h, k) = \langle Ah, k \rangle = \langle h, Bk \rangle$$

for all h in $\mathcal{H}$ and k in $\mathcal{K}$ and $\|A\|, \|B\| \leqslant M$.

PROOF. Only the existence of A will be shown. For each h in $\mathcal{H}$, define L_h: $\mathcal{K} \to \mathbb{F}$ by $L_h(k) = \overline{u(h, k)}$. Then L_h is linear and $|L_h(k)| \leqslant M \|h\| \|k\|$. By the Riesz Representation Theorem there is a unique vector f in $\mathcal{K}$ such that $\langle k, f \rangle = L_h(k) = \overline{u(h, k)}$ and $\|f\| \leqslant M \|h\|$. Let $Ah = f$. It is left as an exercise to show that A is linear (use the uniqueness part of the Riesz Theorem). Also, $\langle Ah, k \rangle = \overline{\langle k, Ah \rangle} = \overline{\langle k, f \rangle} = u(h, k)$.

If $A_1 \in \mathcal{B}(\mathcal{H}, \mathcal{K})$ and $u(h, k) = \langle A_1 h, k \rangle$, then $\langle Ah - A_1 h, k \rangle = 0$ for all k; thus $Ah - A_1 h = 0$ for all h. Thus, A is unique. ∎

2.4. Definition. If $A \in \mathcal{B}(\mathcal{H}, \mathcal{K})$, then the unique operator B in $\mathcal{B}(\mathcal{K}, \mathcal{H})$ satisfying (2.3) is called the *adjoint* of A and is denoted by $B = A^*$.

The adjoint of an operator will usually be used for operators in $\mathcal{B}(\mathcal{H})$, rather than $\mathcal{B}(\mathcal{H}, \mathcal{K})$. There is one notable exception.

2.5. Proposition. *If $U \in \mathcal{B}(\mathcal{H}, \mathcal{K})$, then U is an isomorphism if and only if U is invertible and $U^{-1} = U^*$.*

PROOF. Exercise.

From now on we will examine and prove results for the adjoint of operators in $\mathcal{B}(\mathcal{H})$. Often, as in the next proposition, there are analogous results for the adjoint of operators in $\mathcal{B}(\mathcal{H}, \mathcal{K})$. This simplification is justified, however, by the cleaner statements that result. Also, the interested reader will have no trouble formulating the more general statement when it is needed.

2.6. Proposition. *If $A, B \in \mathcal{B}(\mathcal{H})$ and $\alpha \in \mathbb{F}$, then:*

(a) $(\alpha A + B)^* = \bar{\alpha} A^* + B^*$.
(b) $(AB)^* = B^* A^*$.
(c) $A^{**} \equiv (A^*)^* = A$.
(d) *If A is invertible in $\mathcal{B}(\mathcal{H})$ and A^{-1} is its inverse, then A^* is invertible and $(A^*)^{-1} = (A^{-1})^*$.*

The proof of the preceding proposition is left as an exercise, but a word about part (d) might be helpful. The hypothesis that A is invertible in $\mathcal{B}(\mathcal{H})$ means that there is an operator A^{-1} in $\mathcal{B}(\mathcal{H})$ such that $AA^{-1} = A^{-1}A = I$. It is a remarkable fact that if A is only assumed to be bijective, then A is invertible in $\mathcal{B}(\mathcal{H})$. This is a consequence of the Open Mapping Theorem, which will be proved later.

2.7. Proposition. *If $A \in \mathcal{B}(\mathcal{H})$, $\|A\| = \|A^*\| = \|A^*A\|^{1/2}$.*

PROOF. For h in $\mathcal{H}$, $\|h\| \leq 1$, $\|Ah\|^2 = \langle Ah, Ah \rangle = \langle A^*Ah, h \rangle \leq \|A^*Ah\| \|h\| \leq \|A^*A\| \leq \|A^*\| \|A\|$. Hence $\|A\|^2 \leq \|A^*A\| \leq \|A^*\| \|A\|$. Using the two ends of this string of inequalities gives $\|A\| \leq \|A^*\|$ when $\|A\|$ is cancelled. But $A = A^{**}$ and so if A^* is substituted for A, we get $\|A^*\| \leq \|A^{**}\| = \|A\|$. Hence $\|A\| = \|A^*\|$. Thus the string of inequalities becomes a string of equalities and the proof is complete. ∎

2.8. Example. Let (X, Ω, μ) be a σ-finite measure space and let M_ϕ be the multiplication operator with symbol ϕ (1.5). Then M_ϕ^* is $M_{\bar\phi}$, the multiplication operator with symbol $\bar\phi$.

If an operator on $\mathbb{F}^d$ is presented by a matrix, then its adjoint is represented by the conjuagate transpose of the matrix.

2.9. Example. If K is the integral operator with kernel k as in (1.6), then K^* is the integral operator with kernel $k^*(x, y) \equiv \overline{k(y, x)}$.

2.10. Proposition. *If $S: l^2 \to l^2$ is defined by $S(\alpha_1, \alpha_2, \ldots) = (0, \alpha_1, \alpha_2, \ldots)$, then S is an isometry and $S^*(\alpha_1, \alpha_2, \ldots) = (\alpha_2, \alpha_3, \ldots)$.*

PROOF. It has already been mentioned that S is an isometry (I.5.3). For (α_n) and (β_n) in l^2, $\langle S^*(\alpha_n), (\beta_n) \rangle = \langle (\alpha_n), S(\beta_n) \rangle = \langle (\alpha_1, \alpha_2, \dots), (0, \beta_1, \beta_2, \dots) \rangle = \alpha_2 \bar{\beta}_1 + \alpha_3 \bar{\beta}_2 + \dots = \langle (\alpha_2, \alpha_3, \dots), (\beta_1, \beta_2, \dots) \rangle$. Since this holds for every (β_n), the result is proved. ∎

The operator S in (2.10) is called the *unilateral shift* and the operator S^* is called the *backward shift*.

The operation of taking the adjoint of an operator is, as the reader may have seen from the examples above, analogous to taking the conjugate of a complex number. It is good to keep the analogy in mind, but do not become too religious about it.

2.11. Definition. If $A \in \mathscr{B}(\mathscr{H})$, then: (a) is *hermitian* or *self-adjoint* if $A^* = A$; (b) A *is normal* if $AA^* = A^*A$.

In the analogy between the adjoint and the complex conjugate, hermitian operators become the analogues of real numbers and, by (2.5), unitaries are the analogues of complex numbers of modulus 1. Normal operators, as we shall see, are the true analogues of complex numbers. Notice that hermitian and unitary operators are normal.

In light of (2.8), every multiplication operator M_ϕ is normal; M_ϕ is hermitian if and only if ϕ is real-valued; M_ϕ is unitary if and only if $|\phi| \equiv 1$ a.e. $[\mu]$. By (2.9), an integral operator K with kernel k is hermitian if and only if $k(x, y) = \overline{k(y, x)}$ a.e. $[\mu \times \mu]$. The unilateral shift is not normal (Exercise 6).

2.12. Proposition. *If $\mathscr{H}$ is a $\mathbb{C}$-Hilbert space and $A \in \mathscr{B}(\mathscr{H})$, then A is hermitian if and only if $\langle Ah, h \rangle \in \mathbb{R}$ for all h in $\mathscr{H}$.*

PROOF. If $A = A^*$, then $\langle Ah, h \rangle = \langle h, Ah \rangle = \overline{\langle Ah, h \rangle}$; hence $\langle Ah, h \rangle \in \mathbb{R}$.

For the converse, assume $\langle Ah, h \rangle$ is real for every h in $\mathscr{H}$. If $\alpha \in \mathbb{C}$ and $h, g \in \mathscr{H}$, then $\langle A(h + \alpha g), h + \alpha g \rangle = \langle Ah, h \rangle + \bar{\alpha} \langle Ah, g \rangle + \alpha \langle Ag, h \rangle + |\alpha|^2 \langle Ag, g \rangle \in \mathbb{R}$. So this expression equals its complex conjugate. Using the fact that $\langle Ah, h \rangle$ and $\langle Ag, g \rangle \in \mathbb{R}$ yields

$$\alpha \langle Ag, h \rangle + \bar{\alpha} \langle Ah, g \rangle = \bar{\alpha} \langle h, Ag \rangle + \alpha \langle g, Ah \rangle$$
$$= \bar{\alpha} \langle A^*h, g \rangle + \alpha \langle A^*g, h \rangle.$$

By first taking $\alpha = 1$ and then $\alpha = i$, we obtain the two equations

$$\langle Ag, h \rangle + \langle Ah, g \rangle = \langle A^*h, g \rangle + \langle A^*g, h \rangle,$$
$$i\langle Ag, h \rangle - i\langle Ah, g \rangle = -i\langle A^*h, g \rangle + i\langle A^*g, h \rangle.$$

A little arithmetic implies $\langle Ag, h \rangle = \langle A^*g, h \rangle$, so $A = A^*$. ∎

The preceding proposition is false if it is only assumed that $\mathscr{H}$ is an $\mathbb{R}$-Hilbert space. For example, if $A = \begin{bmatrix} 0 & 1 \\ -1 & 0 \end{bmatrix}$ on $\mathbb{R}^2$, then $\langle Ah, h \rangle = 0$ for all h in $\mathbb{R}^2$. However, A^* is the transpose of A and so $A^* \neq A$. Indeed, for any operator A on an $\mathbb{R}$-Hilbert space, $\langle Ah, g \rangle \in \mathbb{R}$.

2.13. Proposition. *If* $A = A^*$, *then*

$$\|A\| = \sup\{|\langle Ah, h \rangle|: \|h\| = 1\}.$$

PROOF. Put $M = \sup\{|\langle Ah, h \rangle|: \|h\| = 1\}$. If $\|h\| = 1$, then $|\langle Ah, h \rangle| \leqslant \|A\|$; hence $M \leqslant \|A\|$. On the other hand, if $\|h\| = \|g\| = 1$, then

$$\langle A(h \pm g), h \pm g \rangle = \langle Ah, h \rangle \pm \langle Ah, g \rangle \pm \langle Ag, h \rangle + \langle Ag, g \rangle$$
$$= \langle Ah, h \rangle \pm \langle Ah, g \rangle \pm \langle g, A^*h \rangle + \langle Ag, g \rangle.$$

Since $A = A^*$, this implies

$$\langle A(h \pm g), h \pm g \rangle = \langle Ah, h \rangle \pm 2 \operatorname{Re} \langle Ah, g \rangle + \langle Ag, g \rangle.$$

Subtracting one of these two equations from the other gives

$$4 \operatorname{Re} \langle Ah, g \rangle = \langle A(h + g), h + g \rangle - \langle A(h - g), h - g \rangle.$$

Now it is easy to verify that $|\langle Af, f \rangle| \leqslant M \|f\|^2$ for any f in $\mathscr{H}$. Hence using the parallelogram law we get

$$4 \operatorname{Re} \langle Ah, g \rangle \leqslant M(\|h + g\|^2 + \|h - g\|^2)$$
$$= 2M(\|h\|^2 + \|g\|^2)$$
$$= 4M$$

since h and g are unit vectors. Now suppose $\langle Ah, g \rangle = e^{i\theta}|\langle Ah, g \rangle|$. Replacing h in the inequality above with $e^{-i\theta}h$ gives $|\langle Ah, g \rangle| \leqslant M$ if $\|h\| = \|g\| = 1$. Taking the supremum over all g gives $\|Ah\| \leqslant M$ when $\|h\| = 1$. Thus $\|A\| \leqslant M$. ∎

2.14. Corollary. *If* $A = A^*$ *and* $\langle Ah, h \rangle = 0$ *for all* h, *then* $A = 0$.

The preceding corollary is not true unless $A = A^*$, as the example given after Proposition 2.12 shows. However, if a complex Hilbert space is present, this hypothesis can be deleted.

2.15. Proposition. *If* $\mathscr{H}$ *is a* $\mathbb{C}$-*Hilbert space and* $A \in \mathscr{B}(\mathscr{H})$ *such that* $\langle Ah, h \rangle = 0$ *for all* h *in* $\mathscr{H}$, *then* $A = 0$.

The proof of (2.15) is left to the reader.

If $\mathscr{H}$ is a $\mathbb{C}$-Hilbert space and $A \in \mathscr{B}(\mathscr{H})$, then $B = (A + A^*)/2$ and $C = (A - A^*)/2i$ are self-adjoint and $A = B + iC$. The operators B and C are called, respectively, the *real and imaginary parts* of A.

2.16. Proposition. *If* $A \in \mathscr{B}(\mathscr{H})$, *the following statement are equivalent.*

(a) *A is normal.*

(b) $\|Ah\| = \|A^*h\|$ *for all h.*

If $\mathcal{H}$ is a $\mathbb{C}$-Hilbert space, then these statements are also equivalent to:
(c) *The real and imaginary parts of A commute.*

PROOF. If $h \in \mathcal{H}$, then $\|Ah\|^2 - \|A^*h\|^2 = \langle Ah, Ah \rangle - \langle A^*h, A^*h \rangle = \langle (A^*A - AA^*)h, h \rangle$. Since $A^*A - AA^*$ is hermitian, the equivalence of (a) and (b) follows from Corollary 2.14.

If B, C are real and imaginary parts of A, then a calculation yields

$$A^*A = B^2 - iCB + iBC + C^2,$$
$$AA^* = B^2 + iCB - iBC + C^2.$$

Hence $A^*A = AA^*$ if and only if $CB = BC$, and so (a) and (c) are equivalent. ∎

2.17. Proposition. *If $A \in \mathcal{B}(\mathcal{H})$, the following statements are equivalent.*

(a) *A is an isometry.*
(b) $A^*A = I.$
(c) $\langle Ah, Ag \rangle = \langle h, g \rangle$ *for all h, g in $\mathcal{H}$.*

PROOF. The proof that (a) and (c) are equivalent was seen in Proposition I.5.2. Note that if $h, g \in \mathcal{H}$, then $\langle A^*Ah, g \rangle = \langle Ah, Ag \rangle$. Hence (b) and (c) are easily seen to be equivalent. ∎

2.18. Proposition. *If $A \in \mathcal{B}(\mathcal{H})$, then the following statements are equivalent.*

(a) $A^*A = AA^* = I.$
(b) *A is unitary. (That is, A is a surjective isometry.)*
(c) *A is a normal isometry.*

PROOF. (a)⇒(b): Proposition I.5.2.
(b)⇒(c): By (2.17), $A^*A = I$. But it is easy to see that the fact that A is a surjective isometry implies that A^{-1} is also. Hence by (2.17) $I = (A^{-1})^*A^{-1} = (A^*)^{-1}A^{-1} = (AA^*)^{-1}$; this implies that $A^*A = AA^* = I$.
(c)⇒(a): By (2.17), $A^*A = I$. Since A is also normal, $AA^* = A^*A = I$ and so A is surjective. ∎

We conclude with a very important, though easily proved, result.

2.19. Theorem. *If $A \in \mathcal{B}(\mathcal{H})$, then $\ker A = (\operatorname{ran} A^*)^\perp$.*

PROOF. If $h \in \ker A$ and $g \in \mathcal{H}$, then $\langle h, A^*g \rangle = \langle Ah, g \rangle = 0$, so $\ker A \subseteq (\operatorname{ran} A^*)^\perp$. On the other hand, if $h \perp \operatorname{ran} A^*$ and $g \in \mathcal{H}$, then $\langle Ah, g \rangle = \langle h, A^*g \rangle = 0$; so $(\operatorname{ran} A^*)^\perp \subseteq \ker A$. ∎

Two facts should be noted. Since $A^{**} = A$, it also holds that $\ker A^* = (\operatorname{ran} A)^\perp$. Second, it is not true that $(\ker A)^\perp = \operatorname{ran} A^*$ since $\operatorname{ran} A^*$ may not be closed. All that can be said is that $(\ker A)^{\perp} = \mathrm{cl}(\operatorname{ran} A^*)$ and $(\ker A^*)^{\perp} = \mathrm{cl}(\operatorname{ran} A)$.

EXERCISES

1. Prove Proposition 2.5.

2. Prove Proposition 2.6.

3. Verify the statement in Example 2.8.

4. Verify the statement in Example 2.9.

5. Find the adjoint of a diagonal operator (Exercise 1.8).

6. Let S be the unilateral shift and compute SS^* and S^*S. Also compute $S^n S^{*n}$ and $S^{*n}S^n$.

7. Compute the adjoint of the Volterra operator V (1.7) and $V + V^*$. What is $\operatorname{ran}(V + V^*)$?

8. Where was the hypothesis that $\mathcal{H}$ is a Hilbert space over $\mathbb{C}$ used in the proof of Proposition 2.12?

9. Suppose $A = B + iC$, where B and C are hermitian and prove that $B = (A + A^*)/2$, $C = (A - A^*)/2i$.

10. Prove Proposition 2.15.

11. If A and B are self-adjoint, show that AB is self-adjoint if and only if $AB = BA$.

12. Let $\sum_{n=0}^{\infty} \alpha_n z^n$ be a power series with radius of convergence $R, 0 < R \leqslant \infty$. If $A \in \mathcal{B}(\mathcal{H})$ and $\|A\| < R$, show that there is an operator T in $\mathcal{B}(\mathcal{H})$ such that for any h, g in $\mathcal{H}$, $\langle Th, g \rangle = \sum_{n=0}^{\infty} \alpha_n \langle A^n h, g \rangle$. [If $f(z) = \sum \alpha_n z^n$, the operator T is usually denoted by $f(A)$.]

13. Let A and T be as in Exercise 12 and show that $\|T - \sum_{k=0}^{n} \alpha_k A^k\| \to 0$ as $n \to \infty$. If $BA = AB$, show that $BT = TB$.

14. If $f(z) = \exp z = \sum_{n=0}^{\infty} z^n/n!$ and A is hermitian, show that $f(iA)$ is unitary.

15. If A is a normal operator on $\mathcal{H}$, show that A is injective if and only if A has dense range. Give an example of an operator B such that $\ker B = (0)$ but $\operatorname{ran} B$ is not dense. Give an example of an operator C such that C is surjective but $\ker C \neq (0)$.

16. Let M_ϕ be a multiplication operator (1.5) and show that $\ker M_\phi = 0$ if and only if $\mu(\{x: \phi(x) = 0\}) = 0$. Give necessary and sufficient conditions on ϕ that $\operatorname{ran} M_\phi$ be closed.

§3. Projections and Idempotents; Invariant and Reducing Subspaces

3.1. Definition. An *idempotent* on $\mathcal{H}$ is a bounded linear operator E on $\mathcal{H}$ such that $E^2 = E$. A *projection* is an idempotent P such that $\ker P = (\operatorname{ran} P)^{\perp}$.

If $\mathscr{M} \leqslant \mathscr{H}$, then $P_{\mathscr{M}}$ is a projection (Theorem I.2.7). It is not difficult to construct an idempotent that is not a projection (Exercise 1).

Let E be any idempotent and set $\mathscr{M} = \operatorname{ran} E$ and $\mathscr{N} = \ker E$. Since E is continuous, $\mathscr{N}$ is a closed subspace of $\mathscr{H}$. Notice that $(I - E)^2 = I - 2E + E^2 = I - 2E + E = I - E$; thus $I - E$ is also an idempotent. Also, $0 = (I - E)h = h - Eh$, if and only if $Eh = h$. So $\operatorname{ran} E \supseteq \ker(I - E)$. On the other hand, if $h \in \operatorname{ran} E$, $h = Eg$ and so $Eh = E^2 g = Eg = h$; hence $\operatorname{ran} E = \ker(I - E)$. Similarly, $\operatorname{ran}(I - E) = \ker E$. These facts are recorded here.

3.2. Proposition. (a) E is an idempotent if and only if $I - E$ is an idempotent. (b) $\operatorname{ran} E = \ker(I - E)$, $\ker E = \operatorname{ran}(I - E)$, and both $\operatorname{ran} E$ and $\ker E$ are closed linear subspaces of $\mathscr{H}$. (c) If $\mathscr{M} = \operatorname{ran} E$ and $\mathscr{N} = \ker E$, then $\mathscr{M} \cap \mathscr{N} = (0)$ and $\mathscr{M} + \mathscr{N} = \mathscr{H}$.

The proof of part (c) is left as an exercise. There is also a converse to (c). If $\mathscr{M}, \mathscr{N} \leqslant \mathscr{H}$, $\mathscr{M} \cap \mathscr{N} = (0)$, and $\mathscr{M} + \mathscr{N} = \mathscr{H}$, then there is an idempotent E such that $\mathscr{M} = \operatorname{ran} E$ and $\mathscr{N} = \ker E$; moreover, E is unique. The difficult part in proving this converse is to show that E is bounded. The same fact is true in more generality (for Banach spaces) and so this proof will be postponed.

Now we turn our attention to projections, which are peculiar to Hilbert space.

3.3. Proposition. If E is an idempotent on $\mathscr{H}$ and $E \neq 0$, the following statements are equivalent.

(a) E is a projection.
(b) E is the orthogonal projection of $\mathscr{H}$ onto $\operatorname{ran} E$.
(c) $\|E\| = 1$.
(d) E is hermitian.
(e) E is normal.
(f) $\langle Eh, h \rangle \geqslant 0$ for all h in $\mathscr{H}$.

PROOF. (a) $\Rightarrow$ (b): Let $\mathscr{M} = \operatorname{ran} E$ and $P = P_{\mathscr{M}}$. If $h \in \mathscr{H}$, $Ph =$ the unique vector in $\mathscr{M}$ such that $h - Ph \in \mathscr{M}^{\perp} = (\operatorname{ran} E)^{\perp} = \ker E$ by (a). But $h - Eh = (I - E)h \in \ker E$. Hence $Eh = Ph$ by uniqueness.

(b) $\Rightarrow$ (c): By (I.2.7), $\|E\| \leqslant 1$. But $Eh = h$ for h in $\operatorname{ran} E$, so $\|E\| = 1$.

(c) $\Rightarrow$ (a): Let $h \in (\ker E)^{\perp}$. Now $\operatorname{ran}(I - E) = \ker E$, so $h - Eh \in \ker E$. Hence $0 = \langle h - Eh, h \rangle = \|h\|^2 - \langle Eh, h \rangle$. Hence $\|h\|^2 = \langle Eh, h \rangle \leqslant \|Eh\| \|h\| \leqslant \|h\|^2$. So for h in $(\ker E)^{\perp}$, $\|Eh\| = \|h\| = \langle Eh, h \rangle^{1/2}$. But then for h in $(\ker E)^{\perp}$,

$$\|h - Eh\|^2 = \|h\|^2 - 2\operatorname{Re}\langle Eh, h \rangle + \|Eh\|^2 = 0.$$

That is, $(\ker E)^{\perp} \subseteq \ker(I - E) = \operatorname{ran} E$. On the other hand, if $g \in \operatorname{ran} E$, $g = g_1 + g_2$, where $g_1 \in \ker E$ and $g_2 \in (\ker E)^{\perp}$. Thus $g = Eg = Eg_2 = g_2$; that is, $\operatorname{ran} E \subseteq (\ker E)^{\perp}$. Therefore $\operatorname{ran} E = (\ker E)^{\perp}$ and E is a projection.

(b) $\Rightarrow$ (f): If $h \in \mathcal{H}$, write $h = h_1 + h_2$, $h_1 \in \operatorname{ran} E$, $h_2 \in \ker E = (\operatorname{ran} E)^\perp$. Hence $\langle Eh, h \rangle = \langle E(h_1 + h_2), h_1 + h_2 \rangle = \langle Eh_1, h_1 \rangle = \langle h_1, h_1 \rangle = \|h_1\|^2 \geqslant 0$.

(f) $\Rightarrow$ (a): Let $h_1 \in \operatorname{ran} E$ and $h_2 \in \ker E$. Then by (f), $0 \leqslant \langle E(h_1 + h_2), h_1 + h_2 \rangle = \langle h_1, h_1 \rangle + \langle h_1, h_2 \rangle$. Hence $-\|h_1\|^2 \leqslant \langle h_1, h_2 \rangle$ for all h_1 in $\operatorname{ran} E$ and h_2 in $\ker E$. If there are such h_1 and h_2 with $\langle h_1, h_2 \rangle = \bar{\alpha} \neq 0$, then substituting $k_2 = -2\alpha^{-1}\|h_1\|^2 h_2$ for h_2 in this inequality, we obtain $-\|h_1\|^2 \leqslant -2\|h_1\|^2$, a contradiction. Hence $\langle h_1, h_2 \rangle = 0$ whenever $h_1 \in \operatorname{ran} E$ and $h_2 \in \ker E$. That is, E is a projection.

(a) $\Rightarrow$ (d): Let h, $g \in \mathcal{H}$ and put $h = h_1 + h_2$ and $g = g_1 + g_2$, where $h_1, g_1 \in \operatorname{ran} E$ and $h_2, g_2 \in \ker E = (\operatorname{ran} E)^\perp$. Hence $\langle Eh, g \rangle = \langle h_1, g_1 \rangle$. Also, $\langle E^*h, g \rangle = \langle h, Eg \rangle = \langle h_1, g_1 \rangle = \langle Eh, g \rangle$. Thus $E = E^*$.

(d) $\Rightarrow$ (e): clear.

(e) $\Rightarrow$ (a): By (2.16), $\|Eh\| = \|E^*h\|$ for every h. Hence $\ker E = \ker E^*$. But by (2.19), $\ker E^* = (\operatorname{ran} E)^\perp$, so E is a projection.   ∎

Note that by part (b) of the preceding proposition, if E is a projection and $\mathcal{M} = \operatorname{ran} E$, then $E = P_{\mathcal{M}}$.

Let P be a projection with $\operatorname{ran} P = \mathcal{M}$ and $\ker P = \mathcal{N}$. So both $\mathcal{M}$ and $\mathcal{N}$ are closed subspaces of $\mathcal{H}$ and, hence, are also Hilbert spaces. As in (I.6.1), we can form $\mathcal{M} \oplus \mathcal{N}$. If $U: \mathcal{M} \oplus \mathcal{N} \to \mathcal{H}$ is defined by $U(h \oplus g) = h + g$ for h in $\mathcal{M}$ and g in $\mathcal{N}$, then it is easy to see that U is an isomorphism. Making this identification, we will often write $\mathcal{H} = \mathcal{M} \oplus \mathcal{N}$.

More generally, the following will be used.

3.4. Definition. If $\{\mathcal{M}_i\}$ is a collection of pairwise orthogonal subspaces of $\mathcal{H}$, then
$$\oplus_i \mathcal{M}_i \equiv \vee_i \mathcal{M}_i.$$

If $\mathcal{M}$ and $\mathcal{N}$ are two closed linear subspaces of $\mathcal{H}$, then
$$\mathcal{M} \ominus \mathcal{N} \equiv \mathcal{M} \cap \mathcal{N}^\perp.$$

This is called the *orthogonal difference* of $\mathcal{M}$ and $\mathcal{N}$.

Note that if $\mathcal{M}$, $\mathcal{N} \leqslant \mathcal{H}$ and $\mathcal{M} \perp \mathcal{N}$, then $\mathcal{M} + \mathcal{N}$ is closed. (Why?) Hence $\mathcal{M} \oplus \mathcal{N} = \mathcal{M} + \mathcal{N}$. The same is true, of course, for any finite collection of pairwise orthogonal subspaces but not for infinite collections.

3.5. Definition. If $A \in \mathcal{B}(\mathcal{H})$ and $\mathcal{M} \leqslant \mathcal{H}$, say that $\mathcal{M}$ is an *invariant subspace* for A if $Ah \in \mathcal{M}$ whenever $h \in \mathcal{M}$. In other words, if $A\mathcal{M} \subseteq \mathcal{M}$. Say that $\mathcal{M}$ is a *reducing subspace* for A if $A\mathcal{M} \subseteq \mathcal{M}$ and $A\mathcal{M}^\perp \subseteq \mathcal{M}^\perp$.

If $\mathcal{M} \leqslant \mathcal{H}$, then $\mathcal{H} = \mathcal{M} \oplus \mathcal{M}^\perp$. If $A \in \mathcal{B}(\mathcal{H})$, then A can be written as a 2×2 matrix with operator entries,

3.6
$$A = \begin{bmatrix} W & X \\ Y & Z \end{bmatrix},$$

where $W \in \mathcal{B}(\mathcal{M})$, $X \in \mathcal{B}(\mathcal{M}^\perp, \mathcal{M})$, $Y \in \mathcal{B}(\mathcal{M}, \mathcal{M}^\perp)$, and $Z \in \mathcal{B}(\mathcal{M}^\perp)$.

3.7. Proposition. *If* $A \in \mathcal{B}(\mathcal{H})$, $\mathcal{M} \leqslant \mathcal{H}$, *and* $P = P_{\mathcal{M}}$, *then statements* (a) *through* (c) *are equivalent.*

(a) $\mathcal{M}$ *is invariant for* A.
(b) $PAP = AP$.
(c) *In* (3.6), $Y = 0$.

   Also, statements (d) *through* (g) *are equivalent.*

(d) $\mathcal{M}$ *reduces* A.
(e) $PA = AP$.
(f) *In* (3.6), Y *and* X *are* 0.
(g) $\mathcal{M}$ *is invariant for both* A *and* A^*.

PROOF. (a)$\Rightarrow$(b): If $h \in \mathcal{H}$, $Ph \in \mathcal{M}$. So $APh \in \mathcal{M}$. Hence, $P(APh) = APh$. That is, $PAP = AP$.

   (b)$\Rightarrow$(c): If P is represented as a 2×2 operator matrix relative to $\mathcal{H} = \mathcal{M} \oplus \mathcal{M}^{\perp}$, then

$$P = \begin{bmatrix} I & 0 \\ 0 & 0 \end{bmatrix}.$$

Hence,

$$PAP = \begin{bmatrix} W & 0 \\ 0 & 0 \end{bmatrix} = AP = \begin{bmatrix} W & 0 \\ Y & 0 \end{bmatrix}.$$

So $Y = 0$.

   (c)$\Rightarrow$(a): If $Y = 0$ and $h \in \mathcal{M}$, then

$$Ah = \begin{bmatrix} W & X \\ 0 & Z \end{bmatrix} \begin{bmatrix} h \\ 0 \end{bmatrix} = \begin{bmatrix} Wh \\ 0 \end{bmatrix} \in \mathcal{M}.$$

   (d)$\Rightarrow$(e): Since both $\mathcal{M}$ and $\mathcal{M}^{\perp}$ are invariant for A, (b) implies that $AP = PAP$ and $A(I - P) = (I - P)A(I - P)$. Multiplying this second equation gives $A - AP = A - AP - PA + PAP$. Thus $PA = PAP = AP$.
   (e)$\Rightarrow$(f): Exercise.
   (f)$\Rightarrow$(g): If $X = Y = 0$, then

$$A = \begin{bmatrix} W & 0 \\ 0 & Z \end{bmatrix} \quad \text{and} \quad A^* = \begin{bmatrix} W^* & 0 \\ 0 & Z^* \end{bmatrix}.$$

By (c), $\mathcal{M}$ is invariant for both A and A^*.
   (g)$\Rightarrow$(d): If $h \in \mathcal{M}^{\perp}$ and $g \in \mathcal{M}$, then $\langle g, Ah \rangle = \langle A^*g, h \rangle = 0$ since $A^*g \in \mathcal{M}$. Since g was an arbitrary vector in $\mathcal{M}$, $Ah \in \mathcal{M}^{\perp}$. That is, $A\mathcal{M}^{\perp} \subseteq \mathcal{M}^{\perp}$. ∎

   If $\mathcal{M}$ reduces A, then $X = Y = 0$ in (3.6). This says that a study of A is reduced to the study of the smaller operators W and Z. This is the reason for the terminology.
   If $A \in \mathcal{B}(\mathcal{H})$ and $\mathcal{M}$ is an invariant subspace for A, then $A|\mathcal{M}$ is used to denote the restriction of A to $\mathcal{M}$. That is, $A|\mathcal{M}$ is the operator on $\mathcal{M}$ defined

by $(A|\mathcal{M})h = Ah$ whenever $h \in \mathcal{M}$. Note that $A|\mathcal{M} \in \mathcal{B}(\mathcal{M})$ and $\|A|\mathcal{M}\| \leq \|A\|$. Also, if $\mathcal{M}$ is invariant for A and A has the representation (3.6) with $Y = 0$, then $W = A|\mathcal{M}$.

EXERCISES

1. Let $\mathcal{H}$ be the two-dimensional real Hilbert space $\mathbb{R}^2$, let $\mathcal{M} \equiv \{(x,0) \in \mathbb{R}^2 : x \in \mathbb{R}\}$ and let $\mathcal{N} \equiv \{(x, x \tan\theta) : x \in \mathbb{R}\}$, where $0 < \theta < \frac{1}{2}\pi$. Find a formula for the idempotent E_θ with ran $E_\theta = \mathcal{M}$ and ker $E_\theta = \mathcal{N}$. Show that $\|E_\theta\| = (\sin\theta)^{-1}$.

2. Prove Proposition 3.2 (c).

3. Let $\{\mathcal{M}_i : i \in I\}$ be a collection of closed subspaces of $\mathcal{H}$ and show that $\cap\{\mathcal{M}_i^\perp : i \in I\} = [\vee\{\mathcal{M}_i : i \in I\}]^\perp$ and $[\cap\{\mathcal{M}_i : i \in I\}]^\perp = \vee\{\mathcal{M}_i^\perp : i \in I\}$.

4. Let P and Q be projections. Show: (a) $P+Q$ is a projection if and only if ran $P \perp$ ran Q. If $P+Q$ is a projection, then ran$(P+Q) =$ ran $P +$ ran Q and ker$(P+Q) =$ ker $P \cap$ ker Q. (b) PQ is a projection if and only if $PQ = QP$. If PQ is a projection, then ran $PQ =$ ran $P \cap$ ran Q and ker $PQ =$ ker $P +$ ker Q.

5. Generalize Exercise 4 as follows. Suppose $\{\mathcal{M}_i : i \in I\}$ is a collection of subspaces of $\mathcal{H}$ such that $\mathcal{M}_i \perp \mathcal{M}_j$ if $i \neq j$. Let P_i be the projection of $\mathcal{H}$ onto $\mathcal{M}_i$ and show that for all h in $\mathcal{H}$, $\sum\{P_ih : i \in I\}$ converges to Ph, where P is the projection of $\mathcal{H}$ onto $\vee\{\mathcal{M}_i : i \in I\}$.

6. If P and Q are projections, then the following statements are equivalent. (a) $P-Q$ is a projection. (b) ran $Q \subseteq$ ran P. (c) $PQ = Q$. (d) $QP = Q$. If $P-Q$ is a projection, then ran$(P-Q) = ($ran $P) \ominus ($ran $Q)$ and ker$(P-Q) =$ ran $Q +$ ker P.

7. Let P and Q be projections. Show that $PQ = QP$ if and only if $P+Q-PQ$ is a projection. If this is the case, then ran$(P+Q-PQ) =$ ran $P +$ ran Q and ker$(P+Q-PQ) =$ ker $P \cap$ ker Q.

8. Give an example of two noncommuting projections.

9. Let $A \in \mathcal{B}(\mathcal{H})$ and let $\mathcal{N} =$ graph $A \subseteq \mathcal{H} \oplus \mathcal{H}$. That is, $\mathcal{N} = \{h \oplus Ah : h \in \mathcal{H}\}$. Because A is continuous and linear, $\mathcal{N} \leq \mathcal{H} \oplus \mathcal{H}$. Let $\mathcal{M} = \mathcal{H} \oplus (0) \leq \mathcal{H} \oplus \mathcal{H}$. Prove the following statements. (a) $\mathcal{M} \cap \mathcal{N} = (0)$ if and only if ker $A = (0)$. (b) $\mathcal{M} + \mathcal{N}$ is dense in $\mathcal{H} \oplus \mathcal{H}$ if and only if ran A is dense in $\mathcal{H}$. (c) $\mathcal{M} + \mathcal{N} = \mathcal{H} \oplus \mathcal{H}$ if and only if A is surjective.

10. Find two closed linear subspaces $\mathcal{M}, \mathcal{N}$ of an infinite dimensional Hilbert space $\mathcal{H}$ such that $\mathcal{M} \cap \mathcal{N} = (0)$ and $\mathcal{M} + \mathcal{N}$ is dense in $\mathcal{H}$, but $\mathcal{M} + \mathcal{N} \neq \mathcal{H}$.

11. Define $A: l^2(\mathbb{Z}) \to l^2(\mathbb{Z})$ by $A(\ldots, \alpha_{-1}, \overset{\wedge}{\alpha_0}, \alpha_1, \ldots) = (\ldots, \overset{\wedge}{\alpha_{-1}}, \alpha_0, \alpha_1, \ldots)$, where $\wedge$ sits above the coefficient in the 0-place. Find an invariant subspace of A that does not reduce A. This operator is called the bilateral shift.

12. Let $\mu =$ Area measure on $\mathbb{D} \equiv \{z \in \mathbb{C} : |z| < 1\}$ and define $A: L^2(\mu) \to L^2(\mu)$ by $(Af)(z) = zf(z)$ for $|z| < 1$ and f in $L^2(\mu)$. Find a nontrivial reducing subspace for A and an invariant subspace that does not reduce A.

§4. Compact Operators

It turns out that most of the statements about linear transformations on finite dimensional spaces have nice generalizations to a certain class of operators on infinite dimensional spaces—namely, to the compact operators, Let ball $\mathcal{H}$ denote the closed unit ball in $\mathcal{H}$.

4.1. Definition. A linear transformation $T: \mathcal{H} \to \mathcal{K}$ is *compact* if $T(\text{ball } \mathcal{H})$ has compact closure in $\mathcal{K}$. The set of compact operators from $\mathcal{H}$ into $\mathcal{K}$ is denoted by $\mathcal{B}_0(\mathcal{H}, \mathcal{K})$, and $\mathcal{B}_0(\mathcal{H}) = \mathcal{B}_0(\mathcal{H}, \mathcal{H})$.

4.2. Proposition. (a) $\mathcal{B}_0(\mathcal{H}, \mathcal{K}) \subseteq \mathcal{B}(\mathcal{H}, \mathcal{K})$.
(b) $\mathcal{B}_0(\mathcal{H}, \mathcal{K})$ *is a linear space and if* $\{T_n\} \subseteq \mathcal{B}_0(\mathcal{H}, \mathcal{K})$ *and* $T \in \mathcal{B}(\mathcal{H}, \mathcal{K})$ *such that* $\| T_n - T \| \to 0$, *then* $T \in \mathcal{B}_0(\mathcal{H}, \mathcal{K})$.
(c) *If* $A \in \mathcal{B}(\mathcal{H})$, $B \in \mathcal{B}(\mathcal{K})$, *and* $T \in \mathcal{B}_0(\mathcal{H}, \mathcal{K})$, *then* TA *and* $BT \in \mathcal{B}_0(\mathcal{H}, \mathcal{K})$.

PROOF. (a) If $T \in \mathcal{B}_0(\mathcal{H}, \mathcal{K})$, then $\text{cl}[T(\text{ball } \mathcal{H})]$ is compact in $\mathcal{K}$. Hence there is a constant $C > 0$ such that $T(\text{ball } \mathcal{H}) \subseteq \{k \in \mathcal{K} : \|k\| \leqslant C\}$. Thus $\|T\| \leqslant C$.

(b) It is left to the reader to show that $\mathcal{B}_0(\mathcal{H}, \mathcal{K})$ is a linear space. For the second part of (b), it will be shown that $T(\text{ball } \mathcal{H})$ is totally bounded. Since $\mathcal{K}$ is a complete metric space, this is equivalent to showing that $T(\text{ball } \mathcal{H})$ has compact closure. Let $\varepsilon > 0$ and choose n such that $\| T - T_n \| < \varepsilon/3$. Since T_n is compact, there are vectors $h_1, \ldots, h_m$ in ball $\mathcal{H}$ such that $T_n(\text{ball } \mathcal{H}) \subseteq \bigcup_{j=1}^{m} B(T_n h_j; \varepsilon/3)$. So if $\|h\| \leqslant 1$, there is an h_j with $\| T_n h_j - T_n h \| < \varepsilon/3$. Thus

$$\| Th_j - Th \| \leqslant \| Th_j - T_n h_j \| + \| T_n h_j - T_n h \| + \| T_n h - Th \|$$
$$< 2 \| T - T_n \| + \varepsilon/3$$
$$< \varepsilon.$$

Hence $T(\text{ball } \mathcal{H}) \subseteq \bigcup_{j=1}^{m} B(Th_j; \varepsilon)$.
The proof of (c) is left to the reader. ∎

4.3. Definition. An operator T on $\mathcal{H}$ has *finite rank* if ran T is finite dimensional. The set of continuous finite rank operators is denoted by $\mathcal{B}_{00}(\mathcal{H}, \mathcal{K})$; $\mathcal{B}_{00}(\mathcal{H}) = \mathcal{B}_{00}(\mathcal{H}, \mathcal{H})$.

It is easy to see that $\mathcal{B}_{00}(\mathcal{H}, \mathcal{K})$ is a linear space and $\mathcal{B}_{00}(\mathcal{H}, \mathcal{K}) \subseteq \mathcal{B}_0(\mathcal{H}, \mathcal{K})$ (Exercise 2). Before giving other examples of compact operators, however, the next result should be proved.

4.4. Theorem. *If* $T \in \mathcal{B}(\mathcal{H}, \mathcal{K})$, *the following statements are equivalent.*

(a) T *is compact.*
(b) T^* *is compact.*
(c) *There is a sequence* $\{T_n\}$ *of operators of finite rank such that* $\| T - T_n \| \to 0$.

PROOF. (c)⇒(a): This is immediate from (4.2b) and the fact that $\mathscr{B}_{00}(\mathscr{H},\mathscr{K})\subseteq\mathscr{B}_0(\mathscr{H},\mathscr{K})$.

(a)⇒(c): Since cl[T(ball $\mathscr{H}$)] is compact, it is separable. Therefore cl(ran T) = $\mathscr{L}$ is a separable subspace of $\mathscr{K}$. Let $\{e_1,e_2,\ldots\}$ be a basis for $\mathscr{L}$ and let P_n be the orthogonal projection of $\mathscr{K}$ onto $\bigvee\{e_j: 1\leqslant j\leqslant n\}$. Put $T_n = P_nT$; note that each T_n has finite rank. It will be shown that $\|T_n - T\|\to 0$, but first we prove the following:

Claim. If $h\in\mathscr{H}$, $\|T_nh - Th\|\to 0$.

In fact, $k = Th\in\mathscr{L}$, so $\|P_nk - k\|\to 0$ by (I.4.13d) and (I.4.7). That is, $\|P_nTh - Th\|\to 0$ and the claim is proved.

Since T is compact, if $\varepsilon > 0$, there are vectors $h_1,\ldots,h_m$ in ball $\mathscr{H}$ such that T(ball $\mathscr{H}$)$\subseteq\bigcup_{j=1}^m B(Th_j;\varepsilon/3)$. So if $\|h\|\leqslant 1$, choose h_j with $\|Th - Th_j\| < \varepsilon/3$. Thus for any integer n,

$$\|Th - T_nh\| \leqslant \|Th - Th_j\| + \|Th_j - T_nh_j\| + \|P_n(Th_j - Th)\|$$
$$\leqslant 2\|Th - Th_j\| + \|Th_j - T_nh_j\|$$
$$\leqslant 2\varepsilon/3 + \|Th_j - T_nh_j\|.$$

Using the claim we can find an integer n_0 such that $\|Th_j - T_nh_j\| < \varepsilon/3$ for $1\leqslant j\leqslant m$ and $n\geqslant n_0$. So $\|Th - T_nh\| < \varepsilon$ uniformly for h in ball $\mathscr{H}$. Therefore $\|T - T_n\| < \varepsilon$ for $n\geqslant n_0$.

(c)⇒(b): If $\{T_n\}$ is a sequence in $\mathscr{B}_{00}(\mathscr{H},\mathscr{K})$ such that $\|T_n - T\|\to 0$, then $\|T_n^* - T^*\| = \|T_n - T\|\to 0$. But $T_n^*\in\mathscr{B}_{00}(\mathscr{H},\mathscr{K})$ (Exercise 3). Since (c) implies (a), T^* is compact.

(b)⇒(a): Exercise. ■

A fact emerged in the proof that (a) implies (c) in the preceding theorem that is worth recording.

4.5. Corollary. *If $T\in\mathscr{B}_0(\mathscr{H},\mathscr{K})$, then cl(ran T) is separable and if $\{e_n\}$ is a basis for cl(ran T) and P_n is the projection of $\mathscr{K}$ onto $\bigvee\{e_j: 1\leqslant j\leqslant n\}$, then $\|P_nT - T\|\to 0$.*

4.6. Proposition. *Let $\mathscr{H}$ be a separable Hilbert space with basis $\{e_n\}$; let $\{\alpha_n\}\subseteq\mathbb{F}$ with $M = \sup\{|\alpha_n|: n\geqslant 1\} < \infty$. If $Ae_n = \alpha_ne_n$ for all n, then A extends by linearity to a bounded operator on $\mathscr{H}$ with $\|A\| = M$. The operator A is compact if and only if $\alpha_n\to 0$ as $n\to\infty$.*

PROOF. The fact that A is bounded and $\|A\| = M$ is an exercise; such an operator is said to be diagonalizable (see Exercise 1.8). Let P_n be the projection of $\mathscr{H}$ onto $\bigvee\{e_1,\ldots,e_n\}$. Then $A_n = A - AP_n$ is seen to be diagonalizable with $A_ne_j = \alpha_je_j$ if $j > n$ and $A_ne_j = 0$ if $j\leqslant n$. So $AP_n\in\mathscr{B}_{00}(\mathscr{H})$ and $\|A_n\| = \sup\{|\alpha_j|: j > n\}$. If $\alpha_n\to 0$, then $\|A_n\|\to 0$ and so A is compact since it is the limit of a sequence of finite-rank operators. Conversely, if A is compact, then Corollary 4.5 implies $\|A_n\| \to 0$; hence $\alpha_n \to 0$.   ■

4.7. Proposition. *If (X, Ω, μ) is a measure space and $k \in L^2(X \times X, \Omega \times \Omega, \mu \times \mu)$, then*

$$(Kf)(x) = \int k(x, y) f(y) d\mu(y)$$

is a compact operator and $\|K\| \leqslant \|k\|_2$.

The following lemma is useful for proving this proposition. The proof is left to the reader.

4.8. Lemma. *If $\{e_i \colon i \in I\}$ is a basis for $L^2(X, \Omega, \mu)$ and*

$$\phi_{ij}(x, y) = e_j(x)\overline{e_i(y)}$$

for i, j in I and x, y in X, then $\{\phi_{ij} \colon i, j \in I\}$ is an orthonormal set in $L^2(X \times X, \Omega \times \Omega, \mu \times \mu)$. If k and K are as in the preceding proposition, then $\langle k, \phi_{ij} \rangle = \langle Ke_j, e_i \rangle$.

PROOF OF PROPOSITION 4.7. First we show that K defines a bounded operator. In fact, if $f \in L^2(\mu)$, $\|Kf\|^2 = \int |\int k(x, y)f(y)d\mu(y)|^2 d\mu(x) \leqslant \int (\int |k(x, y)|^2 d\mu(y)) \cdot (\int |f(y)|^2 d\mu(y)) d\mu(x) = \|k\|^2 \|f\|^2$. Hence K is bounded and $\|K\| \leqslant \|k\|_2$.

Now let $\{e_i\}$ be a basis for $L^2(\mu)$ and define ϕ_{ij} as in Lemma 4.8. Thus

$$\|k\|^2 \geqslant \sum_{i,j} |\langle k, \phi_{ij} \rangle|^2 = \sum_{i,j} |\langle Ke_j, e_i \rangle|^2.$$

Since $k \in L^2(\mu \times \mu)$, there are at most a countable number of i and j such that $\langle k, \phi_{ij} \rangle \neq 0$; denote these by $\{\psi_{km} \colon 1 \leqslant k, m < \infty\}$. Note that $\langle Ke_j, e_i \rangle = 0$ unless $\phi_{ij} \in \{\psi_{km}\}$. Let $\psi_{km}(x, y) = e_k(x)\overline{e_m(y)}$, let P_n be the orthogonal projection onto $\bigvee \{e_k \colon 1 \leqslant k \leqslant n\}$, and put $K_n = KP_n + P_n K - P_n K P_n$; so K_n is a finite rank operator. We will show that $\|K - K_n\| \to 0$ as $n \to \infty$, thus showing that K is compact.

Let $f \in L^2(\mu)$ with $\|f\|^2 \leqslant 1$; so $f = \sum_j \alpha_j e_j$. Hence

$$\|Kf - K_n f\|^2 = \sum_i |\langle Kf - K_n f, e_i \rangle|^2$$

$$= \sum_i \left| \sum_j \alpha_j \langle (K - K_n)e_j, e_i \rangle \right|^2$$

$$= \sum_k \left| \sum_m \alpha_m \langle (K - K_n)e_m, e_k \rangle \right|^2$$

$$\leqslant \sum_k \left[\sum_m |\alpha_m|^2 \right] \left[\sum_m |\langle (K - K_n)e_m, e_k \rangle|^2 \right]$$

$$\leqslant \|f\|^2 \sum_k \sum_m |\langle Ke_m, e_k\rangle - \langle KP_n e_m, P_n e_k\rangle$$

$$-\langle KP_n e_m, P_n e_k\rangle + \langle KP_n e_m, P_n e_k\rangle|^2$$

$$= \sum_{k=n+1}^{\infty} \sum_{m=n+1}^{\infty} |\langle Ke_m, e_k\rangle|^2$$

$$= \sum_{k=n+1}^{\infty} \sum_{m=n+1}^{\infty} |\langle k, \psi_{km}\rangle|^2.$$

Since $\sum_{k,m} |\langle k, \psi_{km}\rangle|^2 < \infty$, n can be chosen sufficiently large such that for any $\varepsilon > 0$ this last sum will be smaller than ε^2. Thus $\|K - K_n\| \to 0$. ∎

In particular, note that the preceding proposition shows that the Volterra operator (1.7) is compact.

One of the dominant tools in the study of linear transformation on finite dimensional spaces is the concept of eigenvalue.

4.9. Definition. If $A \in \mathcal{B}(\mathcal{H})$, a scalar α is an *eigenvalue* of A if $\ker(A - \alpha) \neq (0)$. If h is a nonzero vector in $\ker(A - \alpha)$, h is called an *eigenvector* for α; thus $Ah = \alpha h$. Let $\sigma_p(A)$ denote the set of eigenvalues of A.

4.10. Example. Let A be the diagonalizable operator in Proposition 4.6. Then $\sigma_p(A) = \{\alpha_1, \alpha_2, \dots\}$. If $\alpha \in \sigma_p(A)$, let $J_\alpha = \{j \in \mathbb{N}: \alpha_j = \alpha\}$. Then h is an eigenvector for α if and only if $h \in \vee \{e_j: j \in J_\alpha\}$.

4.11. Example. The Volterra operator has no eigenvalues.

4.12. Example. Let $h \in \mathcal{H} = L^2_{\mathbb{C}}(-\pi, \pi)$ and define $K: \mathcal{H} \to \mathcal{H}$ by $(Kf)(x) = (2\pi)^{-1/2} \int_{-\pi}^{\pi} h(x - y)f(y)dy$. If $\lambda_n = (2\pi)^{-1/2} \int_{-\pi}^{\pi} h(x)\exp(-inx)dx = \hat{h}(n)$, the nth Fourier coefficient of h, then $Ke_n = \lambda_n e_n$, where $e_n(x) = (2\pi)^{-1/2} \exp(-inx)$.

The way to see this is to extend functions in $L^2_{\mathbb{C}}(-\pi, \pi)$ to $\mathbb{R}$ by periodicity and perform a change of variables in the formula for $(Ke_n)(x)$. The details are left to the reader.

Operators on finite dimensional spaces over $\mathbb{C}$ always have eigenvalues. As the Volterra operator illustrates, the analogy between operators on finite dimensional spaces and compact operators breaks down here. If, however, a compact operator has an eigenvalue, several nice things can be said if the eigenvalue is not zero.

4.13. Proposition. *If* $T \in \mathcal{B}_0(\mathcal{H})$, $\lambda \in \sigma_p(T)$, *and* $\lambda \neq 0$, *then the eigenspace* $\ker(T - \lambda)$ *is finite dimensional.*

PROOF. Suppose there is an infinite orthonormal sequence $\{e_n\}$ in $\ker(T - \lambda)$. Since T is compact, there is a subsequence $\{e_{n_k}\}$ such that $\{Te_{n_k}\}$ converges.

Thus, $\{Te_{n_k}\}$ is a Cauchy sequence. But for $n_k \neq n_j$, $\|Te_{n_k} - Te_{n_j}\|^2 = \|\lambda e_{n_k} - \lambda e_{n_j}\|^2 = 2|\lambda|^2 > 0$ since $\lambda \neq 0$. This contradiction shows that $\ker(T - \lambda)$ must be finite dimensional. ∎

The next result on the existence of eigenvalues is not a practical way to show that a specific example has a nonzero eigenvalue, but it is a good theoretical tool that will be used later in this book (in particular, in the next section).

4.14. Proposition. *If T is a compact operator on $\mathcal{H}$, $\lambda \neq 0$, and $\inf\{\|(T - \lambda)h\| : \|h\| = 1\} = 0$, then $\lambda \in \sigma_p(T)$.*

PROOF. By hypothesis, there is a sequence of unit vectors $\{h_n\}$ such that $\|(T - \lambda)h_n\| \to 0$. Since T is compact, there is a vector f in $\mathcal{H}$ and a subsequence $\{h_{n_k}\}$ such that $\|Th_{n_k} - f\| \to 0$. But $h_{n_k} = \lambda^{-1}[(\lambda - T)h_{n_k} + Th_{n_k}] \to \lambda^{-1}f$. So $1 = \|\lambda^{-1}f\| = |\lambda|^{-1}\|f\|$ and $f \neq 0$. Also, it must be that $Th_{n_k} \to \lambda^{-1}Tf$. Since $Th_{n_k} \to f$, $f = \lambda^{-1}Tf$, or $Tf = \lambda f$. That is, $f \in \ker(T - \lambda)$ and $f \neq 0$, so $\lambda \in \sigma_p(T)$. ∎

4.15. Corollary. *If T is a compact operator on $\mathcal{H}$, $\lambda \neq 0$, $\lambda \notin \sigma_p(T)$, and $\bar{\lambda} \notin \sigma_p(T^*)$, then $\operatorname{ran}(T - \lambda) = \mathcal{H}$ and $(T - \lambda)^{-1}$ is a bounded operator on $\mathcal{H}$.*

PROOF. Since $\lambda \notin \sigma_p(T)$, the preceding proposition implies that there is a constant $c > 0$ such that $\|(T - \lambda)h\| \geq c\|h\|$ for all h in $\mathcal{H}$. If $f \in \operatorname{cl} \operatorname{ran}(T - \lambda)$, then there is a sequence $\{h_n\}$ in $\mathcal{H}$ such that $(T - \lambda)h_n \to f$. Thus $\|h_n - h_m\| \leq c^{-1}\|(T - \lambda)h_n - (T - \lambda)h_m\|$ and so $\{h_n\}$ is a Cauchy sequence. Hence $h_n \to h$ for some h in $\mathcal{H}$. Thus $(T - \lambda)h = f$. So $\operatorname{ran}(T - \lambda)$ is closed and, by (2.19), $\operatorname{ran}(T - \lambda) = [\ker(T - \lambda)^*]^\perp = \mathcal{H}$, by hypothesis.

So for f in $\mathcal{H}$ let $Af =$ the unique vector h such that $(T - \lambda)h = f$. Thus $(T - \lambda)Af = f$ for all f in $\mathcal{H}$. From the inequality above, $c\|Af\| \leq \|(T - \lambda)Af\| = \|f\|$. So $\|Af\| \leq c^{-1}\|f\|$ and A is bounded. Also, $(T - \lambda)A(T - \lambda)h = (T - \lambda)h$, so $0 = (T - \lambda)[A(T - \lambda)h - h]$. Since $\lambda \notin \sigma_p(T)$, $A(T - \lambda)h = h$. That is, $A = (T - \lambda)^{-1}$. ∎

It will be proved in a later chapter that if $\lambda \notin \sigma_p(T)$ and $\lambda \neq 0$, then $\bar{\lambda} \notin \sigma_p(T^*)$.

More will be shown about arbitrary compact operators in Chapter VI. In the next section the theory of compact self-adjoint operators will be explored.

EXERCISES

1. Prove Proposition 4.2(c).

2. Show that every operator of finite rank is compact.

3. If $T \in \mathcal{B}_{00}(\mathcal{H}, \mathcal{K})$, show that $T^* \in \mathcal{B}_{00}(\mathcal{K}, \mathcal{H})$ and $\dim(\operatorname{ran} T) = \dim(\operatorname{ran} T^*)$.

4. Show that an idempotent is compact if and only if it has finite rank.

5. Show that no nonzero multiplication operator on $L^2(0, 1)$ is compact.

6. Show that if $T: \mathcal{H} \to \mathcal{H}$ is a compact operator and $\{e_n\}$ is any orthonormal sequence in $\mathcal{H}$, then $\| Te_n \| \to 0$. Is the converse true?

7. If T is compact and $\mathcal{M}$ is an invariant subspace for T, show that $T | \mathcal{M}$ is compact.

8. If $h, g \in \mathcal{H}$, define $T: \mathcal{H} \to \mathcal{H}$ by $Tf = \langle f, h \rangle g$. Show that T has rank 1 [that is, $\dim(\operatorname{ran} T) = 1$]. Moreover, every rank 1 operator can be so represented. Show that if T is a finite rank operator, then there are orthonormal vectors $e_1, \ldots, e_n$ and vectors $g_1, \ldots, g_n$ such that $Th = \sum_{j=1}^{n} \langle h, e_j \rangle g_j$ for all h in $\mathcal{H}$. In this case show that T is normal if $g_j = \lambda_j e_j$ for some scalars $\lambda_1, \ldots, \lambda_n$. Find $\sigma_p(T)$.

9. Show that a diagonalizable operator is normal.

10. Verify the statements in Example 4.10.

11. Verify the statement in Example 4.11.

12. Verify the statement in Example 4.12. (Note that the operator K in this example is diagonalizable.)

13. If $T_n \in \mathcal{B}(\mathcal{H}_n)$, $n \geqslant 1$, with $\sup_n \| T_n \| < \infty$ and $T = \bigoplus_{n=1}^{\infty} T_n$ on $\mathcal{H} = \bigoplus_{n=1}^{\infty} \mathcal{H}_n$, show that T is compact if and only if each T_n is compact and $\| T_n \| \to 0$.

14. In Lemma 4.8, show that if $L^2(X, \Omega, \mu)$ is separable, then $\{\varphi_{ij}\}$ is a basis for $L^2(X \times X, \Omega \times \Omega, \mu \times \mu)$. What if $L^2(X, \Omega, \mu)$ is not separable?

§5*. The Diagonalization of Compact Self-Adjoint Operators

This section and the remaining ones in this chapter may be omitted if the reader intends to continue through to the end of this book, as the material in these sections (save for Section 6) will be obtained in greater generality in Chapter IX. It is worthwhile, however, to examine this material even if Chapter IX is to be read, since the intuition provided by this special case is valuable.

The main result of this section is the following.

5.1. Theorem. *If T is a compact self-adjoint operator on $\mathcal{H}$, then T has only a countable number of distinct eigenvalues. If $\{\lambda_1, \lambda_2, \ldots\}$ are the distinct nonzero eigenvalues of T, and P_n is the projection of $\mathcal{H}$ onto $\ker(T - \lambda_n)$, then $P_n P_m = P_m P_n = 0$ if $n \neq m$, each λ_n is real, and*

5.2
$$T = \sum_{n=1}^{\infty} \lambda_n P_n,$$

where the series converges to T in the metric defined by the norm of $\mathcal{B}(\mathcal{H})$. [Of course, (5.2) may be only a finite sum.]

The proof of Theorem 5.1 requires a few preliminary results. Before beginning this process, let's look at a few consequences.

5.3. Corollary. *With the notation of* (5.1):

(a) $\ker T = [\vee \{ P_n \mathcal{H} : n \geq 1 \}]^\perp = (\operatorname{ran} T)^\perp$;
(b) *each P_n has finite rank*;
(c) $\| T \| = \sup \{ |\lambda_n| : n \geq 1 \}$ *and* $\lambda_n \to 0$ *as* $n \to \infty$.

PROOF. Since $P_n \perp P_m$ for $n \neq m$, if $h \in \mathcal{H}$, then (5.2) implies $\| Th \|^2 = \sum_{n=1}^\infty \| \lambda_n P_n h \|^2 = \sum_{n=1}^\infty |\lambda_n|^2 \| P_n h \|^2$. Hence $Th = 0$ if and only if $P_n h = 0$ for all n. That is, $h \in \ker T$ if and only if $h \perp P_n \mathcal{H}$ for all n, whence (a).

Part (b) follows by Proposition 4.13.

For part (c), if $\mathcal{L} = \operatorname{cl}[\operatorname{ran} T]$, $\mathcal{L}$ is invariant for T. Since $T = T^*$, $\mathcal{L} = (\ker T)^\perp$ and $\mathcal{L}$ reduces T. So we can consider the restriction of T to $\mathcal{L}$, $T|\mathcal{L}$. Now $\mathcal{L} = \vee \{ P_n \mathcal{H} : n \geq 1 \}$ by (a). Let $\{ e_j^{(n)} : 1 \leq j \leq N_n \}$ be a basis for $P_n \mathcal{H} = \ker(T - \lambda_n)$, so $Te_j^{(n)} = \lambda_n e_j^{(n)}$ for $1 \leq j \leq N_n$. Thus $\{ e_j^{(n)} : 1 \leq j \leq N_n, n \geq 1 \}$ is a basis for $\mathcal{L}$ and $T|\mathcal{L}$ is diagonalizable with respect to this basis. Part (c) now follows by (4.6). ∎

The proof of (c) in the preceding corollary revealed an interesting fact that deserves a statement of its own.

5.4. Corollary. *If T is a compact self-adjoint operator, then there is a sequence $\{\mu_n\}$ of real numbers and an orthonormal basis $\{e_n\}$ for $(\ker T)^\perp$ such that for all h,*

$$Th = \sum_{n=1}^\infty \mu_n \langle h, e_n \rangle e_n.$$

Note that there may be repetitions in the sequence $\{\mu_n\}$ in (5.4). How many repetitions?

5.5. Corollary. *If $T \in \mathcal{B}_0(\mathcal{H})$, $T = T^*$, and $\ker T = (0)$, then $\mathcal{H}$ is separable.*

Also note that by (4.6), if (5.2) holds, $T \in \mathcal{B}_0(\mathcal{H})$.

To begin the proof of Theorem 5.1, we prove a few results about not necessarily compact operators.

5.6. Proposition. *If A is a normal operator and $\lambda \in \mathbb{F}$, then $\ker(A - \lambda) = \ker(A - \lambda)^*$ and $\ker(A - \lambda)$ is a reducing subspace for A.*

PROOF. Since A is normal, so is $A - \lambda$. Hence $\| (A - \lambda)h \| = \| (A - \lambda)^* h \|$ (2.16). Thus $\ker(A - \lambda) = \ker(A - \lambda)^*$. If $h \in \ker(A - \lambda)$, $Ah = \lambda h \in \ker(A - \lambda)$. Also $A^* h = \bar{\lambda} h \in \ker(A - \lambda)$. Therefore $\ker(A - \lambda)$ reduces A. ∎

5.7. Proposition. *If A is a normal operator and λ, μ are distinct eigenvalues of A, then $\ker(A - \lambda) \perp \ker(A - \mu)$.*

PROOF. If $h \in \ker(A - \lambda)$ and $g \in \ker(A - \mu)$, then the fact (5.6) that $A^*g = \bar{\mu}g$ implies that $\lambda \langle h, g \rangle = \langle Ah, g \rangle = \langle h, A^*g \rangle = \langle h, \bar{\mu}g \rangle = \mu \langle h, g \rangle$. Thus $(\lambda - \mu)\langle h, g \rangle = 0$. Since $\lambda - \mu \neq 0$, $h \perp g$.   ∎

5.8. Proposition. *If $A = A^*$ and $\lambda \in \sigma_p(A)$, then λ is a real number.*

PROOF. If $Ah = \lambda h$, then $Ah = A^*h = \bar{\lambda}h$ by (5.6). So $(\lambda - \bar{\lambda})h = 0$. Since h can be chosen different from 0, $\lambda = \bar{\lambda}$.   ∎

The main result prior to entering the proof of Theorem 5.1 is to show that a compact self-adjoint operator has nonzero eigenvalues. If (5.3c) is examined, we see that there is a λ_n in $\sigma_p(T)$ with $|\lambda_n| = \|T\|$. Since the preceding proposition says that $\lambda_n \in \mathbb{R}$, it must be that $\lambda_n = \pm \|T\|$. That is, either $\pm \|T\| \in \sigma_p(T)$. This is the key to showing that $\sigma_p(T)$ is nonvoid.

5.9. Lemma. *If T is a compact self-adjoint operator, then either $\pm \|T\|$ is an eigenvalue of T.*

PROOF. If $T = 0$, the result is clear. So suppose $T \neq 0$. By Proposition 2.13 there is a sequence $\{h_n\}$ of unit vectors such that $|\langle Th_n, h_n \rangle| \to \|T\|$. By passing to a subsequence if necessary, we may assume that $\langle Th_n, h_n \rangle \to \lambda$, where $|\lambda| = \|T\|$. It will be shown that $\lambda \in \sigma_p(T)$. Since $|\lambda| = \|T\|$, $0 \leqslant \|(T - \lambda)h_n\|^2 = \|Th_n\|^2 - 2\lambda \langle Th_n, h_n \rangle + \lambda^2 \leqslant 2\lambda^2 - 2\lambda \langle Th_n, h_n \rangle \to 0$. Hence $\|(T - \lambda)h_n\| \to 0$. By (4.14), $\lambda \in \sigma_p(T)$.   ∎

PROOF OF THEOREM 5.1. By Lemma 5.9 there is a real number λ_1 in $\sigma_p(T)$ with $|\lambda_1| = \|T\|$. Let $\mathscr{E}_1 = \ker(T - \lambda_1)$, $P_1 =$ the projection onto $\mathscr{E}_1$, $\mathscr{H}_2 = \mathscr{E}_1^\perp$. By (5.6) $\mathscr{E}_1$ reduces T, so $\mathscr{H}_2$ reduces T. Let $T_2 = T|\mathscr{H}_2$; then T_2 is a self-adjoint compact operator on $\mathscr{H}_2$. (Why?)

By (5.9) there is an eigenvalue λ_2 for T_2 such that $|\lambda_2| = \|T_2\|$. Let $\mathscr{E}_2 = \ker(T_2 - \lambda_2)$. Note that $(0) \neq \mathscr{E}_2 \subseteq \ker(T - \lambda_2)$. If it were the case that $\lambda_1 = \lambda_2$, then $\mathscr{E}_2 \subseteq \ker(T - \lambda_1) = \mathscr{E}_1$. Since $\mathscr{E}_1 \perp \mathscr{E}_2$, it must be that $\lambda_1 \neq \lambda_2$. Let $P_2 =$ the projection of $\mathscr{H}$ onto $\mathscr{E}_2$ and put $\mathscr{H}_3 = (\mathscr{E}_1 \oplus \mathscr{E}_2)^\perp$. Note that $\|T_2\| \leqslant \|T\|$ so that $|\lambda_2| \leqslant |\lambda_1|$.

Using induction (give the details) we obtain a sequence $\{\lambda_n\}$ of real eigenvalues of T such that

(i) $|\lambda_1| \geqslant |\lambda_2| \geqslant \cdots$;

(ii) If $\mathscr{E}_n = \ker(T - \lambda_n)$, $|\lambda_{n+1}| = \|T|(\mathscr{E}_1 \oplus \cdots \oplus \mathscr{E}_n)^\perp\|$.

By (i) there is a nonnegative number α such that $|\lambda_n| \to \alpha$.

Claim. $\alpha = 0$; that is, $\lim \lambda_n = 0$.

In fact, let $e_n \in \mathscr{E}_n$, $\|e_n\| = 1$. Since T is compact, there is an h in $\mathscr{H}$ and a subsequence $\{e_{n_j}\}$ such that $\|Te_{n_j} - h\| \to 0$. But $e_n \perp e_m$ for $n \neq m$ and $Te_{n_j} = \lambda_{n_j}e_{n_j}$. Hence $\|Te_{n_j} - Te_{n_i}\|^2 = \lambda_{n_j}^2 + \lambda_{n_i}^2 \geqslant 2\alpha^2$. Since $\{Te_{n_j}\}$ is a Cauchy sequence, $\alpha = 0$.

Now put P_n = the projection of $\mathscr{H}$ onto $\mathscr{E}_n$ and examine $T - \sum_{j=1}^{n} \lambda_j P_j$. If $h \in \mathscr{E}_k$, $1 \leqslant k \leqslant n$, then $(T - \sum_{j=1}^{n} \lambda_j P_j)h = Th - \lambda_k h = 0$. Hence $\mathscr{E}_1 \oplus \cdots \oplus \mathscr{E}_n \subseteq \ker(T - \sum_{j=1}^{n} \lambda_j P_j)$. If $h \in (\mathscr{E}_1 \oplus \cdots \oplus \mathscr{E}_n)^{\perp}$, then $P_j h = 0$ for $1 \leqslant j \leqslant n$; so $(T - \sum_{j=1}^{n} \lambda_j P_j)h = Th$. These two statements, together with the fact that $(\mathscr{E}_1 \oplus \cdots \oplus \mathscr{E}_n)^{\perp}$ reduces T, imply that

$$\left\| T - \sum_{j=1}^{n} \lambda_j P_j \right\| = \| T|(\mathscr{E}_1 \oplus \cdots \oplus \mathscr{E}_n)^{\perp} \|$$

$$= |\lambda_{n+1}| \to 0.$$

Therefore the series $\sum_{n=1}^{\infty} \lambda_n P_n$ converges in the metric of $\mathscr{B}(\mathscr{H})$ to T. ∎

Theorem 5.1 is called the *Spectral Theorem* for compact self-adjoint operators. Using it, one can answer virtually every question about compact hermitian operators, as will be seen before the end of this chapter.

If in Theorem 5.1 it is assumed that T is normal and compact, then the same conclusion, except for the statement that each λ_n is real, is true provided that $\mathscr{H}$ is a $\mathbb{C}$-Hilbert space. The proof of this will be given in Section 7.

EXERCISES

1. Prove Corollary 5.4.

2. Prove Corollary 5.5.

3. Let K and k be as in Proposition 4.7 and suppose that $k(x, y) = \overline{k(y, x)}$. Show that K is self-adjoint and if $\{\mu_n\}$ are the eigenvalues of K, each repeated $\dim \ker(K - \mu_n)$ times, then $\sum_{1}^{\infty} |\mu_n|^2 < \infty$.

4. If T is a compact self-adjoint operator and $\{e_n\}$ and $\{\mu_n\}$ are as in (5.4) and if h is a given vector in $\mathscr{H}$, show that there is a vector f in $\mathscr{H}$ such that $Tf = h$ if and only if $h \perp \ker T$ and $\sum_n \mu_n^{-2} |\langle h, e_n \rangle|^2 < \infty$. Find the form of the general vector f such that $Tf = h$.

5. Let T, $\{\mu_n\}$, and $\{e_n\}$ be as in (5.4). If $\lambda \neq 0$ and $\lambda \neq \mu_n$ for any μ_n, then for every h in $\mathscr{H}$ there is a unique f in $\mathscr{H}$ such that $(\lambda - T)f = h$. Moreover, $f = \lambda^{-1}[h + \sum_{n=1}^{\infty} \lambda_n(\lambda - \lambda_n)^{-1} \langle h, e_n \rangle e_n]$. Interpret this when T is an integral operator.

§6*. An Application: Sturm–Liouville Systems

In this section, $[a, b]$ will be a proper interval with $-\infty < a < b < \infty$. $C[a, b]$ denotes the continuous functions $f : [a, b] \to \mathbb{R}$ and for $n \geqslant 1$, $C^{(n)}[a, b]$ denotes those functions in $C[a, b]$ that have n continuous derivatives. $C_{\mathbb{C}}^{(n)}[a, b]$ denotes the corresponding spaces of complex-valued functions. We want to consider the differential equation

6.1 $$-h'' + qh - \lambda h = f,$$

where λ is a given complex number, $q \in C[a,b]$, and $f \in L^2[a,b]$, together with the boundary conditions

6.2 $\begin{cases} \text{(a) } \alpha h(a) + \alpha_1 h'(a) = 0 \\ \text{(b) } \beta h(b) + \beta_1 h'(b) = 0 \end{cases}$,

where α, α_1, β, and β_1 are real numbers and $\alpha^2 + \alpha_1^2 > 0$, $\beta^2 + \beta_1^2 > 0$.

Equation (6.1) together with the boundary conditions (6.2) is called a (*regular*) *Sturm–Liouville system*. Such systems arise in a number of physical problems, including the description of the motion of a vibrating string. In this section we will discuss solutions of the Sturm–Liouville system by relating the system to a certain compact self-adjoint integral operator.

Recall that an absolutely continuous function h on $[a,b]$ has a derivative a.e. and $h(x) = \int_a^x h'(t)\,dt + h(a)$ for all x.

Define

$$\mathscr{D}_a \equiv \{h \in C_{\mathbb{C}}^{(1)}[a,b]: h' \text{ is absolutely continuous,}$$
$$h'' \in L^2[a,b], \text{ and } h \text{ satisfies (6.2a)}\}.$$

$\mathscr{D}_b$ is defined similarly but each h in $\mathscr{D}_b$ satisfies (6.2b) instead of (6.2a). The space $\mathscr{D} = \mathscr{D}_a \cap \mathscr{D}_b$.

Define $L: \mathscr{D} \to L^2[a,b]$ by

6.3 $Lh = -h'' + qh.$

L is called a *Sturm–Liouville operator*.

Note that $\mathscr{D}$ is a linear space and L is a linear transformation. The Sturm–Liouville problem thus becomes: if $\lambda \in \mathbb{C}$ and $f \in L^2[a,b]$, is there an h in $\mathscr{D}$ with $(L - \lambda)h = f$. Equivalently, for which λ is f in $\operatorname{ran}(L - \lambda)$?

By placing a suitable norm on $\mathscr{D}$, L can be made into a bounded operator. This does not help much. The best procedure is to consider $(L - \lambda)^{-1}$. Integration is the inverse of differentiation, and it turns out that $(L - \lambda)^{-1}$ (when we can define it) is an integral operator.

Begin by considering the case when $\lambda = 0$. (Equivalently, replace q by $q - \lambda$.) To define L^{-1} (even if only on the range of L), we need that L is injective. Thus we make an assumption;

6.4   if $h \in \mathscr{D}$ and $Lh = 0$,   then $h = 0$.

The first lemma is from ordinary differential equations and says that certain initial-value problems have nontrivial (nonzero) solutions.

6.5. Lemma. *If* $\alpha, \alpha_1, \beta, \beta_1 \in \mathbb{R}$, $\alpha^2 + \alpha_1^2 > 0$, *and* $\beta^2 + \beta_1^2 > 0$, *then there are functions* h_a, h_b *in* $\mathscr{D}_a$, $\mathscr{D}_b$, *respectively, such that* $L(h_a) = 0$ *and* $L(h_b) = 0$ *and* h_a, h_b *are real-valued and not identically zero.*

The *Wronskian* of h_a and h_b is the function

$$W = \det \begin{bmatrix} h_a & h_b \\ h_a' & h_b' \end{bmatrix} = h_a h_b' - h_a' h_b.$$

Note that $W' = h_a h_b'' - h_a'' h_b = h_a(qh_b) - (qh_a)h_b = 0$. Hence $W(x) \equiv W(a)$ for all x.

6.6. Lemma. *Assuming* (6.4), $W(a) \neq 0$ *and so* h_a *and* h_b *are linearly independent.*

PROOF. If $W(a) = 0$, then linear algebra tells us that the column vectors in the matrix used to define $W(a)$ are linearly dependent. Thus there is a λ in $\mathbb{R}$ such that $h_b(a) = \lambda h_a(a)$ and $h_b'(a) = \lambda h_a'(a)$. Thus $h_b \in \mathscr{D}$ and $L(h_b) = 0$. By (6.4), $h_b \equiv 0$, contradiction. ∎

Put $c = - W(a)$ and define $g: [a, b] \times [a, b] \to \mathbb{R}$ by

6.7
$$g(x, y) = \begin{cases} c^{-1}h_a(x)h_b(y) & \text{if } a \leqslant x \leqslant y \leqslant b \\ c^{-1}h_a(y)h_b(x) & \text{if } a \leqslant y \leqslant x \leqslant b. \end{cases}$$

The function g is the *Green function* for L.

6.8. Lemma. *The function* g *defined in* (6.7) *is real-valued, continuous, and* $g(x, y) = g(y, x)$.

PROOF. Exercise.

6.9. Theorem. *Assume* (6.4). *If* g *is the Green function for* L *defined in* (6.7) *and* $G: L^2[a, b] \to L^2[a, b]$ *is the integral operator defined by*

$$(Gf)(x) = \int_a^b g(x, y) f(y) \, dy,$$

then G *is a compact self-adjoint operator,* ran $G = \mathscr{D}$, $LGf = f$ *for all* f *in* $L^2[a, b]$, *and* $GLh = h$ *for all* h *in* $\mathscr{D}$.

PROOF. That G is self-adjoint follows from the fact that g is real-valued and $g(x, y) = g(y, x)$; G is compact by (4.7). Fix f in $L^2[a, b]$ and put $h = Gf$. It must be shown that $h \in \mathscr{D}$.
  Put

$$H_a(x) = c^{-1} \int_a^x h_a(y) f(y) \, dy \quad \text{and} \quad H_b(x) = c^{-1} \int_x^b h_b(y) f(y) \, dy.$$

Then

$$h(x) = \int_a^b g(x, y) f(y) \, dy$$

$$= c^{-1} \int_a^x h_a(y) h_b(x) f(y) \, dy + c^{-1} \int_x^b h_a(x) h_b(y) f(y) \, dy.$$

That is, $h = H_a h_b + h_a H_b$. Differentiating this equation gives $h' = (c^{-1} h_a f)h_b + H_a h_b' + h_a' H_b + h_a(-c^{-1} h_b f) = H_a h_b' + h_a' H_b$ a.e. Since $H_a h_b' + h_a' H_b$ is absolutely continuous, as part of showing that $h \in \mathscr{D}$ we want to show the following.

Claim. $h' = H_a h'_b + h'_a H_b$ everywhere.

Put $\phi = H_a h'_b + h'_a H_b$ and put $\psi(x) = h(a) + \int_a^x \phi(y)\,dy$. So ϕ and ψ are absolutely continuous, $h(a) = \psi(a)$, and $h' = \psi'$ a.e. Thus $h = \psi$ everywhere. But ψ has a continuous derivative ϕ, so h does too. That is, the claim is proved.

Differentiating $h' = H_a h'_b + h'_a H_b$ gives that a.e., $h'' = (c^{-1} h_a f)h'_b + H_a h''_b + h''_a H_b + h'_a(-c^{-1} h_b f)$; since each of these summands belongs to $L^2[a,b]$, $h'' \in L^2[a,b]$.

Because $H_a(a) = 0$ and $h_a \in \mathcal{D}_a$, $\alpha h(a) + \alpha_1 h'(a) = \alpha h_a(a)H_b(a) + \alpha_1 h'_a(a)H_b(a) = [\alpha h_a(a) + \alpha_1 h'_a(a)]H_b(a) = 0$. Hence $h \in \mathcal{D}_a$. Similarly, $h \in \mathcal{D}_b$. Thus $h \in \mathcal{D}$. Hence ran $G \subseteq \mathcal{D}$.

Now to show that $LGf = f$. If $h = Gf$, $L(h) = -h'' + qh = -(c^{-1} h_a h'_b f + H_a h''_b + h''_a H_b - c^{-1} h'_a h_b f] + q(H_a h_b + h_a H_b) = (-h''_b + qh_b)H_a + (-h''_a + qh_a)H_b + c^{-1}(h'_a h_b - h_a h'_b)f = f$ since $L(h_a) = L(h_b) = 0$ and $h'_a h_b - h_a h'_b = W = c$.

If $h \in \mathcal{D}$, then $Lh \in L^2[a,b]$. So by the first part of the proof, $LGLh = Lh$. Thus $0 = L(GLh - h)$. Since $\ker L = (0)$, $h = GLh$ and so $h \in$ ran G. ∎

6.10. Corollary. *Assume* (6.4). *If* $h \in \mathcal{D}$, $\lambda \in \mathbb{C}\backslash\{0\}$, *and* $Lh = \lambda h$, *then* $Gh = \lambda^{-1} h$. *If* $h \in L^2[a,b]$ *and* $Gh = \lambda^{-1} h$, *then* $h \in \mathcal{D}$ *and* $Lh = \lambda h$.

PROOF. This is immediate from the theorem. ∎

6.11. Lemma. *Assume* (6.4). *If* $\alpha \in \sigma_p(G)$ *and* $\alpha \neq 0$, *then* $\dim \ker(G - \alpha) = 1$.

PROOF. Suppose there are linearly independent functions h_1, h_2 in $\ker(G - \alpha)$. By (6.10), h_1, h_2 are solutions of the equation

$$-h'' + (q - \alpha^{-1})h = 0.$$

Since this is a second-order linear differential equation, every solution of it must be a linear combination of h_1 and h_2. But $h_1, h_2 \in \mathcal{D}$ so they satisfy (6.2). But a solution can be found to this equation satisfying any initial conditions at a—and thus not satisfying (6.2). This contradiction shows that linearly independent h_1, h_2 in $\ker(G - \alpha)$ cannot be found. ∎

6.12. Theorem. *Assume* (6.4). *Then there is a sequence* $\{\lambda_1, \lambda_2, \ldots\}$ *of real numbers and a basis* $\{e_1, e_2, \ldots\}$ *for* $L^2[a,b]$ *such that*

(a) $0 < |\lambda_1| < |\lambda_2| < \cdots$ *and* $|\lambda_n| \to \infty$.
(b) $e_n \in \mathcal{D}$ *and* $Le_n = \lambda_n e_n$ *for all* n.
(c) *If* $\lambda \neq \lambda_n$ *for any* λ_n *and* $f \in L^2[a,b]$, *then there is a unique* h *in* $\mathcal{D}$ *with* $Lh - \lambda h = f$.
(d) *If* $\lambda = \lambda_n$ *for some* n *and* $f \in L^2[a,b]$, *then there is an* h *in* $\mathcal{D}$ *with* $Lh - \lambda h = f$ *if and only if* $\langle f, e_n \rangle = 0$. *If* $\langle f, e_n \rangle = 0$, *any two solutions of* $Lh - \lambda h = f$ *differ by a multiple of* e_n.

PROOF. Parts (a) and (b) follow by Theorem 5.1, Corollary 6.10, and

Lemma 6.1. For parts (c) and (d), first note that

6.13 $Lh - \lambda h = f$ if and only if $h - \lambda Gh = Gf$.

This is, in fact, a straightforward consequence of Theorem 6.9.

(c) The case where $\lambda = 0$ is left to the reader. If $\lambda \neq \lambda_n$ for any n, $\lambda^{-1} \notin \sigma_p(G)$. Since $G = G^*$, Corollary 4.15 implies $G - \lambda^{-1}$ is bijective. So if $f \in L^2[a,b]$, there is a unique h in $L^2[a,b]$ with $Gf = (\lambda^{-1} - G)h$. Thus $h \in \mathscr{D}$ and (6.13) implies $L(h/\lambda) - \lambda(h/\lambda) = f$.

(d) Suppose $\lambda = \lambda_n$ for some n. If $Lh - \lambda_n h = f$, then $h - \lambda_n Gh = Gf$. Hence $\langle Gf, e_n \rangle = \langle h, e_n \rangle - \lambda_n \langle Gh, e_n \rangle = \langle h, e_n \rangle - \lambda_n \langle h, Ge_n \rangle = \langle h, e_n \rangle - \lambda_n \lambda_n^{-1} \langle h, e_n \rangle = 0$. So $0 = \langle Gf, e_n \rangle = \langle f, Ge_n \rangle = \lambda_n \langle f, e_n \rangle$. Hence $f \perp e_n$.

Since $\mathbb{C}\, e_n = \ker(G - \lambda_n^{-1})$, $[e_n]^\perp \equiv \mathscr{N}$ reduces G. Let $G_1 = G | \mathscr{N}$. So G_1 is a compact self-adjoint operator on $\mathscr{N}$ and $\lambda_n^{-1} \notin \sigma_p(G_1)$. By (4.15), $\operatorname{ran}(G_1 - \lambda_n^{-1}) = \mathscr{N}$. As in the proof of (c), if $f \perp e_n$, there is a unique h and $\mathscr{N}$ such that $Lh - \lambda_n h = f$. Note that $h + \alpha e_n$ is also a solution. If h_1, h_2 are two solutions, $h_1 - h_2 \in \ker(L - \lambda_n)$, so $h_1 - h_2 = \alpha e_n$. ∎

What happens if $\ker L \neq (0)$? In this case it is possible to find a real number μ such that $\ker(L - \mu) = (0)$ (Exercise 6). Replacing q by $q - \mu$, Theorem 6.12 now applies. More information on this problem can be found in Exercises 2 through 5.

EXERCISES

1. Consider the Sturm–Liouville operator $Lh = -h''$ with $a = 0$, $b = 1$, and for each of the following boundary conditions find the eigenvalues $\{\lambda_n\}$, the eigenvectors $\{e_n\}$, and the Green function $g(x, y)$: (a) $h(0) = h(1) = 0$; (b) $h'(0) = h'(1) = 0$; (c) $h(0) = 0$ and $h'(1) = 0$; (d) $h(0) = h'(0)$ and $h(1) = -h'(1)$.

2. In Theorem 6.12 show that $\sum_{n=1}^{\infty} \lambda_n^{-2} < \infty$ (see Exercise 5.3).

3. In Theorem 6.12 show that $h \in \mathscr{D}$ if and only if $h \in L^2[a,b]$ and $\sum_{n=1}^{\infty} \lambda_n^2 |\langle h, e_n \rangle|^2 < \infty$. If $h \in \mathscr{D}$, show that $h(x) = \sum_{n=1}^{\infty} \langle h, e_n \rangle e_n(x)$, where this series converges uniformly and absolutely on $[a,b]$.

4. In Theorem 6.12(c), show that $h(x) = \sum_{n=1}^{\infty} (\lambda_n - \lambda)^{-1} \langle f, e_n \rangle e_n(x)$ and this series converges uniformly and absolutely on $[a,b]$.

5. In Theorem 6.12(d), show that if $f \perp e_n$ and $Lh - \lambda_n h = f$, then $h(x) = \sum_{j \neq n} (\lambda_j - \lambda_n)^{-1} \langle f, e_j \rangle e_j(x) + \alpha e_n(x)$ for some α, where the series converges uniformly and absolutely on $[a,b]$.

6. This exercise demonstrates how to handle the case in which $\ker L \neq (0)$. (a) If $h, g \in C^{(1)}[a,b]$ with h', g' absolutely continuous and $h'', g'' \in L^2[a,b]$, show that

$$\int_a^b (h''g - hg'') = [h'(b)g(b) - h(b)g'(b)] - [h'(a)g(a) - h(a)g'(a)].$$

(b) If $h, g \in \mathscr{D}$, show that $\langle Lh, g \rangle = \langle h, Lg \rangle$. (The inner product is in $L^2[a,b]$.)
(c) If $h, g \in \mathscr{D}$ and $\lambda, \mu \in \mathbb{R}$, $\lambda \neq \mu$, and if $h \in \ker(L - \lambda)$, $g \in \ker(L - \mu)$, then $h \perp g$.
(d) Show that there is a real number μ with $\ker(L - \mu) = (0)$.

§7*. The Spectral Theorem and Functional Calculus for Compact Normal Operators

We begin by characterizing the operators that commute with a diagonalizable operator. If one considers the definition of a diagonalizable operator (4.6), it is possible to reformulate it in a way that is more tractable for the present purpose and closer to the form of a compact self-adjoint operator given in (5.2). Unlike (4.6), it will not be assumed that the underlying Hilbert space is separable.

7.1. Proposition. *Let* $\{P_i: i \in I\}$ *be a family of pairwise orthogonal projections in* $\mathscr{B}(\mathscr{H})$. *(That is,* $P_i P_j = P_j P_i = 0$ *for* $i \neq j$.) *If* $h \in \mathscr{H}$, *then* $\sum_i \{P_i h: i \in I\}$ *converges in* $\mathscr{H}$ *to* Ph, *where* P *is the projection of* $\mathscr{H}$ *onto* $\vee \{P_i \mathscr{H}: i \in I\}$.

This appeared as Exercise 3.5 and its proof is left to the reader.

If $\{P_i: i \in I\}$ is as in the preceding proposition and $\mathscr{M}_i = P_i \mathscr{H}$, then with the notation of Definition 3.4, P is the projection of $\mathscr{H}$ onto $\oplus_i \mathscr{M}_i$. Write $P = \sum_i P_i$. A word of caution here: $Ph = \sum_i P_i h$, where the convergence is in the norm of $\mathscr{H}$. However, $\sum_i P_i$ does not converge to P in the norm of $\mathscr{B}(\mathscr{H})$. In fact, it never does unless I is finite (Exercise 1).

7.2. Definition. A *partition of the identity* on $\mathscr{H}$ is a family $\{P_i: i \in I\}$ of pairwise orthogonal projections on $\mathscr{H}$ such that $\vee_i P_i \mathscr{H} = \mathscr{H}$. This might be indicated by $1 = \sum_i P_i$ or $1 = \oplus_i P_i$. [Note that 1 is often used to denote the operator on $\mathscr{H}$ defined by $1(h) = h$ for all h. Similarly if $\alpha \in \mathbb{F}$, α is the operator defined by $\alpha(h) = \alpha h$ for all h.]

7.3. Definition. An operator A on $\mathscr{H}$ is *diagonalizable* if there is a partition of the identity on $\mathscr{H}$, $\{P_i: i \in I\}$, and a family of scalars $\{\alpha_i: i \in I\}$ such that $\sup_i |\alpha_i| < \infty$ and $Ah = \alpha_i h$ whenever $h \in \operatorname{ran} P_i$.

It is easy to see that this is equivalent to the definition given in (4.6) when $\mathscr{H}$ is separable (Exercise 2). Also, $\|A\| = \sup_i |\alpha_i|$.

To denote a diagonalizable operator satisfying the conditions of (7.3), write

$$A = \sum_i \alpha_i P_i \quad \text{or} \quad A = \oplus_i \alpha_i P_i.$$

Note that it was not assumed that the scalars α_i in (7.3) are distinct. There is no loss in generality in assuming this, however. In fact, if $\alpha_i = \alpha_j$, then we can replace P_i and P_j with $P_i + P_j$.

7.4. Proposition. *An operator* A *on* $\mathscr{H}$ *is diagonalizable if and only if there is an orthonormal basis for* $\mathscr{H}$ *consisting of eigenvectors for* A.

PROOF. Exercise.

Also note that if $A = \oplus_i \alpha_i P_i$, then $A^* = \oplus_i \bar{\alpha}_i P_i$ and A is normal (Exercise 5).

7.5. Theorem. *If $A = \oplus_i \alpha_i P_i$ is diagonalizable and all the α_i are distinct, then an operator B in $\mathcal{B}(\mathcal{H})$ satisfies $AB = BA$ if and only if for each i, ran P_i reduces B.*

PROOF. If all the α_i are distinct, then ran $P_i = \ker(A - \alpha_i)$. If $AB = BA$ and $Ah = \alpha_i h$, then $ABh = BAh = B(\alpha_i h) = \alpha_i Bh$; hence $Bh \in$ ran P_i whenever $h \in$ ran P_i. Thus ran P_i is left invariant by B. Therefore B leaves $\bigvee \{$ran $P_j: j \neq i\} = \mathcal{N}_i$ invariant. But since $\oplus_i P_i = 1$, $\mathcal{N}_i = ($ran $P_i)^\perp$. Thus ran P_i reduces B.

Now assume that B is reduced by each ran P_i. Thus $BP_i = P_iB$ for all i. If $h \in \mathcal{H}$, then $Ah = \sum_i \alpha_i P_i h$. Hence $BAh = \sum_i \alpha_i BP_i h = \sum_i \alpha_i P_i Bh = ABh$. (Why is the first equality valid?) ∎

Using the notation of the preceding theorem, if $AB = BA$, let $B_i = B|$ran P_i. Then it is appropriate to write $B = \oplus_i B_i$ on $\mathcal{H} = \oplus_i (P_i \mathcal{H})$. One might paraphrase Theorem 7.5 by saying that B commutes with a diagonalizable operator if and only if B can be "diagonalized with operator entries."

7.6. Spectral Theorem for Compact Normal Operators. *If T is a compact normal operator on the complex Hilbert space $\mathcal{H}$, then T has only a countable number of distinct eigenvalues. If $\{\lambda_1, \lambda_2, \ldots\}$ are the distinct nonzero eigenvalues of T, and P_n is the projection of $\mathcal{H}$ onto $\ker(T - \lambda_n)$, then $P_nP_m = P_mP_n = 0$ if $n \neq m$ and*

7.7
$$T = \sum_{n=1}^{\infty} \lambda_n P_n,$$

where this series converges to T in the metric defined by the norm on $\mathcal{B}(\mathcal{H})$.

PROOF. Let $A = (T + T^*)/2$, $B = (T - T^*)/2i$. So A, B are compact self-adjoint operators, $T = A + iB$, and $AB = BA$ since T is normal. The idea of the proof is rather simple. We'll get started in this proof together but the reader will have to complete the details.

By Theorem 5.1, $A = \sum_1^\infty \alpha_n E_n$, where $\alpha_n \in \mathbb{R}$, $\alpha_n \neq \alpha_m$ if $n \neq m$, and E_n is the projection of $\mathcal{H}$ onto $\ker(A - \alpha_n)$. Since $AB = BA$, the idea is to use Theorem 7.5 and Theorem 5.1 applied to B to diagonalize A and B simultaneously; that is, to find an orthonormal basis for $\mathcal{H}$ consisting of vectors that are simultaneously eigenvectors of A and B.

Since $BA = AB$, $E_n\mathcal{H} = \mathscr{L}_n$ reduces B for every n (7.5). Let $B_n = B|\mathscr{L}_n$; then $B_n = B_n^*$ and dim $\mathscr{L}_n < \infty$. Applying (5.1) (or, rather, the corresponding theorem from linear algebra) to B_n, there is a basis $\{e^{(n)}: 1 \leq j \leq d_n\}$ for $\mathscr{L}_n$ and real numbers $\{\beta_j^{(n)}: 1 \leq j \leq d_n\}$ such that $B_n e_j^{(n)} = \beta_j^{(n)} e_j^{(n)}$. Thus $Te_j^{(n)} = Ae_j^{(n)} + iBe_j^{(n)} = (\alpha_n + i\beta_j^{(n)})e_j^{(n)}$.

Therefore $\{e_j^{(n)}: 1 \leq j \leq d_n, n \geq 1\}$ is a basis for cl(ran A) consisting of eigenvectors for T. It may be that cl(ran A) $\neq$ cl(ran T). Since B is reduced by $\ker A = ($ran $A)^\perp$ and $B_0 = B|\ker A$ is a compact self-adjoint operator there is an orthonormal basis $\{e_j^{(0)}: j \geqslant 1\}$ for cl(ran B_0) and scalars $\{\beta_j^{(0)}: j \geqslant 1\}$ such that $Be_j^{(0)} = \beta_j^{(0)} e_j^{(0)}$. It follows that $Te_j^{(0)} = i\beta_j^{(0)} e_j^{(0)}$. Moreover, ker $T^* \supseteq$ ker $A \cap$ ker B_0 so cl(ran T) $\subseteq$ cl(ran A) $\oplus$ cl(ran B_0).

The remainder of the proof now consists in a certain amount of bookeeping to gather together the eigenvectors belonging to the same eigenvalues of T and the performing of some light housekeeping chores to obtain the convergence of the series (7.7)   ∎

7.8. Corollary. *With the notation of* (7.6):

(a) ker $T = [\vee \{P_n \mathcal{H}: n \geqslant 1\}]^\perp$;
(b) *each P_n has finite rank;*
(c) $\|T\| = \sup\{|\lambda_n|: n \geqslant 1\}$ *and either $\{\lambda_n\}$ is finite or $\lambda_n \to 0$ as $n \to \infty$.*

The proof of (7.8) is similar to the proof of (5.3).

7.9. Corollary. *If T is a compact operator on a complex Hilbert space, then T is normal if and only if T is diagonalizable.*

If T is a normal operator which is not necessarily compact, there is a spectral theorem for T which has a somewhat different form. This theorem states that T can be represented as an integral with respect to a measure whose values are not numbers but projections on a Hilbert space. Theorem 7.6 will be a consequence of this more general theorem and correspond to the case in which this projection-valued measure is "atomic."

The approach to this more general spectral theorem will be to develop a functional calculus for normal operators T. That is, an operator $\phi(T)$ will be defined for every bounded Borel function ϕ on $\mathbb{C}$ and certain properties of the map $\phi \mapsto \phi(T)$ will be deduced. The projection-valued measure will then be obtained by letting $\mu(\Delta) = \chi_\Delta(T)$, where χ_Δ is the characteristic function of the set Δ. These matters are taken up in Chapter IX.

At this point, Theorem 7.6 will be used to develop a functional calculus for compact normal operators. For the remainder of this section $\mathcal{H}$ is a complex Hilbert space.

7.10. Definition. Denote by $l^\infty(\mathbb{C})$ all the bounded functions $\phi: \mathbb{C} \to \mathbb{C}$. If T is a compact normal operator satisfying (7.7), define $\phi(T): \mathcal{H} \to \mathcal{H}$ by

$$\phi(T) = \sum_{n=1}^{\infty} \phi(\lambda_n) P_n + \phi(0) P_0,$$

where P_0 = the projection of $\mathcal{H}$ onto ker T.

Note that $\phi(T)$ is a diagonalizable operator and $\|\phi(T)\| = \sup\{|\phi(0)|, |\phi(\lambda_1)|, \ldots\}$ (4.6). Much more can be said.

7.11. Functional Calculus for Compact Normal Operators. *If T is a compact normal operator on a $\mathbb{C}$-Hilbert space $\mathcal{H}$, then the map $\phi \mapsto \phi(T)$ of*

$l^\infty(\mathbb{C}) \to \mathcal{B}(\mathcal{H})$ *has the following properties:*

(a) $\phi \mapsto \phi(T)$ *is a multiplicative linear map of* $l^\infty(\mathbb{C})$ *into* $\mathcal{B}(\mathcal{H})$. *If* $\phi \equiv 1$,
   $\phi(T) = 1$; *if* $\phi(z) = z$ *on* $\sigma_p(T) \cup \{0\}$, *then* $\phi(T) = T$.
(b) $\|\phi(T)\| = \sup\{|\phi(\lambda)|: \lambda \in \sigma_p(T)\}$.
(c) $\phi(T)^* = \phi^*(T)$, *where* ϕ^* *is the function defined by* $\phi^*(z) = \overline{\phi(z)}$.
(d) *If* $A \in \mathcal{B}(\mathcal{H})$ *and* $AT = TA$, *then* $A\phi(T) = \phi(T)A$ *for all* ϕ *in* $l^\infty(\mathbb{C})$.

PROOF. Adopt the notation of Theorem 7.6 and (7.10).

   (a) If $\phi, \psi \in l^\infty(\mathbb{C})$, then $(\phi\psi)(z) = \phi(z)\psi(z)$ for z in $\mathbb{C}$. Also, $\phi(T)\psi(T)h = [\phi(0)P_0 + \sum \phi(\lambda_n)P_n][\psi(0)P_0 + \sum \psi(\lambda_m)P_m]h = [\phi(0)P_0 + \sum_n \phi(\lambda_n)P_n]$ $[\psi(0)P_0 h + \sum_m \psi(\lambda_m)P_m h]$. Since $P_n P_m = 0$ when $n \neq m$, this gives that $\phi(T)\psi(T)h = \phi(0)\psi(0)P_0 h + \sum_n \phi(\lambda_n)\psi(\lambda_n)P_n h = (\phi\psi)(T)h$. Thus $\phi \mapsto \phi(T)$ is multiplicative. The linearity of the map is left to the reader. If $\phi(z) = 1$, then $\phi(T) = 1(T) = P_0 + \sum_{n=1}^\infty P_n = 1$ since $\{P_0, P_1, \ldots\}$ is a partition of the identity. If $\phi(z) = z$, $\phi(\lambda_n) = \lambda_n$ and so $\phi(T) = T$.

   Parts (b) and (c) follow from Exercise 5.

   (d) If $AT = TA$, Theorem 7.5 implies that $P_0\mathcal{H}, P_1\mathcal{H}, \ldots$ all reduce A. Fix h_n in $P_n\mathcal{H}$, $n \geq 0$. If $\phi \in l^\infty(\mathbb{C})$, then $Ah_n \in P_n\mathcal{H}$ and so $\phi(T)Ah_n = \phi(\lambda_n)Ah_n = A(\phi(\lambda_n)h_n) = A\phi(T)h_n$. If $h \in \mathcal{H}$, then $h = \sum_{n=0}^\infty h_n$, where $h_n \in P_n$. Hence $\phi(T)Ah = \sum_{n=0}^\infty \phi(T)Ah_n = \sum_{n=0}^\infty A\phi(T)h_n = A\phi(T)h$. (Justify the first equality.) ∎

   Which operators on $\mathcal{H}$ can be expressed as $\phi(T)$ for some ϕ in $l^\infty(\mathbb{C})$? Part (d) of the preceding theorem provides the answer.

7.12. Theorem. *If* T *is a compact normal operator on a* $\mathbb{C}$-*Hilbert space, then* $\{\phi(T): \phi \in l^\infty(\mathbb{C})\}$ *is equal to*

$$\{B \in \mathcal{B}(\mathcal{H}): BA = AB \text{ whenever } AT = TA\}.$$

PROOF. Half of the desired equality is obtained from (7.11d). So let $B \in \mathcal{B}(\mathcal{H})$ and assume that $BA = AB$ whenever $AT = TA$. Thus, B must commute with T itself. By (7.5), B is reduced by each $P_n\mathcal{H} \equiv \mathcal{H}_n$, $n \geq 0$; put $B_n = B|\mathcal{H}_n$. Fix $n \geq 0$ for the moment and let A_n be any bounded operator in $\mathcal{B}(\mathcal{H}_n)$. Define $Ah = A_n h$ if $h \in \mathcal{H}_n$ and $Ah = 0$ if $h \in \mathcal{H}_m$, $m \neq n$, and extend A to $\mathcal{H}$ by linearity; so $A = \bigoplus_{m=0}^\infty A_m$ where $A_m = 0$ if $m \neq n$. By (7.5), $AT = TA$; hence $BA = AB$. This implies that $B_n A_n = A_n B_n$. Since A_n was arbitrarily chosen from $\mathcal{B}(\mathcal{H}_n)$, $B_n = \beta_n$ for some β_n (Exercise 7). If $\phi: \mathbb{C} \to \mathbb{C}$ is defined by $\phi(0) = \beta_0$ and $\phi(\lambda_n) = \beta_n$ for $n \geq 1$, then $B = \phi(T)$. ∎

7.13. Definition. *If* $A \in \mathcal{B}(\mathcal{H})$, *then* A *is positive if* $\langle Ah, h \rangle \geq 0$ *for all* h *in* $\mathcal{H}$. *In symbols this is denoted by* $A \geq 0$.

   Note that by Proposition 2.12 every positive operator on a complex Hilbert space is self-adjoint.

7.14. Proposition. *If T is a compact normal operator, then T is positive if and only if all its eigenvalues are non-negative real numbers.*

PROOF. Let $T = \sum_1^\infty \lambda_n P_n$. If $T \geqslant 0$ and $h \in P_n \mathcal{H}$ with $\|h\| = 1$, then $Th = \lambda_n h$. Hence $\lambda_n = \langle Th, h \rangle \geqslant 0$. Conversely, assume each $\lambda_n \geqslant 0$. If $h \in \mathcal{H}$, $h = h_0 + \sum_{n=1}^\infty h_n$, where $h_0 \in \ker T$ and $h_n \in P_n \mathcal{H}$ for $n \geqslant 1$. Then $Th = \sum_1^\infty \lambda_n h_n$. Hence $\langle Th, h \rangle = \langle \sum_{n=1}^\infty \lambda_n h_n, h_0 + \sum_{m=1}^\infty h_m \rangle = \sum_{n=1}^\infty \sum_{m=0}^\infty \lambda_n \langle h_n, h_m \rangle = \sum_{n=1}^\infty \lambda_n \|h_n\|^2 \geqslant 0$ since $\langle h_n, h_m \rangle = 0$ when $n \neq m$. ∎

7.15. Theorem. *If T is a compact self-adjoint operator, then there are unique positive compact operators A, B such that $T = A - B$ and $AB = BA = 0$.*

PROOF. Let $T = \sum_{n=1}^\infty \lambda_n P_n$ as in (7.6). Define $\phi, \psi \colon \mathbb{C} \to \mathbb{C}$ by $\phi(\lambda_n) = \lambda_n$ if $\lambda_n > 0$, $\phi(z) = 0$ otherwise; $\psi(\lambda_n) = -\lambda_n$ if $\lambda_n < 0$, $\psi(z) = 0$ otherwise. Put $A = \phi(T)$ and $B = \psi(T)$. Then $A = \sum \{\lambda_n P_n \colon \lambda_n > 0\}$ and $B = \sum \{-\lambda_n P_n \colon \lambda_n < 0\}$. Thus $T = A - B$. Since $\phi\psi = 0$. $AB = BA = 0$ by (7.11a). Since $\phi, \psi \geqslant 0$, $A, B \geqslant 0$ by the preceding proposition. It remains to show that A, B are unique.

Suppose $T = C - D$ where C, D are compact positive operators and $CD = DC = 0$. It is easy to check that C and D commute with T. Put $\lambda_0 = 0$ and $P_0 = $ the projection of $\mathcal{H}$ onto $\ker T$. Thus C and D are reduced by $P_n \mathcal{H} \equiv \mathcal{H}_n$ for all $n \geqslant 0$. Let $C_n = C | \mathcal{H}_n$ and $D_n = D | \mathcal{H}_n$. So $C_n D_n = D_n C_n = 0$, $\lambda_n P_n = T | \mathcal{H}_n = C_n - D_n$, and C_n, D_n are positive. Suppose $\lambda_n > 0$ and let $h \in \mathcal{H}_n$. Since $C_n D_n = 0$, $\ker C_n \supseteq \mathrm{cl}[\mathrm{ran}\, D_n] = (\ker D_n)^\perp$. So if $h \in (\ker D_n)^\perp$, then $\lambda_n h = -D_n h$. Hence $\lambda_n \|h\|^2 = -\langle D_n h, h \rangle \leqslant 0$. Thus $h = 0$ since $\lambda_n > 0$. That is, $\ker D_n = \mathcal{H}_n$. Thus $D_n = 0 = B | \mathcal{H}_n$ and $C_n = \lambda_n P_n = A | \mathcal{H}_n$. Similarly, if $\lambda_n < 0$, $C_n = 0 = A | \mathcal{H}_n$ and $D_n = -\lambda_n P_n = B | \mathcal{H}_n$. On $\mathcal{H}_0$, $T | \mathcal{H}_0 = 0 = C_0 - D_0$. Thus $C_0 = D_0$. But $0 = C_0 D_0 = C_0^2$. Thus $0 = \langle C_0^2 h, h \rangle = \|C_0 h\|^2$, so $C_0 = 0 = A | \mathcal{H}_0$ and $D_0 = 0 = B | \mathcal{H}_0$. Therefore $C = A$ and $D = B$. ∎

Positive operators are analogous to positive numbers. With this in mind, the next result seems reasonable.

7.16. Theorem. *If T is a positive compact operator, then there is a unique positive compact operator A such that $A^2 = T$.*

PROOF. Let $T = \sum_{n=1}^\infty \lambda_n P_n$ as in the Spectral Theorem. Since $T \geqslant 0$, $\lambda_n > 0$ for all n (7.14). Let $\phi(\lambda_n) = \lambda_n^{1/2}$ and $\phi(z) = 0$ otherwise; put $A = \phi(T)$. It is easy to check that $A \geqslant 0$; $A = \sum_1^\infty \lambda_n^{1/2} P_n$ so that A is compact; and $A^2 = T$. The proof of uniqueness is left to the reader. ∎

EXERCISES

1. If $\{P_n\}$ is a sequence of pairwise orthogonal nonzero projections and $P = \sum P_n$, show that $\|P - \sum_{j=1}^n P_j\| = 1$ for all n.

2. If $\mathcal{H}$ is separable, show that the definitions of a diagonalizable operator in (4.6) and (7.3) are equivalent.

3. If $A = \sum \alpha_i P_i$ as in (7.3), show that A is compact if and only if: (a) $\alpha_i = 0$ for all but a countable number of i; (b) P_i has finite rank whenever $\alpha_i \neq 0$; (c) if $\{\alpha_1, \alpha_2, \ldots\} = \{\alpha_i : \alpha_i \neq 0\}$, then $\alpha_n \to 0$ as $n \to \infty$.

4. Prove Proposition 7.4.

5. If $A = \bigoplus_i \alpha_i P_i$, show that $A^* = \bigoplus_i \bar{\alpha}_i P_i$, A is normal, and $\|A\| = \sup\{|\alpha_i| : i \in I\}$.

6. Give the remaining details in the proof of (7.6).

7. If $A \in \mathcal{B}(\mathcal{H})$ and $AT = TA$ for every compact operator T, show that A is a multiple of the identity operator.

8. Suppose T is a compact normal operator on a $\mathbb{C}$-Hilbert space such that $\dim \ker(T - \lambda) \leq 1$ for all λ in $\mathbb{C}$. Show that if $A \in \mathcal{B}(\mathcal{H})$ and $AT = TA$, then $A = \phi(T)$ for some ϕ in $l^\infty(\mathbb{C})$.

9. Prove a converse to Exercise 8: if T is a compact normal operator such that $\{A \in \mathcal{B}(\mathcal{H}) : AT = TA\} = \{\phi(T) : \phi \in l^\infty(\mathbb{C})\}$, then $\dim \ker(T - \lambda) \leq 1$ for all λ in $\mathbb{C}$.

10. Let T be a compact normal operator and show that $\dim \ker(T - \lambda) \leq 1$ for all λ in $\mathbb{C}$ if and only if there is a vector h in $\mathcal{H}$ such that $\{p(T)h : p$ is a polynomial in one variable$\}$ is dense in $\mathcal{H}$. (Such a vector h is called a *cyclic vector* for T.)

11. If $\lambda \in \mathbb{C}$, let δ_λ be the unit point mass at λ; that is, δ_λ is the measure on $\mathbb{C}$ such that $\delta_\lambda(\Delta) = 1$ if $\lambda \in \Delta$ and $\delta_\lambda(\Delta) = 0$ if $\lambda \notin \Delta$. If $\{\lambda_1, \lambda_2, \ldots\}$ is a bounded sequence of distinct complex numbers and $\{\alpha_n\}$ is a sequence of real numbers with $\alpha_n > 0$ and $\sum_n \alpha_n < \infty$, let $\mu = \sum_{n=1}^\infty \alpha_n \delta_{\lambda_n}$; so μ is a finite measure. If $\phi \in l^\infty(\mathbb{C})$, let M_ϕ be the multiplication operator on $L^2(\mu)$. Define $T : L^2(\mu) \to L^2(\mu)$ by $(Tf)(\lambda_n) = \lambda_n f(\lambda_n)$. Prove: (a) T is a normal operator; (b) T has a cyclic vector (see Exercise 10); (c) if $A \in \mathcal{B}(\mathcal{H})$ and $AT = TA$, then $A = M_\phi$ for some ϕ in $l^\infty(\mathbb{C})$; (d) T is compact if and only if $\lambda_n \to 0$. (e) If T is compact, find all of the cyclic vectors for T. (f) If T is compact, find the decomposition (7.7) for T.

12. Using the notation of Theorem 7.11, give necessary and sufficient conditions on T and ϕ that $\phi(T)$ be compact. (Hint: consider separately the cases where $\ker T$ is finite or infinite dimensional.)

13. Prove the uniqueness part of Theorem 7.16.

14. If $T \in \mathcal{B}(\mathcal{H})$, show that $T^*T \geq 0$.

15. Let T be a compact normal operator and show that there is a compact positive operator A and a unitary operator U such that $T = UA = AU$. discuss the uniqueness of A and U.

16. (*Polar decomposition* of compact operators.) Let $T \in \mathcal{B}_0(\mathcal{H})$ and let A be the unique positive square root of T^*T [(7.16) and Exercise 14]. (a) Show that $\|Ah\| = \|Th\|$ for all h in $\mathcal{H}$. (b) Show that there is a unique oprator U such that $\|Uh\| = \|h\|$ when $h \perp \ker T$, $Uh = 0$ when $h \in \ker T$, and $UA = T$. (c) If U and A are as in (a) and (b), show that $T = AU$ if and only if T is normal.

17. Prove the following uniqueness statement for the functional calculus (7.11). If T is a compact normal operator on a $\mathbb{C}$-Hilbert space $\mathcal{H}$ and $\tau : l^\infty(\mathbb{C}) \to \mathcal{B}(\mathcal{H})$ is

a multiplicative linear map such that $\|\tau(\phi)\| = \sup\{|\phi(\lambda)|: \lambda \in \sigma_p(T)\}$, $\tau(1) = 1$, and $\tau(\psi) = T$ whenever $\psi(z) = z$ on $\sigma_p(T) \cup \{0\}$, then $\tau(\phi) = \phi(T)$ for every ϕ in $l^\infty(\mathbb{C})$.

§8*. Unitary Equivalence for Compact Normal Operators

In Section I.5 the concept of an isomorphism between Hilbert spaces was defined as the natural equivalence relation on Hilbert spaces. This equivalence relation between the spaces induces a natural equivalence relation between the operators on the spaces.

8.1. Definition. If A, B are bounded operators on Hilbert spaces $\mathcal{H}, \mathcal{K}$, then A and B are *unitarily equivalent* if there is an isomorphism $U: \mathcal{H} \to \mathcal{K}$ such that $UAU^{-1} = B$. In symbols this is denoted by $A \cong B$.

Some of the elementary properties of unitary equivalence are contained in Exercises 1 and 2. Note that if $UAU^{-1} = B$, then $UA = BU$.

The purpose of this section is to give necessary and sufficient conditions that two compact normal operators be unitarily equivalent. Later, in Section IX.10, necessary and sufficient conditions that any two normal operators be unitarily equivalent are given and the results of this section are subsumed by those of that section.

8.2. Definition. If T is a compact operator, the *multiplicity function* for T is the cardinal number valued function m_T defined for every complex number λ by $m_T(\lambda) = \dim \ker(T - \lambda)$.

Hence $m_T(\lambda) \geqslant 0$ for all λ and $m_T(\lambda) > 0$ if and only if λ is an eigenvalue for T. Note that by Proposition 4.13, $m_T(\lambda) < \infty$ if $\lambda \neq 0$.

If, T, S are compact operators on Hilbert spaces and $U: \mathcal{H} \to \mathcal{K}$ is an isomorphism with $UTU^{-1} = S$, then $U \ker(T - \lambda) = \ker(S - \lambda)$ for every λ in $\mathbb{C}$. In fact, if $Th = \lambda h$, then $SUh = UTh = \lambda Uh$ and so $Uh \in \ker(S - \lambda)$. Conversely, if $k \in \ker(S - \lambda)$ and $h = U^{-1}k$, then $Th = TU^{-1}k = U^{-1}Sk = \lambda h$. In particular, it must be that $m_T = m_S$. If S and T are normal, this condition is also sufficient for unitary equivalence.

8.3. Theorem. *Two compact normal operators are unitarily equivalent if and only if they have the same multiplicity function.*

PROOF. Let T, S be compact normal operators on Hilbert spaces $\mathcal{H}, \mathcal{K}$. If $T \cong S$, then it has already been shown that $m_T = m_S$. Suppose now that $m_T = m_S$. We must manufacture a unitary operator $U: \mathcal{H} \to \mathcal{K}$ such that $UTU^{-1} = S$.

Let $T = \sum_{n=1}^\infty \lambda_n P_n$ and let $S = \sum_{n=1}^\infty \mu_n Q_n$ as in the Spectral Theorem (7.6). So if $n \neq m$, then $\lambda_n \neq \lambda_m$ and $\mu_n \neq \mu_m$, and each of the projections P_n and Q_n

has finite rank. Let P_0, Q_0 be the projections of $\mathcal{H}, \mathcal{K}$ onto $\ker T$, $\ker S$; so $P_0 = (\sum_1^\infty P_n)^\perp$ and $Q_0 = (\sum_1^\infty Q_n)^\perp$. Put $\lambda_0 = \mu_0 = 0$.

Since $m_T = m_S$, $0 < m_T(\lambda_n) = m_S(\lambda_n)$. Hence there is a unique μ_j such that $\mu_j = \lambda_n$. Define $\pi: \mathbb{N} \to \mathbb{N}$ by letting $\mu_{\pi(n)} = \lambda_n$. Let $\pi(0) = 0$. Note that π is one-to-one. Also, since $0 < m_S(\mu_n) = m_T(\mu_n)$, for every n there is a j such that $\pi(j) = n$. Thus $\pi: \mathbb{N} \cup \{0\} \to \mathbb{N} \cup \{0\}$ is a bijection or permutation. Since $\dim P_n = m_T(\lambda_n) = m_S(\mu_{\pi(n)}) = \dim Q_{\pi(n)}$, there is an isomorphism $U_n: P_n\mathcal{H} \to Q_{\pi(n)}\mathcal{K}$ for $n \geq 0$. Define $U: \mathcal{H} \to \mathcal{K}$ by letting $U = U_n$ on $P_n\mathcal{H}$ and extending by linearity. Hence $U = \bigoplus_{n=0}^\infty U_n$. It is easy to check that U is an isomorphism. Also, if $h \in P_n\mathcal{H}$, $n \geq 0$, then $UTh = \lambda_n Uh = \mu_{\pi(n)} Uh = SUh$. Hence $UTU^{-1} = S$. ∎

If V is the Volterra operator, then $m_V \equiv 0$ (4.11) and V and the zero operator are definitely not unitarily equivalent, so the preceding theorem only applies to compact normal operators. There are no known necessary and sufficient conditions for two arbitrary compact operators to be unitarily equivalent. In fact, there are no known necessary and sufficient conditions that two arbitrary operators on a finite-dimensional space be unitarily equivalent.

EXERCISES

1. Show that "unitary equivalence" is an equivalence relation on $\mathcal{B}(\mathcal{H})$.

2. Let $U: \mathcal{H} \to \mathcal{K}$ be an isomorphism and define $\rho: \mathcal{B}(\mathcal{H}) \to \mathcal{B}(\mathcal{K})$ by $\rho(A) = UAU^{-1}$. Prove: (a) $\|\rho(A)\| = \|A\|$, $\rho(A^*) = \rho(A^*)$, and ρ is an isomorphism between the two algebras $\mathcal{B}(\mathcal{H})$ and $\mathcal{B}(\mathcal{K})$. (b) $\rho(A) \in \mathcal{B}_0(\mathcal{K})$ if and only if $A \in \mathcal{B}_0(\mathcal{H})$. (c) If $T \in \mathcal{B}(\mathcal{H})$, then $AT = TA$ if and only if $\rho(T)\rho(A) = \rho(A)\rho(T)$. (d) If $A \in \mathcal{B}(\mathcal{H})$ and $\mathcal{M} \leq \mathcal{H}$, then $\mathcal{M}$ is invariant (reducing) for A if and only if $U\mathcal{M}$ is invariant (reducing) for $\rho(A)$.

3. Say that an operator A on $\mathcal{H}$ is *irreducible* if the only reducing subspaces for A are (0) and $\mathcal{H}$. Prove: (a) The Volterra operator is irreducible. (b) The unilateral shift is irreducible.

4. Suppose $A = \oplus\{A_i: i \in I\}$ and $B = \oplus\{B_i: i \in I\}$ where each A_i and B_i is irreducible (Exercise 3). Show that $A \cong B$ if and only if there is a bijection $\pi: I \to I$ such that $A_i \cong B_{\pi(i)}$.

5. If T is a compact normal operator and $m_T = m$ is its multiplicity function, prove: (a) $\{\lambda: m(\lambda) > 0\}$ is countable and 0 is its only possible cluster point; (b) $m(\lambda) < \infty$ if $\lambda \neq 0$. Show that if $m: \mathbb{C} \to \mathbb{N} \cup \{0, \infty\}$ is any function satisfying (a) and (b), then there is a compact normal operator T such that $m_T = m$.

6. Show that two projections P and Q are unitarily equivalent if and only if $\dim(\operatorname{ran} P) = \dim(\operatorname{ran} Q)$ and $\dim(\ker P) = \dim(\ker Q)$.

7. Let $A: L^2(0, 1) \to L^2(0, 1)$ be defined by $(Af)(x) = xf(x)$ for f in $L^2(0, 1)$ and x in $(0, 1)$. Show that $A \cong A^2$.

8. Say that a compact normal operator T is *simple* if $m_T \leq 1$. (See Exercises 7.10 and 7.11.) Show that every compact normal operator T on a separable Hilbert

space is unitarily equivalent to $\bigoplus_{n=1}^{\infty} T_n$, where each T_n is a simple compact normal operator and $m_{T_n} \geqslant m_{T_{n+1}}$ for all n. Show that $\| T_n \| \to 0$. (Of course, there may only be a finite number of T_n.)

9. Using the notation of Exercise 8, suppose also that S is a compact normal operator and $S \cong \bigoplus_{n=1}^{\infty} S_n$, where S_n is a simple compact normal operator and $m_{S_n} \geqslant m_{S_{n+1}}$ for all n. Show that $T \cong S$ if and only if $T_n \cong S_n$ for all n.

10. If T is a compact normal operator on a separable Hilbert space, show that there are simple compact normal operators $T_1, T_2, \ldots$ such that $T \cong 0 \oplus T_1 \oplus T_2^{(2)} \oplus T_2^{(3)} \oplus \cdots$, where: (a) for any operator A, $A^{(n)} \equiv A \oplus \cdots \oplus A$ (n times); (b) 0 is the zero operator on an infinite dimensional space; (c) for $n \neq k$, $m_{T_n} m_{T_k} \equiv 0$; and (d) if $\ker T$ is infinite dimensional, then $\ker T_n = (0)$ for all n. (Of course not all of the summands need be present.) Show that $\| T_n \| \to 0$.

11. Using the notation of Exercise 10, let S be a compact normal operator and let $0 \oplus S_1 \oplus S_2^{(2)} \oplus \cdots$ be the corresponding decomposition. Show that $T \cong S$ if and only if $T_n \cong S_n$ and $\ker T$ and $\ker S$ have the same dimension.

12. If T is a non-zero compact normal operator, show that T and $T \oplus T$ are not unitarily equivalent.

13. Give an example of a nontrivial operator T such that $T \cong T \oplus T$. Show that if $T \cong T \oplus T$, then $T \cong T \oplus T \oplus \cdots$. Characterize the diagonalizable normal operators T such that $T \cong T \oplus T$.

14. Let $\mathcal{H}$ be the space defined in Example I.1.8 and let $U \colon \mathcal{H} \to L^2(0, 1)$ be the isomorphism defined by $Uf = f'$ (Exercise I.1.4). If $(Af)(x) = xf(x)$ for f in $\mathcal{H}$, what is UAU^{-1}?

CHAPTER III
Banach Spaces

The concept of a Banach space is a generalization of Hilbert space. A Banach space assumes that there is a norm on the space relative to which the space is complete, but it is not assumed that the norm is defined in terms of an inner product. There are many examples of Banach spaces that are not Hilbert spaces, so that the generalization is quite useful.

§1. Elementary Properties and Examples

1.1. Definition. If $\mathscr{X}$ is a vector space over $\mathbb{F}$, a *seminorm* is a function $p\colon \mathscr{X} \to [0, \infty)$ having the properties:

(a) $p(x + y) \leqslant p(x) + p(y)$ for all x, y in $\mathscr{X}$.
(b) $p(\alpha x) = |\alpha| p(x)$ for all α in $\mathbb{F}$ and x in $\mathscr{X}$.

It follows from (b) that $p(0) = 0$. A *norm* is a seminorm p such that

(c) $x = 0$ if $p(x) = 0$.

Usually a norm is denoted by $\|\cdot\|$.
The norm on a Hilbert space is a norm. Also, the norm on $\mathscr{B}(\mathscr{H})$ is a norm.
If $\mathscr{X}$ has a norm, then $d(x, y) = \|x - y\|$ defines a metric on $\mathscr{X}$.

1.2. Definition. A *normed space* is a pair $(\mathscr{X}, \|\cdot\|)$, where $\mathscr{X}$ is a vector space and $\|\cdot\|$ is a norm on $\mathscr{X}$. A *Banach space* is a normed space that is complete with respect to the metric defined by the norm.

1.3. Proposition. *If $\mathscr{X}$ is a normed space, then*

(a) *the function $\mathscr{X} \times \mathscr{X} \to \mathscr{X}$ defined by $(x, y) \mapsto x + y$ is continuous;*
(b) *the function $\mathbb{F} \times \mathscr{X} \to \mathscr{X}$ defined by $(\alpha, x) \mapsto \alpha x$ is continuous.*

PROOF. If $x_n \to x$ and $y_n \to y$, then $\|(x_n + y_n) - (x + y)\| = \|(x_n - x) + (y_n - y)\| \le$ $\|x_n - x\| + \|y_n - y\| \to 0$ as $n \to \infty$. This proves (a). The proof of (b) is left to the reader. ∎

The next lemma is quite useful.

1.4. Lemma. *If p and q are seminorms on a vector space $\mathscr{X}$, then the following statements are equivalent.*

(a) $p(x) \le q(x)$ *for all* x. *(That is, $p \le q$.)*
(b) $\{x \in \mathscr{X}: q(x) < 1)\} \subseteq \{x \in \mathscr{X}: p(x) < 1\}$.
(b') $p(x) < 1$ *whenever* $q(x) < 1$.
(c) $\{x: q(x) \le 1\} \subseteq \{x: p(x) \le 1\}$.
(c') $p(x) \le 1$ *whenever* $q(x) \le 1$.
(d) $\{x: q(x) < 1\} \subseteq \{x: p(x) \le 1\}$.
(d') $p(x) \le 1$ *whenever* $q(x) < 1$.

PROOF. It is clear that (b) and (b'), (c) and (c'), and (d) and (d') are equivalent. It is also clear that (a) implies all of the remaining conditions and that both (b) and (c) imply (d). It remains to show that (d) implies (a).

Assume that (d) holds and put $q(x) = \alpha$. If $\varepsilon > 0$, then $q((\alpha + \varepsilon)^{-1} x) = (\alpha + \varepsilon)^{-1} \alpha < 1$. By (d), $1 \ge p((\alpha + \varepsilon)^{-1} x) = (\alpha + \varepsilon)^{-1} p(x)$, so $p(x) \le \alpha + \varepsilon = q(x) + \varepsilon$. Letting $\varepsilon \to 0$ shows (a). ∎

If $\|\cdot\|_1$ and $\|\cdot\|_2$ are two norms on $\mathscr{X}$, they are said to be *equivalent norms* if they define the same topology on $\mathscr{X}$.

1.5. Proposition. *If $\|\cdot\|_1$ and $\|\cdot\|_2$ are two norms on $\mathscr{X}$, then these norms are equivalent if and only if there are positive constants c and C such that*

$$c\|x\|_1 \le \|x\|_2 \le C\|x\|_1$$

for all x in $\mathscr{X}$.

PROOF. Suppose there are constants c and C such that $c\|x\|_1 \le \|x\|_2 \le C\|x\|_1$ for all x in $\mathscr{X}$. Fix x_0 in $\mathscr{X}$, $\varepsilon > 0$. Then

$$\{x \in \mathscr{X}: \|x - x_0\|_1 < \varepsilon/C\} \subseteq \{x \in \mathscr{X}: \|x - x_0\|_2 < \varepsilon\},$$
$$\{x \in \mathscr{X}: \|x - x_0\|_2 < c\varepsilon\} \subseteq \{x \in \mathscr{X}: \|x - x_0\|_1 < \varepsilon\}.$$

This shows that the two topologies are the same. Now assume that the two norms are equivalent. Hence $\{x: \|x\|_1 < 1\}$ is an open neighborhood of 0 in the topology defined by $\|\cdot\|_2$. Therefore there is an $r > 0$ such that $\{x: \|x\|_2 < r\} \subseteq \{x: \|x\|_1 < 1\}$. If $q(x) = r^{-1}\|x\|_2$ and $p(x) = \|x\|_1$, the preceding lemma implies $\|x\|_1 \le r^{-1}\|x\|_2$ or $c\|x\|_1 \le \|x\|_2$, where $c = r$. The other inequality is left to the reader. ∎

There are two types of properties of a Banach space: those that are topological and those that are metric. The metric properties depend on the

precise norm; the topological ones depend only on the equivalence class of
norms (see Exercise 4).

1.6. Example. Let X be any Hausdorff space (all spaces in this book are
assumed to be Hausdorff unless the contrary is specified) and let $C_b(X) = $ all
continuous functions $f: X \to \mathbb{F}$ such that $\|f\| \equiv \sup\{|f(x)|: x \in X\} < \infty$. For
f, g in $C_b(X)$, define $(f + g): X \to \mathbb{F}$ by $(f + g)(x) = f(x) + g(x)$; for α in $\mathbb{F}$
define $(\alpha f)(x) = \alpha f(x)$. Then $C_b(X)$ is a Banach space.

The proofs of the statements in (1.6) are all routine except, perhaps,
for the fact that $C_b(X)$ is complete. To see this, let $\{f_n\}$ be a Cauchy
sequence in $C_b(X)$. So if $\varepsilon > 0$, there is an integer N_ε such that for $n, m \geq N_\varepsilon$,
$\varepsilon > \|f_n - f_m\| = \sup\{|f_n(x) - f_m(x)|: x \in X\}$. In particular, for any x in
$X, |f_n(x) - f_m(x)| \leq \|f_n - f_m\| < \varepsilon$ when $n, m \geq N_\varepsilon$. So $\{f_n(x)\}$ is a Cauchy
sequence in $\mathbb{F}$. Let $f(x) = \lim f_n(x)$ if $x \in X$. Now fix x in X. If $n, m \geq N_\varepsilon$, then
$|f(x) - f_n(x)| \leq |f(x) - f_m(x)| + \|f_m - f_n\| < |f(x) - f_m(x)| + \varepsilon$. Letting $m \to \infty$
gives that $|f(x) - f_n(x)| \leq \varepsilon$ when $n \geq N_\varepsilon$. This is independent of x. Hence
$\|f - f_n\| \leq \varepsilon$ for $n \geq N_\varepsilon$.

What has been just shown is that $\|f - f_n\| \to 0$ as $n \to \infty$. Note that this
implies that $f_n(x) \to f(x)$ uniformly on X. It is standard that f is continuous.
Also, $\|f\| \leq \|f - f_n\| + \|f_n\| < \infty$. Hence $f \in C_b(X)$ and so $C_b(X)$ is complete.

Note that a linear subspace $\mathcal{Y}$ of a Banach space $\mathcal{X}$ that is topologically
closed is also a Banach space if it has the norm of $\mathcal{X}$.

1.7. Proposition. *If X is a locally compact space and $C_0(X) = $ all continuous
functions $f: X \to \mathbb{F}$ such that for all $\varepsilon > 0$, $\{x \in X: |f(x)| \geq \varepsilon\}$ is compact, then
$C_0(X)$ is a closed subspace of $C_b(X)$ and hence is a Banach space.*

PROOF. That $C_0(X)$ is a linear manifold in $C_b(X)$ is left as an exercise. It will
only be shown that $C_0(X)$ is closed in $C_b(X)$. Let $\{f_n\} \subseteq C_0(X)$ and suppose
$f_n \to f$ in $C_b(X)$. If $\varepsilon > 0$, there is an integer N such that $\|f_n - f\| < \varepsilon/2$; that
is, $|f_n(x) - f(x)| < \varepsilon/2$ for all $n \geq N$ and x in X. If $|f(x)| \geq \varepsilon$, then
$\varepsilon \leq |f(x) - f_n(x) + f_n(x)| \leq \varepsilon/2 + |f_n(x)|$ for $n \geq N$; so $|f_n(x)| \geq \varepsilon/2$ for $n \geq N$.
Thus, $\{x \in X: |f(x)| \geq \varepsilon\} \subseteq \{x \in X: |f_N(x)| \geq \varepsilon/2\}$ so that $f \in C_0(X)$. ∎

The space $C_0(X)$ is the set of continuous functions on X that *vanish at
infinity*. If $X = \mathbb{R}$, then $C_0(\mathbb{R}) = $ all of the continuous functions $f: \mathbb{R} \to \mathbb{F}$ such
that $\lim_{x \to \pm\infty} f(x) = 0$. If X is compact, $C_0(X) = C_b(X) \equiv C(X)$.

If I is any set, then give I the discrete topology. Hence I becomes locally
compact. Also any function on I is continuous. Rather than $C_b(I)$, the
customary notation is $l^\infty(I)$. That is, $l^\infty(I) = $ all bounded functions $f: I \to \mathbb{F}$
with $\|f\| = \sup\{|f(i)|: i \in I\}$. $c_0(I)$ consists of all functions $f: I \to \mathbb{F}$ such that
for every $\varepsilon > 0$, $\{i \in I: |f(i)| \geq \varepsilon\}$ is finite. If $I = \mathbb{N}$, the usual notation for these
spaces is l^∞ and c_0. Note that l^∞ consists of all bounded sequences of scalars
and c_0 consists of all sequences that converge to 0.

1.8. Example. If (X, Ω, μ) is a measure space and $1 \leqslant p \leqslant \infty$, then $L^p(X, \Omega, \mu)$ is a Banach space.

The preceding example is usually proved in courses on integration and no proof is given here.

1.9. Example. Let I be a set and $1 \leqslant p < \infty$. Define $l^p(I)$ to be the set of all functions $f: I \to \mathbb{F}$ such that $\sum \{|f(i)|^p : i \in I\} < \infty$; and define $\|f\|_p = (\sum \{|f(i)|^p : i \in I\})^{1/p}$. Then $l^p(I)$ is a Banach space. If $I = \mathbb{N}$, then $l^p(\mathbb{N}) = l^p$.

If $\Omega =$ all subsets of I and for each Δ in Ω, $\mu(\Delta) =$ the number of points in Δ if Δ is finite and $\mu(\Delta) = \infty$ otherwise, then $l^p(I) = L^p(I, \Omega, \mu)$. So the statement in (1.9) is a consequence of the one in (1.8).

1.10. Example. Let $n \geqslant 1$ and let $C^{(n)}[0, 1] =$ the collection of functions $f: [0, 1] \to \mathbb{F}$ such that f has n continuous derivatives. Define $\|f\| = \sup_{0 \leqslant k \leqslant n} \{\sup\{|f^{(k)}(x)| : 0 \leqslant x \leqslant 1\}\}$. Then $C^{(n)}[0, 1]$ is a Banach space.

1.11. Example. Let $1 \leqslant p < \infty$ and $n \geqslant 1$ and let $W_p^n[0, 1] =$ the functions $f: [0, 1] \to \mathbb{F}$ such that f has $n - 1$ continuous derivatives, $f^{(n-1)}$ is absolutely continuous, and $f^{(n)} \in L^p[0, 1]$. For f in $W_p^n[0, 1]$, define

$$\|f\| = \sum_{k=0}^{n} \left[\int_0^1 |f^{(k)}(x)|^p \, dx \right]^{1/p}.$$

Then $W_p^n[0, 1]$ is a Banach space.

The following is a useful fact about seminorms.

1.12. Proposition. *If p is a seminorm on $\mathcal{X}$, $|p(x) - p(y)| \leqslant p(x - y)$ for all x, y in $\mathcal{X}$. If $\|\cdot\|$ is a norm, then $|\, \|x\| - \|y\| \,| \leqslant \|x - y\|$ for all x, y in $\mathcal{X}$.*

PROOF. Of course, the inequality for norms is a consequence of the one for seminorms. Note that if $x, y \in \mathcal{X}$, $p(x) = p(x - y + y) \leqslant p(x - y) + p(y)$, so $p(x) - p(y) \leqslant p(x - y)$. Similarly, $p(y) - p(x) \leqslant p(x - y)$. ∎

There is the concept of "isomorphism" for the category of Banach spaces.

1.13. Definition. If $\mathcal{X}$ and $\mathcal{Y}$ are normed spaces, $\mathcal{X}$ and $\mathcal{Y}$ are *isometrically isomorphic* if there is a surjective linear isometry from $\mathcal{X}$ onto $\mathcal{Y}$.

The term *isomorphism* in Banach space theory is reserved for linear bijections $T: \mathcal{X} \to \mathcal{Y}$ that are homeomorphisms.

EXERCISES

1. Complete the proof of Proposition 1.3.

2. Complete the proof of Proposition 1.5.

3. For $1 \leqslant p < \infty$ and $x = (x_1, \ldots, x_d)$ in $\mathbb{F}^d$, define $\|x\|_p \equiv [\sum_{j=1}^{d} |x_j|^p]^{1/p}$; define $\|x\|_\infty \equiv \sup\{|x_j|: 1 \leqslant j \leqslant d\}$. Show that all of these norms are equivalent. For $1 \leqslant p, q \leqslant \infty$, what are the best constants c and C such that $c\|x\|_p \leqslant \|x\|_q \leqslant C\|x\|_p$ for all x in $\mathbb{F}^d$?

4. If $1 \leqslant p \leqslant \infty$ and $\|\cdot\|_p$ is defined on $\mathbb{R}^2$ as in Exercise 3, graph $\{x \in \mathbb{R}^2: \|x\|_p = 1\}$. Note that if $1 < p < \infty$, $\|x\|_p = \|y\|_p = 1$, and $x \neq y$, then for $0 < t < 1$, $\|tx + (1-t)y\|_p < 1$. The same cannot be said for $p = 1, \infty$.

5. Let c = the set of all sequences $\{\alpha_n\}_1^\infty, \alpha_n$ in $\mathbb{F}$, such that $\lim \alpha_n$ exists. Show that c is a closed subspace of l^∞ and hence is a Banach space.

6. Let $X = \{n^{-1}: n \geqslant 1\} \cup \{0\}$. Show that $C(X)$ and the space of c of Exercise 5 are isometrically isomorphic.

   (a) Show that if $1 \leqslant p < \infty$ and I is an infinite set, then $l^p(I)$ has a dense set of the same cardinality as I.

   (b) Show that if $1 \leqslant p < \infty$, $l^p(I)$ and $l^p(J)$ are isometrically isomorphic if and only if I and J have the same cardinality.

7. If $l^\infty(I)$ and $l^\infty(J)$ are isometrically isomorphic, do I and J have the same cardinality?

8. Show that l^∞ is not separable.

9. Complete the proof of Proposition 1.7.

10. Verify the statements in Example 1.10.

11. Verify the statements in Example 1.11.

12. Let X be locally compact and let $X_\infty = X \cup \{\infty\}$ be the one-point compactification of X. Show that $C_0(X)$ and $\{f \in C(X_\infty): f(\infty) = 0\}$, with the norm it inherits as a subspace of $C(X_\infty)$, are isometrically isomorphic Banach spaces.

13. Let X be locally compact and define $C_c(X)$ to be the continuous functions $f: X \to \mathbb{F}$ such that spt $f \equiv \mathrm{cl}\{x \in X: f(x) \neq 0\}$ is compact (spt f is the *support* of f). Show that $C_c(X)$ is dense in $C_0(X)$.

14. If $W_p^n[0, 1]$ is defined as in Example 1.11 and $f \in W_p^n[0, 1]$, let $\|\|f\|\| \equiv [\int |f(x)|^p\, dx]^{1/p} + [\int |f^{(n)}(x)|^p\, dx]^{1/p}$. Show that $\|\|\cdot\|\|$ is equivalent to the norm defined on $W_p^n[0, 1]$.

15. Let $\mathcal{X}$ be a normed space and let $\hat{\mathcal{X}}$ be its completion as a metric space. Show that $\hat{\mathcal{X}}$ is a Banach space.

16. Show that the norm on $C([0, 1]) = C_b([0, 1])$ does not come from an inner product by showing that it does not satisfy the parallelogram law.

§2. Linear Operators on Normed Spaces

This section gathers together a few pertinent facts and examples concerning linear operators on normed spaces. A fuller study of operators on Banach spaces will be pursued later.

The proof of the first result is similar to that of Proposition I.3.1 and is left to the reader. [Also see (II.1.1).] $\mathscr{B}(\mathscr{X}, \mathscr{Y}) = $ all continuous linear transformations $A: \mathscr{X} \to \mathscr{Y}$.

2.1. Proposition. *If $\mathscr{X}$ and $\mathscr{Y}$ are normed spaces and $A: \mathscr{X} \to \mathscr{Y}$ is a linear transformation, the following statements are equivalent.*

(a) $A \in \mathscr{B}(\mathscr{X}, \mathscr{Y})$.
(b) *A is continuous at 0.*
(c) *A is continuous at some point.*
(d) *There is a positive constant c such that $\| Ax \| \leqslant c \| x \|$ for all x in $\mathscr{X}$.*

If $A \in \mathscr{B}(\mathscr{X}, \mathscr{Y})$ and

$$\| A \| = \sup\{ \| Ax \|: \| x \| \leqslant 1 \},$$

then

$$\| A \| = \sup\{ \| Ax \|: \| x \| = 1 \}$$
$$= \sup\{ \| Ax \| / \| x \|: x \neq 0 \}$$
$$= \inf\{ c > 0: \| Ax \| \leqslant c \| x \| \text{ for } x \text{ in } \mathscr{X} \}.$$

$\| A \|$ is called the *norm* of A and $\mathscr{B}(\mathscr{X}, \mathscr{Y})$ becomes a normed space if addition and scalar multiplication are defined pointwise. $\mathscr{B}(\mathscr{X}, \mathscr{Y})$ is a Banach space if $\mathscr{Y}$ is a Banach space (Exercise 1). A continuous linear operator is also called a *bounded linear operator*.

The following examples are reminiscent of those that were given in Section II.1.

2.2. Example. If (X, Ω, μ) is a σ-finite measure space and $\phi \in L^\infty(X, \Omega, \mu)$, define $M_\phi: L^p(X, \Omega, \mu) \to L^p(X, \Omega, \mu)$, $1 \leqslant p \leqslant \infty$, by $M_\phi f = \phi f$ for all f in $L^p(X, \Omega, \mu)$. Then $M_\phi \in \mathscr{B}(L^p(X, \Omega, \mu))$ and $\| M_\phi \| = \| \phi \|_\infty$.

2.3. Example. If (X, Ω, μ), k, c_1, and c_2 are as in Example II.1.6 and $1 \leqslant p \leqslant \infty$, then $K: L^p(\mu) \to L^p(\mu)$, defined by

$$(Kf)(x) = \int k(x, y) f(y) \, d\mu(y)$$

for all f in $L^p(\mu)$ and x in X, is a bounded operator on $L^p(\mu)$ and $\| K \| \leqslant c_1^{1/q} c_2^{1/p}$, where $1/p + 1/q = 1$.

2.4. Example. If X and Y are compact spaces and $\tau: Y \to X$ is a continuous map, define $A: C(X) \to C(Y)$ by $(Af)(y) = f(\tau(y))$. Then $A \in \mathscr{B}(C(X), C(Y))$ and $\| A \| = 1$.

EXERCISES

1. Show that for $\mathscr{B}(\mathscr{X}, \mathbb{F}) \neq (0)$, $\mathscr{B}(\mathscr{X}, \mathscr{Y})$ is a Banach space if and only if $\mathscr{Y}$ is a Banach space.

2. Let $\mathscr{X}$ be a normed space, let $\mathscr{Y}$ be a Banach space, and let $\hat{\mathscr{X}}$ be the completion of $\mathscr{X}$. Show that if $\rho\colon \mathscr{B}(\hat{\mathscr{X}}, \mathscr{Y}) \to \mathscr{B}(\mathscr{X}, \mathscr{Y})$ is defined by $\rho(A) = A|\mathscr{X}$, then ρ is an isometric isomorphism.

3. If (X, Ω, μ) is a σ-finite measure space, $\phi\colon X \to \mathbb{F}$ is an Ω-measurable function, $1 \leqslant p \leqslant \infty$, and $\phi f \in L^p(\mu)$ whenever $f \in L^p(\mu)$, then show that $\phi \in L^\infty(\mu)$.

4. Verify the statements in Example 2.2.

5. Verify the statements in Example 2.3.

6. Verify the statements in Example 2.4.

7. Let A and τ be as in Example 2.4. (a) Give necessary and sufficient conditions on τ that A be injective. (b) Give such a condition that A be surjective. (c) Give such a condition that A be an isometry. (d) If $X = Y$, show that $A^2 = A$ if and only if τ is a retraction.

8. (Wilansky [1951]) Assume that $A\colon \mathscr{X} \to \mathscr{Y}$ is an additive mapping (that is, $A(x_1 + x_2) = A(x_1) + A(x_2)$ for all x_1 and x_2 in $\mathscr{X}$) and show that conditions (b), (c), and (d) in Proposition 2.1 are equivalent to the continuity of A.

§3. Finite Dimensional Normed Spaces

In functional analysis it is always good to see what significance a concept has for finite dimensional spaces.

3.1. Theorem. *If $\mathscr{X}$ is a finite dimensional vector space over $\mathbb{F}$, then any two norms on $\mathscr{X}$ are equivalent.*

PROOF. Let $\{e_1, \ldots, e_d\}$ be a Hamel basis for $\mathscr{X}$. For $x = \sum_{j=1}^d x_j e_j$, define $\|x\|_\infty \equiv \max\{|x_j|\colon 1 \leqslant j \leqslant d\}$. It is left to the reader to verify that $\|\cdot\|_\infty$ is a norm. Let $\|\cdot\|$ be any norm on $\mathscr{X}$. It will be shown that $\|\cdot\|$ and $\|\cdot\|_\infty$ are equivalent.

If $x = \sum_j x_j e_j$, then $\|x\| \leqslant \sum_j |x_j| \|e_j\| \leqslant C \|x\|_\infty$, when $C = \sum_j \|e_j\|$. To show the other inequality, let $\mathscr{T}$ be the topology defined on $\mathscr{X}$ by $\|\cdot\|_\infty$ and let $\mathscr{U}$ be the topology defined on $\mathscr{X}$ by $\|\cdot\|$. Put $B = \{x \in \mathscr{X}\colon \|x\|_\infty \leqslant 1\}$. The first part of the proof implies $\mathscr{T} \supseteq \mathscr{U}$. Since B is $\mathscr{T}$-compact and $\mathscr{T} \supseteq \mathscr{U}$, B is $\mathscr{U}$-compact and the relativizations of the two topologies to B agree. Let $A = \{x \in \mathscr{X}\colon \|x\|_\infty < 1\}$. Since A is $\mathscr{T}$-open, it is open in $(B, \mathscr{U})$. Hence there is a set U in $\mathscr{U}$ such that $U \cap B = A$. Thus $0 \in U$ and there is an $r > 0$ such that $\{x \in \mathscr{X}\colon \|x\| < r\} \subseteq U$. Hence

3.2 $\qquad\qquad \|x\| < r$ and $\|x\|_\infty \leqslant 1$ implies $\|x\|_\infty < 1$.

Claim. $\|x\| < r$ implies $\|x\|_\infty < 1$.

Let $\|x\| < r$ and put $x = \sum x_j e_j$, $\alpha = \|x\|_\infty$. So $\|x/\alpha\|_\infty = 1$ and $x/\alpha \in B$. If

$\alpha \geq 1$, then $\|x/\alpha\| < r/\alpha \leq r$, and hence $\|x/\alpha\|_\infty < 1$ by (3.2), a contradiction. Thus $\|x\|_\infty = \alpha < 1$ and the claim is established.

By Lemma 1.4, $\|x\|_\infty \leq r^{-1}\|x\|$ for all x and so the proof is complete. ∎

3.3. Proposition. *If $\mathscr{X}$ is a normed space and $\mathscr{M}$ is a finite dimensional linear manifold in $\mathscr{X}$, then $\mathscr{M}$ is closed.*

PROOF. Using a Hamel basis $\{e_1,\ldots,e_n\}$ for $\mathscr{M}$, define a norm $\|\cdot\|_\infty$ on $\mathscr{M}$ as in the proof of Theorem 3.1. It is easy to see that $\mathscr{M}$ is complete with respect to this new norm. But then Theorem 3.1 implies that $\mathscr{M}$ is complete with respect to its original norm and hence must be a closed subspace of $\mathscr{X}$. ∎

3.4. Proposition. *Let $\mathscr{X}$ be a finite dimensional normed space and let $\mathscr{Y}$ be any normed space. If $T: \mathscr{X} \to \mathscr{Y}$ is a linear transformation, then T is continuous.*

PROOF. Since all norms on $\mathscr{X}$ are equivalent and $T: \mathscr{X} \to \mathscr{Y}$ is continuous with respect to one norm on $\mathscr{X}$ precisely when it is continuous with respect to any equivalent norm, we may assume that $\|\sum_{j=1}^d \xi_j e_j\| = \max\{|\xi_j|: 1 \leq j \leq d\}$, where $\{e_j\}$ is a Hamel basis for $\mathscr{X}$. Thus, for $x = \sum \xi_j e_j$, $\|Tx\| = \|\sum_j \xi_j Te_j\| \leq \sum_j |\xi_j| \|Te_j\| \leq C\|x\|$, where $C = \sum_j \|Te_j\|$. By (2.1), T is continuous. ∎

EXERCISES

1. Show that if $\mathscr{X}$ is a locally compact normed space, then $\mathscr{X}$ is finite dimensional. (This same result, due to F Riesz, is valid in the more general topological vector spaces-see IV.1.1 for the definition. For a nice proof of this look at Pitcairn [1966].)

2. Show that $\|\cdot\|_\infty$ defined in the proof of Theorem 3.1 is a norm.

§4. Quotients and Products of Normed Spaces

Let $\mathscr{X}$ be a normed space, let $\mathscr{M}$ be a linear manifold in $\mathscr{X}$, and let $Q: \mathscr{X} \to \mathscr{X}/\mathscr{M}$ be the natural map $Qx = x + \mathscr{M}$. We want to make $\mathscr{X}/\mathscr{M}$ into a normed space, so define

4.1 $$\|x + \mathscr{M}\| = \inf\{\|x + y\|: y \in \mathscr{M}\}.$$

Note that because $\mathscr{M}$ is a linear space, $\|x + \mathscr{M}\| = \inf\{\|x - y\|: y \in \mathscr{M}\} = \text{dist}(x, \mathscr{M})$, the distance from x to $\mathscr{M}$. It is left to the reader to show that (4.1) defines a seminorm on $\mathscr{X}/\mathscr{M}$. But if $\mathscr{M}$ is not closed in $\mathscr{X}$, (4.1) cannot define a norm. (Why?) If, however, $\mathscr{M}$ is closed, then (4.1) does define a norm.

4.2. Theorem. *If $\mathscr{M} \leq \mathscr{X}$ and $\|x + \mathscr{M}\|$ is defined as in (4.1), then $\|\cdot\|$ is a norm on $\mathscr{X}/\mathscr{M}$. Also:*

(a) *$\|Q(x)\| \leq \|x\|$ for all x in $\mathscr{X}$ and hence Q is continuous.*
(b) *If $\mathscr{X}$ is a Banach space, then so is $\mathscr{X}/\mathscr{M}$.*

(c) *A subset W of $\mathscr{X}/\mathscr{M}$ is open relative to the norm if and only if $Q^{-1}(W)$ is open in $\mathscr{X}$.*

(d) *If U is open in $\mathscr{X}$, then $Q(U)$ is open in $\mathscr{X}/\mathscr{M}$.*

PROOF. It is left as an exercise to show that (4.1) defines a norm on $\mathscr{X}/\mathscr{M}$. To show (a), $\| Q(x) \| = \| x + \mathscr{M} \| \leqslant \| x \|$ since $0 \in \mathscr{M}$; Q is therefore continuous by (2.1).

(b) Let $\{x_n + \mathscr{M}\}$ be a Cauchy sequence in $\mathscr{X}/\mathscr{M}$. There is a subsequence $\{x_{n_k} + \mathscr{M}\}$ such that

$$\|(x_{n_k} + \mathscr{M}) - (x_{n_{k+1}} + \mathscr{M})\| = \| x_{n_k} - x_{n_{k+1}} + \mathscr{M} \| < 2^{-k}.$$

Let $y_1 = 0$. Choose y_2 in $\mathscr{M}$ such that

$$\| x_{n_1} - x_{n_2} + y_2 \| \leqslant \| x_{n_1} - x_{n_2} + \mathscr{M} \| + 2^{-1} < 2 \cdot 2^{-1}.$$

Choose y_3 in $\mathscr{M}$ such that

$$\|(x_{n_2} + y_2) - (x_{n_3} + y_3)\| \leqslant \| x_{n_2} - x_{n_3} + \mathscr{M} \| + 2^{-2} < 2 \cdot 2^{-2}.$$

Continuing, there is a sequence $\{y_k\}$ in $\mathscr{M}$ such that

$$\|(x_{n_k} + y_k) - (x_{n_{k+1}} + y_{k+1})\| < 2 \cdot 2^{-k}.$$

Thus $\{x_{n_k} + y_k\}$ is a Cauchy sequence in $\mathscr{X}$ (Why?). Since $\mathscr{X}$ is complete, there is an x_0 in $\mathscr{X}$ such that $x_{n_k} + y_k \to x_0$ in $\mathscr{X}$. By (a), $x_{n_k} + \mathscr{M} = Q(x_{n_k} + y_k) \to Qx_0 = x_0 + \mathscr{M}$. Since $\{x_n + \mathscr{M}\}$ is a Cauchy sequence, $x_n + \mathscr{M} \to x_0 + \mathscr{M}$ and $\mathscr{X}/\mathscr{M}$ is complete (Exercise 3).

(c) If W is open in $\mathscr{X}/\mathscr{M}$, then $Q^{-1}(W)$ is open in $\mathscr{X}$ because Q is continuous. Now assume that $W \subseteq \mathscr{X}/\mathscr{M}$ and $Q^{-1}(W)$ is open in $\mathscr{X}$. Let $r > 0$ and put $B_r \equiv \{x \in \mathscr{X}: \| x \| < r\}$. It will be shown that $Q(B_r) = \{x + \mathscr{M}: \| x + \mathscr{M} \| < r\}$. In fact, if $\| x \| < r$, then $\| x + \mathscr{M} \| \leqslant \| x \| < r$. On the other hand, if $\| x + \mathscr{M} \| < r$, then there is a y in $\mathscr{M}$ such that $\| x + y \| < r$. Thus $x + \mathscr{M} = Q(x + y) \in Q(B_r)$. If $x_0 + \mathscr{M} \in W$, then $x_0 \in Q^{-1}(W)$. Since $Q^{-1}(W)$ is open, there is an $r > 0$ such that $x_0 + B_r = \{x: \| x - x_0 \| < r\} \subseteq Q^{-1}(W)$. The preceding argument now implies that $W = QQ^{-1}(W) \supseteq Q(x_0 + B_r) = \{x + \mathscr{M}: \| x - x_0 + \mathscr{M} \| < r\}$. Hence W is open.

(d) If U is open in $\mathscr{X}$, then $Q^{-1}(Q(U)) = U + \mathscr{M} \equiv \{u + y: u \in U, y \in \mathscr{M}\} = \cup \{U + y: y \in \mathscr{M}\}$. Each $U + y$ is open, so $Q^{-1}(Q(U))$ is open in $\mathscr{X}$. By (c), $Q(U)$ is open in $\mathscr{X}/\mathscr{M}$. ∎

Because Q is an open map [part (d)], it does not follow that Q is a closed map (Exercise 4).

4.3. Proposition. *If $\mathscr{X}$ is a normed space, $\mathscr{M} \leqslant \mathscr{X}$, and $\mathscr{N}$ is a finite dimensional subspace of $\mathscr{X}$, then $\mathscr{M} + \mathscr{N}$ is a closed subspace of $\mathscr{X}$.*

PROOF. Consider $\mathscr{X}/\mathscr{M}$ and the quotient map $Q: \mathscr{X} \to \mathscr{X}/\mathscr{M}$. Since $\dim Q(\mathscr{N}) \leqslant \dim \mathscr{N} < \infty$, $Q(\mathscr{N})$ is closed in $\mathscr{X}/\mathscr{M}$. Since Q is continuous $Q^{-1}(Q(\mathscr{N}))$ is closed in $\mathscr{X}$; but $Q^{-1}(Q(\mathscr{N})) = \mathscr{M} + \mathscr{N}$. ∎

Now for the product or direct sum of normed spaces. Here there is a difficulty because, unlike Hilbert space, there is no canonical way to proceed. Suppose $\{\mathscr{X}_i : i \in I\}$ is a collection of normed spaces. Then $\prod \{\mathscr{X}_i : i \in I\}$ is a vector space if the linear operations are defined coordinatewise. The idea is to put a norm on a linear subspace of this product.

Let $\|\cdot\|$ denote the norm on each $\mathscr{X}_i$. For $1 \leqslant p < \infty$, define

$$\oplus_p \mathscr{X}_i \equiv \left\{ x \in \prod_i \mathscr{X}_i : \|x\| \equiv \left[\sum_i \|x(i)\|^p \right]^{1/p} < \infty \right\}.$$

Define

$$\oplus_\infty \mathscr{X}_i \equiv \left\{ x \in \prod_i \mathscr{X}_i : \|x\| \equiv \sup_i \|x(i)\| < \infty \right\}.$$

If $\{\mathscr{X}_1, \mathscr{X}_2, \ldots\}$ is a sequence of normed spaces, define

$$\oplus_0 \mathscr{X}_n \equiv \left\{ x \in \prod_{n=1}^{\infty} \mathscr{X}_n : \|x(n)\| \to 0 \right\};$$

give $\oplus_0 \mathscr{X}_n$ the norm it has as a subspace of $\oplus_\infty \mathscr{X}_n$.

The proof of the next proposition is left as an exercise.

4.4. Proposition. *Let $\{\mathscr{X}_i : i \in I\}$ be a collection of normed spaces and let $\mathscr{X} = \oplus_p \mathscr{X}_i$, $1 \leqslant p \leqslant \infty$.*

(a) *$\mathscr{X}$ is a normed space and the projection $P_i : \mathscr{X} \to \mathscr{X}_i$ is a continuous linear map with $\|P_i(x)\| \leqslant \|x\|$ for each x in $\mathscr{X}$.*
(b) *$\mathscr{X}$ is a Banach space if and only if each $\mathscr{X}_i$ is a Banach space.*
(c) *Each projection P_i is an open map of $\mathscr{X}$ onto $\mathscr{X}_i$.*

A similar result holds for $\oplus_0 \mathscr{X}_n$, but the formulation and proof of this is left to the reader.

EXERCISES

1. Show that if $\mathscr{M} \leqslant \mathscr{X}$, then (4.1) defines a norm on $\mathscr{X}/\mathscr{M}$.

2. Prove that $\mathscr{X}$ is a Banach space if and only if whenever $\{x_n\}$ is a sequence in $\mathscr{X}$ such that $\sum \|x_n\| < \infty$, then $\sum_{n=1}^{\infty} x_n$ converges in $\mathscr{X}$.

3. Show that if (X, d) is a metric space and $\{x_n\}$ is a Cauchy sequence such that there is a subsequence $\{x_{n_k}\}$ that converges to x_0, then $x_n \to x_0$.

4. Find a Banach space $\mathscr{X}$ and a closed subspace $\mathscr{M}$ such that the natural map $Q: \mathscr{X} \to \mathscr{X}/\mathscr{M}$ is not a closed map. Can the natural map ever be a closed map?

5. Prove the converse of (4.2b): If $\mathscr{X}$ is a normed space, $\mathscr{M} \leqslant \mathscr{H}$, and both $\mathscr{M}$ and $\mathscr{X}/\mathscr{M}$ are complete, then $\mathscr{X}$ is complete. (This is an example of what is called a "two-out-of-three" result. If any two of $\mathscr{X}, \mathscr{M}$, and $\mathscr{X}/\mathscr{M}$ are complete, so is the third.)

6. Let $\mathscr{M} = \{x \in l^p : x(2n) = 0 \text{ for all } n\}$, $1 \leqslant p \leqslant \infty$. Show that $l^p/\mathscr{M}$ is isometrically isomorphic to l^p.

7. Let X be a normal locally compact space and F a closed subset of X. If $\mathcal{M} \equiv \{f \in C_0(X): f(x) = 0 \text{ for all } x \text{ in } F\}$, then $C_0(X)/\mathcal{M}$ is isometrically isomorphic to $C_0(F)$.

8. Prove Proposition 4.4.

9. Formulate and prove a version of Proposition 4.4 for $\oplus_0 \mathcal{X}_n$.

10. If $\{\mathcal{X}_1, \ldots, \mathcal{X}_n\}$ is a finite collection of normed spaces and $1 \leqslant p \leqslant \infty$, show that the norms on $\oplus_p \mathcal{X}_k$ are all equivalent.

11. Here is an abstraction of Proposition 4.4. Suppose $\{\mathcal{X}_i: i \in I\}$ is a collection of normed spaces and Y is a normed space contained in $\mathbb{F}^I$. Define $\mathcal{X} \equiv \{x \in \prod_i \mathcal{X}_i: \text{there is a } y \text{ in } Y \text{ with } \|x(i)\| \leqslant y(i) \text{ for all } i\}$. If $x \in \mathcal{X}$, define $\|x\| \equiv \inf\{\|y\|: \|x(i)\| \leqslant y(i) \text{ for all } i\}$. Then $(\mathcal{X}, \|\cdot\|)$ is a normed space. Give necessary and sufficient conditions on Y that each of the parts of (4.4) be valid for $\mathcal{X}$.

12. Let $\mathcal{X}$ be a normed space and $\mathcal{M} \leqslant \mathcal{X}$. (a) If $\mathcal{X}$ is separable, so is $\mathcal{X}/\mathcal{M}$. (b) If $\mathcal{X}/\mathcal{M}$ and $\mathcal{M}$ are separable, then $\mathcal{X}$ is separable. (c) Give an example such that $\mathcal{X}/\mathcal{M}$ is separable but $\mathcal{X}$ is not.

13. Let $\{\mathcal{X}_i: i \in I\}$ be a collection of non-zero normed spaces. For $1 \leqslant p < \infty$, put $\mathcal{X} = \oplus_p \mathcal{X}_i$. Show that $\mathcal{X}$ is separable if and only if I is countable and each $\mathcal{X}_i$ is separable. Show that $\oplus_\infty \mathcal{X}_i$ is separable if and only if I is finite and each $\mathcal{X}_i$ is separable.

14. Show that $\oplus_0 \mathcal{X}_n$ is separable if and only if each $\mathcal{X}_n$ is separable.

15. Let $J \subseteq I$, and $\mathcal{X} \equiv \oplus_p \{\mathcal{X}_i: i \in I\}$, $\mathcal{M} \equiv \{x \in \mathcal{X}: x(j) = 0 \text{ for } j \text{ in } J\}$. Show that $\mathcal{X}/\mathcal{M}$ is isometrically isomorphic to $\oplus_p \{\mathcal{X}_j: j \in J\}$.

16. Let $\mathcal{H}$ be a Hilbert space and suppose $\mathcal{M} \leqslant \mathcal{H}$. Show that if $Q: \mathcal{H} \to \mathcal{H}/\mathcal{M}$ is the natural map, then $Q: \mathcal{M}^\perp \to \mathcal{H}/\mathcal{M}$ is an isometric isomorphism.

§5. Linear Functionals

Let $\mathcal{X}$ be a vector space over $\mathbb{F}$. A *hyperplane* in $\mathcal{X}$ is a linear manifold $\mathcal{M}$ in $\mathcal{X}$ such that $\dim(\mathcal{X}/\mathcal{M}) = 1$. If $f: \mathcal{X} \to \mathbb{F}$ is a linear functional and $f \not\equiv 0$, then $\ker f$ is a hyperplane. In fact, f induces an isomorphism between $\mathcal{X}/\ker f$ and $\mathbb{F}$. Conversely, if $\mathcal{M}$ is a hyperplane, let $Q: \mathcal{X} \to \mathcal{X}/\mathcal{M}$ be the natural map and let $T: \mathcal{X}/\mathcal{M} \to \mathbb{F}$ be an isomorphism. Then $f \equiv T \circ Q$ is a linear functional on $\mathcal{X}$ and $\ker f = \mathcal{M}$.

Suppose now that f and g are linear functionals on $\mathcal{X}$ such that $\ker f = \ker g$. Let $x_0 \in \mathcal{X}$ such that $f(x_0) = 1$; so $g(x_0) \neq 0$. If $x \in \mathcal{X}$ and $\alpha = f(x)$, then $x - \alpha x_0 \in \ker f = \ker g$. So $0 = g(x) - \alpha g(x_0)$, or $g(x) = (g(x_0))\alpha = (g(x_0))f(x)$. Thus $g = \beta f$ for a scalar β. This is summarized as follows.

5.1. Proposition. *A linear manifold in $\mathcal{X}$ is a hyperplane if and only if it is the kernel of a non-zero linear functional. Two linear functionals have the same kernel if and only if one is a non-zero multiple of the other.*

Hyperplanes in a normed space fall into one of two categories.

5.2. Proposition. *If $\mathscr{X}$ is a normed space and $\mathscr{M}$ is a hyperplane in $\mathscr{X}$, then either $\mathscr{M}$ is closed or $\mathscr{M}$ is dense.*

PROOF. Consider cl $\mathscr{M}$, the closure of $\mathscr{M}$. By Proposition 1.3, cl $\mathscr{M}$ is a linear manifold in $\mathscr{X}$. Since $\mathscr{M} \subseteq$ cl $\mathscr{M}$ and dim $\mathscr{X}/\mathscr{M} = 1$, either cl $\mathscr{M} = \mathscr{M}$ or cl $\mathscr{M} = \mathscr{X}$. ∎

If $\mathscr{X} = c_0$ and $f: \mathscr{X} \to \mathbb{F}$ is defined by $f(\alpha_1, \alpha_2, \ldots) = \alpha_1$, then ker $f = \{(\alpha_n) \in c_0: \alpha_1 = 0\}$ is closed in c_0. To get an example of a dense hyperplane, let $\mathscr{X} = c_0$ and let e_n be the element of c_0 such that $e_n(k) = 0$ if $k \neq n$ and $e_n(n) = 1$. (It is best to think of c_0 as a collection of functions on $\mathbb{N}$.) Let $x_0(n) = 1/n$ for all n; so $x_0 \in c_0$ and $\{x_0, e_1, e_2, \ldots\}$ is a linearly independent set in c_0. Let $\mathscr{B} = $ a Hamel basis in c_0 which contains $\{x_0, e_1, e_2, \ldots\}$. Put $\mathscr{B} = \{x_0, e_1, e_2, \ldots\} \cup \{b_i: i \in I\}$, $b_i \neq x_0$ or e_n for any i or n. Define $f: c_0 \to \mathbb{F}$ by $f(\alpha_0 x_0 + \sum_{n=1}^{\infty} \alpha_n e_n + \sum_i \beta_i b_i) = \alpha_0$. (Remember that in the preceding expression at most a finite number of the α_n and β_i are not zero.) Since $e_n \in$ ker f for all $n \geq 1$, ker f is dense but clearly ker $f \neq c_0$.

The dichotomy that exists for hyperplanes should be reflected in a dichotomy for linear functionals.

5.3. Theorem. *If $\mathscr{X}$ is a normed space and $f: \mathscr{X} \to \mathbb{F}$ is a linear functional, then f is continuous if and only if ker f is closed.*

PROOF. If f is continuous, ker $f = f^{-1}(\{0\})$ and so ker f must be closed. Assume now that ker f is closed and let $Q: \mathscr{X} \to \mathscr{X}/\text{ker } f$ be the natural map. By (4.2), Q is continuous. Let $T: \mathscr{X}/\text{ker } f \to \mathbb{F}$ be an isomorphism; by (3.4), T is continuous. Thus, if $g = T \circ Q: \mathscr{X} \to \mathbb{F}$, g is continuous and ker $f = $ ker g. Hence (5.1) $f = \alpha g$ for some α in $\mathbb{F}$ and so f is continuous. ∎

If $f: \mathscr{X} \to \mathbb{F}$ is a linear functional, then f is a linear transformation and so Proposition 2.1 applies. Continuous linear functionals are also called *bounded linear functionals* and

$$\|f\| \equiv \sup\{|f(x)|: \|x\| \leq 1\}.$$

The other formulas for $\|f\|$ given in (2.1) are also valid here. Let $\mathscr{X}^* \equiv$ the collection of all bounded linear functionals on $\mathscr{X}$. If $f, g \in \mathscr{X}^*$ and $\alpha \in F$, define $(\alpha f + g)(x) = \alpha f(x) + g(x)$; $\mathscr{X}^*$ is called the *dual space* of $\mathscr{X}$. Note that $\mathscr{X}^* = \mathscr{B}(\mathscr{X}, \mathbb{F})$.

5.4. Proposition. *If $\mathscr{X}$ is a normed space, $\mathscr{X}^*$ is a Banach space.*

PROOF. It is left as an exercise for the reader to show that $\mathscr{X}^*$ is a normed space. To show that $\mathscr{X}^*$ is complete, let $B = \{x \in \mathscr{X}: \|x\| \leq 1\}$. If $f \in \mathscr{X}^*$, define $\rho(f): B \to \mathbb{F}$ by $\rho(f)(x) = f(x)$; that is, $\rho(f)$ is the restriction of f to B. Note that $\rho: \mathscr{X}^* \to C_b(B)$ is a linear isometry. Thus to show that $\mathscr{X}^*$ is complete, it suffices, since $C_b(B)$ is complete (1.6), to show that $\rho(\mathscr{X}^*)$ is closed. Let $\{f_n\} \subseteq \mathscr{X}^*$ and suppose $g \in C_b(B)$ such that $\|\rho(f_n) - g\| \to 0$ as $n \to \infty$. Let $x \in \mathscr{X}$. If $\alpha, \beta \in \mathbb{F}$, $\alpha, \beta \neq 0$, such that $\alpha x, \beta x \in B$, then $\alpha^{-1} g(\alpha x) = \lim \alpha^{-1} f_n(\alpha x) = \lim \beta^{-1} f_n(\beta x) = \beta^{-1} g(\beta x)$. Define $f: \mathscr{X} \to \mathbb{F}$ by letting $f(x) = \alpha^{-1} g(\alpha x)$ for any $\alpha \neq 0$ such that $\alpha x \in B$. It is left as an exercise for the reader to show that $f \in \mathscr{X}^*$ and $\rho(f) = g$. ∎

Compare the preceding result with Exercise 2.1.

It should be emphasized that it is not assumed in the preceding proposition that $\mathscr{X}$ is complete. In fact, if $\mathscr{X}$ is a normed space and $\hat{\mathscr{X}}$ is its completion (Exercise 1.16), then $\mathscr{X}^*$ and $\hat{\mathscr{X}}^*$ are isometrically isomorphic (Exercise 2.2).

5.5. Theorem. *Let (X, Ω, μ) be a measure space and let $1 < p < \infty$. If $1/p + 1/q = 1$ and $g \in L^q(X, \Omega, \mu)$, define $F_g: L^p(\mu) \to \mathbb{F}$ by*

$$F_g(f) = \int fg \, d\mu.$$

Then $F_g \in L^p(\mu)^*$ and the map $g \mapsto F_g$ defines an isometric isomorphism of $L^q(\mu)$ onto $L^p(\mu)^*$.

Since this theorem is often proved in courses in measure and integration, the proof of this result, as well as the next two, is contained in the Appendix. See Appendix B for the proofs of (5.5) and (5.6).

5.6. Theorem. *If (X, Ω, μ) is a σ-finite measure space and $g \in L^\infty(X, \Omega, \mu)$, define $F_g: L^1(\mu) \to \mathbb{F}$ by*

$$F_g(f) = \int fg \, d\mu.$$

Then $F_g \in L^1(\mu)^*$ and the map $g \mapsto F_g$ defines an isometric isomorphism of $L^\infty(\mu)$ onto $L^1(\mu)^*$.

Note that when $p = 2$ in Theorem 5.5, there is a little difference between (5.5) and (I.3.5) owing to the absence of a complex conjugate in (5.5). Also, note that (5.6) is false if the measure space is not assumed to be σ-finite (Exercise 3).

If X is a locally compact space, $M(X)$ denotes the space of all $\mathbb{F}$-valued regular Borel measures on X with the total variation norm. See Appendix C for the definitions as well as the proof of the next theorem.

5.7. Riesz Representation Theorem. *If X is a locally compact space and $\mu \in M(X)$, define $F_\mu: C_0(X) \to \mathbb{F}$ by*

$$F_\mu(f) = \int f \, d\mu.$$

Then $F_\mu \in C_0(X)^*$ and the map $\mu \to F_\mu$ is an isometric isomorphism of $M(X)$ onto $C_0(X)^*$.

There are special cases of these theorems that deserve to be pointed out.

5.8. Example. The dual of c_0 is isometrically isomorphic to l^1. In fact, $c_0 = C_0(\mathbb{N})$, if $\mathbb{N}$ is given the discrete topology, and $l^1 = M(\mathbb{N})$.

5.9. Example. The dual of l^1 is isometrically isomorphic to l^∞. In fact, $l^1 = L^1(\mathbb{N}, 2^\mathbb{N}, \mu)$, where $\mu(\Delta) =$ the number of points in Δ. Also, $l^\infty = L^\infty(\mathbb{N}, 2^\mathbb{N}, \mu)$.

5.10. Example. If $1 < p < \infty$, the dual of l^p is l^q, where $1 = 1/p + 1/q$.

What is the dual of $L^\infty(X, \Omega, \mu)$? There are two possible representations. One is to identify $L^\infty(X, \Omega, \mu)^*$ with the space of finitely additive measures defined on Ω that are "absolutely continuous" with respect to μ and have finite total variation (see Dunford and Schwartz [1958], p. 296). Another representation is to obtain a compact space Z such that $L^\infty(X, \Omega, \mu)$ is isometrically isomorphic to $C(Z)$ and then use the Riesz Representation Theorem. This will be done later in this book (VIII.2.1).

What is the dual of $M(X)$? For this, define $L^\infty(M(X))$ as the set of all F in $\prod \{L^\infty(\mu): \mu \in M(X)\}$ such that if $\mu \ll \nu$, then $F(\mu) = F(\nu)$ a.e. $[\mu]$. This is an inverse limit of the spaces $L^\infty(\mu), \mu$ in $M(X)$.

5.11. Lemma. If $F \in L^\infty(M(X))$, then

$$\|F\| \equiv \sup_\mu \|F(\mu)\|_\infty < \infty.$$

PROOF. If $\|F\| = \infty$, then there is a sequence $\{\mu_n\}$ in $M(X)$ such that $\|F(\mu_n)\|_\infty \geq n$. Let $\mu = \sum_{n=1}^\infty 2^{-n}|\mu_n|/\|\mu_n\|$. Then $\mu_n \ll \mu$ for all n, so $F(\mu_n) = F(\mu)$ a.e. $[\mu_n]$ for each n. Hence $\|F(\mu)\|_\infty \geq \|F(\mu_n)\|_\infty \geq n$ for each n, a contradiction. ∎

5.12. Theorem. If X is locally compact and $F \in L^\infty(M(X))$, define $\Phi_F: M(X) \to \mathbb{F}$ by

$$\Phi_F(\mu) = \int F(\mu) \, d\mu.$$

Then $\Phi_F \in M(X)^*$ and the map $F \mapsto \Phi_F$ is an isometric isomorphism of $L^\infty(M(X))$ onto $M(X)^*$.

PROOF. It is easy to see that Φ_F is linear. Also, $|\Phi_F(\mu)| \leq \int |F(\mu)| \, d|\mu| \leq \|F(\mu)\|_\infty \|\mu\| \leq \|F\| \|\mu\|$. Thus $\Phi_F \in M(X)^*$ and $\|\Phi_F\| \leq \|F\|$.

Now fix Φ in $M(X)^*$. If $\mu \in M(X)$ and $f \in L^1(|\mu|)$, then $\nu = f\mu \in M(X)$. (That is, $\nu(\Delta) = \int_\Delta f \, d\mu$ for every Borel set Δ.) Also $\|\nu\| = \int |f| \, d|\mu|$. In fact, the

Radon–Nikodym Theorem can be interpreted as an identification (isometrically isomorphic) of $L^1(|\mu|)$ with $\{\eta \in M(X): \eta \ll |\mu|\}$. Thus $f \mapsto \Phi(f\mu)$ is a linear functional on $L^1(|\mu|)$ and $|\Phi(f\mu)| \leqslant \|\Phi\| \int |f| \, d|\mu|$. Hence there is an $F(\mu)$ in $L^\infty(|\mu|)$ such that $\Phi(f\mu) = \int f F(\mu) \, d\mu$ for every f in $L^1(|\mu|)$ and $\|F(\mu)\|_\infty \leqslant \|\Phi\|$. (We have been a little nonchalant about using μ or $|\mu|$, but what was said is perfectly correct. Fill in the details.) In particular, taking $f = 1$ gives $\Phi(\mu) = \int F(\mu) \, d\mu$. It must be shown that $F \in L^\infty(M(X))$; it then follows that $\Phi = \Phi_F$ and $\|\Phi_F\| \geqslant \|F\|_\infty$.

To show that $F \in L^\infty(M(X))$, let μ and v be measures such that $v \ll \mu$. By the Radon–Nikodym Theorem, there is an f in $L^1(|\mu|)$ such that $v = f\mu$. Hence if $g \in L^1(|v|)$, then $gf \in L^1(|\mu|)$ and $\int g \, dv = \int gf \, d\mu$. Thus, $\int g F(v) \, dv = \Phi(gv) = \Phi(gf\mu) = \int gf F(\mu) \, d\mu = \int g F(\mu) \, dv$. So $F(v) = F(\mu)$ a.e. $[v]$ and $F \in L^\infty(M(X))$. ∎

EXERCISES

1. Complete the proof of Proposition 5.4.

2. Show that $\mathscr{X}^*$ is a normed space.

3. Give an example of a measure space (X, Ω, μ) that is not σ-finite for which the conclusion of Theorem 5.6 is false.

4. Let $\{\mathscr{X}_i: i \in I\}$ be a collection of normed spaces. If $1 \leqslant p < \infty$, show that the dual space of $\oplus_p \mathscr{X}_i$ is isometrically isomorphic to $\oplus_q \mathscr{X}_i^*$, where $1/p + 1/q = 1$.

5. If $\mathscr{X}_1, \mathscr{X}_2, \ldots$ are normed spaces, show that $(\oplus_0 \mathscr{X}_n)^*$ is isometrically isomorphic to $\oplus_1 \mathscr{X}_n^*$.

6. Let $n \geqslant 1$ and let $C^{(n)}[0, 1]$ be defined as in Example 1.10. Show that $\|f\| = \sum_{k=0}^{n-1} |f^{(k)}(0)| + \sup\{|f^{(n)}(x)|: 0 \leqslant x \leqslant 1\}$ is an equivalent norm on $C^{(n)}[0, 1]$. Show that $L \in (C^{(n)}[0, 1])^*$ if and only if there are scalars $\alpha_0, \alpha_1, \ldots, \alpha_{n-1}$ and a measure μ on $[0, 1]$ such that $L(f) = \sum_{k=0}^{n-1} \alpha_k f^{(k)}(0) + \int f^{(n)} \, d\mu$. If $C^{(n)}[0, 1]$ is given this new norm, find a formula for $\|L\|$ in terms of $|\alpha_0|, |\alpha_1|, \ldots, |\alpha_{n-1}|$, and $\|\mu\|$?

7. Give $\mathscr{X} = C([0, 1])$ the norm $\|f\| = \int |f(t)| \, dt$ and define $L: \mathscr{X} \to F$ by $L(f) = f(\frac{1}{2})$. Show directly (without using Theorem 5.6) that L is not bounded. Now prove this as a consequence of (5.6).

§6. The Hahn–Banach Theorem

The Hahn–Banach Theorem is one of the most important results in mathematics. It is used so often it is rightly considered as a cornerstone of functional analysis. It is one of those theorems that when it or one of its immediate consequences is used, it is used without quotation or reference and the reader is assumed to realize that it is being invoked.

6.1. Definition. If $\mathscr{X}$ is a vector space, a *sublinear functional* is a function

$q: \mathscr{X} \to \mathbb{R}$ such that

(a) $q(x + y) \leqslant q(x) + q(y)$ for all x, y in $\mathscr{X}$;
(b) $q(\alpha x) = \alpha q(x)$ for x in $\mathscr{X}$ and $\alpha \geqslant 0$.

Note that every seminorm is a sublinear functional, but not conversely. In fact, it should be emphasized that a sublinear functional is allowed to assume negative values and that (b) in the definition only holds for $\alpha \geqslant 0$.

6.2. The Hahn–Banach Theorem. *Let $\mathscr{X}$ be a vector space over $\mathbb{R}$ and let q be a sublinear functional on $\mathscr{X}$. If $\mathscr{M}$ is a linear manifold in $\mathscr{X}$ and $f: \mathscr{M} \to \mathbb{R}$ is a linear functional such that $f(x) \leqslant q(x)$ for all x in $\mathscr{M}$, then there is a linear functional $F: \mathscr{X} \to \mathbb{R}$ such that $F|\mathscr{M} = f$ and $F(x) \leqslant q(x)$ for all x in $\mathscr{X}$.*

Note that the substance of the theorem is not that the extension exists but that an extension can be found that remains dominated by q. Just to find an extension, let $\{e_i\}$ be a Hamel basis for $\mathscr{M}$ and let $\{y_j\}$ be vectors in $\mathscr{X}$ such that $\{e_i\} \cup \{y_j\}$ is a Hamel basis for $\mathscr{X}$. Now define $F: \mathscr{X} \to \mathbb{R}$ by $F(\sum_i \alpha_i e_i + \sum_j \beta_j y_j) = \sum_i \alpha_i f(e_i) = f(\sum_i \alpha_i e_i)$. This extends f. If $\{\gamma_j\}$ is any collection of real numbers, then $F(\sum_i \alpha_i e_i + \sum_j \beta_j y_j) = f(\sum_i \alpha_i e_i) + \sum_j \beta_j \gamma_j$ is also an extension of f. Moreover, any extension of f has this form. The difficulty is that we must find one of these extensions that is dominated by q.

Before proving the theorem, let's see some of its immediate corollaries. The first is an extension of the theorem to complex spaces. For this a lemma is needed. Note that if $\mathscr{X}$ is a vector space over $\mathbb{C}$, it is also a vector space over $\mathbb{R}$. Also, if $f: \mathscr{X} \to \mathbb{C}$ is $\mathbb{C}$-linear, then $\operatorname{Re} f: \mathscr{X} \to \mathbb{R}$ is $\mathbb{R}$-linear. The following lemma is the converse of this.

6.3. Lemma. *Let $\mathscr{X}$ be a vector space over $\mathbb{C}$.*

(a) *If $f: \mathscr{X} \to \mathbb{R}$ is an $\mathbb{R}$-linear functional, then $\tilde{f}(x) = f(x) - if(ix)$ is a $\mathbb{C}$-linear functional and $f = \operatorname{Re} \tilde{f}$.*
(b) *If $g: \mathscr{X} \to \mathbb{C}$ is $\mathbb{C}$-linear, $f = \operatorname{Re} g$, and $\tilde{f}$ is defined as in (a), then $\tilde{f} = g$.*
(c) *If p is a seminorm on $\mathscr{X}$ and f and $\tilde{f}$ are as in (a), then $|f(x)| \leqslant p(x)$ for all x if and only if $|\tilde{f}(x)| \leqslant p(x)$ for all x.*
(d) *If $\mathscr{X}$ is a normed space and f and $\tilde{f}$ are as in (a), then $\|f\| = \|\tilde{f}\|$.*

PROOF. The proofs of (a) and (b) are left as an exercise. To prove (c), suppose $|\tilde{f}(x)| \leqslant p(x)$. Then $f(x) = \operatorname{Re} \tilde{f}(x) \leqslant |\tilde{f}(x)| \leqslant p(x)$. Also, $-f(x) = \operatorname{Re} \tilde{f}(-x) \leqslant |\tilde{f}(-x)| \leqslant p(x)$. Hence $|f(x)| \leqslant p(x)$. Now assume that $|f(x)| \leqslant p(x)$. Choose θ such that $\tilde{f}(x) = e^{i\theta} |\tilde{f}(x)|$. Hence $|\tilde{f}(x)| = \tilde{f}(e^{-i\theta}x) = \operatorname{Re} \tilde{f}(e^{-i\theta}x) = f(e^{-i\theta}x) \leqslant p(e^{-i\theta}x) = p(x)$.

Part (d) is an easy application of (c). ∎

6.4. Corollary. *Let $\mathscr{X}$ be a vector space, let $\mathscr{M}$ be a linear manifold in $\mathscr{X}$, and let $p: \mathscr{X} \to [0, \infty)$ be a seminorm. If $f: \mathscr{M} \to \mathbb{F}$ is a linear functional such that*

$|f(x)| \leqslant p(x)$ for all x in $\mathcal{M}$, then there is a linear functional $F: \mathcal{X} \to \mathbb{F}$ such that $F|\mathcal{M} = f$ and $|F(x)| \leqslant p(x)$ for all x in $\mathcal{X}$.

PROOF. *Case 1*: $\mathbb{F} = \mathbb{R}$. Note that $f(x) \leqslant |f(x)| \leqslant p(x)$ for x in $\mathcal{M}$. By (6.2) there is an extension $F: \mathcal{X} \to \mathbb{R}$ of f such that $F(x) \leqslant p(x)$ for all x. Hence $-F(x) = F(-x) \leqslant p(-x) = p(x)$. Thus $|F| \leqslant p$.

Case 2: $\mathbb{F} = \mathbb{C}$. Let $f_1 = \operatorname{Re} f$. By (6.3c), $|f_1| \leqslant p$. By Case 1, there is an $\mathbb{R}$-linear functional $F_1: \mathcal{X} \to \mathbb{R}$ such that $F_1|\mathcal{M} = f_1$ and $|F_1| \leqslant p$. Let $F(x) = F_1(x) - iF_1(ix)$ for all x in $\mathcal{X}$. By (6.3c), $|F| \leqslant p$. Clearly, $F|\mathcal{M} = f$. ∎

6.5. Corollary. *If $\mathcal{X}$ is a normed space, $\mathcal{M}$ is a linear manifold in $\mathcal{X}$, and $f: \mathcal{M} \to \mathbb{F}$ is a bounded linear functional, then there is an F in $\mathcal{X}^*$ such that $F|\mathcal{M} = f$ and $\|F\| = \|f\|$.*

PROOF. Use Corollary 6.4 with $p(x) = \|f\| \, \|x\|$. ∎

6.6. Corollary. *If $\mathcal{X}$ is a normed space, $\{x_1, x_2, \ldots, x_d\}$ is a linearly independent subset of $\mathcal{X}$, and $\alpha_1, \alpha_2, \ldots, \alpha_d$ are arbitrary scalars, then there is an f in $\mathcal{X}^*$ such that $f(x_j) = \alpha_j$ for $1 \leqslant j \leqslant d$.*

PROOF. Let $\mathcal{M}$ = the linear span of $x_1, \ldots, x_d$ and define $g: \mathcal{M} \to \mathbb{F}$ by $g(\sum_j \beta_j x_j) = \sum_j \beta_j \alpha_j$. So g is linear. Since $\mathcal{M}$ is finite dimensional, g is continuous. Let f be a continuous extension of g to $\mathcal{X}$. ∎

6.7. Corollary. *If $\mathcal{X}$ is a normed space and $x \in \mathcal{X}$, then*

$$\|x\| = \sup\{|f(x)|: f \in \mathcal{X}^* \text{ and } \|f\| \leqslant 1\}.$$

Moreover, this supremum is attained.

PROOF. Let $\alpha = \sup\{|f(x)|: f \in \mathcal{X}^* \text{ and } \|f\| \leqslant 1\}$. If $f \in \mathcal{X}^*$ and $\|f\| \leqslant 1$, then $|f(x)| \leqslant \|f\| \, \|x\| \leqslant \|x\|$; hence $\alpha \leqslant \|x\|$. Now let $\mathcal{M} = \{\beta x: \beta \in \mathbb{F}\}$ define $g: \mathcal{M} \to \mathbb{F}$ by $g(\beta x) = \beta \|x\|$. Then $g \in \mathcal{M}^*$ and $\|g\| = 1$. By Corollary 6.5, there is an f in $\mathcal{X}^*$ such that $\|f\| = 1$ and $f(x) = g(x) = \|x\|$. ∎

This introduces a certain symmetry in the definitions of the norms in $\mathcal{X}$ and $\mathcal{X}^*$ that will be explored later (§11).

6.8. Corollary. *If $\mathcal{X}$ is a normed space, $\mathcal{M} \leqslant \mathcal{X}$, $x_0 \in \mathcal{X} \setminus \mathcal{M}$, and $d = \operatorname{dist}(x_0, \mathcal{M})$, then there is an f in $\mathcal{X}^*$ such that $f(x_0) = 1$, $f(x) = 0$ for all x in $\mathcal{M}$, and $\|f\| = d^{-1}$.*

PROOF. Let $Q: \mathcal{X} \to \mathcal{X}/\mathcal{M}$ be the natural map. Since $\|x_0 + \mathcal{M}\| = d$, by the preceding corollary there is a g in $(\mathcal{X}/\mathcal{M})^*$ such that $g(x_0 + \mathcal{M}) = d$ and $\|g\| = 1$. Let $f = d^{-1} g \circ Q: \mathcal{X} \to \mathbb{F}$. Then f is continuous, $f(x) = 0$ for x in $\mathcal{M}$, and $f(x_0) = 1$. Also, $|f(x)| = d^{-1}|g(Q(x))| \leqslant d^{-1}\|Q(x)\| \leqslant d^{-1}\|x\|$; hence $\|f\| \leqslant d^{-1}$. On the other hand, $\|g\| = 1$ so there is a sequence $\{x_n\}$ such that $|g(x_n + \mathcal{M})| \to 1$ and $\|x_n + \mathcal{M}\| < 1$ for all n. Let $y_n \in \mathcal{M}$ such that $\|x_n + y_n\| < 1$. Then $|f(x_n + y_n)| = d^{-1}|g(x_n + \mathcal{M})| \to d^{-1}$, so $\|f\| = d^{-1}$. ∎

To prove the Hahn–Banach Theorem, we first show that we can extend the functional to a space of one dimension more.

6.9. Lemma. *Suppose the hypothesis of (6.2) is satisfied and, in addition, $\dim \mathcal{X}/\mathcal{M} = 1$. Then the conclusion of (6.2) is valid.*

PROOF. Fix x_0 in $\mathcal{X} \backslash \mathcal{M}$; so $\mathcal{X} = \mathcal{M} \vee \{x_0\} = \{tx_0 + y: t \in \mathbb{R}, y \in \mathcal{M}\}$. For the moment assume that the extension $F: \mathcal{X} \to \mathbb{R}$ of f exists with $F \leqslant q$. Let's see what F must look like. Put $\alpha_0 = F(x_0)$. If $t > 0$ and $y_1 \in \mathcal{M}$, then $F(tx_0 + y_1) = t\alpha_0 + f(y_1) \leqslant q(tx_0 + y_1)$. Hence $\alpha_0 \leqslant -t^{-1}f(y_1) + t^{-1}q(tx_0 + y_1) = -f(y_1/t) + q(x_0 + y_1/t)$ for every y_1 in $\mathcal{M}$. Since $y_1/t \in \mathcal{M}$, this gives that.

6.10 $$\alpha_0 \leqslant -f(y_1) + q(x_0 + y_1)$$

for all y_1 in $\mathcal{M}$. Also note that if α_0 satisfies (6.10), then by reversing the preceding argument, it follows that $t\alpha_0 + f(y_1) \leqslant q(tx_0 + y_1)$ whenever $t \geqslant 0$.

If $t \geqslant 0$ and $y_2 \in \mathcal{M}$ and if F exists, then $F(-tx_0 + y_2) = -t\alpha_0 + f(y_2) \leqslant q(-tx_0 + y_2)$. As above, this implies that

6.11 $$\alpha_0 \geqslant f(y_2) - q(-x_0 + y_2)$$

for all y_2 in $\mathcal{M}$. Moreover, (6.11) is sufficient that $-t\alpha_0 + f(y_2) \leqslant q(-tx_0 + y_2)$ for all $t \geqslant 0$ and y_2 in $\mathcal{M}$.

Combining (6.10) and (6.11) we see that we must show that α_0 can be chosen satisfying (6.10) and (6.11) simultaneously. Thus we must show that

6.12 $$f(y_2) - q(-x_0 + y_2) \leqslant -f(y_1) + q(x_0 + y_1)$$

for all y_1, y_2 in $\mathcal{M}$. But this means we want to show that $f(y_1 + y_2) \leqslant q(x_0 + y_1) + q(-x_0 + y_2)$. But

$$f(y_1 + y_2) \leqslant q(y_1 + y_2) = q((y_1 + x_0) + (-x_0 + y_2))$$
$$\leqslant q(y_1 + x_0) + q(-x_0 + y_2),$$

so (6.12) is satisfied. If α_0 is chosen with $\sup\{f(y_2) - q(-x_0 + y_2): y_2 \in \mathcal{M}\} \leqslant \alpha_0 \leqslant \inf\{-f(y_1) + q(x_0 + y_1): y_1 \in \mathcal{M}\}$ and $F(tx_0 + y) \equiv t\alpha_0 + f(y_1)$, F satisfies the conclusion of (6.2). ∎

PROOF OF THE HAHN–BANACH THEOREM. Let $\mathcal{S}$ be the collection of all pairs $(\mathcal{M}_1, f_1)$, where $\mathcal{M}_1$ is a linear manifold in $\mathcal{X}$ such that $\mathcal{M}_1 \supseteq \mathcal{M}$ and $f_1: \mathcal{M}_1 \to \mathbb{R}$ is a linear functional with $f_1|\mathcal{M} = f$ and $f_1 \leqslant q$ on $\mathcal{M}_1$. If $(\mathcal{M}_1, f_1)$ and $(\mathcal{M}_2, f_2) \in \mathcal{S}$, define $(\mathcal{M}_1, f_1) \lesssim (\mathcal{M}_2, f_2)$ to mean that $\mathcal{M}_1 \subseteq \mathcal{M}_2$ and $f_2|\mathcal{M}_1 = f_1$. So $(\mathcal{S}, \lesssim)$ is a partially ordered set. Suppose $\mathcal{C} = \{(\mathcal{M}_i, f_i): i \in I\}$ is a chain in $\mathcal{S}$. If $\mathcal{N} \equiv \cup\{\mathcal{M}_i: i \in I\}$, then the fact that $\mathcal{C}$ is a chain implies that $\mathcal{N}$ is a linear manifold. Define $F: \mathcal{N} \to \mathbb{R}$ by setting $F(x) = f_i(x)$ if $x \in \mathcal{M}_i$.

It is easily checked that F is well defined, linear, and satisfies $F \leqslant q$ on $\mathcal{N}$. So $(\mathcal{N}, F) \in \mathcal{S}$ and $(\mathcal{N}, F)$ is an upper bound for $\mathscr{C}$. By Zorn's Lemma, $\mathcal{S}$ has a maximal element $(\mathcal{Y}, F)$. But the preceding lemma implies that $\mathcal{Y} = \mathcal{X}$. Hence F is the desired extension. ∎

This section concludes with one important consequence of the Hahn–Banach Theorem. It will be generalized later (IV.3.11), but it is used so often it is worth singling out for consideration.

6.13. Theorem. *If $\mathcal{X}$ is a normed space and $\mathcal{M}$ is a linear manifold in $\mathcal{X}$, then*

$$\mathrm{cl}\, \mathcal{M} = \cap \{\ker f : f \in \mathcal{X}^* \text{ and } \mathcal{M} \subseteq \ker f\}.$$

PROOF. Let $\mathcal{N} = \cap \{\ker f : f \in \mathcal{X}^* \text{ and } \mathcal{M} \subseteq \ker f\}$. If $f \in \mathcal{X}^*$ and $\mathcal{M} \subseteq \ker f$, then the continuity of f implies that $\mathrm{cl}\, \mathcal{M} \subseteq \ker f$. Hence $\mathrm{cl}\, \mathcal{M} \subseteq \mathcal{N}$. If $x_0 \notin \mathrm{cl}\, \mathcal{M}$, then $d = \mathrm{dist}(x_0, \mathcal{M}) > 0$. By Corollary 6.8 there is an f in $\mathcal{X}^*$ such that $f(x_0) = 1$ and $f(x) = 0$ for every x in $\mathcal{M}$. Hence $x_0 \notin \mathcal{N}$. Thus $\mathcal{N} \subseteq \mathrm{cl}\, \mathcal{M}$ and the proof is complete. ∎

6.14. Corollary. *If $\mathcal{X}$ is a normed space and $\mathcal{M}$ is a linear manifold in $\mathcal{X}$, then $\mathcal{M}$ is dense in $\mathcal{X}$ if and only if the only bounded linear functional on $\mathcal{X}$ that annihilates $\mathcal{M}$ is the zero functional.*

EXERCISES

1. Complete the proof of Lemma 6.3.

2. Give the details of the proof of Corollary 6.5.

3. Show that c^* is isometrically isomorphic to l^1. Are c and c_0 isometrically isomorphic?

4. If μ is a Borel measure on $[0, 1]$ and $\int x^n \, d\mu(x) = 0$ for all $n \geqslant 0$, show that $\mu = 0$.

5. If $n \geqslant 1$, show that there is a measure μ on $[0, 1]$ such that for every polynomial p of degree at most n,

$$\int p \, d\mu = \sum_{k=1}^{n} p^{(k)}(k/n).$$

6. If $n \geqslant 1$, does there exist a measure μ on $[0, 1]$ such that $p'(0) = \int p \, d\mu$ for every polynomial of degree at most n?

7. Does there exist a measure μ on $[0, 1]$ such that $\int p \, d\mu = p'(0)$ for every polynomial p?

8. Let K be a compact subset of $\mathbb{C}$ and define $A(K)$ to be $\{f \in C(K) : f \text{ is analytic on int } K\}$. (Functions here are complex valued.) Show that if $a \in K$, then there is a probability measure μ supported on ∂K such that $f(a) = \int f \, d\mu$ for every f in $A(K)$. (A *probability measure* is a nonnegative measure μ such that $\|\mu\| = 1$.)

9. If $K = \mathrm{cl}\, \mathbb{D}$ ($\mathbb{D} = \{|z| < 1\}$) and $a \in K$, find the measure μ whose existence was proved in Exercise 8.

10. Let $P = \{p|\partial\mathbb{D}: p = $ an analytic polynomial$\}$ and consider P as a manifold in $C(\partial\mathbb{D})$. Show that if μ is a real-valued measure on $\partial\mathbb{D}$ such that $\int p\,d\mu = 0$ for every p in P, then $\mu = 0$. Give an example of a complex-valued measure μ such that $\mu \neq 0$ but $\int p\,d\mu = 0$ for every p in P.

§7*. An Application: Banach Limits

If $x = \{x(n)\} \in c$, define $L(x) = \lim x(n)$. Then L is a linear functional, $\|L\| = 1$, and, if for x in c, x' is defined by $x' = (x(2), x(3),\ldots)$, then $L(x) = L(x')$. Also, if $x \geqslant 0$ [that is, $x(n) \geqslant 0$ for all n], then $L(x) \geqslant 0$. In this section it will be shown that these properties of the limit functional can be extended to l^∞. The proof uses the Hahn–Banach Theorem.

7.1. Theorem. *There is a linear functional $L: l^\infty \to \mathbb{F}$ such that*

(a) $\|L\| = 1$.
(b) *If $x \in c$, $L(x) = \lim x(n)$.*
(c) *If $x \in l^\infty$ and $x(n) \geqslant 0$ for all n, then $L(x) \geqslant 0$.*
(d) *If $x \in l^\infty$ and $x' \equiv (x(2), x(3),\ldots)$, then $L(x) = L(x')$.*

PROOF. First assume $\mathbb{F} = \mathbb{R}$; that is, $l^\infty = l^\infty_{\mathbb{R}}$. If $x \in l^\infty$, let x' denote the element of l^∞ defined in part (d) above. Put $\mathcal{M} = \{x - x': x \in l^\infty\}$. Note that $(x + \alpha y)' = x' + \alpha y'$ for any x, y in l^∞ and α in $\mathbb{R}$; hence $\mathcal{M}$ is a linear manifold in l^∞. Let 1 denote the sequence $(1, 1, 1,\ldots)$ in l^∞.

7.2. Claim. $\text{dist}(1, \mathcal{M}) = 1$.

Since $0 \in \mathcal{M}$, $\text{dist}(1, \mathcal{M}) \leqslant 1$. Let $x \in l^\infty$; if $(x - x')(n) \leqslant 0$ for any n, then $\|1 - (x - x')\|_\infty \geqslant |1 - (x(n) - x'(n))| \geqslant 1$. Suppose $0 \leqslant (x - x')(n) = x(n) - x'(n) = x(n) - x(n+1)$ for all n. Thus $x(n+1) \leqslant x(n)$ for all n. Since $x \in l^\infty$, $\alpha = \lim x(n)$ exists. Thus $\lim(x - x')(n) = 0$ and $\|1 - (x - x')\|_\infty \geqslant 1$. This proves the claim.

By Corollary 6.8 there is a linear functional $L: l^\infty \to \mathbb{R}$ such that $\|L\| = 1$, $L(1) = 1$, and $L(\mathcal{M}) = 0$. So this functional satisfies (a) and (d) of the theorem. To prove (b), we establish the following.

7.3. Claim. $c_0 \subseteq \ker L$.

If $x \in c_0$, let $x^{(1)} = x'$ and let $x^{(n+1)} = (x^{(n)})'$ for $n \geqslant 1$. Note that $x^{(n+1)} - x = [x^{(n+1)} - x^{(n)}] + \cdots + [x' - x] \in \mathcal{M}$. Hence $L(x) = L(x^{(n)})$ for all $n \geqslant 1$. If $\varepsilon > 0$, then let n be such that $|x(m)| < \varepsilon$ for $m > n$. Hence $|L(x)| = |L(x^{(n)})| \leqslant \|x^{(n)}\|_\infty = \sup\{|x(m)|: m > n\} < \varepsilon$. Thus $x \in \ker L$. Condition (b) is now clear.

To show (c), suppose there is an x in l^∞ such that $x(n) \geqslant 0$ for all n and $L(x) < 0$. If x is replaced by $x/\|x\|_\infty$, it remains true that $L(x) < 0$ and it is also true that $1 \geqslant x(n) \geqslant 0$ for all n. But then $\|1 - x\|_\infty \leqslant 1$ and $L(1 - x) = 1 - L(x) > 1$, contradicting (a). Thus (c) holds.

Now assume that $\mathbb{F} = \mathbb{C}$. Let L_1 be the functional obtained on $l_\mathbb{R}^\infty$. If $x \in l_\mathbb{C}^\infty$, then $x = x_1 + ix_2$ where $x_1, x_2 \in l_\mathbb{R}^\infty$. Define $L(x) = L_1(x_1) + iL_1(x_2)$. It is left as an exercise to show that L is $\mathbb{C}$-linear. It's clear that (b), (c), and (d) hold. It remains to show that $\|L\| = 1$.

Let $E_1, \ldots, E_m$ be pairwise disjoint subsets of $\mathbb{N}$ and let $\alpha_1, \ldots, \alpha_m \in \mathbb{C}$ with $|\alpha_k| \leqslant 1$ for all k. Put $x = \sum_{k=1}^m \alpha_k \chi_{E_k}$; so $x \in l^\infty$ and $\|x\|_\infty \leqslant 1$. Then $L(x) = \sum_k \alpha_k L(\chi_{E_k}) = \sum_k \alpha_k L_1(\chi_{E_k})$. But $L_1(\chi_{E_k}) \geqslant 0$ and $\sum_k L_1(\chi_{E_k}) = L_1(\chi_E)$, where $E = \bigcup_k E_k$. Hence $\sum_k L_1(\chi_{E_k}) \leqslant 1$. Because $|\alpha_k| \leqslant 1$ for all k, $|L(x)| \leqslant 1$. If x is an arbitrary element of l^∞, $\|x\|_\infty \leqslant 1$, then there is a sequence $\{x_n\}$ of elements of l^∞ such that $\|x_n - x\|_\infty \to 0$, $\|x_n\|_\infty \leqslant 1$, and each x_n is the type of element of l^∞ just discussed that takes on only a finite number of values (Exercise 3). Clearly, $\|L\| \leqslant 2$, so $L(x_n) \to L(x)$. Since $|L(x_n)| \leqslant 1$ for all n, $|L(x)| \leqslant 1$. Hence $\|L\| \leqslant 1$. Since $L(1) = 1$, $\|L\| = 1$. ∎

A linear functional of the type described in Theorem 7.1 is called a *Banach limit*. They are useful for a variety of things, among which is the construction of representations of the algebra of bounded operators on a Hilbert space.

EXERCISES

1. If L is a Banach limit, show that there are x and y in l^∞ such that $L(xy) \neq L(x)L(y)$.

2. Let X be a set and Ω a σ-algebra of subsets of X. Suppose μ is a complex-valued countably additive measure defined on Ω such that $\|\mu\| = \mu(X) < \infty$. Show that $\mu(\Delta) \geqslant 0$ for every Δ in Ω. (Though it is difficult to see at this moment, this fact is related to the proof of (c) in Theorem 7.1 for the complex case.)

3. Show that if $x \in l^\infty$, $\|x\|_\infty \leqslant 1$, then there is a sequence $\{x_n\}$, x_n in l^∞ such that $\|x_n\|_\infty \leqslant 1$, $\|x_n - x\|_\infty \to 0$, and each x_n takes on only a finite number of values.

§8*. An Application: Runge's Theorem

The symbol $\mathbb{C}_\infty$ denotes the extended complex plane.

8.1. Runge's Theorem. *Let K be a compact subset of $\mathbb{C}$ and let E be a subset of $\mathbb{C}_\infty \backslash K$ that meets each component of $\mathbb{C}_\infty \backslash K$. If f is analytic in a neighborhood of K, then there are rational functions f_n whose only poles lie in E such that $f_n \to f$ uniformly on K.*

The main tool in proving Runge's Theorem is Theorem 6.13. (A proof that does not use functional analysis can be found on p. 189 of Conway [1978].) To do this, let $R(K, E)$ be the closure in the space $C(K)$ of the rational functions with poles in E. By (6.13) and the Riesz Representation Theorem, it suffices to show that if $\mu \in M(K)$ and $\int g\, d\mu = 0$ for each g in $R(K, E)$, then $\int f\, d\mu = 0$.

Let $R > 0$ and let λ be area measure. Pick $\rho > 0$ such that $B(0; R) \subseteq B(z; \rho)$ for every z in K. Then for z in K,

$$\int_{B(0;R)} |z - w|^{-1}\, d\lambda(w) \leqslant \int_{B(z;\rho)} |z - w|^{-1}\, d\lambda(w)$$
$$= \int_0^{2\pi} \int_0^\rho dr\, d\theta = 2\pi\rho.$$

If $\mu \in M(K)$, define $\tilde{\mu}: \mathbb{C} \to [0, \infty]$ by

$$\tilde{\mu}(w) = \int \frac{d|\mu|(z)}{|z - w|}$$

when the integral is finite, and $\tilde{\mu}(w) = \infty$ otherwise. The inequality above implies

$$\int_{B(0;R)} \tilde{\mu}(w)\, d\lambda(w) = \int_{B(0;R)} \int_K \frac{d|\mu|(z)}{|z - w|}\, d\lambda(w)$$
$$= \int_K \int_{B(0;R)} \frac{d\lambda(w)}{|z - w|}\, d|\mu|(z)$$
$$\leqslant 2\pi\rho \|\mu\|.$$

Thus $\tilde{\mu}(w) < \infty$ a.e. $[\lambda]$.

8.2. Lemma. *If $\mu \in M(K)$, then*

$$\hat{\mu}(w) = \int \frac{d\mu(z)}{z - w}$$

is in $L^1(B(0; R), \lambda)$ for any $R > 0$, $\hat{\mu}$ is analytic on $\mathbb{C}_\infty \setminus K$, and $\hat{\mu}(\infty) = 0$.

PROOF. The first statement follows from what came before the statement of this lemma. To show that $\hat{\mu}$ is analytic on $\mathbb{C}_\infty \setminus K$, let $w, w_0 \in \mathbb{C} \setminus K$ and note that

$$\frac{\hat{\mu}(w) - \hat{\mu}(w_0)}{w - w_0} = \int_K \frac{d\mu(z)}{(z - w)(z - w_0)}.$$

As $w \to w_0$, $[(z - w)(z - w_0)]^{-1} \to (z - w_0)^{-2}$ uniformly for z in K, so that $\hat{\mu}$ has a derivative at w_0 and

$$\frac{d\hat{\mu}}{dw}(w_0) = \int_K (z - w_0)^{-2}\, d\mu(z).$$

So $\hat{\mu}$ is analytic on $\mathbb{C} \setminus K$. To show that it is analytic at infinity, note that $\hat{\mu}(z) \to 0$ as $z \to \infty$, so infinity is a removable singularity. ∎

It is not difficult to see that for w_0 in $\mathbb{C}\backslash K$,

8.3
$$\left(\frac{d}{dw}\right)^n \hat{\mu}(w_0) = n! \int (z - w_0)^{-n-1} d\mu(z)$$

Also, we can easily find the power series expansion of $\hat{\mu}$ at infinity. Indeed,

$$\hat{\mu}(w) = \int \frac{1}{z - w} d\mu(z) = -\frac{1}{w} \int \left(1 - \frac{z}{w}\right)^{-1} d\mu(z).$$

Choose w near enough to infinity that $|z/w| < 1$ for all z in K. Then

$$\hat{\mu}(w) = -\frac{1}{w} \sum_{n=0}^{\infty} \int \left(\frac{z}{w}\right)^n d\mu(z)$$

8.4
$$= -\sum_{n=0}^{\infty} \frac{a_n}{w^{n+1}},$$

where $a_n = \int z^n d\mu(z)$.

Now assume $\mu \in M(K)$ and $\int g \, d\mu = 0$ for every rational function g with poles in E. Let U be a component of $\mathbb{C}_\infty \backslash K$, and let $w_0 \in E \cap U$. If $w_0 \neq \infty$, then the hypothesis and (8.3) imply that each derivative of $\hat{\mu}$ at w_0 vanishes. Hence $\hat{\mu} \equiv 0$ on U. If $w_0 = \infty$, then (8.4) implies $\hat{\mu} \equiv 0$ on U. Thus $\hat{\mu} \equiv 0$ on $\mathbb{C}_\infty \backslash K$.

If f is analytic on an open set G containing K, let $\gamma_1, \ldots, \gamma_n$ be straight-line segments in $G \backslash K$ such that

$$f(z) = \sum_{k=1}^{n} \frac{1}{2\pi i} \int_{\gamma_k} \frac{f(w)}{w - z} dw$$

for all z in K. (See p. 195 of Conway [1978].) Thus

$$\int_K f(z) d\mu(z) = \sum_{k=1}^{n} \frac{1}{2\pi i} \int_K \int_{\gamma_k} \frac{f(w)}{w - z} dw \, d\mu(z)$$

$$= -\sum_{k=1}^{n} \frac{1}{2\pi i} \int_{\gamma_k} f(w) \hat{\mu}(w) dw$$

by Fubini's Theorem. But $\hat{\mu}(w) = 0$ on γ_k ($\subseteq \mathbb{C}\backslash K$), so $\int f \, d\mu = 0$. By (6.13), $f \in R(K, E)$. This proves Runge's Theorem. ∎

8.5. Corollary. *If K is compact and $\mathbb{C}\backslash K$ is connected and if f is analytic in a neighborhood of K, then there is a sequence of polynomials that converges to f uniformly on K.*

EXERCISES

1. Let μ be a compactly supported measure on $\mathbb{C}$ that is boundedly absolutely continuous with respect to area measure. Show that $\hat{\mu}$ is continuous on $\mathbb{C}_\infty$.

2. Let $m =$ Lebesgue measure on $[0, 1]$. Show that $\hat{m}$ is not continuous at any point of $[0, 1]$.

§9*. An Application: Ordered Vector Spaces

In this section only vector spaces over $\mathbb{R}$ are considered.

There are numerous spaces in which there is a notion of $\leqslant$ in addition to the vector space structure. The L^p spaces and $C(X)$ are some that spring to mind. The concept of an ordered vector space is an attempt to study such spaces in an abstract setting. The first step is to abstract the notion of the positive elements.

9.1. Definition. An *ordered vector space* is a pair $(\mathscr{X}, \leqslant)$ where $\mathscr{X}$ is a vector space over $\mathbb{R}$ and $\leqslant$ is a relation on $\mathscr{X}$ satisfying

(a) $x \leqslant x$ for all x;
(b) if $x \leqslant y$ and $y \leqslant z$, then $x \leqslant z$;
(c) if $x \leqslant y$ and $z \in \mathscr{X}$, then $x + z \leqslant y + z$;
(d) if $x \leqslant y$ and $\alpha \in [0, \infty)$, then $\alpha x \leqslant \alpha y$.

Note that it is not assumed that $\leqslant$ is *antisymmetric*. That is, it is not assumed that if $x \leqslant y$ and $y \leqslant x$, then $x = y$.

9.2. Definition. If $\mathscr{X}$ is a real vector space, a *wedge* is a nonempty subset P of $\mathscr{X}$ such that

(a) if $x, y \in P$, then $x + y \in P$;
(b) if $x \in P$ and $\alpha \in [0, \infty)$, then $\alpha x \in P$.

9.3. Proposition. (a) *If* $(\mathscr{X}, \leqslant)$ *is an ordered vector space and* $P = \{x \in \mathscr{X}: x \geqslant 0\}$, *then* P *is a wedge.* (b) *If* P *is a wedge in the real vector space* $\mathscr{X}$ *and* $\leqslant$ *is defined on* $\mathscr{X}$ *by declaring* $x \leqslant y$ *if and only if* $y - x \in P$, *then* $(\mathscr{X}, \leqslant)$ *is an ordered vector space.*

PROOF. Exercise.

If $(\mathscr{X}, \leqslant)$ is an ordered vector space, $P = \{x \in \mathscr{X}: x \geqslant 0\}$ is called the *wedge of positive elements*. The next result is also left as an exercise.

9.4. Proposition. *If* $(\mathscr{X}, \leqslant)$ *is an ordered vector space and* P *is the wedge of positive elements,* $\leqslant$ *is antisymmetric if and only if* $P \cap (-P) = (0)$.

9.5. Definition. A *cone* in $\mathscr{X}$ is a wedge P such that $P \cap (-P) = (0)$.

9.6. Definition. If $(\mathscr{X}, \leqslant)$ is an ordered vector space, a subset A of $\mathscr{X}$ is *cofinal* if for every $x \geqslant 0$ in $\mathscr{X}$ there is an a in A such that $a \geqslant x$. An element e of $\mathscr{X}$ is an *order unit* if for every x in $\mathscr{X}$ there is a positive integer n such that $-ne \leqslant x \leqslant ne$.

If X is a compact space $\mathscr{X} = C(X)$, then any non-zero constant function is an order unit. ($f \leqslant g$ if and only if $f(x) \leqslant g(x)$ for all x.) If $\mathscr{X} = C(\mathbb{R})$, all real-valued continuous functions on $\mathbb{R}$, then $\mathscr{X}$ has no order unit (Exercise 4). If e is an order unit, then $\{ne: n \geqslant 1\}$ is cofinal.

9.7. Definition. If $(\mathscr{X}, \leqslant)$ and $(\mathscr{Y}, \leqslant)$ are ordered vector spaces and $T: \mathscr{X} \to \mathscr{Y}$ is a linear map, then T is *positive* (in symbols $T \geqslant 0$) if $Tx \geqslant 0$ whenever $x \geqslant 0$.

The principal result of this section is the following.

9.8. Theorem. *Let $(\mathscr{X}, \leqslant)$ be an ordered vector space and let $\mathscr{Y}$ be a linear manifold in $\mathscr{X}$ that is confinal. If $f: \mathscr{Y} \to \mathbb{R}$ is a positive linear functional, then there is a positive linear functional $\tilde{f}: \mathscr{X} \to \mathbb{R}$ such that $\tilde{f}|\mathscr{Y} = f$.*

PROOF. Let $P = \{x \in \mathscr{X}: x \geqslant 0\}$ and put $\mathscr{X}_1 = \mathscr{Y} + P - P$. It is easy to see that $\mathscr{X}_1$ is a linear manifold in $\mathscr{X}$. If there is a positive linear functional $g: \mathscr{X}_1 \to \mathbb{R}$ that extends f, let $\tilde{f}$ be any linear functional on $\mathscr{X}$ that extends g (use a Hamel basis). If $x \geqslant 0$, then $x \in P \subseteq \mathscr{X}_1$ so that $\tilde{f}(x) = g(x) \geqslant 0$. Hence $\tilde{f}$ is positive. Thus, we may assume that $\mathscr{X} = \mathscr{Y} + P - P$.

9.9. Claim. $\mathscr{X} = \mathscr{Y} + P = \mathscr{Y} - P$.

Let $x \in \mathscr{X}$; so $x = y + p_1 - p_2$, y in $\mathscr{Y}$, p_1, p_2 in P. Since $\mathscr{Y}$ is confinal there is a y_1 in $\mathscr{Y}$ such that $y_1 \geqslant p_1$. Hence $p_1 = y_1 - (y_1 - p_1) \in \mathscr{Y} - P$. Thus $x = y - p_2 + p_1 \in (\mathscr{Y} - P) + (\mathscr{Y} - P) \subseteq \mathscr{Y} - P$. So $\mathscr{X} = \mathscr{Y} - P$. Also, $\mathscr{X} = -\mathscr{X} = -\mathscr{Y} + P = \mathscr{Y} + P$.

9.10. Claim. *If $x \in \mathscr{X}$, there are y_1, y_2 in $\mathscr{Y}$ such that $y_2 \leqslant x \leqslant y_1$.*

In fact, Claim 9.9 states that we can write $x = y_1 - p_1 = y_2 + p_2$, $p_1, p_2 \in P$ and $y_1, y_2 \in \mathscr{Y}$. Thus $y_2 \leqslant x \leqslant y_1$.

By Claim 9.10, it is possible to define for each x in $\mathscr{X}$,

$$q(x) = \inf \{f(y): y \in \mathscr{Y} \text{ and } y \geqq x\}.$$

9.11. Claim. *The function q is a sublinear functional on $\mathscr{X}$.*

The proof of (9.11) is left as an exercise.

For y in $\mathscr{Y}$, let $y_1 \in \mathscr{Y}$ such that $y_1 \geqslant y$. Because f is positive, $f(y) \leqslant f(y_1)$. Hence $f(y) \leqslant q(y)$ for all y in $\mathscr{Y}$. The Hahn–Banach Theorem implies that there is a linear functional $\tilde{f}: \mathscr{X} \to \mathbb{R}$ such that $\tilde{f}|\mathscr{Y} = f$ and $\tilde{f} \leqslant q$ on $\mathscr{X}$. If $x \in P$, then $-x \leqslant 0$ (and $0 \in \mathscr{Y}$). Hence $q(-x) \leqslant f(0)$. Thus $-\tilde{f}(x) = \tilde{f}(-x) \leqslant q(-x) \leqslant 0$, or $\tilde{f}(x) \geqslant 0$. Therefore $\tilde{f}$ is positive. ∎

9.12. Corollary. *Let $(\mathscr{X}, \leqslant)$ be an ordered vector space with an order unit e. If $\mathscr{Y}$ is a linear manifold in $\mathscr{X}$ and $e \in \mathscr{Y}$, then any positive linear functional defined on $\mathscr{Y}$ has an extension to a positive linear functional defined on $\mathscr{X}$.*

EXERCISES

1. Prove Proposition 9.3.

2. Prove Proposition 9.4.

3. Show that e is an order unit for $(\mathcal{X}, \leq)$ if and only if for every x in $\mathcal{X}$ there is a $\delta > 0$ such that $e \pm tx \geq 0$ for $0 \leq t \leq \delta$.

4. Show that $C(\mathbb{R})$, the space of all continuous real-valued functions on $\mathbb{R}$, has no order unit.

5. Prove (9.11).

6. Characterize the order units of $C_b(X)$. Does $C_b(X)$ always have an order unit?

7. Characterize the order units of $C_0(X)$ if X is locally compact. Does $C_0(X)$ always have an order unit?

8. Let $\mathcal{X} = M_2(\mathbb{R})$, the 2×2 matrices over $\mathbb{R}$. Define A in $M_2(\mathbb{R})$ to be positive if $A = A^*$ and $\langle Ax, x \rangle \geq 0$ for all x in $\mathbb{R}^2$. Characterize the order units of $M_2(\mathbb{R})$.

9. If $1 \leq p < \infty$ and $\mathcal{X} = L^p(0,1)$, define $f \leq g$ to mean that $f(x) \leq g(x)$ a.e. Show that $\mathcal{X}$ is an ordered vector space that has no order unit.

§10. The Dual of a Quotient Space and a Subspace

Let $\mathcal{X}$ be a normed space and $\mathcal{M} \leq \mathcal{X}$. If $f \in \mathcal{X}^*$, then $f|\mathcal{M}$, the restriction of f to $\mathcal{M}$, belongs to $\mathcal{M}^*$ and $\|f|\mathcal{M}\| \leq \|f\|$. According to the Hahn–Banach Theorem, every bounded linear functional on $\mathcal{M}$ is obtainable as the restriction of a functional from $\mathcal{X}^*$. In fact, more can be said.

Note that if $\mathcal{M}^\perp \equiv \{g \in \mathcal{X}^*: g(\mathcal{M}) = 0\}$ (note the analogy with Hilbert space notation); then $\mathcal{M}^\perp$ is a closed subspace of the Banach space $\mathcal{X}^*$. Hence $\mathcal{X}^*/\mathcal{M}^\perp$ is a Banach space. Moreover, if $f + \mathcal{M}^\perp \in \mathcal{X}^*/\mathcal{M}^\perp$, then $f + \mathcal{M}^\perp$ induces a linear functional on $\mathcal{M}$, namely $f|\mathcal{M}$.

10.1. Theorem. If $\mathcal{M} \leq \mathcal{X}$ and $\mathcal{M}^\perp \equiv \{g \in \mathcal{X}^*: g(\mathcal{M}) = 0\}$, then the map $\rho: \mathcal{X}^*/\mathcal{M}^\perp \to \mathcal{M}^*$ defined by

$$\rho(f + \mathcal{M}^\perp) = f|\mathcal{M}$$

is an isometric isomorphism.

PROOF. It is easy to see that ρ is linear and injective. If $f \in \mathcal{X}^*$ and $g \in \mathcal{M}^\perp$, then $\|f|\mathcal{M}\| = \|(f + g)|\mathcal{M}\| \leq \|f + g\|$. Taking the infimum over all g we get that $\|f|\mathcal{M}\| \leq \|f + \mathcal{M}^\perp\|$. Suppose $\phi \in \mathcal{M}^*$. The Hahn–Banach Theorem implies that there is an f in $\mathcal{X}^*$ such that $f|\mathcal{M} = \phi$ and $\|f\| = \|\phi\|$. Hence $\phi = \rho(f + \mathcal{M}^\perp)$ and $\|\phi\| = \|f\| \geq \|f + \mathcal{M}^\perp\|$. ∎

Now consider $\mathcal{X}/\mathcal{M}$; what is $(\mathcal{X}/\mathcal{M})^*$? Let $Q: \mathcal{X} \to \mathcal{X}/\mathcal{M}$ be the natural map. If $f \in (\mathcal{X}/\mathcal{M})^*$, then $f \circ Q \in \mathcal{X}^*$ and $\|f \circ Q\| \leq \|f\|$. (Why?) This gives a way of mapping $(\mathcal{X}/\mathcal{M})^* \to \mathcal{X}^*$. What is its image? Is it an isometry?

10.2. Theorem. *If $\mathcal{M} \leqslant \mathcal{X}$ and $Q: \mathcal{X} \to \mathcal{X}/\mathcal{M}$ is the natural map, then $\rho(f) = f \circ Q$ defines an isometric isomorphism of $(\mathcal{X}/\mathcal{M})^*$ onto $\mathcal{M}^\perp$.*

PROOF. If $f \in (\mathcal{X}/\mathcal{M})^*$ and $y \in \mathcal{M}$, then $f \circ Q(y) = 0$, so $f \circ Q \in \mathcal{M}^\perp$. Again, it is easy to see that $\rho: (\mathcal{X}/\mathcal{M})^* \to \mathcal{M}^\perp$ is linear and, as was seen earlier, $\|\rho(f)\| \leqslant \|f\|$. Let $\{x_n + \mathcal{M}\}$ be a sequence in $\mathcal{X}/\mathcal{M}$ such that $\|x_n + \mathcal{M}\| < 1$ and $|f(x_n + \mathcal{M})| \to \|f\|$. For each n there is a y_n in $\mathcal{M}$ such that $\|x_n + y_n\| < 1$. Thus $\|\rho(f)\| \geqslant |\rho(f)(x_n + y_n)| = |f(x_n + \mathcal{M})| \to \|f\|$, so ρ is an isometry.

   To see that ρ is surjective, let $g \in \mathcal{M}^\perp$; then $g \in \mathcal{X}^*$ and $g(\mathcal{M}) = 0$. Define f: $\mathcal{X}/\mathcal{M} \to \mathbb{F}$ by $f(x + \mathcal{M}) = g(x)$. Because $g(\mathcal{M}) = 0$, f is well defined. Also, if $x \in \mathcal{X}$ and $y \in \mathcal{M}$, $|f(x + \mathcal{M})| = |g(x)| = |g(x + y)| \leqslant \|g\| \, \|x + y\|$. Taking the infimum over all y gives $|f(x + \mathcal{M})| \leqslant \|g\| \, \|x + \mathcal{M}\|$. Hence $f \in (\mathcal{X}/\mathcal{M})^*$, $\rho(f) = g$, and $\|f\| \leqslant \|\rho(f)\|$. ∎

§11. Reflexive Spaces

If $\mathcal{X}$ is a normed space, then we have seen that $\mathcal{X}^*$ is a Banach space (5.4). Because $\mathcal{X}^*$ is a Banach space, it too has a dual space $(\mathcal{X}^*)^* \equiv \mathcal{X}^{**}$ and $\mathcal{X}^{**}$ is a Banach space. Hence $\mathcal{X}^{**}$ has a dual. Can this be kept up?

   Before answering this question, let's examine a curious phenomenon. If $x \in \mathcal{X}$, then x defines an element $\hat{x}$ of $\mathcal{X}^{**}$; namely, define $\hat{x}: \mathcal{X}^* \to \mathbb{F}$ by

11.1 $$\hat{x}(x^*) = x^*(x)$$

for every x^* in $\mathcal{X}^*$. Note that Corollary 6.7 implies that $\|\hat{x}\| = \|x\|$ for all x in $\mathcal{X}$. The map $x \to \hat{x}$ of $\mathcal{X} \to \mathcal{X}^{**}$ is called the *natural map* of $\mathcal{X}$ into its *second dual*.

11.2. Definition. A normed space $\mathcal{X}$ is *reflexive* if $\mathcal{X}^{**} = \{\hat{x}: x \in \mathcal{X}\}$, where $\hat{x}$ is defined in (11.1).

   First note that a reflexive space $\mathcal{X}$ is isometrically isomorphic to $\mathcal{X}^{**}$, and hence must be a Banach space. It is not true however, that a Banach space $\mathcal{X}$ that is isometric to $\mathcal{X}^{**}$ is reflexive. The definition of reflexivity stipulates that the isometry be the natural embedding of $\mathcal{X}$ into $\mathcal{X}^{**}$. In fact, James [1951] gives an example of a nonreflexive space $\mathcal{X}$ that is isometric to $\mathcal{X}^{**}$.

11.3. Example. If $1 < p < \infty$, $L^p(X, \Omega, \mu)$ is reflexive.

11.4. Example. c_0 is not reflexive. We know that $c_0^* = l^1$, so $c_0^{**} = (l^1)^* = l^\infty$. With these identifications, the natural map $c_0 \to c_0^{**}$ is precisely the inclusion map $c_0 \to l^\infty$.

   A discussion of reflexivity is best pursued after the weak topology is understood (Chapter V). Until that time, we will say *adieu* to reflexivity.

EXERCISES

1. Show that $(\mathscr{X}^*)^{**}$ and $(\mathscr{X}^{**})^*$ are equal.

2. Show that for a locally compact space X, $C_b(X)$ is reflexive if and only if X is finite.

3. Let $\mathscr{M} \leq \mathscr{X}$ and let $\rho_{\mathscr{X}}\colon \mathscr{X} \to \mathscr{X}^{**}$ and $\rho_{\mathscr{M}}\colon \mathscr{M} \to \mathscr{M}^{**}$ be the natural maps. If $i\colon \mathscr{M} \to \mathscr{X}$ is the inclusion map, show that there is an isometry $\phi\colon \mathscr{M}^{**} \to \mathscr{X}^{**}$ such that the diagram

$$
\begin{array}{ccc}
\mathscr{X} & \xrightarrow{\ \rho_{\mathscr{X}}\ } & \mathscr{X}^{**} \\
\Big\uparrow{\scriptstyle i} & & \Big\downarrow{\scriptstyle \phi} \\
\mathscr{M} & \xrightarrow[\ \rho_{\mathscr{M}}\ ]{} & \mathscr{M}^{**}
\end{array}
$$

commutes. Prove that $\phi(\mathscr{M}^{**}) = (\mathscr{M}^{\perp})^{\perp} \equiv \{x^{**} \in \mathscr{X}^{**}\colon x^{**}(\mathscr{M}^{\perp}) = 0\}$.

4. Use Exercise 3 to show that if $\mathscr{X}$ is reflexive, then any closed subspace of $\mathscr{X}$ is also reflexive. See Yang [1967].

§12. The Open Mapping and Closed Graph Theorems

12.1. The Open Mapping Theorem. *If $\mathscr{X}$, $\mathscr{Y}$ are Banach spaces and $A\colon \mathscr{X} \to \mathscr{Y}$ is a continuous linear surjection, then $A(G)$ is open in $\mathscr{Y}$ whenever G is open in $\mathscr{X}$.*

PROOF. For $r > 0$, let $B(r) = \{x \in \mathscr{X}\colon \|x\| < r\}$.

12.2. Claim. $0 \in \mathrm{int}\ \mathrm{cl}\ A(B(r))$.

Note that because A is surjective, $\mathscr{Y} = \bigcup_{k=1}^{\infty} \mathrm{cl}\,[A(B(kr/2))] = \bigcup_{k=1}^{\infty} k\,\mathrm{cl}\,[A(B(r/2))]$. By the Baire Category Theorem, there is a $k \geq 1$ such that $k\,\mathrm{cl}\,[A(B(r/2))]$ has nonempty interior. Thus $V = \mathrm{int}\,\{\mathrm{cl}\,[A(B(r/2))]\} \neq \square$. If $y_0 \in V$, let $s > 0$ such that $\{y \in \mathscr{Y}\colon \|y - y_0\| < s\} \subseteq V \subseteq \mathrm{cl}\,A(B(r/2))$. Let $y \in \mathscr{Y}$, $\|y\| < s$. Since $y_0 \in \mathrm{cl}\,A(B(r/2))$, there is a sequence $\{x_n\}$ in $B(r/2)$ such that $A(x_n) \to y_0$. There is also a sequence $\{z_n\}$ in $B(r/2)$ such that $A(z_n) \to y_0 + y$. Thus $A(z_n - x_n) \to y$ and $\{z_n - x_n\} \subseteq B(r)$; that is, $\{y \in \mathscr{Y}\colon \|y\| < s\} \subseteq \mathrm{cl}\,A(B(r))$. This establishes Claim 12.2.

It will now be shown that

12.3 $$\mathrm{cl}\,A(B(r/2)) \subseteq A(B(r)).$$

Note that if (12.3) is proved, then Claim 12.2 implies that $0 \in \mathrm{int}\,A(B(r))$ for any $r > 0$. From here the theorem is easily proved. Indeed, if G is an open subset of $\mathscr{X}$, then for every x in G let $r_x > 0$ such that $B(x; r_x) \subseteq G$. But $0 \in \mathrm{int}\,A(B(r_x))$ and so $A(x) \in \mathrm{int}\,A(B(x; r_x))$. Thus there is an $s_x > 0$ such that $U_x \equiv \{y \in \mathscr{Y}\colon \|y - A(x)\| < s_x\} \subseteq A(B(x; r_x))$. Therefore $A(G) \supseteq \cup \{U_x\colon x \in G\}$. But $A(x) \in U_x$, so $A(G) = \cup \{U_x\colon x \in G\}$ and hence $A(G)$ is open.

To prove (12.3), fix y_1 in cl $A(B(r/2))$. By (12.2), $0 \in \mathrm{int}\,[\mathrm{cl}\,A(B(2^{-2}r))]$. Hence $[y_1 - \mathrm{cl}\,A(B(2^{-2}r))] \cap A(B(r/2)) \neq \square$. Let $x_1 \in B(r/2)$ such that $A(x_1) \in [y_1 - \mathrm{cl}\,A(B(2^{-2}r))]$; now $A(x_1) = y_1 - y_2$, where $y_2 \in \mathrm{cl}\,A(B(2^{-2}r))$. Using induction, we obtain a sequence $\{x_n\}$ in $\mathcal{X}$ and a sequence $\{y_n\}$ in $\mathcal{Y}$ such that

12.4 $\begin{cases} \text{(i)} & x_n \in B(2^{-n}r), \\[2mm] \text{(ii)} & y_n \in \mathrm{cl}\,A(B(2^{-n}r)), \\[2mm] \text{(iii)} & y_{n+1} = y_n - A(x_n). \end{cases}$

But $\|x_n\| < 2^{-n}r$, so $\sum_1^\infty \|x_n\| < \infty$; hence $x = \sum_{n=1}^\infty x_n$ exists in $\mathcal{X}$ and $\|x\| < r$. Also,

$$\sum_{k=1}^n A(x_k) = \sum_{k=1}^n (y_k - y_{k+1}) = y_1 - y_{n+1}.$$

But (12.4ii) implies $\|y_n\| \leqslant \|A\|\,2^{-n}r$; hence $y_n \to 0$. Therefore $y_1 = \sum_{k=1}^\infty A(x_k) = A(x) \in A(B(r))$, proving (12.3) and completing the proof of the theorem. ∎

The same method used to prove the Open Mapping Theorem can also be used to prove the Tietze Extension Theorem. See Grabiner [1986].

The Open Mapping Theorem has several applications. Here are two important ones.

12.5. The Inverse Mapping Theorem. *If $\mathcal{X}$ and $\mathcal{Y}$ are Banach spaces and $A: \mathcal{X} \to \mathcal{Y}$ is a bounded linear transformation that is bijective, then A^{-1} is bounded.*

PROOF. Because A is continuous, bijective, and open by Theorem 12.1, A is a homeomorphism. ∎

12.6. The Closed Graph Theorem. *If $\mathcal{X}$ and $\mathcal{Y}$ are Banach spaces and $A: \mathcal{X} \to \mathcal{Y}$ is a linear transformation such that the graph of A,*

$$\mathrm{gra}\,A \equiv \{x \oplus Ax \in \mathcal{X} \oplus_1 \mathcal{Y}: x \in \mathcal{X}\}$$

is closed, then A is continuous.

PROOF. Let $\mathcal{G} = \mathrm{gra}\,A$. Since $\mathcal{X} \oplus_1 \mathcal{Y}$ is a Banach space and $\mathcal{G}$ is closed, $\mathcal{G}$ is a Banach space. Define $P: \mathcal{G} \to \mathcal{X}$ by $P(x \oplus Ax) = x$. It is easy to check that P is bounded and bijective. (Do it). By the Inverse Mapping Theorem, $P^{-1}: \mathcal{X} \to \mathcal{G}$ is continuous. Thus $A: \mathcal{X} \to \mathcal{Y}$ is the composition of the continuous map $P^{-1}: \mathcal{X} \to \mathcal{G}$ and the continuous map of $\mathcal{G} \to \mathcal{Y}$ defined by $x \oplus Ax \mapsto Ax$; A is therefore continuous. ∎

Let $\mathcal{X}$ = all functions $f: [0,1] \to \mathbb{F}$ such that the derivative f' exists and is continuous on $[0,1]$. Let $\mathcal{Y} = C[0,1]$ and give both $\mathcal{X}$ and $\mathcal{Y}$ the supremum norm: $\|f\| = \sup\{|f(t)|: t \in [0,1]\}$. So $\mathcal{X}$ is not a Banach space, though $\mathcal{Y}$ is. Define $A: \mathcal{X} \to \mathcal{Y}$ by $Af = f'$. Clearly, A is linear. If $\{f_n\} \subseteq \mathcal{X}$ and

$(f_n, f'_n) \to (f, g)$ in $\mathscr{X} \times \mathscr{Y}$, then $f'_n \to g$ uniformly on $[0,1]$. Hence

$$f_n(t) - f_n(0) = \int_0^t f'_n(s)\, ds \to \int_0^t g(s)\, ds.$$

But $f_n(t) - f_n(0) \to f(t) - f(0)$, so

$$f(t) = f(0) + \int_0^t g(s)\, ds.$$

Thus $f' = g$ and gra A is closed. However, A is not bounded. (Why?)

The preceding example shows that the domain of the operator in the Closed Graph Theorem must be assumed to be complete. The next example (due to Alp Eden) shows that the range must also be assumed to be complete.

Let $\mathscr{X}$ be a separable infinite-dimensional Banach space and let $\{e_i : i \in I\}$ be a Hamel basis for $\mathscr{X}$ with $\|e_i\| = 1$ for all i. Note that a Baire Category argument shows that I is uncountable. If $x \in \mathscr{X}$, then $x = \sum_i \alpha_i e_i, \alpha_i \in \mathbb{F}$, and $\alpha_i = 0$ for all but a finite number of i in I. Define $\|x\|_1 \equiv \sum_i |\alpha_i|$. It is left as an exercise for the reader to show that $\|\cdot\|_1$ is a norm on $\mathscr{X}$. Since $\|e_i\| = 1$ for all i, $\|x\| \le \sum_i |\alpha_i| = \|x\|_1$. Let $\mathscr{Y} = \mathscr{X}$ with the norm $\|\cdot\|_1$ and let $T: \mathscr{Y} \to \mathscr{X}$ be defined by $T(x) = x$. Note that it was just shown that $T: \mathscr{Y} \to \mathscr{X}$ is a contraction. Therefore gra T is closed and hence so is gra T^{-1}. But T^{-1} is not continuous because if it were, then T would be a homeomorphism. Since $\mathscr{X}$ is separable, it would follow that $\mathscr{Y}$ is separable. But $\mathscr{Y}$ is not separable. To see this, note that $\|e_i - e_j\|_1 = 2$ for $i \ne j$ and since I is uncountable, $\mathscr{Y}$ cannot be separable.

When applying the Closed Graph Theorem, the following result is useful.

12.7. Proposition. *If $\mathscr{X}$ and $\mathscr{Y}$ are normed spaces and $A: \mathscr{X} \to \mathscr{Y}$ is a linear transformation, then gra A is closed if and only if whenever $x_n \to 0$ and $Ax_n \to y$, it must be that $y = 0$.*

PROOF. Exercise 3.

Note that (12.7) underlines the advantage of the Closed Graph Theorem. To show that A is continuous, it suffices to show that if $x_n \to 0$, then $Ax_n \to 0$. By (12.7) this is eased by allowing us to assume that $\{Ax_n\}$ is convergent.

It is possible to give a measure-theoretic solution to Exercise 2.3, but here is one using the Closed Graph Theorem. Let (X, Ω, μ) be a σ-finite measure space, $1 \le p \le \infty$, and $\phi: X \to \mathbb{F}$ an Ω-measurable function such that $\phi f \in L^p(\mu)$ whenever $f \in L^p(\mu)$. Define $A: L^p(\mu) \to L^p(\mu)$ by $Af = \phi f$. Thus A is linear and well defined. Suppose $f_n \to 0$ and $\phi f_n \to g$ in $L^p(\mu)$. If $1 \le p < \infty$, then $f_n \to 0$ in measure. By a theorem of Riesz, there is a subsequence $\{f_{n_k}\}$ such that $f_{n_k}(x) \to 0$ a.e. $[\mu]$. Hence $\phi(x) f_{n_k}(x) \to 0$ a.e. $[\mu]$. This implies $g = 0$ and so gra A is closed. If $p = \infty$, then $f_n(x) \to 0$ a.e. $[\mu]$ and the same argument implies gra A is closed. By the Closed Graph Theorem, A is bounded. Clearly, it may be assumed that $\|A\| = 1$. If $\delta > 0$, let E be a measurable subset of

$\{x: |\phi(x)| \geqslant 1 + \delta\}$ with $\mu(E) < \infty$. We want to show that $\mu(E) = 0$. But if $f = \chi_E$, then $\mu(E) = \|f\|_p^p \geqslant \|Af\|_p^p = \|\phi f\|_p^p = \int_E |\phi|^p \, d\mu \geqslant (1 + \delta)^p \mu(E)$. Hence $\mu(E) = 0$. Since E was arbitrary it follows that ϕ is an essentially bounded function and $\|\phi\|_\infty \leqslant 1$. ∎

12.8. Definition. If $\mathscr{X}, \mathscr{Y}$ are Banach spaces, an *isomorphism* of $\mathscr{X}$ and $\mathscr{Y}$ is a linear bijection $T: \mathscr{X} \to \mathscr{Y}$ that is a homeomorphism. Say that $\mathscr{X}$ and $\mathscr{Y}$ are *isomorphic* if there is an isomorphism of $\mathscr{X}$ onto $\mathscr{Y}$.

Note that the Inverse Mapping Theorem says that a continuous bijection is an isomorphism.

The use of the word "isomorphism" is counter to the spirit of category theory, but it is traditional in Banach space theory.

EXERCISES

1. Suppose $\mathscr{X}$ and $\mathscr{Y}$ are Banach spaces. If $A \in \mathscr{B}(\mathscr{X}, \mathscr{Y})$ and ran A is a second category space, show that ran A is closed.

2. Give both $C^{(1)}[0, 1]$ and $C[0, 1]$ the supremum norm. If $A: C^{(1)}[0, 1] \to C[0, 1]$ is defined by $Af = f'$, show that A is not bounded.

3. Prove Proposition 12.7.

4. Let $\mathscr{X}$ be a vector space and suppose $\|\cdot\|_1$ and $\|\cdot\|_2$ are two norms on $\mathscr{X}$ and that $\mathscr{I}_1$ and $\mathscr{I}_2$ are the corresponding topologies. Show that if $\mathscr{X}$ is complete in both norms and $\mathscr{I}_1 \supseteq \mathscr{I}_2$, then $\mathscr{I}_1 = \mathscr{I}_2$.

5. Let $\mathscr{X}$ and $\mathscr{Y}$ be Banach spaces and let $A \in \mathscr{B}(\mathscr{X}, \mathscr{Y})$. Show that there is a constant $c > 0$ such that $\|Ax\| \geqslant c\|x\|$ for all x in $\mathscr{X}$ if and only if ker $A = (0)$ and ran A is closed.

6. Let X be compact and suppose that $\mathscr{X}$ is a Banach subspace of $C(X)$. If E is a closed subset of X such that for every g in $C(E)$ there is an f in $\mathscr{X}$ with $f|E = g$, show that there is a constant $c > 0$ such that for each g in $C(E)$ there is an f in $\mathscr{X}$ with $f|E = g$ and $\max\{|f(x)|: x \in X\} \leqslant c \max\{|g(x)|: x \in E\}$.

7. Let $1 \leqslant p \leqslant \infty$ and suppose (α_{ij}) is a matrix such that $(Af)(i) = \sum_{j=1}^{\infty} \alpha_{ij} f(j)$ defines an element Af of l^p for every f in l^p. Show that $A \in \mathscr{B}(l^p)$.

8. Let (X, Ω, μ) be a σ-finite measure space, $1 \leqslant p < \infty$, and suppose that $k: X \times X \to \mathbb{F}$ is an $\Omega \times \Omega$ measurable function such that for f in $L^p(\mu)$ and a.e. x, $k(x, \cdot) f(\cdot) \in L^1(\mu)$ and $(Kf)(x) = \int k(x, y) f(y) \, d\mu(y)$ defines an element Kf of $L^p(\mu)$. Show that $K: L^p(\mu) \to L^p(\mu)$ is a bounded operator.

§13. Complemented Subspaces of a Banach Space

If $\mathscr{X}$ is a Banach space and $\mathscr{M} \leqslant \mathscr{X}$, say that $\mathscr{M}$ is *algebraically complemented* in $\mathscr{X}$ if there is an $\mathscr{N} \leqslant \mathscr{X}$ such that $\mathscr{M} \cap \mathscr{N} = (0)$ and $\mathscr{M} + \mathscr{N} = \mathscr{X}$. Of course, the definition makes sense in a purely algebraic setting, so the requirement that $\mathscr{M}$ and $\mathscr{N}$ be closed seems fatuous. Why is it made?

If $\mathcal{M}$ is a linear manifold in a vector space $\mathcal{X}$ (a Banach space or not), then a Hamel-basis argument can be fashioned to produce a linear manifold $\mathcal{N}$ such that $\mathcal{M} \cap \mathcal{N} = (0)$ and $\mathcal{M} + \mathcal{N} = \mathcal{X}$. So the requirement in the definition that $\mathcal{M}$ and $\mathcal{N}$ be closed subspaces of the Banach space $\mathcal{X}$ makes the existence problem more interesting. Also, since we are dealing with the category of Banach spaces, all definitions should involve only objects in that category.

If $\mathcal{M}$ and $\mathcal{N}$ are algebraically complemented closed subspaces of a normed space $\mathcal{X}$, then $A: \mathcal{M} \oplus_1 \mathcal{N} \to \mathcal{X}$ defined by $A(m \oplus n) = m + n$ is a linear bijection. Also, $\|A(m \oplus n)\| = \|m + n\| \leq \|m\| + \|n\| = \|m \oplus n\|$. Hence A is bounded. Say that $\mathcal{M}$ and $\mathcal{N}$ are *topologically complemented* if A is a homeomorphism; equivalently, if $\|\!\|\!\|m + n\|\!\|\!\| \equiv \|m\| + \|n\|$ is an equivalent norm. If $\mathcal{X}$ is a Banach space, then the Inverse Mapping Theorem implies A is a homeomorphism. This proves the following.

13.1. Theorem. *If two subspaces of a Banach space are algebraically complementary, then they are topologically complementary.*

This permits us to speak of *complementary subspaces* of a Banach space without modifying the term. The proof of the next result is left to the reader.

13.2. Theorem. (a) *If $\mathcal{M}$ and $\mathcal{N}$ are complementary subspaces of a Banach space $\mathcal{X}$ and $E: \mathcal{X} \to \mathcal{X}$ is defined by $E(m + n) = m$ for m in $\mathcal{M}$ and n in $\mathcal{N}$, then E is a continuous linear operator such that $E^2 = E$, $\operatorname{ran} E = \mathcal{M}$, and $\ker E = \mathcal{N}$. (b) If $E \in \mathcal{B}(\mathcal{X})$ and $E^2 = E$, then $\mathcal{M} = \operatorname{ran} E$ and $\mathcal{N} = \ker E$ are complementary subspaces of $\mathcal{X}$.*

If $\mathcal{M} \leq \mathcal{X}$ and $\mathcal{M}$ is complemented in $\mathcal{X}$, its complementary subspace may not be unique. Indeed, finite dimensional spaces furnish the necessary examples.

A result due to R.S. Phillips [1940] is that c_0 is not complemented in l^∞. A straightforward proof of this can be found in Whitley [1966]. Murray [1937] showed that $l^p, p \neq 2, p > 1$ has *uncomplemented* subspaces. This seems to be the first paper to exhibit uncomplemented subspaces of a Banach space.

Lindenstrauss [1967] showed that if $\mathcal{M}$ is an infinite dimensional subspace of l^∞ that is complemented in l^∞, then $\mathcal{M}$ is isomorphic to l^∞. This same result holds if l^∞ is replaced by l^p, $1 \leq p < \infty$, c, or c_0.

Does there exist a Banach space $\mathcal{X}$ such that every closed subspace of $\mathcal{X}$ is complemented? Of course, if $\mathcal{X}$ is a Hilbert space, then this is true. But are there any Banach spaces that have this property and are not Hilbert spaces? Lindenstrauss and Tzafriri [1971] proved that if $\mathcal{X}$ is a Banach space and every subspace of $\mathcal{X}$ is complemented, then $\mathcal{X}$ is isomorphic to a Hilbert space.

EXERCISES

1. If $\mathscr{X}$ is a vector space and $\mathscr{M}$ is a linear manifold in $\mathscr{X}$, show that there is a linear manifold $\mathscr{N}$ in $\mathscr{X}$ such that $\mathscr{M} \cap \mathscr{N} = (0)$ and $\mathscr{M} + \mathscr{N} = \mathscr{X}$.

2. Let $\mathscr{X}$ be a Banach space and let $E: \mathscr{X} \to \mathscr{X}$ be a linear map such that $E^2 = E$ and both ran E and ker E are closed. Show that E is continuous.

3. Prove Theorem 13.2.

4. Let $\mathscr{X}$ be a Banach space and show that if $\mathscr{M}$ is a complemented subspace of $\mathscr{X}$, then every complementary subspace is isomorphic to $\mathscr{X}/\mathscr{M}$.

5. Let X be a compact set and let Y be a closed subset of X. A *simultaneous extension* for Y is a bounded linear map $T: C(Y) \to C(X)$ such that for each g in $C(Y)$, $T(g)|Y = g$. Let $C_0(X \setminus Y) = \{f \in C(X): f(y) = 0 \text{ for all } y \text{ in } Y\}$. Show that if there is a simultaneous extension for Y, then $C_0(X \setminus Y)$ is complemented in $C(X)$.

6. Show that if Y is a closed subset of $[0, 1]$, then there is a simultaneous extension for Y (see Exercise 5). (Hint: Write $[0, 1] \setminus Y$ as the union of disjoint intervals.)

7. Using the notation of Exercise 5, show that if Y is a retract of X, then $C_0(X \setminus Y)$ is complemented in $C(X)$.

§14. The Principle of Uniform Boundedness

There are several results that may be called the Principle of Uniform Boundedness (PUB) and all of these are called the PUB by various mathematicians. In this book the PUB will refer to any of the results of this section, though in a formal way the next result plays the role of the founder of the family.

14.1. Principle of Uniform Boundedness (PUB). *Let $\mathscr{X}$ be a Banach space and $\mathscr{Y}$ a normed space. If $\mathscr{A} \subseteq \mathscr{B}(\mathscr{X}, \mathscr{Y})$ such that for each x in $\mathscr{X}$, $\sup\{\|Ax\|: A \in \mathscr{A}\} < \infty$, then $\sup\{\|A\|: A \in \mathscr{A}\} < \infty$.*

PROOF. (Due to William R. Zame, 1978. Also see J. Hennefeld [1980].) For each x in $\mathscr{X}$ let $M(x) = \sup\{\|Ax\|: A \in \mathscr{A}\}$, so $\|Ax\| \leqslant M(x)$ for all x in $\mathscr{X}$. Suppose $\sup\{\|A\|: A \in \mathscr{A}\} = \infty$. Then there is a sequence $\{A_n\} \subseteq \mathscr{A}$ and a sequence $\{x_n\}$ of vectors in $\mathscr{X}$ such that $\|x_n\| = 1$ and $\|A_n x_n\| > 4^n$. Let $y_n = 2^{-n} x_n$; thus $\|y_n\| = 2^{-n}$ and $\|A_n y_n\| > 2^n$.

14.2. Claim. There is a subsequence $\{y_{n_k}\}$ such that for $k \geqslant 1$:

(a) $\|A_{n_{k+1}} y_{n_{k+1}}\| > 1 + k + \sum_{j=1}^{k} M(y_{n_j})$;
(b) $\|y_{n_{k+1}}\| < 2^{-k-1} [\sup\{\|A_{n_j}\|: 1 \leqslant j \leqslant k\}]^{-1}$.

The proof of (14.2) is by induction. Let $n_1 = 1$. The induction step is valid since $\|y_n\| \to 0$ and $\|A_n y_n\| \to \infty$. The details are left to the reader.

Since $\sum_k \|y_{n_k}\| < \infty$, $\sum_k y_{n_k} = y$ in $\mathscr{X}$ (here is where the completeness of $\mathscr{X}$ is used.) Now for any $k \geq 1$,

$$\|A_{n_{k+1}} y\| = \left\| \sum_{j=1}^{k} A_{n_{k+1}} y_{n_j} + A_{n_{k+1}} y_{n_{k+1}} + \sum_{j=k+2}^{\infty} A_{n_{k+1}} y_{n_j} \right\|$$

$$= \left\| A_{n_{k+1}} y_{n_{k+1}} - \left[- \sum_{j=1}^{k} A_{n_{k+1}} y_{n_j} - \sum_{j=k+2}^{\infty} A_{n_{k+1}} y_{n_j} \right] \right\|$$

$$\geq \|A_{n_{k+1}} y_{n_{k+1}}\| - \left\| \sum_{j=1}^{k} A_{n_{k+1}} y_{n_j} + \sum_{j=k+2}^{\infty} A_{n_{k+1}} y_{n_j} \right\|$$

$$\geq 1 + k + \sum_{j=1}^{k} M(y_{n_j}) - \left[\sum_{j=1}^{k} M(y_{n_j}) + \sum_{j=k+2}^{\infty} \|A_{n_{k+1}}\| \, \|y_{n_j}\| \right]$$

$$\geq 1 + k - \sum_{j=k+2}^{\infty} 2^{-j}$$

$$\geq k.$$

That is, $M(y) \geq k$ for all k, a contradiction. ∎

14.3. Corollary. *If $\mathscr{X}$ is a normed space and $A \subseteq \mathscr{X}$, then A is a bounded set if and only if for every f in $\mathscr{X}^*$, $\sup\{|f(a)|: a \in A\} < \infty$.*

PROOF. Consider $\mathscr{X}$ as a subset of $\mathscr{B}(\mathscr{X}^*, \mathbb{F})$ ($= \mathscr{X}^{**}$) by letting $\hat{x}(f) = f(x)$ for every f in $\mathscr{X}^*$. Since $\mathscr{X}^*$ is a Banach space and $\|x\| = \|\hat{x}\|$ for all x, the corollary is a special case of the PUB. ∎

14.4. Corollary. *If $\mathscr{X}$ is a Banach space and $A \subseteq \mathscr{X}^*$, then A is a bounded set if and only if for every x in $\mathscr{X}$, $\sup\{|f(x)|: f \in A\} < \infty$.*

PROOF. Consider $\mathscr{X}^*$ as $\mathscr{B}(\mathscr{X}, \mathbb{F})$. ∎

Using Corollary 14.3, it is possible to prove the following improvement of (14.1).

14.5. Corollary. *If $\mathscr{X}$ is a Banach space and $\mathscr{Y}$ is a normed space and if $\mathscr{A} \subseteq \mathscr{B}(\mathscr{X}, \mathscr{Y})$ such that for every x in $\mathscr{X}$ and g in $\mathscr{Y}^*$,*

$$\sup\{|g(A(x))|: A \in \mathscr{A}\} < \infty,$$

then $\sup\{\|A\|: A \in \mathscr{A}\} < \infty$.

PROOF. Fix x in $\mathscr{X}$. By the hypothesis and Corollary 14.3, $\sup\{\|A(x)\|: A \in \mathscr{A}\} < \infty$. By (14.1), $\sup\{\|A\|: A \in \mathscr{A}\} < \infty$. ∎

A special form of the PUB that is quite useful is the following.

14.6. The Banach–Steinhaus Theorem. *If $\mathscr{X}$ and $\mathscr{Y}$ are Banach spaces and*

$\{A_n\}$ is a sequence in $\mathscr{B}(\mathscr{X}, \mathscr{Y})$ with the property that for every x in $\mathscr{X}$ there is a y in $\mathscr{Y}$ such that $\|A_n x - y\| \to 0$, then there is an A in $\mathscr{B}(\mathscr{X}, \mathscr{Y})$ such that $\|A_n x - Ax\| \to 0$ for every x in $\mathscr{X}$ and $\sup\|A_n\| < \infty$.

PROOF. If $x \in \mathscr{X}$, let $Ax = \lim A_n x$. By hypothesis $A: \mathscr{X} \to \mathscr{Y}$ is defined and it is easy to see that it is linear. To show that A is bounded, note that the PUB implies that there is a constant $M > 0$ such that $\|A_n\| \leqslant M$ for all n. If $x \in \mathscr{X}$ and $\|x\| \leqslant 1$, then for any $n \geqslant 1$, $\|Ax\| \leqslant \|Ax - A_n x\| + \|A_n x\| \leqslant \|Ax - A_n x\| + M$. Letting $n \to \infty$ shows that $\|Ax\| \leqslant M$ whenever $\|x\| \leqslant 1$. ∎

The Banach–Steinhaus Theorem is a result about sequences, not nets. Note that if I is the identity operator on $\mathscr{X}$ and for each $n \geqslant 1$, $A_n = n^{-1}I$ and for $n \leqslant 0$, $A_n = nI$, then $\{A_n : n \in \mathbb{Z}\}$ is a countable net that converges in norm to 0, but the net is not bounded.

14.7. Proposition. Let X be locally compact and let $\{f_n\}$ be a sequence in $C_0(X)$. Then $\int f_n d\mu \to \int f d\mu$ for every μ in $M(X)$ if and only if $\sup_n \|f_n\| < \infty$ and $f_n(x) \to f(x)$ for every x in X.

PROOF. Suppose $\int f_n d\mu \to \int f d\mu$ for every μ in $M(X)$. Since $M(X) = C_0(X)^*$, (14.3) implies that $\sup_n \|f_n\| < \infty$. By letting $\mu = \delta_x$, the unit point mass at x, we see that $\int f_n d\delta_x = f_n(x) \to f(x)$. The converse follows by the Lebesgue Dominated Convergence Theorem. ∎

EXERCISES

1. Here is another proof of the PUB using the Baire Category Theorem. With the notation of (14.1), let $B_n \equiv \{x \in \mathscr{X} : \|Ax\| \leqslant n$ for all A in $\mathscr{A}\}$. By hypothesis, $\bigcup_{n=1}^{\infty} B_n = \mathscr{X}$. Now apply the Baire Category Theorem.

2. If $1 < p < \infty$ and $\{x_n\} \subseteq l^p$, then $\sum_{j=1}^{\infty} x_n(j) y(j) \to 0$ for every y in l^q, $1/p + 1/q = 1$, if and only if $\sup_n \|x_n\|_p < \infty$ and $x_n(j) \to 0$ for every $j \geqslant 1$.

3. If $\{x_n\} \subseteq l^1$, then $\sum_{j=1}^{\infty} x_n(j) y(j) \to 0$ for every y in c_0 if and only if $\sup_n \|x_n\|_1 < \infty$ and $x_n(j) \to 0$ for every $j \geqslant 1$.

4. If (X, Ω, μ) is a measure space, $1 < p < \infty$, and $\{f_n\} \subseteq L^p(X, \Omega, \mu)$, then $\int f_n g \, d\mu \to 0$ for every g in $L^q(\mu)$, $1/p + 1/q = 1$, if and only if $\sup\{\|f_n\|_p : n \geqslant 1\} < \infty$ and for every set E in Ω with $\mu(E) < \infty$, $\int_E f_n d\mu \to 0$ as $n \to \infty$.

5. If (X, Ω, μ) is a σ-finite measure space and $\{f_n\}$ is a sequence in $L^1(X, \Omega, \mu)$, then $\int f_n g \, d\mu \to 0$ for every g in $L^\infty(\mu)$ if and only if $\sup\{\|f_n\|_1 : n \geqslant 1\} < \infty$ and $\int_E f_n d\mu \to 0$ for every E in Ω.

6. Let $\mathscr{H}$ be a Hilbert space and let $\mathscr{E}$ be an orthonormal basis for $\mathscr{H}$. Show that a sequence $\{h_n\}$ in $\mathscr{H}$ satisfies $\langle h_n, h \rangle \to 0$ for every h in $\mathscr{H}$ if and only if $\sup\{\|h_n\| : n \geqslant 1\} < \infty$ and $\langle h_n, e \rangle \to 0$ for every e in $\mathscr{E}$.

7. If X is locally compact and $\{\mu_n\}$ is a sequence in $M(X)$, then $L(\mu_n) \to 0$ for every L in $M(X)^*$ if and only if $\sup\{\|\mu_n\| : n \geqslant 1\} < \infty$ and $\mu_n(E) \to 0$ for every Borel set E.

8. In (14.6), show that $\|A\| \leqslant \liminf \|A_n\|$.

9. If (S, d) is a metric space and $\mathscr{X}$ is a normed space, say that a function $f: S \to \mathscr{X}$ is a *Lipschitz function* if there is a constant $M > 0$ such that $\| f(s) - f(t) \| \leqslant M d(s, t)$ for all s, t in S. Show that if $f: S \to \mathscr{X}$ is a function such that for all L in $\mathscr{X}^*$, $L \circ f: S \to \mathbb{F}$ is Lipschitz, then $f: S \to \mathscr{X}$ is a Lipschitz function.

10. Let $\mathscr{X}$ be a Banach space and suppose $\{x_n\}$ is a sequence in $\mathscr{X}$ such that for each x in $\mathscr{X}$ there are unique scalars $\{\alpha_n\}$ such that $\lim_{n \to \infty} \| x - \sum_{k=1}^{n} \alpha_k x_k \| = 0$. Such a sequence is called a *Schauder basis*. (a) Prove that $\mathscr{X}$ is separable. (b) Let $\mathscr{Y} = \{ \{\alpha_n\} \in \mathbb{F}^{\mathbb{N}} : \sum_{n=1}^{\infty} \alpha_n x_n \text{ converges in } \mathscr{X} \}$ and for $y = \{\alpha_n\}$ in $\mathscr{Y}$ define $\| y \| = \sup_n \| \sum_{k=1}^{n} \alpha_k x_k \|$. Show that $\mathscr{Y}$ is a Banach space. (c) Show that there is a bounded bijection $T: \mathscr{X} \to \mathscr{Y}$. (d) If $n \geqslant 1$ and $f_n: \mathscr{X} \to \mathbb{F}$ is defined by $f_n(\sum_{k=1}^{\infty} \alpha_k x_k) = \alpha_n$, show that $f_n \in \mathscr{X}^*$. (e) Show that $x_n \notin$ the closed linear span of $\{x_k : k \neq n\}$.

CHAPTER IV

Locally Convex Spaces

A topological vector space is a generalization of the concept of a Banach space. The locally convex spaces are encountered repeatedly when discussing weak topologies on a Banach space, sets of operators on Hilbert space, or the theory of distributions. This book will only skim the surface of this theory, but it will treat locally convex spaces in sufficient detail as to enable the reader to understand the use of these spaces in the three areas of analysis just mentioned. For more details on this theory, see Bourbaki [1967], Robertson and Robertson [1966], or Schaefer [1971].

§1. Elementary Properties and Examples

A topological vector space is a vector space that is also a topological space such that the linear structure and the topological structure are vitally connected.

1.1. Definition. A *topological vector space* (TVS) is a vector space $\mathscr{X}$ together with a topology such that with respect to this topology

(a) the map of $\mathscr{X} \times \mathscr{X} \to \mathscr{X}$ defined by $(x, y) \mapsto x + y$ is continuous;
(b) the map of $\mathbb{F} \times \mathscr{X} \to \mathscr{X}$ defined by $(\alpha, x) \mapsto \alpha x$ is continuous.

It is easy to see that a normed space is a TVS (Proposition III.1.3).

Suppose $\mathscr{X}$ is a vector space and $\mathscr{P}$ is a family of seminorms on $\mathscr{X}$. Let $\mathscr{T}$ be the topology on $\mathscr{X}$ that has as a subbase the sets $\{x: p(x - x_0) < \varepsilon\}$, where $p \in \mathscr{P}$, $x_0 \in \mathscr{X}$, and $\varepsilon > 0$. Thus a subset U of $\mathscr{X}$ is open if and only if for every x_0 in U there are $p_1, \ldots, p_n$ in $\mathscr{P}$ and $\varepsilon_1, \ldots, \varepsilon_n > 0$ such that

$\bigcap_{j=1}^{n} \{x \in \mathcal{X}: p_j(x - x_0) < \varepsilon_j\} \subseteq U$. It is not difficult to show that $\mathcal{X}$ with this topology is a TVS (Exercise 2).

1.2. Definition. A *locally convex space* (LCS) is a TVS whose topology is defined by a family of seminorms $\mathscr{P}$ such that $\bigcap_{p \in \mathscr{P}} \{x: p(x) = 0\} = (0)$.

The attitude that has been adopted in this book is that all topological spaces are Hausdorff. The condition in Definition 1.2 that $\bigcap_{p \in \mathscr{P}} \{x: p(x) = 0\} = (0)$ is imposed precisely so that the topology defined by $\mathscr{P}$ be Hausdorff. In fact, suppose that $x \neq y$. Then there is a p in $\mathscr{P}$ such that $p(x - y) \neq 0$; let $p(x - y) > \varepsilon > 0$. If $U = \{z: p(x - z) < \frac{1}{2}\varepsilon\}$ and $V = \{z: p(y - z) < \frac{1}{2}\varepsilon\}$, then $U \cap V = \square$ and U and V are neighborhoods of x and y, respectively.

If $\mathcal{X}$ is a TVS and $x_0 \in \mathcal{X}$, then $x \mapsto x + x_0$ is a homeomorphism of $\mathcal{X}$; also, if $\alpha \in \mathbb{F}$ and $\alpha \neq 0$, $x \mapsto \alpha x$ is a homeomorphism of $\mathcal{X}$ (Exercise 4). Thus the topology of $\mathcal{X}$ looks the same at any point. This might make the next statement less surprising.

1.3. Proposition. *Let $\mathcal{X}$ be a TVS and let p be a seminorm on $\mathcal{X}$. The following statements are equivalent.*

(a) *p is continuous.*
(b) *$\{x \in \mathcal{X}: p(x) < 1\}$ is open.*
(c) *$0 \in \mathrm{int}\{x \in \mathcal{X}: p(x) < 1\}$.*
(d) *$0 \in \mathrm{int}\{x \in \mathcal{X}: p(x) \leq 1\}$.*
(e) *p is continuous at 0.*
(f) *There is a continuous seminorm q on $\mathcal{X}$ such that $p \leq q$.*

PROOF. It is clear that (a) $\Rightarrow$ (b) $\Rightarrow$ (c) $\Rightarrow$ (d).

(d) *implies* (e): Clearly (d) implies that for every $\varepsilon > 0$, $0 \in \mathrm{int}\{x \in \mathcal{X}: p(x) \leq \varepsilon\}$; so if $\{x_i\}$ is a net in $\mathcal{X}$ that converges to 0 and $\varepsilon > 0$, there is an i_0 such that $x_i \in \{x: p(x) \leq \varepsilon\}$ for $i \geq i_0$; that is, $p(x_i) \leq \varepsilon$ for $i \geq i_0$. So p is continuous at 0.

(e) *implies* (a): If $x_i \to x$, then $|p(x) - p(x_i)| \leq p(x - x_i)$. Since $x - x_i \to 0$, (e) implies that $p(x - x_i) \to 0$. Hence $p(x_i) \to p(x)$.

Clearly (a) implies (f). So it remains to show that (f) implies (e). If $x_i \to 0$ in $\mathcal{X}$, then $q(x_i) \to 0$. But $0 \leq p(x_i) \leq q(x_i)$, so $p(x_i) \to 0$. ∎

1.4. Proposition. *If $\mathcal{X}$ is a TVS and $p_1, \ldots, p_n$ are continuous seminorms, then $p_1 + \cdots + p_n$ and $\max_i(p_i(x))$ are continuous seminorms. If $\{p_i\}$ is a family of continuous seminorms such that there is a continuous seminorm q with $p_i \leq q$ for all i, then $x \mapsto \sup_i\{p_i(x)\}$ defines a continuous seminorm.*

PROOF. Exercise.

If $\mathscr{P}$ is a family of seminorms of $\mathcal{X}$ that makes $\mathcal{X}$ into a LCS, it is often convenient to enlarge $\mathscr{P}$ by assuming that $\mathscr{P}$ is closed under the formation of finite sums and supremums of bounded families [as in (1.4)]. Sometimes it is convenient to assume that $\mathscr{P}$ consists of all continuous seminorms. In either case the resulting topology on $\mathscr{X}$ remains unchanged.

1.5. Example. Let X be completely regular and let $C(X) = $ all continuous functions from X into $\mathbb{F}$. If K is a compact subset of X, define $p_K(f) = \sup\{|f(x)|: x \in K\}$. Then $\{p_K: K$ compact in $X\}$ is a family of seminorms that makes $C(X)$ into a LCS.

1.6. Example. Let G be an open subset of $\mathbb{C}$ and let $H(G)$ be the subset of $C_{\mathbb{C}}(G)$ consisting of all analytic functions on G. Define the seminorms of (1.5) on $H(G)$. Then $H(G)$ is a LCS. Also, the topology defined on $H(G)$ by these seminorms is the topology of uniform convergence on compact subsets—the usual topology for discussing analytic functions.

1.7. Example. Let $\mathscr{X}$ be a normed space. For each x^* in $\mathscr{X}^*$, define $p_{x^*}(x) = |x^*(x)|$. Then p_{x^*} is a seminorm and if $\mathscr{P} = \{p_{x^*}: x^* \in \mathscr{X}^*\}$, $\mathscr{P}$ makes $\mathscr{X}$ into a LCS. The topology defined on $\mathscr{X}$ by these seminorms is called the *weak topology* and is often denoted by $\sigma(\mathscr{X}, \mathscr{X}^*)$.

1.8. Example. Let $\mathscr{X}$ be a normed space and for each x in $\mathscr{X}$ define $p_x: \mathscr{X}^* \to [0, \infty)$ by $p_x(x^*) = |x^*(x)|$. Then p_x is a seminorm and $\mathscr{P} = \{p_x: x \in \mathscr{X}\}$ makes $\mathscr{X}^*$ into a LCS. The topology defined by these seminorms is called the *weak-star* (or *weak** or wk*) *topology* on $\mathscr{X}^*$. It is often denoted by $\sigma(\mathscr{X}^*, \mathscr{X})$.

The spaces $\mathscr{X}$ with its weak topology and $\mathscr{X}^*$ with its weak* topology are very important and will be explored in depth in Chapter V.

Recall the definition of convex set from (I.2.4). If $a, b \in \mathscr{X}$, then the *line segment* from a to b is defined as $[a, b] \equiv \{tb + (1 - t)a: 0 \leqslant t \leqslant 1\}$. So a set A is convex if and only if $[a, b] \subseteq A$ whenever $a, b \in A$. The proof of the next result is left to the reader.

1.9. Proposition. (a) *A set A is convex if and only if whenever $x_1, \ldots, x_n \in A$ and $t_1, \ldots, t_n \in [0, 1]$ with $\sum_j t_j = 1$, then $\sum_j t_j x_j \in A$.* (b) *If $\{A_i: i \in I\}$ is a collection of convex sets, then $\bigcap_i A_i$ is convex.*

1.10. Definition. If $A \subseteq \mathscr{X}$, the *convex hull* of A, denoted by co(A), is the intersection of all convex sets that contain A. If $\mathscr{X}$ is a TVS, then the *closed convex hull* of A is the intersection of all closed convex subsets of $\mathscr{X}$ that contain A; it is denoted by $\overline{\text{co}}(A)$.

Since a vector space is itself convex, each subset of $\mathscr{X}$ is contained in a convex set. This fact and Proposition 1.9(b) imply that co(A) is well defined and convex. Also, $\overline{\text{co}}(A)$ is a closed convex set.

If $\mathscr{X}$ is a normed space, then $\{x: \|x\| \leqslant 1\}$ and $\{x: \|x\| < 1\}$ are both convex sets. If $f \in \mathscr{X}^*$, $\{x: |f(x)| \leqslant 1\}$, $\{x: \text{Re } f(x) \leqslant 1\}$, $\{x: \text{Re } f(x) > 1\}$ are all convex. In fact, if $T: \mathcal{X} \to \mathcal{Y}$ is a real linear map and C is a convex subset of $\mathcal{Y}$, then $T^{-1}(C)$ is convex in $\mathcal{X}$.

1.11. Proposition. *Let $\mathcal{X}$ be a TVS and let A be a convex subset of $\mathcal{X}$. Then (a) cl A is convex; (b) if $a \in \text{int } A$ and $b \in \text{cl } A$, then $[a, b) \equiv \{tb + (1-t) a : 0 \le t < 1\} \subseteq \text{int } A$.*

PROOF. Let $a \in A$, $b \in \text{cl } A$, and $0 \le t \le 1$. Let $\{x_i\}$ be a net in A such that $x_i \to b$. Then $tx_i + (1-t)a \to tb + (1-t)a$. This shows that

1.12 b in cl A and a in A imply $[a, b] \subseteq \text{cl } A$.

Using (1.12) it is easy to show that cl A is convex. To prove (b), fix t, $0 < t < 1$, and put $c = tb + (1-t)a$, where $a \in \text{int } A$ and $b \in \text{cl } A$. There is an open set V in $\mathcal{X}$ such that $0 \in V$ and $a + V \subseteq A$. (Why?) Hence for any d in A

$$A \supseteq td + (1-t)(a+V)$$
$$= t(d - b) + tb + (1-t)(a+V)$$
$$= [t(d-b) + (1-t)V] + c.$$

If it can be shown that there is an element d in A such that $0 \in t(d-b) + (1-t)V = U$, then the preceding inclusion shows that $c \in \text{int } A$ since U is open (Exercise 4). Note that the finding of such a d in A is equivalent to finding a d such that $0 \in t^{-1}(1-t)V + (d-b)$ or $d \in b - t^{-1}(1-t)V$. But $0 \in -t^{-1}(1-t)V$ and this set is open. Since $b \in \text{cl } A$, d can be found in A. ∎

1.13. Corollary. *If $A \subseteq \mathcal{X}$, then $\overline{\text{co}}(A)$ is the closure of $\text{co}(A)$.*

A set $A \subseteq \mathcal{X}$ is *balanced* if $\alpha x \in A$ whenever $x \in A$ and $|\alpha| \le 1$. A set A is *absorbing* if for each x in $\mathcal{X}$ there is an $\varepsilon > 0$ such that $tx \in A$ for $0 \le t < \varepsilon$. Note that an absorbing set must contain the origin. If $a \in A$, then A is *absorbing at a* if the set $A - a$ is absorbing. Equivalently, A is absorbing at a if for every x in $\mathcal{X}$ there is an $\varepsilon > 0$ such that $a + tx \in A$ for $0 \le t < \varepsilon$.

If $\mathcal{X}$ is a vector space and p is a seminorm, then $V = \{x : p(x) < 1\}$ is a convex balanced set that is absorbing at each of its points. It is rather remarkable that the converse of this is true. This fact will be used to give an abstract formulation of a LCS and also to explore some geometric consequences of the Hahn–Banach Theorem.

1.14. Proposition. *If $\mathcal{X}$ is a vector space over $\mathbb{F}$ and V is a nonempty convex, balanced set that is absorbing at each of its points, then there is a unique seminorm p on $\mathcal{X}$ such that $V = \{x \in \mathcal{X} : p(x) < 1\}$.*

PROOF. Define $p(x)$ by

$$p(x) = \inf\{t : t \ge 0 \text{ and } x \in tV\}.$$

Since V is absorbing, $\mathcal{X} = \bigcup_{n=1}^{\infty} nV$, so that the set whose infimum is $p(x)$ is nonempty. Clearly $p(0) = 0$. To see that $p(\alpha x) = |\alpha| p(x)$, we can suppose that $\alpha \neq 0$. Hence, because V is balanced,

$$p(\alpha x) = \inf\{t \geqslant 0: \alpha x \in tV\}$$

$$= \inf\left\{t \geqslant 0: x \in t\left(\frac{1}{\alpha}V\right)\right\}$$

$$= \inf\left\{t \geqslant 0: x \in t\left(\frac{1}{|\alpha|}V\right)\right\}$$

$$= |\alpha|\inf\left\{\frac{t}{|\alpha|}: x \in \frac{t}{|\alpha|}V\right\}$$

$$= |\alpha| p(x).$$

To complete the proof that p is a seminorm, note that if $\alpha, \beta \geqslant 0$ and $a, b \in V$, then

$$\alpha a + \beta b = (\alpha + \beta)\left(\frac{\alpha}{\alpha + \beta}a + \frac{\beta}{\alpha + \beta}b\right) \in (\alpha + \beta)V$$

by the convexity of V. If $x, y \in \mathcal{X}$, $p(x) = \alpha$, and $p(y) = \beta$, let $\delta > 0$. Then $x \in (\alpha + \delta)V$ and $y \in (\beta + \delta)V$. (Why?) Hence $x + y \in (\alpha + \delta)V + (\beta + \delta)V = (\alpha + \beta + 2\delta)V$ (Exercise 11). Letting $\delta \to 0$ shows that $p(x + y) \leqslant \alpha + \beta = p(x) + p(y)$.

It remains to show that $V = \{x: p(x) < 1\}$. If $p(x) = \alpha < 1$, then $\alpha < \beta < 1$ implies $x \in \beta V \subseteq V$ since V is balanced. Thus $V \supseteq \{x: p(x) < 1\}$. If $x \in V$, then $p(x) \leqslant 1$. Since V is absorbing at x, there is an $\varepsilon > 0$ such that for $0 < t < \varepsilon$, $x + tx = y \in V$. But $x = (1 + t)^{-1}y$, $y \in V$. Hence $p(x) = (1 + t)^{-1}p(y) \leqslant (1 + t)^{-1} < 1$.

Uniqueness follows by (III.1.4). ∎

The seminorm p defined in the preceding proposition is called the *Minkowski function* of V or the *gauge of V*.

Note that if $\mathcal{X}$ is a TVS space and V is an open set in $\mathcal{X}$, then V is absorbing at each of its points.

Using Proposition 1.14, the following characterization of a LCS can be obtained. The proof is left to the reader.

1.15. Proposition. *Let $\mathcal{X}$ be a TVS and let $\mathcal{U}$ be the collection of all open convex balanced subsets of $\mathcal{X}$. Then $\mathcal{X}$ is locally convex if and only if $\mathcal{U}$ is a basis for the neighborhood system at 0.*

EXERCISES

1. Let $\mathcal{X}$ be a TVS and let $\mathcal{U}$ be all the open sets containing 0. Prove the following.
   (a) If $U \in \mathcal{U}$, there is a V in $\mathcal{U}$ such that $V + V \subseteq U$. (b) If $U \in \mathcal{U}$, there is a V in $\mathcal{U}$ such that $V \subseteq U$ and $\alpha V \subseteq V$ for all $|\alpha| \leqslant 1$. (V is balanced.)
   (Hint: If $W \in \mathcal{U}$ and $\alpha W \subseteq U$ for $|\alpha| \leqslant \varepsilon$, then $\varepsilon W \subseteq \beta U$ for $|\beta| \geqslant 1$.)

2. Show that a LCS is a TVS.

3. Suppose that $\mathscr{X}$ is a TVS but do not assume that $\mathscr{X}$ is Hausdorff. (a) Show that $\mathscr{X}$ is Hausdorff if and only if the singleton set $\{0\}$ is closed. (b) If $\mathscr{X}$ is Hausdorff, show that $\mathscr{X}$ is a regular topological space.

4. Let $\mathscr{X}$ be a TVS. Show: (a) if $x_0 \in \mathscr{X}$, the map $x \mapsto x + x_0$ is a homeomorphism of $\mathscr{X}$ onto $\mathscr{X}$; (b) if $\alpha \in \mathbb{F}$ and $\alpha \neq 0$, the map $x \mapsto \alpha x$ is a homeomorphism.

5. Prove Proposition 1.4.

6. Verify the statements made in Example 1.5. Show that a net $\{f_i\}$ in $C(X)$ converges to f if and only if $f_i \to f$ uniformly on compact subsets of X.

7. Show that the space $H(G)$ defined in (1.6) is complete. (Every Cauchy net converges.)

8. Verify the statements made in Example 1.7. Give a basis for the neighborhood system at 0.

9. Verify the statements made in Example 1.8.

10. Prove Proposition 1.9.

11. Show that if A is a convex set and $\alpha, \beta > 0$, then $\alpha A + \beta A = (\alpha + \beta)A$, Give an example of a nonconvex set A for which this is untrue.

12. If $\mathscr{X}$ is a TVS and A is closed, show that A is convex if and only if $\frac{1}{2}(x + y) \in A$ whenever x and $y \in A$.

13. Let $s = $ the space of all sequences of scalars. Thus $s = $ all functions $x: \mathbb{N} \to \mathbb{F}$. Define addition and scalar multiplication in the usual way. If $x, y \in s$, define

$$d(x, y) = \sum_{n=1}^{\infty} 2^{-n} \frac{|x(n) - y(n)|}{1 + |x(n) - y(n)|}.$$

Show that d is a metric on s and that with this topology s is a TVS. Also show that s is complete.

14. Let (X, Ω, μ) be a finite measure space, let $\mathscr{M}$ be the space of Ω-measurable functions, and identify two functions that agree a.e. $[\mu]$. If $f, g \in \mathscr{M}$, define

$$d(f, g) = \int \frac{|f - g|}{1 + |f - g|} d\mu.$$

Then d is a metric on $\mathscr{M}$ and $(\mathscr{M}, d)$ is a complete TVS. Is there a relationship between this example and the space s of Exercise 13?

15. If $\mathscr{X}$ is a TVS and $A \subseteq \mathscr{X}$, then cl $A = \cap \{A + V: 0 \in V$ and V is open$\}$.

16. If $\mathscr{X}$ is a TVS and $\mathscr{M}$ is a closed linear space, then $\mathscr{X}/\mathscr{M}$ with the quotient topology is a TVS. If p is a seminorm on $\mathscr{X}$, define $\bar{p}$ on $\mathscr{X}/\mathscr{M}$ by $\bar{p}(x + \mathscr{M}) = \inf\{p(x + y): y \in \mathscr{M}\}$ Show that $\bar{p}$ is a seminorm on $\mathscr{X}/\mathscr{M}$. Show that if $\mathscr{X}$ is a LCS, then so is $\mathscr{X}/\mathscr{M}$.

17. If $\{\mathscr{X}_i: i \in I\}$ is a family of TVS's, then $\mathscr{X} = \Pi\{\mathscr{X}_i: i \in I\}$ with the product topology is a TVS. If each $\mathscr{X}_i$ is a LCS, then so is $\mathscr{X}$. If $\mathscr{X}$ is a LCS, must each $\mathscr{X}_i$ be a LCS?

18. If $\mathscr{X}$ is a finite-dimensional vector space and $\mathscr{T}_1, \mathscr{T}_2$ are two topologies on $\mathscr{X}$ that make $\mathscr{X}$ into a TVS, then $\mathscr{T}_1 = \mathscr{T}_2$.

19. If $\mathscr{X}$ is a TVS and $\mathscr{M}$ is a finite dimensional linear manifold in $\mathscr{X}$, then $\mathscr{M}$ is closed and $\mathscr{Y} + \mathscr{M}$ is closed for any closed subspace $\mathscr{Y}$ of $\mathscr{X}$.

20. Let $\mathscr{X}$ be any infinite dimensional vector space and let $\mathscr{T}$ be the collection of all subsets W of $\mathscr{X}$ such that if $x \in W$, then there is a convex balanced set U with $x + U \subseteq W$ and $U \cap \mathscr{M}$ open in $\mathscr{M}$ for every finite dimensional linear manifold $\mathscr{M}$ in $\mathscr{X}$. (Each such $\mathscr{M}$ is given its usual topology.) Show: (a) $(\mathscr{X}, \mathscr{T})$ is a LCS; (b) a set F is closed in $\mathscr{X}$ if and only if $F \cap \mathscr{M}$ is closed for every finite dimensional subspace $\mathscr{M}$ of $\mathscr{X}$; (c) if Y is a topological space and $f: \mathscr{X} \to Y$ (not necessarily linear), then f is continuous if and only if $f | \mathscr{M}$ is continuous for every finite dimensional space $\mathscr{M}$; (d) if $\mathscr{Y}$ is a TVS and $T: \mathscr{X} \to \mathscr{Y}$ is a linear map, then T is continuous.

21. Let X be a locally compact space and for each ϕ in $C_0(X)$, define $p_\phi(f) = \|\phi f\|_\infty$ for f in $C_b(X)$. Show that p_ϕ is a seminorm on $C_b(X)$. Let β = the topology defined by these seminorms. Show that $(C_b(X), \beta)$ is a LCS that is complete. β is called the *strict topology*.

22. For $0 < p < 1$, let l^p = all sequences x such that $\sum_{n=1}^{\infty} |x(n)|^p < \infty$. Define $d(x, y) = \sum_{n=1}^{\infty} |x(n) - y(n)|^p$ (no pth root). Then d is a metric and (l^p, d) is a TVS that is not locally convex.

23. Let $\mathscr{X}$ and $\mathscr{Y}$ be locally convex spaces and let $T: \mathscr{X} \to \mathscr{Y}$ be a linear transformation. Show that T is continuous if and only if for every continuous seminorm p on $\mathscr{Y}$, $p \circ T$ is a continuous seminorm on $\mathscr{X}$.

24. Let $\mathscr{X}$ be a LCS and let G be an open connected subset of $\mathscr{X}$. Show that G is arcwise connected.

§2. Metrizable and Normable Locally Convex Spaces

Which LCS's are metrizable? That is, which have a topology which is defined by a metric? Which LCS's have a topology that is defined by a norm? Both are interesting questions and both answers could be useful.

If $\mathscr{P}$ is a family of seminorms on $\mathscr{X}$ and $\mathscr{X}$ is a TVS, say that $\mathscr{P}$ *determines* the topology on $\mathscr{X}$ if the topology of $\mathscr{X}$ is the same as the topology induced by $\mathscr{P}$.

2.1. Proposition. *Let* $\{p_1, p_2, \ldots\}$ *be a sequence of seminorms on* $\mathscr{X}$ *such that* $\bigcap_{n=1}^{\infty} \{x: p_n(x) = 0\} = (0)$. *For* x *and* y *in* $\mathscr{X}$, *define*

$$d(x, y) = \sum_{n=1}^{\infty} 2^{-n} \frac{p_n(x - y)}{1 + p_n(x - y)}.$$

Then d *is a metric on* $\mathscr{X}$ *and the topology on* $\mathscr{X}$ *defined by* d *is the topology* on $\mathscr{X}$ defined by the seminorms $\{p_1. p_2, \ldots\}$. *Thus a LCS is metrizable if and only if its topology is determined by a countable family of seminorms.*

PROOF. It is left as an exercise for the reader to show that d is a metric and induces the same topology as $\{p_n\}$. If $\mathscr{X}$ is a LCS and its topology is determined by a countable family of seminorms, it is immediate that $\mathscr{X}$ is metrizable. For the converse, assume that $\mathscr{X}$ is metrizable and its metric is ρ. Let $U_n = \{x: \rho(x, 0) < 1/n\}$. Because $\mathscr{X}$ is locally convex, there are continuous seminorms $q_1, \ldots, q_k$ and positive numbers $\varepsilon_1, \ldots, \varepsilon_k$ such that $\bigcap_{j=1}^{k} \{x: q_j(x) < \varepsilon_j\} \subseteq U_n$. If $p_n = \varepsilon_1^{-1} q_1 + \cdots + \varepsilon_k^{-1} q_k$, then $x \in U_n$ whenever $p_n(x) < 1$. Clearly, p_n is continuous for each n. Thus if $x_j \to 0$ in $\mathscr{X}$, then for each n, $p_n(x_j) \to 0$ as $j \to \infty$. Conversely, suppose that for each n, $p_n(x_j) \to 0$ as $j \to \infty$. If $\varepsilon > 0$, let $n > \varepsilon^{-1}$. Then there is a j_0 such that for $j \geq j_0$, $p_n(x_j) < 1$. Thus, for $j \geq j_0$, $x_j \in U_n \subseteq \{x: \rho(x, 0) < \varepsilon\}$. That is, $\rho(x_j, 0) < \varepsilon$ for $j \geq j_0$ and so $x_j \to 0$ in $\mathscr{X}$. This shows that $\{p_n\}$ determines the topology on $\mathscr{X}$. (Why?)   ∎

2.2. Example. If $C(X)$ is as in Example 1.5, then $C(X)$ is metrizable if and only if $X = \bigcup_{n=1}^{\infty} K_n$, where each K_n is compact, $K_1 \subseteq K_2 \subseteq \cdots$, and if K is any compact subset of X, then $K \subseteq K_n$ for some n.

2.3. Example. If X is locally compact and $C(X)$ is as in Example 1.5, then $C(X)$ is metrizable if and only if X is σ-*compact* (that is, X is the union of a sequence of compact sets). If $H(G)$ is as in Example 1.6, then $H(G)$ is metrizable.

If $\mathscr{X}$ is a vector space and d is a metric on $\mathscr{X}$, say that d is *translation invariant* if $d(x + z, y + z) = d(x, y)$ for all x, y, z in $\mathscr{X}$. Note that the metric defined by a norm as well as the metric defined in (2.1) are translation invariant.

2.4. Definition. A *Fréchet space* is a TVS $\mathscr{X}$ whose topology is defined by a translation invariant metric d and such that $(\mathscr{X}, d)$ is complete.

It should be pointed out that some authors include in the definition of a Fréchet space the assumption that $\mathscr{X}$ is locally convex.

2.5. Definition. If $\mathscr{X}$ is a TVS and $B \subseteq \mathscr{X}$, then B is *bounded* if for every open set U containing 0, there is an $\varepsilon > 0$ such that $\varepsilon B \subseteq U$.

If $\mathscr{X}$ is a normed space, then it is easy to see that a set B is bounded if and only if $\sup\{\|b\|: b \in B\} < \infty$, so the definition is intuitively correct.

Also, notice that if $\|\cdot\|$ is a norm, $\{x: \|x\| < 1\}$ is itself bounded. This is not true for seminorms. For example, if $C(\mathbb{R})$ is topologized as in (1.5), let $p(f) = \sup\{|f(t)|: 0 \leq t \leq 1\}$. Then p is a continuous seminorm. However,

$\{f: p(f) < 1\}$ is not bounded. In fact, if f_0 is any function in $C(\mathbb{R})$ that vanishs on $[0, 1]$, $\{\alpha f_0: \alpha \in \mathbb{R}\} \subseteq \{f: p(f) < 1\}$. The fact that a normed space posseses a bounded open set is characteristic.

2.6. Proposition. *If $\mathcal{X}$ is a LCS, then $\mathcal{X}$ is normable if and only if $\mathcal{X}$ has a nonempty bounded open set.*

PROOF. It has already been shown that a normed space has a bounded open set. So assume that $\mathcal{X}$ is a LCS that has a bounded open set U. It must be shown that there is norm on $\mathcal{X}$ that defines the same topology. By translation, it may be assumed that $0 \in U$ (see Exercise 4i). By local convexity, there is a continuous seminorm p such that $\{x: p(x) < 1\} \equiv V \subseteq U$(Why?). It will be shown that p is a norm and defines the topology on $\mathcal{X}$.

To see that p is a norm, suppose that $x \in \mathcal{X}$, $x \neq 0$. Let W_0, W_x be disjoint open sets such that $0 \in W_0$ and $x \in W_x$. Then there is an $\varepsilon > 0$ such that $W_0 \supseteq \varepsilon U \supseteq \varepsilon V$. But $\varepsilon V = \{y: p(y) < \varepsilon\}$. Since $x \notin W_0$, $p(x) \geqslant \varepsilon$. Hence p is a norm.

Because p is continuous on $\mathcal{X}$, to show that p defines the topology of $\mathcal{X}$ it suffices to show that if q is any continuous seminorm on $\mathcal{X}$, there is an $\alpha > 0$ such that $q \leqslant \alpha p$ (Why?). But because q is continuous, there is an $\varepsilon > 0$ such that $\{x: q(x) < 1\} \supseteq \varepsilon U \supseteq \varepsilon V$. That is, $p(x) < \varepsilon$ implies $q(x) < 1$. By Lemma III.1.4, $q \leqslant \varepsilon^{-1} p$. ∎

EXERCISES

1. Supply the missing details in the proof of Proposition 2.1.

2. Verify the statements in Example 2.2.

3. Verify the statements in Example 2.3.

4. Let $\mathcal{X}$ be a TVS and prove the following: (a) If B is a bounded subset of $\mathcal{X}$, then so is cl B. (b) The finite union of bounded sets is bounded. (c) Every compact set is bounded. (d) If $B \subseteq \mathcal{X}$, then B is bounded if and only if for every sequence $\{x_n\}$ contained in B and for every $\{\alpha_n\}$ in c_0, $\alpha_n x_n \to 0$ in $\mathcal{X}$. (e) If $\mathcal{Y}$ is a TVS, $T: \mathcal{X} \to \mathcal{Y}$ is a continuous linear transformation, and B is a bounded subset of $\mathcal{X}$, then $T(B)$ is a bounded subset of $\mathcal{Y}$. (f) If $\mathcal{X}$ is a LCS and $B \subseteq \mathcal{X}$, then B is bounded if and only if for every continuous seminorm p, $\sup\{p(b): b \in B\} < \infty$. (g) If $\mathcal{X}$ is a normed space and $B \subseteq \mathcal{X}$, then B is bounded if and only if $\sup\{\|b\|: b \in B\} < \infty$. (h) If $\mathcal{X}$ is a Fréchet space, then bounded sets have finite diameter, but not conversely. (i) The translate of a bounded set is bounded.

5. If $\mathcal{X}$ is a LCS, show that $\mathcal{X}$ is metrizable if and only if $\mathcal{X}$ is first countable. Is this equivalent to saying that $\{0\}$ is a G_δ set?

6. Let X be a locally compact space and give $C_b(X)$ the strict topology defined in Exercise 1.21. Show that a subset of $C_b(X)$ is β-bounded if and only if it is norm bounded.

7. With the notation of Exercise 6, show that $(C_b(X), \beta)$ is metrizable if and only if X is compact.

8. Prove the Open Mapping Theorem for Fréchet spaces.

§3. Some Geometric Consequences of the Hahn–Banach Theorem

In order to exploit the Hahn–Banach Theorem in the setting of a LCS, it is necessary to establish some properties of continuous linear functionals. The proofs of the relevant propositions are similar to the proofs of the corresponding facts about linear functionals on normed spaces given in §III.5. For example, a hyperplane in a TVS is either closed or dense (see III.5.2). The proof of the next fact is similar to the proof of (III.2.1) and (III.5.3) and will not be given.

3.1. Theorem. *If $\mathscr{X}$ is a TVS and $f: \mathscr{X} \to \mathbb{F}$ is a linear functional, then the following statements are equivalent.*

(a) *f is continuous.*
(b) *f is continuous at 0.*
(c) *f is continuous at some point.*
(d) *$\ker f$ is closed.*
(e) *$x \mapsto |f(x)|$ is a continuous seminorm.*

If $\mathscr{X}$ is a LCS and $\mathscr{P}$ is a family of seminorms that defines the topology on $\mathscr{X}$, then the statements above are equivalent to the following:

(f) *There are $p_1, \ldots, p_n$ in $\mathscr{P}$ and positive scalars $\alpha_1, \ldots, \alpha_n$ such that $|f(x)| \leqslant \sum_{k=1}^{n} \alpha_k p_k(x)$ for all x.*

The proof of the next proposition is similar to the proof of Proposition 1.14 and will not be given.

3.2. Proposition. *Let $\mathscr{X}$ be a TVS and suppose that G is an open convex subset of $\mathscr{X}$ that contains the origin. If*

$$q(x) = \inf\{t: t \geqslant 0 \text{ and } x \in tG\},$$

then q is a non-negative continuous sublinear functional and $G = \{x: q(x) < 1\}$.

Note that the difference between the preceding proposition and (1.14) is that here G is not assumed to be balanced and the consequence is a sublinear functional ($q(\alpha x) = \alpha q(x)$ if $\alpha \geqslant 0$) that is not necessarily a seminorm.

The geometric consequences of the Hahn–Banach Theorem are achieved by interpreting that theorem in light of the correspondence between linear functionals and hyperplanes and between sublinear functionals and open convex neighborhoods of the origin. The next result is typical.

3.3. Theorem. *If $\mathscr{X}$ is a TVS and G is an open convex nonempty subset of $\mathscr{X}$ that does not contain the origin, then there is a closed hyperplane $\mathscr{M}$ such that $\mathscr{M} \cap G = \square$.*

PROOF. *Case 1.* $\mathscr{X}$ is an $\mathbb{R}$-linear space. Pick any x_0 in G and let $H = x_0 - G$. Then H is an open convex set containing 0. (Verify). By (3.2) there is a continuous nonnegative sublinear functional q: $\mathscr{X} \to \mathbb{R}$ such that $H = \{x: q(x) < 1\}$. Since $x_0 \notin H$, $q(x_0) \geq 1$.

Let $\mathscr{Y} \equiv \{\alpha x_0: \alpha \in \mathbb{R}\}$ and define f_0: $\mathscr{Y} \to \mathbb{R}$ by $f_0(\alpha x_0) = \alpha q(x_0)$. If $\alpha \geq 0$, then $f_0(\alpha x_0) = \alpha q(x_0) = q(\alpha x_0)$; if $\alpha < 0$, then $f_0(\alpha x_0) = \alpha q(x_0) \leq \alpha < 0 \leq q(\alpha x_0)$. So $f_0 \leq q$ on $\mathscr{Y}$. Let f: $\mathscr{X} \to \mathbb{R}$ be a linear functional such that $f|\mathscr{Y} = f_0$ and $f \leq q$ on $\mathscr{X}$. Put $\mathscr{M} = \ker f$.

Now if $x \in G$, then $x_0 - x \in H$ and so $f(x_0) - f(x) = f(x_0 - x) \leq q(x_0 - x) < 1$. Therefore $f(x) > f(x_0) - 1 = q(x_0) - 1 \geq 0$ for all x in G. Thus $\mathscr{M} \cap G = \square$.

Case 2. $\mathscr{X}$ is a $\mathbb{C}$-linear space. Lemma III.6.3 will be used here. Using Case 1 and the fact that $\mathscr{X}$ is also an $\mathbb{R}$-linear space, there is a continuous $\mathbb{R}$-linear functional f: $\mathscr{X} \to \mathbb{R}$ such that $G \cap \ker f = \square$. If $F(x) = f(x) - if(ix)$, then F is a $\mathbb{C}$-linear functional and $f = \operatorname{Re} F$ (III.6.3). Hence $F(x) = 0$ if and only if $f(x) = f(ix) = 0$; that is, $\mathscr{M} = \ker F = \ker f \cap [i \ker f]$. So $\mathscr{M}$ is a closed hyperplane and $\mathscr{M} \cap G = \square$. ∎

An *affine hyperplane* in $\mathscr{X}$ is a set $\mathscr{M}$ such that for every x_0 in $\mathscr{M}$, $\mathscr{M} - x_0$ is a hyperplane. (See Exercise 3.) An *affine manifold* in $\mathscr{X}$ is a set $\mathscr{Y}$ such that for every x_0 in $\mathscr{Y}$, $\mathscr{Y} - x_0$ is a linear manifold in $\mathscr{X}$. An *affine subspace* of a TVS $\mathscr{X}$ is a closed affine manifold.

3.4. Corollary. *Let $\mathscr{X}$ be a TVS and let G be an open convex nonempty subset of $\mathscr{X}$. If $\mathscr{Y}$ is an affine subspace of $\mathscr{X}$ such that $\mathscr{Y} \cap G = \square$, then there is a closed affine hyperplane $\mathscr{M}$ in $\mathscr{X}$ such that $\mathscr{Y} \subseteq \mathscr{M}$ and $\mathscr{M} \cap G = \square$.*

PROOF. By considering $G - x_0$ and $\mathscr{Y} - x_0$ for any x_0 in $\mathscr{Y}$, it may be assumed that $\mathscr{Y}$ is a linear subspace of $\mathscr{X}$. Let Q: $\mathscr{X} \to \mathscr{X}/\mathscr{Y}$ be the natural map. Since $Q^{-1}(Q(G)) = \{y + G: y \in \mathscr{Y}\}$, $Q(G)$ is open in $\mathscr{X}/\mathscr{Y}$. It is easy to see that $Q(G)$ is also convex. Since $\mathscr{Y} \cap G = \square$, $0 \notin Q(G)$. By the preceding theorem, there is a closed hyperplane $\mathscr{N}$ in $\mathscr{X}/\mathscr{Y}$ such that $\mathscr{N} \cap Q(G) = \square$. Let $\mathscr{M} = Q^{-1}(\mathscr{N})$. It is easy to check that $\mathscr{M}$ has the desired properties. ∎

There is a great advantage inherent in a geometric discussion of real TVS's. Namely, if f: $\mathscr{X} \to \mathbb{R}$ is a nonzero continuous $\mathbb{R}$-linear functional, then the hyperplane $\ker f$ disconnects the space. That is, $\mathscr{X} \backslash \ker f$ has two components (see Exercises 4 and 5). It thus becomes convenient to make the following definitions.

3.5. Definition. Let $\mathscr{X}$ be a real TVS. A subset S of $\mathscr{X}$ is called an *open half-space* if there is a continuous linear functional f: $\mathscr{X} \to \mathbb{R}$ such that $S = \{x \in \mathscr{X}: f(x) > \alpha\}$ for some α. S is a *closed half-space* if there is a continuous linear functional f: $\mathscr{X} \to \mathbb{R}$ such that $S = \{x \in \mathscr{X}: f(x) \geq \alpha\}$ for some α.

Two subsets A and B of $\mathscr{X}$ are said to be *strictly separated* if they are contained in disjoint open half-spaces; they are *separated* if they are contained in two closed half-spaces whose intersection is a closed affine hyperplane.

3.6. Proposition. *Let $\mathscr{X}$ be a real* TVS.

(a) *The closure of an open half-space is a closed half-space and the interior of a closed half-space is an open half-space.*
(b) *If $A, B \subseteq \mathscr{X}$, then A and B are strictly separated (separated) if and only if there is a continuous linear functional $f: \mathscr{X} \to \mathbb{R}$ and a real scalar α such that $f(a) > \alpha$ for all a in A and $f(b) < \alpha$ for all b in B ($f(a) \geqslant \alpha$ for all a in A and $f(b) \leqslant \alpha$ for all b in B).*

PROOF. Exercise 6.

In many ways, the next result is the most important "separation" theorem as the other separation theorems follow from this one. However, the most used separation theorem is Theorem 3.9 below.

3.7. Theorem. *If $\mathscr{X}$ is a real* TVS *and A and B are disjoint convex sets with A open, then there is a continuous linear functional $f: \mathscr{X} \to \mathbb{R}$ and a real scalar α such that $f(a) < \alpha$ for all a in A and $f(b) \geqslant \alpha$ for all b in B. If B is also open, then A and B are strictly separated.*

PROOF. Let $G = A - B \equiv \{a - b : a \in A, b \in B\}$; it is easy to verify that G is convex (do it!). Also, $G = \cup \{A - b : b \in B\}$, so G is open. Moreover, because $A \cap B = \square$, $0 \notin G$. By Theorem 3.3 there is a closed hyperplane $\mathscr{M}$ in $\mathscr{X}$ such that $\mathscr{M} \cap G = \square$. Let $f: \mathscr{X} \to \mathbb{R}$ be a linear functional such that $\mathscr{M} = \ker f$. Now $f(G)$ is a convex subset of $\mathbb{R}$ and $0 \notin f(G)$. Hence either $f(x) > 0$ for all x in G or $f(x) < 0$ for all x in G; suppose $f(x) > 0$ for all x in G. Thus if $a \in A$ and $b \in B$, $0 < f(a - b) = f(a) - f(b)$; that is, $f(a) > f(b)$. Therefore there is a real number α such that

$$\sup\{f(b) : b \in B\} \leqslant \alpha \leqslant \inf\{f(a) : a \in A\}.$$

But $f(A)$ and $f(B)$ are open intervals if B is open (Exercise 7), so $f < \alpha$ on B and $f > \alpha$ on A. ∎

3.8. Lemma. *If $\mathscr{X}$ is a* TVS, *K is a compact subset of $\mathscr{X}$, and V is an open subset of $\mathscr{X}$ such that $K \subseteq V$, then there is an open neighborhood of 0, U, such that $K + U \subseteq V$.*

PROOF. Let $\mathscr{U}_0 =$ all of the open neighborhoods of 0. Suppose that for each U in $\mathscr{U}_0$, $K + U$ is not contained in V. Thus, for each U in $\mathscr{U}_0$ there is a vector x_U in K and a y_U in U such that $x_U + y_U \in \mathscr{X} \backslash V$. Order $\mathscr{U}_0$ by reverse inclusion; that is, $U_1 \geqslant U_2$ if $U_1 \subseteq U_2$. Then $\mathscr{U}_0$ is a directed set and $\{x_U\}$ and $\{y_U\}$ are nets. Now $y_U \to 0$ in $\mathscr{X}$. Because K is compact there is an x in K such that $x_U \underset{\text{cl}}{\longrightarrow} x$ ($\{x_U\}$ cluster at x). Hence $x_U + y_U \underset{\text{cl}}{\longrightarrow} x + 0 = x$. (Why?) Hence $x \in \text{cl}(\mathscr{X} \backslash V) = \mathscr{X} \backslash V$, a contradiction. ∎

The condition that K be compact in the preceding lemma is necessary; it is not enough to assume that K is closed. (What is a counterexample?)

3.9. Theorem. *Let $\mathscr{X}$ be a real LCS and let A and B be two disjoint closed convex subsets of $\mathscr{X}$. If B is compact, then A and B are strictly separated.*

PROOF. By hypothesis, B is a compact subset of the open set $\mathscr{X}\backslash A$. The preceding lemma implies there is an open neighborhood U_1 of 0 such that $B + U_1 \subseteq \mathscr{X}\backslash A$. Because $\mathscr{X}$ is locally convex, there is a continuous seminorm p on $\mathscr{X}$ such that $\{x\colon p(x) < 1\} \subseteq U_1$. Put $U = \{x\colon p(x) < \frac{1}{2}\}$. Then $(B + U)\cap(A + U) = \square$ (Verify!), and $A + U$ and $B + U$ are open convex subsets of $\mathscr{X}$ that contain A and B, respectively. (Why?) So the result follows from Theorem 3.7. ∎

The fact that one of the two closed convex sets in the preceding theorem is assumed to be compact is necessary. In fact, if $\mathscr{X} = \mathbb{R}^2$, $A = \{(x, y) \in \mathbb{R}^2\colon y \leqslant 0\}$, and $B = \{(x, y) \in \mathbb{R}^2\colon y \geqslant x^{-1} > 0\}$, then A and B are disjoint closed convex subsets of $\mathbb{R}^2$ that cannot be strictly separated.

The next result generalizes Corollary III.6.8, though, of course, the metric content of (III.6.8) is missing.

3.10. Corollary. *If $\mathscr{X}$ is a real LCS, A is a closed convex subset of $\mathscr{X}$, and $x \notin A$, then x is strictly separated from A.*

3.11. Corollary. *If $\mathscr{X}$ is a real LCS and $A \subseteq \mathscr{X}$, then $\overline{co}(A)$ is the intersection of the closed half-spaces containing A.*

PROOF. Let $\mathscr{H}$ be the collection of all closed half-spaces containing A. Since each set in $\mathscr{H}$ is closed and convex, $\overline{co}(A) \subseteq \cap\{H\colon H \in \mathscr{H}\}$. On the other hand, if $x_0 \notin \overline{co}(A)$, then (3.10) implies there is a continuous linear functional $f\colon \mathscr{X} \to \mathbb{R}$ and an α in $\mathbb{R}$ such that $f(x_0) > \alpha$ and $f(x) < \alpha$ for all x in $\overline{co}(A)$. Thus $H = \{x\colon f(x) \leqslant \alpha\}$ belongs to $\mathscr{H}$ and $x_0 \notin H$. ∎

The next result generalizes Theorem III.6.13.

3.12. Corollary. *If $\mathscr{X}$ is a real LCS and $A \subseteq \mathscr{X}$, then the closed linear span of A is the intersection of all closed hyperplanes containing A.*

If $\mathscr{X}$ is a complex LCS, it is also a real LCS. This can be used to formulate and prove versions of the preceding results. As a sample, the following complex version of Theorem 3.9 is presented.

3.13. Theorem. *Let $\mathscr{X}$ be a complex LCS and let A and B be two disjoint closed convex subsets of $\mathscr{X}$. If B is compact, then there is a continuous linear functional $f\colon \mathscr{X} \to \mathbb{C}$, an α in $\mathbb{R}$, and an $\varepsilon > 0$ such that for a in A and b in B,*

$$\operatorname{Re} f(a) \leqslant \alpha < \alpha + \varepsilon \leqslant \operatorname{Re} f(b).$$

3.14. Corollary. *If $\mathscr{X}$ is a LCS and $\mathscr{Y}$ is a linear manifold in $\mathscr{X}$, then $\mathscr{Y}$ is* dense in $\mathscr{X}$ if and only if the only continuous linear functional on $\mathscr{X}$ that vanishes on $\mathscr{Y}$ is the identically zero functional.

3.15. Corollary. *If $\mathscr{X}$ is a LCS, $\mathscr{Y}$ is a closed linear subspace of $\mathscr{X}$, and $x_0 \in \mathscr{X} \setminus \mathscr{Y}$, then there is a continuous linear functional $f: \mathscr{X} \to \mathbb{F}$ such that $f(y) = 0$ for all y in $\mathscr{Y}$ and $f(x_0) = 1$.*

These results imply that on a LCS there are many continuous linear functionals. Compare the results of this section with those of §III.6.

The hypothesis that $\mathscr{X}$ is locally convex does not appear in the results prior to Theorem 3.9. The reason for this is that in the preceding results the existence of an open convex subset of $\mathscr{X}$ is assumed. In Theorem 3.9 such a set must be manufactured. Without the hypothesis of local convexity it may be that the only open convex sets are the whole space itself and the empty set.

3.16. Example. For $0 < p < 1$, let $L^p(0, 1)$ be the collection of equivalence classes of measurable functions $f: (0, 1) \to \mathbb{R}$ such that

$$((f))_p = \int_0^1 |f(x)|^p \, dx < \infty.$$

It will be shown that $d(f, g) = ((f - g))_p$ is a metric on $L^p(0, 1)$ and that with this metric $L^p(0, 1)$ is a Fréchet space. It will also be shown, however, that $L^p(0, 1)$ has only one nonempty open convex set, namely itself. So $L^p(0, 1)$, $0 < p < 1$, is most emphatically not locally convex. The proof of these facts begins with the following inequality.

3.17   For s, t in $[0, \infty)$ and $0 < p < 1$, $(s + t)^p \leqslant s^p + t^p$.

To see this, let $f(t) = s^p + t^p - (s + t)^p$ for $t \geqslant 0$, s fixed. Then $f'(t) = pt^{p-1} - p(s + t)^{p-1}$. Since $p - 1 < 0$ and $s + t \geqslant t$, $f'(t) \geqslant 0$. Thus $0 = f(0) \leqslant f(t)$. This proves (3.17)

If $d(f, g) = ((f - g))_p$ for f, g in $L^p(0, 1)$, then (3.17) implies that $d(f, g) \leqslant d(f, h) + d(h, g)$ for all f, g, h in $L^p(0, 1)$. It follows that d is a metric on $L^p(0, 1)$. Clearly d is translation invariant.

3.18 $L^p(0, 1)$, $0 < p < 1$, is complete.

The proof of this is left as an exercise.

3.19 $L^p(0, 1)$ is a TVS.

The continuity of addition is a direct consequence of the translation invariance of d. If $f_n \to f$ and $\alpha_n \to \alpha$, α_n in $\mathbb{R}$, $d(\alpha_n f_n, \alpha f) = ((\alpha_n f_n - \alpha f))_p \leqslant ((\alpha_n f_n - \alpha_n f))_p + ((\alpha_n f - \alpha f))_p = |\alpha_n|^p((f_n - f))_p + |\alpha_n - \alpha|^p((f))_p \leqslant C((f_n - f))_p + |\alpha_n - \alpha|^p((f))_p$, where C is a constant independent of n. Hence $\alpha_n f_n \to \alpha f$. Thus $L^p(0, 1)$ is a Fréchet space.

If G is a nonempty open convex subset of $L^p(0,1)$, then

3.20 $$G = L^p(0,1).$$

To see this, first suppose $f \in L^p(0,1)$ and $((f))_p = r < R$. As a function of t, $\int_0^t |f(x)|^p dx$ is continuous, assumes the value 0 at $t = 0$, and assumes the value r at $t = 1$. Let $0 < t < 1$ such that $\int_0^t |f(x)|^p dx = r/2$. Define $g, h: (0,1) \to \mathbb{R}$ by $g(x) = f(x)$ for $x \le t$ and 0 otherwise; $h(x) = f(x)$ for $x \ge t$ and 0 otherwise. Now $f = g + h = \frac{1}{2}(2g + 2h)$ and $((2g))_p = ((2h))_p = 2^p(r/2) = r/2^{1-p}$. Hence $f \in \operatorname{co} B(0; R/2^{1-p})$. This implies that $B(0; R) \subseteq \operatorname{co} B(0; R/2^{1-p})$, or, equivalently, $B(0; 2^{1-p}R) \subseteq \operatorname{co} B(0, R)$. Hence $B(0; 4^{1-p}R) \subseteq \operatorname{co} B(0; 2^{1-p}R) \subseteq \operatorname{co} B(0; R)$. Continuing we see that for all n, $B(0; 2^{n(1-p)}R) \subseteq \operatorname{co} B(0; R)$.

So if G is a nonempty open convex subset of $L^p(0,1)$, then by translation it may be assumed that $0 \in G$. Thus there is an $R > 0$ with $B(0; R) \subseteq G$. By the preceding paragraph, $B(0; 2^{n(1-p)}R) \subseteq \operatorname{co} B(0; R) \subseteq G$ for all $n \ge 1$. Therefore $L^p(0,1) \subseteq G$.

Also note that this says that the only continuous linear functional on $L^p(0,1)$, $0 < p < 1$, is the identically zero functional.

EXERCISES

1. Prove Theorem 3.1.

2. Let p be a sublinear functional, $G \equiv \{x: p(x) < 1\}$, and define the sublinear functional q for the set G as in Proposition 3.2. Show that $q(x) = \max(p(x)), 0)$ for all x in $\mathcal{X}$.

3. Let $\mathcal{M} \subseteq \mathcal{X}$, a TVS, and show that the following statements are equivalent : (a) $\mathcal{M}$ is an affine hyperplane; (b) there exists an x_0 in $\mathcal{M}$ such that $\mathcal{M} - x_0$ is a hyperplane; (c) there is a non-zero linear function $f: \mathcal{X} \to \mathbb{F}$ and an α in $\mathbb{F}$ such that $\mathcal{M} = \{x \in \mathcal{X}: f(x) = \alpha\}$.

4. Let $\mathcal{X}$ be a real TVS. Show: (a) if G is an open connected subset of $\mathcal{X}$, then G is arcwise connected; (b) if $f: \mathcal{X} \to \mathbb{R}$ is a continuous non-zero linear functional, then $\mathcal{X} \backslash \ker f$ has two components, $\{x: f(x) > 0\}$ and $\{x: f(x) < 0\}$.

5. If $\mathcal{X}$ is a complex TVS and $f: \mathcal{X} \to \mathbb{C}$ is a nonzero continuous linear function, show that $\mathcal{X} \backslash \ker f$ is connected.

6. Prove Proposition 3.6.

7. If $f: \mathcal{X} \to \mathbb{R}$ is a continuous $\mathbb{R}$-linear functional and A is an open convex subset of $\mathcal{X}$, then $f(A)$ is an open interval.

8. Prove Corollary 3.12.

9. Prove Theorem 3.13.

10. State and prove a version of Theorem 3.7 for a complex TVS.

11. State and prove a version of Corollary 3.11 for a complex LCS.

12. State and prove a version of Corollary 3.12 for a complex LCS.

13. Prove (3.18).

14. Give an example of a TVS $\mathscr{X}$ that is not locally convex and a subspace $\mathscr{Y}$ of $\mathscr{X}$ such that there is a continuous linear functional f on $\mathscr{Y}$ with no continuous extension to $\mathscr{X}$.

15. Let $\mathscr{X}$ be a real LCS and let A and B be disjoint compact convex subsets of $\mathscr{X}$. Suppose $\mathscr{Y}$ is a subspace of $\mathscr{X}$ and $f_0: \mathscr{Y} \to \mathbb{R}$ is a continuous linear functional such that $f_0(a) < 0$ for a in $A \cap \mathscr{Y}$ and $f_0(b) > 0$ for b in $B \cap \mathscr{Y}$. Show by an example that it is not always possible to extend f_0 to a continuous linear functional on $\mathscr{X}$ such that $f(a) < 0$ or a in A and $f(b) > 0$ for b in B. (Hint: Let $\mathscr{X} = \mathbb{R}^3$ and let $\mathscr{Y}$ be the plane.)

§4*. Some Examples of the Dual Space of a Locally Convex Space

As with a normed space, if $\mathscr{X}$ is a LCS, $\mathscr{X}^*$ denotes the space of all continuous linear functionals $f: \mathscr{X} \to \mathbb{F}$. $\mathscr{X}^*$ is called the *dual space* of $\mathscr{X}$.

4.1. Proposition. *Let X be completely regular and let $C(X)$ be topologized as in Example 1.5. If $L: C(X) \to \mathbb{F}$ is a continuous linear functional, then there is a compact set K and a regular Borel measure μ on K such that $L(f) = \int_K f \, d\mu$ for every f in $C(X)$. Conversely, each such measure defines an element of $C(X)^*$.*

PROOF. It is easy to see that each measure μ supported on a compact set K defines an element of $C(X)^*$. In fact, if $p_K(f) = \sup\{|f(x)|: x \in K\}$ and $L(f) = \int_K f \, d\mu$, then $|L(f)| \leq \|\mu\| p_K(f)$, and so L is continuous.

Now assume $L \in C(X)^*$. There are compact sets $K_1, \ldots, K_n$ and positive numbers $\alpha_1, \ldots, \alpha_n$ such that $|L(f)| \leq \sum_{j=1}^n \alpha_j p_{K_j}(f)$ (3.1f). Let $K = \bigcup_{j=1}^n K_j$ and $\alpha = \max\{\alpha_j: 1 \leq j \leq n\}$. Then $|L(f)| \leq \alpha p_K(f)$. Hence if $f \in C(X)$ and $f|K \equiv 0$, then $L(f) = 0$.

Define $F: C(K) \to \mathbb{F}$ as follows. If $g \in C(K)$, let $\tilde{g}$ be any continuous extension of g to X and put $F(g) = L(\tilde{g})$. To check that F is well defined, suppose that $\tilde{g}_1$ and $\tilde{g}_2$ are both extensions of g to X. Then $\tilde{g}_1 - \tilde{g}_2 = 0$ on K, and hence $L(\tilde{g}_1) = L(\tilde{g}_2)$. Thus F is well defined. It is left as an exercise for the reader to show that $F: C(K) \to \mathbb{F}$ is linear. If $g \in C(K)$ and $\tilde{g}$ is an extension in $C(X)$, then $|F(g)| = |L(\tilde{g})| \leq \alpha p_K(\tilde{g}) = \alpha \|g\|$, where the norm is the norm of $C(K)$. By (III.5.7) there is a measure μ in $M(K)$ such that $F(g) = \int_K g \, d\mu$. If $f \in C(X)$, then $g = f|K \in C(K)$ and so $L(f) = F(g) = \int_K f \, d\mu$. ∎

If $\gamma: [0,1] \to \mathbb{C}$ is a rectifiable curve and f is a continuous function defined on the trace of γ, $\gamma([0,1])$, then $\int_\gamma f$ is the line integral of f over γ. That is, $\int_\gamma f \equiv \int_0^1 f(\gamma(t)) \, d\gamma(t)$. (See Conway [1978].) The next result generalizes to arbitrary regions in the plane, but for simplicity it is stated only for the disk $\mathbb{D}$. Recall the definition of $H(\mathbb{D})$ from Example 1.6.

4.2. Proposition. *$L \in H(\mathbb{D})^*$ if and only if there is an $r < 1$ and a unique function* *g analytic on* $\mathbb{C}_\infty \backslash \bar{B}(0;r)$ *with* $g(\infty) = 0$ *such that*

4.3
$$L(f) = \frac{1}{2\pi i} \int_\gamma fg$$

for every f in $H(\mathbb{D})$, *where* $\gamma(t) = \rho e^{it}$, $0 \leqslant t \leqslant 2\pi$, *and* $r < \rho < 1$.

PROOF. Let g be given and define L as in (4.3). If $K = \{z: |z| = \rho\}$, then

$$|L(f)| = \frac{1}{2\pi} \left| \int_0^{2\pi} f(\rho e^{it}) g(\rho e^{it}) i \rho e^{it} \, dt \right|$$

$$\leqslant \frac{1}{2\pi} p_K(f) p_K(g) 2\pi \rho.$$

So if $c = \rho p_K(g)$, $|L(f)| \leqslant c p_K(f)$, and $L \in H(\mathbb{D})^*$.

Now assume that $L \in H(\mathbb{D})^*$. The Hahn–Banach Theorem implies there is an F in $C(\mathbb{D})^*$ such that $F|H(\mathbb{D}) = L$. By Proposition 4.1 there is a compact set K contained in $\mathbb{D}$ and a measure μ on K such that $L(f) = \int_K f \, d\mu$ for every f in $H(\mathbb{D})$. Define $g: \mathbb{C}_\infty \backslash K \to \mathbb{C}$ by $g(\infty) = 0$ and $g(z) = -\int_K 1/(w-z) \, d\mu(w)$ for z in $\mathbb{C} \backslash K$. By Lemma III.8.2, g is analytic on $\mathbb{C}_\infty \backslash K$. Let $\rho < 1$ such that $K \subseteq B(0; \rho)$. If $\gamma(t) = \rho e^{it}, 0 \leqslant t \leqslant 2\pi$, then Cauchy's Integral Formula implies

$$f(w) = \frac{1}{2\pi i} \int_\gamma \frac{f(z)}{z - w} \, dz$$

for $|w| < \rho$; in particular, this is true for w in K. Thus,

$$L(f) = \int_K f(w) \, d\mu(w)$$

$$= \int_K \left[\frac{\rho}{2\pi} \int_0^{2\pi} \frac{f(\rho e^{it})}{\rho e^{it} - w} e^{it} \, dt \right] d\mu(w)$$

$$= \frac{\rho}{2\pi} \int_0^{2\pi} f(\rho e^{it}) e^{it} \left[\int_K \frac{1}{\rho e^{it} - w} \, d\mu(w) \right] dt$$

$$= \frac{1}{2\pi i} \int_\gamma f(z) g(z) \, dz.$$

This completes the proof except for the uniqueness of g (Exercise 3). ∎

EXERCISES

1. Let $\{\mathscr{X}_i: i \in I\}$ be a family of LCS's and give $\mathscr{X} = \prod \{\mathscr{X}_i: i \in I\}$ the product topology. (See Exercise 1.17.) Show that $L \in \mathscr{X}^*$ if and only if there is a finite subset F contained in I and there are x_j^* in $\mathscr{X}_j^*$ for j in F such that $L(x) = \sum_{j \in F} x_j^*(x(j))$ for each x in $\mathscr{X}$.

2. Show that the space s (Exercise 1.13) is linearly homeomorphic to $C(\mathbb{N})$ and describe s^*.

3. Show that the function g obtained in Proposition 4.2 is unique.

4. Show that $L \in H(\mathbb{D})^*$ if and only if there are scalars $b_0, b_1, \ldots$ in $\mathbb{C}$ such that $\limsup |b_n|^{1/n} < 1$ and $L(f) = \sum_{n=0}^{\infty} 1/(n!) f^{(n)}(0) b_n$.

5. If G is an annulus, describe $H(G)^*$.

6. (Buck [1958]). Let X be locally compact and let β be the strict topology on $C_b(X)$ defined in Exercise 1.21. (Also see Exercises 2.6 and 2.7.) Prove the following statements: (a) If $\mu \in M(X)$ and $\varepsilon_n \downarrow 0$, then there are compact sets $K_1, K_2, \ldots$ such that for each $n \geqslant 1$, $K_n \subseteq \operatorname{int} K_{n+1}$ and $|\mu|(X \backslash K_n) < \varepsilon_n$. (b) If $\mu \in M(X)$, then there is a ϕ in $C_0(X)$ such that $\phi \geqslant 0$, $|\mu|(X \backslash \{x: \phi(x) > 0\}) = 0$, $1/\phi \in L^1(|\mu|)$, and $\int 1/\phi \, d|\mu| \leqslant 1$. (c) Show that if $\mu \in M(X)$ and $L(f) = \int f \, d\mu$ for f in $C_b(X)$, then $L \in (C_b(X), \beta)^*$. (d) Conversely, if $L \in (C_b(X), \beta)^*$, then there is a μ in $M(X)$ such that $L(f) = \int f \, d\mu$ for f in $C_b(X)$.

7. Let X be completely regular and let $\mathcal{M}$ be a linear manifold in $C(X)$. Show that if for every compact subset K of X, $\mathcal{M}|K \equiv \{f|K: f \in \mathcal{M}\}$ is dense in $C(K)$, then $\mathcal{M}$ is dense in $C(X)$.

§5*. Inductive Limits and the Space of Distributions

In this section the most general definition of an inductive limit will not be presented. Rather one that removes certain technicalities from the arguments and yet covers the most important examples will be given. For the more general definition see Köthe [1969], Robertson and Robertson [1966], or Schaefer [1971].

5.1. Definition. An *inductive system* is a pair $(\mathcal{X}, \{\mathcal{X}_i: i \in I\})$, where $\mathcal{X}$ is a vector space, $\mathcal{X}_i$ is a linear manifold in $\mathcal{X}$ that has a topology $\mathcal{T}_i$ such that $(\mathcal{X}_i, \mathcal{T}_i)$ is a LCS, and, moreover:

(a) I is a directed set and $\mathcal{X}_i \subseteq \mathcal{X}_j$ if $i \leqslant j$;
(b) if $i \leqslant j$ and $U_j \in \mathcal{T}_j$, then $U_j \cap \mathcal{X}_i \in \mathcal{T}_i$;
(c) $\mathcal{X} = \cup \{\mathcal{X}_i: i \in I\}$.

Note that condition (b) is equivalent to the condition that the inclusion map $\mathcal{X}_i \hookrightarrow \mathcal{X}_j$ is continuous.

5.2. Example. Let $d \geqslant 1$ and let Ω be an open subset of $\mathbb{R}^d$. Denote by $C_c^{(\infty)}(\Omega)$ all the functions $\phi: \Omega \to \mathbb{F}$ such that ϕ is infinitely differentiable and has compact support in Ω. (The *support* of ϕ is defined by $\operatorname{spt} \phi \equiv \operatorname{cl}\{x: \phi(x) \neq 0\}$.) If K is a compact subset of Ω, define $\mathscr{D}(K) \equiv \{\phi \in C_c^{(\infty)}(\Omega): \operatorname{spt} \phi \subseteq K\}$. Let $\mathscr{D}(K)$ have the topology defined by the seminorms

$$p_{K,m}(\phi) = \sup\{|\phi^{(k)}(x)|: |k| \leqslant m, x \in K\},$$

where $k = (k_1, \ldots, k_d)$, $k_j \in \mathbb{N} \cup \{0\}$, $|k| = k_1 + \cdots + k_d$, and

$$\phi^{(k)} = \frac{\partial^{|k|} \phi}{\partial x_1^{k_1} \cdots \partial x_d^{k_d}}.$$

Then $(C_c^{(\infty)}(\Omega), \{\mathscr{D}(K): K \text{ is compact in } \Omega\})$ is an inductive system. The space $C_c^{(\infty)}(\Omega)$ is often denoted in the literature by $\mathscr{D}(\Omega)$, as it will be in this book.

This example of an inductive system is the most important one as it is connected with the theory of distributions (below). In fact, this example was the inspiration for the definition of an inductive limit given now.

5.3. Proposition. *If* $(\mathscr{X}, \{\mathscr{X}_i, \mathscr{T}_i\})$ *is an inductive system, let* $\mathscr{B} = $ *all convex balanced sets* V *such that* $V \cap \mathscr{X}_i \in \mathscr{T}_i$ *for all* i. *Let* $\mathscr{T} = $ *the collection of all subsets* U *of* $\mathscr{X}$ *such that for every* x_0 *in* U *there is a* V *in* $\mathscr{B}$ *with* $x_0 + V \subseteq U$. *Then* $(\mathscr{X}, \mathscr{T})$ *is a (not necessarily Hausdorff) LCS.*

Before proving this proposition, it seems appropriate to make the following definition.

5.4. Definition. If $(\mathscr{X}, \{\mathscr{X}_i\})$ is an inductive system and $\mathscr{T}$ is the topology defined in (5.3), $\mathscr{T}$ is called the *inductive limit topology* and $(\mathscr{X}, \mathscr{T})$ is said to be the *inductive limit* of $\{\mathscr{X}_i\}$.

5.5. Lemma. *With the notation as in (5.3),* $\mathscr{B} \subseteq \mathscr{T}$.

PROOF. Fix V is $\mathscr{B}$. It will be shown that V is absorbing at each of its points. Indeed, if $x_0 \in V$ and $x \in \mathscr{X}$, then there is an $\mathscr{X}_i$ and an $\mathscr{X}_j$ such that $x_0 \in \mathscr{X}_i$ and $x \in \mathscr{X}_j$. Since I is directed, there is a k in I with $k \geqslant i, j$. Hence $x_0, x \in \mathscr{X}_k$. But $V \cap \mathscr{X}_k \in \mathscr{T}_k$. Thus there is an $\varepsilon > 0$ such that $x_0 + \alpha x \in V \cap \mathscr{X}_k \subseteq V$ for $|\alpha| < \varepsilon$.

Since V is convex, balanced, and absorbing at each of its points, there is a seminorm p on $\mathscr{X}$ such that $V = \{x \in \mathscr{X}: p(x) < 1\}$ (1.14). So if $x_0 \in V$, $p(x_0) = r_0 < 1$. Let $W = \{x \in \mathscr{X}: p(x) < \frac{1}{2}(1 - r_0)\}$. Then $W = \frac{1}{2}(1 - r_0)V$ and so $W \in \mathscr{B}$. Since $x_0 + W \subseteq V$, $V \in \mathscr{T}$. ∎

PROOF OF PROPOSITION 5.3. The proof that $\mathscr{T}$ is a topology is left as an exercise. To see that $(\mathscr{X}, \mathscr{T})$ is a LCS, note that Lemma 5.5 and Theorem 1.14 imply that $\mathscr{T}$ is defined by a family of seminorms. ∎

For all we know the inductive limit topology may be trivial. However, the fact that this topology has not been shown to be Hausdorff need not concern us, since we will concentrate on a particular type of inductive limit which will be shown to be Hausdorff. But for the moment we will continue at the present level of generality.

5.6. Proposition. *Let* $(\mathscr{X}, \{\mathscr{X}_i\})$ *be an inductive system and let* $\mathscr{T}$ *be the inductive* *limit topology. Then*

(a) *the relative topology on $\mathcal{X}_i$ induced by $\mathcal{T}$ (viz., $\mathcal{T}\,|\,\mathcal{X}_i$) is smaller than $\mathcal{T}_i$;*
(b) *if $\mathcal{U}$ is a locally convex topology on $\mathcal{X}$ such that for every i, $\mathcal{U}\,|\,\mathcal{X}_i \subseteq \mathcal{T}_i$, then $\mathcal{U} \subseteq \mathcal{T}$;*
(c) *a seminorm p on $\mathcal{X}$ is continuous if and only if $p\,|\,\mathcal{X}_i$ is continuous for each i.*

PROOF. Exercise 3.

5.7. Proposition. *Let $(\mathcal{X}, \mathcal{T})$ be the inductive limit of the spaces $\{(\mathcal{X}_i, \mathcal{T}_i): i\in I\}$. If $\mathcal{Y}$ is a LCS and $T: \mathcal{X} \to \mathcal{Y}$ is a linear transformation, then T is continuous if and only if the restriction of T to each $\mathcal{X}_i$ is $\mathcal{T}_i$-continuous.*

PROOF. Suppose $T: \mathcal{X} \to \mathcal{Y}$ is continuous. By (5.6a), the inclusion map $(\mathcal{X}_i, \mathcal{T}_i) \to (\mathcal{X}, \mathcal{T})$ is continuous. Since the restriction of T to $\mathcal{X}_i$ is the composition of the inclusion map $\mathcal{X}_i \to \mathcal{X}$ and T, the restriction is continuous.

Now assume that each restriction is continuous. If p is a continuous seminorm on $\mathcal{Y}$, then $p \circ T\,|\,\mathcal{X}_i$ is a $\mathcal{T}_i$-continuous seminorm for every i. By (5.6c), $p \circ T$ is continuous on $\mathcal{X}$. By Exercise 1.23, T is continuous. ∎

It may have occurred to the reader that the definition of the inductive limit topology depends on the choice of the spaces $\mathcal{X}_i$ in more than the obvious way. That is, if $\mathcal{X} = \bigcup_j \mathcal{Y}_j$ and each $\mathcal{Y}_j$ has a topology that is "compatible" with that of the spaces $\{\mathcal{X}_i\}$, perhaps the inductive limit topology defined by the spaces $\{\mathcal{Y}_j\}$ will differ from that defined by the $\{\mathcal{X}_i\}$. This is not the case.

5.8. Proposition. *Let $(\mathcal{X}, \{(\mathcal{X}_i, \mathcal{T}_i)\})$ and $(\mathcal{X}, \{(\mathcal{Y}_j, \mathcal{U}_j)\})$ be two inductive systems and let $\mathcal{T}$ and $\mathcal{U}$ be the corresponding inductive limit topologies on $\mathcal{X}$. If for every i there is a j such that $\mathcal{X}_i \subseteq \mathcal{Y}_j$ and $\mathcal{U}_j\,|\,\mathcal{X}_i \subseteq \mathcal{T}_i$, then $\mathcal{U} \subseteq \mathcal{T}$.*

PROOF. Let V be a convex balanced subset of $\mathcal{X}$ such that for every j, $V \cap \mathcal{Y}_j \in \mathcal{U}_j$. If $\mathcal{X}_i$ is given, let j be such that $\mathcal{X}_i \subseteq \mathcal{Y}_j$ and $\mathcal{U}_j\,|\,\mathcal{X}_i \subseteq \mathcal{T}_i$. Hence $V \cap \mathcal{X}_i = (V \cap \mathcal{Y}_j) \cap \mathcal{X}_i \in \mathcal{T}_i$. Thus $V \in \mathcal{B}$ [as defined in (5.3)]. It now follows that $\mathcal{U} \subseteq \mathcal{T}$. ∎

5.9. Example. Let $\mathcal{X}$ be any vector space and let $\{\mathcal{X}_i: i\in I\}$ be all of the finite dimensional subspaces of $\mathcal{X}$. Give each $\mathcal{X}_i$ the unique topology from its identification with a Euclidean space. Then $(\mathcal{X}, \{\mathcal{X}_i\})$ is an inductive system. Let $\mathcal{T}$ be the inductive limit topology. If $\mathcal{Y}$ is a LCS and $T: \mathcal{X} \to \mathcal{Y}$ is a linear transformation, then T is $\mathcal{T}$-continuous.

5.10. Example. Let X be a locally compact space and let $\{K_i: i\in I\}$ be the collection of all compact subsets of X. Let $\mathcal{X}_i = $ all f in $C(X)$ such that spt $f \subseteq K_i$. Then $\bigcup_i \mathcal{X}_i = C_c(X)$, the continuous functions on X with compact support. Topologize each $\mathcal{X}_i$ by giving it the supremum norm. Then $(C_c(X), \{\mathcal{X}_i\})$ is an inductive system.

Let $\{U_j\}$ be the open subsets of X such that $\mathrm{cl}\, U_j$ is compact. Let $C_0(U_j)$ be the continuous functions on U_j vanishing at ∞ with the supremum norm. If $f \in C_0(U_j)$ and f is defined on X by letting it be identically 0 on $X \backslash U_j$, then $f \in C_c(X)$. Thus $(C_c(X), \{C_0(U_j)\})$ is an inductive system. Proposition 5.8 implies that these two inductive systems define the same inductive limit topology on $C_c(X)$.

5.11. Example. Let $d \geqslant 1$ and put $K_n = \{x \in \mathbb{R}^d : \|x\| \leqslant n\}$. Then $(\mathscr{D}(\mathbb{R}^d),$ $\{\mathscr{D}(K_n)\}_{n=1}^{\infty})$ is an inductive system. By (5.9), the inductive limit topology defined on $\mathscr{D}(\mathbb{R}^d)$ by this system equals the inductive limit topology defined by the system given in Example 5.2.

If Ω is any subset of $\mathbb{R}^d$, then Ω can be written as the union of a sequence of compact subsets $\{K_n\}$ such that $K_n \subseteq \mathrm{int}\, K_{n+1}$. It follows by (5.9) that $\{\mathscr{D}(K_n)\}$ defines the same topology on $\mathscr{D}(\Omega)$ as was defined in Example 5.2.

The preceding example inspires the following definition.

5.12. Definition. A strict inductive system is an inductive system $(\mathscr{X}, \{\mathscr{X}_n, \mathscr{T}_n\}_{n=1}^{\infty})$ such that for every $n \geqslant 1$, $\mathscr{X}_n \subseteq \mathscr{X}_{n+1}$, $\mathscr{T}_{n+1} | \mathscr{X}_n = \mathscr{T}_n$, and $\mathscr{X}_n$ is closed in $\mathscr{X}_{n+1}$. The inductive limit topology defined on $\mathscr{X}$ by such a system is called a *strict inductive limit topology* and $\mathscr{X}$ is said to be the *strict inductive limit* of $\{\mathscr{X}_n\}$.

Example 5.11 shows that $\mathscr{D}(\mathbb{R}^d)$, indeed $\mathscr{D}(\Omega)$, is a strict inductive limit.

The following lemma is useful in the study of strict inductive limits as well as in other situations.

5.13. Proposition. *If $\mathscr{X}$ is a LCS, $\mathscr{Y} \leqslant \mathscr{X}$, and p is a continuous seminorm on $\mathscr{Y}$, then there is a continuous seminorm $\tilde{p}$ on $\mathscr{X}$ such that $\tilde{p} | \mathscr{Y} = p$.*

PROOF. Let $U = \{y \in \mathscr{Y} : p(y) < 1\}$. So U is open in $\mathscr{Y}$; hence there is an open subset V_1 of $\mathscr{X}$ such that $V_1 \cap \mathscr{Y} = U$. Since $0 \in V_1$ and $\mathscr{X}$ is a LCS, there is an open convex balanced set V in $\mathscr{X}$ such that $V \subseteq V_1$. Let $q =$ the gauge of V. So if $y \in \mathscr{Y}$ and $q(y) < 1$, then $p(y) < 1$. By Lemma III.1.4, $p \leqslant q | \mathscr{Y}$.

Let $W = \mathrm{co}(U \cup V)$; it is easy to see that W is convex and balanced since both U and V are. It will be shown that W is open. First observe that $W = \{tu + (1-t)v : 0 \leqslant t \leqslant 1,\ u \in U,\ v \in V\}$ (verify). Hence $W = \cup\{tU + (1-t)V : 0 \leqslant t \leqslant 1\}$. Put $W_t = tU + (1-t)V$. So $W_0 = V$, which is open. If $0 < t < 1$, $W_t = \cup\{tu + (1-t)V : u \in U\}$, and hence is open. But $W_1 = U$, which is not open. However, if $u \in U$, then there is an $\varepsilon > 0$ such that $\varepsilon u \in V$. For $0 < t < 1$, let $y_t = t^{-1}[1 - \varepsilon + t\varepsilon]u$ $(\in \mathscr{Y})$. As $t \to 1$, $y_t \to u$. Since U is open in $\mathscr{Y}$, there is a $t, 0 < t < 1$, with y_t in U. Thus $u = ty_t + (1-t)(\varepsilon u) \in W_t$. Therefore $W = \cup\{W_t : 0 \leqslant t < 1\}$ and W is open.

5.14. Claim. $W \cap \mathscr{Y} = U$.

In fact, $U \subseteq W$, so $U \subset W \cap \mathscr{Y}$. If $w \in W \cap \mathscr{Y}$, then $w = tu + (1-t)v, u$ in U, v in V, $0 \leqslant t \leqslant 1$; it may be assumed that $0 < t < 1$. (Why?) Hence, $v = (1-t)^{-1}(w - tu) \in \mathscr{Y}$. So $v \in V \cap \mathscr{Y} \subseteq U$; hence $w \in U$.

Let $\tilde{p}$ = the gauge of W. By Claim 5.14, $\{y \in \mathscr{Y}: \tilde{p}(y) < 1\} = \{y \in \mathscr{Y}: p(y) < 1\}$. By the uniqueness of the gauge, $\tilde{p}|\mathscr{Y} = p$. ∎

5.15. Corollary. *If $\mathscr{X}$ is the strict inductive limit of $\{\mathscr{X}_n\}$, k is fixed, and p_k is a continuous seminorm on $\mathscr{X}_k$, then there is a continuous seminorm p on $\mathscr{X}$ such that $p|\mathscr{X}_k = p_k$. In particular, the inductive limit topology is Hausdorff and the topology on $\mathscr{X}$ when restricted to $\mathscr{X}_k$ equals the original topology of $\mathscr{X}_k$.*

PROOF. By (5.13) and induction, for every integer $n > k$, there is a continuous seminorm p_n such that $p_n|\mathscr{X}_{n-1} = p_{n-1}$. If $x \in \mathscr{X}$, define $p(x) = p_n(x)$ when $x \in \mathscr{X}_n$. Since $\mathscr{X}_n \subseteq \mathscr{X}_{n+1}$ for all n, the properties of $\{p_n\}$ insure that p is well defined. Clearly p is a seminorm and by (5.6c) p is continuous.

If $x \in \mathscr{X}$ and $x \neq 0$, there is a $k \geqslant 1$ such that $x \in \mathscr{X}_k$. Thus there is a continuous seminorm p_k on $\mathscr{X}_k$ such that $p_k(x) \neq 0$. Using the first part of the corollary, we get a continuous seminorm p on $\mathscr{X}$ such that $p(x) \neq 0$. Thus $(\mathscr{X}, \mathscr{T})$ is Hausdorff. The proof that the topology on $\mathscr{X}$ when relativized to $\mathscr{X}_k$ equals the original topology is an easy application of (5.13). ∎

5.16. Proposition. *Let $\mathscr{X}$ be the strict inductive limit of $\{\mathscr{X}_n\}$. A subset B of $\mathscr{X}$ is bounded if and only if there is an $n \geqslant 1$ such that $B \subseteq \mathscr{X}_n$ and B is bounded in $\mathscr{X}_n$.*

The proof will be accomplished only after a few preliminaries are settled. Before doing this, here are a few consequences of (5.16).

5.17. Corollary. *If $\mathscr{X}$ is the strict inductive limit of $\{\mathscr{X}_n\}$, then a subset K of $\mathscr{X}$ is compact if and only if there is an $n \geqslant 1$ such that $K \subseteq \mathscr{X}_n$ and K is compact in $\mathscr{X}_n$.*

5.18. Corollary. *If $\mathscr{X}$ is the strict inductive limit of Fréchet spaces $\{\mathscr{X}_n\}$, $\mathscr{Y}$ is a LCS, and $T: \mathscr{X} \to \mathscr{Y}$ is a linear transformation, then T is continuous if and only if T is sequentially continuous.*

PROOF. By Proposition 5.7, T is continuous if and only if $T|\mathscr{X}_n$ is continuous for every n. Since each $\mathscr{X}_n$ is metrizable, the result follows. ∎

Note that using Example 5.11 it follows that for an open subset Ω of $\mathbb{R}^d$, $\mathscr{D}(\Omega)$ is the strict inductive limit of Fréchet spaces [each $\mathscr{D}(K_n)$ is a Fréchet space by Proposition 2.1]. So (5.18) applies.

5.19. Definition. If Ω is an open subset of $\mathbb{R}^d$, a *distribution* on Ω is a continuous linear functional on $\mathscr{D}(\Omega)$.

Distributions are, in a certain sense, generalizations of the concept of function as the following example illustrates.

5.20. Example. Let f be a Lebesgue measurable function on Ω that is locally integrable (that is, $\int_K |f| d\lambda < \infty$ for every compact subset K of Ω—here λ is d-dimensional Lebesgue measure). If $L_f \colon \mathcal{D}(\Omega) \to \mathbb{F}$ is defined by $L_f(\phi) = \int f\phi \, d\lambda$, L_f is a distribution.

From Corollary 5.18 we arrive at the following.

5.21. Proposition. *A linear functional* $L \colon \mathcal{D}(\Omega) \to \mathbb{F}$ *is a distribution if and only if for every sequence* $\{\phi_n\}$ *in* $\mathcal{D}(\Omega)$ *such that* $\mathrm{cl}[\bigcup_{n=1}^{\infty} \mathrm{spt}\, \phi_n] = K$ *is compact in* Ω *and* $\phi_n^{(k)}(x) \to 0$ *uniformly on* K *as* $n \to \infty$ *for every* $k = (k_1, \ldots, k_d)$, *it follows that* $L(\phi_n) \to 0$.

Proposition 5.21 is usually taken as the definition of a distribution in books on differential equations. There is the advantage that (5.21) can be understood with no knowledge of locally convex spaces and inductive limits. Moreover, most theorems on distributions can be proved by using (5.21). However, the realization that a distribution is precisely a continuous linear functional on a LCS contributes more than cultural edification. This knowledge brings power as it enables you to apply the theory of LCS's (including the Hahn–Banach Theorem).

The exercises contain more results on distributions, but now we must return to the proof of Proposition 5.16. To do this the idea of a topological complement is needed. We have seen this idea in Section III.13.

5.22. Proposition. *If* $\mathcal{X}$ *is a TVS and* $\mathcal{Y} \leqslant \mathcal{X}$, *the following statements are equivalent.*

(a) *There is a closed linear subspace* $\mathcal{Z}$ *of* $\mathcal{X}$ *such that* $\mathcal{Y} \cap \mathcal{Z} = (0)$, $\mathcal{Y} + \mathcal{Z} = \mathcal{X}$, *and the map of* $\mathcal{Y} \times \mathcal{Z} \to \mathcal{X}$ *given by* $(y, z) \mapsto y + z$ *is a homeomorphism.*
(b) *There is a continuous linear map* $P \colon \mathcal{X} \to \mathcal{X}$ *such that* $P\mathcal{X} = \mathcal{Y}$ *and* $P^2 = P$.

PROOF. (a) $\Rightarrow$ (b): Define $P \colon \mathcal{X} \to \mathcal{X}$ by $P(y + z) = y$, for y in $\mathcal{Y}$ and z in $\mathcal{Z}$. It is easy to verify that P is linear and $P\mathcal{X} = \mathcal{Y}$. Also, $P^2(y + z) = PP(y + z) = Py = y = P(y + z)$; so $P^2 = P$. If $\{y_i + z_i\}$ is a net in $\mathcal{X}$ such that $y_i + z_i \to y + z$, then (a) implies that $y_i \to y$ (and $z_i \to z$). Hence $P(y_i + z_i) \to P(y + z)$ and P is continuous.

(b) $\Rightarrow$ (a): If P is given, let $\mathcal{Z} = \ker P$. So $\mathcal{Z} \leqslant \mathcal{X}$. Also, $x = Px + (x - Px)$ and $y = Px \in \mathcal{Y}$, and $z = x - Px$ has $Pz = Px - P^2x = Px - Px = 0$, so $z \in \mathcal{Z}$. Thus, $\mathcal{Y} + \mathcal{Z} = \mathcal{X}$. If $x \in \mathcal{Y} \cap \mathcal{Z}$, then $Px = 0$ since $x \in \mathcal{Z}$; but also $x = Pw$ for some w in $\mathcal{X}$ since $x \in \mathcal{Y} = P\mathcal{X}$. Therefore $0 = Px = P^2w = Pw = x$; that is, $\mathcal{Y} \cap \mathcal{Z} = (0)$. Now suppose that $\{y_i\}$ and $\{z_i\}$ are nets in $\mathcal{Y}$ and $\mathcal{Z}$. If $y_i \to y$ and $z_i \to z$, then $y_i + z_i \to y + z$ because addition is continuous. If, on the other hand, it is assumed that $y_i + z_i \to y + z$, then $y = P(y + z) = \lim P(y_i + z_i) = \lim y_i$ and $z_i = (y_i + z_i) - y_i \to z$. This proves (a). ∎

5.23. Definition. If $\mathscr{X}$ is a TVS and $\mathscr{Y} \leqslant \mathscr{X}, \mathscr{Y}$ is *topologically complemented* in $\mathscr{X}$ if either (a) or (b) of (5.22) is satisfied.

5.24. Proposition. *If $\mathscr{X}$ is a LCS and $\mathscr{Y} \leqslant \mathscr{X}$ such that either $\dim \mathscr{Y} < \infty$ or $\dim \mathscr{X}/\mathscr{Y} < \infty$, then $\mathscr{Y}$ is topologically complemented in $\mathscr{X}$.*

PROOF. The proof will only be sketched. The reader is asked to supply the details (Exercise 9).

(a) Suppose $d = \dim \mathscr{Y} < \infty$ and let $y_1, \ldots, y_d$ be a basis for $\mathscr{Y}$. By the Hahn–Banach Theorem (III.6.6), there are $f_1, \ldots, f_d$ in $\mathscr{X}^*$ such that $f_i(y_j) = 1$ if $i = j$ and 0 otherwise. Define $Px = \sum_{j=1}^{d} f_j(x) y_j$.

(b) Suppose $d = \dim \mathscr{X}/\mathscr{Y} < \infty$, $Q: \mathscr{X} \to \mathscr{X}/\mathscr{Y}$ is the natural map, and $z_1, \ldots, z_d \in \mathscr{X}$ such that $Q(z_1), \ldots, Q(z_d)$ is a basis for $\mathscr{X}/\mathscr{Y}$. Let $\mathscr{Z} = \vee \{z_1, \ldots, z_d\}$. ∎

PROOF OF PROPOSITION 5.16. Suppose $\mathscr{X}$ is the strict inductive limit of $\{(\mathscr{X}_n, \mathscr{T}_n)\}$ and B is a bounded subset of $\mathscr{X}$. It must be shown that there is an n such that $B \subseteq \mathscr{X}_n$ (the rest of the proof is easy). Suppose this is not the case. By replacing $\{\mathscr{X}_n\}$ by a subsequence if necessary, it follows that for each n there is an x_n in $(B \cap \mathscr{X}_{n+1}) \backslash \mathscr{X}_n$. Let p_1 be a continuous seminorm on $\mathscr{X}_1$ such that $p_1(x_1) = 1$.

5.25. Claim. *For every $n \geqslant 2$ there is a continuous seminorm p_n on $\mathscr{X}_n$ such that $p_n(x_n) = n$ and $p_n|\mathscr{X}_{n-1} = p_{n-1}$.*

The proof of (5.25) is by induction. Suppose p_n is given and let $\mathscr{Y} = \mathscr{X}_n \vee \{x_{n+1}\}$. By (5.24), $\mathscr{X}_n$ and $\vee \{x_{n+1}\}$ are topologically complementary in $\mathscr{Y}$. Define $q: \mathscr{Y} \to [0, \infty)$ by $q(x + \alpha x_{n+1}) = p_n(x) + (n+1)|\alpha|$, where $x \in \mathscr{X}_n$ and $\alpha \in \mathbb{F}$. Then q is a continuous seminorm on $(\mathscr{Y}, \mathscr{T}_{n+1}|\mathscr{Y})$. (Verify!) By Proposition 5.13 there is a continuous seminorm p_{n+1} on $\mathscr{X}_{n+1}$ such that $p_{n+1}|\mathscr{Y} = q$. Thus $p_{n+1}|\mathscr{X}_n = p_n$ and $p_{n+1}(x_{n+1}) = n + 1$. This proves the claim.

Now define $p: \mathscr{X} \to [0, \infty)$ by $p(x) = p_n(x)$ if $x \in \mathscr{X}_n$. By (5.25), p is well defined. It is easy to see that p is a continuous seminorm. However, $\sup\{p(x): x \in B\} = \infty$, so B is not bounded (Exercise 2.4f). ∎

EXERCISES

1. Verify the statements made in Example 5.2.

2. Fill in the details of the proof of Proposition 5.3.

3. Prove Proposition 5.6.

4. Verify the statements made in Example 5.9.

5. Verify the statements made in Example 5.10.

6. Verify the statements made in Example 5.11.

7. With the notation of (5.10), show that if X is σ-compact, then the dual of $C_c(X)$ is the space of all extended $\mathbb{F}$-valued measures.

8. Is the inductive limit topology on $C_c(X)$ (5.10) different from the topology of uniform convergence on compact subsets of X (1.5)?

9. Prove Proposition 5.24.

10. Verify the statements made in Example 5.20.

For the remaining exercises, Ω is always an open subset of $\mathbb{R}^d$, $d \geqslant 1$.

11. If μ is a measure on Ω, $\phi \mapsto \int \phi \, d\mu$ is a distribution Ω.

12. Let $f : \Omega \to \mathbb{F}$ be a function with continuous partial derivatives and let L_f be defined as in (5.20). Show that for every ϕ in $\mathscr{D}(\Omega)$ and $1 \leqslant j \leqslant d$, $L_f(\partial \phi / \partial x_j) = -L_g(\phi)$, where $g = \partial f / \partial x_j$. (Hint: Use integration by parts.)

13. Exercise 12 motivates the following definition. If L is a distribution on Ω, define $\partial L / \partial x_j : \mathscr{D}(\Omega) \to \mathbb{F}$ by $\partial L / \partial x_j(\phi) = -L(\partial \phi / \partial x_j)$ for all ϕ in $\mathscr{D}(\Omega)$. Show that $\partial L / \partial x_j$ is a distribution.

14. Using Example 5.20 and Exercise 13, one is justified to talk of the derivative of any locally integrable function as a *distribution*. By Exercise 11 we can differentiate measures. Let $f : \mathbb{R} \to \mathbb{R}$ be the characteristic function of $[0, \infty)$ and show that its derivative as a distribution is δ_0, the unit point mass at 0. [That is, δ_0 is the measure such that $\delta_0(\Delta) = 1$ if $0 \in \Delta$ and $\delta_0(\Delta) = 0$ if $0 \notin \Delta$.]

15. Let f be an absolutely continuous function on $\mathbb{R}$ and show that $(L_f)' = L_{f'}$.

16. Let f be a left continuous nondecreasing function on $\mathbb{R}$ and show that $(L_f)'$ is the distribution defined by the measure μ such that $\mu[a, b) = f(b) - f(a)$ for all $a < b$.

17. Let f be a C^∞ function on Ω and let L be a distribution on $\mathscr{D}(\Omega)$. Show that $M(\phi) \equiv L(\phi f)$, ϕ in $\mathscr{D}(\Omega)$, is a distribution. State and prove a product rule for finding the derivative of M.

CHAPTER V

Weak Topologies

The principal objects of study in this chapter are the weak topology on a Banach space and the weak-star topology on its dual. In order to carry out this study efficiently, the first two sections are devoted to the study of the weak topology on a locally convex space.

§1. Duality

As in §IV.4, for a LCS $\mathscr{X}$, let $\mathscr{X}^*$ denote the space of continuous linear functionals on $\mathscr{X}$. If $x^*, y^* \in \mathscr{X}^*$ and $\alpha \in \mathbb{F}$, then $(\alpha x^* + y^*)(x) \equiv \alpha x^*(x) + y^*(x)$, x in $\mathscr{X}$, defines an element $\alpha x^* + y^*$ in $\mathscr{X}^*$. Thus $\mathscr{X}^*$ has a natural vector-space structure.

It is convenient and, more importantly, helpful to introduce the notation

$$\langle x, x^* \rangle$$

to stand for $x^*(x)$, for x in $\mathscr{X}$ and x^* in $\mathscr{X}^*$. Also, because of a certain symmetry, we will use $\langle x^*, x \rangle$ to stand for $x^*(x)$. Thus

$$x^*(x) = \langle x, x^* \rangle = \langle x^*, x \rangle.$$

We begin by recalling two definitions (IV.1.7 and IV.1.8).

1.1. Definition. If $\mathscr{X}$ is a LCS, the *weak topology* on $\mathscr{X}$, denoted by "wk" or $\sigma(\mathscr{X}, \mathscr{X}^*)$, is the topology defined by the family of seminorms $\{p_{x^*} : x^* \in \mathscr{X}^*\}$, where

$$p_{x^*}(x) = |\langle x, x^* \rangle|.$$

The *weak-star topology* on $\mathscr{X}^*$, denoted by "wk*" or $\sigma(\mathscr{X}^*, \mathscr{X})$, is the topology defined by the seminorms $\{p_x: x \in \mathscr{X}\}$, where

$$p_x(x^*) = |\langle x, x^* \rangle|.$$

So a subset U of $\mathscr{X}$ is weakly open if and only if for every x_0 in U there is an $\varepsilon > 0$ and there are $x_1^*, \ldots, x_n^*$ in $\mathscr{X}^*$ such that

$$\bigcap_{k=1}^{n} \{x \in \mathscr{X}: |\langle x - x_0, x_k^* \rangle| < \varepsilon\} \subseteq U.$$

A net $\{x_i\}$ in $\mathscr{X}$ converges weakly to x_0 if and only if $\langle x_i, x^* \rangle \to \langle x_0, x^* \rangle$ for every x^* in $\mathscr{X}^*$. (What are the analogous statements for the weak-star topology?)

Note that both $(\mathscr{X}, \text{wk})$ and $(\mathscr{X}^*, \text{wk}^*)$ are LCS's. Also, $\mathscr{X}$ already possesses a topology so that wk is a second topology on $\mathscr{X}$. However, $\mathscr{X}^*$ has no topology to begin with so that wk* is the only topology on $\mathscr{X}^*$. Of course if $\mathscr{X}$ is a normed space, this last statement is not correct since $\mathscr{X}^*$ is a Banach space (III.5.4). The reader should also be cautioned that some authors abuse the language and use the term weak topology to designate both the weak and weak-star topologies. Finally, pay attention to the positions of $\mathscr{X}$ and $\mathscr{X}^*$ in the notation $\sigma(\mathscr{X}, \mathscr{X}^*) = \text{wk}$ and $\sigma(\mathscr{X}^*, \mathscr{X}) = \text{wk}^*$.

If $\{x_i\}$ is a net in $\mathscr{X}$ and $x_i \to 0$ in $\mathscr{X}$, then for every x^* in $\mathscr{X}^*$, $\langle x_i, x^* \rangle \to 0$. So if $\mathscr{T}$ is the topology on $\mathscr{X}$, wk $\subseteq \mathscr{T}$ (A.2.9) and each x^* in $\mathscr{X}^*$ is weakly continuous. The first result gives the converse of this.

1.2. Theorem. *If $\mathscr{X}$ is a LCS, $(\mathscr{X}, \text{wk})^* = \mathscr{X}^*$.*

PROOF. Since every weakly open set is open in the original topology, each f in $(\mathscr{X}, \text{wk})^*$ belongs to $\mathscr{X}^*$. The converse is even easier. ∎

1.3. Theorem. *If $\mathscr{X}$ is a LCS, $(\mathscr{X}^*, \text{wk}^*)^* = \mathscr{X}$.*

PROOF. Clearly if $x \in \mathscr{X}$, $x^* \to \langle x, x^* \rangle$ is a wk* continuous functional on $\mathscr{X}^*$. Hence $\mathscr{X} \subseteq (\mathscr{X}^*, \text{wk}^*)^*$. Conversely, if $f \in (\mathscr{X}^*, \text{wk}^*)^*$, then (IV.3.1) implies there are vectors $x_1, \ldots, x_n$ in $\mathscr{X}$ such that $|f(x^*)| \leqslant \sum_{k=1}^{n} |\langle x_k, x^* \rangle|$ for all x^* in $\mathscr{X}^*$. This implies that $\cap \{\ker x_k: 1 \leqslant k \leqslant n\} \subseteq \ker f$. By (A.1.4) there are scalars $\alpha_1, \ldots, \alpha_n$ such that $f = \sum_{k=1}^{n} \alpha_k x_k$; hence $f \in \mathscr{X}$. ∎

So $\mathscr{X}$ is the dual of a LCS—$(\mathscr{X}^*, \text{wk}^*)$—and hence has a weak-star topology—$\sigma((\mathscr{X}, \text{wk}^*), \mathscr{X}^*)$. As an exercise in notational juggling, note that $\sigma((\mathscr{X}, \text{wk}^*), \mathscr{X}^*) = \sigma(\mathscr{X}, \mathscr{X}^*)$.

All unmodified topological statements about $\mathscr{X}$ refer to its original topology. So if $A \subseteq \mathscr{X}$ and we say that it is closed, we mean that A is closed in the original topology of $\mathscr{X}$. To say that A is closed in the weak topology of $\mathscr{X}$ we say that A is weakly closed or wk-closed. Also cl A means the closure of A in the original topology while wk $-$ cl A means the closure of A in the weak topology. The next result shows that under certain circumstances this distinction in unnecessary.

1.4. Theorem. *If $\mathscr{X}$ is a LCS and A is a convex subset of $\mathscr{X}$, then* cl $A = $ wk $-$ cl A.

PROOF. If $\mathscr{T}$ is the original topology of $\mathscr{X}$, then wk $\subseteq \mathscr{T}$, hence cl $A \subseteq$ wk $-$ cl A. Conversely, if $x \in \mathscr{X} \backslash$ cl A, then (IV.3.13) implies that there is an x^* in $\mathscr{X}^*$, an α in $\mathbb{R}$, and an $\varepsilon > 0$ such that

$$\operatorname{Re}\langle a, x^* \rangle \leqslant \alpha < \alpha + \varepsilon \leqslant \operatorname{Re}\langle x, x^* \rangle$$

for all a in cl A. Hence cl $A \subseteq B \equiv \{ y \in \mathscr{X} : \operatorname{Re}\langle y, x^* \rangle \leqslant \alpha \}$. But B is clearly wk-closed since x^* is wk-continuous. Thus wk $-$ cl $A \subseteq B$. Since $x \notin B$, $x \notin$ wk $-$ cl A. ∎

1.5. Corollary. *A convex subset of $\mathscr{X}$ is closed if and only if it is weakly closed.*

There is a useful observation that can be made here. Because of (III.6.3) it can be shown that if $\mathscr{X}$ is a complex linear space, then the weak topology on $\mathscr{X}$ is the same as the weak topology it has if it is considered as a real linear space (Exercise 4). This will be used in the future.

1.6. Definition. If $A \subseteq \mathscr{X}$, the *polar* of A, denoted by A°, is the subset of $\mathscr{X}^*$ defined by

$$A^\circ \equiv \{ x^* \in \mathscr{X}^* : |\langle a, x^* \rangle| \leqslant 1 \text{ for all } a \text{ in } A \}.$$

If $B \subseteq \mathscr{X}^*$, the *prepolar* of B, denoted by $^\circ B$, is the subset of $\mathscr{X}$ defined by

$$^\circ B \equiv \{ x \in \mathscr{X} : |\langle x, b^* \rangle| \leqslant 1 \text{ for all } b^* \text{ in } B \}.$$

If $A \subseteq \mathscr{X}$ the *bipolar* of A is the set $^\circ(A^\circ)$. If there is no confusion, then it is also denoted by $^\circ A^\circ$.

The prototype for this idea is that if A is the unit ball in a normed space, A° is the unit ball in the dual space.

1.7. Proposition. *If $A \subseteq \mathscr{X}$, then*

(a) A° *is convex and balanced.*
(b) *If $A_1 \subseteq A$, then $A^\circ \subseteq A_1^\circ$.*
(c) *If $\alpha \in \mathbb{F}$ and $\alpha \neq 0$, $(\alpha A)^\circ = \alpha^{-1} A^\circ$.*
(d) $A \subseteq {}^\circ A^\circ$.
(e) $A^\circ = ({}^\circ A^\circ)^\circ$.

PROOF. The proofs of parts (a) through (d) are left as an exercise. To prove (e) note that $A \subseteq {}^\circ A^\circ$ by (d), so $({}^\circ A^\circ)^\circ \subseteq A^\circ$ by (b). But $A^\circ \subseteq {}^\circ(A^\circ)^\circ$ by an analog of (d) for prepolars. Also, $^\circ(A^\circ)^\circ = ({}^\circ A^\circ)^\circ$. ∎

There is an analogous result for prepolars. In fact, it is more than analogy that is at work here. By Theorem 1.3, $(\mathscr{X}^*, \text{wk}^*)^* = \mathscr{X}$. Thus the result for prepolars is a consequence of the preceding proposition.

If A is a linear manifold in $\mathscr{X}$ and $x^* \in A^\circ$, then $ta \in A$ for all $t > 0$ and a in A. So $1 \geqslant |\langle ta, x^* \rangle| = t|\langle a, x^* \rangle|$. Letting $t \to \infty$ show that $A^\circ = A^\perp$, where

$$A^\perp \equiv \{x^* \text{ in } \mathscr{X}^* : \langle a, x^* \rangle = 0 \text{ for all } a \text{ in } A\}.$$

Similarly, if B is a linear manifold in $\mathscr{X}^*$, $^\circ B = {}^\perp B$, where

$$^\perp B \equiv \{x \text{ in } \mathscr{X} : \langle x, b^* \rangle = 0 \text{ for all } b^* \text{ in } B\}.$$

The next result is a slight generalization of Corollary IV.3.12.

1.8. Bipolar Theorem. *If $\mathscr{X}$ is a LCS and $A \subseteq \mathscr{X}$, then $^\circ A^\circ$ is the closed convex balanced hull of A.*

PROOF. Let A_1 be the intersection of all closed convex balanced subsets of $\mathscr{X}$ that contain A. It must be shown that $A_1 = {}^\circ A^\circ$. Since $^\circ A^\circ$ is closed, convex, and balanced and $A \subseteq {}^\circ A^\circ$, it follows that $A_1 \subseteq {}^\circ A^\circ$.

Now assume that $x_0 \in \mathscr{X} \backslash A_1$. A_1 is a closed convex balanced set so by (IV.3.13) there is an x^* in $\mathscr{X}^*$, an α in $\mathbb{R}$, and an $\varepsilon > 0$ such that

$$\operatorname{Re}\langle a_1, x^* \rangle < \alpha < \alpha + \varepsilon < \operatorname{Re}\langle x_0, x^* \rangle$$

for all a_1 in A_1. Since $0 \in A_1$, $0 = \langle 0, x^* \rangle < \alpha$. By replacing x^* with $\alpha^{-1}x^*$ it follows that there is an $\varepsilon > 0$ (not the same as the first ε) such that

$$\operatorname{Re}\langle a_1, x^* \rangle < 1 < 1 + \varepsilon < \operatorname{Re}\langle x_0, x^* \rangle$$

for all a_1 in A_1. If $a_1 \in A_1$ and $\langle a_1, x^* \rangle = |\langle a_1, x^* \rangle| e^{-i\theta}$, then $e^{-i\theta} a_1 \in A_1$ and so

$$|\langle a_1, x^* \rangle| = \operatorname{Re}\langle e^{-i\theta} a_1, x^* \rangle < 1 < \operatorname{Re}\langle x_0, x^* \rangle$$

for all a_1 in A_1. Hence $x^* \in A_1^\circ$, and $x_0 \notin {}^\circ A^\circ$. That is, $\mathscr{X} \backslash A_1 \subseteq \mathscr{X} \backslash {}^\circ A^\circ$. ∎

1.9. Corollary. *If $\mathscr{X}$ is a LCS and $B \subseteq \mathscr{X}^*$, then $(^\circ B)^\circ$ is the wk* closed convex balanced hull of B.*

Using the weak and weak* topologies and the concept of a bounded subset of a LCS (IV.2.5), it is possible to rephrase the results associated with the Principle of Uniform Boundedness (§III.14). As an example we offer the following reformulation of Corollary III.14.5 (which is, in fact, the most general form of the result).

1.10. Theorem. *If $\mathscr{X}$ is a Banach space, $\mathscr{Y}$ is a normed space, and $\mathscr{A} \subseteq \mathscr{B}(\mathscr{X}, \mathscr{Y})$ such that for every x in $\mathscr{X}$, $\{Ax : A \in \mathscr{A}\}$ is weakly bounded in $\mathscr{Y}$, then $\mathscr{A}$ is norm bounded in $\mathscr{B}(\mathscr{X}, \mathscr{Y})$.*

EXERCISES

1. Show that wk is the smallest topology on $\mathscr{X}$ such that each x^* in $\mathscr{X}^*$ is continuous.

2. Show that wk* is the smallest topology on $\mathscr{X}^*$ such that for each x in $\mathscr{X}$, $x^* \mapsto \langle x, x^* \rangle$ is continuous.

3. Prove Theorem 1.3.

4. Let $\mathscr{X}$ be a complex LCS and let $\mathscr{X}_{\mathbb{R}}^*$ denote the collection of all continuous real linear functionals on $\mathscr{X}$. Use the elements of $\mathscr{X}_{\mathbb{R}}^*$ to define seminorms on $\mathscr{X}$ and let $\sigma(\mathscr{X}, \mathscr{X}_{\mathbb{R}}^*)$ be the corresponding topology. Show that $\sigma(\mathscr{X}, \mathscr{X}^*) = \sigma(\mathscr{X}, \mathscr{X}_{\mathbb{R}}^*)$.

5. Prove the remainder of Proposition 1.7.

6. If $A \subseteq \mathscr{X}$, show that A is weakly bounded if and only if A° is absorbing in $\mathscr{X}^*$.

7. Let $\mathscr{X}$ be a normed space and let $\{x_n\}$ be a sequence in $\mathscr{X}$ such that $x_n \to x$ weakly. Show that there is a sequence $\{y_n\}$ such that $y_n \in \operatorname{co}\{x_1, x_2, \ldots, x_n\}$ and $\|y_n - x\| \to 0$. (Hint: use Theorem 1.4.)

8. If $\mathscr{H}$ is a Hilbert space and $\{h_n\}$ is a sequence in $\mathscr{H}$ such that $h_n \to h$ weakly and $\|h_n\| \to \|h\|$, then $\|h_n - h\| \to 0$. (The same type of result is true for L^p-spaces if $1 < p < \infty$. See W.P. Novinger [1972].)

9. If $\mathscr{X}$ is a normed space show that the norm on $\mathscr{X}$ is lower semicontinuous for the weak topology and the norm of $\mathscr{X}^*$ is lower semicontinuous for the weak-star topology.

10. Suppose $\mathscr{X}$ is an infinite-dimensional normed space. If $S = \{x \in \mathscr{X}: \|x\| = 1\}$, then the weak closure of S is $\{x: \|x\| \leqslant 1\}$.

§2. The Dual of a Subspace and a Quotient Space

In §III.4 the quotient of a normed space $\mathscr{X}$ by a closed subspace $\mathscr{M}$ was defined and in (III.10.2) it was shown that the dual of a quotient space $\mathscr{X}/\mathscr{M}$ is isometrically isomorphic to $\mathscr{M}^\perp$. These results are generalized in this section to the setting of a LCS and, moreover, it is shown that when $(\mathscr{X}/\mathscr{M})^*$ and $\mathscr{M}^\perp$ are identified, the weak-star topology on $(\mathscr{X}/\mathscr{M})^*$ is precisely the relative weak-star topology that $\mathscr{M}^\perp$ receives as a subspace of $\mathscr{X}^*$.

The first result was presented in abbreviated form as Exercise IV.1.16.

2.1. Proposition. *If p is a seminorm on $\mathscr{X}$, $\mathscr{M}$ is a linear manifold in $\mathscr{X}$, and $\bar{p}: \mathscr{X}/\mathscr{M} \to [0, \infty)$ is defined by*

$$\bar{p}(x + \mathscr{M}) = \inf\{p(x + y): y \in \mathscr{M}\},$$

then $\bar{p}$ is a seminorm on $\mathscr{X}/\mathscr{M}$. If $\mathscr{X}$ is a locally convex space and $\mathscr{P}$ is the family of all continuous seminorms on $\mathscr{X}$, then the family $\bar{\mathscr{P}} \equiv \{\bar{p}: p \in \mathscr{P}\}$ defines the quotient topology on $\mathscr{X}/\mathscr{M}$.

PROOF. Exercise.

Thus if $\mathscr{X}$ is a LCS and $\mathscr{M} \leqslant \mathscr{X}$, then $\mathscr{X}/\mathscr{M}$ is a LCS. Let $f \in (\mathscr{X}/\mathscr{M})^*$. If $Q: \mathscr{X} \to \mathscr{X}/\mathscr{M}$ is the natural map, then $f \circ Q \in \mathscr{X}^*$. Moreover, $f \circ Q \in \mathscr{M}^\perp$. Hence $f \mapsto f \circ Q$ is a map of $(\mathscr{X}/\mathscr{M})^* \to \mathscr{M}^\perp \subseteq \mathscr{X}^*$.

2.2. Theorem. *If $\mathcal{X}$ is a LCS, $\mathcal{M} \leqslant \mathcal{X}$, and $Q: \mathcal{X} \to \mathcal{X}/\mathcal{M}$ is the natural map, then $f \mapsto f \circ Q$ defines a linear bijection between $(\mathcal{X}/\mathcal{M})^*$ and $\mathcal{M}^\perp$. If $(\mathcal{X}/\mathcal{M})^*$ has its weak-star topology $\sigma((\mathcal{X}/\mathcal{M})^*, \mathcal{X}/\mathcal{M})$ and $\mathcal{M}^\perp$ has the relative weak-star topology $\sigma(\mathcal{X}^*, \mathcal{X})|\mathcal{M}^\perp$, then this bijection is a homeomorphism. If $\mathcal{X}$ is a normed space, then this bijection is an isometry.*

PROOF. Let $\rho: (\mathcal{X}/\mathcal{M})^* \to \mathcal{M}^\perp$ be defined by $\rho(f) = f \circ Q$. It was shown prior to the statement of the theorem that ρ is well defined and maps $(\mathcal{X}/\mathcal{M})^*$ into $\mathcal{M}^\perp$. It is easy to see that ρ is linear and if $0 = \rho(f) = f \circ Q$, then $f = 0$ since Q is surjective. So ρ is injective. Now let $x^* \in \mathcal{M}^\perp$ and define $f: \mathcal{X}/\mathcal{M} \to \mathbb{F}$ by $f(x + \mathcal{M}) = \langle x, x^* \rangle$. Because $\mathcal{M} \subseteq \ker x^*$, f is well defined and linear. Also, $Q^{-1}\{x + \mathcal{M}: |f(x + \mathcal{M})| < 1\} = \{x \in \mathcal{X}: |\langle x, x^* \rangle| < 1\}$ and this is open in $\mathcal{X}$ since x^* is continuous. Thus $\{x + \mathcal{M}: |f(x + \mathcal{M})| < 1\}$ is open in $\mathcal{X}/\mathcal{M}$ and so f is continuous. Clearly $\rho(f) = x^*$, so ρ is a bijection.

If $\mathcal{X}$ is a normed space, it was shown in (III.10.2) that ρ is an isometry. It remains to show that ρ is a weak-star homeomorphism. Let $wk^* = \sigma(\mathcal{X}^*, \mathcal{X})$ and let $\sigma^* = \sigma((\mathcal{X}/\mathcal{M})^*, \mathcal{X}/\mathcal{M})$. If $\{f_i\}$ is a net in $(\mathcal{X}/\mathcal{M})^*$ and $f_i \to 0(\sigma^*)$, then for each x in $\mathcal{X}$, $\langle x, \rho(f_i) \rangle = f_i(Q(x)) \to 0$. Hence $\rho(f_i) \to 0(wk^*)$. Conversely, if $\rho(f_i) \to 0(wk^*)$, then for each x in $\mathcal{X}$, $f_i(x + \mathcal{M}) = \langle x, \rho(f_i) \rangle \to 0$; hence $f_i \to 0(\sigma^*)$. ∎

Once again let $\mathcal{M} \leqslant \mathcal{X}$. If $x^* \in \mathcal{X}^*$, then the restriction of $\mathcal{X}^*$ to $\mathcal{M}$, $x^*|\mathcal{M}$, belongs to $\mathcal{M}^*$. Also, the Hahn–Banach Theorem implies that the map $x^* \mapsto x^*|\mathcal{M}$ is surjective. If $\rho(x^*) = x^*|\mathcal{M}$, then $\rho: \mathcal{X}^* \to \mathcal{M}^*$ is clearly linear as well as surjective. It fails, however, to be injective. How does it fail? It's easy to see that $\ker \rho = \mathcal{M}^\perp$. Thus ρ induces a linear bijection $\tilde{\rho}: \mathcal{X}^*/\mathcal{M}^\perp \to \mathcal{M}^*$.

2.3. Theorem. *If $\mathcal{X}$ is a LCS, $\mathcal{M} \leqslant \mathcal{X}$, and $\rho: \mathcal{X}^* \to \mathcal{M}^*$ is the restriction map, then ρ induces a linear bijection $\tilde{\rho}: \mathcal{X}^*/\mathcal{M}^\perp \to \mathcal{M}^*$. If $\mathcal{X}^*/\mathcal{M}^\perp$ has the quotient topology induced by $\sigma(\mathcal{X}^*, \mathcal{X})$ and $\mathcal{M}^*$ has its weak-star topology $\sigma(\mathcal{M}^*, \mathcal{M})$, then $\tilde{\rho}$ is a homeomorphism. If $\mathcal{X}$ is a normed space, then $\tilde{\rho}$ is an isometry.*

PROOF. The fact that $\tilde{\rho}$ is an isometry when $\mathcal{X}$ is a normed space was shown in (III.10.1). Let $wk^* = \sigma(\mathcal{M}^*, \mathcal{M})$ and let η^* be the quotient topology on $\mathcal{X}^*/\mathcal{M}^\perp$ defined by $\sigma(\mathcal{X}^*, \mathcal{X})$. Let $Q: \mathcal{X}^* \to \mathcal{X}^*/\mathcal{M}^\perp$ be the natural map. Therefore the diagram

commutes. If $y \in \mathcal{M}$, then the commutativity of the diagram implies that

$$Q^{-1}(\tilde{\rho}^{-1}\{y^* \in \mathcal{M}^*: |\langle y, y^* \rangle| < 1\}) = Q^{-1}\{x^* + \mathcal{M}^\perp: |\langle y, x^* \rangle| < 1\}$$
$$= \{x^* \in \mathcal{X}^*: |\langle y, x^* \rangle| < 1\},$$

which is weak-star open in $\mathscr{X}^*$. Hence $\tilde{\rho}$: $(\mathscr{X}^*/\mathscr{M}^{\perp}, \eta^*) \to (\mathscr{M}^*, \mathrm{wk}^*)$ is continuous.

How is the topology on $\mathscr{X}^*/\mathscr{M}^{\perp}$ defined? If $x \in \mathscr{X}$, $p_x(x^*) = |\langle x, x^* \rangle|$ is a typical seminorm on $\mathscr{X}^*$. By Proposition 2.1, the topology on $\mathscr{X}^*/\mathscr{M}^{\perp}$ is defined by the seminorms $\{\bar{p}_x \colon x \in \mathscr{X}\}$, where

$$\bar{p}_x(x^* + \mathscr{M}^{\perp}) = \inf\{|\langle x, x^* + z^* \rangle| \colon z^* \in \mathscr{M}^{\perp}\}.$$

2.4. Claim. If $x \notin \mathscr{M}$, then $\bar{p}_x = 0$.

In fact, let $\mathscr{L} = \{\alpha x \colon \alpha \in \mathbb{F}\}$. If $x \notin \mathscr{M}$, then $\mathscr{L} \cap \mathscr{M} = (0)$. Since $\dim \mathscr{L} < \infty$, $\mathscr{M}$ is topologically complemented in $\mathscr{L} + \mathscr{M}$. Let $x^* \in \mathscr{X}^*$ and define f: $\mathscr{L} + \mathscr{M} \to \mathbb{F}$ by $f(\alpha x + y) = \langle y, x^* \rangle$ for y in $\mathscr{M}$ and α in $\mathbb{F}$. Because $\mathscr{M}$ is topologically complemented in $\mathscr{L} + \mathscr{M}$, if $\alpha_i x + y_i \to 0$, then $y_i \to 0$. Hence $f(\alpha_i x + y_i) = \langle y_i, x^* \rangle \to 0$. Thus f is continuous. By the Hahn–Banach Theorem, there is an x_1^* in $\mathscr{X}^*$ that extends f. Note that $x^* - x_1^* \in \mathscr{M}^{\perp}$. Thus $\bar{p}_x(x^* + \mathscr{M}^{\perp}) = \bar{p}_x(x_1^* + \mathscr{M}^{\perp}) \leqslant p_x(x_1^*) = |\langle x, x_1^* \rangle| = 0$. This proves (2.4).

Now suppose that $\{x_i^* + \mathscr{M}^{\perp}\}$ is a net in $\mathscr{X}^*/\mathscr{M}^{\perp}$ such that $\tilde{\rho}(x_i^* + \mathscr{M}^{\perp}) = x_i^* | \mathscr{M} \to 0(\mathrm{wk}^*)$ in $\mathscr{M}^*$. If $x \in \mathscr{X}$ and $x \notin \mathscr{M}$, the Claim (2.4) implies that $\bar{p}_x(x_i^* + \mathscr{M}^{\perp}) = 0$. If $x \in \mathscr{M}$, then $\bar{p}_x(x_i^* + \mathscr{M}^{\perp}) \leqslant |\langle x, x_i^* \rangle| \to 0$. Thus $x_i^* + \mathscr{M}^{\perp} \to 0(\eta^*)$ and $\tilde{\rho}$ is a weak-star homeomorphism. ∎

EXERCISES

1. In relation to Claim 2.4, show that if $\mathscr{L} \leqslant \mathscr{X}$, $\dim \mathscr{L} < \infty$, and $\mathscr{M} \leqslant \mathscr{X}$, then $\mathscr{L} + \mathscr{M}$ is closed.

2. Show that if $\mathscr{M} \leqslant \mathscr{X}$ and $\mathscr{M}$ is topologically complemented in $\mathscr{X}$, then $\mathscr{M}^{\perp}$ is topologically complemented in $\mathscr{X}^*$ and that its complement is weak-star and linearly homeomorphic to $\mathscr{X}^*/\mathscr{M}^{\perp}$.

§3. Alaoglu's Theorem

If $\mathscr{X}$ is any normed space, let's agree to denote by ball $\mathscr{X}$ the closed unit ball in $\mathscr{X}$. So ball $\mathscr{X} \equiv \{x \in \mathscr{X} \colon \|x\| \leqslant 1\}$.

3.1. Alaoglu's Theorem. *If $\mathscr{X}$ is a normed space, then ball $\mathscr{X}^*$ is weak-star compact.*

PROOF. For each x in ball $\mathscr{X}$, let $D_x \equiv \{\alpha \in \mathbb{F} \colon |\alpha| \leqslant 1\}$ and put $D = \prod\{D_x \colon x \in \mathrm{ball}\,\mathscr{X}\}$. By Tychonoff's Theorem, D is compact. Define τ: ball $\mathscr{X}^* \to D$ by

$$\tau(x^*)(x) = \langle x, x^* \rangle.$$

That is, $\tau(x^*)$ is the element of the product space D whose x coordinate is $\langle x, x^* \rangle$. It will be shown that τ is a homeomorphism from (ball $\mathscr{X}^*$, wk*)

onto $\tau(\text{ball } \mathscr{X}^*)$ with the relative topology from D, and that $\tau(\text{ball } \mathscr{X}^*)$ is closed in D. Thus it will follow that $\tau(\text{ball } \mathscr{X}^*)$, and hence ball $\mathscr{X}^*$, is compact.

To see that τ is injective, suppose that $\tau(x_1^*) = \tau(x_2^*)$. Then for each x in ball $\mathscr{X}$, $\langle x, x_1^* \rangle = \langle x, x_2^* \rangle$. It follows by definition that $x_1^* = x_2^*$.

Now let $\{x_i^*\}$ be a net in ball $\mathscr{X}^*$ such that $x_i^* \to x^*$. Then for each x in ball $\mathscr{X}$, $\tau(x_i^*)(x) = \langle x, x_i^* \rangle \to \langle x, x^* \rangle = \tau(x^*)(x)$. That is, each coordinate of $\{\tau(x_i^*)\}$ converges to $\tau(x^*)$. Hence $\tau(x_i^*) \to \tau(x^*)$ and τ is continuous.

Let x_i^* be a net in ball $\mathscr{X}^*$, let $f \in D$, and suppose $\tau(x_i^*) \to f$ in D. So $f(x) = \lim \langle x, x_i^* \rangle$ exists for every x in ball $\mathscr{X}$. If $x \in \mathscr{X}$, let $\alpha > 0$ such that $\| \alpha x \| \leq 1$. Then define $f(x) = \alpha^{-1} f(\alpha x)$. If also $\beta > 0$ such that $\| \beta x \| \leq 1$, then $\alpha^{-1} f(\alpha x) = \alpha^{-1} \lim \langle \alpha x, x_i^* \rangle = \beta^{-1} \lim \langle \beta x, x_i^* \rangle = \beta^{-1} f(\beta x)$. So $f(x)$ is well defined. It is left as an exercise for the reader to show that $f: \mathscr{X} \to \mathbb{F}$ is a linear functional. Also, if $\| x \| \leq 1$, $f(x) \in D_x$ so $|f(x)| \leq 1$. Thus $x^* \in \text{ball }$ $\mathscr{X}^*$ and $\tau(x^*) = f$. Thus $\tau(\text{ball } \mathscr{X}^*)$ is closed in D. This implies that $\tau(\text{ball } \mathscr{X}^*)$ is compact. The proof that τ^{-1} is continuous is left to the reader. ■

EXERCISES

1. Show that the functional f occurring in the proof of Alaoglu's Theorem is linear.

2. Let $\mathscr{X}$ be a LCS and let V be an open neighborhood of 0. Show that V° is weak-star compact in $\mathscr{X}^*$.

3. If $\mathscr{X}$ is a Banach space, show that there is a compact space X such that $\mathscr{X}$ is isometrically isomorphic to a closed subspace of $C(X)$.

§4. Reflexivity Revisited

In §III.11 a Banach space $\mathscr{X}$ was defined to be reflexive if the natural embedding of $\mathscr{X}$ into its double dual, $\mathscr{X}^{**}$, is surjective. Recall that if $x \in \mathscr{X}$, then the image of x in $\mathscr{X}^{**}$, $\hat{x}$, is defined by (using our recent notation)

$$\langle x^*, \hat{x} \rangle = \langle x, x^* \rangle$$

for all x^* in $\mathscr{X}^*$. Also recall that the map $x \mapsto \hat{x}$ is an isometry.

To begin, note that $\mathscr{X}^{**}$, being the dual space of $\mathscr{X}^*$, has its weak-star topology $\sigma(\mathscr{X}^{**}, \mathscr{X}^*)$. Also note that if $\mathscr{X}$ is considered as a subspace of $\mathscr{X}^{**}$, then the topology $\sigma(\mathscr{X}^{**}, \mathscr{X}^*)$ when relativized to $\mathscr{X}$ is $\sigma(\mathscr{X}, \mathscr{X}^*)$, the weak topology on $\mathscr{X}$. This will be important later when it is combined with Alaoglu's Theorem applied to $\mathscr{X}^{**}$ in the discussion of reflexivity. But now the next result must occupy us.

4.1. Proposition. *If $\mathscr{X}$ is a normed space, then ball $\mathscr{X}$ is $\sigma(\mathscr{X}^{**}, \mathscr{X}^*)$ dense in ball $\mathscr{X}^{**}$.*

PROOF. Let $B = $ the $\sigma(\mathscr{X}^{**}, \mathscr{X}^*)$ closure of ball $\mathscr{X}$ in $\mathscr{X}^{**}$; clearly, $B \subseteq $ ball $\mathscr{X}^{**}$. If there is an x_0^{**} in ball $\mathscr{X}^{**} \backslash B$, then the Hahn–Banach Theorem implies there is an x^* in $\mathscr{X}^*$, an α in $\mathbb{R}$, and an $\varepsilon > 0$ such that

$$\operatorname{Re}\langle x, x^* \rangle < \alpha < \alpha + \varepsilon < \operatorname{Re}\langle x^*, x_0^{**} \rangle$$

for all x in ball $\mathscr{X}$. (Exactly how does the Hahn–Banach Theorem imply this?) Since $0 \in$ ball $\mathscr{X}$, $0 < \alpha$. Dividing by α and replacing x^* by $\alpha^{-1} x^*$, it may be assumed that there is an x^* in $\mathscr{X}^*$ and an $\varepsilon > 0$ such that

$$\operatorname{Re}\langle x, x^* \rangle < 1 < 1 + \varepsilon < \operatorname{Re}\langle x^*, x_0^{**} \rangle$$

for all x in ball $\mathscr{X}$. Since $e^{i\theta} x \in$ ball $\mathscr{X}$ whenever $x \in$ ball $\mathscr{X}$, this implies that $|\langle x, x^* \rangle| \leqslant 1$ if $\|x\| \leqslant 1$. Hence $x^* \in$ ball $\mathscr{X}^*$. But then $1 + \varepsilon < \operatorname{Re}\langle x^*, x_0^{**} \rangle \leqslant |\langle x^*, x_0^{**} \rangle| \leqslant \|x_0^{**}\| \leqslant 1$, a contradiction. ∎

4.2. Theorem. *If $\mathscr{X}$ is a Banach space, the following statements are equivalent.*

(a) $\mathscr{X}$ *is reflexive.*
(b) $\mathscr{X}^*$ *is reflexive.*
(c) $\sigma(\mathscr{X}^*, \mathscr{X}) = \sigma(\mathscr{X}^*, \mathscr{X}^{**})$.
(d) ball $\mathscr{X}$ *is weakly compact.*

PROOF. (a)$\Rightarrow$(c): This is clear since $\mathscr{X} = \mathscr{X}^{**}$.

(d)$\Rightarrow$(a): Note that $\sigma(\mathscr{X}^{**}, \mathscr{X}^*) | \mathscr{X} = \sigma(\mathscr{X}, \mathscr{X}^*)$. By (d), ball $\mathscr{X}$ is $\sigma(\mathscr{X}^{**}, \mathscr{X}^*)$ closed in ball $\mathscr{X}^{**}$. But the preceding proposition implies ball $\mathscr{X}$ is $\sigma(\mathscr{X}^{**}, \mathscr{X}^*)$ dense in ball $\mathscr{X}^{**}$. Hence ball $\mathscr{X} =$ ball $\mathscr{X}^{**}$ and so $\mathscr{X}$ is reflexive.

(c)$\Rightarrow$(b): By Alaoglu's Theorem, ball $\mathscr{X}^*$ is $\sigma(\mathscr{X}^*, \mathscr{X})$-compact. By (c), ball $\mathscr{X}^*$ is $\sigma(\mathscr{X}^*, \mathscr{X}^{**})$ compact. Since it has already been shown that (d) implies (a), this implies that $\mathscr{X}^*$ is reflexive.

(b)$\Rightarrow$(a): Now ball $\mathscr{X}$ is norm closed in $\mathscr{X}^{**}$; hence ball $\mathscr{X}$ is $\sigma(\mathscr{X}^{**}, \mathscr{X}^{***})$ closed in $\mathscr{X}^{**}$ (Corollary 1.5). Since $\mathscr{X}^* = \mathscr{X}^{***}$ by (b), this says that ball $\mathscr{X}$ is $\sigma(\mathscr{X}^{**}, \mathscr{X}^*)$ closed in $\mathscr{X}^{**}$. But, according to (4.1), ball $\mathscr{X}$ is $\sigma(\mathscr{X}^{**}, \mathscr{X}^*)$ dense in ball $\mathscr{X}^{**}$. Hence ball $\mathscr{X} =$ ball $\mathscr{X}^{**}$ and $\mathscr{X}$ is reflexive.

(a)$\Rightarrow$(d): By Alaoglu's Theorem, ball $\mathscr{X}^{**}$ is $\sigma(\mathscr{X}^{**}, \mathscr{X}^*)$ compact. Since $\mathscr{X} = \mathscr{X}^{**}$, this says that ball $\mathscr{X}$ is $\sigma(\mathscr{X}, \mathscr{X}^*)$ compact. ∎

4.3. Corollary. *If $\mathscr{X}$ is a reflexive Banach space and $\mathscr{M} \leqslant \mathscr{X}$, then $\mathscr{M}$ is a reflexive Banach space.*

PROOF. Note that ball $\mathscr{M} = \mathscr{M} \cap [\text{ball } \mathscr{X}]$, so ball $\mathscr{M}$ is $\sigma(\mathscr{X}, \mathscr{X}^*)$ compact. It remains to show that $\sigma(\mathscr{X}, \mathscr{X}^*) | \mathscr{M} = \sigma(\mathscr{M}, \mathscr{M}^*)$. But this follows by (2.3). (How?) ∎

Call a sequence $\{x_n\}$ in $\mathscr{X}$ a *weakly Cauchy sequence* if for every x^* in $\mathscr{X}^*$, $\{\langle x_n, x^* \rangle\}$ is a Cauchy sequence in $\mathbb{F}$.

4.4. Corollary. *If $\mathscr{X}$ is reflexive, then every weakly Cauchy sequence in $\mathscr{X}$ converges weakly. That is, $\mathscr{X}$ is weakly sequentially complete.*

PROOF. Since $\{\langle x_n, x^* \rangle\}$ is a Cauchy sequence in $\mathbb{F}$ for each x^* in $\mathscr{X}^*$, $\{x_n\}$

is weakly bounded. By the PUB there is a constant M such that $\|x_n\| \leqslant M$ for all $n \geqslant 1$. But $\{x \in \mathscr{X}: \|x\| \leqslant M\}$ is weakly compact since $\mathscr{X}$ is reflexive. Thus there is an x in $\mathscr{X}$ such that $x_n \xrightarrow{\text{cl}} x$ weakly. But for each x^* in $\mathscr{X}^*$, $\lim \langle x_n, x^* \rangle$ exists. Hence $\langle x_n, x^* \rangle \to \langle x, x^* \rangle$, so $x_n \to x$ weakly. ∎

Not all Banach spaces are weakly sequentially complete.

4.5. Example. $C[0,1]$ is not weakly sequentially complete. In fact, let $f_n(t) = (1 - nt)$ if $0 \leqslant t \leqslant 1/n$ and $f_n(t) = 0$ if $1/n \leqslant t \leqslant 1$. If $\mu \in M[0,1]$, then $\int f_n d\mu \to \mu(\{0\})$ by the Monotone Convergence Theorem. Hence $\{f_n\}$ is a weakly Cauchy sequence. However, $\{f_n\}$ does not converge weakly to any continuous function on $[0,1]$.

4.6. Corollary. *If $\mathscr{X}$ is a reflexive Banach space, $\mathcal{M} \leqslant \mathscr{X}$, and $x_0 \in \mathscr{X} \setminus \mathcal{M}$, then there is a point y_0 in $\mathcal{M}$ such that $\|x_0 - y_0\| = \text{dist}(x_0, \mathcal{M})$.*

PROOF. $x \mapsto \|x - x_0\|$ is weakly lower semicontinuous (Exercise 1.9). If $d = \text{dist}(x_0, \mathcal{M})$, then $\mathcal{M} \cap \{x: \|x - x_0\| \leqslant 2d\}$ is weakly compact and a lower semicontinuous function attains its minimum on a compact set. ∎

It is not generally true that the distance from a point to a linear subspace is attained. If $\mathcal{M} \subseteq \mathscr{X}$, call $\mathcal{M}$ *proximinal* if for every x in $\mathscr{X}$ there is a y in $\mathcal{M}$ such that $\|x - y\| = \text{dist}(x, \mathcal{M})$. So if $\mathscr{X}$ is reflexive, Corollary 4.6 implies that every closed linear subspace of $\mathscr{X}$ is proximinal. If $\mathscr{X}$ is any Banach space and $\mathcal{M}$ is a finite dimensional subspace, then it is easy to see that $\mathcal{M}$ is proximinal. How about if $\dim(\mathscr{X}/\mathcal{M}) < \infty$?

4.7. Proposition. *If $\mathscr{X}$ is a Banach space and $x^* \in \mathscr{X}^*$, then $\ker x^*$ is proximinal if and only if there is an x in $\mathscr{X}$, $\|x\| = 1$, such that $\langle x, x^* \rangle = \|x^*\|$.*

PROOF. Let $\mathcal{M} = \ker x^*$ and suppose that $\mathcal{M}$ is proximinal. If $f: \mathscr{X}/\mathcal{M} \to \mathbb{F}$ is defined by $f(x + \mathcal{M}) = \langle x, x^* \rangle$, then f is a linear functional and $\|f\| = \|x^*\|$. Since $\dim \mathscr{X}/\mathcal{M} = 1$, there is an x in $\mathscr{X}$ such that $\|x + \mathcal{M}\| = 1$ and $f(x + \mathcal{M}) = \|f\|$. Because $\mathcal{M}$ is proximinal, there is a y in $\mathcal{M}$ such that $1 = \|x + \mathcal{M}\| = \|x + y\|$. Thus $\langle x + y, x^* \rangle = \langle x, x^* \rangle = f(x + \mathcal{M}) = \|f\| = \|x^*\|$.

Now assume that there is an x_0 in $\mathscr{X}$ such that $\|x_0\| = 1$ and $\langle x_0, x^* \rangle = \|x^*\|$. If $x \in \mathscr{X}$ and $\|x + \mathcal{M}\| = \alpha > 0$, then $\|\alpha^{-1} x + \mathcal{M}\| = 1$. But also $\|x_0 + \mathcal{M}\| = 1$. (Why?) Since $\dim \mathscr{X}/\mathcal{M} = 1$, there is a β in $\mathbb{F}$, $|\beta| = 1$, such that $\alpha^{-1} x + \mathcal{M} = \beta(x_0 + \mathcal{M})$. Hence $\alpha^{-1} x - \beta x_0 \in \mathcal{M}$, or, equivalently, $x - \alpha\beta x_0 \in \mathcal{M}$. However, $\|x - (x - \alpha\beta x_0)\| = \|\alpha\beta x_0\| = \alpha = \text{dist}(x, \mathcal{M})$. So the distance from x to $\mathcal{M}$ is attained at $x - \alpha\beta x_0$. ∎

4.8. Example. If $L: C[0,1] \to \mathbb{F}$ is defined by

$$L(f) = \int_0^{1/2} f(x)dx - \int_{1/2}^1 f(x)dx,$$

then ker L is not proximinal.

There is a result in James [1964b] that states that a Banach space is reflexive if and only if every closed hyperplane is proximinal. This result is very deep. A nice reference on reflexivity is Yang [1967].

EXERCISES

1. Show that if $\mathscr{X}$ is reflexive and $\mathscr{M} \leqslant \mathscr{X}$, then $\mathscr{X}/\mathscr{M}$ is reflexive.

2. If $\mathscr{X}$ is a Banach space, $\mathscr{M} \leqslant \mathscr{X}$, and both $\mathscr{M}$ and $\mathscr{X}/\mathscr{M}$ are reflexive, must $\mathscr{X}$ be reflexive?

3. If (X, Ω, μ) is a σ-finite measure space, show that $L^1(X, \Omega, \mu)$ is reflexive if and only if it is finite dimensional.

4. Give the details of the proofs of the statements made in Example 4.5.

5. Verify the statement made in Example 4.8.

6. If (X, Ω, μ) is a σ-finite measure space, show that $L^\infty(\mu)$ is weak-star sequentially complete but is reflexive if and only if it is finite dimensional.

7. Let X be compact and suppose there is a norm on $C(X)$ that is given by an inner product making $C(X)$ into a Hilbert space such that for every x in X the functional $f \mapsto f(x)$ on $C(X)$ is continuous with respect to the Hilbert space norm. Show that X is finite.

§5. Separability and Metrizability

The weak and weak-star topologies on an infinite dimensional Banach space are never metrizable. It is possible, however, to show that under certain conditions these topologies are metrizable when restricted to bounded sets. In applications this is often sufficient.

5.1. Theorem. *If $\mathscr{X}$ is a Banach space, then ball $\mathscr{X}^*$ is weak-star metrizable if and only if $\mathscr{X}$ is separable.*

PROOF. Assume that $\mathscr{X}$ is separable and let $\{x_n\}$ be a countable dense subset of ball $\mathscr{X}$. For each n let $D_n = \{\alpha \in \mathbb{F} : |\alpha| \leqslant 1\}$. Put $X = \prod_{n=1}^\infty D_n$; X is a compact metric space. So if (ball $\mathscr{X}^*$, wk*) is homeomorphic to a subset of X, ball $\mathscr{X}^*$ is weak-star metrizable.

Define τ: ball $\mathscr{X}^* \to X$ by $\tau(x^*) = \{\langle x_n, x^* \rangle\}$. If $\{x_i^*\}$ is a net in ball $\mathscr{X}^*$ and $x_i^* \to x^*$ (wk*), then for each $n \geqslant 1$, $\langle x_n, x_i^* \rangle \to \langle x_n, x^* \rangle$; hence $\tau(x_i^*) \to \tau(x^*)$ and τ is continuous. If $\tau(x^*) = \tau(y^*)$, $\langle x_n, x^* - y^* \rangle = 0$ for all n. Since $\{x_n\}$ is dense, $x^* - y^* = 0$. Thus τ is injective. Since ball $\mathscr{X}^*$ is wk* compact, τ is a homeomorphism onto its image (A.2.8) and ball $\mathscr{X}^*$ is wk* metrizable.

Now assume that (ball $\mathscr{X}^*$, wk*) is metrizable. Thus there are open sets $\{U_n : n \geqslant 1\}$ in (ball $\mathscr{X}^*$, wk*) such that $0 \in U_n$ and $\bigcap_{n=1}^\infty U_n = (0)$. By the definition of the relative weak-star topology on ball $\mathscr{X}^*$, for each n there is a finite set F_n contained in $\mathscr{X}$ such that $\{x^* \in \text{ball } \mathscr{X}^*: |\langle x, x^* \rangle| < 1 \text{ for all } x \text{ in } F_n\} \subseteq U_n$. Let $F = \bigcup_{n=1}^{\infty} F_n$; so F is countable. Also, $^{\perp}(F^{\perp})$ is the closed linear span of F and this subspace of $\mathscr{X}$ is separable. But if $x^* \in F^{\perp}$, then for each $n \geq 1$ and for each x in F_n, $|\langle x, x^*/\|x^*\| \rangle| = 0 < 1$. Hence $x^*/\|x^*\| \in U_n$ for all $n \geq 1$; thus $x^* = 0$. Since $F^{\perp} = (0)$, $^{\perp}(F^{\perp}) = \mathscr{X}$ and $\mathscr{X}$ must be separable. ∎

Is there a corresponding result for the weak topology? If $\mathscr{X}^*$ is separable, then the weak topology on ball $\mathscr{X}$ is metrizable. In fact, this follows from Theorem 5.1 if the embedding of $\mathscr{X}$ into $\mathscr{X}^{**}$ is considered. This result is not very useful since there are few examples of Banach spaces $\mathscr{X}$ such that $\mathscr{X}^*$ is separable. Of course if $\mathscr{X}$ is separable and reflexive, then $\mathscr{X}^*$ is separable (Exercise 3), but in this case the weak topology on $\mathscr{X}$ is the same as its weak-star topology when $\mathscr{X}$ is identified with $\mathscr{X}^{**}$. Thus (5.1) is adequate for a discussion of the weak topology on the unit ball of a separable reflexive space. If $\mathscr{X} = c_0$, then $\mathscr{X}^* = l^1$ and this is separable but not reflexive. This is one of the few nonreflexive spaces with a separable dual space.

If $\mathscr{X}$ is separable, is (ball $\mathscr{X}$, wk) metrizable? The answer is no, as the following result of Schur demonstrates.

5.2. Proposition. *If a sequence in l^1 converges weakly, it converges in norm.*

PROOF. Recall that $l^{\infty} = (l^1)^*$. Since l^1 is separable, Theorem 5.1 implies that ball l^{∞} is wk* metrizable. By Alaoglu's Theorem, ball l^{∞} is wk* compact. Hence (ball l^{∞}, wk*) is a complete metric space and the Baire Category Theorem is applicable.

Let $\{f_n\}$ be a sequence of elements in l^1 such that $f_n \to 0$ weakly and let $\varepsilon > 0$. For each positive integer m let

$$F_m = \{\phi \in \text{ball } l^{\infty}: |\langle f_n, \phi \rangle| \leq \varepsilon/3 \text{ for } n \geq m\}.$$

It is easy to see that F_m is wk* closed in ball l^{∞} and, because $f_n \to 0 \,(\text{wk})$, $\bigcup_{m=1}^{\infty} F_m = \text{ball } l^{\infty}$. By the theorem of Baire, there is an F_m with non-empty weak-star interior.

An equivalent metric on (ball l^{∞}, wk*) is given by

$$d(\phi, \psi) = \sum_{j=1}^{\infty} 2^{-j} |\phi(j) - \psi(j)|$$

(see Exercise 4). Since F_m has a nonempty wk* interior, there is a ϕ in F_m and a $\delta > 0$ such that $\{\psi \in \text{ball } l^{\infty}: d(\phi, \psi) < \delta\} \subseteq F_m$. Let $J \geq 1$ such that $2^{-(J-1)} < \delta$. Fix $n \geq m$ and define ψ in l^{∞} by $\psi(j) = \phi(j)$ for $1 \leq j \leq J$ and $\psi(j) = \text{sign}(f_n(j))$ for $j > J$. Thus $\psi(j) f_n(j) = |f_n(j)|$ for $j > J$. It is easy to see that $\psi \in \text{ball } l^{\infty}$. Also, $d(\phi, \psi) = \sum_{j=J+1}^{\infty} 2^{-j} |\phi(j) - \psi(j)| \leq 2 \cdot 2^{-J} = 2^{-(J-1)} < \delta$.

So $\psi \in F_m$ and hence $|\langle \psi, f_m \rangle| \leqslant \varepsilon/3$ for $n \geqslant m$. Thus

5.3
$$\left| \sum_{j=1}^{J} \phi(j) f_n(j) + \sum_{j=J+1}^{\infty} |f_n(j)| \right| \leqslant \frac{\varepsilon}{3}$$

for $n \geqslant m$. But there is an $m_1 \geqslant m$ such that for $n \geqslant m_1$, $\sum_{j=1}^{J} |f_n(j)| < \varepsilon/3$. (Why?) Combining this with (5.3) gives that

$$\| f_n \| = \sum_{j=1}^{\infty} |f_n(j)|$$

$$< \frac{\varepsilon}{3} + \left| \sum_{j=J+1}^{\infty} |f_n(j)| + \sum_{j=1}^{J} \phi(j) f_n(j) \right| + \left| \sum_{j=1}^{J} \phi(j) f_n(j) \right|$$

$$< \frac{2\varepsilon}{3} + \sum_{j=1}^{J} |f_n(j)|$$

$$< \varepsilon$$

whenever $n \geqslant m_1$.   ∎

So if (ball l^1, wk) were metrizable, the preceding proposition would say that the weak and norm topologies on l^1 agree. But this is not the case (Exercise 1.10).

Also, note that the preceding result demonstrates in a dramatic way that in discussions concerning the weak topology it is essential to consider nets and not just sequences.

A proof of (5.2) that avoids the Baire Category Theorem can be found in Banach [1955], p. 218.

EXERCISES

1. Let $B =$ ball $M[0, 1]$ and for μ, v in $M[0, 1]$ define $d(\mu, v) = \sum_{n=0}^{\infty} 2^{-n} |\int_0^1 x^n d\mu - \int_0^1 x^n dv|$. Show that d is a metric on $M[0, 1]$ that defines the weak-star topology on B but not on $M[0, 1]$.

2. Let X be a compact space and let $\mathscr{U} = \{(U, V): U, V \text{ are open subsets of } X \text{ and } \text{cl } U \subseteq V\}$. For $u = (U, V)$ in $\mathscr{U}$, let $f_u: X \to [0, 1]$ be a continuous function such that $f_u \equiv 1$ on cl U and $f_u \equiv 0$ on $X \backslash V$. Show: (a) the linear span of $\{f_u: u \in \mathscr{U}\}$ is dense in $C(X)$; (b) if X is a metric space, then $C(X)$ is separable; (c) if X is a σ-compact metrizable locally compact space, then $C_0(X)$ is separable. (X is σ-compact if X is the union of a countable number of compact subsets.)

3. If $\mathscr{X}$ is a Banach space and $\mathscr{X}^*$ is separable, show that (a) $\mathscr{X}$ is separable; (b) if K is a weakly compact subset of $\mathscr{X}$, then K with the relative weak topology is metrizable.

4. If $B =$ ball l^{∞}, show that $d(\phi, \psi) = \sum_{j=1}^{\infty} 2^{-j} |\phi(j) - \psi(j)|$ defines a metric on B and that this metric defines the weak-star topology on B.

5. Use the type of argument used in the proof of the Principle of Uniform Boundedness to obtain a proof of Proposition 5.2 that does not need the Baire Category Theorem.

6. Show that Proposition 5.2 fails for l^p if $1 < p < \infty$.

§6*. An Application: The Stone–Čech Compactification

Let X be any topological space and consider the Banach space $C_b(X)$. Unless some assumption is made regarding X, it may be that $C_b(X)$ is "very small." If, for example, it is assumed that X is completely regular, then $C_b(X)$ has many elements. The next result says that this assumption is also necessary in order for $C_b(X)$ to be "large." But first, here is some notation.

If $x \in X$, let $\delta_x \colon C_b(X) \to \mathbb{F}$ be defined by $\delta_x(f) = f(x)$ for every f in $C_b(X)$. It is easy to see that $\delta_x \in C_b(X)^*$ and $\|\delta_x\| = 1$. Let $\Delta \colon X \to C_b(X)^*$ be defined by $\Delta(x) = \delta_x$. If $\{x_i\}$ is a net in X and $x_i \to x$, then $f(x_i) \to f(x)$ for every f in $C_b(X)$. This says that $\delta_{x_i} \to \delta_x$ (wk*) in $C_b(X)^*$. Hence $\Delta \colon X \to (C_b(X)^*, \text{wk*})$ is continuous. Is Δ a homeomorphism of X onto $\Delta(X)$?

6.1. Proposition. *The map* $\Delta \colon X \to (\Delta(X), \text{wk*})$ *is a homeomorphism if and only if* X *is completely regular.*

PROOF. Assume X is completely regular. If $x_1 \neq x_2$, then there is an f in $C_b(X)$ such that $f(x_1) = 1$ and $f(x_2) = 0$; thus $\delta_{x_1}(f) \neq \delta_{x_2}(f)$. Hence Δ is injective. To show that $\Delta \colon X \to (\Delta(X), \text{wk*})$ is an open map, let U be an open subset of X and let $x_0 \in U$. Since X is completely regular, there is an f in $C_b(X)$ such that $f(x_0) = 1$ and $f \equiv 0$ on $X \backslash U$. Let $V_1 = \{\mu \in C_b(X)^* \colon \langle f, \mu \rangle > 0\}$. Then V_1 is wk* open in $C_b(X)^*$ and $V_1 \cap \Delta(X) = \{\delta_x \colon f(x) > 0\}$. So if $V = V_1 \cap \Delta(X)$, V is wk* open in $\Delta(X)$ and $\delta_{x_0} \in V \subseteq \Delta(U)$. Since x_0 was arbitrary, $\Delta(U)$ is open in $\Delta(X)$. Therefore $\Delta \colon X \to (\Delta(X), \text{wk*})$ is a homeomorphism.

Now assume that Δ is a homeomorphism onto its image. Since $(\text{ball } C_b(X)^*, \text{wk*})$ is a compact space, it is completely regular. Since $\Delta(X) \subseteq \text{ball } C_b(X)^*$, $\Delta(X)$ is completely regular (Exercise 2). Thus X is completely regular. ∎

6.2. Stone–Čech Compactification. *If* X *is completely regular, then there is a compact space* βX *such that:*

(a) *there is a continuous map* $\Delta \colon X \to \beta X$ *with the property that* $\Delta \colon X \to \Delta(X)$ *is a homeomorphism;*

(b) $\Delta(X)$ *is dense in* βX;

(c) *if* $f \in C_b(X)$, *then there is a continuous map* $f^\beta \colon \beta X \to \mathbb{F}$ *such that* $f^\beta \circ \Delta = f$.

Moreover, if Ω *is a compact space having these properties, then* Ω *is homeomorphic to* βX.

PROOF. Let $\Delta \colon X \to C_b(X)^*$ be the map defined by $\Delta(x) = \delta_x$ and let $\beta X =$ the weak-star closure of $\Delta(X)$ in $C_b(X)^*$. By Alaoglu's Theorem and the fact that $\|\delta_x\| = 1$ for all x, βX is compact. By the preceding proposition, (a) holds. Part (b) is true by definition. It remains to show (c).

Fix f in $C_b(X)$ and define $f^\beta\colon \beta X \to \mathbb{F}$ by $f^\beta(\tau) = \langle f, \tau \rangle$ for every τ in βX. [Remember that $\beta X \subseteq C_b(X)^*$, so that this makes sense.] Clearly f^β is continuous and $f^\beta \circ \Delta(x) = f^\beta(\delta_x) = \langle f, \delta_x \rangle = f(x)$. So $f^\beta \circ \Delta = f$ and (c) holds.

To show that βX is unique, assume that Ω is a compact space and $\pi\colon X \to \Omega$ is a continuous map such that:

(a') $\pi\colon X \to \pi(X)$ is a homeomorphism;
(b') $\pi(X)$ is dense in Ω;
(c') if $f \in C_b(X)$, there is an $\tilde{f}$ in $C(\Omega)$ such that $\tilde{f} \circ \pi = f$.

Define $g\colon \Delta(X) \to \Omega$ by $g(\Delta(x)) = \pi(x)$. In other words, $g = \pi \circ \Delta^{-1}$. The idea is to extend g to a homeomorphism of βX onto Ω. If $\tau_0 \in \beta X$, then (b) implies that there is a net $\{x_i\}$ in X such that $\Delta(x_i) \to \tau_0$ in βX. Now $\{\pi(x_i)\}$ is a net in Ω and since Ω is compact, there is an ω_0 in Ω such that $\pi(x_i) \xrightarrow{\text{cl}} \omega_0$. If $F \in C(\Omega)$, let $f = F \circ \pi$; so $f \in C_b(X)$ (and $F = \tilde{f}$). Also, $f(x_i) = \langle f, \delta_{x_i} \rangle \to \langle f, \tau_0 \rangle = f^\beta(\tau_0)$. But it is also true that $f(x_i) = F(\pi(x_i)) \xrightarrow{\text{cl}} F(\omega_0)$. Hence $F(\omega_0) = f^\beta(\tau_0)$ for any F in $C(\Omega)$. This implies that ω_0 is the unique cluster point of $\{\pi(x_i)\}$; thus $\pi(x_i) \to \omega_0$ (A.2.7). Let $g(\tau_0) = \omega_0$. It must be shown that the definition of $g(\tau_0)$ does not depend on the net $\{x_i\}$ in X such that $\Delta(x_i) \to \tau_0$. This is left as an exercise. To summarize, it has been shown that

6.3   There is a function $g\colon \beta X \to \Omega$
such that if $f \in C_b(X)$, then $f^\beta = \tilde{f} \circ g$.

To show that $g\colon \beta X \to \Omega$ is continuous, let $\{\tau_i\}$ be a net in βX such that $\tau_i \to \tau$. If $F \in C(\Omega)$, let $f = F \circ \pi$; so $f \in C_b(X)$ and $\tilde{f} = F$. Also, $f^\beta(\tau_i) \to f^\beta(\tau)$. But $F(g(\tau_i)) = f^\beta(\tau_i) \to f^\beta(\tau) = F(g(\tau))$. It follows (6.1) that $g(\tau_i) \to g(\tau)$ in Ω. Thus g is continuous.

It is left as an exercise for the reader to show that g is injective. Since $g(\beta X) \supseteq g(\Delta(X)) = \pi(X)$, $g(\beta X)$ is dense in Ω. But $g(\beta X)$ is compact, so g is bijective. By (A.2.8), g is a homeomorphism. ∎

The compact set βX obtained in the preceding theorem is called the *Stone–Čech compactification* of X. By properties (a) and (b), X can be considered as a dense subset of βX and the map Δ can be taken to be the inclusion map. With this convention, (c) can be interpreted as saying that every bounded continuous function on X has a continuous extension to βX.

The space βX is usually very much larger than X. In particular, it is almost never true that βX is the one-point compactification of X. For example, if $X = (0, 1]$, then the one-point compactification of X is $[0, 1]$. However, $\sin(1/x) \in C_b(X)$ but it has no continuous extension to $[0, 1]$, so $\beta X \neq [0, 1]$.

To obtain an idea of how large $\beta X \setminus X$ is, see Exercise 6, which indicates how to show that if $\mathbb{N}$ has the diserete topology, then $\beta \mathbb{N} \setminus \mathbb{N}$ has $2^{\aleph_0}$ pairwise disjoint open sets. The best source of information on the

Stone–Čech compactification is the book by Gillman and Jerison [1960], though the approach to βX is somewhat different there than here. Two recent works on the Stone–Čech compactification are Johnstone [1982] and Walker [1974].

6.4. Corollary. *if X is completely regular and $\mu \in M(\beta X)$, define $L_\mu: C_b(X) \to \mathbb{F}$ by*

$$L_\mu(f) = \int_{\beta X} f^\beta d\mu$$

for each f in $C_b(X)$. Then the map $\mu \mapsto L_\mu$ is an isometric isomorphism of $M(\beta X)$ onto $C_b(X)^*$.

PROOF. Define $V: C_b(X) \to C(\beta X)$ by $Vf = f^\beta$. It is easy to see that V is linear. Considering X as a subset of βX, the fact that $\beta X = \operatorname{cl} X$ implies that V is an isometry. If $g \in C(\beta X)$ and $f = g|X$, then $g = f^\beta = Vf$; hence V is surjective.

If $\mu \in M(\beta X) = C(\beta X)^*$, it is easy to check that $L_\mu \in C_b(X)^*$ and $\|L_\mu\| = \|\mu\|$ since V is an isometry. Conversely, if $L \in C_b(X)^*$, then $L \circ V^{-1} \in C(\beta X)^*$ and $\|L \circ V^{-1}\| = \|L\|$. Hence there is a μ in $M(\beta X)$ such that $\int fg\, d\mu = L \circ V^{-1}(g)$ for every g in $C(\beta X)$. Since $V^{-1}g = |X$, it follows that $L = L_\mu$. ∎

The next result is from topology. It may be known to the reader, but it is presented here for the convenience of those to whom it is not.

6.5. Partition of Unity. *If X is normal and $\{U_1, \ldots, U_n\}$ is an open covering of X, then there are continuous functions $f_1, \ldots, f_n$ from X into $[0, 1]$ such that*

(a) $\sum_{k=1}^n f_k(x) = 1$ *for all x in X;*
(b) $f_k(x) = 0$ *for x in $X \backslash U_k$ and $1 \leqslant k \leqslant n$.*

PROOF. First observe that it may be assumed that $\{U_1, \ldots, U_n\}$ has no proper subcover. The proof now proceeds by induction.

If $n = 1$, let $f_1 \equiv 1$. Suppose $n = 2$. Then $X \backslash U_1$ and $X \backslash U_2$ are disjoint closed subsets of X. By Urysohn's Lemma there is a continuous function $f_1: X \to [0, 1]$ such that $f_1(x) = 0$ for x in $X \backslash U_1$ and $f_1(x) = 1$ for x in $X \backslash U_2$. Let $f_2 = 1 - f_1$ and the proof of this case is complete.

Now suppose the theorem has been proved for some $n \geqslant 2$ and $\{U_1, \ldots, U_{n+1}\}$ is an open cover of X that is minimal. Let $F = X \backslash U_{n+1}$; then F is closed, nonempty, and $F \subseteq \bigcup_{k=1}^n U_k$. Let V be an open subset of X such that $F \subseteq V \subseteq \operatorname{cl} V \subseteq \bigcup_{k=1}^n U_k$. Since $\operatorname{cl} V$ is normal and $\{U_1 \cap \operatorname{cl} V, \ldots, U_n \cap \operatorname{cl} V\}$ is an open cover of $\operatorname{cl} V$, the induction hypothesis implies that there are continuous functions $g_1, \ldots, g_n$ on $\operatorname{cl} V$ such that $\sum_{k=1}^n g_k = 1$ and for $1 \leqslant k \leqslant n$, $0 \leqslant g_k \leqslant 1$, and $g_k(\operatorname{cl} V \backslash U_k) = 0$. By Tietze's Extension Theorem there are continuous functions $\tilde{g}_1, \ldots, \tilde{g}_n$ on X such that $\tilde{g}_k = g_k$ on $\operatorname{cl} V$ and $0 \leqslant \tilde{g}_k \leqslant 1$ for $1 \leqslant k \leqslant n$.

Also, there is a continuous function $h: X \to [0, 1]$ such that $h = 0$ on $X \backslash V$

and $h = 1$ on F. Put $f_k = \tilde{g}_k h$ for $1 \leqslant k \leqslant n$ and let $f_{n+1} = 1 - \sum_{k=1}^{n} f_k$. Clearly $0 \leqslant f_k \leqslant 1$ if $1 \leqslant k \leqslant n$. If $x \in \text{cl } V$, then $f_{n+1}(x) = 1 - (\sum_{k=1}^{n} g_k(x))h(x) = 1 - h(x)$; so $0 \leqslant f_{n+1}(x) \leqslant 1$ on cl V. If $x \in X \setminus V$, then $f_{n+1}(x) = 1$ since $h(x) = 0$. Hence $0 \leqslant f_{n+1} \leqslant 1$.

Clearly (a) holds. Let $1 \leqslant k \leqslant n$; if $x \in X \setminus U_k$, then either $x \in (\text{cl } V) \setminus U_k$ or $x \in (X \setminus \text{cl } V) \setminus U_k$. If the first alternative is the case, then $g_k(x) = 0$, so $f_k(x) = 0$. If the second alternative is true, then $h(x) = 0$ so that $f_k(x) = 0$. If $x \in X \setminus U_{n+1} = F$, then $h(x) = 1$ and so $f_{n+1}(x) = 1 - \sum_{k=1}^{n} g_k(x) = 0$. ∎

Partitions of unity are a standard way to put together local results to obtain global results. If $\{f_k\}$ is related to $\{U_k\}$ as in the statement of (6.5), then $\{f_k\}$ is said to be a partition of unity *subordinate to the cover* $\{U_k\}$.

6.6. Theorem. *If X is completely regular, then $C_b(X)$ is separable if and only if X is a compact metric space.*

PROOF. Suppose X is a compact metric space with metric d. For each n, let $\{U_k^{(n)}: 1 \leqslant k \leqslant N_n\}$ be an open cover of X by balls of radius $1/n$. Let $\{f_k^{(n)}: 1 \leqslant k \leqslant N_n\}$ be a partition of unity subordinate to $\{U_k^{(n)}: 1 \leqslant k \leqslant N_n\}$. Let $\mathscr{Y}$ be the rational (or complex–rational) linear span of $\{f_k^{(n)}: n \geqslant 1, 1 \leqslant k \leqslant N_n\}$; thus $\mathscr{Y}$ is countable. It will be shown that $\mathscr{Y}$ is dense in $C(X)$.

Fix f in $C(X)$ and $\varepsilon > 0$. Since f is uniformly continuous there is a $\delta > 0$ such that $|f(x_1) - f(x_2)| < \varepsilon/2$ whenever $d(x_1, x_2) < \delta$. Choose $n > 2/\delta$ and consider the cover $\{U_k^{(n)}: 1 \leqslant k \leqslant N_n\}$. If $x_1, x_2 \in U_k^{(n)}, d(x_1, x_2) < 2/n < \delta$; hence $|f(x_1) - f(x_2)| < \varepsilon/2$. Pick x_k in $U_k^{(n)}$ and let $\alpha_k \in \mathbb{Q} + i\mathbb{Q}$ such that $|\alpha_k - f(x_k)| < \varepsilon/2$. Let $g = \sum_k \alpha_k f_k^{(n)}$, so $g \in \mathscr{Y}$. Therefore for every x in X,

$$|f(x) - g(x)| = \left| \sum_k f(x) f_k^{(n)}(x) - \sum_k \alpha_k f_k^{(n)}(x) \right|$$
$$\leqslant \sum_k |f(x) - \alpha_k| f_k^{(n)}(x).$$

Examine each of these summands. If $x \in U_k^{(n)}$, then $|f(x) - \alpha_k| \leqslant |f(x) - f(x_k)| + |f(x_k) - \alpha_k| < \varepsilon$. If $x \notin U_k^{(n)}$, then $f_k^{(n)}(x) = 0$. Hence $|f(x) - g(x)| < \sum_k \varepsilon f_k^{(n)}(x) = \varepsilon$. Thus $\|f - g\| < \varepsilon$ and $\mathscr{Y}$ is dense in $C(X)$. This shows that $C(X)$ is separable.

Now assume that $C_b(X)$ is separable. Thus (ball $C_b(X)^*$, wk*) is metrizable (5.1). Since X is homeomorphic to a subset of ball $C_b(X)^*$ (6.1), X is metrizable. It also follows that βX is metrizable. It must be shown that $X = \beta X$.

Suppose there is a τ in $\beta X \setminus X$. Let $\{x_n\}$ be a sequence in X such that $x_n \to \tau$. It can be assumed that $x_n \neq x_m$ for $n \neq m$. Let $A = \{x_n: n \text{ is even}\}$ and $B = \{x_n: n \text{ is odd}\}$. Then A and B are disjoint closed subsets of X (not closed in βX, but in X) since A and B contain all of their limit points in X. Since X is normal, there is a continuous function $f: X \to [0, 1]$ such that $f = 0$ on A and $f = 1$ on B. But then $f^\beta(\tau) = \lim f(x_{2n}) = 0$ and $f^\beta(\tau) = \lim f(x_{2n+1}) = 1$, a contradiction. Thus $\beta X \setminus X = \square$. ∎

EXERCISES

1. If $x \in X$ and $\delta_x(f) = f(x)$ for all f in $C_b(X)$, show that $\|\delta_x\| = 1$.

2. Prove that a subset of a completely regular space is completely regular.

3. Fill in the details of the proof of Theorem 6.2.

4. If X is completely regular, Ω is compact, and $f\colon X \to \Omega$ is continuous, show that there is a continuous map $f^\beta\colon \beta X \to \Omega$ such that $f^\beta | X = f$.

5. If X is completely regular, show that X is open in βX if and only if X is locally compact.

6. Let $\mathbb{N}$ have the discrete topology. Let $\{r_n\colon n \in \mathbb{N}\}$ be an enumeration of the rational numbers in $[0, 1]$. Let $S =$ the irrational numbers in $[0, 1]$ and for each s in S let $\{r_n\colon n \in N_s\}$ be a subsequence of $\{r_n\}$ such that $s = \lim\{r_n\colon n \in N_s\}$. Show: (a) if $s, t \in S$ and $s \neq t$, $N_s \cap N_t$ is finite; (b) if for each s in S, $\operatorname{cl} N_s =$ the closure of N_s in $\beta \mathbb{N}$ and $A_s = (\operatorname{cl} N_s) \setminus \mathbb{N}$, then $\{A_s\colon s \in S\}$ are pairwise disjoint subsets of $\beta \mathbb{N} \setminus \mathbb{N}$ that are both open and closed.

7. Show that if X is normal, $\tau \in \beta X$, and there is a sequence $\{x_n\}$ in X such that $x_n \to \tau$ in βX, then $\tau \in X$. If X is not normal, is the result still true?

8. Let X be the space of all ordinals less than the first uncountable ordinal and give X the order topology. Show that $\beta X =$ the one point compactification of X. (You can find the pertinent definitions in Kelley [1955].)

§7. The Krein–Milman Theorem

7.1. Definition. If K is a convex subset of a vector space $\mathcal{X}$, then a point a in K is an *extreme point* of K if there is no proper open line segment that contains a and lies entirely in K. Let $\operatorname{ext} K$ be the set of extreme points of K.

Recall that an open line segment is a set of the form $(x_1, x_2) \equiv \{tx_2 + (1 - t)x_1\colon 0 < t < 1\}$, and to say that this line segment is proper is to say that $x_1 \neq x_2$.

7.2. Examples.

(a) If $\mathcal{X} = \mathbb{R}^2$ and $K = \{(x, y) \in \mathbb{R}^2\colon x^2 + y^2 \leqslant 1\}$, then $\operatorname{ext} K = \{(x, y)\colon x^2 + y^2 = 1\}$.

(b) If $\mathcal{X} = \mathbb{R}^2$ and $K = \{(x, y) \in \mathbb{R}^2\colon x \leqslant 0\}$, then $\operatorname{ext} K = \square$.

(c) If $\mathcal{X} = \mathbb{R}^2$ and $K = \{(x, y) \in \mathbb{R}^2\colon x < 0\} \cup \{(0, 0)\}$, then $\operatorname{ext} K = \{(0, 0)\}$.

(d) If $K =$ the closed region in $\mathbb{R}^2$ bordered by a regular polygon, then $\operatorname{ext} K =$ the vertices of the polygon.

(e) If $\mathcal{X}$ is any normed space and $K = \{x \in \mathcal{X}\colon \|x\| \leqslant 1\}$, then $\operatorname{ext} K \subseteq \{x\colon \|x\| = 1\}$, though for all we know it may be that $\operatorname{ext} K = \square$.

(f) If $\mathcal{X} = L^1[0, 1]$ and $K = \{f \in L^1[0, 1]\colon \|f\|_1 \leqslant 1\}$, then $\operatorname{ext} K = \square$. This last statement requires a bit of proof. Let $f \in L^1[0,1]$ such that $\|f\|_1 = 1$. Choose x in $[0,1]$ such that $\int_0^x |f(t)| dt = \frac{1}{2}$. Let $h(t) = 2f(t)$ if $t \leq x$ and 0 otherwise; let $g(t) = 2f(t)$ if $t \geq x$ and 0 otherwise. Then $\|h\|_1 = \|g\|_1 = 1$ and $f = \frac{1}{2}(h + g)$. So ball $L^1[0,1]$ has no extreme points.

The next proposition is left as an exercise.

7.3. Proposition. *If K is a convex subset of a vector space $\mathscr{X}$ and $a \in K$, then the following statements are equivalent.*

(a) $a \in \text{ext } K$.
(b) *If $x_1, x_2 \in \mathscr{X}$ and $a = \frac{1}{2}(x_1 + x_2)$, then either $x_1 \notin K$ or $x_2 \notin K$ or $x_1 = x_2 = a$.*
(c) *If $x_1, x_2 \in \mathscr{X}$, $0 < t < 1$, and $a = tx_1 + (1 - t)x_2$, then either $x_1 \notin K$, $x_2 \notin K$, or $x_1 = x_2 = a$.*
(d) *If $x_1, \ldots, x_n \in K$ and $a \in \text{co}\{x_1, \ldots, x_n\}$, then $a = x_k$ for some k.*
(e) *$K \setminus \{a\}$ is a convex set.*

7.4. The Krein–Milman Theorem. *If K is a nonempty compact convex subset of a LCS $\mathscr{X}$, then $\text{ext } K \neq \square$ and $K = \overline{\text{co}}(\text{ext } K)$.*

PROOF. (Léger [1968].) Note that (7.3e) says that a point a is an extreme point if and only if $K \setminus \{a\}$ is a relatively open convex subset. We thus look for a maximal proper relatively open convex subset of K. Let $\mathscr{U}$ be all the proper relatively open convex subsets of K. Since $\mathscr{X}$ is a LCS and $K \neq \square$ (and let's assume that K is not a singleton), $\mathscr{U} \neq \square$. Let $\mathscr{U}_0$ be a chain in $\mathscr{U}$ and put $U_0 = \cup\{U : U \in \mathscr{U}_0\}$. Clearly U_0 is open, and since $\mathscr{U}_0$ is a chain, U_0 is convex. If $U_0 = K$, then the compactness of K implies that there is a U in $\mathscr{U}_0$ with $U = K$, a contradiction to the property of U. Thus $U_0 \in \mathscr{U}$. By Zorn's Lemma, $\mathscr{U}$ has a maximal element U.

If $x \in K$ and $0 \leq \lambda \leq 1$, let $T_{x,\lambda} : K \to K$ be defined by $T_{x,\lambda}(y) = \lambda y + (1 - \lambda)x$. Note that $T_{x,\lambda}$ is continuous and $T_{x,\lambda}(\sum_{j=1}^n \alpha_j y_j) = \sum_{j=1}^n \alpha_j T_{x,\lambda}(y_j)$ whenever $y_1, \ldots, y_n \in K$, $\alpha_1, \ldots, \alpha_n \geq 0$, and $\sum_{j=1}^n \alpha_j = 1$. (This means that $T_{x,\lambda}$ is an *affine map* of K into K.) If $x \in U$ and $0 \leq \lambda < 1$, then $T_{x,\lambda}(U) \subseteq U$. Thus $U \subseteq T_{x,\lambda}^{-1}(U)$ and $T_{x,\lambda}^{-1}(U)$ is an open convex subset of K. If $y \in (\text{cl } U) \setminus U$, $T_{x,\lambda}(y) \in [x, y) \subseteq U$ by Proposition IV.1.11. So $\text{cl } U \subseteq T_{x,\lambda}^{-1}(U)$ and hence the maximality of U implies $T_{x,\lambda}^{-1}(U) = K$. That is,

7.5 $T_{x,\lambda}(K) \subseteq U$ if $x \in U$ and $0 \leq \lambda < 1$.

Claim. If V is any open convex subset of K, then either $V \cup U = U$ or $V \cup U = K$.

In fact, (7.5) implies that $V \cup U$ is convex so that the claim follows from the maximality of U.

It now follows from the claim that $K \setminus U$ is a singleton. In fact, if $a, b \in K \setminus U$ and $a \neq b$, let V_a, V_b be disjoint open convex subsets of K such that $a \in V_a$

and $b \in V_b$. By the claim $V_a \cup U = K$ since $a \notin U$. But $b \notin V_a \cup U$, a contradiction. Thus $K \backslash U = \{a\}$ and $a \in \text{ext } K$ by (7.3e). Hence $\text{ext } K \neq \square$.

Note that we have actually proved the following.

7.6 If V is an open convex subset of $\mathcal{X}$ and $\text{ext } K \subseteq V$, then $K \subseteq V$.

Assume (7.6) is false. That is, assume there is an open convex subset V of $\mathcal{X}$ such that $\text{ext } K \subseteq V$ but $V \cap K \neq K$. Then $V \cap K \in \mathcal{U}$ and is contained in a maximal element U of $\mathcal{U}$. Since $K \backslash U = \{a\}$ for some a in $\text{ext } K$, this is a contradiction. Thus (7.6) holds.

Let $E = \overline{\text{co}}(\text{ext } K)$. If $x^* \in \mathcal{X}^*$, $\alpha \in \mathbb{R}$, and $E \subseteq \{x \in \mathcal{X}: \text{Re}\langle x, x^* \rangle < \alpha\} = V$, then $K \subseteq V$ by (7.6). Thus the Hahn–Banach Theorem (IV.3.13) implies $E = K$. ∎

The Krein–Milman Theorem seems innocent enough, but it has widespread application. Two such applications will be seen in Sections 8 and 10; another will occur later when C^*-algebras are studied. Here a small application is given.

If $\mathcal{X}$ is a Banach space, then ball $\mathcal{X}^*$ is weak* compact by Alaoglu's Theorem. By the Krein–Milman Theorem, ball $\mathcal{X}^*$ has many extreme points. Keep this in mind.

7.7. Example. c_0 is not the dual of a Banach space. That is, c_0 is not isometrically isomorphic to the dual of a Banach space. In light of the preceding comments, in order to prove this statement, it suffices to show that ball c_0 has few extreme points. In fact, ball c_0 has no extreme points. Let $x \in$ ball c_0. It must be that $0 = \lim x(n)$. Let N be such that $|x(n)| < \frac{1}{2}$ for $n \geqslant N$. Define y_1, y_2 in c_0 by letting $y_1(n) = y_2(n) = x(n)$ for $n \leqslant N$, and for $n > N$ let $y_1(n) = x(n) + 2^{-n}$ and $y_2(n) = x(n) - 2^{-n}$. It is easy to check that y_1 and $y_2 \in$ ball c_0, $\frac{1}{2}(y_1 + y_2) = x$, and $y_1 \neq x$.

In light of Example 7.2(f), $L^1[0, 1]$ is not the dual of a Banach space.

The next two results are often useful in applying the Krein–Milman Theorem. Indeed, the first is often taken as part of that result.

7.8. Theorem. If $\mathcal{X}$ is a LCS, K is a compact convex subset of $\mathcal{X}$, and $F \subseteq K$ such that $K = \overline{\text{co}}(F)$, then $\text{ext } K \subseteq \text{cl } F$.

PROOF. Clearly it suffices to assume that F is closed. Suppose that there is an extreme point x_0 of K such that $x_0 \notin F$. Let p be a continuous seminorm on $\mathcal{X}$ such that $F \cap \{x \in \mathcal{X}: p(x - x_0) < 1\} = \square$. Let $U_0 = \{x \in \mathcal{X}: p(x) < \frac{1}{3}\}$. So $(x_0 + U_0) \cap (F + U_0) = \square$; hence $x_0 \notin \text{cl}(F + U_0)$.

Because F is compact, there are $y_1, \ldots, y_n$ in F such that $F \subseteq \bigcup_{k=1}^{n}(y_k + U_0)$. Let $K_k = \overline{\text{co}}(F \cap (y_k + U_0))$. Thus $K_k \subseteq y_k + \text{cl } U_0$ (Why?), and $K_k \subseteq K$. Now that fact that $K_1, \ldots, K_n$ are compact and convex implies that $\overline{\text{co}}(K_1 \cup \cdots \cup K_n) = \text{co}(K_1 \cup \cdots \cup K_n)$ (Exercise 8). Therefore

$$K = \overline{\text{co}}(F) = \text{co}(K_1 \cup \cdots \cup K_n).$$

Since $x_0 \in K$, $x_0 = \sum_{k=1}^{n} \alpha_k x_k$, $x_k \in K_k$, $\alpha_k \geqslant 0$, $\alpha_1 + \cdots + \alpha_n = 1$. But x_0 is an extreme point of K. Thus, $x_0 = x_k \in K_k$ for some k. But this implies that $x_0 \in K_k \subseteq y_k + \mathrm{cl}\, U_0 \subseteq cl(F + U_0)$, a contradiction. ∎

You might think that the set of extreme points of a compact convex subset would have to be closed. This is untrue even if the LCS is finite dimensional, as Figure V-1 illustrates.

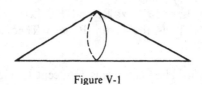

Figure V-1

7.9. Proposition. *If K is a compact convex subset of a LCS $\mathcal{X}$, $\mathcal{Y}$ is a LCS, and $T: K \to \mathcal{Y}$ is a continuous affine map, then $T(K)$ is a compact convex subset of $\mathcal{Y}$ and if y is an extreme point of $T(K)$, then there is an extreme point x of K such that $T(x) = y$.*

PROOF. Because T is affine, $T(K)$ is convex and it is compact by the continuity of T. Let y be an extreme point of $T(K)$. It is easy to see that $T^{-1}(y)$ is compact and convex. Let x be an extreme point of $T^{-1}(y)$. It now follows that $x \in \mathrm{ext}\, K$ (Exercise 9). ∎

Note that it is possible that there are extreme points x of K such that $T(x)$ is not an extreme point of $T(K)$. For example, let T be the orthogonal projection of $\mathbb{R}^3$ onto $\mathbb{R}^2$ and let $K = \mathrm{ball}\, \mathbb{R}^3$.

EXERCISES

1. If (X, Ω, μ) is a σ-finite measure space and $1 < p < \infty$, then the set of extreme points of ball $L^p(\mu)$ is $\{f \in L^p(\mu): \|f\|_p = 1\}$.

2. If (X, Ω, μ) is a σ-finite measure space, the set of extreme points of ball $L^1(\mu)$ is $\{\alpha \chi_E : E \text{ is an atom of } \mu, \ \alpha \in \mathbb{F}, \text{ and } |\alpha| = \mu(E)^{-1}\}$.

3. If (X, Ω, μ) is a σ-finite measure space, the set of extreme points of ball $L^\infty(\mu)$ is $\{f \in L^\infty(\mu): |f(x)| = 1 \text{ a.e. } [\mu]\}$.

4. If X is completely regular, the set of extreme points of ball $C_b(X)$ is $\{f \in C_b(X): |f(x)| = 1 \text{ for all } x\}$. So ball $C_{\mathbb{R}}[0, 1]$ has only two extreme points.

5. Let X be a totally disconnected compact space. (That is, X is compact and if $x \in X$ and U is an neighborhood of x, then there is a subset V of X that is both open and closed and such that $x \in V \subseteq U$. The Cantor set is an example of such a space.) Show that ball $C(X)$ is the norm closure of the convex hull of its extreme points. (If $\mathbb{F} = \mathbb{C}$, the result is true for all compact Hausdorff spaces X (Phelps

[1965]). If $\mathbb{F} = \mathbb{R}$, then this characterizes totally disconnected compact spaces (Goodner [1964]).)

6. Show that ball l^1 is the norm closure of the convex hull of its extreme points.

7. Show that if X is locally compact but not compact, then ball $C_0(X)$ has no extreme points.

8. If $\mathscr{X}$ is a LCS and $K_1, \ldots, K_n$ are compact convex subsets of $\mathscr{X}$, then $\overline{\mathrm{co}}(K_1 \cup \cdots \cup K_n) = \mathrm{co}(K_1 \cup \cdots \cup K_n)$ and this convex hull is compact.

9. Let K be convex and let $T: K \to \mathscr{Y}$ be an affine map. If y is an extreme point of $T(K)$ and x is an extreme point of $T^{-1}(y)$, then x is an extreme point of K.

10. If $\mathscr{H}$ is a Hilbert space and either T or T^* is an isometry, show that T is an extreme point of the closed unit ball of $\mathscr{B}(\mathscr{H})$. (The converse of this is also true, but it may be hard unless you use the Polar Decomposition of operators (VIII.3.11).)

§8. An Application: The Stone–Weierstrass Theorem

If $f: X \to \mathbb{C}$ is a function, then $\bar{f}$ denotes the function from X into $\mathbb{C}$ whose value at each x is the complex conjugate of $f(x)$, $\overline{f(x)}$.

8.1. The Stone–Weierstrass Theorem. *If X is compact and $\mathscr{A}$ is a closed subalgebra of $C(X)$ such that:*

(a) $1 \in \mathscr{A}$;
(b) *if $x, y \in X$ and $x \neq y$, then there is an f in $\mathscr{A}$ such that $f(x) \neq f(y)$;*
(c) *if $f \in \mathscr{A}$, then $\bar{f} \in \mathscr{A}$;*
    then $\mathscr{A} = C(X)$.

If $C(X)$ is the algebra of continuous functions from X into $\mathbb{R}$, then condition (c) is not needed. Also, an algebra in $C(X)$ that has property (b) is said to *separate the points* of X (see Exercise 1).

The proof of this result that will be presented here makes use of the Krein–Milman Theorem and is due to L. de Branges [1959].

PROOF OF THE STONE–WEIERSTRASS THEOREM. To prove the theorem it suffices to show that $\mathscr{A}^\perp = (0)$ (III.6.14). Suppose $\mathscr{A}^\perp \neq (0)$. By Alaoglu's Theorem, ball $\mathscr{A}^\perp$ is weak* compact. By the Krein–Milman Theorem, there is an extreme point μ of ball $\mathscr{A}^\perp$. Let $K = $ the support of μ. That is,

$$ K = X \backslash \bigcup \{V: V \text{ is open and } |\mu|(V) = 0\}. $$

Hence $|\mu|(X \backslash K) = 0$ and $\int f \, d\mu = \int_K f \, d\mu$ for all continuous functions f on X. Since $\mathscr{A}^\perp \neq (0)$, $\|\mu\| = 1$ and $K \neq \square$. Fix x_0 in K. It will be shown that $K = \{x_0\}$.

Let $x \in X$, $x \neq x_0$. By (b) there is an f_1 in $\mathscr{A}$ such that $f_1(x_0) \neq f_1(x) = \beta$. By (a), the function $\beta \in \mathscr{A}$. Hence $f_2 = f_1 - \beta \in \mathscr{A}$, $f_2(x_0) \neq 0 = f_2(x)$. By (c), $f_3 = |f_2|^2 = f_2 \bar{f}_2 \in \mathscr{A}$. Also, $f_3(x) = 0 < f_3(x_0)$ and $f_3 \geqslant 0$. Put $f = (\|f_3\| + 1)^{-1} f_3$. Then $f \in \mathscr{A}$, $f(x) = 0$, $f(x_0) > 0$, and $0 \leqslant f < 1$ on X. Moreover, because $\mathscr{A}$ is an algebra, gf and $g(1 - f) \in \mathscr{A}$ for every g in $\mathscr{A}$. Because $\mu \in \mathscr{A}^\perp$, $0 = \int g f d\mu = \int g(1 - f) d\mu$ for every g in $\mathscr{A}$. Therefore $f\mu$ and $(1 - f)\mu \in \mathscr{A}^\perp$.

(For any bounded Borel function h on X, $h\mu$ denotes the measure whose value at a Borel set Δ is $\int_\Delta h d\mu$. Note that $\|h\mu\| = \int |h| d|\mu|$.)

Put $\alpha = \|f\mu\| = \int f d|\mu|$. Since $f(x_0) > 0$, there is an open neighborhood U of x_0 and an $\varepsilon > 0$ such that $f(y) > \varepsilon$ for y in U. Thus, $\alpha = \int f d|\mu| \geqslant \int_U f d|\mu| \geqslant \varepsilon |\mu|(U) > 0$ since $U \cap K \neq \square$. Similarly, since $f(x_0) < 1$, $\alpha < 1$. Therefore, $0 < \alpha < 1$. Also, $1 - \alpha = 1 - \int f d|\mu| = \int (1 - f) d|\mu| = \|(1 - f)\mu\|$. Since

$$\mu = \alpha \left[\frac{f\mu}{\|f\mu\|} \right] + (1 - \alpha) \left[\frac{(1 - f)\mu}{\|(1 - f)\mu\|} \right]$$

and μ is an extreme point of ball $\mathscr{A}^\perp$, $\mu = f\mu \|f\mu\|^{-1} = \alpha^{-1} f\mu$. But the only way that the measures μ and $\alpha^{-1} f\mu$ can be equal is if $\alpha^{-1} f = 1$ a.e. $[\mu]$. Since f is continuous, it must be that $f \equiv \alpha$ on K. Since $x_0 \in K$, $f(x_0) = \alpha$. But $f(x_0) > f(x) = 0$. Hence $x \notin K$. This establishes that $K = \{x_0\}$ and so $\mu = \gamma \delta_{x_0}$ where $|\gamma| = 1$. But $\mu \in \mathscr{A}^\perp$ and $1 \in \mathscr{A}$, so $0 = \int 1 d\mu = \gamma$, a contradiction. Therefore $\mathscr{A}^\perp = (0)$ and $\mathscr{A} = C(X)$. ∎

With an important theorem it is good to ask what happends if part of the hypothesis is deleted. If $x_0 \in X$ and $\mathscr{A} = \{f \in C(X): f(x_0) = 0\}$, then $\mathscr{A}$ is a closed subalgebra of $C(X)$ that satisfies (b) and (c) but $\mathscr{A} \neq C(X)$. This is the worst that can happen.

8.2. Corollary. *If X is compact and $\mathscr{A}$ is a closed subalgebra of $C(X)$ that separates the points of X and is closed under complex conjugation, then either $\mathscr{A} = C(X)$ or there is a point x_0 in X such that $\mathscr{A} = \{f \in C(X): f(x_0) = 0\}$.*

PROOF. Identify $\mathbb{F}$ and the one-dimensional subspace of $C(X)$ consisting of the constant functions. Since $\mathscr{A}$ is closed, $\mathscr{A} + \mathbb{F}$ is closed (III.4.3). It is easy to see that $\mathscr{A} + \mathbb{F}$ is an algebra and satisfies the hypothesis of the Stone–Weierstrass Theorem; hence $\mathscr{A} + \mathbb{F} = C(X)$. Suppose $\mathscr{A} \neq C(X)$. Then $C(X)/\mathscr{A}$ is one dimensional; thus $\mathscr{A}^\perp$ is one dimensional (Theorem 2.2). Let $\mu \in \mathscr{A}^\perp$, $\|\mu\| = 1$. If $f \in \mathscr{A}$, then $f\mu \in \mathscr{A}^\perp$; hence there is an α in $\mathbb{F}$ such that $f\mu = \alpha\mu$. This implies that each f in $\mathscr{A}$ is constant on the support of μ. But the functions in $\mathscr{A}$ separate the points of X. Hence the support of μ is a single point x_0 and so $\mathscr{A}^\perp = \{\beta \delta_{x_0}: \beta \in \mathbb{F}\}$. Thus $\mathscr{A} = \mathscr{A}^{\perp\perp} = \{f \in C(X): f(x_0) = 0\}$. ∎

There are many examples of subalgebras of $C(X)$ that separate the points of X, contain the constants, but are not necessarily closed under complex conjugation. Indeed, a subalgebra of $C(X)$ having these properties is called a *uniform algebra* or *function algebra* and their study forms a separate area of mathematics (Gamelin [1969]). One example (the most famous) is obtained by letting X be a subset of $\mathbb{C}$ and letting $\mathscr{A} = R(X) \equiv$ the uniform closure of rational functions with poles off X.

Let $x_0, x_1 \in X$, $x_0 \neq x_1$, and let $\mathscr{A} \equiv \{f \in C(X): f(x_0) = f(x_1)\}$. Then $\mathscr{A}$ is a uniformly closed subalgebra of $C(X)$, contains the constant functions, and is closed under conjugation. In a certain sense this is the worst that can happen if the only hypothesis of the Stone–Weierstrass Theorem that does not hold is that $\mathscr{A}$ fails to separate the points of X (see Exercise 4).

If X is only assumed to be locally compact, then the story is similar.

8.3. Corollary. *If X is locally compact and $\mathscr{A}$ is a closed subalgebra of $C_0(X)$ such that*

(a) *for each x in X there is an f in $\mathscr{A}$ such that $f(x) \neq 0$;*
(b) *$\mathscr{A}$ separates the points of X;*
(c) *$\bar{f} \in \mathscr{A}$ whenever $f \in \mathscr{A}$;*

then $\mathscr{A} = C_0(X)$.

PROOF. Let $X_\infty =$ the one point compactification of X and identify $C_0(X)$ with $\{f \in C(X_\infty): f(\infty) = 0\}$. So $\mathscr{A}$ becomes a subalgebra of $C(X_\infty)$. Now apply Corollary 8.2. The details are left to the reader. ∎

What are the extreme points of the unit ball of $M(X)$? The characterization of these extreme points as well as the extreme points of the set $P(X)$ of probability measures on X is given in the next theorem. [A *probability measure* is a positive measure μ such that $\mu(X) = 1$.]

8.4. Theorem. *If X is compact, then the set of extreme points of ball $M(X)$ is*

$$\{\alpha\delta_x: |\alpha| = 1 \text{ and } x \in X\}.$$

The set of extreme points of $P(X)$, the probability measures on X, is

$$\{\delta_x: x \in X\}.$$

PROOF. It is left as an exercise for the reader to show that if $x \in X$, δ_x is an extreme point of $P(X)$ and $\alpha\delta_x$ is an extreme point of ball $M(X)$ (Exercise 3).

It will now be shown that if μ is an extreme point of $P(X)$, then μ is an extreme point of ball $M(X)$. Thus the first part of the theorem implies the second. Suppose μ is an extreme point of $P(X)$ and $v_1, v_2 \in$ ball $M(X)$ such that $\mu = \frac{1}{2}(v_1 + v_2)$. Then $1 = \|\mu\| \leqslant \frac{1}{2}(\|v_1\| + \|v_2\|) \leqslant 1$; hence $\|v_1\| + \|v_2\| = 2$ and so $\|v_1\| = \|v_2\| = 1$. Also, $1 = \mu(X) = \frac{1}{2}(v_1(X) + v_2(X))$. Now $|v_1(X)|, |v_2(X)| \leqslant 1$ and 1 is an extreme point of $\{\alpha \in \mathbb{F}: |\alpha| \leqslant 1\}$. Hence for $k = 1, 2$, $\|v_k\| = v_k(X) = 1$. By Exercise III.7.2, $v_k \in P(X)$ for $k = 1, 2$. Since $\mu \in \text{ext } P(X)$, $\mu = v_1 = v_2$. So μ is an extreme point of ball $M(X)$. Thus it suffices to prove the first part of the theorem.

Suppose that μ is an extreme point of ball $M(X)$ and let K be the support of μ. It will be shown that K is a singleton set.

Fix x_0 in K and suppose there is a second point x in $K, x \neq x_0$. Let U and V be open subsets of X such that $x_0 \in U$, $x \in V$, and cl $U \cap$ cl $V = \square$. By Urysohn's Lemma there is an f in $C(X)$ such that $0 \leqslant f \leqslant 1$, $f(y) = 1$ for y in cl U, and $f(y) = 0$ for y in cl V. Consider the measures $f\mu$ and $(1 - f)\mu$. Put $\alpha = \|f\mu\| = \int |f| d|\mu| = \int f d|\mu|$. Then $\alpha = \int f d|\mu| \leqslant \|\mu\| = 1$ and $\alpha = \int f d|\mu| \geqslant |\mu|(U) > 0$ since U is open and $U \cap K \neq \square$. Also, $1 - \alpha = 1 - \int f d|\mu| = \int (1 - f) d|\mu| = \|(1 - f)\mu\|$ and so $1 - \alpha \geqslant \int_V (1 - f) d|\mu| = |\mu|(V) > 0$ since $x \in K$. Hence $0 < \alpha < 1$.

But $f\mu/\alpha$ and $(1 - f)\mu/(1 - \alpha) \in$ ball $M(X)$ and

$$\mu = \alpha \left[\frac{f\mu}{\alpha} \right] + (1 - \alpha) \left[\frac{(1 - f)\mu}{1 - \alpha} \right].$$

Since μ is an extreme point of ball $M(X)$ and $\alpha \neq 0$, $\mu = f\mu/\alpha$. This can only happen if $f \equiv \alpha < 1$ a.e. $[\mu]$. But $f \equiv 1$ on U and $|\mu|(U) > 0$, a contradiction. Hence $K = \{x_0\}$.

Since the only measures whose support can be the singleton set $\{x_0\}$ have the form $\alpha \delta_{x_0}, \alpha$ in $\mathbb{F}$, the theorem is proved.   ∎

EXERCISES

1. Suppose that $\mathscr{A}$ is a subalgebra of $C(X)$ that separates the points of X and $1 \in \mathscr{A}$. Show that if $x_1, \ldots, x_n$ are distinct points in X and $\alpha_1, \ldots, \alpha_n \in \mathbb{F}$, there is an f in $\mathscr{A}$ such that $f(x_j) = \alpha_j$ for $1 \leqslant j \leqslant n$.

2. Give the details of the proof of Corollary 8.3.

3. If X is compact, show that for each x in X, δ_x is an extreme point of $P(X)$ and $\alpha \delta_x, |\alpha| = 1$, is an extreme point of ball $M(X)$.

4. Let X be compact and let $\mathscr{A}$ be a closed subalgebra of $C(X)$ such that $1 \in \mathscr{A}$ and $\mathscr{A}$ is closed under conjugation. Define an equivalence relation $\sim$ on X by declaring $x \sim y$ if and only if $f(x) = f(y)$ for all f in $\mathscr{A}$. Let $X/\sim$ be the corresponding quotient space and let $\pi: X \to X/\sim$ be the natural map. Give $X/\sim$ the quotient topology. (a) Show that if $f \in \mathscr{A}$, then there is a unique function $\pi^*(f)$ in $C(X/\sim)$ such that $\pi^*(f) \circ \pi = f$. (b) Show that $\pi^*: \mathscr{A} \to C(X/\sim)$ is an isometry. (c) Show that π^* is surjective. (d) Show that $\mathscr{A} = \{f \in C(X): f(x) = f(y) \text{ whenever } x \sim y\}$.

5. (This exercise requires Exercise IV.4.7.) Let X be completely regular and topologize $C(X)$ as in Example IV.1.5. If $\mathscr{A}$ is a closed subalgebra of $C(X)$ such that $1 \in \mathscr{A}$, $\mathscr{A}$ separates the points of X, and $\bar{f} \in \mathscr{A}$ whenever $f \in \mathscr{A}$, then $\mathscr{A} = C(X)$.

6. Let X, Y be compact spaces and show that if $f \in C(X \times Y)$ and $\varepsilon > 0$, then there are functions $g_1, \ldots, g_n$ in $C(X)$ and $h_1, \ldots, h_n$ in $C(Y)$ such that $|f(x, y) - \sum_{k=1}^n g_k(x) h_k(y)| < \varepsilon$ for all (x, y) in $X \times Y$.

7. Let $\mathscr{A}$ be the uniformly closed subalgebra of $C_b(\mathbb{R})$ generated by $\sin x$ and $\cos x$. Show that $\mathscr{A} = \{f \in C_b(\mathbb{R}): f(t) = f(t + 2\pi) \text{ for all } t \text{ in } \mathbb{R}\}$.

8. If K is a compact subset of $\mathbb{C}$, $f \in C(K)$, and $\varepsilon > 0$, show that there is a polynomial $p(z, \bar{z})$ in z and $\bar{z}$ such that $|f(z) - p(z, \bar{z})| < \varepsilon$ for all z in K.

§9*. The Schauder Fixed Point Theorem

Fixed-point theorems hold a fascination for mathematicians and they are very applicable to a variety of mathematical and physical situations. In this section and the next two such theorems are presented.

The results of this section are different from the rest of this book in an essential way. Although we will continue to look at convex subsets of Banach spaces, the functions will not be assumed to be linear or affine. This is a small part of nonlinear functional analysis.

To begin with, recall the following classical result whose proof can be found in any algebraic topology book. (Also see Dugundji [1966].)

9.1. Brouwer's Fixed Point Theorem. *If* $1 \leqslant d < \infty$, $B =$ *the closed unit ball of* $\mathbb{R}^d$, *and* $f\colon B \to B$ *is a continuous map, then there is a point* x *in* B *such that* $f(x) = x$.

9.2. Corollary. *If* K *is a nonempty compact convex subset of a finite dimensional normed space* $\mathscr{X}$ *and* $f\colon K \to K$ *is a continuous function, then there is a point* x *in* K *such that* $f(x) = x$.

PROOF. Since $\mathscr{X}$ is isomorphic to either $\mathbb{C}^d$ or $\mathbb{R}^d$, it is homeomorphic to either $\mathbb{R}^{2d}$ or $\mathbb{R}^d$. So it suffices to assume that $\mathscr{X} = \mathbb{R}^d$, $1 \leqslant d < \infty$. If $K = \{x \in \mathbb{R}^d \colon \|x\| \leqslant r\}$, then the result is immediate from Brouwer's Theorem (Exercise). If K is any compact convex subset of $\mathbb{R}^d$, let $r > 0$ such that $K \subseteq B \equiv \{x \in \mathbb{R}^d \colon \|x\| \leqslant r\}$. Let $\phi\colon B \to K$ be the function defined by $\phi(x) =$ the unique point y in K such that $\|x - y\| = \mathrm{dist}(x, K)$ (I.2.5). Then ϕ is continuous (Exercise) and $\phi(x) = x$ for each x in K. (In topological parlance, K is a retract of B.) Hence $f \circ \phi\colon B \to K \subseteq B$ is continuous. By Brouwer's Theorem, there is an x in B such that $f(\phi(x)) = x$. Since $f \circ \phi(B) \subseteq K$, $x \in K$. Hence $\phi(x) = x$ and $f(x) = x$. ∎

Schauder's Fixed Point Theorem is a generalization of the preceding corollary to infinite dimensional spaces.

9.3. Definition. If $\mathscr{X}$ is a normed space and $E \subseteq \mathscr{X}$, a function $f\colon E \to \mathscr{X}$ is said to be *compact* if f is continuous and $\mathrm{cl}\, f(A)$ is compact whenever A is a bounded subset of E.

If E is itself a compact subset of $\mathscr{X}$, then every continuous function from E into $\mathscr{X}$ is compact.

The following lemma will be needed in the proof of Schauder's Theorem.

9.4. Lemma. *If* K *is a compact subset of the normed space* $\mathscr{X}$, $\varepsilon > 0$, *and* A *is a finite subset of* K *such that* $K \subseteq \bigcup \{B(a; \varepsilon) \colon a \in A\}$, *define* $\phi_A \colon K \to \mathscr{X}$ *by*

$$\phi_A(x) = \frac{\sum \{m_a(x)a \colon a \in A\}}{\sum \{m_a(x) \colon a \in A\}},$$

where $m_a(x) = 0$ if $\|x - a\| \geqslant \varepsilon$ and $m_a(x) = \varepsilon - \|x - a\|$ if $\|x - a\| \leqslant \varepsilon$. Then ϕ_A is a continuous function and

$$\|\phi_A(x) - x\| < \varepsilon$$

for all x in K.

PROOF. Note that for each a in A, $m_a(x) \geqslant 0$ and $\sum\{m_a(x): a \in A\} > 0$ for all x in K. So ϕ_A is well defined on K. The fact that ϕ_A is continuous follows from the fact that for each a in A, $m_a: K \to [0, \varepsilon]$ is continuous. (Verify!)
If $x \in K$, then

$$\phi_A(x) - x = \frac{\sum\{m_a(x)[a - x]: a \in A\}}{\sum\{m_a(x): a \in A\}}.$$

If $m_a(x) > 0$, then $\|x - a\| < \varepsilon$. Hence

$$\|\phi_A(x) - x\| \leqslant \frac{\sum\{m_a(x)\|a - x\|: a \in A\}}{\sum\{m_a(x): a \in A\}} < \varepsilon.$$

This concludes the proof. ∎

9.5. The Schauder Fixed Point Theorem. *Let E be a closed bounded convex subset of a normed space $\mathcal{X}$. If $f: E \to \mathcal{X}$ is a compact map such that $f(E) \subseteq E$, then there is an x in E such that $f(x) = x$.*

PROOF. Let $K = \operatorname{cl} f(E)$, so $K \subseteq E$. For each positive integer n let A_n be a finite subset of K such that $K \subseteq \bigcup\{B(a; 1/n): a \in A_n\}$. For each n let $\phi_n = \phi_{A_n}$ as in the preceding lemma. Now the definition of ϕ_n clearly implies that $\phi_n(K) \subseteq \operatorname{co}(K) \subseteq E$ since E is convex; thus $f_n \equiv \phi_n \circ f$ maps E into E. Also, Lemma 9.4 implies

9.6 $\|f_n(x) - f(x)\| < 1/n$ for x in E.

Let $\mathcal{X}_n$ be the linear span of the set A_n and put $E_n = E \cap \mathcal{X}_n$. So $\mathcal{X}_n$ is a finite dimensional normed space, E_n is a compact convex subset of $\mathcal{X}_n$, and $f_n: E_n \to E_n$ (Why?) is continuous. By Corollary 9.2, there is a point x_n in E_n such that $f_n(x_n) = x_n$.
Now $\{f(x_n)\}$ is a sequence in the compact set K, so there is a point x_0 and a subsequence $\{f(x_{n_j})\}$ such that $f(x_{n_j}) \to x_0$. Since $f_{n_j}(x_{n_j}) = x_{n_j}$, (9.6) implies

$$\|x_{n_j} - x_0\| \leqslant \|f_{n_j}(x_{n_j}) - f(x_{n_j})\| + \|f(x_{n_j}) - x_0\|$$

$$\leqslant \frac{1}{n_j} + \|f(x_{n_j}) - x_0\|.$$

Thus $x_{n_j} \to x_0$. Since f is continuous, $f(x_0) = \lim f(x_{n_j}) = x_0$. ∎

There is a generalization of Schauder's Theorem where $\mathcal{X}$ is only assumed to be a LCS. See Dunford and Schwartz [1958], p. 456.

EXERCISE
1. Let $E = \{x \in l^2(\mathbb{N}): \|x\| \leqslant 1\}$ and for x in E define $f(x) = ((1 - \|x\|^2)^{1/2}, x(1), x(2), \ldots)$. Show that $f(E) \subseteq E$, f is continuous, and f has no fixed points.

§10*. The Ryll–Nardzewski Fixed Point Theorem

This section begins by proving a fixed point theorem that in addition to being used to prove the result in the title of this section has some interest of its own. Recall that a map T defined from a convex set K into a vector space is said to be *affine* if $T(\sum \alpha_j x_j) = \sum \alpha_j T(x_j)$ when $x_j \in K$, $\alpha_j \geqslant 0$, and $\sum \alpha_j = 1$.

10.1. The Markov–Kakutani Fixed Point Theorem. *If K is a nonempty compact convex subset of a LCS $\mathscr{X}$ and $\mathscr{F}$ is a family of continuous affine maps of K into itself that is abelian, then there is an x_0 in K such that $T(x_0) = x_0$ for all T in $\mathscr{F}$.*

PROOF. If $T \in \mathscr{F}$ and $n \geqslant 1$, define $T^{(n)}: K \to K$ by

$$T^{(n)} = \frac{1}{n} \sum_{k=0}^{n-1} T^k.$$

If S and $T \in \mathscr{F}$ and $n, m \geqslant 1$, then it is easy to check that $S^{(n)} T^{(m)} = T^{(m)} S^{(n)}$. Let $\mathscr{K} = \{T^{(n)}(K): T \in \mathscr{F}, n \geqslant 1\}$. Each set in $\mathscr{K}$ is compact and convex. If $T_1, \ldots, T_p \in \mathscr{F}$ and $n_1, \ldots, n_p \geqslant 1$, then the commutativity of $\mathscr{F}$ implies that $T_1^{(n_1)} \cdots T_p^{(n_p)}(K) \subseteq \bigcap_{j=1}^{p} T_j^{(n_j)}(K)$. This says that $\mathscr{K}$ has the finite intersection property and hence there is an x_0 in $\bigcap \{B: B \in \mathscr{K}\}$. It is claimed that x_0 is the desired common fixed point for the maps in $\mathscr{F}$.

If $T \in \mathscr{F}$ and $n \geqslant 1$, then $x_0 \in T^{(n)}(K)$. Thus there is an x in K such that

$$x_0 = T^{(n)}(x) = \frac{1}{n}[x + T(x) + \cdots + T^{n-1}(x)].$$

Using this equation for x_0, it follows that

$$T(x_0) - x_0 = \frac{1}{n}[T(x) + \cdots + T^n(x)]$$

$$- \frac{1}{n}[x + T(x) + \cdots + T^{n-1}(x)]$$

$$= \frac{1}{n}[T^n(x) - x]$$

$$\in \frac{1}{n}[K - K].$$

Now K is compact and so $K - K$ is also. If U is an open neighborhood of 0 in $\mathscr{X}$, there is an integer $n \geqslant 1$ such that $n^{-1}[K - K] \subseteq U$. Therefore $T(x_0) - x_0 \in U$ for every open neighborhood U of 0. This implies that $T(x_0) - x_0 = 0$. ∎

If p is a seminorm on $\mathscr{X}$ and $A \subseteq \mathscr{X}$, define the *p-diameter* of A to be the number
$$p\text{-diam } A \equiv \sup\{p(x - y)\colon x, y \in A\}.$$

10.2. Lemma. *If $\mathscr{X}$ is a LCS, K is a nonempty separable weakly compact convex subset of $\mathscr{X}$, and p is a continuous seminorm on $\mathscr{X}$, then for every $\varepsilon > 0$ there is a closed convex subset C of K such that:*

(a) $C \neq K$;
(b) $p\text{-diam}(K \backslash C) \leqslant \varepsilon$.

PROOF. Let $S = \{x \in \mathscr{X}\colon p(x) \leqslant \varepsilon/4\}$ and let $D = $ the weak closure of the set of extreme points of K. Note that $D \subseteq K$. By hypothesis there is a countable subset A of K such that $D \subseteq K \subseteq \bigcup\{a + S\colon a \in A\}$. Now each $a + S$ is weakly closed. (Why?) Since D is weakly compact, there is an a in A such that $(a + S) \cap D$ has interior in the relative weak topology of D (Exercise 2). Thus, there is a weakly open subset W of $\mathscr{X}$ such that

10.3 $(a + S) \cap D \supseteq W \cap D \neq \square.$

Let $K_1 = \overline{\text{co}}(D \backslash W)$ and $K_2 = \overline{\text{co}}(D \cap W)$. Because K_1 and K_2 are compact and convex and $K_1 \cup K_2$ contains the extreme points of K, the Krein–Milman Theorem and Exercise 7.8 imply $K = \text{co}(K_1 \cup K_2)$.

10.4. Claim. $K_1 \neq K$.

In fact, if $K_1 = K$, then $K = \overline{\text{co}}(D \backslash W)$ so that $\text{ext } K \subseteq D \backslash W$ (Theorem 7.8). This implies that $D \subseteq D \backslash W$, or that $W \cap D = \square$, a contradiction to (10.3).

Now (10.3) implies that $K_2 \subseteq a + S$; so the definition of S implies that $p\text{-diam } K_2 \leqslant \varepsilon/2$. Let $0 < r \leqslant 1$ and define $f_r\colon K_1 \times K_2 \times [r, 1] \to K$ by $f_r(x_1, x_2, t) = t x_1 + (1 - t)x_2$. So f_r is continuous and $C_r \equiv f_r(K_1 \times K_2 \times [r, 1])$ is weakly compact and convex. (Verify!)

10.5. Claim. $C_r \neq K$ for $0 < r \leqslant 1$.

In fact, if $C_r = K$ and $e \in \text{ext } K$, then $e = t x_1 + (1 - t)x_2$ for some $t, r \leqslant t \leqslant 1$, x_j in K_j. Because e is an extreme point and $t \neq 0$, $e = x_1$. Thus $\text{ext } K \subseteq K_1$ and $K = K_1$, contradicting (10.4).

Let $y \in K \backslash C_r$. The definition of C_r and the fact that $K = \text{co}(K_1 \cup K_2)$ imply $y = t x_1 + (1 - t)x_2$ with x_j in K_j and $0 \leqslant t < r$. Hence $p(y - x_2) = p(t(x_1 - x_2)) = t p(x_1 - x_2) \leqslant rd$, where $d = p\text{-diam } K$. Therefore, if $y' = t' x_1' + (1 - t')x_2' \in K \backslash C_r$, then $p(y - y') \leqslant p(y - x_2) + p(x_2 - x_2') + p(x_2' - y') \leqslant 2rd + p\text{-diam } K_2 \leqslant 2rd + \varepsilon/2$. Choosing $r = \varepsilon/4d$ and putting $C = C_r$, we have proved the lemma. ∎

10.6. Definition. Let $\mathscr{X}$ be a LCS and let Q be a nonempty subset of $\mathscr{X}$. If $\mathscr{S}$ is a family of maps (not necessarily linear) of Q into Q, then $\mathscr{S}$ is said to be a *noncontracting family of maps* if for two distinct points x and y in Q,

$$0 \notin \mathrm{cl}\{T(x) - T(y): T \in \mathscr{S}\}.$$

The next lemma has a straightforward proof whose discovery is left to the reader.

10.7. Lemma. *If $\mathscr{X}$ is a LCS, $Q \subseteq \mathscr{X}$, and $\mathscr{S}$ is a family of maps of Q into Q, then $\mathscr{S}$ is a noncontracting family if and only if for every pair of distinct points x and y in Q there is a continuous seminorm p such that*

$$\inf\{p(T(x) - T(y)): T \in \mathscr{S}\} > 0.$$

10.8. The Ryll–Nardzewski Fixed Point Theorem. *If $\mathscr{X}$ is a LCS, Q is a weakly compact convex subset of $\mathscr{X}$, and $\mathscr{S}$ is a noncontracting semigroup of weakly continuous affine maps of Q into Q, then there is a point x_0 in Q such that $T(x_0) = x_0$ for every T in $\mathscr{S}$.*

PROOF. The proof begins by showing that every finite subset of $\mathscr{S}$ has a common fixed point.

10.9. Claim. *If $\{T_1, \ldots, T_n\} \subseteq \mathscr{S}$, then there is an x_0 in Q such that $T_k x_0 = x_0$ for $1 \leqslant k \leqslant n$.*

Put $T_0 = (T_1 + \cdots + T_n)/n$; so $T_0 : Q \to Q$ and T_0 is weakly continuous and affine. By (10.1), there is an x_0 in Q such that $T_0(x_0) = x_0$. It will be shown that $T_k(x_0) = x_0$ for $1 \leqslant k \leqslant n$. In fact, if $T_k(x_0) \neq x_0$ for some k, then by renumbering the T_k, it can be assumed that there is an integer m such that $T_k(x_0) \neq x_0$ for $1 \leqslant k \leqslant m$ and $T_k(x_0) = x_0$ for $m < k \leqslant n$. Let $T_0' = (T_1 + \cdots + T_m)/m$. Then

$$x_0 = T_0(x_0)$$
$$= \frac{1}{n}[T_1(x_0) + \cdots + T_m(x_0)] + \left(\frac{n-m}{n}\right)x_0.$$

Hence

$$T_0'(x_0) = \frac{1}{m}[T_1(x_0) + \cdots + T_m(x_0)]$$
$$= \frac{n}{m}\frac{1}{n}[T_1(x_0) + \cdots + T_m(x_0)]$$
$$= \frac{n}{m}\left[x_0 - \left(\frac{n-m}{n}\right)x_0\right]$$
$$= x_0.$$

Thus it may be assumed that $T_k(x_0) \neq x_0$ for all k, but $T_0(x_0) = x_0$. Make this assumption.

By Lemma 10.7, there is an $\varepsilon > 0$ and there is a continuous seminorm p on $\mathscr{X}$ such that for every T in $\mathscr{S}$ and $1 \leqslant k \leqslant n$,

10.10 $p(T(T_k(x_0)) - T(x_0)) > \varepsilon.$

Let $\mathscr{L}_1 =$ the semigroup generated by $\{T_1, T_2, \ldots, T_n\}$. So $\mathscr{L}_1 \subseteq \mathscr{L}$ and $\mathscr{S}_1 = \{T_{l_1} \cdots T_{l_m} : m \geqslant 1, 1 \leqslant l_j \leqslant n\}$. Thus $\mathscr{S}_1$ is a countable subsemigroup of $\mathscr{S}$. Put $K = \overline{\text{co}}\{T(x_0) : T \in \mathscr{S}_1\}$. Therefore K is a weakly compact convex subset of Q and K is separable. By Lemma 10.2, there is a closed convex subset C of K such that $C \neq K$ and p-diam$(K \backslash C) \leqslant \varepsilon$.

Since $C \neq K$, there is an S in $\mathscr{S}_1$ such that $S(x_0) \in K \backslash C$. Hence

$$S(x_0) = ST_0(x_0) = \frac{1}{n}[ST_1(x_0) + \cdots + ST_n(x_0)] \in K \backslash C.$$

Since C is convex, there must be a k, $1 \leqslant k \leqslant n$, such that $ST_k(x_0) \in K \backslash C$. But this implies that $p(S(T_k(x_0)) - S(x_0)) \leqslant p$-diam$(K \backslash C) \leqslant \varepsilon$, contradicting (10.10). This establishes Claim 10.9.

Let $\mathscr{F} = $ all finite nonempty subsets of $\mathscr{S}$. If $F \in \mathscr{F}$, let $Q_F = \{x \in Q : T(x) = x$ for all T in $F\}$. By Claim 10.9, $Q_F \neq \square$ for every F in $\mathscr{F}$. Also, since each T in $\mathscr{S}$ is weakly continuous and affine, Q_F is convex and weakly compact. It is easy to see that $\{Q_F : F \in \mathscr{F}\}$ has the finite intersection property. Therefore, there is an x_0 in $\bigcap\{Q_F : F \in \mathscr{F}\}$. The point x_0 is the desired common fixed point for $\mathscr{S}$.   ■

The original reference for this theorem is Ryll–Nardzewski [1967]; the treatment here is from Namioka and Asplund [1967]. Another proof can be found in Hansel and Troallic [1976]. An application of this theorem is given in the next section.

EXERCISES

1. Was local convexity used in the proof of Theorem 10.1?
2. Show that if X is locally compact and $X = \bigcup_{n=1}^{\infty} F_n$, where each F_n is closed in X, then there is an integer n such that int $F_n \neq \square$. (Hint: Look at the proof of the Baire Category Theorem.)

§11*. An Application: Haar Measure on a Compact Group

In this section the operation in all semigroups and groups is multiplication and is denoted by juxtaposition.

11.1. Definition. A *topological semigroup* is a semigroup G that also is a topological space and such that the map $G \times G \to G$ defined by $(x, y) \mapsto xy$ is continuous. A *topological group* is a topological semigroup that is also a group such that the map $G \to G$ defined by $x \mapsto x^{-1}$ is continuous.

So a topological group is both a group and a topological space with a property that ties these two structures together.

$\mathbb{R}_{\geq 0}$ means the set of non-negative real numbers.

11.2. Examples

(a) $\mathbb{N}$ and $\mathbb{R}_{\geq 0}$ are topological semigroups under addition.
(b) $\mathbb{Z}$, $\mathbb{R}$, and $\mathbb{C}$ are topological groups under addition.
(c) $\partial \mathbb{D}$ is a topological group under multiplication.
(d) If X is a topological space and $G = \{f \in C(X): f(X) \subset \partial \mathbb{D}\}$, define $(fg)(x) = f(x)g(x)$ for f, g in G and x in X. Then G is a group. If G is given the topology of uniform convergence on X, G is a topological group.
(e) For $n \geq 1$, let $M_n(\mathbb{C}) =$ the $n \times n$ matrices with entries in $\mathbb{C}$; $O(n) \equiv \{A \in M_n(\mathbb{C}): A$ is invertible and $A^{-1} = A^*\}$; $SO(n) \equiv \{A \in O(n): \det A = 1\}$. If $M_n(\mathbb{C})$ is given the usual topology, $O(n)$ and $SO(n)$ are compact topological groups under multiplication.

There are many more examples and the subject is a self-sustaining area of research. Some good references are Hewitt and Ross [1963] and Rudin [1962].

11.3. Definition. If S is a semigroup and $f: S \to \mathbb{F}$, then for every x in S define $f_x: S \to \mathbb{F}$ and $_xf: S \to \mathbb{F}$ by $f_x(s) = f(sx)$ and $_xf(s) = f(xs)$ for all s in S. If S is also a group, let $f^{\#}(s) = f(s^{-1})$ for all s in S.

11.4. Theorem. *If G is a compact topological group, then there is a unique positive regular Borel measure m on G such that*

(a) $m(G) = 1$;
(b) *if U is a nonempty open subset of G, then $m(U) > 0$;*
(c) *if Δ is any Borel subset of G and $x \in G$, then $m(\Delta) = m(\Delta x) = m(x\Delta) = m(\Delta^{-1})$, where $\Delta x \equiv \{ax: a \in \Delta\}$, $x\Delta \equiv \{xa: a \in \Delta\}$, and $\Delta^{-1} \equiv \{a^{-1}: a \in \Delta\}$.*

The measure m is called the *Haar measure* for G. If G is locally compact, then it is also true that there is a positive Borel measure m on G satisfying (b) and such that $m(\Delta x) = m(\Delta)$ for all x in G and every Borel subset Δ of G. It is not necessarily true that $m(\Delta) = m(x\Delta)$, let alone that $m(\Delta) = m(\Delta^{-1})$ (see Exercise 4). The measure m is necessarily unbounded if G is not compact, so that (a) is not possible. Uniqueness, however, is still true in a modified form: if m_1, m_2 are two such measures, then $m_1 = \alpha m_2$ for some $\alpha > 0$.

By using the Riesz Representation Theorem for representing bounded linear functionals on $C(G)$, Theorem 11.4 is equivalent to the following.

11.5. Theorem. *If G is a compact topological group, then there exists a unique positive linear functional $I: C(G) \to \mathbb{F}$ such that*

(a) $I(1) = 1$;
(b) *if $f \in C(G)$, $f \geq 0$, and $f \neq 0$, then $I(f) > 0$;*
(c) *if $f \in C(G)$ and $x \in G$, then $I(f) = I(f_x) = I(_xf) = I(f^\#)$.*

Before proving Theorem 11.5, we need the following lemma. For a compact topological group G, if $x \in G$, define $L_x: M(G) \to M(G)$ and $R_x: M(G) \to M(G)$ by

$$\langle f, L_x(\mu) \rangle = \int {}_xf\, d\mu,$$

$$\langle f, R_x(\mu) \rangle = \int f_x\, d\mu$$

for f in $C(G)$ and μ in $M(G)$. Define $S_0: M(G) \to M(G)$ by

$$\langle f, S_0(\mu) \rangle = \int f^\#\, d\mu$$

for f in $C(G)$ and μ in $M(G)$. It is easy to check that L_x, R_x, and S_0 are linear isometries of $M(G)$ onto $M(G)$ (Exercise 5).

11.6. Lemma. *If G is a compact topological group, $\mu \in M(G)$, and $\rho: G \times G \to (M(G), \text{wk*})$ is defined by $\rho(x, y) = L_x R_y(\mu)$, then ρ is continuous. Similarly, if $\rho_0: G \times G \to (M(G), \text{wk*})$ is defined by $\rho_0(x, y) = S_0 L_x R_y(\mu)$, then ρ is continuous.*

PROOF. Let $f \in C(G)$ and let $\varepsilon > 0$. Then (Exercise 10) there is a neighborhood U of e (the identity of G) such that $|f(x) - f(y)| < \varepsilon$ whenever $xy^{-1} \in U$ or $x^{-1}y \in U$. Suppose $\{(x_i, y_i)\}$ is a net in $G \times G$ such that $(x_i, y_i) \to (x, y)$. Let i_0 be such that for $i \geq i_0, x_i x^{-1} \in U$ and $y_i^{-1} y \in U$. If $x \in G$, then $|f(x_i z y_i) - f(xzy)| \leq |f(x_i z y_i) - f(xzy_i)| + |f(xzy_i) - f(xzy)|$. But if $i \geq i_0$ and $z \in G$, $(x_i z y_i)(x z y_i)^{-1} = x_i x^{-1} \in U$ and $(xzy_i)^{-1}(xzy) = y_i^{-1} y \in U$. Hence $|f(x_i z y_i - f(xzy)| < 2\varepsilon$ for $i \geq i_0$ and for all z in G. Thus $\lim_i \int f(x_i z y_i) d\mu(z) = \int f(xzy) d\mu(z)$. Since f was arbitrary, this implies that $\rho(x_i, y_i) \to \rho(x, y) \text{wk*}$ in $M(G)$. The proof for ρ_0 is similar. ∎

PROOF OF THEOREM 11.5. If $e = $ the identity of G, then

11.7

$$\begin{cases} L_x R_y = R_y L_x \\ L_x L_y = L_{yx} \\ R_x R_y = R_{xy} \\ S_0^2 = L_e = R_e = \text{the identity on } M(G) \\ S_0 L_x R_y = L_{y^{-1}} R_{x^{-1}} S_0 \end{cases}$$

for x, y in G. Hence

$$(S_0 L_x R_y)(S_0 L_u R_v) = (L_{y^{-1}} R_{x^{-1}} S_0)(S_0 L_u R_v)$$
$$= L_{y^{-1}} R_{x^{-1}} L_u R_v$$
$$= L_{y^{-1}} L_u R_{x^{-1}} R_v$$
$$= L_{uy^{-1}} R_{x^{-1}v}.$$

Hence if $S_1 =$ the identity on $M(G)$,

$$\mathscr{S} = \{S_i L_x R_y : i = 0, 1; \, x, y \in G\}$$

is a group of surjective linear isometries of $M(G)$. Let $Q =$ the probability measures on G; that is, $Q = \{\mu \in M(G): \mu \geq 0 \text{ and } \mu(G) = 1\}$. So Q is a convex subset of $M(G)$ that is wk* compact. Furthermore, $T(Q) \subseteq Q$ for every T in $\mathscr{S}$.

11.8. Claim. If $\mu \in M(G)$ and $\mu \neq 0$, then $0 \notin$ the weak* closure of $\{T(\mu): T \in \mathscr{S}\}$.

In fact, Lemma 11.6 implies that $\{T(\mu): T \in \mathscr{S}\}$ is weak* closed. Since each T in $\mathscr{S}$ is an isometry, $T(\mu) \neq 0$ for every T in $\mathscr{S}$.

By Claim 11.8, $\mathscr{S}$ is a noncontracting family of affine maps of Q into itself. Moreover, if $T = S_0 L_x R_y$ and $\{\mu_i\}$ is a net in Q such that $\mu_i \to \mu(\text{wk*})$, then for every f in $C(G)$, $\langle f, T(\mu_i) \rangle = \int f(xs^{-1}y) d\mu_i(s) \to \int f(xs^{-1}y) d\mu = \langle f, T(\mu) \rangle$. So each T in $\mathscr{S}$ in wk* continuous on Q. By the Ryll–Nardzewski Fixed Point Theorem, there is a measure m in Q such that $T(m) = m$ for all T in $\mathscr{S}$.

By definition, (a) holds. Also, for any x in G and f in $C(G)$, $\int f(xs) dm(s) = \langle f, L_x(m) \rangle = \int f dm$. By similar equations, (c) holds. Now suppose $f \in C(G)$, $f \geq 0$, and $f \neq 0$. Then there is an $\varepsilon > 0$ such that $U = \{x \in G: f(x) > \varepsilon\}$ is nonempty. Since U is open, $G = \cup \{Ux: x \in G\}$, and G is compact, there are $x_1, x_2, \ldots, x_n$ in G such that $G \subseteq \bigcup_{k=1}^n Ux_k$. (Why is Ux open?) Define $g_k(x) = f(xx_k^{-1})$ and put $g = \sum_{k=1}^n g_k$. Then $g \in C(G)$ and $\int g\, dm = \sum_{k=1}^n \int g_k dm = n \int f dm$ by (c). But for any x in G there is an x_k such that $xx_k^{-1} \in U$; hence $g(x) \geq g_k(x) = f(xx_k^{-1}) > \varepsilon$. Thus

$$\int f dm = \frac{1}{n} \int g\, dm \geq \varepsilon/n > 0.$$

This proves (b).

To prove uniqueness, let μ be a probability measure on G having properties (a), (b), (c). If $f \in C(G)$ and $x \in G$, then $\int f d\mu = \int_x f d\mu$. Hence

$$\int f d\mu = \int \left[\int f(y) d\mu(y) \right] dm(x)$$

$$= \int \left[\int f(xy) d\mu(y) \right] dm(x)$$

$$= \int \left[\int f(xy)dm(x) \right] d\mu(y)$$

$$= \int \left[\int f(x)dm(x) \right] d\mu(y)$$

$$= \int f dm.$$

Hence $\mu = m$. ∎

For further information on Haar measure see Nachbin [1965].

What happens if G is only a semigroup? In this case L_x and R_x may not be isometries, so $\{L_x R_y : x, y \in G\}$ may not be noncontractive. However, there are measures for some semigroups that are invariant (see Exercise 7). For further reading see Greenleaf [1969].

EXERCISES

1. Let G be a group and a topological space. Show that G is a topological group if and only if the map of $G \times G \to G$ defined by $(x, y) \mapsto x^{-1}y$ is continuous.

2. Verify the statements in (11.2).

3. Show that Theorems (11.4) and (11.5) are equivalent.

4. Let G be a locally compact group. If m is a regular Borel measure on G, show that any two of the following properties imply the third: (a) $m(\Delta x) = m(\Delta)$ for every Borel set Δ and every x in G; (b) $m(x\Delta) = m(\Delta)$ for every Borel set Δ and every x in G; (c) $m(\Delta) = m(\Delta^{-1})$ for every Borel set Δ.

5. Show that the maps S_0, L_x, R_x are linear isometries of $M(G)$ onto $M(G)$.

6. Prove (11.7).

7. Let S be an abelian semigroup and show that there is a positive linear functional $L: l^\infty(S) \to \mathbb{F}$ such that (a) $L(1) = 1$, (b) $L(f_x) = L(f)$ for every f in $l^\infty(S)$.

8. If $S = \mathbb{N}$, what does Exercise 7 say about Banach limits?

9. If G is a compact group, $f: G \to \mathbb{F}$ is a continuous function, and $\varepsilon > 0$, show that there is a neighborhood U of the identity in G such that $|f(x) - f(y)| < \varepsilon$ whenever $xy^{-1} \in U$. (Note that this say that every continuous function on a compact group is uniformly continuous.)

10. If G is a locally compact group and $f \in C_b(G)$, let $\mathcal{O}(f) \equiv$ the closure of $\{f_x : x \in G\}$ in $C_b(G)$. Let $AP(G) = \{f \in C_b(G) : \mathcal{O}(f) \text{ is compact}\}$. Functions in $AP(G)$ are called *almost periodic*. (a) Show that every periodic function in $C_b(\mathbb{R})$ belongs to $AP(\mathbb{R})$. (b) If G is compact, show that $AP(G) = C(G)$. (c) Show that if $f \in C_b(\mathbb{R})$, then $f \in AP(\mathbb{R})$ if and only if for every $\varepsilon > 0$ there is a positive number T such that in every interval of length T there is a number p such that $|f(x) - f(x + p)| < \varepsilon$ for all x in $\mathbb{R}$. (d) If G is not compact, then the only function in $AP(G)$ having compact support is the zero function. For more information on this topic, see Exercise 13.5 below.

§12*. The Krein–Smulian Theorem

Let A be a convex subset of a Banach space $\mathscr{X}$. If A is weakly closed, then for every $r > 0$, $A \cap \{x \in \mathscr{X}: \|x\| \leq r\}$ is weakly closed; this is clear since each of the sets in the intersection is weakly closed. But the converse of this is also true: if A is convex and $A \cap \{X \in \mathscr{X}: \|x\| \leq r\}$ is weakly closed for every $r > 0$, then A is weakly closed. In fact, because A is convex it suffices to prove that A is norm closed (Corollary 1.5). If $\{x_n\} \subseteq A$ and $\|x_n - x_0\| \to 0$, then there is a constant r such that $\|x_n\| \leq r$ for all n. By hypothesis, $A \cap \{x \in \mathscr{X}: \|x\| \leq r\}$ is weakly closed and hence norm closed. Thus $x_0 \in A$.

Now let A be a convex subset of $\mathscr{X}^*, \mathscr{X}$ a Banach space. If $A \cap \{x^* \in \mathscr{X}^*: \|x^*\| \leq r\}$ is weak-star closed for every $r > 0$, is A weak-star closed? If $\mathscr{X}$ is reflexive, then this is the same question that was asked and answered affirmatively in the preceding paragraph. If $\mathscr{X}$ is not reflexive, then the preceding argument fails since there are norm closed convex subsets of $\mathscr{X}^*$ that are not weak-star closed. (Example: let $x^{**} \in \mathscr{X}^{**} \setminus \mathscr{X}$ and consider $A = \ker x^{**}$.) Nevertheless, even though the argument fails, the statement is true.

12.1. The Krein–Smulian Theorem. *If $\mathscr{X}$ is a Banach space and A is a convex subset of $\mathscr{X}^*$ such that $A \cap \{x^* \in \mathscr{X}^*: \|x^*\| \leq r\}$ is weak-star closed for every $r > 0$, then A is weak-star closed.*

To prove this theorem, two lemmas are needed.

12.2. Lemma. *If $\mathscr{X}$ is a Banach space, $r > 0$, and $\mathscr{F}_r$ is the collection of all finite subsets of $\{x \in \mathscr{X}: \|x\| \leq r^{-1}\}$, then*

$$\bigcap \{F^\circ: F \in \mathscr{F}_r\} = \{x^* \in \mathscr{X}^*: \|x^*\| \leq r\}.$$

PROOF. Let $E = \bigcap \{F^\circ: F \in \mathscr{F}_r\}$; it is easy to see that $r(\text{ball } \mathscr{X}^*) \subseteq E$. If $x^* \notin r(\text{ball } \mathscr{X}^*)$, then there is an x in ball $\mathscr{X}$ such that $|\langle x, x^* \rangle| > r$. Hence $|\langle r^{-1}x, x^* \rangle| > 1$ and $x^* \notin E$. ∎

12.3. Lemma. *If A and $\mathscr{X}$ satisfy the hypothesis of the Krein–Smulian Theorem and, moreover, $A \cap \text{ball } \mathscr{X}^* = \square$, then there is an x in $\mathscr{X}$ such that*

$$\text{Re} \langle x, x^* \rangle \geq 1$$

for all x^* in A.

PROOF. The proof begins by showing that there are finite subsets $F_0, F_1, \ldots$ of $\mathscr{X}$ such that

12.4 $\begin{cases} \text{(i) } nF_n \subseteq \text{ball } \mathscr{X}; \\ \text{(ii) } n(\text{ball } \mathscr{X}^*) \cap \bigcap_{k=0}^{n-1} F_k^\circ \cap A = \square. \end{cases}$

To establish (12.4) use induction as follows. Let $F_0 = (0)$. Suppose that $F_0, \ldots, F_{n-1}$ have been chosen satisfying (12.4) and set $Q = [(n+1)\text{ball } \mathscr{X}^*] \cap$

$\bigcap_{k=0}^{n-1}F_k^{\circ}\cap A$. Note that Q is wk* compact. So if $Q\cap F^{\circ}\neq\square$ for every finite subset F of n^{-1}ball $\mathcal{X}$, then $\square\neq Q\cap\bigcap\{F^{\circ}:\ F$ is a finite subset of $n^{-1}(\text{ball }\mathcal{X})\}=Q\cap[n(\text{ball}\mathcal{X}^*)]$ by the preceding lemma. This contradicts (12.4ii). Therefore there is a finite subset F_n of $n^{-1}(\text{ball }\mathcal{X})$ such that $Q\cap F_n^{\circ}=\square$. This proves (12.4).

If $\{F_n\}_{n=1}^{\infty}$ satisfies (12.4), then $A\cap\bigcap_{n=1}^{\infty}F_n^{\circ}=\square$. Arrange the elements of $\bigcup_{n=1}^{\infty}F_n$ in a sequence and denote this sequence by $\{x_n\}$. Note that $\lim\|x_n\|=0$. Thus if $x^*\in\mathcal{X}^*$, $\{\langle x_n,x^*\rangle\}\in c_0$. Define $T:\ \mathcal{X}^*\to c_0$ by $T(x^*)=\{\langle x_n,x^*\rangle\}$. It is easy to see that T is linear (and bounded, though this fact is unnecessary). Hence $T(A)$ is a convex subset of c_0. Also, from the construction of $\{x_n\}=\bigcup_{n=1}^{\infty}F_n$, for each x^* in A, $\|T(x^*)\|=\sup_n|\langle x_n,x^*\rangle|>1$. That is, $T(A)\cap\text{ball }c_0=\square$. Thus Theorem IV.3.7 applies to the sets $T(A)$ and $\text{int}[\text{ball }c_0]$ and there is an f in $l^1=c_0^*$ and an α in $\mathbb{R}$ such that $\text{Re}\langle\phi,f\rangle<\alpha\leqslant\text{Re}\langle T(x^*),f\rangle$ for every ϕ in int $[\text{ball }c_0]$ and x^* in A. That is

12.5
$$\text{Re}\sum_{n=1}^{\infty}\phi(n)f(n)<\alpha\leqslant\text{Re}\sum_{n=1}^{\infty}\langle x_n,x^*\rangle f(n)$$

for every ϕ in c_0 with $\|\phi\|<1$ and for every x^* in A. Replacing f by $f/\|f\|$ and α by $\alpha/\|f\|$, it is clear that it may be assumed that (12.5) holds with $\|f\|=1$. If $\phi\in c_0$, $\|\phi\|<1$, let $\mu\in\mathbb{F}$ such that $|\mu|=1$ and $\langle\mu\phi,f\rangle=|\langle\phi,f\rangle|$. Applying this to (12.5) and taking the supremum over all ϕ in int$[\text{ball }c_0]$ gives that $1\leqslant\text{Re}\sum_{n=1}^{\infty}\langle x_n,x^*\rangle f(n)$ for all x^* in A. But $f\in l^1$ so $x=\sum_{n=1}^{\infty}f(n)x_n\in\mathcal{X}$ and $1\leqslant\text{Re}\langle x,x^*\rangle$ for all x^* in A. ∎

Where was the completeness of $\mathcal{X}$ used in the preceding proof?

PROOF OF THE KREIN–SMULIAN THEOREM. Let $x_0^*\in\mathcal{X}^*\backslash A$; it will be shown that $x_0^*\notin$wk* $-$ cl A. It is easy to see that A is norm closed. So there is an $r>0$ such that $\{x^*\in\mathcal{X}^*:\ \|x^*-x_0^*\|\leqslant r\}\cap A=\square$. But this implies that ball $\mathcal{X}^*\cap[r^{-1}(A-x_0^*)]=\square$. With this it is easy to see that $r^{-1}(A-x_0^*)$ satisfies the hypothesis of the preceding lemma. Therefore there is an x in $\mathcal{X}$ such that $\text{Re}\langle x,x^*\rangle\geqslant1$ for all x^* in $r^{-1}(A-x_0^*)$. In particular, $0\notin$wk* $-$ cl$[r^{-1}(A-x_0^*)]$ and hence $x_0^*\notin$wk* $-$ cl A. ∎

12.6. Corollary. *If $\mathcal{X}$ is a Banach space and $\mathcal{Y}$ is a linear manifold in $\mathcal{X}^*$, then $\mathcal{Y}$ is weak-star closed if and only if $\mathcal{Y}\cap$ ball $\mathcal{X}^*$ is weak-star closed.*

12.7. Corollary. *If $\mathcal{X}$ is a separable Banach space and A is a convex subset of $\mathcal{X}^*$ that is weak-star sequentially closed, then A is weak-star closed.*

PROOF. Because $\mathcal{X}$ is separable, $r(\text{ball }\mathcal{X}^*)$ is weak-star metrizable for every $r>0$ (Theorem 5.1). So if A is weak-star sequentially closed, $A\cap[r(\text{ball }\mathcal{X}^*)]$ is weak-star closed for every $r>0$. Hence the Krein–Smulian Theorem applies. ∎

This last corollary is one of the most useful forms of the Krein–Smulian Theorem. To show that a convex subset A of $\mathscr{X}^*$ is weak-star closed it is not necessary to show that every weak-star convergent net from A has its limit in A; it suffices to prove this for sequences.

12.8. Corollary. *If $\mathscr{X}$ is a separable Banach space and $F: \mathscr{X}^* \to \mathbb{F}$ is a linear functional, then F is weak-star continuous if and only if F is weak-star sequentially continuous.*

PROOF. By Theorem IV.3.1, F is wk* continuous if and only if $\ker F$ is wk* closed. This corollary is, therefore, a direct consequence of the preceding one. ∎

A proof of Corollary 12.8 that is independent of The Krein–Smulian Theorem can be found as a lemma in Whitley [1986].

There is a misinterpretation of the Krein–Smulian Theorem that the reader should be warned about. If A is a weak-star closed convex balanced subset of ball $\mathscr{X}^*$, let $\mathscr{M} = \bigcup \{rA: r > 0\}$. It is easy to see that $\mathscr{M}$ is a linear manifold, but it does not follow that $\mathscr{M}$ is weak-star closed. What is true is the following.

12.9. Theorem. *Let $\mathscr{X}$ be a Banach space and let A be a weak-star closed subset of $\mathscr{X}^*$. If $\mathscr{Y} =$ the linear span of A, then $\mathscr{Y}$ is norm closed in $\mathscr{X}^*$ if and only if $\mathscr{Y}$ is weak-star closed.*

The proof will not be presented here. The interested reader can consult Dunford and Schwartz [1958], p. 429.

There is a method for finding the weak-star closure of a linear manifold that is quite useful despite its seemingly bizarre appearance. Let $\mathscr{X}$ be a Banach space and let $\mathscr{M}$ be a linear manifold in $\mathscr{X}^*$. For each ordinal number α define a linear manifold $\mathscr{M}_\alpha$ as follows. Let $\mathscr{M}_1 = \mathscr{M}$. Suppose α is an ordinal number and $\mathscr{M}_\beta$ has been defined for each ordinal $\beta < \alpha$. If α has an immediate predecessor, $\alpha - 1$, let $\mathscr{M}_\alpha$ be the weak-star sequential closure of $\mathscr{M}_{\alpha-1}$. If α is a limit ordinal and has no immediate predecessor, let $\mathscr{M}_\alpha = \bigcup \{\mathscr{M}_\beta: \beta < \alpha\}$. In each case $\mathscr{M}_\alpha$ is a linear manifold in $\mathscr{X}^*$ and $\mathscr{M}_\beta \subseteq \mathscr{M}_\alpha$ if $\beta \leqslant \alpha$.

12.10. Theorem. *If $\mathscr{X}$ is a separable Banach space, $\mathscr{M}$ is a linear manifold in $\mathscr{X}^*$, and $\mathscr{M}_\alpha$ is defined as above for every ordinal number α, then $\mathscr{M}_\Omega$ is the weak-star closure of $\mathscr{M}$, where Ω is the first uncountable ordinal. Moreover, there is an ordinal number $\alpha < \Omega$ such that $\mathscr{M}_\alpha = \mathscr{M}_\Omega$.*

PROOF. By Corollary 12.7 it suffices to show that $\mathscr{M}_\Omega$ is weak-star sequentially closed. Let $\{x_n^*\}$ be a sequence in $\mathscr{M}_\Omega$ such that $x_n^* \to x^*$ (wk*). Since $\mathscr{M}_\Omega = \bigcup \{\mathscr{M}_\alpha: \alpha < \Omega\}$, for each n there is an $\alpha_n < \Omega$ such that $x_n^* \in \mathscr{M}_{\alpha_n}$. But $\alpha = \sup_n \alpha_n < \Omega$. Hence $x_n^* \in \mathscr{M}_\alpha$ for all n; thus $x^* \in \mathscr{M}_{\alpha+1} \subseteq \mathscr{M}_\Omega$ and $\mathscr{M}_\Omega$ is weak-star closed.

To see that $\mathcal{M}_\Omega = \mathcal{M}_\alpha$ for some $\alpha < \Omega$, let $\{x_n^*\}$ be a countable wk* dense subset of ball $\mathcal{M}_\Omega$. For each n there is an α_n such that $x_n^* \in \mathcal{M}_{\alpha_n}$. But $\alpha = \sup_n \alpha_n$. So $\{x_n^*\} \subset$ ball $\mathcal{M}_\alpha$. But ball $\mathcal{M}_\Omega$ is a compact metric space in the weak-star topology, so $\{x_n^*\}$ is wk* sequentially dense in ball $\mathcal{M}_\Omega$. Therefore ball $\mathcal{M}_\Omega \subseteq$ ball $\mathcal{M}_{\alpha+1}$ and $\mathcal{M}_\Omega = \mathcal{M}_{\alpha+1}$. ∎

When is $\mathcal{M}$ weak-star sequentially dense in $\mathcal{X}^*$? The following result of Banach answers this question.

12.11. Theorem. *If $\mathcal{X}$ is a separable Banach space and $\mathcal{M}$ is a linear manifold in $\mathcal{X}^*$, then the following statements are equivalent.*

(a) *$\mathcal{M}$ is weak-star sequentially dense in $\mathcal{X}^*$.*

(b) *There is a positive constant c such that for every x in $\mathcal{X}$,*

$$\|x\| \leqslant \sup\{|\langle x, x^*\rangle|: x^* \in \mathcal{M}, \|x^*\| \leqslant c\}.$$

(c) *There is a positive constant c such that if $x^* \in$ ball $\mathcal{X}^*$, there is a sequence $\{x_k^*\}$ in $\mathcal{M}$, $\|x_k^*\| \leqslant c$, such that $x_k^* \to x^*$ (wk*).*

PROOF. It is clear that (c) implies (a). The proof will consist in showing that (a) implies (c) and that (b) and (c) are equivalent.

(a)$\Rightarrow$(c): For each positive integer n, let $A_n =$ the wk* closure of n(ball $\mathcal{M}$). If $x^* \in \mathcal{X}^*$, let $\{x_k^*\}$ be a sequence in $\mathcal{M}$ such that $x_k^* \to x^*$ (wk*). By the PUB, there is an n such that $\|x_k^*\| \leqslant n$ for all k. Hence $x^* \in A_n$. That is, $\bigcup_{n=1}^\infty A_n = \mathcal{X}^*$. Clearly each A_n is norm closed, so the Baire Category Theorem implies that there is an A_n that has interior in the norm topology. Thus there is an x_0^* in A_n and an $r > 0$ such that $A_n \supseteq \{x^* \in \mathcal{X}^*: \|x^* - x_0^*\| \leqslant r\}$. Let $\{x_k^*\} \subseteq n$(ball $\mathcal{M}$) such that $x_k^* \to x_0^*$(wk*). If $x^* \in$ ball $\mathcal{X}^*$, then $x_0^* + rx^* \in A_n$; hence there is a sequence $\{y_k^*\}$ in n(ball $\mathcal{M}$) such that $y_k^* \to x_0^* + rx^*$ (wk*). Thus $r^{-1}(y_k^* - x_k^*) \to x^*$ (wk*) and $r^{-1}(y_k^* - x_k^*) \in c$(ball $\mathcal{M}$), where $c = 2n/r$ is independent of x^*.

(c)$\Rightarrow$(b): If $x \in \mathcal{X}$, then Alaoglu's Theorem implies there is an x^* in ball $\mathcal{X}^*$ such that $\langle x, x^*\rangle = \|x\|$. By (c), there is a sequence $\{x_k^*\}$ in c(ball $\mathcal{M}$) such that $x_k^* \to x^*$ (wk*). Thus $\langle x_k^*, x\rangle \to \|x\|$ and (b) holds.

(b)$\Rightarrow$(c): According to (b), ball $\mathcal{X} \supseteq {}^\circ[c$(ball $\mathcal{M}$)]. Hence ball $\mathcal{X}^* =$ (ball $\mathcal{X})^\circ \subseteq {}^\circ[c$(ball $\mathcal{M}$)]$^\circ$. By (1.8), ${}^\circ[c$(ball $\mathcal{M}$)]$^\circ =$ the weak-star closure of c(ball $\mathcal{M}$). But bounded subsets of $\mathcal{X}^*$ are weak-star metrizable (5.1) and hence (c) follows. ∎

EXERCISES

1. Suppose $\mathcal{X}$ is a normed space and that the only hyperplanes $\mathcal{M}$ in $\mathcal{X}^*$ such that $\mathcal{M} \cap$ ball $\mathcal{X}^*$ is weak-star closed are those that are weak-star closed. Prove that $\mathcal{X}$ is a Banach space.

2. (von Neumann) Let A be the subset of l^2 consisting of all vectors $\{x_{mn}: 1 \leqslant m < n < \infty\}$ where $x_{mn}(m) = 1$, $x_{mn}(n) = m$, and $x_{mn}(k) = 0$ if $k \neq m, n$. Show that $0 \in$ wk $-$ cl A but no sequence in A converges weakly to 0.

3. Where were the hypotheses of the separability and completeness of $\mathscr{X}$ used in the proof of Theorem 12.11?

4. Let $\mathscr{X}$ be a separable Banach space. If $\mathscr{M}$ is a linear manifold in $\mathscr{X}^*$ give necessary and sufficient conditions that every functional in wk* $-$ cl $\mathscr{M}$ be the wk* limit of a sequence from $\mathscr{M}$.

5. Let $\mathscr{X}$ be a normed space and let $\mathscr{T}$ be a locally convex topology on $\mathscr{X}$ such that ball $\mathscr{X}$ is $\mathscr{T}$-compact. Show that there is a Banach space $\mathscr{Y}$ such that $\mathscr{X}$ is isometrically isomorphic to $\mathscr{Y}^*$. (Hint: Let $\mathscr{Y} = \{x^* \in \mathscr{X}^*: x^* | \text{ball } \mathscr{X} \text{ is } \mathscr{T}\text{-continuous}\}$.)

6. If X is a compact connected topological space that is not a singleton, show that $C(X)$ is not the dual of a Banach space. (For $C_{\mathbf{R}}(X)$ this is a consequence of Exercise 7.4. For the complex case, show that if $C(X)$ is a dual space, then $C_{\mathbf{R}}(X)$ is a weak* closed real linear subspace of $C(X)$. Hint (S. Axler): Suppose $\{u_j\}$ is a net in ball $C_{\mathbf{R}}(X)$ such that $u_j \to u + iv$ weak*. If $v(x) = t > 0$, choose n such that $1 + n^2 < (n + t)^2$ and examine the net $\{u_j + in\}$.)

§13*. Weak Compactness

In this section, two results are stated without proof. These results are among the deepest in the study of weak topologies.

13.1. The Eberlein–Smulian Theorem. *If $\mathscr{X}$ is a Banach space and $A \subseteq \mathscr{X}$, then the following statements are equivalent.*

(a) *Each sequence of elements of A has a subsequence that is weakly convergent.*
(b) *Each sequence of elements of A has a weak cluster point.*
(c) *The weak closure of A is weakly compact.*

An elementary proof of the Eberlein–Smulian Theorem can be found in Whitley [1967] and Kremp [1986]. Another proof can be found in Dunford and Schwartz [1958], p. 430. The serious student should examine Chapter V of Dunford and Schwartz [1958] for several results not presented here as well as for some of the history behind the material of this chapter.

The following is a consequence of the Eberlein–Smulian Theorem.

13.2. Corollary. *If $\mathscr{X}$ is a Banach space and $A \subseteq \mathscr{X}$, then A is weakly compact if and only if $A \cap \mathscr{M}$ is weakly compact for every separable subspace $\mathscr{M}$ of $\mathscr{X}$.*

If $\mathscr{X}$ is Banach space and A is a weakly compact subset of $\mathscr{X}$, then for each x^* in $\mathscr{X}^*$ there is an x_0 in A such that $|\langle x_0, x^* \rangle| = \sup\{|\langle x, x^* \rangle|: x \in A\}$. It is a rather deep fact due to R.C. James [1964a] that the converse is true.

13.3. James's Theorem. *If $\mathscr{X}$ is a Banach space and A is a closed convex subset of $\mathscr{X}$ such that for each x^* in $\mathscr{X}^*$ there is an x_0 in A with*

$$|\langle x_0, x^* \rangle| = \sup\{|\langle x, x^* \rangle|: x \in A\},$$

then A is weakly compact.

A proof of James's Theorem can be found in Pyrce [1966]. Another reference for a proof of this theorem as well as a number of other equivalent formulations of weak compactness and reflexivity is James [1964b]. Also, if $\mathscr{X}$ is only assumed to be a normed space in Theorem 13.2, the conclusion is false (see James [1971]).

The next result, presented with proof, is also called the Krein–Smulian Theorem and must not be confused with the theorem of the preceding section.

13.4. Krein–Smulian Theorem. *If $\mathscr{X}$ is a Banach space and K is a weakly compact subset of $\mathscr{X}$, then $\overline{\mathrm{co}}(K)$ is weakly compact.*

PROOF. *Case 1*: $\mathscr{X}$ *is separable.* Endow K with the relative weak topology; so $M(K) = C(K)^*$. If $\mu \in M(K)$, define $F_\mu : \mathscr{X}^* \to \mathbb{F}$ by

$$F_\mu(x^*) = \int_K \langle x, x^* \rangle \, d\mu(x).$$

It is easy to see that F_μ is a bounded linear functional on $\mathscr{X}^*$ and $\|F_\mu\| \leqslant \|\mu\| \sup\{\|x\|: x \in K\}$.

13.5. Claim. $F_\mu : \mathscr{X}^* \to \mathbb{F}$ *is weak-star continuous.*

By (12.8) it suffices to show that F_μ is weak* sequentially continuous. Let $\{x_n^*\}$ be a sequence in $\mathscr{X}^*$ such that $x_n^* \to x^*$ (wk*). By the PUB, $M = \sup_n \|x_n^*\| < \infty$. Also, $\langle x, x_n^* \rangle \to \langle x, x^* \rangle$ for every x in K. By the Lebesgue Dominated Convergence Theorem, $F_\mu(x_n^*) = \int \langle x, x_n^* \rangle \, d\mu(x) \to F_\mu(x^*)$. So (13.5) is established.

By (1.3), $F_\mu \in \mathscr{X}$. That is, there is an x_μ in $\mathscr{X}$ such that $F_\mu(x^*) = \langle x_\mu, x^* \rangle$. Define $T: M(K) \to \mathscr{X}$ by $T(\mu) = x_\mu$.

13.6. Claim. $T: (M(K), \text{wk*}) \to (\mathscr{X}, \text{wk})$ *is continuous.*

In fact, this is clear. If $\mu_i \to 0$ weak* in $M(K)$, then for each x^* in $\mathscr{X}^*$, $x^* | K \in C(K)$. Hence $\langle T(\mu_i), x^* \rangle = \int \langle x, x^* \rangle \, d\mu_i(x) \to 0$.

Let $\mathscr{P}$ = the probability measures on K. By Alaoglu's Theorem $\mathscr{P}$ is weak* compact. Thus $T(\mathscr{P})$ is weakly compact and convex. However, if $x \in K$, $\langle T(\delta_x), x^* \rangle = \langle x, x^* \rangle$; that is, $T(\delta_x) = x$. So $T(\mathscr{P}) \supseteq K$. Hence $T(\mathscr{P}) \supseteq \overline{\mathrm{co}}(K)$ and $\overline{\mathrm{co}}(K)$ must be compact.

Case 2: $\mathscr{X}$ *is arbitrary.* Let $\{x_n\}$ be a sequence in $\mathrm{co}(K)$. So for each n there is a finite subset F_n of K such that $x_n \in \mathrm{co}(F_n)$. Let $F = \bigcup_{n=1}^\infty F_n$ and let $\mathscr{M} = \vee F$. Then $K_1 = K \cap \mathscr{M}$ is weakly compact and $\{x_n\} \subseteq \mathrm{co}(K_1)$. Since $\mathscr{M}$ is separable, Case 1 implies that $\overline{\mathrm{co}}(K_1)$ is weakly compact. By the Eberlein–Smulian Theorem, there is a subsequence $\{x_{n_k}\}$ and an x in $\overline{\mathrm{co}}(K_1) \subseteq \overline{\mathrm{co}}(K)$ such that $x_{n_k} \to x$. Thus $\overline{\mathrm{co}}(K)$ is weakly compact. ∎

Another proof of the Krein–Smulian Theorem that avoids 13.1 and 12.8 can be found in Simons [1967].

EXERCISES

1. Prove Corollary 13.2.

2. If $\mathscr{X}$ is a Banach space and K is a compact subset of $\mathscr{X}$, prove that $\overline{\mathrm{co}}(K)$ is compact. (This will be proved in Theorem VI.4.8 below.)

3. In the proof of (13.4), if $\mathscr{P}$ = the probability measures on K, show that $T(\mathscr{P}) = \overline{\mathrm{co}}(K)$.

4. Prove the Eberlein–Smulian Theorem in the setting of Hilbert space.

5. Refer to the notation of Exercise 11.10. Prove that there is a bounded linear functional $L: AP(G) \to \mathbb{F}$ such that $L(1) = 1$, $L(f) \geqslant 0$ if $f \geqslant 0$, and $L(f_x) = L(f)$ for all f in $AP(G)$ and x in G.

CHAPTER VI

Linear Operators on a Banach Space

As has been said before in this book, the theory of bounded linear operators on a Banach space has seen relatively little activity owing to the difficult geometric problems inherent in the concept of a Banach space. In this chapter several of the general concepts of this theory are presented. When combined with the few results from the next chapter, they constitute essentially the whole of the general theory of these operators.

   We begin with a study of the adjoint of a Banach space operator. Unlike the adjoint of an operator on a Hilbert space (Section II.2), the adjoint of a bounded linear operator on a Banach space does not operate on the space but on the dual space.

§1. The Adjoint of a Linear Operator

Suppose $\mathscr{X}$ and $\mathscr{Y}$ are vector spaces and $T: \mathscr{X} \to \mathscr{Y}$ is a linear transformation. Let $\mathscr{Y}' =$ all of the linear functionals of $\mathscr{Y} \to \mathbb{F}$. If $y' \in \mathscr{Y}'$, then $y' \circ T: \mathscr{X} \to \mathbb{F}$ is easily seen to be a linear functional on $\mathscr{X}$. That is, $y' \circ T \in \mathscr{X}'$. This defines a map

$$T': \mathscr{Y}' \to \mathscr{X}'$$

by $T'(y') = y' \circ T$. The first result shows that if $\mathscr{X}$ and $\mathscr{Y}$ are Banach spaces, then the map T' can be used to determine when T is bounded. Another equivalent formulation of boundedness is given by means of the weak topology.

1.1. Theorem. *If $\mathscr{X}$ and $\mathscr{Y}$ are Banach spaces and $T: \mathscr{X} \to \mathscr{Y}$ is a linear transformation, then the following statements are equivalent.*

(a) T is bounded.

(b) $T'(\mathscr{Y}^*) \subseteq \mathscr{X}^*$.

(c) $T: (\mathscr{X}, \text{weak}) \to (\mathscr{Y}, \text{weak})$ is continuous.

PROOF. (a)$\Rightarrow$(b): If $y^* \in \mathscr{Y}^*$, then $T'(y^*) \in \mathscr{X}'$; it must be shown that $T'(y^*) \in \mathscr{X}^*$. But $|T'(y^*)(x)| = |y^* \circ T(x)| = |\langle T(x), y^* \rangle| \leqslant \|T(x)\| \|y^*\| \leqslant \|T\| \|y^*\| \|x\|$. So $T'(y^*) \in \mathscr{X}^*$.

(b)$\Rightarrow$(c): If $\{x_i\}$ is a net in $\mathscr{X}$ and $x_i \to 0$ weakly, then for y^* in $\mathscr{Y}^*$, $\langle T(x_i), y^* \rangle = T'(y^*)(x_i) \to 0$ since $T'(y^*) \in \mathscr{X}^*$. Hence $T(x_i) \to 0$ weakly in $\mathscr{Y}$.

(c)$\Rightarrow$(b): If $y^* \in \mathscr{Y}^*$, then $y^* \circ T: \mathscr{X} \to \mathbb{F}$ is weakly continuous by (c). Hence $T'(y^*) = y^* \circ T \in \mathscr{X}^*$ by (V.1.2).

(b)$\Rightarrow$(a): Let $y^* \in \mathscr{Y}^*$ and put $x^* = T'(y^*)$. So $x^* \in \mathscr{X}^*$ by (b). So if $x \in$ ball $\mathscr{X}$, $|\langle T(x), y^* \rangle| = |\langle x, x^* \rangle| \leqslant \|x^*\|$. That is, $\sup\{|\langle T(x), y^* \rangle|: x \in \text{ball } \mathscr{X}\} < \infty$. Hence $T(\text{ball } \mathscr{X})$ is weakly bounded; by the PUB, $T(\text{ball } \mathscr{X})$ is norm bounded and so $\|T\| < \infty$. ∎

The preceding result is useful, though strictly speaking it is not necessary for the purpose of defining the adjoint of an operator A in $\mathscr{B}(\mathscr{X}, \mathscr{Y})$, which we now turn to. If $A \in \mathscr{B}(\mathscr{X}, \mathscr{Y})$ and $y^* \in \mathscr{Y}^*$, then $y^* \circ A = A'(y^*) \in \mathscr{X}^*$. This defines a map $A^*: \mathscr{Y}^* \to \mathscr{X}^*$, where $A^* = A'|\mathscr{Y}^*$. Hence

1.2
$$\langle x, A^*(y^*) \rangle = \langle A(x), y^* \rangle$$

for x in $\mathscr{X}$ and y^* in $\mathscr{Y}^*$. A^* is called the *adjoint* of A.

Before exploring the concept let's see how this compares with the definition of the adjoint of an operator on Hilbert space given in §II.2. There is a difference, but only a small one. When $\mathscr{H}$ is identified with $\mathscr{H}^*$, the dual space of $\mathscr{H}$, the identification is not linear but conjugate linear (if $\mathbb{F} = \mathbb{C}$). The isometry $h \mapsto L_h$ of $\mathscr{H}$ onto $\mathscr{H}^*$, where $L_h(f) = \langle f, h \rangle$, satisfies $L_{\alpha h} = \bar{\alpha} L_h$. Thus the definition of A^* given in (1.2) above is not the same as the adjoint of an operator on Hilbert space, since in (1.2) A^* is defined on $\mathscr{Y}^*$ and not some conjugate-linear isomorphic image of it. In particular, if the definition (1.2) is applied to a matrix A acting on $\mathbb{C}^d$ considered as a Banach space, its adjoint corresponds to the transpose of A. If $\mathbb{C}^d$ is considered as a Hilbert space, then the matrix of A^* is the conjugate transpose of the matrix of A. This difference will not confuse us but it will serve to explain minor differences that will appear in the treatment of the two types of adjoints. The first of these occurs in the next result.

1.3. Proposition. *If $\mathscr{X}$ and $\mathscr{Y}$ are Banach spaces, $A, B \in \mathscr{B}(\mathscr{X}, \mathscr{Y})$, and $\alpha, \beta \in \mathbb{F}$, then $(\alpha A + \beta B)^* = \alpha A^* + \beta B^*$. Moreover, $A^*: (\mathscr{Y}^*, \text{wk}^*) \to (\mathscr{X}^*, \text{wk}^*)$ is continuous.*

Note the absence of conjugates. The proof is left to the reader.

If $A \in \mathscr{B}(\mathscr{X}, \mathscr{Y})$, then it is easy to see that $A^* \in \mathscr{B}(\mathscr{Y}^*, \mathscr{X}^*)$. In fact, if $y^* \in$ ball $\mathscr{Y}^*$ and $x \in$ ball $\mathscr{X}$, then $|\langle x, A^* y^* \rangle| = |\langle Ax, y^* \rangle| \leqslant \|Ax\| \leqslant \|A\|$. Hence $\|A^* y^*\| \leqslant \|A\|$ if $y^* \in$ ball $\mathscr{Y}^*$, so that $\|A^*\| \leqslant \|A\|$. This implies that

$(A^*)^* \equiv A^{**}$ can be defined,

$$A^{**}: \mathcal{X}^{**} \to \mathcal{Y}^{**},$$
$$\langle A^{**}x^{**}, y^* \rangle = \langle x^{**}, A^*y^* \rangle$$

for x^{**} in $\mathcal{X}^{**}$ and y^* in $\mathcal{Y}^*$.

Suppose $x \in \mathcal{X}$ and consider x as an element of $\mathcal{X}^{**}$ via the natural embedding of $\mathcal{X}$ into its double dual. What is $A^{**}(x)$? For y^* in $\mathcal{Y}^*$,

$$\langle A^{**}(x), y^* \rangle = \langle x, A^*y^* \rangle$$
$$= \langle Ax, y^* \rangle.$$

That is, $A^{**}|\mathcal{X} = A$. This is the first part of the next proposition.

1.4. Proposition. *If $\mathcal{X}$ and $\mathcal{Y}$ are Banach spaces and $A \in \mathcal{B}(\mathcal{X}, \mathcal{Y})$, then:*

(a) $A^{**}|\mathcal{X} = A$;
(b) $\|A^*\| = \|A\|$;
(c) *if A is invertible, then A^* is invertible and $(A^*)^{-1} = (A^{-1})^*$;*
(d) *if $\mathcal{Z}$ is a Banach space and $B \in \mathcal{B}(\mathcal{Y}, \mathcal{Z})$, then $(BA)^* = A^*B^*$.*

PROOF. Part (a) was proved above. It was shown that $\|A^*\| \leq \|A\|$. Thus $\|A^{**}\| \leq \|A^*\|$. So if $x \in$ ball $\mathcal{X}$, then (a) implies that $\|Ax\| = \|A^{**}x\| \leq \|A^{**}\| \leq \|A^*\|$. Hence $\|A\| \leq \|A^*\|$.

The remainder of the proof is left to the reader. ∎

1.5. Example. Let (X, Ω, μ) and $M_\phi: L^p(\mu) \to L^p(\mu)$ be as in Example III.2.2. If $1 \leq p < \infty$ and $1/p + 1/q = 1$, then $M_\phi^*: L^q(\mu) \to L^q(\mu)$ is given by $M_\phi^* f = \phi f$. That is, $M_\phi^* = M_\phi$.

1.6. Example. Let K and k be as in Example III.2.3. If $1 \leq p < \infty$ and $1/p + 1/q = 1$, then $K^*: L^q(\mu) \to L^q(\mu)$ is the integral operator with kernel $k^*(x, y) \equiv k(y, x)$.

1.7. Example. Let X, Y, τ, and A be as in Example III.2.4. Then $A^*: M(Y) \to M(X)$ is given by

$$(A^*\mu)(\Delta) = \mu(\tau^{-1}(\Delta))$$

for every Borel subset Δ of X and every μ in $M(Y)$.

Compare (1.5) and (1.6) with (II.2.8) and (II.2.9) to see the contrast between the adjoint of an operator on a Banach space with the adjoint of a Hilbert space operator.

1.8. Proposition. *If $A \in \mathcal{B}(\mathcal{X}, \mathcal{Y})$, then $\ker A^* = (\operatorname{ran} A)^\perp$ and $\ker A = {}^\perp(\operatorname{ran} A^*)$.*

The proof of this useful result is similar to that of Proposition II.2.19 and is left to the reader.

This enables us to prove the converse of Proposition 1.4c.

1.9. Proposition. *If $A \in \mathscr{B}(\mathscr{X}, \mathscr{Y})$, then A is invertible if and only if A^* is invertible.*

PROOF. In light of (1.4c) it suffices to assume that A^* is invertible and show that A is invertible. Since A^* is an open mapping, there is a constant $c > 0$ such that $A^*(\text{ball } \mathscr{Y}^*) \supseteq \{x^* \in \mathscr{X}^* : \|x^*\| \leqslant c\}$. So if $x \in \mathscr{X}$, then

$$
\begin{aligned}
\|Ax\| &= \sup\{|\langle Ax, y^* \rangle| : y^* \in \text{ball } \mathscr{Y}^*\} \\
&= \sup\{|\langle x, A^*y^* \rangle| : y^* \in \text{ball } \mathscr{Y}^*\} \\
&\geqslant \sup\{|\langle x, x^* \rangle| : x^* \in \mathscr{X}^* \text{ and } \|x^*\| \leqslant c\} \\
&= c^{-1}\|x\|.
\end{aligned}
$$

Thus $\ker A = (0)$ and $\operatorname{ran} A$ is closed. (Why?) On the other hand, $(\operatorname{ran} A)^{\perp} = \ker A^* = (0)$ since A^* is invertible. Thus $\operatorname{ran} A$ is also dense. This implies that A is surjective and thus invertible. ∎

This section concludes with the following useful result that seems to be somewhat unfamiliar to parts of the mathematical community.

1.10. Theorem. *If $\mathscr{X}$ and $\mathscr{Y}$ are Banach spaces and $A \in \mathscr{B}(\mathscr{X}, \mathscr{Y})$, then the following statements are equivalent.*

(a) $\operatorname{ran} A$ *is closed.*
(b) $\operatorname{ran} A^*$ *is weak* closed.*
(c) $\operatorname{ran} A^*$ *is norm closed.*

PROOF. It is clear that (b) implies (c), so it will be shown that (a) implies (b) and (c) implies (a). Before this is done, it will be shown that it suffices to prove the theorem under the additional hypothesis that A is injective and has dense range.

Let $\mathscr{Z} = \operatorname{cl}(\operatorname{ran} A)$. Thus $A: \mathscr{X} \to \mathscr{Z}$ induces a bounded linear map $B: \mathscr{X}/\ker A \to \mathscr{Z}$ defined by $B(x + \ker A) = Ax$. If $Q: \mathscr{X} \to \mathscr{X}/\ker A$ is the natural map, the diagram

$$
\begin{array}{ccc}
\mathscr{X} & \xrightarrow{\ A\ } & \mathscr{Z} \hookrightarrow \mathscr{Y} \\
{\scriptstyle Q} \searrow & & \nearrow {\scriptstyle B} \\
& \mathscr{X}/\ker A &
\end{array}
$$

commutes. (Why is B bounded?) It is easy to see that B is injective and that B has dense range. In fact, $\operatorname{ran} B = \operatorname{ran} A$, so $\operatorname{ran} A$ is closed if and only if $\operatorname{ran} B$ is closed. Let's examine $B^*: \mathscr{Z}^* \to (\mathscr{X}/\ker A)^*$. By (V.2.2), $(\mathscr{X}/\ker A)^* = (\ker A)^{\perp} = \text{wk*cl}(\operatorname{ran} A^*) \subseteq \mathscr{X}^*$ by (1.8). Also by (V.2.3), since $\mathscr{Z} \leqslant \mathscr{Y}$, $\mathscr{Z}^* = \mathscr{Y}^*/\mathscr{Z}^{\perp} = \mathscr{Y}^*/(\operatorname{ran} A)^{\perp} = \mathscr{Y}^*/\ker A^*$ by (1.8). Thus,

$$
B^*: \mathscr{Y}^*/\ker A^* \to (\ker A)^{\perp}.
$$

1.11. Claim. $B^*(y^* + \ker A^*) = A^*y^*$ for all y^* in $\mathcal{Y}^*$.

To see this, let $x \in \mathcal{X}$ and $y^* \in \mathcal{Y}^*$. Making the appropriate identifications as in (V.2.2) and (V.2.3) gives $\langle x + \ker A, B^*(y^* + \ker A^*) \rangle = \langle B(x + \ker A), y^* + \ker A^* \rangle = \langle Ax, y^* + (\operatorname{ran} A)^{\perp} \rangle = \langle Ax, y^* \rangle = \langle x, A^*y^* \rangle = \langle x + {}^{\perp}(\operatorname{ran} A^*), A^*y^* \rangle = \langle x + \ker A, A^*y^* \rangle$. Since x was arbitrary. (1.11) is established.

Note that Claim 1.11 implies that $\operatorname{ran} B^* = \operatorname{ran} A^*$. Hence $\operatorname{ran} A^*$ is weak* (resp., norm) closed if and only if $\operatorname{ran} B^*$ is weak* (resp., norm) closed.

This discussion shows that the theorem is equivalent to the analogous theorem in which there is the additional hypothesis that A is injective and has dense range. It is assumed, therefore, that $\ker A = (0)$ and $\operatorname{cl}(\operatorname{ran} A) = \mathcal{Y}$.

(a) $\Rightarrow$ (b): Since $\operatorname{ran} A$ is closed, the additional hypothesis implies that A is bijective. By the Inverse Mapping Theorem, $A^{-1} \in \mathcal{B}(\mathcal{Y}, \mathcal{X})$. Hence A^* is invertible (1.4c). Since A^* is invertible, $\operatorname{ran} A^* = \mathcal{X}^*$ and hence is weak* closed.

(c) $\Rightarrow$ (b): Since $\operatorname{ran} A$ is dense in $\mathcal{Y}$, $\ker A^* = (\operatorname{ran} A)^{\perp}$ (1.8) $= (0)$. Thus $A^*: \mathcal{Y}^* \to \operatorname{ran} A^*$ is a bijection. Since $\operatorname{ran} A^*$ is norm closed, it is a Banach space. By the Inverse Mapping Theorem, there is a constant $c > 0$ such that $\|A^*y^*\| \geq c\|y^*\|$ for all y^* in $\mathcal{Y}^*$.

To show that $\operatorname{ran} A^*$ is weak* closed, the Krein–Smulian Theorem (V.12.6) will be used. Thus suppose $\{A^*y_i^*\}$ is a net in $\operatorname{ran} A^*$ with $\|A^*y_i^*\| \leq 1$ such that $A^*y_i^* \to x^* \sigma(\mathcal{X}^*, \mathcal{X})$ for some x^* in $\mathcal{X}^*$. Thus $\|y_i^*\| \leq c^{-1}$ for all y_i^*. By Alaoglu's Theorem there is a y^* in $\mathcal{Y}^*$ such that $y_i^* \xrightarrow{\text{cl}} y^* \sigma(\mathcal{Y}^*, \mathcal{Y})$. Thus (1.3), $A^*y_i^* \xrightarrow{\text{cl}} A^*y^* \sigma(\mathcal{X}^*, \mathcal{X})$, and so $x^* = A^*y^* \in \operatorname{ran} A^*$. By (V.12.6), $\operatorname{ran} A^*$ is weak* closed.

(b) $\Rightarrow$ (a): Since $\operatorname{ran} A^*$ is weak* closed, $\operatorname{ran} A^* = (\ker A)^{\perp} = \mathcal{X}^*$. Also $\ker A^* = (\operatorname{ran} A)^{\perp} = (0)$ since A has dense range. Thus A^* is a bijection and is thus invertible. By Proposition 1.9, A is invertible and thus has closed range. $\blacksquare$

A proof that condition (c) in the preceding theorem implies (a) which avoids the weak* topology can be found in Kaufman [1966].

EXERCISES

1. Prove Proposition 1.3.

2. Complete the proof of Proposition 1.4.

3. Verify the statement made in (1.5).

4. Verify the statement made in (1.6).

5. Verify the statement made in (1.7).

6. Let $1 \leq p < \infty$ and define $S: l^p \to l^p$ by $S(\alpha_1, \alpha_2, \ldots) = (0, \alpha_1, \alpha_2, \ldots)$. Compute S^*.

7. Let $A \in \mathcal{B}(c_0)$ and for $n \geq 1$, define e_n in c_0 by $e_n(n) = 1$ and $e_n(m) = 0$ for $m \neq n$. Put $\alpha_{mn} = (Ae_n)(m)$ for $m, n \geq 1$. Prove: (a) $M \equiv \sup_m \sum_{n=1}^{\infty} |\alpha_{mn}| < \infty$; (b) for every

n, $\alpha_{mn} \to 0$ as $\overset{\bullet}{m} \to \infty$. Conversely, if $\{\alpha_{mn}: m, n \geqslant 1\}$ are scalars satisfying (a) and (b), then

$$(Ax)(m) = \sum_{n=1}^{\infty} \alpha_{mn} x(n)$$

defines a bounded operator A on c_0 and $\|A\| = M$. Find A^*.

8. Let $A \in \mathcal{B}(l^1)$ and for $n \geqslant 1$ define e_n in l^1 by $e_n(n) = 1$, $e_n(m) = 0$ for $m \neq n$. Put $\alpha_{mn} = (Ae_n)(m)$ for $m, n \geqslant 1$. Prove: (a) $M \equiv \sup_n \sum_{m=1}^{\infty} |\alpha_{mn}| < \infty$; (b) for every m, $\sup_n |\alpha_{mn}| < \infty$. Conversely, if $\{\alpha_{mn}: m, n \geqslant 1\}$ are scalars satisfying (a) and (b), then

$$(Af)(m) = \sum_{n=1}^{\infty} \alpha_{mn} f(n)$$

defines a bounded operator A on l^1 and $\|A\| = M$. Find A^*.

9. (Bonsall [1986]) Let $\mathcal{X}$ be a Banach space, Z a nonempty set, and $u: Z \to \mathcal{X}$. If there are positive constants M_1 and M_2 such that (i) $\|u(z)\| \leqslant M_1$ for all z in Z and (ii) for every x^* in $\mathcal{X}$, $\sup\{|\langle u(z), x^* \rangle|: z \in Z\} \geqslant M_2 \|x^*\|$; then for every x in $\mathcal{X}$ there is an f in $l^1(Z)$ such that $(*) x = \sum \{f(z)u(z): z \in Z\}$ and $M_2 \inf \|f\|_1 \leqslant \|x\| \leqslant M_1 \inf \|f\|_1$, where the infimum is taken over all f in $l^1(Z)$ such that $(*)$ holds. (Hint: define $T: l^1(Z) \to \mathcal{X}$ by $Tf = \sum \{f(z)u(z): z \in Z\}$.)

10. (Bonsall [1986]) Let m be normalized Lebesgue measure on $\partial \mathbb{D}$ and for $|z| < 1$ and $|w| = 1$ let $p_z(w) = (1 - |z|^2)/|1 - \bar{z}w|^2$. So p_z is the Poisson kernel. Show that if $f \in L^1(m)$, then there is a sequence $\{z_n\} \subseteq \mathbb{D}$ and a sequence $\{\lambda_n\}$ in l^1 such that $(*) f = \sum_{n=1}^{\infty} \lambda_n p_{z_n}$. Moreover, $\|f\|_1 = \inf \sum_{n=1}^{\infty} |\lambda_n|$, where the infimum is taken over all $\{\lambda_n\}$ in l^1 such that $(*)$ holds. (Hint: use Exercise 9.)

11. If $\mathcal{X}$ and $\mathcal{Y}$ are Banach spaces and $B \in \mathcal{B}(\mathcal{Y}^*, \mathcal{X}^*)$, then there is an operator A in $\mathcal{B}(\mathcal{X}, \mathcal{Y})$ such that $B = A^*$ if and only if B is wk*-continuous.

12. If $\mathcal{X}$ is a Banach space and $\mathcal{M}$ and $\mathcal{N}$ are closed subspaces, show that the following statements are equivalent. (a) $\mathcal{M} + \mathcal{N}$ is closed. (b) the range of the linear transformation $x \to (x + \mathcal{M}) \oplus (x + \mathcal{N})$ from $\mathcal{X}$ into $\mathcal{X}/\mathcal{M} \oplus \mathcal{X}/\mathcal{N}$ is closed. (c) $\mathcal{M}^\perp + \mathcal{N}^\perp$ is norm closed in $\mathcal{X}^*$. (d) $\mathcal{M}^\perp + \mathcal{N}^\perp$ is weak* closed in $\mathcal{X}^*$.

§2*. The Banach–Stone Theorem

As an application of the adjoint of a linear map, the isometries between spaces of the form $C(X)$ and $C(Y)$ will be characterized. Note that if X and Y are compact spaces, $\tau: Y \to X$ is continuous map, and $Af = f \circ \tau$ for f in $C(X)$, then (III.2.4) A is a bounded linear map and $\|A\| = 1$. Moreover, A is an isometry if and only if τ is surjective. If A is a surjective isometry, then τ must be a homeomorphism. Indeed, suppose A is a surjective isometry; it must be shown that τ is injective. If $y_0, y_1 \in Y$ and $y_0 \neq y_1$, then there is a g

in $C(Y)$ such that $g(y_0) = 0$ and $g(y_1) = 1$. Let $f \in C(X)$ such that $Af = g$. Thus $f(\tau(y_0)) = g(y_0) = 0$ and $f(\tau(y_1)) = 1$. Hence $\tau(y_0) \neq \tau(y_1)$.

So if $\tau: Y \to X$ is a homeomorphism and $\alpha: Y \to \mathbb{F}$ is a continuous function, with $|\alpha(y)| \equiv 1$, then $T: C(X) \to C(Y)$ defined by $(Tf)(y) = \alpha(y)f(\tau(y))$ is a surjective isometry. The next result gives a converse to this.

2.1. The Banach–Stone Theorem. *If X and Y are compact and $T: C(X) \to C(Y)$ is a surjective isometry, then there is a homeomorphism $\tau: Y \to X$ and a function α in $C(Y)$ such that $|\alpha(y)| = 1$ for all y and*

$$(Tf)(y) = \alpha(y)f(\tau(y))$$

for all f in $C(X)$ and y in Y.

PROOF. Consider $T^*: M(Y) \to M(X)$. Because T is a surjective isometry, T^* is also. (Verify.) Thus T^* is a weak* homeomorphism of ball $M(Y)$ onto ball $M(X)$ that distributes over convex combinations. Hence (Why?)

$$T^*(\text{ext}[\text{ball } M(Y)]) = \text{ext}[\text{ball } M(X)].$$

By Theorem V.8.4 this implies that for every y in Y there is a unique $\tau(y)$ in X and a unique scalar $\alpha(y)$ such that $|\alpha(y)| = 1$ and

$$T^*(\delta_y) = \alpha(y)\delta_{\tau(y)}.$$

By the uniqueness, $\alpha: Y \to \mathbb{F}$ and $\tau: Y \to X$ are well-defined functions.

2.2. Claim. $\alpha: Y \to \mathbb{F}$ is continuous.

If $\{y_i\}$ is a net in Y and $y_i \to y$, then $\delta_{y_i} \to \delta_y$ weak* in $M(Y)$. Hence $\alpha(y_i)\delta_{\tau(y_i)} = T^*(\delta_{y_i}) \to T^*(\delta_y) = \alpha(y)\delta_{\tau(y)}$ weak* in $M(X)$. In particular, $\alpha(y_i) = \langle 1, T^*(\delta_{y_i}) \rangle \to \langle 1, T^*(\delta_y) \rangle = \alpha(y)$, proving (2.2).

2.3. Claim. $\tau: Y \to X$ is a homeomorphism.

As in the proof of (2.2), if $y_i \to y$ in Y, then $\alpha(y_i)\delta_{\tau(y_i)} \to \alpha(y)\delta_{\tau(y)}$ weak* in $M(X)$. Also, $\alpha(y_i) \to \alpha(y)$ in $\mathbb{F}$ by (2.2). Thus $\delta_{\tau(y_i)} = \alpha(y_i)^{-1}[\alpha(y_i)\delta_{\tau(y_i)}] \to \delta_{\tau(y)}$. By (V.6.1) this implies that $\tau(y_i) \to \tau(y)$, so that $\tau: Y \to X$ is continuous.

If $y_1, y_2 \in Y$ and $y_1 \neq y_2$, then $\overline{\alpha(y_1)\delta_{y_1}} \neq \alpha(y_2)\delta_{y_2}$. Since T^* is injective, it is easy to see that $\tau(y_1) \neq \tau(y_2)$ and so τ is one-to-one. If $x \in X$, then the fact that T^* is surjective implies that there is a μ in $M(Y)$ such that $T^*\mu = \delta_x$. It must be that $\mu \in \text{ext}[\text{ball } M(Y)]$ (Why?), so that $\mu = \beta\delta_y$ for some y in Y and β in $\mathbb{F}$ with $|\beta| = 1$. Thus $\delta_x = T^*(\beta\delta_y) = \beta\alpha(y)\delta_{\tau(y)}$. Hence $\beta = \overline{\alpha(y)}$ and $\tau(y) = x$. Therefore $\tau: Y \to X$ is a continuous bijection and hence must be a homeomorphism (A.2.8). This establishes (2.3).

If $f \in C(X)$ and $y \in Y$, then $T(f)(y) = \langle Tf, \delta_y \rangle = \langle f, T^*\delta_y \rangle = \langle f, \alpha(y)\delta_{\tau(y)} \rangle = \alpha(y)f(\tau(y))$. ■

§3. Compact Operators

The following definition generalizes the concept of a compact operator from a Hilbert space to a Banach space.

3.1. Definition. If $\mathscr{X}$ and $\mathscr{Y}$ are Banach spaces and $A\colon \mathscr{X}\to\mathscr{Y}$ is a linear transformation, then A is *compact* if cl $A(\text{ball }\mathscr{X})$ is compact in $\mathscr{Y}$.

The reader should become reacquainted with Section II.4.

It is easy to see that compact operators are bounded.

For operators on a Hilbert space the following concept is equivalent to compactness, as will be seen.

3.2. Definition. If $\mathscr{X}$ and $\mathscr{Y}$ are Banach spaces and $A\in\mathscr{B}(\mathscr{X},\mathscr{Y})$, then A is *completely continuous* if for any sequence $\{x_n\}$ in $\mathscr{X}$ such that $x_n\to x$ weakly it follows that $\|Ax_n - Ax\|\to 0$.

3.3. Proposition. *Let $\mathscr{X}$ and $\mathscr{Y}$ be Banach spaces and let $A\in\mathscr{B}(\mathscr{X},\mathscr{Y})$.*

(a) *If A is a compact operator, then A is completely continuous.*

(b) *If $\mathscr{X}$ is reflexive and A is completely continuous, then A is compact.*

PROOF. (a) Let $\{x_n\}$ be a sequence in $\mathscr{X}$ such that $x_n\to 0$ weakly. By the PUB, $M = \sup_n\|x_n\| < \infty$. Without loss of generality, it may be assumed that $M \leqslant 1$. Hence $\{Ax_n\}\subseteq \text{cl } A(\text{ball }\mathscr{X})$. Since A is compact, there is a subsequence $\{x_{n_k}\}$ and a y in $\mathscr{Y}$ such that $\|Ax_{n_k} - y\|\to 0$. But $x_{n_k}\to 0$ (wk) and $A\colon (\mathscr{X},\text{wk})\to(\mathscr{Y},\text{wk})$ is continuous (1.1c). Hence $Ax_{n_k}\to A(0) = 0$ (wk). Thus $y = 0$. Since 0 is the unique cluster point of $\{Ax_n\}$ and this sequence is contained in a compact set, $\|Ax_n\|\to 0$.

(b) First assume that $\mathscr{X}$ is separable; so (ball $\mathscr{X}$, wk) is a compact metric space. So if $\{x_n\}$ is a sequence in ball $\mathscr{X}$ there is an x in $\mathscr{X}$ and a subsequence $\{x_{n_k}\}$ such that $x_{n_k}\to x$ weakly. Since A is completely continuous, $\|Ax_{n_k} - Ax\|\to 0$. Thus $A(\text{ball }\mathscr{X})$ is sequentially compact; that is, A is a compact operator.

Now let $\mathscr{X}$ be arbitrary and let $\{x_n\}\subseteq \text{ball }\mathscr{X}$. If $\mathscr{X}_1 = $ the closed linear span of $\{x_n\}$, then $\mathscr{X}_1$ is separable and reflexive. If $A_1 = A|\mathscr{X}_1$, then $A_1\colon \mathscr{X}_1\to\mathscr{Y}$ is easily seen to be completely continuous. By the first paragraph, A_1 is compact. Thus $\{Ax_n\} = \{A_1 x_n\}$ has a convergent subsequence. Since $\{x_n\}$ was arbitrary, A is a compact operator. ■

In the proof of (3.3b), the fact that $A(\text{ball }\mathscr{X})$ is compact, and hence closed, is a consequence of the reflexivity of $\mathscr{X}$.

By Proposition V.5.2, every operator in $\mathscr{B}(l^1)$ is completely continuous. However, there are noncompact operators in $\mathscr{B}(l^1)$ (for example, the identity operator).

There has been relatively little study of completely continuous operators that I am aware of. Most of the effort has been devoted to the study of compact operators and this is the direction we now pursue.

3.4. Schauder's Theorem. *If $A \in \mathcal{B}(\mathcal{X}, \mathcal{Y})$, then A is compact if and only if A^* is compact.*

PROOF. Assume A is a compact operator and let $\{y_n^*\}$ be a sequence in ball $\mathcal{Y}^*$. It must be shown that $\{A^* y_n^*\}$ has a norm convergent subsequence or, equivalently, a cluster point in the norm topology. By Alaoglu's Theorem, there is a y^* in ball $\mathcal{Y}^*$ such that $y_n^* \xrightarrow{\text{cl}} y^*$ (weak*). It will be shown that $A^* y_n^* \xrightarrow{\text{cl}} A^* y^*$ in norm.

Let $\varepsilon > 0$ and fix $N \geqslant 1$. Because $A(\text{ball } \mathcal{X})$ has compact closure, there are vectors $y_1, \ldots, y_m$ in $\mathcal{Y}$ such that $A(\text{ball } \mathcal{X}) \subseteq \bigcup_{k=1}^m \{y \in \mathcal{Y}: \|y - y_k\| < \varepsilon/3\}$. Since $y_n^* \xrightarrow{\text{cl}} y^*$ (weak*), there is an $n \geqslant N$ such that $|\langle y_k, y^* - y_n^* \rangle| < \varepsilon/3$ for $1 \leqslant k \leqslant m$. Let x be an arbitrary element in ball $\mathcal{X}$ and choose y_k such that $\|Ax - y_k\| < \varepsilon/3$. Then

$$
\begin{aligned}
|\langle x, A^* y^* - A^* y_n^* \rangle| &= |\langle Ax, y^* - y_n^* \rangle| \\
&\leqslant |\langle Ax - y_k, y^* - y_n^* \rangle| + |\langle y_k, y^* - y_n^* \rangle| \\
&\leqslant 2\|Ax - y_k\| + \varepsilon/3 < \varepsilon.
\end{aligned}
$$

Thus $\|A^* y - A^* y_n^*\| \leqslant \varepsilon$.

For the converse, assume A^* is compact. By the first half of the proof, $A^{**}: \mathcal{X}^{**} \to \mathcal{Y}^{**}$ is compact. It is easy to check that $A = A^{**}|\mathcal{X}$ is compact. ∎

For Banach spaces $\mathcal{X}$ and $\mathcal{Y}$, $\mathcal{B}_0(\mathcal{X}, \mathcal{Y})$ denotes the set of all compact operators from $\mathcal{X}$ into $\mathcal{Y}$; $\mathcal{B}_0(\mathcal{X}) = \mathcal{B}_0(\mathcal{X}, \mathcal{X})$.

3.5. Proposition. *Let $\mathcal{X}, \mathcal{Y},$ and $\mathcal{Z}$ be Banach spaces.*

(a) $\mathcal{B}_0(\mathcal{X}, \mathcal{Y})$ *is a closed linear subspace of $\mathcal{B}(\mathcal{X}, \mathcal{Y})$.*
(b) *If $K \in \mathcal{B}_0(\mathcal{X}, \mathcal{Y})$ and $A \in \mathcal{B}(\mathcal{Y}, \mathcal{Z})$, then $AK \in \mathcal{B}_0(\mathcal{X}, \mathcal{Z})$.*
(c) *If $K \in \mathcal{B}_0(\mathcal{X}, \mathcal{Y})$ and $A \in \mathcal{B}(\mathcal{Z}, \mathcal{X})$, then $KA \in \mathcal{B}_0(\mathcal{Z}, \mathcal{Y})$.*

The proof of (3.5) is left as an exercise.

3.6. Corollary. *If $\mathcal{X}$ is a Banach space, $\mathcal{B}_0(\mathcal{X})$ is a closed two-sided ideal in the algebra $\mathcal{B}(\mathcal{X})$.*

Let $\mathcal{B}_{00}(\mathcal{X}, \mathcal{Y}) =$ the bounded operators $T: \mathcal{X} \to \mathcal{Y}$ for which ran T is finite dimensional. Operators in $\mathcal{B}_{00}(\mathcal{X}, \mathcal{Y})$ are called operators with *finite rank*. It is easy to see that $\mathcal{B}_{00}(\mathcal{X}, \mathcal{Y}) \subseteq \mathcal{B}_0(\mathcal{X}, \mathcal{Y})$ and by (3.5a) the closure of $\mathcal{B}_{00}(\mathcal{X}, \mathcal{Y})$ is contained in $\mathcal{B}_0(\mathcal{X}, \mathcal{Y})$. Is $\mathcal{B}_{00}(\mathcal{X}, \mathcal{Y})$ dense in $\mathcal{B}_0(\mathcal{X}, \mathcal{Y})$?

It was shown in (II.4.4) that if $\mathscr{H}$ is a Hilbert space, then $\mathscr{B}_0(\mathscr{H})$ is indeed the closure of $\mathscr{B}_{00}(\mathscr{H})$. Note that the availability of an orthonormal basis in a Hilbert space played a significant role in the proof of this theorem. There is a concept of a basis for a Banach space called a Schauder basis. Any Banach space $\mathscr{X}$ with a Schauder basis has the property that $\mathscr{B}_{00}(\mathscr{X})$ is dense in $\mathscr{B}_0(\mathscr{X})$. Enflo [1973] gave an example of a separable reflexive Banach space $\mathscr{X}$ for which $\mathscr{B}_{00}(\mathscr{X})$ is not dense in $\mathscr{B}_0(\mathscr{X})$, and, hence, $\mathscr{H}$ has no Schauder basis. Davie [1973] and [1975] have simplifications of Enflo's proof. For the classical Banach spaces, however, every compact operator is the limit of a sequence of finite-rank operators.

The remainder of this section is devoted to proving that for X compact, $\mathscr{B}_{00}(C(X))$ is dense in $\mathscr{B}_0(C(X))$. This begins with material that may be familiar to many readers but will be presented for those who are unacquainted with it.

3.7. Definition. If X is completely regular and $\mathscr{F} \subseteq C(X)$, then $\mathscr{F}$ is *equicontinuous* if for every $\varepsilon > 0$ and for every x_0 in X there is a neighborhood U of x_0 such that $|f(x) - f(x_0)| < \varepsilon$ for all x in U and for all f in $\mathscr{F}$.

Note that for a single function f in $C(X)$, $\mathscr{F} = \{f\}$ is equicontinuous. The concept of equicontinuity states that one neighborhood works for all f in $\mathscr{F}$.

3.8. The Arzela–Ascoli Theorem. *If X is compact and $\mathscr{F} \subseteq C(X)$, then $\mathscr{F}$ is totally bounded if and only if $\mathscr{F}$ is bounded and equicontinuous.*

PROOF. Suppose $\mathscr{F}$ is totally bounded. It is easy to see that $\mathscr{F}$ is bounded. If $\varepsilon > 0$, then there are $f_1, \ldots, f_n$ in $\mathscr{F}$ such that $\mathscr{F} \subseteq \bigcup_{k=1}^{n} \{f \in C(X): \|f - f_k\| < \varepsilon/3\}$. If $x_0 \in X$, let U be an open neighborhood of x_0 such that for $1 \leqslant k \leqslant n$ and x in U, $|f_k(x) - f_k(x_0)| < \varepsilon/3$. If $f \in \mathscr{F}$, let f_k be such that $\|f - f_k\| < \varepsilon/3$. Then for x in U,

$$
\begin{aligned}
|f(x) - f(x_0)| &\leqslant |f(x) - f_k(x)| + |f_k(x) - f_k(x_0)| \\
&\quad + |f_k(x_0) - f(x_0)| \\
&< \varepsilon.
\end{aligned}
$$

Hence $\mathscr{F}$ is equicontinuous.

Now assume that $\mathscr{F}$ is equicontinuous and $\mathscr{F} \subseteq \text{ball } C(X)$. Let $\varepsilon > 0$. For each x in X, let U_x be an open neighborhood of x such that $|f(x) - f(y)| < \varepsilon/3$ for f in $\mathscr{F}$ and y in U_x. Now $\{U_x: x \in X\}$ is an open covering of X. Since X is compact, there are points $x_1, \ldots, x_n$ in X such that $X = \bigcup_{j=1}^{n} U_x$.

Let $\{\alpha_1, \ldots, \alpha_m\} \subseteq \mathbb{D}$ such that $\text{cl } \mathbb{D} \subseteq \bigcup_{k=1}^{m} \{\alpha: |\alpha - \alpha_k| < \varepsilon/6\}$. Consider the collection B of those ordered n-tuples $b = (\beta_1, \ldots, \beta_n)$ for which there is a function f_b in $\mathscr{F}$ such that $|f_b(x_j) - \beta_j| < \varepsilon/6$ for $1 \leqslant j \leqslant n$. Note that B is not empty since $f(x) \subseteq \text{cl } \mathbb{D}$ for every f in $\mathscr{F}$. In fact, each f in $\mathscr{F}$ gives rise to such a b in B. Moreover B is finite. Fix one function f_b in $\mathscr{F}$ associated as above with b in B.

3.9. Claim. $\mathscr{F} \subseteq \bigcup_{b \in B} \{f : \|f - f_b\| < \varepsilon\}$.

Note that (3.9) implies that $\mathscr{F}$ is totally bounded.

If $f \in \mathscr{F}$, there is a b in B such that $|f(x_j) - f_b(x_j)| < \varepsilon/3$ for $1 \leq j \leq n$. Therefore if $x \in X$, let x_j be chosen such that $x \in U_{x_j}$. Thus $|f(x) - f_b(x)| \leq |f(x) - f(x_j)| + |f(x_j) - f_b(x_j)| + |f_b(x_j) - f_b(x)| < \varepsilon$. Since x was arbitrary, $\|f - f_b\| < \varepsilon$. ∎

3.10. Corollary. *If X is compact and $\mathscr{F} \subseteq C(X)$, then $\mathscr{F}$ is compact if and only if $\mathscr{F}$ is closed, bounded, and equicontinuous.*

3.11. Theorem. *If X is compact, then $\mathscr{B}_{00}(C(X))$ is dense in $\mathscr{B}_0(C(X))$.*

PROOF. Let $T \in \mathscr{B}_0(C(X))$. Thus $T(\text{ball } C(X))$ is bounded and equicontinuous by the Arzela–Ascoli Theorem. If $\varepsilon > 0$ and $x \in X$, let U_x be an open neighborhood of x such that $|(Tf)(x) - (Tf)(y)| < \varepsilon$ for all f in ball $C(X)$ and y in U_x. Let $\{x_1, \ldots, x_n\} \subseteq X$ such that $X \subseteq \bigcup_{j=1}^n U_{x_j}$. Let $\{\phi_1, \ldots, \phi_n\}$ be a partition of unity subordinate to $\{U_{x_1}, \ldots, U_{x_n}\}$. Define $T_\varepsilon : C(X) \to C(X)$ by

$$T_\varepsilon f = \sum_{j=1}^n (Tf)(x_j)\phi_j.$$

Since ran $T_\varepsilon \subseteq \bigvee \{\phi_1, \ldots, \phi_n\}$, $T_\varepsilon \in \mathscr{B}_{00}(C(X))$.

If $f \in \text{ball } C(X)$ and $x \in X$, then

$$|(T_\varepsilon f)(x) - (Tf)(x)| = \left| \sum_{j=1}^n [(Tf)(x_j) - (Tf)(x)]\phi_j(x) \right|$$

$$\leq \sum_{j=1}^n |(Tf)(x_j) - (Tf)(x)|\phi_j(x)$$

$$< \varepsilon. \qquad \blacksquare$$

If X is locally compact, then the operators on $C_0(X)$ of finite rank are dense in $\mathscr{B}_0(C_0(X))$. See Exercise 18.

EXERCISES

1. If $\mathscr{X}$ is reflexive and $A \in \mathscr{B}(\mathscr{X}, \mathscr{Y})$, show that $A(\text{ball } \mathscr{X})$ is closed in $\mathscr{Y}$.

2. Prove Proposition 3.5.

3. If $A \in \mathscr{B}_0(\mathscr{X}, \mathscr{Y})$, show that $\text{cl}[\text{ran } A]$ is separable.

4. If $A \in \mathcal{B}_0(\mathcal{X}, \mathcal{Y})$ and ran A is closed, show that ran A is finite dimensional.

5. If $A \in \mathcal{B}_0(\mathcal{X})$ and A is invertible, show that dim $\mathcal{X} < \infty$.

6. Let (X, Ω, μ) be a finite measure space, $1 < p < \infty$, and $1/p + 1/q = 1$. If $k: X \times X \to \mathbb{F}$ is an $\Omega \times \Omega$-measurable function such that $\sup \{ \int |k(x, y)|^q d\mu(y): x \in X \} < \infty$, then $(Kf)(x) = \int k(x, y)f(y)d\mu(y)$ defines a compact operator on $L^p(\mu)$.

7. Let (X, Ω, μ) be an arbitrary measure space, $1 < p < \infty$, and $1/p + 1/q = 1$. If $k: X \times X \to \mathbb{F}$ is an $\Omega \times \Omega$-measurable function such that $M = [\int (\int |k(x, y)|^p d\mu(x))^{q/p} d\mu(y)]^{1/q} < \infty$ and if $(Kf)(x) = \int k(x, y)f(y)d\mu(y)$, then $K \in \mathcal{B}_0(L^p(\mu))$ and $\|K\| \leqslant M$.

8. Let X be a compact space and let μ be a positive Borel measure on X. Let $T \in \mathcal{B}(L^p(\mu), C(X))$ where $1 < p < \infty$. Show that if $A: L^p(\mu) \to L^p(\mu)$ is defined by $Af = Tf$, then A is compact.

9. (B.J. Pettis) If $\mathcal{X}$ is reflexive and $T \in \mathcal{B}(\mathcal{X}, l^1)$, then T is a compact operator. Also, if $\mathcal{Y}$ is reflexive and $T \in \mathcal{B}(c_0, \mathcal{Y})$, T is compact.

10. If X is compact and $\{f_1, \ldots, f_n, g_1, \ldots, g_n\} \subseteq C(X)$, define $k(x, y) = \sum_{j=1}^n f_j(x)g_j(y)$ for $x, y \in X$. Let μ be a regular Borel measure on X and put $Kf(x) = \int k(x, y)f(y)d\mu(y)$. Show that $K \in \mathcal{B}(C(X))$ and K has finite rank.

11. If X is compact, $k \in C(X \times X)$, and μ is a regular Borel measure on X, show that $Kf(x) = \int k(x, y)f(y)d\mu(y)$ defines a compact operator on $C(X)$.

12. Let (X, Ω, μ) be a σ-finite measure space and for ϕ in $L^\infty(\mu)$ let $M_\phi: L^p(\mu) \to L^p(\mu)$ be the multiplication operator defined in Example III.2.2. Give necessary and sufficient conditions on (X, Ω, μ) and ϕ for M_ϕ to be compact.

13. Let $\tau: [0, 1] \to [0, 1]$ be continuous and define $A: C[0, 1] \to C[0, 1]$ by $Af = f \circ \tau$. Give necessary and sufficient conditions on τ for A to be compact.

14. Let $A \in \mathcal{B}(c_0)$ and let (α_{mn}) be the corresponding matrix as in Exercise 1.7. Give necessary and sufficient conditions on (α_{mn}) for A to be compact.

15. Let $A \in \mathcal{B}(l^1)$ and let (α_{mn}) be the corresponding matrix as in Exercise 1.8. Give a necessary and sufficient condition on (α_{mn}) for A to be compact.

16. If (X, d) is a compact metric space and $\mathcal{F} \subseteq C(X)$, show that $\mathcal{F}$ is equicontinuous if and only if for every $\varepsilon > 0$ there is a $\delta > 0$ such that $|f(x) - f(y)| < \varepsilon$ whenever $d(x, y) < \delta$ and $f \in \mathcal{F}$.

17. If X is locally compact and $\mathcal{F} \subseteq C_0(X)$, show that $\mathcal{F}$ is totally bounded if and only if (a) $\mathcal{F}$ is bounded; (b) $\mathcal{F}$ is equicontinuous; (c) for every $\varepsilon > 0$ there is a compact subset K of X such that $|f(x)| < \varepsilon$ for all f in $\mathcal{F}$ and x in $X \backslash K$.

18. If X is locally compact and $A \in \mathcal{B}_0(C_0(X))$, then there is a sequence $\{A_n\}$ of finite-rank operators such that $\|A_n - A\| \to 0$.

19. Let $\mathcal{X}$ be a Banach space and suppose there is a net $\{F_i\}$ of finite-rank operators on $\mathcal{X}$ such that (a) $\sup_i \|F_i\| < \infty$; (b) $\|F_i x - x\| \to 0$ for all x in $\mathcal{X}$. Show that if $A \in \mathcal{B}_0(\mathcal{X})$, then $\|F_i A - A\| \to 0$ and hence there is a sequence $\{A_n\}$ of finite-rank operators on $\mathcal{X}$ such that $\|A_n - A\| \to 0$.

20. Let $1 \leqslant p \leqslant \infty$ and let (X, Ω, μ) be a σ-finite measure space. If $A \in \mathcal{B}_0(L^p(\mu))$, show that there is a sequence $\{A_n\}$ of finite-rank operators such that $\|A_n - A\| \to 0$. (Hint: Use Exercise 19.)

21. Let X be compact and let $\mathcal{U}$ be the collection of all pairs (C, F) where $C = \{U_1, \ldots, U_n\}$ is a finite open cover of X and $F = \{x_1, \ldots, x_n\} \subseteq X$ such that $x_j \in U_j$ for $1 \leqslant j \leqslant n$. If (C_1, F_1) and $(C_2, F_2) \in \mathcal{U}$, define $(C_1, F_1) \leqslant (C_2, F_2)$ to mean: (a) C_2 is a refinement of C_1; that is, each member of C_2 is contained in some member of C_1. (b) $F_1 \subseteq F_2$. If $\alpha = (C, F) \in \mathcal{U}$ let $\{\phi_1, \ldots, \phi_n\}$ be a partition of unity subordinate to C. If $F = \{x_1, \ldots, x_n\}$, define $T_\alpha : C(X) \to C(X)$ by

$$(T_\alpha f)(x) = \sum_{j=1}^{n} f(x_j)\phi_j(x).$$

Then: (a) $T_\alpha \in \mathcal{B}_{00}(C(X))$; (b) $\|T_\alpha\| = 1$; (c) $(\mathcal{U}, \leqslant)$ is a directed set and $\{T_\alpha : \alpha \in \mathcal{U}\}$ is a net; (d) $\|T_\alpha f - f\| \to 0$ for each f. Now apply Exercise 19 to obtain a new proof of Theorem 3.11.

§4. Invariant Subspaces

4.1. Definition. If $\mathcal{X}$ is a Banach space and $T \in \mathcal{B}(\mathcal{X})$, an *invariant subspace* for T is a closed linear subspace $\mathcal{M}$ of $\mathcal{X}$ such that $Tx \in \mathcal{M}$ whenever $x \in \mathcal{M}$. $\mathcal{M}$ is nontrivial if $\mathcal{M} \neq (0)$ or $\mathcal{X}$. Lat $T =$ the collection of all invariant subspaces for T. If $\mathcal{A} \subseteq \mathcal{B}(\mathcal{X})$, then Lat $\mathcal{A} = \bigcap \{\text{Lat } T : T \in \mathcal{A}\}$.

This generalizes the corresponding concept of invariant subspace for an operator on Hilbert space (II.3.5). Note that the idea of a reducing subspace for an operator on a Hilbert space has no generalization to Banach spaces since there is no concept of an orthogonal complement in Banach spaces.

4.2. Proposition.

(a) *If* $\mathcal{M}_1, \mathcal{M}_2 \in \text{Lat } T$, *then* $\mathcal{M}_1 \vee \mathcal{M}_2 \equiv \text{cl}(\mathcal{M}_1 + \mathcal{M}_2) \in \text{Lat } T$ *and* $\mathcal{M}_1 \wedge \mathcal{M}_2 \equiv \mathcal{M}_1 \cap \mathcal{M}_2 \in \text{Lat } T$.

(b) *If* $\{\mathcal{M}_i : i \in I\} \subseteq \text{Lat } T$, *then* $\vee \{\mathcal{M}_i : i \in I\}$, *the closed linear span of* $\bigcup_i \mathcal{M}_i$, *and* $\wedge \{\mathcal{M}_i : i \in I\} \equiv \bigcap_i \mathcal{M}_i$ *belong to* Lat T.

The proof of this proposition is left as an exercise. The proposition, however, does justify the use of the symbol "Lat" to denote the collection of invariant subspaces. With the operations $\vee$ and $\wedge$, Lat T is a lattice (a) that is complete (b). Moreover, Lat T has a largest element, $\mathcal{X}$, and a smallest element, (0).

The main question is: does Lat T have any elements besides (0) and $\mathcal{X}$? In other words, does T have a nontrivial invariant subspace? C.J. Read [1984] showed the existence of a Banach space and an operator on the Banach space having no non-trivial invariant subspaces. This was preceded by some work of P Enflo (not published, but circulated) which did the same thing. Later B Beauzamy [1985] sorted out Enflo's ideas and gave an exposition and simplification of Enflo's construction. Enflo's work eventually appeared in Enflo [1987]. Read [1986] gave a self contained exposition showing that there is a bounded operator on l^1 having no nontrivial invariant subspace. This deep work does not completely settle the matter. Which Banach spaces $\mathscr{X}$ have the property that there is a bounded operator on $\mathscr{X}$ with no nontrivial invariant subspaces? If $\mathscr{X}$ is reflexive, is Lat T nontrivial for every T in $\mathscr{B}(\mathscr{X})$? The question is unanswered even if $\mathscr{X}$ is a Hilbert space. However, for certain specific operators and classes of operators it has been shown that the lattice of invariant subspaces is not trivial. In this section it will be shown that any compact operator has a nontrivial invariant subspace. This will be obtained as a corollary of a more general result of V. Lomonosov. But first some examples.

4.3. Example. If $\mathscr{X}$ is a finite dimensional space over $\mathbb{C}$ and $T \in \mathscr{B}(\mathscr{X})$, then Lat T is not trivial. In fact, let $\mathscr{X} = \mathbb{C}^d$ and let $T = $ a matrix. Then $p(z) = \det(T - zI)$ is a polynomial of degree d. Hence it has a zero, say α. If $\det(T - \alpha I) = 0$, then $(T - \alpha I)$ is not invertible. But in finite dimensional spaces this means that $T - \alpha I$ is not injective. Thus $\ker(T - \alpha I) \neq (0)$. Let $\mathscr{M} \leqslant \ker(T - \alpha I)$ such that $\mathscr{M} \neq (0)$. If $x \in \mathscr{M}$, then $Tx = \alpha x \in \mathscr{M}$, so $\mathscr{M} \in \text{Lat } T$.

4.4. Example. If $T = \begin{bmatrix} 0 & -1 \\ 1 & 0 \end{bmatrix}$ on $\mathbb{R}^2$, then Lat T is trivial. Indeed, if Lat T is not trivial, there is a one-dimensional space $\mathscr{M}$ in Lat T. Let $\mathscr{M} = \{\alpha e : \alpha \in \mathbb{R}\}$. Since $\mathscr{M} \in \text{Lat } T$, $Te = \lambda e$ for some λ in $\mathbb{R}$. Hence $T^2 e = T(Te) = \lambda Te = \lambda^2 e$. But $T^2 = -I$, so $-e = \lambda^2 e$ and it must be that $\lambda^2 = -1$ if $e \neq 0$. But this cannot be if λ is real.

If $d \geqslant 3$, however, and $T \in \mathscr{B}(\mathbb{R}^d)$, then Lat T is not trivial (Exercise 6).

4.5. Example. If $V: L^2[0,1] \to L^2[0,1]$ is the Volterra operator, $Vf(x) = \int_0^x f(t)\,dt$, and $0 \leqslant \alpha \leqslant 1$, put $\mathscr{M}_\alpha = \{f \in L^2[0,1]: f(t) = 0 \text{ for } 0 \leqslant t \leqslant \alpha\}$. Then $\mathscr{M}_\alpha \in \text{Lat } V$. Moreover, it can be shown Lat $V = \{\mathscr{M}_\alpha: 0 \leqslant \alpha \leqslant 1\}$. (See Donoghue [1957], and Radjavi and Rosenthal [1973], p. 68.)

4.6. Example. If $S: l^p \to l^p$ is defined by $S(\alpha_1, \alpha_2, \ldots) = (0, \alpha_1, \alpha_2, \ldots)$, and $\mathscr{M}_n = \{x \in l^p: x(k) = 0 \text{ for } 1 \leqslant k \leqslant n\}$, then $\mathscr{M}_n \in \text{Lat } S$.

4.7. Example. Let (X, Ω, μ) be a σ-finite measure space and for ϕ in $L^\infty(\mu)$ let M_ϕ denote the multiplication operator on $L^p(\mu)$, $1 \leqslant p \leqslant \infty$. If $\Delta \in \Omega$, let $\mathscr{M}_\Delta = \{f \in L^p(\mu): f = 0 \text{ a.e. } [\mu] \text{ off } \Delta\}$. Then for each ϕ in $L^\infty(\mu)$, $\mathscr{M}_\Delta \in \text{Lat } M_\phi$.

It is a difficult if not impossible task to determine all the invariant subspaces of a specific operator. The Volterra operator and the shift operator are examples where all the invariant subspaces have been determined. But there

are multiplication operators M_ϕ for which there is no characterization of
Lat M_ϕ as well as some M_ϕ for which such a characterization has been
achieved. One such example follows: let $\mu = $ Lebesgue area measure on $\mathbb{D}$
and let $(Af)(z) = zf(z)$ for f in $L^2(\mu)$. There is no known characterization of
Lat A.

It is necessary at this point to return to the geometry of Banach spaces
to prove the following classical theorem, which appeared as Exercise V.13.2.

4.8. Mazur's Theorem. *If $\mathscr{X}$ is a Banach space and K is a compact subset of
$\mathscr{X}$, then $\overline{\mathrm{co}}(K)$ is compact.*

PROOF. It suffices to show that $\overline{\mathrm{co}}(K)$ is totally bounded. Let $\varepsilon > 0$ and
choose $x_1, \ldots, x_n$ in K such that $K \subseteq \bigcup_{j=1}^n B(x_j; \varepsilon/4)$. Put $C = \mathrm{co}\{x_1, \ldots, x_n\}$.
It is easy to see that C is compact (Exercise V.7.8). Hence there are vectors
$y_1, \ldots, y_m$ in C such that $C \subseteq \bigcup_{i=1}^m B(y_i; \varepsilon/4)$. If $w \in \overline{\mathrm{co}}(K)$, there is a z in $\mathrm{co}(K)$
with $\|w - z\| < \varepsilon/4$. Thus $z = \sum_{p=1}^l \alpha_p k_p$, where $k_p \in K$, $\alpha_p \geq 0$, and $\sum \alpha_p = 1$.
Now for each k_p there is an $x_{j(p)}$ with $\|k_p - x_{j(p)}\| < \varepsilon/4$. Therefore

$$\left\| z - \sum_{p=1}^l \alpha_p x_{j(p)} \right\| = \left\| \sum_{p=1}^l \alpha_p (k_p - x_{j(p)}) \right\|$$

$$\leq \sum_{p=1}^l \alpha_p \| k_p - x_{j(p)} \|$$

$$< \varepsilon/4.$$

But $\sum_p \alpha_p x_{j(p)} \in C$ so there is a y_i with $\|\sum_p \alpha_p x_{j(p)} - y_i\| < \varepsilon/4$. The triangle
inequality now shows that $\overline{\mathrm{co}}(K) \subseteq \bigcup_{i=1}^m B(y_i; \varepsilon)$ and so $\overline{\mathrm{co}}(K)$ is totally
bounded. ∎

The next result is from Lomonosov [1973]. When it appeared it caused
great excitement, both for the strength of its conclusion and for the simplicity
of its proof. The proof uses Schauder's Fixed-Point Theorem (V.9.5).

4.9. Lomonosov's Lemma. *If $\mathscr{A}$ is a subalgebra of $\mathscr{B}(\mathscr{X})$ such that $1 \in \mathscr{A}$ and
Lat $\mathscr{A} = \{(0), \mathscr{X}\}$ and if K is a nonzero compact operator on $\mathscr{X}$, then there is
an A in $\mathscr{A}$ such that $\ker(AK - 1) \neq (0)$.*

PROOF. It may be assumed that $\|K\| = 1$. Fix x_0 in $\mathscr{X}$ such that $\|Kx_0\| > 1$
and put $S = \{x \in \mathscr{X} : \|x - x_0\| \leq 1\}$. It is easy to check that

4.10 $0 \notin S$ and $0 \notin \mathrm{cl}\, K(S)$.

Now if $x \in \mathscr{X}$ and $x \neq 0$, $\mathrm{cl}\{Tx : T \in \mathscr{A}\}$ is an invariant subspace for $\mathscr{A}$ (because
$\mathscr{A}$ is an algebra) that contains the nonzero vector x (because $1 \in \mathscr{A}$). By
hypothesis, $\mathrm{cl}\{Tx : T \in \mathscr{A}\} = \mathscr{X}$. By (4.10) this says that for every y in $\mathrm{cl}\, K(S)$
there is a T in $\mathscr{A}$ with $\|Ty - x_0\| < 1$. Equivalently,

$$\operatorname{cl} K(S) \subseteq \bigcup_{T \in \mathscr{A}} \{y \colon \| Ty - x_0 \| < 1\}.$$

Because $\operatorname{cl} K(S)$ is compact, there are $T_1, \ldots, T_n$ in $\mathscr{A}$ such that

4.11 $$\operatorname{cl} K(S) \subseteq \bigcup_{j=1}^{n} \{y \colon \| T_j y - x_0 \| < 1\}.$$

For y in $\operatorname{cl} K(S)$ and $1 \leqslant j \leqslant n$, let $a_j(y) = \max\{0, 1 - \| T_j y - x_0 \|\}$. By (4.11), $\sum_{j=1}^{n} a_j(y) > 0$ for all y in $\operatorname{cl} K(S)$. Define $b_j \colon \operatorname{cl} K(S) \to \mathbb{R}$ by

$$b_j(y) = \frac{a_j(y)}{\sum\limits_{i=1}^{n} a_i(y)},$$

and define $\psi \colon S \to \mathscr{X}$ by

$$\psi(x) = \sum_{j=1}^{n} b_j(Kx) T_j Kx.$$

It is easy to see that $a_j \colon \operatorname{cl} K(S) \to [0, 1]$ is a continuous function. Hence b_j and ψ are continuous.

If $x \in S$, then $Kx \in K(S)$. If $b_j(Kx) > 0$, then $a_j(Kx) > 0$ and so $\| T_j Kx - x_0 \| < 1$. That is, $T_j Kx \in S$ whenever $b_j(Kx) > 0$. Since S is a convex set and $\sum_{j=1}^{n} b_j(Kx) = 1$ for x in S,

$$\psi(S) \subseteq S.$$

Note that $T_j K \in \mathscr{B}_0(\mathscr{X})$ for each j so that $\bigcup_{j=1}^{n} T_j K(S)$ has compact closure. By Mazur's Theorem, $\overline{\operatorname{co}}\,(\bigcup_{j=1}^{n} T_j K(S))$ is compact. But this convex set contains $\psi(S)$ so that $\operatorname{cl} \psi(S)$ is compact. This is, ψ is a compact map. By the Schauder Fixed-Point Theorem, there is a vector x_1 in S such that $\psi(x_1) = x_1$.

Let $\beta_j = b_j(Kx_1)$ and put $A = \sum_{j=1}^{n} \beta_j T_j$. So $A \in \mathscr{A}$ and $AKx_1 = \psi(x_1) = x_1$. Since $x_1 \neq 0$ (Why?), $\ker(AK - 1) \neq 0$. ∎

4.12. Definition. If $T \in \mathscr{B}(\mathscr{X})$, then a *hyperinvariant subspace* for T is a subspace $\mathscr{M}$ of $\mathscr{X}$ such that $A\mathscr{M} \subseteq \mathscr{M}$ for every operator A in the commutant of T, $\{T\}'$; that is, $A\mathscr{M} \subseteq \mathscr{M}$ whenever $AT = TA$.

Note that every hyperinvariant subspace for T is invariant.

4.13. Lomonosov's Theorem. *If $\mathscr{X}$ is a Banach space over $\mathbb{C}$, $T \in \mathscr{B}(\mathscr{X})$, T is not a multiple of the identity, and $TK = KT$ for some nonzero compact operator K, then T has a nontrivial hyperinvariant subspace.*

PROOF. Let $\mathscr{A} = \{T\}'$. We want to show that $\operatorname{Lat} \mathscr{A} \neq \{(0), \mathscr{X}\}$. If this is not the case, then Lomonosov's Lemma implies that there is an operator A in $\mathscr{A}$ such that $\mathscr{N} = \ker(AK - 1) \neq (0)$. But $\mathscr{N} \in \operatorname{Lat}(AK)$ and $AK|\mathscr{N}$ is the identity operator. Since $AK \in \mathcal{B}_0(\mathcal{X})$, $AK|\mathcal{N} \in \mathcal{B}_0(\mathcal{N})$. Thus dim $\mathcal{N} < \infty$. Since $AK \in \mathcal{A} = \{T\}'$, for any x in $\mathcal{N}$, $AK(Tx) = T(AKx) = Tx$; hence $T\mathcal{N} \subseteq \mathcal{N}$. But dim $\mathcal{N} < \infty$ so that $T|\mathcal{N}$ must have an eigenvalue λ. Thus ker$(T - \lambda) = \mathcal{M} \neq (0)$. But $\mathcal{M} \neq \mathcal{X}$ since T is not a multiple of the identity. It is easy to check that $\mathcal{M}$ is hyperinvariant for T. ∎

A proof of a slightly weaker version of Lomonosov's Theorem that avoids Schauder's Fixed Point Theorem can be found in Michaels [1977].

4.14. Corollary. (Aronszajn–Smith [1954].) *If $K \in \mathcal{B}_0(\mathcal{X})$, then* Lat K *is nontrivial.*

The next result appeared in Bernstein and Robinson [1966], where it is proved using nonstandard analysis. Halmos [1966] gave a proof using standard analysis. Now it is an easy consequence of Lomonosov's Theorem.

4.15. Corollary. *If $\mathcal{X}$ is infinite dimensional, $A \in \mathcal{B}(\mathcal{X})$, and there is a polynomial in one variable, p, such that $p(A) \in \mathcal{B}_0(\mathcal{X})$, then* Lat A *is nontrivial.*

PROOF. If $p(A) \neq 0$, then Lomonosov's Theorem applies. If $p(A) = 0$, let $p(z) = \alpha_0 + \alpha_1 z + \cdots + \alpha_n z^n$, $\alpha_n \neq 0$. For $x \neq 0$, let $\mathcal{M} = \bigvee\{x, Ax, \ldots, A^{n-1}x\}$. Since $A^n = -\alpha_n^{-1}[\alpha_0 + \alpha_1 A + \cdots + \alpha_{n-1}A^{n-1}]$, $\mathcal{M} \in$ Lat A. Since $x \in \mathcal{M}$, $\mathcal{M} \neq (0)$; since dim $\mathcal{M} < \infty$, $\mathcal{M} \neq \mathcal{X}$. ∎

4.16. Corollary. *If $K_1, K_2 \in \mathcal{B}_0(\mathcal{X})$ and $K_1K_2 = K_2K_1$, then K_1 and K_2 have a common nontrivial invariant subspace.*

EXERCISES

1. Let $A, B, T \in \mathcal{B}(\mathcal{X})$ such that $TA = BT$. Show that graph $(T) \in$ Lat$(A \oplus B)$.

2. Prove that $\mathcal{M} \in$ Lat T if and only if $\mathcal{M}^\perp \in$ Lat T^*. What does the map $\mathcal{M} \mapsto \mathcal{M}^\perp$ of Lat T into Lat T^* do to the lattice operations?

3. Let $\{e_1, e_2, e_3\}$ be the usual basis for $\mathbb{F}^3$ and let $\alpha_1, \alpha_2, \alpha_3 \in \mathbb{F}$. Define $T: \mathbb{F}^3 \to \mathbb{F}^3$ by $Te_j = \alpha_j e_j$, $1 \leq j \leq 3$. (a) If $\alpha_1, \alpha_2, \alpha_3$ are all distinct, show that $\mathcal{M} \in$ Lat T if and only if $\mathcal{M} = \bigvee E$, where $E \subseteq \{e_1, e_2, e_3\}$. (b) If $\alpha_1 = \alpha_2 \neq \alpha_3$, show that $\mathcal{M} \in$ Lat T if and only if $\mathcal{M} = \mathcal{N} + \mathcal{L}$, where $\mathcal{N} \leq \bigvee\{e_1, e_2\}$ and $\mathcal{L} \leq \{\alpha e_3 : \alpha \in \mathbb{F}\}$.

4. Generalize Exercise 3 by characterizing Lat T, where T is defined by $Te_j = \alpha_j e_j$, $1 \leq j \leq d$, for any choice of scalars $\alpha_1, \ldots, \alpha_d$ and where $\{e_1, \ldots, e_d\}$ is the usual basis for $\mathbb{F}^d$.

5. Let $\{e_1, \ldots, e_d\}$ be the usual basis for $\mathbb{F}^d$, let $\{\alpha_1, \ldots, \alpha_{d-1}\} \subseteq \mathbb{F}$ with no $\alpha_j = 0$. If $Te_j = \alpha_j e_{j+1}$ for $1 \leq j \leq d - 1$ and $Te_d = 0$, find Lat T.

6. If $T \in \mathcal{B}(\mathbb{R}^d)$ and $d \geq 3$, show that T has a nontrivial subspace.

7. Show that if $T \in \mathcal{B}(\mathcal{X})$ and $\mathcal{X}$ is not separable, then T has a nontrivial invariant subspace.

8. Give an example of an invertible operator T on a Banach space $\mathscr{X}$ and an invariant subspace $\mathscr{M}$ for T such that $\mathscr{M}$ is not invariant for T^{-1}.

9. Let $\mathscr{X}$ be a Banach space over $\mathbb{C}$, let $K \in \mathscr{B}_0(\mathscr{X})$ and show that if $\mathscr{C}$ is a maximal chain in Lat K, then $\mathscr{C}$ is a maximal chain in the lattice of all subspaces of $\mathscr{X}$.

§5. Weakly Compact Operators

5.1. Definition. If $\mathscr{X}$ and $\mathscr{Y}$ are Banach spaces, an operator T in $\mathscr{B}(\mathscr{X}, \mathscr{Y})$ is *weakly compact* if the closure of $T(\text{ball } \mathscr{X})$ is weakly compact.

Weakly compact operators are generalizations of compact operators, but the hypothesis is not sufficiently strong to yield good information about their structure.

Recall that in a reflexive Banach space the weak closure of any bounded set is weakly compact. Also, a bounded operator $T: \mathscr{X} \to \mathscr{Y}$ is continuous if both $\mathscr{X}$ and $\mathscr{Y}$ have their weak topologies (1.1). With these facts in mind, the proof of the next result becomes an easy exercise for the reader.

5.2. Proposition.

(a) *If either $\mathscr{X}$ or $\mathscr{Y}$ is reflexive, then every operator in $\mathscr{B}(\mathscr{X}, \mathscr{Y})$ is weakly compact.*

(b) *If $T: \mathscr{X} \to \mathscr{Y}$ is weakly compact and $A \in \mathscr{B}(\mathscr{Y}, \mathscr{Z})$, then AT is weakly compact.*

(c) *If $T: \mathscr{X} \to \mathscr{Y}$ is weakly compact and $B \in \mathscr{B}(\mathscr{Z}, \mathscr{X})$, then TB is weakly compact.*

This proposition shows that assuming that an operator is weakly compact is not that strong an assumption. For example, if $\mathscr{X}$ is reflexive, every operator in $\mathscr{B}(\mathscr{X})$ is weakly compact. In particular, every operator on a Hilbert space is weakly compact. So any theorem about weakly compact operators is a theorem about all operators on a reflexive space.

In fact, there is a degree of validity for the converse of this statement. In a certain sense, theorems about operators on reflexive spaces are also theorems about weakly compact operators. The precise meaning of this statement is the content of Theorem 5.4 below. But before we begin to prove this, a lemma is needed.

Let $\mathscr{Y}$ be a Banach space and let W be a bounded convex balanced subset of $\mathscr{Y}$. For $n > 1$ put $U_n = 2^n W + 2^{-n} \text{int}[\text{ball } \mathscr{Y}]$. Let $p_n = $ the gauge of U_n (IV.1.14). Because $U_n \supseteq 2^{-n} \text{int}[\text{ball } \mathscr{Y}]$, it is easy to check that p_n is a norm on $\mathscr{Y}$. In fact, p_n and $\|\cdot\|$ are equivalent norms. To see this note that if $\|y\| < 1$, then $2^{-n}y \in U_n$ so that $p_n(y) < 2^n$. Hence $p_n(y) \leqslant 2^n \|y\|$. Also, because W is bounded, U_n must be bounded; let $M > \sup\{\|y\|: y \in U_n\}$. So if $p_n(y) < 1$, $\|y\| < M$. Thus $\|y\| \leqslant M p_n(y)$, and $\|\cdot\|$ and p_n are equivalent norms.

5.3. Lemma. *For a Banach space $\mathcal{Y}$ let W, U_n, and p_n be as above. Let $\mathcal{R} = $ the set of all y in $\mathcal{Y}$ such that $\|\|y\|\| \equiv [\sum_{n=1}^{\infty} p_n(y)^2]^{1/2} < \infty$. Then*

(a) $W \subseteq \{y\colon \|\|y\|\| < 1\}$;
(b) $(\mathcal{R}, \|\|\cdot\|\|)$ *is a Banach space and the inclusion map* $A\colon \mathcal{R} \to \mathcal{Y}$ *is continuous;*
(c) $A^{**}\colon \mathcal{R}^{**} \to \mathcal{Y}^{**}$ *is injective and* $(A^{**})^{-1}(\mathcal{Y}) = \mathcal{R}$;
(d) $\mathcal{R}$ *is reflexive if and only if* cl W *is weakly compact.*

PROOF. (a) If $w \in W$, then $2^n w \in U_n$. Hence $1 > p_n(2^n w) = 2^n p_n(w)$, so $p_n(w) < 2^{-n}$. Thus $\|\|w\|\|^2 < \sum_n (2^{-n})^2 < 1$.

(b) Let $\mathcal{Y}_n = \mathcal{Y}$ with the norm p_n and put $\mathcal{X} = \oplus_2 \mathcal{Y}_n$ (III.4.4). Define $\Phi\colon \mathcal{R} \to \mathcal{X}$ by $\Phi(y) = (y, y, \ldots)$. It is easy to see that Φ is an isometry though it is clearly not surjective. In fact, ran $\Phi = \{(y_n) \in \mathcal{X}\colon y_n = y_m \text{ for all } n, m\}$. Thus $\mathcal{R}$ is a Banach space. Let $P_1 = $ the projection of $\mathcal{X}$ onto the first coordinate. Then $A = P_1 \circ \Phi$ and hence A is continuous.

(c) With the notation from the proof of (b), it follows that $\mathcal{X}^{**} = \oplus_2 \mathcal{Y}_n^{**}$ and $\Phi^{**}\colon \mathcal{R}^{**} \to \mathcal{X}^{**}$ is given by $\Phi^{**}(y^{**}) = (A^{**}y^{**}, A^{**}y^{**}, \ldots)$. Now the fact that Φ is an isometry implies that Φ^* is surjective. (This follows in two ways. One is by a direct argument (see Exercise 2). Also, ran Φ^* is closed since ran Φ is closed (1.10), and ran Φ^* is dense since $^\perp(\text{ran } \Phi^*) = \ker \Phi = (0)$.) Hence $\ker \Phi^{**} = (\text{ran } \Phi^*)^\perp = (0)$; that is, Φ^{**} is injective. Therefore A^{**} is injective.

Now let $y^{**} \in A^{**-1}(\mathcal{Y})$. It follows that $\Phi^{**}y^{**} = x \in \mathcal{X}$. Let $\{y_i\}$ be a net in $\mathcal{R}$ such that $\|\|y_i\|\| \le \|y^{**}\|$ for all i and $y_i \to y^{**}$ $\sigma(\mathcal{R}^{**}, \mathcal{R}^*)$ (V.4.1). Thus $\Phi^{**}(y_i) \to \Phi^{**}(y^{**})$ $\sigma(\mathcal{X}^{**}, \mathcal{X}^*)$. But $\Phi^{**}(y_i) = \Phi(y_i) \in \mathcal{X}$ and $\Phi^{**}(y^{**}) = x$. Hence $\Phi(y_i) \to x$ $\sigma(\mathcal{X}, \mathcal{X}^*)$. Since ran Φ is closed, $x \in$ ran Φ; let $\Phi(y) = x$. Then $0 = \Phi^{**}(y^{**} - y)$. Since Φ^{**} is injective, $y^{**} = y \in \mathcal{R}$.

(d) An argument using Alaoglu's Theorem shows that $A^{**}(\text{ball }\mathcal{R}^{**}) = $ the $\sigma(\mathcal{Y}^{**}, \mathcal{Y}^*)$ closure of $A(\text{ball }\mathcal{R})$. Put $C = A(\text{ball }\mathcal{R})$. Suppose cl W is weakly compact. Now $C \subseteq 2^n$ cl $W + 2^{-n}$ ball $\mathcal{Y}^{**}$ and this set is $\sigma(\mathcal{Y}^{**}, \mathcal{Y}^*)$ compact. From the preceding comments, $A^{**}(\text{ball }\mathcal{R}^{**}) \subseteq 2^n$ cl $W + 2^{-n}$ ball $\mathcal{Y}^{**}$. Thus,

$$A^{**}(\text{ball }\mathcal{R}^{**}) \subseteq \bigcap_{n=1}^{\infty} [2^n \text{ cl } W + 2^{-n} \text{ ball } \mathcal{Y}^{**}]$$

$$\subseteq \bigcap_{n=1}^{\infty} [\mathcal{Y} + 2^{-n} \text{ ball } \mathcal{Y}^{**}]$$

$$= \mathcal{Y}.$$

By (c), $\mathcal{R}^{**} = \mathcal{R}$ and $\mathcal{R}$ is reflexive.

Now assume $\mathcal{R}$ is reflexive; thus ball $\mathcal{R}$ is $\sigma(\mathcal{R}, \mathcal{R}^*)$-compact. Therefore $C = A(\text{ball }\mathcal{R})$ is weakly compact in $\mathcal{Y}$. By (a), cl W is weakly compact. ∎

The next theorem, as well as the preceding lemma, are from Davis, Figiel, Johnson, and Pelczynski [1974].

5.4. Theorem. *If* $\mathcal{X}$, $\mathcal{Y}$ *are Banach spaces and* $T \in \mathcal{B}(\mathcal{X}, \mathcal{Y})$, *then* T *is weakly compact if and only if there is a reflexive space* $\mathcal{R}$ *and operators* A *in* $\mathcal{B}(\mathcal{R}, \mathcal{Y})$ *and* B *in* $\mathcal{B}(\mathcal{X}, \mathcal{R})$ *such that* $T = AB$.

PROOF. If $T = AB$, where A, B have the described form, then T is weakly compact by Proposition 5.2.

Now assume that T is weakly compact and put $W = T(\text{ball } \mathcal{X})$. Define $\mathcal{R}$ as in Lemma 5.3. By (5.3d), $\mathcal{R}$ is reflexive. Let $A: \mathcal{R} \to \mathcal{Y}$ be the inclusion map. Note that if $x \in \text{ball } \mathcal{X}$, then $Tx \in W$. By (5.3a), $\|\|Tx\|\| < 1$ whenever $\|x\| \leqslant 1$. So $B: \mathcal{X} \to \mathcal{R}$ defined by $Bx = Tx$ is a bounded operator. Clearly $AB = T$. ∎

The preceding result can be used to prove several standard results from antiquity.

5.5. Theorem. *If* $\mathcal{X}$, $\mathcal{Y}$ *are Banach spaces and* $T \in \mathcal{B}(\mathcal{X}, \mathcal{Y})$, *the following statements are equivalent.*

(a) T *is weakly compact.*
(b) $T^{**}(\mathcal{X}^{**}) \subseteq \mathcal{Y}$.
(c) T^* *is weakly compact.*

PROOF. (a) ⇒ (b): Let $\mathcal{R}$ be a reflexive space, $A \in \mathcal{B}(\mathcal{R}, \mathcal{Y})$, and $B \in \mathcal{B}(\mathcal{X}, \mathcal{R})$ such that $T = AB$. So $T^{**} = A^{**}B^{**}$. But $A^{**}: \mathcal{R} \to \mathcal{Y}^{**}$ since $\mathcal{R}^{**} = \mathcal{R}$. Hence $A^{**} = A$. Thus $T^{**} = AB^{**}$, and so ran $T^{**} \subseteq$ ran $A \subseteq \mathcal{Y}$.

(b) ⇒ (a): $T^{**}(\text{ball } \mathcal{X}^{**})$ is $\sigma(\mathcal{Y}^{**}, \mathcal{Y}^*)$ compact by Alaoglu's Theorem and the weak* continuity of T^{**}. By (b), $T^{**}(\text{ball } \mathcal{X}^{**}) = C$ is $\sigma(\mathcal{Y}, \mathcal{Y}^*)$ compact in $\mathcal{Y}$. Hence $T(\text{ball } \mathcal{X}) \subseteq C$ and must have weakly compact closure.

(c) ⇒ (a): Let $\mathcal{S}$ be a reflexive space, $C \in \mathcal{B}(\mathcal{Y}^*, \mathcal{S})$, $D \in \mathcal{B}(\mathcal{S}, \mathcal{X}^*)$ such that $T^* = DC$. So $T^{**} = C^*D^*$, $D^*: \mathcal{X}^{**} \to \mathcal{S}^*$, and $C^*: \mathcal{S}^* \to \mathcal{Y}^{**}$. Put $\mathcal{R} = \text{cl } D^*(\mathcal{X})$ and $B = D^*|\mathcal{X}$; then $B: \mathcal{X} \to \mathcal{R}$ and $\mathcal{R}$ is reflexive. Let $A = C^*|\mathcal{R}$; so $A: \mathcal{R} \to \mathcal{Y}^{**}$. But if $x \in \mathcal{X}$, $ABx = C^*D^*x = T^{**}x = Tx \in \mathcal{Y}$. Thus $A: \mathcal{R} \to \mathcal{Y}$. Clearly $AB = T$.

(a) ⇒ (c): Exercise. ∎

EXERCISES

1. Prove Proposition 5.2.

2. If $\mathcal{R}$ and $\mathcal{X}$ are Banach spaces and $\Phi: \mathcal{R} \to \mathcal{X}$ is an isometry, give an elementary proof that Φ^* is surjective.

3. Let $\mathcal{X}$ be a Banach space and recall the definition of a weakly Cauchy sequence (V.4.4). (a) Show that every bounded sequence in c_0 has a weakly Cauchy subsequence, but not every weakly Cauchy sequence in c_0 converges. (b) Show that if $T \in \mathcal{B}(c_0)$ and T is weakly compact, then T is compact.

4. Say that a Banach space $\mathcal{X}$ is *weakly compactly generated* (WCG) if there is a weakly compact subset K of $\mathcal{X}$ such that $\mathcal{X}$ is the closed linear span of K. Prove

(Davis, Figiel, Johnson, and Pelczynski, [1974]) that $\mathscr{X}$ is WCG if and only if there is a reflexive space and an injective bounded operator $T:\mathscr{R} \to \mathscr{X}$ such that ran T is dense. (Hint: The Krein–Smulian Theorem (V.13.4) may be useful.)

5. If (X,Ω,μ) is a finite measure space, $k \in L^{\infty}(X \times X, \Omega \times \Omega, \mu \times \mu)$, and K: $L^1(\mu) \to L^1(\mu)$ is defined by $(Kf)(x) = \int k(x,y)f(y)d\mu(y)$, show that K is weakly compact and K^2 is compact.

6. Let $\mathscr{Y}$ be a weakly sequentially complete Banach space. That is, if $\{y_n\}$ is a sequence in $\mathscr{Y}$ such that $\{\langle y_n, y^* \rangle\}$ is a Cauchy sequence in $\mathbb{F}$ for every y^* in $\mathscr{Y}^*$, then there is a y in $\mathscr{Y}$ such that $y_n \to y$ weakly [see (V.4.4)]. (a) If $T \in \mathscr{B}(\mathscr{X}, \mathscr{Y})$ and $x^{**} \in \mathscr{X}^{**}$ such that x^{**} is the $\sigma(\mathscr{X}^{**}, \mathscr{X}^*)$ limit of a sequence $\{x_n^{**}\}$ from $\mathscr{X}^{**}$ such that $T^{**}(x_n^{**}) \in \mathscr{Y}$ for every n, show that $T^{**}(x^{**}) \in \mathscr{Y}$. Let X be a compact space and put $\mathscr{F} = $ all subsets of X that are the union of a countable number of compact G_δ sets. Let $\mathscr{L} = $ the linear span of $\{\chi_F: F \in \mathscr{F}\}$ considered as a subset of $M(X)^* = C(X)^{**}$. (b) Show that if $T \in \mathscr{B}(C(X), \mathscr{Y})$, then $T^{**}(\mathscr{L}) \subseteq \mathscr{Y}$. (c) (Grothendieck [1953].) If $T \in \mathscr{B}(C(X), \mathscr{Y})$, then T is weakly compact. [Hint (Spain [1976]): Use James's Theorem V.13.3).]

CHAPTER VII

Banach Algebras and Spectral Theory for Operators on a Banach Space

The theory of Banach algebras is a large area in functional analysis with several subdivisions and applications to diverse areas of analysis and the rest of mathematics. Some monographs on this subject are by Bonsall and Duncan [1973] and C.E. Rickart [1960].

A significant change occurs in this chapter that will affect the remainder of this book. In order to prove that the spectrum of an element of a Banach algebra is nonvoid (Section 3), it is necessary to assume that the underlying field of scalars $\mathbb{F}$ is the field of complex numbers $\mathbb{C}$. It will be assumed from Section 3 until the end of this book that all vector spaces are over $\mathbb{C}$. This will also enable us to apply the theory of analytic functions to the study of Banach algebras and linear operators.

In this chapter only the rudiments of this subject are discussed. Enough, however, is presented to allow a treatment of the basics of spectral theory for operators on a Banach space.

§1. Elementary Properties and Examples

An *algebra* over $\mathbb{F}$ is a vector space $\mathscr{A}$ over $\mathbb{F}$ that also has a multiplication defined on it that makes $\mathscr{A}$ into a ring such that if $\alpha \in \mathbb{F}$ and $a, b \in \mathscr{A}$, $\alpha(ab) = (\alpha a)b = a(\alpha b)$.

1.1. Definition. A *Banach algebra* is an algebra $\mathscr{A}$ over $\mathbb{F}$ that has a norm $\|\cdot\|$ relative to which $\mathscr{A}$ is a Banach space and such that for all a, b in $\mathscr{A}$,

1.2 $$\|ab\| \leqslant \|a\| \, \|b\|.$$

If $\mathscr{A}$ has an identity, e, then it is assumed that $\|e\| = 1$.

The fact that (1.2) is satisfied is not essential. If $\mathscr{A}$ is an algebra and has a norm relative to which $\mathscr{A}$ is a Banach space and is such that the map of $\mathscr{A} \times \mathscr{A} \to \mathscr{A}$ defined by $(a, b) \mapsto ab$ is continuous, then there is an equivalent norm on $\mathscr{A}$ that satisfies (1.2) (Exercise 1).

If $\mathscr{A}$ has an identity e, then the map $\alpha \mapsto \alpha e$ is an isomorphism of $\mathbb{F}$ into $\mathscr{A}$ and $\|\alpha e\| = |\alpha|$. So it will be assumed that $\mathbb{F} \subseteq \mathscr{A}$ via this identification. Thus the identity will be denoted by 1.

The content of the next proposition is that if $\mathscr{A}$ does not have an identity, it is possible to find a Banach algebra $\mathscr{A}_1$ that contains $\mathscr{A}$, that has an identity, and is such that $\dim \mathscr{A}_1 / \mathscr{A} = 1$.

1.3. Proposition. *If $\mathscr{A}$ is a Banach algebra without an identity, let $\mathscr{A}_1 = \mathscr{A} \times \mathbb{F}$. Define algebraic operations on $\mathscr{A}_1$ by*

(i) $(a, \alpha) + (b, \beta) = (a + b, \alpha + \beta)$;

(ii) $\beta(a, \alpha) = (\beta a, \beta \alpha)$;

(iii) $(a, \alpha)(b, \beta) = (ab + \alpha b + \beta a, \alpha \beta)$.

Define $\|(a, \alpha)\| = \|a\| + |\alpha|$. Then $\mathscr{A}_1$ with this norm and the algebraic operations defined in (i), (ii), and (iii) is a Banach algebra with identity $(0, 1)$ and $a \mapsto (a, 0)$ is an isometric isomorphism of $\mathscr{A}$ into $\mathscr{A}_1$.

PROOF. Only (1.2) will be verified here; the remaining details are left to the reader. If $(a, \alpha), (b, \beta) \in \mathscr{A}_1$, then $\|(a, \alpha)(b, \beta)\| = \|(ab + \beta a + \alpha b, \alpha \beta)\| = \|ab + \beta a + \alpha b\| + |\alpha \beta| \leq \|a\| \|b\| + |\beta| \|a\| + |\alpha| \|b\| + |\alpha| |\beta| = \|(a, \alpha)\| \|(b, \beta)\|$. ∎

1.4. Example. If X is a compact space, then $\mathscr{A} = C(X)$ is a Banach algebra if $(fg)(x) = f(x)g(x)$ whenever $f, g \in \mathscr{A}$ and $x \in X$. Note that $\mathscr{A}$ is abelian and has an identity (the constantly 1 function).

If X is completely regular and $\mathscr{A} = C_b(X)$, then $\mathscr{A}$ is also a Banach algebra. In fact, $C_b(X) \cong C(\beta X)$ (V.6) so that this is a special case of Example 1.4. Another special case is l^∞.

1.5. Example. If X is a locally compact space, $\mathscr{A} = C_0(X)$ is a Banach algebra when the multiplication is defined pointwise as in the preceding example. $\mathscr{A}$ is abelian, but if X is not compact, $\mathscr{A}$ does not have an identity. If X_∞ is the one-point compactification of X, then $C(X_\infty) \supseteq C_0(X)$ and $C(X_\infty)$ is a Banach algebra with identity.

Note that c_0 is a special case of Example 1.5.

1.6. Example. If (X, Ω, μ) is a σ-finite measure space and $\mathscr{A} = L^\infty(X, \Omega, \mu)$, then $\mathscr{A}$ is an abelian Banach algebra with identity if the operations are defined pointwise.

1.7. Example. Let $\mathscr{X}$ be a Banach space and put $\mathscr{A} = \mathscr{B}(\mathscr{X})$. If multiplication is defined by composition, then $\mathscr{A}$ is a Banach algebra with identity, 1. If $\dim \mathscr{X} \geqslant 2$, $\mathscr{A}$ is not abelian.

1.8. Example. If $\mathscr{X}$ is a Banach space and $\mathscr{A} = \mathscr{B}_0(\mathscr{X})$, the compact operators on $\mathscr{X}$, then $\mathscr{A}$ is a Banach algebra without identity if $\dim \mathscr{X} = \infty$. In fact, $\mathscr{B}_0(\mathscr{X})$ is an ideal of $\mathscr{B}(\mathscr{X})$.

Note that a special case of Example 1.7 occurs when $\mathscr{A} = M_n(\mathbb{F})$, the $n \times n$ matrices, where $\mathscr{A}$ is given the norm resulting when $M_n(\mathbb{F})$ is identified with $\mathscr{B}(\mathbb{F}^n)$.

1.9. Example. Let G be a locally compact topological group and let $M(G) = $ all finite regular Borel measures on G. If $\mu, \nu \in M(G)$, define $L: C_0(G) \to \mathbb{F}$ by

$$L(f) = \int\int f(xy) \, d\mu(x) \, d\nu(y) = \int\int f(xy) \, d\nu(y) \, d\mu(x).$$

Then L is a linear functional on $C_0(G)$ and

$$|L(f)| \leqslant \int\int |f(xy)| \, d|\mu|(x) \, d|\nu|(y)$$

$$\leqslant \|f\| \, \|\mu\| \, \|\nu\|.$$

So $L \in C_0(G)^* = M(G)$. Define $\mu * \nu$ by $L(f) = \int f \, d\mu * \nu$ for f in $C_0(G)$. That is,

1.10
$$\int f \, d\mu * \nu = \int\int f(xy) \, d\mu(x) \, d\nu(y).$$

Note that $\|\mu * \nu\| = \|L\| \leqslant \|\mu\| \, \|\nu\|$. If follows that $M(G)$ is a Banach algebra with this definition of multiplication. The product $\mu * \nu$ is called the *convolution* of μ and ν.

Let $e = $ the identity of G and let $\delta_e = $ the unit point mass at e. If $f \in C_0(G)$, then

$$\int f \, d\mu * \delta_e = \int\int f(xy) \, d\mu(x) \, d\delta_e(y)$$

$$= \int f(xe) \, d\mu(x)$$

$$= \int f \, d\mu.$$

So $\mu * \delta_e = \mu$; similarly, $\delta_e * \mu = \mu$. Hence δ_e is the identity for $M(G)$.

If $x, y \in G$, then it is easy to check that $\delta_x * \delta_y = \delta_{xy}$ and $M(G)$ is abelian if and only if G is abelian.

1.11. Example. Let G be a σ-compact locally compact group and let $m = $ right Haar measure on G. That is, m is a non-negative regular Borel measure on G such that $m(U) > 0$ for every nonempty open subset U of G and $\int f(xy)dm(x) = \int f(x)dm(x)$ for every f in $C_c(G)$ (the continuous functions f: $G \to \mathbb{F}$ with compact support). If G is compact, the existence of m was established in Section V.11. If G is not compact, m exists but its existence must be established by nonfunctional analytic methods. If G is not assumed to be σ-compact, then a Haar measure exists but it is not regular in the sense defined in this book. (See Nachbin [1965].)

If $f, g \in L^1(m)$, let $\mu = fm$ and $\nu = gm$ as in the proof of (V.8.1). Then $\mu, \nu \in M(G)$ and $\|\mu\| = \|f\|_1$, $\|\nu\| = \|g\|_1$. In fact, the Radon–Nikodym Theorem makes it possible to identify $L^1(m)$ with a closed subspace of $M(G)$. Is it a closed subalgebra?

Let $\phi \in C_c(G)$. Then

$$\int \phi\, d\mu * \nu = \int\int \phi(xy)f(x)g(y)dm(x)dm(y)$$

$$= \int g(y)\left[\int \phi(xy)f(x)dm(x)\right]dm(y)$$

$$= \int g(y)\left[\int \phi(x)f(xy^{-1})dm(x)\right]dm(y)$$

$$= \int \phi(x)\left[\int f(xy^{-1})g(y)dm(y)\right]dm(x)$$

$$= \int \phi(x)h(x)dm(x),$$

where $h(x) = \int f(xy^{-1})g(y)dm(y)$, x in G. It follows that $h \in L^1(m)$ (see Exercise 4). Thus $\mu * \nu = hm$, so $L^1(m)$ is a Banach subalgebra of $M(G)$. In fact, the preceding discussion enables us to define $f * g$ in $L^1(m)$ for f, g in $L^1(m)$ by

$$f * g(x) = \int f(xy^{-1})g(y)dm(y).$$

The algebra $L^1(m)$ is denoted by $L^1(G)$.

It can be shown that $L^1(G)$ is abelian if and only if G is abelian and $L^1(G)$ has an identity if and only if G is discrete (in which case $L^1(G) = M(G)$—what is m?). This algebra is examined more closely in Section 9.

If $\{\mathscr{A}_i\}$ is a collection of Banach algebras, let $\oplus_0\mathscr{A}_i \equiv \{a \in \prod_i\mathscr{A}_i$: for all $\varepsilon > 0$, $\{i: \|a(i)\| \geq \varepsilon\}$ is finite$\}$.

1.12. Proposition. *If $\{\mathscr{A}_i\}$ is a collection of Banach algebras, $\oplus_0\mathscr{A}_i$ and $\oplus_\infty\mathscr{A}_i$ are Banach algebras.*

PROOF. Exercise.

EXERCISES

1. Let $\mathscr{A}$ be an algebra that is also a Banach space and such that if $a \in \mathscr{A}$, the maps $x \mapsto ax$ and $x \mapsto xa$ of $\mathscr{A} \to \mathscr{A}$ are continuous. Let $\mathscr{A}_1 = \mathscr{A} \times \mathbb{F}$ as in Proposition 1.3. If $a \in \mathscr{A}$, define $L_a \colon \mathscr{A}_1 \to \mathscr{A}_1$ by $L_a(x, \xi) = (ax + \xi a, 0)$. Show that $L_a \in \mathscr{B}(\mathscr{A}_1)$ and if $\|\!|a|\!\| = \|L_a\|$, then $\|\!|\cdot|\!\|$ is equivalent to the norm of $\mathscr{A}$ and $\mathscr{A}$ with $\|\!|\cdot|\!\|$ is a Banach algebra.

2. Complete the proof of Proposition 1.3.

3. Verify the statements made in Examples (1.4) through (1.9) and (1.11).

4. Let G be a locally compact group. (a) If $\phi \in C_c(G)$ and $\varepsilon > 0$, show that there is an open neighborhood U of e in G such that $\|\phi_x - \phi_y\| < \varepsilon$ whenever $xy^{-1} \in U$. [Here $\phi_x(z) = \phi(xz)$.] (b) Show that if $f \in L^p(G)$, $1 \le p < \infty$, and $\varepsilon > 0$, there is an open neighborhood U of e in G such that $\|f_x - f_y\|_p < \varepsilon$ whenever $xy^{-1} \in U$. (c) Show that if $f \in L^1(G)$ and $g \in L^\infty(G)$, $h(x) = \int f(xy^{-1})g(y)dm(y)$ defines a bounded continuous function $h \colon G \to \mathbb{F}$. (d) If f, $g \in L^1(G)$ and h is defined as in (c), show that $h \in L^1(G)$.

5. Prove Proposition 1.12.

6. Let $\{\mathscr{A}_i \colon i \in I\}$ be a collection of Banach algebras. (a) Show that $\oplus_0 \mathscr{A}_i$ is a closed ideal of $\oplus_\infty \mathscr{A}_i$. (b) Show that $\oplus_\infty \mathscr{A}_i$ has an identity if and only if each $\mathscr{A}_i$ has an identity. (c) Show that $\oplus_0 \mathscr{A}_i$ has an identity if I is finite and each $\mathscr{A}_i$ has an identity.

7. If X, Y are completely regular, show that $C_b(X) \oplus_\infty C_b(Y)$ is isometrically isomorphic to $C_b(X \oplus Y)$, where $X \oplus Y$ is the disjoint union of X and Y.

8. If X and Y are locally compact, show that $C_0(X) \oplus_\infty C_0(Y)$ is isometrically isomorphic to $C_0(X \oplus Y)$.

9. Let $\{X_i \colon i \in I\}$ be a collection of locally compact spaces and let $X = $ the disjoint union of these spaces furnished with the topology $\{U \subseteq X \colon U \cap X_i \text{ is open in } X_i \text{ for all } i\}$. Show that X is locally compact and $\oplus_0 C_0(X_i)$ is isometrically isomorphic to $C_0(X)$.

§2. Ideals and Quotients

If $\mathscr{A}$ is an algebra, a *left ideal* of $\mathscr{A}$ is a subalgebra $\mathscr{M}$ of $\mathscr{A}$ such that $ax \in \mathscr{M}$ whenever $a \in \mathscr{A}$, $x \in \mathscr{M}$. A *right ideal* of $\mathscr{A}$ is a subalgebra $\mathscr{M}$ such that $xa \in \mathscr{M}$ whenever $a \in \mathscr{A}$, $x \in \mathscr{M}$. A *(bilateral) ideal* is a subalgebra of $\mathscr{A}$ that is both a left ideal and a right ideal.

If $a \in \mathscr{A}$ and $\mathscr{A}$ has an identity 1, say that a is *left invertible* if there is an x in $\mathscr{A}$ with $xa = 1$. Similarly, define *right invertible* and *invertible* elements. If a is invertible and x, $y \in \mathscr{A}$ such that $xa = 1 = ay$, then $y = 1y = (xa)y = x(ay) = x1 = x$. So if a is invertible, there is a unique element a^{-1} such that $aa^{-1} = a^{-1}a = 1$.

If $\mathscr{M}$ is a left ideal in $\mathscr{A}$, $a \in \mathscr{M}$, and a is left invertible, then $\mathscr{M} = \mathscr{A}$. In fact, if $xa = 1$, then $1 \in \mathcal{M}$ since $\mathcal{M}$ is a left ideal. Thus for y in $\mathcal{A}$, $y = y1 \in \mathcal{M}$. This forms a link between ideals and invertibility.

In the case of a Banach algebra some bonuses occur due to the interplay of the norm and the algebra. The results of this section will be for Banach algebras with an identity. To discuss invertibility this is, of course, the only feasible setting. For Banach algebras without an identity some analogous results can be obtained, however, by a consideration of the algebra obtained by adjoining an identity (1.3). The concept of a modular ideal and a modular unit can also be employed (see Exercise 6).

The next proof is based on the geometric series.

2.1. Lemma. *If $\mathcal{A}$ is a Banach algebra with identity and $x \in \mathcal{A}$ such that $\|x - 1\| < 1$, then x is invertible.*

PROOF. Let $y = 1 - x$; so $\|y\| = r < 1$. Since $\|y^n\| \leqslant \|y\|^n = r^n$ (Why?), $\sum_{n=0}^{\infty} \|y^n\| < \infty$. Hence $z = \sum_{n=0}^{\infty} y^n$ converges in $\mathcal{A}$. If $z_n = 1 + y + y^2 + \cdots + y^n$,

$$z_n(1 - y) = (1 + y + \cdots + y^n) - (y + y^2 + \cdots + y^{n+1}) = 1 - y^{n+1}.$$

But $\|y^{n+1}\| \leqslant r^{n+1}$, so $y^{n+1} \to 0$ as $n \to \infty$. Hence $z(1 - y) = \lim z_n(1 - y) = 1$. Similarly, $(1 - y)z = 1$. So $(1 - y)$ is invertible and $(1 - y)^{-1} = z = \sum_0^{\infty} y^n$. But $1 - y = 1 - (1 - x) = x$. ∎

Note that completeness was used to show that $\sum y^n$ converges.

2.2. Theorem. *If $\mathcal{A}$ is a Banach algebra with identity, $G_l = \{a \in \mathcal{A}: a \text{ is left invertible}\}$, $G_r = \{a \in \mathcal{A}: a \text{ is right invertible}\}$, and $G = \{a \in \mathcal{A}: a \text{ is invertible}\}$, then G_l, G_r, and G are open subsets of $\mathcal{A}$. Also, the map $a \mapsto a^{-1}$ of $G \to G$ is continuous.*

PROOF. Let $a_0 \in G_l$ and let $b_0 \in \mathcal{A}$ such that $b_0 a_0 = 1$. If $\|a - a_0\| < \|b_0\|^{-1}$, then $\|b_0 a - 1\| = \|b_0(a - a_0)\| < 1$. By the preceding lemma, $x = b_0 a$ is invertible. If $b = x^{-1}b_0$, then $ba = 1$. Hence $G_l \supseteq \{a \in \mathcal{A}: \|a - a_0\| < \|b_0\|^{-1}\}$ and G_l must be open. Similarly, G_r is open. Since $G = G_l \cap G_r$ (Why?), G is open.

To prove that $a \mapsto a^{-1}$ is a continuous map of $G \to G$, first assume that $\{a_n\}$ is a sequence in G such that $a_n \to 1$. Let $0 < \delta < 1$ and suppose $\|a_n - 1\| < \delta$. From the preceding lemma, $a_n^{-1} = (1 - (1 - a_n))^{-1} = \sum_{k=0}^{\infty}(1 - a_n)^k = 1 + \sum_{k=1}^{\infty}(1 - a_n)^k$. Hence

$$\|a_n^{-1} - 1\| = \left\|\sum_{k=1}^{\infty}(1 - a_n)^k\right\|$$

$$\leqslant \sum_{k=1}^{\infty}\|1 - a_n\|^k$$

$$< \delta/(1 - \delta).$$

If $\varepsilon > 0$ is given, then δ can be chosen such that $\delta/(1 - \delta) < \varepsilon$. So $\|a_n - 1\| < \delta$ implies $\|a_n^{-1} - 1\| < \varepsilon$. Hence $\lim a_n^{-1} = 1$.

Now let $a \in G$ and suppose $\{a_n\}$ is a sequence in G such that $a_n \to a$. Hence $a^{-1}a_n \to 1$. By the preceding paragraph, $a_n^{-1}a = (a^{-1}a_n)^{-1} \to 1$. Hence $a_n^{-1} = a_n^{-1}aa^{-1} \to a^{-1}$. ∎

Two facts surfaced in the preceding proofs that are worth recording for the future.

2.3. Corollary. *Let $\mathscr{A}$ be a Banach algebra with identity.*

(a) *If $\|a - 1\| < 1$, then $a^{-1} = \sum_{k=0}^{\infty}(1 - a)^k$.*
(b) *If $b_0 a_0 = 1$ and $\|a - a_0\| < \|b_0\|^{-1}$, then a is left invertible.*

A *maximal ideal* is a proper ideal that is contained in no larger proper ideal.

2.4. Corollary. *If $\mathscr{A}$ is a Banach algebra with identity, then*

(a) *the closure of a proper left, right, or bilateral ideal is a proper left, right, or bilateral ideal;*
(b) *a maximal left, right, or bilateral ideal is closed.*

PROOF. (a) Let $\mathscr{M}$ be a proper left ideal and let G_l be the set of left-invertible elements in $\mathscr{A}$. It follows that $\mathscr{M} \cap G_l = \square$. (See the introduction to this section.) Thus $\mathscr{M} \subseteq \mathscr{A} \backslash G_l$. By the preceding theorem, $\mathscr{A} \backslash G_l$ is closed. Hence $\mathrm{cl}\, \mathscr{M} \subseteq \mathscr{A} \backslash G_l$; and thus $\mathrm{cl}\, \mathscr{M} \neq \mathscr{A}$. It is easy to check that $\mathrm{cl}\, \mathscr{M}$ is an ideal. The proof of the remainder of (a) is similar.

(b) If $\mathscr{M}$ is a maximal left ideal, $\mathrm{cl}\, \mathscr{M}$ is a proper left ideal by (a). Hence $\mathscr{M} = \mathrm{cl}\, \mathscr{M}$ by maximality. ∎

If $\mathscr{A}$ does not have an identity, then $\mathscr{A}$ may contain some proper, dense ideals. For example, let $\mathscr{A} = C_0(\mathbb{R})$. Then $C_c(\mathbb{R})$, the continuous functions with compact support, is a dense ideal in $C_0(\mathbb{R})$. There is something that can be said, however (see Exercise 6).

2.5. Proposition. *If $\mathscr{A}$ is a Banach algebra with identity, then every proper left, right, or bilateral ideal is contained in a maximal ideal of the same type.*

The proof of the preceding proposition is an exercise in the application of Zorn's Lemma and is left to the reader. Actually, this is a theorem from algebra and it is not necessary to assume that $\mathscr{A}$ is a Banach algebra.

Let $\mathscr{A}$ be a Banach algebra and let $\mathscr{M}$ be a proper closed ideal. Note that $\mathscr{A}/\mathscr{M}$ becomes an algebra. Indeed, $(x + \mathscr{M})(y + \mathscr{M}) = xy + \mathscr{M}$ is a well-defined multiplication on $\mathscr{A}/\mathscr{M}$. (Why?)

2.6. Theorem. *If $\mathscr{A}$ is a Banach algebra and $\mathscr{M}$ is a proper closed ideal in $\mathscr{A}$, then $\mathscr{A}/\mathscr{M}$ is a Banach algebra. If $\mathscr{A}$ has an identity, so does $\mathscr{A}/\mathscr{M}$.*

PROOF. We have already seen that $\mathscr{A}/\mathscr{M}$ is a Banach space and, as was mentioned prior to the statement of the theorem, $\mathscr{A}/\mathscr{M}$ is an algebra. If $x, y \in \mathscr{A}$ and $u, v \in \mathscr{M}$, then $(x + u)(y + v) = xy + (xv + uy + uv) \in xy + \mathscr{M}$. Hence $\|(x + \mathscr{M})(y + \mathscr{M})\| = \|xy + \mathscr{M}\| \leqslant \|(x + u)(y + v)\| \leqslant \|x + u\| \|y + v\|$. Taking the infimum over all u, v in $\mathscr{M}$ gives that $\|(x + \mathscr{M})(y + \mathscr{M})\| \leqslant \|x + \mathscr{M}\|$ $\|y + \mathscr{M}\|$. The remainder of the proof is left to the reader. ∎

It may be that $\mathscr{A}/\mathscr{M}$ has an identity even if $\mathscr{A}$ does not. For example, let $\mathscr{A} = C_0(\mathbb{R})$ and let $\mathscr{M} = \{\phi \in C_0(\mathbb{R}): \phi(x) = 0 \text{ when } |x| \leqslant 1\}$. If $\phi_0 \in C_0(\mathbb{R})$ such that $\phi_0(x) = 1$ for $|x| \leqslant 1$, then $\phi_0 + \mathscr{M}$ is an identity for $\mathscr{A}/\mathscr{M}$. In fact, if $\phi \in C_0(\mathbb{R})$, $(\phi\phi_0 - \phi)(x) = 0$ if $|x| \leqslant 1$. Hence $(\phi + \mathscr{M})(\phi_0 + \mathscr{M}) = \phi + \mathscr{M}$ (see Exercises 6 through 9).

EXERCISES

1. Let $\mathscr{A}$ be a Banach algebra and let $\mathscr{L}$ be all of the closed left ideals in $\mathscr{A}$. If I_1, $I_2 \in \mathscr{L}$, define $I_1 \vee I_2 \equiv \text{cl}(I_1 + I_2)$ and $I_1 \wedge I_2 = I_1 \cap I_2$. Show that with these definitions $\mathscr{L}$ is a complete lattice with a largest and a smallest element.

2. Let X be locally compact. For every open subset U of X, let $I(U) = \{\phi \in C_0(X):$ $\phi = 0$ on $X \backslash U\}$. Show that $U \mapsto I(U)$ is a lattice monomorphism of the collection of open subsets of X into the lattice of close ideals of $C_0(X)$. (It is, in fact, surjective, but the proof of that should wait.)

3. Let (X, Ω, μ) be a σ-finite measure space and let I be an ideal in $L^\infty(X, \Omega, \mu)$ that is weak* closed. Show that there is a set Δ in Ω such that $I = \{\phi \in L^\infty(X, \Omega, \mu):$ $\phi = 0$ a.e. on $\Delta\}$.

4. Let $\mathscr{A} = \left\{ \begin{bmatrix} \alpha & 0 \\ \beta & \alpha \end{bmatrix} : \alpha, \beta \in \mathbb{F} \right\}$ and let $\mathscr{M} = \left\{ \begin{bmatrix} 0 & 0 \\ \beta & 0 \end{bmatrix} : \beta \in \mathbb{F} \right\}$. Show that $\mathscr{A}$ is a Banach algebra and $\mathscr{M}$ is a maximal ideal in $\mathscr{A}$.

5. Show that for $n \geqslant 1$, $M_n(\mathbb{C})$ has no nontrivial ideals. How about $M_n(\mathbb{R})$?

6. Let $\mathscr{A}$ be a Banach algebra but do not assume that $\mathscr{A}$ has an identity. If I is a left ideal of $\mathscr{A}$, say that I is a *modular left ideal* if there is a u in $\mathscr{A}$ such that $\mathscr{A}(1 - u) \equiv \{a - au: a \in \mathscr{A}\} \subseteq I$; call such an element u of $\mathscr{A}$ a *right modular unit* for I. Similarly, define *right modular ideals* and *left modular units*. Prove the following. (a) If u is a right modular unit for the left ideal I and $u \in I$, then $I = \mathscr{A}$. (b) Maximal modular left ideals are maximal left ideals. (c) If I is a proper modular left ideal, then I is contained in a maximal left ideal. (d) If I is a proper modular left ideal and u is a modular right unit for I, then $\|u - x\| \geqslant 1$ for all x in I and cl I is a proper modular left ideal. (e) Every maximal modular left ideal of $\mathscr{A}$ is closed.

7. Using the terminology of Exercise 6, let I be an ideal of $\mathscr{A}$. Show: (a) if u is a right modular unit for I and v is a left modular unit for I, then $u - v \in I$. (b) If I is closed, $\mathscr{A}/I$ has an identity if and only if there is a right modular unit and a left modular unit for I. Call an ideal I such that $\mathscr{A}/I$ has an identity a *modular ideal*. An element u such that $u + I$ is an identity for $\mathscr{A}/I$ is called a *modular identity* for I.

8. If $\mathscr{A}$ is a Banach algebra, a net $\{e_i\}$ in $\mathscr{A}$ is called an *approximate identity* for $\mathscr{A}$ if $\sup_i \|e_i\| < \infty$ and for each a in $\mathscr{A}$, $e_i a \to a$ and $a e_i \to a$. Show that $\mathscr{A}$ has an approximate identity if and only if there is a bounded subset E of $\mathscr{A}$ such that for every $\varepsilon > 0$ and for every a in $\mathscr{A}$ there is an e in E with $\|ae - a\| + \|ea - a\| < \varepsilon$. See Wichmann [1973] for more information.

9. Show that if X is locally compact, then $C_0(X)$ has an approximate identity.

10. If $\mathscr{H}$ is a Hilbert space, show that $\mathscr{B}_0(\mathscr{H})$ has an approximate identity.

11. If G is a locally compact group, show that $L^1(G)$ (1.11) has an approximate identity. [Hint: Let $\mathscr{U} =$ all neighborhoods U of the identity e of G such that cl U is compact. Order $\mathscr{U}$ by reverse inclusion. For U in $\mathscr{U}$, let $f_U = m(U)^{-1} \chi_U$. Then $\{f_U: U \in \mathscr{U}\}$ is an approximate identity for $L^1(G)$.]

12. For $0 < r < 1$, let $P_r: \partial \mathbb{D} \to [0, \infty)$ defined by $P_r(z) = \sum_{n=-\infty}^{\infty} r^{|n|} z^n$ (the *Poisson kernel*). Show that $\{P_r\}$ is an approximate identity for $L^1(\partial \mathbb{D})$ (under convolution).

13. If $\mathscr{H}$ is a Hilbert space and P is a projection, show that $\mathscr{B}_0(\mathscr{H})P$ is a closed modular left ideal of $\mathscr{B}_0(\mathscr{H})$. What is the associated right modular unit?

14. Find the minimal proper left ideals of $M_n(\mathbb{F})$.

15. Find the minimal closed proper left ideals of $\mathscr{B}_0(\mathscr{H})$, $\mathscr{H}$ a Hilbert space. How about for $\mathscr{B}_0(\mathscr{X})$, $\mathscr{X}$ a Banach space?

16. What are the maximal modular left ideals of $\mathscr{B}_0(\mathscr{H})$, $\mathscr{H}$ a Hilbert space?

§3. The Spectrum

3.1. Definition. If $\mathscr{A}$ is a Banach algebra with identity and $a \in \mathscr{A}$, the *spectrum* of a, denoted by $\sigma(a)$, is defined by

$$\sigma(a) = \{\alpha \in \mathbb{F}: a - \alpha \text{ is not invertible}\}.$$

The *left spectrum*, $\sigma_l(a)$, is the set $\{\alpha \in \mathbb{F}: a - \alpha \text{ is not left invertible}\}$; the *right spectrum*, $\sigma_r(a)$, is defined similarly.

The resolvent set of a is defined by $\rho(a) = \mathbb{F} \backslash \sigma(a)$. The *left* and *right* resolvents of a are $\rho_l(a) = \mathbb{F} \backslash \sigma_l(a)$ and $\rho_r(a) = \mathbb{F} \backslash \sigma_r(a)$.

3.2. Example. Let X be compact. If $f \in C(X)$, then $\sigma(f) = f(X)$. In fact, if $\alpha = f(x_0)$, then $f - \alpha$ has a zero and cannot be invertible. So $f(X) \subseteq \sigma(f)$. On the other hand, if $\alpha \notin f(X)$, $f - \alpha$ is a nonvanishing continuous function on X. Hence $(f - \alpha)^{-1} \in C(X)$ and so $f - \alpha$ is invertible. Thus $\alpha \notin \sigma(f)$.

3.3. Example. If $\mathscr{X}$ is a Banach space and $A \in \mathscr{B}(\mathscr{X})$, then $\sigma(A) = \{\alpha \in \mathbb{F}:$ either $\ker(A - \alpha) \neq (0)$ or $\operatorname{ran}(A - \alpha) \neq \mathscr{X}\}$. In fact, this means that $\rho(A) = \mathbb{F} \backslash \sigma(A) = \{\alpha \in \mathbb{F}: A - \alpha \text{ is bijective}\}$. If $\alpha \in \rho(A)$, there is an operator T in $\mathscr{B}(\mathscr{X})$ such that $T(A - \alpha) = (A - \alpha)T = 1$; clearly, $A - \alpha$ is bijective. On the other hand, if $A - \alpha$ is bijective, $(A - \alpha)^{-1} \in \mathscr{B}(\mathscr{X})$ by the Inverse Mapping Theorem.

3.4. Example. If $\mathscr{H}$ is a Hilbert space and $A \in \mathscr{B}(\mathscr{H})$, then $\sigma_l(A) = \{\alpha \in \mathbb{F}:$ $\inf\{\|(A - \alpha)h\|: \|h\| = 1\} = 0\}$. In fact, suppose $B \in \mathscr{B}(\mathscr{H})$ such that $B(A - \alpha) = 1$. If $\|h\| = 1$, then $1 = \|h\| = \|B(A - \alpha)h\| \leqslant \|B\| \|(A - \alpha)h\|$. So $\|(A - \alpha)h\| \geqslant \|B\|^{-1}$ whenever $\|h\| = 1$.

Conversely, suppose $\|(A - \alpha)h\| \geqslant \delta > 0$ whenever $\|h\| = 1$. Note that $\ker(A - \alpha) = (0)$. It will now be shown that $\operatorname{ran}(A - \alpha)$ is closed. In fact, assume that $(A - \alpha)f_n \to g$. Then $\delta\|f_n - f_m\| \leqslant \|(A - \alpha)(f_n - f_m)\| = \|(A - \alpha)f_n - (A - \alpha)f_m\|$. Thus $\{f_n\}$ is a Cauchy sequence. Let $f_n \to f$. Then $g = \lim(A - \alpha)f_n = (A - \alpha)f$; hence $g \in \operatorname{ran}(A - \alpha)$. Let $\mathscr{K} = \operatorname{ran}(A - \alpha)$; so $(A - \alpha): \mathscr{H} \to \mathscr{K}$ is a bijection. Thus $(A - \alpha)^{-1}: \mathscr{K} \to \mathscr{H}$ is bounded. Define $B: \mathscr{H} \to \mathscr{H}$ by letting $B(k + h) = (A - \alpha)^{-1}k$ when $k \in \mathscr{K}$ and $h \in \mathscr{K}^\perp$. Thus $B \in \mathscr{B}(\mathscr{H})$ and $B(A - \alpha) = 1$.

3.5. Example. If $\mathscr{A} = M_2(\mathbb{R})$ and $A = \begin{bmatrix} 0 & -1 \\ 1 & 0 \end{bmatrix}$, then $\sigma(A) = \square$. In fact, $A - \alpha$ is not invertible if and only if $0 = \det(A - \alpha) = \alpha^2 + 1$, which is impossible in $\mathbb{R}$.

The phenomenon of the last example does not occur if $\mathscr{A}$ is a Banach algebra over $\mathbb{C}$.

3.6. Theorem. *If $\mathscr{A}$ is a Banach algebra over $\mathbb{C}$ with an identity, then for each a in $\mathscr{A}$, $\sigma(a)$ is a nonempty compact subset of $\mathbb{C}$. Moreover, if $|\alpha| > \|a\|$, $\alpha \notin \sigma(a)$ and $z \mapsto (z - a)^{-1}$ is an $\mathscr{A}$-valued analytic function defined on $\rho(a)$.*

Before beginning the proof, a few words on vector-valued analytic functions are in order. If G is a region in $\mathbb{C}$ and $\mathscr{X}$ is a Banach space, define the derivative of $f: G \to \mathscr{X}$ at z_0 to be $\lim_{h \to 0} h^{-1}[f(z_0 + h) - f(z_0)]$ if the limit exists. Say that f is analytic if f has a continuous derivative on G. The whole theory of analytic functions transfers to this situation. The statements and proofs of such theorems as Cauchy's Integral Formula, Liouville's Theorem, etc., transfer verbatim. Also, $f: G \to \mathscr{X}$ is analytic if and only if for each z_0 in G there is a sequence $x_0, x_1, x_2, \ldots$ in $\mathscr{X}$ such that $f(z) = \sum_{k=0}^\infty (z - z_0)^k x_k$ whenever $z \in B(z_0; r)$, where $r = \operatorname{dist}(z_0, \partial G)$. Moreover, the convergence is uniform on compact subsets of $B(z_0; r)$.

There is also a way of obtaining the vector-valued case as a consequence of the scalar-valued case (see Exercise 4).

PROOF OF THEOREM 3.6. If $|\alpha| > \|a\|$, then $\alpha - a = \alpha(1 - a/\alpha)$ and $\|a/\alpha\| < 1$. By Corollary 2.3, $(1 - a/\alpha)$ is invertible. Hence $\alpha - a$ is invertible and so $\alpha \notin \sigma(a)$. Thus $\sigma(a) \subseteq \{\alpha \in \mathbb{C}: |\alpha| \leqslant \|a\|\}$ and $\sigma(a)$ is bounded.

Let G be the set of invertible elements of $\mathscr{A}$. The map $\alpha \mapsto (\alpha - a)$ is a continuous function of $\mathbb{C} \to \mathscr{A}$. Since G is open and $\rho(a)$ is the inverse image of G under this map, $\rho(a)$ is open. Thus $\sigma(a) = \mathbb{C} \backslash \rho(a)$ is compact.

Define $F: \rho(a) \to \mathscr{A}$ by $F(z) = (z - a)^{-1}$. In the identity $x^{-1} - y^{-1} =$

$x^{-1}(y-x)y^{-1}$, let $x = (\alpha + h - a)$ and $y = (\alpha - a)$, where $\alpha \in \rho(a)$ and $h \in \mathbb{C}$ such that $h \neq 0$ and $\alpha + h \in \rho(a)$. This gives

$$\frac{F(\alpha + h) - F(\alpha)}{h} = \frac{(\alpha + h - a)^{-1}(-h)(\alpha - a)^{-1}}{h}$$

$$= -(\alpha + h - a)^{-1}(\alpha - a)^{-1}.$$

Since $(\alpha + h - a)^{-1} \to (\alpha - a)^{-1}$ as $h \to 0$, $F'(\alpha)$ exists and

$$F'(\alpha) = -(\alpha - a)^{-2}.$$

Clearly $F': \rho(a) \to \mathscr{A}$ is continuous, so F is analytic on $\rho(a)$.

From the first paragraph of the proof and Corollary 2.3, if $|z| > \|a\|$, then $F(z) = z^{-1}(1 - a/z)^{-1}$. But as $z \to \infty$, $(1 - a/z) \to 1$ and so $(1 - a/z)^{-1} \to 1$. Thus $F(z) \to 0$ as $z \to \infty$.

Therefore if $\rho(a) = \mathbb{C}$, F is an entire function that vanishes at ∞. By Liouville's Theorem F is constant. Since $F' \neq 0$, this is a contradiction. Thus $\rho(a) \neq \mathbb{C}$, or $\sigma(a) \neq \square$. ∎

Because the spectrum of an element of a complex Banach algebra is not empty, the following assumption is made.

Assumption. *Henceforward, all Banach spaces and all Banach algebras are over* $\mathbb{C}$.

3.7. Definition. If $\mathscr{A}$ is a Banach algebra with identity and $a \in \mathscr{A}$, the *spectral radius* of a, $r(a)$, is defined by

$$r(a) = \sup\{|\alpha|: \alpha \in \sigma(a)\}.$$

Because $\sigma(a) \neq \square$ and is bounded, $r(a)$ is well defined and finite; because $\sigma(a)$ is compact, this supremum is attained.

Let $\mathscr{A} = M_2(\mathbb{C})$ and let $A = \begin{bmatrix} 0 & 0 \\ 1 & 0 \end{bmatrix}$. Then $A^2 = 0$ and $\sigma(A) = \{0\}$; so $r(A) = 0$. So it is possible to have $r(A) = 0$ with $A \neq 0$.

3.8. Proposition. *If $\mathscr{A}$ is a Banach algebra with identity and $a \in \mathscr{A}$,* $\lim \|a^n\|^{1/n}$ *exists and*

$$r(a) = \lim \|a^n\|^{1/n}.$$

PROOF. Let $G = \{z \in \mathbb{C}: z = 0 \text{ or } z^{-1} \in \rho(a)\}$. Define $f: G \to \mathscr{A}$ by $f(0) = 0$ and for $z \neq 0$, $f(z) = (z^{-1} - a)^{-1}$. Since $(a - \alpha)^{-1} \to 0$ as $\alpha \to \infty$, f is analytic on G, and so f has a power series expansion. In fact, by Corollary 2.3, for $|z| < \|a\|^{-1}$,

$$f(z) = \sum_{n=0}^{\infty} a^n/(z^{-1})^{n+1} = z \sum_{n=0}^{\infty} z^n a^n.$$

From complex variable theory, this power series converges for $|z| < R \equiv \text{dist}(0, \partial G) = \text{dist}(0, \sigma(a)^{-1})$ (Here $\sigma(a)^{-1} = \{z^{-1}: z \in \sigma(a)\}$). Thus $R = \inf\{|\alpha|: \alpha^{-1} \in \sigma(a)\} = r(a)^{-1}$. Also, from the theory of power series, $R^{-1} = \limsup \|a^n\|^{1/n}$. Thus

$$r(a) = \limsup \|a^n\|^{1/n}.$$

Now if $\alpha \in \mathbb{C}$ and $n \geq 1$, $\alpha^n - a^n = (\alpha - a)(\alpha^{n-1} + \alpha^{n-2}a + \cdots + a^{n-1}) = (\alpha^{n-1} + \alpha^{n-2}a + \cdots + a^{n-1})(\alpha - a)$. So if $\alpha^n - a^n$ is invertible, $\alpha - a$ is invertible and $(\alpha - a)^{-1} = (\alpha^n - a^n)^{-1}(\alpha^{n-1} + \cdots + a^{n-1})$. So for α in $\sigma(a)$, $\alpha^n - a^n$ is not invertible for every $n \geq 1$. By Theorem 3.6, $|\alpha|^n \leq \|a^n\|$. Hence $|\alpha| \leq \|a^n\|^{1/n}$ for all $n \geq 1$ and α in $\sigma(a)$. So if $\alpha \in \sigma(a)$, $|\alpha| \leq \liminf \|a^n\|^{1/n}$. Taking the supremum over all α in $\sigma(a)$ gives that $r(a) \leq \liminf \|a^n\|^{1/n} \leq \limsup \|a^n\|^{1/n} = r(a)$. So $r(a) = \lim \|a^n\|^{1/n}$. ∎

3.9. Proposition. *Let $\mathscr{A}$ be a Banach algebra with identity and let $a \in \mathscr{A}$.*

(a) *If $\alpha \in \rho(a)$, then $\text{dist}(\alpha, \sigma(a)) \geq \|(\alpha - a)^{-1}\|^{-1}$.*
(b) *If $\alpha, \beta \in \rho(a)$, then*

$$(\alpha - a)^{-1} - (\beta - a)^{-1} = (\beta - \alpha)(\alpha - a)^{-1}(\beta - a)^{-1}$$
$$= (\beta - \alpha)(\beta - a)^{-1}(\alpha - a)^{-1}.$$

PROOF. (a) By Corollary 2.3, if $\alpha \in \rho(a)$ and $\|x - (\alpha - a)\| < \|(\alpha - a)^{-1}\|^{-1}$, x is invertible. So if $\beta \in \mathbb{C}$ and $|\beta| < \|(\alpha - a)^{-1}\|^{-1}$, $(\beta + \alpha - a)$ is invertible; that is, $\alpha + \beta \in \rho(a)$. Hence $\text{dist}(\alpha, \sigma(a)) \geq \|(\alpha - a)^{-1}\|^{-1}$. (b) This follows by letting $x = \alpha - a$ and $y = \beta - a$ in the identity $x^{-1} - y^{-1} = x^{-1}(y - x)y^{-1} = y^{-1}(y - x)x^{-1}$. ∎

The identity in part (b) of the preceding proposition is called the *resolvent identity* and the function $\alpha \mapsto (\alpha - a)^{-1}$ of $\rho(a) \to \mathscr{A}$ is called the *resolvent* of a.

EXERCISES

1. Let S be the unilateral shift on l^2 (II.2.10). Show that S is left invertible but not right invertible.

2. If $\mathscr{A}$ is a Banach algebra with identity and $a \in \mathscr{A}$ and is nilpotent (that is, $a^n = 0$ for some n), then $\sigma(a) = \{0\}$.

3. Let (X, Ω, μ) be a σ-finite measure space and let $\mathscr{A} = L^\infty(X, \Omega, \mu)$ (1.6). If $\phi \in \mathscr{A}$, show that the following are equivalent: (a) $\alpha \in \sigma(\phi)$; (b) $0 = \sup\{\inf\{|\phi(x) - \alpha|: x \in X \setminus \Delta\}: \Delta \in \Omega$ and $\mu(\Delta) = 0\}$; (c) if $\varepsilon > 0$, $\mu(\{x \in X: |\phi(x) - \alpha| < \varepsilon\}) > 0$; (d) if ν is the measure defined on the Borel subsets of $\mathbb{C}$ by $\nu(\Delta) = \mu(\phi^{-1}(\Delta))$, then $\alpha \in$ the support of ν.

4. If G is an open subset of $\mathbb{C}$ and $f: G \to \mathscr{X}$ is a function such that for each x^* in $\mathscr{X}^*$, $x^* \circ f: G \to \mathbb{C}$ is analytic, then f is analytic. If the word "continuous" is substituted for both occurrences of the word "analytic", is the preceding statement still true?

5. If $\mathscr{A}$ is a Banach algebra with identity, $\{a_n\} \subseteq \mathscr{A}$, $a_n \to a$, $\alpha_n \in \sigma(a_n)$, and $\alpha_n \to \alpha$, then $\alpha \in \sigma(a)$.

6. If $\mathscr{A}$ is a Banach algebra with identity and $r: \mathscr{A} \to [0, \infty)$ is the spectral radius, show that r is upper semicontinuous. If $a \in \mathscr{A}$ such that $r(a) = 0$, show that r is continuous at a.

7. If $\mathscr{A}$ is a Banach algebra with identity, a, $b \in \mathscr{A}$, and α is a nonzero scalar such that $(\alpha - ab)$ is invertible, show that $(\alpha - ba)$ is invertible and $(\alpha - ba)^{-1} = \alpha^{-1} + \alpha^{-1}b(\alpha - ab)^{-1}a$. Show that $\sigma(ab) \cup \{0\} = \sigma(ba) \cup \{0\}$ and give an example such that $\sigma(ab) \neq \sigma(ba)$.

§4. The Riesz Functional Calculus

Before coming to the main course of this section, it is necessary to have an appetizer from complex analysis. Many of these topics can be found in Conway [1978] with complete proofs. Only a few results are presented here.

If γ is a closed rectifiable curve in $\mathbb{C}$ and $a \notin \{\gamma\} \equiv \{\gamma(t): 0 \leqslant t \leqslant 1\}$, then the *winding number of γ about a* is defined to be the number

$$n(\gamma; a) = \frac{1}{2\pi i} \int_\gamma \frac{1}{z - a} dz.$$

The number $n(\gamma; a)$ is always an integer and is constant on each component of $\mathbb{C}\backslash\{\gamma\}$ and vanishes on the unbounded component of $\mathbb{C}\backslash\{\gamma\}$.

Let G be an open subset of $\mathbb{C}$ and let $\mathscr{X}$ be a Banach space. If $f: G \to \mathscr{X}$ is analytic and $x^* \in \mathscr{X}^*$, then $z \mapsto \langle f(z), x^* \rangle$ is analytic on G and its derivative is $\langle f'(z), x^* \rangle$. By Exercise 4 of the preceding section, if $f: G \to \mathscr{X}$ is a function such that $z \mapsto \langle f(z), x^* \rangle$ is analytic for each x^* in $\mathscr{X}^*$, then $f: G \to \mathscr{X}$ is analytic. These facts will help in discussing and proving many of the results below.

If γ is a rectifiable curve in G and f is a continuous function defined in a neighborhood of $\{\gamma\}$ with values in $\mathscr{X}$, then $\int_\gamma f$ can be defined as for a scalar-valued f as the limit in $\mathscr{X}$ of sums of the form

$$\sum_j [\gamma(t_j) - \gamma(t_{j-1})]f(\gamma(t_j)),$$

where $\{t_0, t_1, \ldots, t_n\}$ is a partition of $[0, 1]$. Hence $\int_\gamma f = \int_0^1 f(\gamma(t))d\gamma(t) \in \mathscr{X}$. It is easy to see that for every x^* in $\mathscr{X}^*$, $\langle \int_\gamma f, x^* \rangle = \int_\gamma \langle f(\cdot), x^* \rangle$.

4.1. Cauchy's Theorem. *If $\mathscr{X}$ is a Banach space, G is an open subset of $\mathbb{C}$, $f: G \to \mathscr{X}$ is an analytic function, and $\gamma_1, \ldots, \gamma_m$ are closed rectifiable curves in G such that $\sum_{j=1}^m n(\gamma_j; a) = 0$ for all a in $\mathbb{C}\backslash G$, then $\sum_{j=1}^m \int_{\gamma_j} f = 0$.*

PROOF. If $x^* \in \mathscr{X}^*$, then $\langle \sum_{j=1}^m \int_{\gamma_j} f, x^* \rangle = \sum_{j=1}^m \int_{\gamma_j} \langle f(\cdot), x^* \rangle = 0$ by the scalar-valued version of Cauchy's Theorem. Hence $\sum_{j=1}^m \int_{\gamma_j} f = 0$. ∎

4.2. Cauchy's Integral Formula. *If $\mathscr{X}$ is a Banach space, G is an open subset of $\mathbb{C}$, $f: G \to \mathscr{X}$ is analytic, γ is a closed rectifiable curve in G such that $n(\gamma; a) = 0$ for every a in $\mathbb{C}\backslash G$, and $\lambda \in G\backslash\{\gamma\}$, then for every integer $k \geqslant 0$,*

$$n(\gamma; \lambda) f^{(k)}(\lambda) = \frac{k!}{2\pi i} \int_\gamma (z - \lambda)^{-(k+1)} f(z) dz.$$

4.3. Definition. A closed rectifiable curve γ is *positively oriented* if for every a in $G\backslash\{\gamma\}$, $n(\gamma; a)$ is either 0 or 1. In this case the *inside* of γ, denoted by ins γ, is defined by

$$\text{ins } \gamma \equiv \{a \in \mathbb{C}\backslash\{\gamma\}: n(\gamma; a) = 1\}.$$

The *outside* of γ, denoted by out γ, is defined by

$$\text{out } \gamma \equiv \{a \in \mathbb{C}\backslash\{\gamma\}: n(\gamma; a) = 0\}.$$

Thus $\mathbb{C} = \{\gamma\} \cup \text{ins } \gamma \cup \text{out } \gamma$.

A curve $\gamma: [0, 1] \to \mathbb{C}$ is *simple* if $\gamma(s) = \gamma(t)$ implies that either $s = t$ or $s = 0$ and $t = 1$. The Jordan Curve Theorem says that if γ is a simple closed rectifiable curve, then $\mathbb{C}\backslash\{\gamma\}$ has two components and $\{\gamma\}$ is the boundary of each. Hence $n(\gamma; a)$ takes on only two values and one of these must be 0; the other must be ± 1.

If $\Gamma = \{\gamma_1, \ldots, \gamma_m\}$ is a collection of closed rectifiable curves, then Γ is *positively oriented* if: (a) $\{\gamma_i\} \cap \{\gamma_j\} = \square$ for $i \neq j$; (b) for a in $\mathbb{C}\backslash\bigcup_{j=1}^m \{\gamma_j\}$, $n(\Gamma; a) \equiv \sum_{j=1}^m n(\gamma_j; a)$ is either 0 or 1; (c) each γ_j is a simple curve. The *inside* of Γ, ins Γ, is defined by

$$\text{ins } \Gamma \equiv \{a: n(\Gamma; a) = 1\}.$$

The *outside* of Γ, out Γ, is defined by

$$\text{out } \Gamma \equiv \{a: n(\Gamma; a) = 0\}.$$

Let $\{\Gamma\} \equiv \cup\{\gamma_j: 1 \leqslant j \leqslant m\}$.

4.4. Proposition. *If G is an open subset of $\mathbb{C}$ and K is a compact subset of G, then there is a positively oriented system of curves $\Gamma = \{\gamma_1, \ldots, \gamma_m\}$ in $G\backslash K$ such that $K \subseteq \text{ins } \Gamma$ and $\mathbb{C}\backslash G \subseteq \text{out } \Gamma$. The curves $\gamma_1, \ldots, \gamma_m$ can be found such that they are infinitely differentiable.*

The proof of this proposition can be found on p. 195 of Conway [1978], though some details are missing.

If $\Gamma = \{\gamma_1, \ldots, \gamma_m\}$ and each γ_j is rectifiable, define

$$\int_\Gamma f = \sum_{j=1}^m \int_{\gamma_j} f$$

whenever f is continuous in a neighborhood of $\{\Gamma\}$.

Let $\mathscr{A}$ be a Banach algebra with identity and let $a \in \mathscr{A}$. One of the principal uses of Proposition 4.4 in this book will occur when $K = \sigma(a)$. If $f: G \to \mathbb{C}$ is analytic and $\sigma(a) \subseteq G$, we will define an element $f(a)$ in $\mathscr{A}$ by

4.5
$$f(a) = \frac{1}{2\pi i} \int_{\Gamma} f(z)(z - a)^{-1} dz$$

where Γ is as in Proposition 4.4 with $K = \sigma(a)$. But first it must be shown that (4.5) does not depend on the choice of Γ. That is, it must be shown that $f(a)$ is well defined.

4.6. Proposition. *Let $\mathscr{A}$ be a Banach algebra with identity, let $a \in \mathscr{A}$, and let G be an open subset of $\mathbb{C}$ such that $\sigma(a) \subseteq G$. If $\Gamma = \{\gamma_1, \ldots, \gamma_m\}$ and $\Lambda = \{\lambda_1, \ldots, \lambda_k\}$ are two positively oriented collections of curves in G such that $\sigma(a) \subseteq \text{ins } \Gamma \subseteq G$ and $\sigma(a) \subseteq \text{ins } \Lambda \subseteq G$ and if $f: G \to \mathbb{C}$ is analytic, then*

$$\int_{\Gamma} f(z)(z - a)^{-1} dz = \int_{\Lambda} f(z)(z - a)^{-1} dz.$$

PROOF. For $1 \leqslant j \leqslant k$, let $\gamma_{m+j} = \lambda_j^{-1}$; that is, $\gamma_{m+j}(t) = \lambda_j(1 - t)$ for $0 \leqslant t \leqslant 1$. If $z \notin G \backslash \sigma(a)$, then either $z \in \mathbb{C} \backslash G$ or $z \in \sigma(a)$. If $z \in \mathbb{C} \backslash G$, then $\sum_{j=1}^{m+k} n(\gamma_j; z) = n(\Gamma; z) - n(\Lambda; z) = 0 - 0 = 0$. If $z \in \sigma(a)$, then $\sum_{j=1}^{m+k} n(\gamma_j; z) = n(\Gamma; z) - n(\Lambda; z) = 1 - 1 = 0$. Thus $\Sigma \equiv \{\gamma_j : 1 \leqslant j \leqslant m + k\}$ is a system of closed curves in $U = G \backslash \sigma(a)$ such that $n(\Sigma; z) = 0$ for all z in $\mathbb{C} \backslash U$. Since $z \mapsto f(z)(z - a)^{-1}$ is analytic on U, Cauchy's Theorem implies

$$0 = \int_{\Sigma} f(z)(z - a)^{-1} dz = \int_{\Gamma} f(z)(z - a)^{-1} dz - \int_{\Lambda} f(z)(z - a)^{-1} dz. \qquad \blacksquare$$

As was pointed out before, Proposition 4.6 implies that (4.5) gives a well-defined element $f(a)$ of $\mathscr{A}$ whenever f is analytic in a neighborhood of $\sigma(a)$. Let $\text{Hol}(a) = $ all of the functions that are analytic in a neighborhood of $\sigma(a)$. Note that $\text{Hol}(a)$ is an algebra where if $f, g \in \text{Hol}(a)$ and f and g have domains $D(f)$ and $D(g)$, then fg and $f + g$ have domain $D(f) \cap D(g)$. $\text{Hol}(a)$ is not, however, a Banach algebra.

4.7. The Riesz Functional Calculus. *Let $\mathscr{A}$ be a Banach algebra with identity and let $a \in \mathscr{A}$.*

(a) *The map $f \mapsto f(a)$ of $\text{Hol}(a) \to \mathscr{A}$ is an algebra homomorphism.*
(b) *If $f(z) = \sum_{k=0}^{\infty} \alpha_k z^k$ has radius of convergence $> r(a)$, then $f \in \text{Hol}(a)$ and $f(a) = \sum_{k=0}^{\infty} \alpha_k a^k$.*
(c) *If $f(z) \equiv 1$, then $f(a) = 1$.*
(d) *If $f(z) = z$ for all z, $f(a) = a$.*
(e) *If $f, f_1, f_2, \ldots$ are all analytic on G, $\sigma(a) \subseteq G$, and $f_n(z) \to f(z)$ uniformly on compact subsets of G, then $\| f_n(a) - f(a) \| \to 0$ as $n \to \infty$.*

PROOF. (a) Let $f, g \in \text{Hol}(a)$ and let G be an open neighborhood of $\sigma(a)$ on which both f and g are analytic. Let Γ be a positively oriented system of closed curves in G such that $\sigma(a) \subseteq \text{ins } \Gamma$. Let Λ be a positively oriented system of closed curves in G such that $(\text{ins } \Gamma) \cup \{\Gamma\} = \text{cl}(\text{ins } \Gamma) \subseteq \text{ins } \Lambda$. Then

$$f(a)g(a) = -\frac{1}{4\pi^2}\left[\int_\Gamma f(z)(z-a)^{-1}dz\right]\left[\int_\Lambda g(\zeta)(\zeta-a)^{-1}d\zeta\right]$$

$$= -\frac{1}{4\pi^2}\int_\Gamma\int_\Lambda f(z)g(\zeta)(z-a)^{-1}(\zeta-a)^{-1}d\zeta\,dz$$

[by (3.9b)]
$$= -\frac{1}{4\pi^2}\int_\Gamma\int_\Lambda f(z)g(\zeta)\left[\frac{(z-a)^{-1}-(\zeta-a)^{-1}}{\zeta-z}\right]d\zeta\,dz$$

$$= -\frac{1}{4\pi^2}\int_\Gamma f(z)\left[\int_\Lambda\frac{g(\zeta)}{\zeta-z}d\zeta\right](z-a)^{-1}dz$$

$$+ \frac{1}{4\pi^2}\int_\Lambda g(\zeta)\left[\int_\Gamma\frac{f(z)}{\zeta-z}dz\right](\zeta-a)^{-1}d\zeta.$$

But for ζ on Λ, $\zeta \in \text{out } \Gamma$ and hence $\int_\Gamma[f(z)/(\zeta-z)]dz = 0$ (Cauchy's Theorem). If $z \in \{\Gamma\}$, then $z \in \text{ins } \Lambda$ and so $\int_\Lambda[g(\zeta)/(\zeta-z)]d\zeta = 2\pi i g(z)$. Hence

$$f(a)g(a) = \frac{1}{2\pi i}\int_\Gamma f(z)g(z)(z-a)^{-1}dz$$

$$= (fg)(a).$$

The proof that $(\alpha f + \beta g)(a) = \alpha f(a) + \beta g(a)$ is left to the reader.

(c) and (d). Let $f(z) = z^k$, $k \geq 0$. Let $\gamma(t) = R\exp(2\pi it)$, $0 \leq t \leq 1$, where $R > \|a\|$. So $\sigma(a) \subset \text{ins } \gamma$, and hence

$$f(a) = \frac{1}{2\pi i}\int_\gamma z^k(z-a)^{-1}dz$$

$$= \frac{1}{2\pi i}\int_\gamma z^{k-1}\left(1-\frac{a}{z}\right)^{-1}dz$$

$$= \frac{1}{2\pi i}\int_\gamma z^{k-1}\sum_{n=0}^\infty a^n/z^n\,dz,$$

since $\|a/z\| < 1$ for $|z| = R$. Since this infinite series converges uniformly for z on γ,

$$f(a) = \sum_{n=0}^\infty\left[\frac{1}{2\pi i}\int_\gamma\frac{1}{z^{n-k+1}}dz\right]a^n.$$

If $n \neq k$, then $z^{-(n-k+1)}$ has a primitive and hence $\int_\gamma z^{-(n-k+1)}dz = 0$. For $n = k$ this integral becomes $\int_\gamma z^{-1}dz = 2\pi i$. Hence $f(a) = a^k$.

(e) Let $\Gamma = \{\gamma_1, \ldots, \gamma_m\}$ be a positively oriented system of closed curves in

G such that $\sigma(a) \subseteq \text{ins } \Gamma$. Fix $1 \leqslant k \leqslant m$; then

$$\left\| \int_{\gamma_k} f_n(z)(z-a)^{-1}dz - \int_{\gamma_k} f(z)(z-a)^{-1}dz \right\|$$

$$= \left\| \int_0^1 [f_n(\gamma_k(t)) - f(\gamma_k(t))][\gamma_k(t) - a]^{-1}d\gamma_k(t) \right\|$$

$$\leqslant \int_0^1 |f_n(\gamma_k(t)) - f(\gamma_k(t))| \, \|[\gamma_k(t) - a]^{-1}\| \, d|\gamma_k|(t).$$

Now $t \mapsto \|[\gamma_k(t) - a]^{-1}\|$ is continuous on $[0, 1]$ and hence bounded by some constant, say M. Thus

$$\left\| \int_{\gamma_k} f_n(z)(z-a)^{-1}dz - \int_{\gamma_k} f(z)(a-a)^{-1}dz \right\|$$

$$\leqslant M \|\gamma_k\| \max \{|f_n(z) - f(z)|: z \in \{\gamma_k\}\},$$

where $\|\gamma_k\|$ is the total variation (length) of γ_k. By hypothesis it follows that $\|f_n(a) - f(a)\| \to 0$ as $n \to \infty$.

(b) If $p(z) = \sum_{k=0}^n \alpha_k z^k$ is a polynomial, then (a), (c), and (d) combine to give that $p(a) = \sum_{k=0}^n \alpha_k a^k$. Now let $f(z) = \sum_{k=0}^\infty \alpha_k z^k$ have radius of convergence $R > r(a)$, the spectral radius of a. If $p_n(z) = \sum_{k=0}^n \alpha_k z^k$, $p_n(z) \to f(z)$ uniformly on compact subsets of $\{z: |z| < R\}$. By (e), $p_n(a) \to f(a)$. So (b) follows. ∎

The Riesz Functional Calculus is used in the study of Banach algebras and is especially useful in the study of linear operators on a Banach space (Sections 6 and 7). Now our attention must focus on the basic properties of this functional calculus. The first such property is its uniqueness.

4.8. Proposition. *Let $\mathscr{A}$ be a Banach algebra with identity and let $a \in \mathscr{A}$. Let $\tau: \text{Hol}(a) \to \mathscr{A}$ be a homomorphism such that (a) $\tau(1) = 1$, (b) $\tau(z) = a$, (c) if $\{f_n\}$ is a sequence of analytic functions on an open set G such that $\sigma(a) \subseteq G$ and $f_n(z) \to f(z)$ uniformly on compact subsets of G, then $\tau(f_n) \to \tau(f)$. Then $\tau(f) = f(a)$ for every f in $\text{Hol}(a)$.*

PROOF. The proof uses Runge's Theorem (III.8.1), but first it must be shown that $\tau(f) = f(a)$ whenever f is a rational function. If $n \geqslant 1$, $\tau(z^n) = \tau(z)^n = a^n$; hence $\tau(p) = p(a)$ for any polynomial p. Let q be a polynomial such that q never vanishes on $\sigma(a)$, so $1/q \in \text{Hol}(a)$. Also, $1 = \tau(1) = \tau(q \cdot q^{-1}) = \tau(q)\tau(q^{-1}) = q(a)\tau(q^{-1})$. Hence $q(a)$ is invertible and $q(a)^{-1} = \tau(q^{-1})$. But using the Riesz Functional Calculus, a similar argument shows that $q(a)^{-1} = (1/q)(a)$. Thus $\tau(q^{-1}) = (1/q)(a)$. Therefore if $f = p/q$, where p and q are polynomials and q never vanishes on $\sigma(a)$, $\tau(f) = \tau(p \cdot q^{-1}) = \tau(p)\tau(q^{-1}) = p(a)(1/q)(a) = f(a)$.

Now let $f \in \text{Hol}(a)$ and suppose f is analytic on an open set G such that $\sigma(a) \subseteq G$. By Runge's theorem there are rational functions $\{f_n\}$ in $\text{Hol}(a)$ such that $f_n(z) \rightarrow f(z)$ uniformly on compact subsets of G. By (c) of the hypothesis, $\tau(f_n) \rightarrow \tau(f)$. But $\tau(f_n) = f_n(a)$ and $f_n(a) \rightarrow f(a)$ by (4.7e). Hence $\tau(f) = f(a)$. ∎

A fact that has been implicit in the manipulations involving the functional calculus is that $f(a)$ and $g(a)$ commute for all f and g in Hol(a). In fact, if τ: Hol(a) → $\mathscr{A}$ is defined by $\tau(f) = f(a)$, then $f(a)g(a) = \tau(fg) = \tau(gf) = g(a)f(a)$. Still more can be said.

4.9. Proposition. *If* $a, b \in \mathscr{A}$, $ab = ba$, *and* $f \in$ Hol(a), *then* $f(a)b = bf(a)$.

PROOF. An algebraic exercise demonstrates that $f(a)b = bf(a)$ if f is a rational function with poles off $\sigma(a)$. The general result now follows by Runge's Theorem. ∎

4.10. The Spectral Mapping Theorem. *If* $a \in \mathscr{A}$ *and* $f \in$ Hol(a), *then*

$$\sigma(f(a)) = f(\sigma(a)).$$

PROOF. If $\alpha \in \sigma(a)$, let $g \in$ Hol(a) such that $f(z) - f(\alpha) = (z - \alpha)g(z)$. If it were the case that $f(\alpha) \notin \sigma(f(a))$, then $(a - \alpha)$ would be invertible with inverse $g(a)[f(a) - f(\alpha)]^{-1}$. Hence $f(\alpha) \in \sigma(f(a))$; that is, $f(\sigma(a)) \subseteq \sigma(f(a))$.
   Conversely, if $\beta \notin f(\sigma(a))$, then $g(z) = [f(z) - \beta]^{-1} \in$ Hol(a) and so $g(a)[f(a) - \beta] = 1$. Thus $\beta \notin \sigma(f(a))$; that is, $\sigma(f(a)) \subseteq f(\sigma(a))$. ∎

This section closes with an application of the functional calculus that is typical.

4.11. Proposition. *Suppose* $a \in \mathscr{A}$ *and* $\sigma(a) = F_1 \cup F_2$, *where* F_1 *and* F_2 *are disjoint nonempty closed sets. Then there is a nontrivial idempotent* e *in* $\mathscr{A}$ *such that*

(a) *if* $ba = ab$, *then* $be = eb$;
(b) *if* $a_1 = ae$ *and* $a_2 = a(1 - e)$, *then* $a = a_1 + a_2$ *and* $a_1 a_2 = a_2 a_1 = 0$;
(c) $\sigma(a_1) = F_1 \cup \{0\}$, $\sigma(a_2) = F_2 \cup \{0\}$.

PROOF. Let G_1, G_2 be disjoint open subsets of $\mathbb{C}$ such that $F_j \subset G_j, j = 1, 2$. Let Γ be a positively oriented system of closed curves in G_1 such that $F_1 \subseteq$ ins Γ, $F_2 \subseteq$ out Γ. If $f = $ the characteristic function of G_1, $f \in$ Hol(a); let $e = f(a)$. Since $f^2 = f$, $e^2 = e$. Part (a) follows from (4.9).
   Note that $e(1 - e) = 0 = (1 - e)e$. Hence (b) is immediate. Let $f_1(z) = zf(z)$, $f_2(z) = z(1 - f(z))$. It follows from (4.7a) that $a_j = f_j(a)$, $j = 1, 2$. Hence the Spectral Mapping Theorem implies that $\sigma(a_j) = f_j(\sigma(a)) = F_j \cup \{0\}$. The proof that e is neither 0 nor 1 is left to the reader. ∎

Part (c) of the preceding proposition has the somewhat unattractive conclusion that $\sigma(a_1) = F_1 \cup \{0\}$. It would be much neater if the conclusion were that $\sigma(a_1) = F_1$. This is, in a sense, the case. Since $a_1(1 - e) = 0$ and

$1 - e \neq 0$, a_1 cannot be invertible. However, consider the algebra $\mathscr{A}_1 \equiv \{b \in \mathscr{A}:$ $ab = ba$ and $be = eb = b\}$. It is left to the reader to show that $\mathscr{A}_1$ is a Banach algebra and e is the identity for $\mathscr{A}_1$. If a_1 is considered as an element of the algebra $\mathscr{A}_1$, then its spectrum as an element of $\mathscr{A}_1$ is F_1. This is an illustration of how the spectrum depends on the Banach algebra (the subject of the next section; also see Exercise 9).

EXERCISES

1. Let $\mathscr{A} = C(X)$, X compact (see Example 3.2). If $g \in C(X)$ and $f \in \mathrm{Hol}(g)$, show that $f(g) = f \circ g$.

2. Let a be a nilpotent element of $\mathscr{A}$. For f, g in $\mathrm{Hol}(a)$, give a necessary and sufficient condition on f and g that $f(a) = g(a)$.

3. Let $d \geqslant 1$ and let $A \in M_d(\mathbb{C})$. Give a necessary and sufficient condition on f in $\mathrm{Hol}(A)$ such that $f(A) = 0$. (Hint: Consider the Jordan canonical form for A.)

4. If $\mathscr{A}$ is a Banach algebra with identity, $a \in \mathscr{A}$, $f \in \mathrm{Hol}(a)$, and g is analytic in a neighborhood of $f(\sigma(a))$, then $g \circ f \in \mathrm{Hol}(a)$ and $g(f(a)) = g \circ f(a)$.

5. If $\mathscr{X}$ is a Banach space, $A \in \mathscr{B}(\mathscr{X})$, and $\mathscr{M} \leqslant \mathscr{X}$ such that $(A - \alpha)^{-1} \mathscr{M} \subseteq \mathscr{M}$ for all α in $\rho(A)$, show that $f(A)\mathscr{M} \subseteq \mathscr{M}$ whenever $f \in \mathrm{Hol}(A)$.

6. If $\mathscr{X}$ is a Banach space, $A \in \mathscr{B}(\mathscr{X})$, and $f \in \mathrm{Hol}(A)$, show that $f(A)^* = f(A^*)$. (See (6.1) below.)

7. If $\mathscr{H}$ is a Hilbert space, $A \in \mathscr{B}(\mathscr{H})$, and $f \in \mathrm{Hol}(A)$, show that $f(A)^* = \tilde{f}(A^*)$, where $\tilde{f}(z) = \overline{f(\bar{z})}$ (See (6.1) below.)

8. If $\mathscr{H}$ is a Hilbert space, A is a normal operator on $\mathscr{H}$, and $f \in \mathrm{Hol}(A)$, show that $f(A)$ is normal.

9. Let $\mathscr{X}$ be a Banach space and let $A \in \mathscr{B}(\mathscr{X})$. Show that if $\sigma(A) = F_1 \cup F_2$ where F_1, F_2 are disjoint closed subsets of $\mathbb{C}$, then there are topologically complementary subspaces $\mathscr{X}_1, \mathscr{X}_2$ of $\mathscr{X}$ such that (a) $B\mathscr{X}_j \subseteq \mathscr{X}_j$ ($j = 1, 2$) whenever $BA = AB$; (b) if $A_j = A|\mathscr{X}_j$, $\sigma(A_j) = F_j$; (c) there is an invertible operator $R: \mathscr{X} \to \mathscr{X}_1 \oplus_1 \mathscr{X}_2$ such that $RAR^{-1} = A_1 \oplus A_2$.

10. Let $A \in M_d(\mathbb{C})$, $\sigma(A) = \{\alpha_1, \ldots, \alpha_n\}$, where $\alpha_i \neq \alpha_j$ for $i \neq j$. Show that for $1 \leqslant j \leqslant n$ there is a matrix A_j in $M_{d_j}(\mathbb{C})$ such that $\sigma(A_j) = \{\alpha_j\}$ and A is similar to $A_1 \oplus \cdots \oplus A_n$.

11. If $\mathscr{A}$ is a Banach algebra, I is an ideal of $\mathscr{A}$ (not necessarily closed), $a \in I$, and $f \in \mathrm{Hol}(a)$ such that $f(0) = 0$, show that $f(a) \in I$.

§5. Dependence of the Spectrum on the Algebra

If $\partial \mathbb{D} = \{z \in \mathbb{C}: |z| = 1\}$, let $\mathscr{B} =$ the uniform closure of the polynomials in $C(\partial \mathbb{D})$. (Here "polynomial" means a polynomial in z.) If $\mathscr{A} = C(\partial \mathbb{D})$, then the spectrum of z as an element of $\mathscr{A}$ is $\partial \mathbb{D}$ (Example 3.2). That is,

$$\sigma_{\mathscr{A}}(z) = \partial \mathbb{D}.$$

Now $z \in \mathscr{B}$ and so it has a spectrum as an element of this algebra; denote this spectrum by $\sigma_{\mathscr{B}}(z)$. There is no reason to believe that $\sigma_{\mathscr{B}}(z) = \sigma_{\mathscr{A}}(z)$. In fact, they are not equal.

5.1. Example. If $\mathscr{B} =$ the closure in $C(\partial \mathbb{D})$ of the polynomials in z, then $\sigma_{\mathscr{B}}(z) = \operatorname{cl} \mathbb{D}$.

To see this first note that $\|z\| = 1$, so that $\sigma_{\mathscr{B}}(z) \subseteq \operatorname{cl} \mathbb{D}$ by Theorem 3.6. If $|\lambda| \leqslant 1$ and $\lambda \notin \sigma_{\mathscr{B}}(z)$, there is an f in $\mathscr{B}$ such that $(z - \lambda)f = 1$. Note that this implies that $|\lambda| < 1$. Because $f \in \mathscr{B}$, there is a sequence of polynomials $\{p_n\}$ such that $p_n \to f$ uniformly on $\partial \mathbb{D}$. Thus for every $\varepsilon > 0$ there is a N such that for $m, n \geqslant N$, $\varepsilon > \|p_n - p_m\|_{\partial \mathbb{D}} \equiv \sup\{|p_n(z) - p_n(z)|: z \in \partial \mathbb{D}\}$. By the Maximum Principle, $\varepsilon > \|p_n - p_m\|_{\operatorname{cl} \mathbb{D}}$ for $m, n \geqslant N$. Thus $g(z) = \lim p_n(z)$ is analytic on $\mathbb{D}$ and continuous on $\operatorname{cl} \mathbb{D}$; also, $g|\partial \mathbb{D} = f$. By the same argument, since $p_n(z)(z - \lambda) \to 1$ uniformly on $\partial \mathbb{D}$, $p_n(z)(z - \lambda) \to 1$ uniformly on $\mathbb{D}$. Thus $g(z)(z - \lambda) = 1$ on $\mathbb{D}$. But $1 = g(\lambda)(\lambda - \lambda) = 0$, a contradiction. Thus, $\operatorname{cl} \mathbb{D} \subseteq \sigma_{\mathscr{B}}(z)$.

Thus the spectrum not only depends on the element of the algebra, but also on the algebra. Precisely how this dependence occurs is given below, but it can be said that the example above is typical, both in its statement and its proof, of the general situation. To phrase these results it is necessary to introduce the polynomially convex hull of a compact subset of $\mathbb{C}$.

5.2. Definition. If A is a set and $f: A \to \mathbb{C}$, define

$$\|f\|_A \equiv \sup\{|f(z)|: z \in A\}.$$

If K is a compact subset of $\mathbb{C}$, define the *polynomially convex hull* of K to be the set $K^{\wedge}$ given by

$$K^{\wedge} \equiv \{z \in \mathbb{C}: |p(z)| \leqslant \|p\|_K \text{ for every polynomial } p\}.$$

The set K is *polynomially convex* if $K = K^{\wedge}$.

Note that the polynomially convex hull of $\partial \mathbb{D}$ is $\operatorname{cl} \mathbb{D}$. This is, again, quite typical. If K is any compact set, then $\mathbb{C} \backslash K$ has a countable number of components, only one of which is unbounded. The bounded components are sometimes called the *holes* of K; a few pictures should convince the reader of the appropriateness of this terminology.

5.3. Proposition. *If K is a compact subset of $\mathbb{C}$, then $\mathbb{C} \backslash K^{\wedge}$ is the unbounded component of $\mathbb{C} \backslash K$. Hence K is polynomially convex if and only if $\mathbb{C} \backslash K$ is connected.*

PROOF. Let $U_0, U_1, \ldots$ be the components of $\mathbb{C} \backslash K$, where U_0 is unbounded. Put $L = \mathbb{C} \backslash U_0$; hence $L = K \cup \bigcup_{n=1}^{\infty} U_n$. Clearly $K \subseteq K^{\wedge}$. If $n \geqslant 1$, then U_n is a bounded open set and a topological argument implies $\partial U_n \subset K$. By the Maximum Principle $U_n \subseteq K^{\wedge}$. Thus, $L \subseteq K^{\wedge}$.

If $\alpha \in U_0$, $(z - \alpha)^{-1}$ is analytic in a neighborhood of L. By (III.8.5), there is a sequence of polynomials $\{p_n\}$ such that $\|p_n - (z - \alpha)^{-1}\|_L \to 0$. If $q_n = (z - \alpha)p_n$, then $\|q_n - 1\|_L \to 0$. Thus for large n, $\|q_n - 1\|_L < 1/2$. Since $K \subset L$ and $|q_n(\alpha) - 1| = 1$, this implies that $\alpha \notin K^{\wedge}$. Thus $K^{\wedge} \subseteq L$. ∎

5.4. Theorem. *If $\mathscr{A}$ and $\mathscr{B}$ are Banach algebras with a common identity such that $\mathscr{B} \subseteq \mathscr{A}$ and $a \in \mathscr{B}$, then*

(a) $\sigma_{\mathscr{A}}(a) \subseteq \sigma_{\mathscr{B}}(a)$ and $\partial\sigma_{\mathscr{B}}(a) \subseteq \partial\sigma_{\mathscr{A}}(a)$.
(b) $\sigma_{\mathscr{A}}(a)^{\wedge} = \sigma_{\mathscr{B}}(a)^{\wedge}$.
(c) *If G is a hole of $\sigma_{\mathscr{A}}(a)$, then either $G \subseteq \sigma_{\mathscr{B}}(a)$ or $G \cap \sigma_{\mathscr{B}}(a) = \square$.*
(d) *If $\mathscr{B}$ is the closure in $\mathscr{A}$ of all polynomials in a, then $\sigma_{\mathscr{B}}(a) = \sigma_{\mathscr{A}}(a)^{\wedge}$.*

PROOF. (a) If $\alpha \notin \sigma_{\mathscr{B}}(a)$, then there is a b in $\mathscr{B}$ such that $b(a - \alpha) = (a - \alpha)b = 1$. Since $\mathscr{B} \subseteq \mathscr{A}$, $\alpha \notin \sigma_{\mathscr{A}}(a)$. Now assume that $\lambda \in \partial\sigma_{\mathscr{B}}(a)$. Since $\text{int } \sigma_{\mathscr{A}}(a) \subseteq \text{int } \sigma_{\mathscr{B}}(a)$, it suffices to show that $\lambda \in \sigma_{\mathscr{A}}(a)$. Suppose $\lambda \notin \sigma_{\mathscr{A}}(a)$; there is thus an x in $\mathscr{A}$ such that $x(a - \lambda) = (a - \lambda)x = 1$. Since $\lambda \in \partial\sigma_{\mathscr{B}}(a)$, there is a sequence $\{\lambda_n\}$ in $\mathbb{C}\backslash\sigma_{\mathscr{B}}(a)$ such that $\lambda_n \to \lambda$. Let $(a - \lambda_n)^{-1}$ be the inverse of $(a - \lambda_n)$ in $\mathscr{B}$; so $(a - \lambda_n)^{-1} \in \mathscr{A}$. Since $\lambda_n \to \lambda$, $(a - \lambda_n) \to (a - \lambda)$. By Theorem 2.2, $(a - \lambda_n)^{-1} \to x$. Thus $x \in \mathscr{B}$ since $\mathscr{B}$ is complete. This contradicts the fact that $\lambda \in \sigma_{\mathscr{B}}(a)$.

(b) This is a consequence of (a) and the Maximum Principle.

(c) Let G be a hole of $\sigma_{\mathscr{A}}(a)$ and put $G_1 = G \cap \sigma_{\mathscr{B}}(a)$ and $G_2 = G\backslash\sigma_{\mathscr{B}}(a)$. So $G = G_1 \cup G_2$ and $G_1 \cap G_2 = \square$. Clearly G_2 is open. On the other hand, the fact that $\partial\sigma_{\mathscr{B}}(a) \subseteq \sigma_{\mathscr{A}}(a)$ and $G \cap \sigma_{\mathscr{A}}(a) = \square$ implies that $G_1 = G \cap \text{int } \sigma_{\mathscr{B}}(a)$, so G_1 is open. Because G is connected, either G_1 or G_2 is empty.

(d) Let $\mathscr{B}$ be as in (d). From (a) and (b) it is known that $\sigma_{\mathscr{A}}(a) \subseteq \sigma_{\mathscr{B}}(a) \subseteq \sigma_{\mathscr{A}}(a)^{\wedge}$. Fix λ in $\sigma_{\mathscr{A}}(a)^{\wedge}$. If $\lambda \notin \sigma_{\mathscr{B}}(a)$, $(a - \lambda)^{-1} \in \mathscr{B} \subseteq \mathscr{A}$. Hence there is a sequence of polynomials $\{p_n\}$ such that $p_n(a) \to (a - \lambda)^{-1}$. Let $q_n(z) = (z - \lambda)p_n(z)$. Thus $\|q_n(a) - 1\| \to 0$. By the Spectral Mapping Theorem, $\sigma_{\mathscr{A}}(q_n(a)) = q_n(\sigma_{\mathscr{A}}(a))$. Thus, because $\lambda \in \sigma_{\mathscr{A}}(a)^{\wedge}$,

$$\|q_n(a) - 1\| \geq r(q_n(a) - 1)$$

$$= \sup\{|z - 1|: z \in \sigma_{\mathscr{A}}(q_n(a))\}$$

$$= \sup\{|q_n(w) - 1|: w \in \sigma_{\mathscr{A}}(a)\}$$

$$\geq |q_n(\lambda) - 1|$$

$$= 1.$$

This is a contradiction. ∎

EXERCISES

1. If K is a compact subset of $\mathbb{C}$, let $P(K)$ be the closure of the polynomials in $C(K)$. Show that the identity map on polynomials extends to an isometric isomorphism of $P(K)$ onto $P(K^{\wedge})$.

2. If K is a compact subset of $\mathbb{C}$, let $R(K)$ be the closure in $C(K)$ of all rational functions with poles off K. If $f \in R(K)$, show that $\sigma_{R(K)}(f) = f(K)$. If $f \in P(K)$, show that $\sigma_{P(K)}(f) = \hat{f}(K^{\wedge})$, where $\hat{f}$ is a natural extension of f to $K^{\wedge}$.

3. Let $\mathscr{A}, \mathscr{B}$ be as in Theorem 5.4. If $a \in \mathscr{B}$ and $\sigma_{\mathscr{A}}(a) \subseteq \mathbb{R}$, show that $\sigma_{\mathscr{A}}(a) = \sigma_{\mathscr{A}}(a)$.

4. Let $\mathscr{A}$ be a Banach algebra with identity and let $a \in \mathscr{A}$. If $G_1, G_2, \ldots$ are the holes of $\sigma_{\mathscr{A}}(a)$ and $1 \leq n_1 \leq n_2, \ldots$, show that there is a subalgebra $\mathscr{B}$ of $\mathscr{A}$ such that $a \in \mathscr{B}$ and $\sigma_{\mathscr{B}}(a) = \sigma_{\mathscr{A}}(a) \cup \bigcup_{k=1}^{\infty} G_{n_k}$.

5. If $\mathscr{A}, \mathscr{B}$, and a are as in Theorem 5.4, $\mathscr{A}$ is not abelian, and $\mathscr{B}$ is a maximal abelian subalgebra of $\mathscr{A}$, show that $\sigma_{\mathscr{B}}(a) = \sigma_{\mathscr{A}}(a)$.

6. If K is a nonempty compact subset of $\mathbb{C}$ that is polynomially convex, show that the components of int K are simply connected.

§6. The Spectrum of a Linear Operator

The proof of the first result is left as an exercise.

6.1. Proposition.

(a) *If $\mathscr{X}$ is a Banach space and $A \in \mathscr{B}(\mathscr{X})$, $\sigma(A^*) = \sigma(A)$.*
(b) *If $\mathscr{H}$ is a Hilbert space and $A \in \mathscr{B}(\mathscr{H})$, $\sigma(A^*) = \sigma(A)^*$, where for any subset Δ of $\mathbb{C}$, $\Delta^* \equiv \{\bar{z} : z \in \Delta\}$.*

In this section only results about operators on Banach spaces will be given. For the corresponding results about operators on a Hilbert space involving the adjoint, the reader is asked to supply the details. The preceding proposition should be kept in mind as a model of the probable differences.

In this section and the next $\mathscr{X}$ always denotes a Banach space over $\mathbb{C}$.

6.2. Definition. If $A \in \mathscr{B}(\mathscr{X})$, the *point spectrum* of $A, \sigma_p(A)$, is defined by

$$\sigma_p(A) \equiv \{\lambda \in \mathbb{C} : \ker(A - \lambda) \neq (0)\}.$$

As in the case of operators on a Hilbert space, elements of $\sigma_p(A)$ are called *eigenvalues*. If $\lambda \in \sigma_p(A)$, non-zero vectors in $\ker(A - \lambda)$ are called *eigenvectors*; $\ker(A - \lambda)$ is called the *eigenspace* of A at λ.

6.3. Definition. If $A \in \mathscr{B}(\mathscr{X})$, the *approximate point spectrum* of $A, \sigma_{ap}(A)$, is defined by

$$\sigma_{ap}(A) \equiv \{\lambda \in \mathbb{C} : \text{there is a sequence } \{x_n\} \text{ in } \mathscr{X}$$
$$\text{such that } \|x_n\| = 1 \text{ for all } n \text{ and } \|(A - \lambda)x_n\| \to 0\}.$$

Note that $\sigma_p(A) \subseteq \sigma_{ap}(A)$.

6.4. Proposition. *If $A \in \mathscr{B}(\mathscr{X})$ and $\lambda \in \mathbb{C}$, the following statements are equivalent.*

(a) $\lambda \notin \sigma_{ap}(A)$.
(b) $\ker(A - \lambda) = (0)$ *and* $\operatorname{ran}(A - \lambda)$ *is closed.*

(c) *There is a constant $c > 0$ such that $\|(A - \lambda)x\| \geqslant c\|x\|$ for all x.*

PROOF. Clearly it may be assumed that $\lambda = 0$.

(a)$\Rightarrow$(c): Suppose (c) fails to hold; then for every n there is a non-zero vector x_n with $\|Ax_n\| \leqslant \|x_n\|/n$. If $y_n = x_n/\|x_n\|$, $\|y_n\| = 1$ and $\|Ay_n\| \to 0$. Hence $0 \in \sigma_{ap}(A)$.

(c)$\Rightarrow$(b): Suppose $\|Ax\| \geqslant c\|x\|$. Clearly $\ker A = (0)$. If $Ax_n \to y$, $\|x_n - x_m\| \leqslant c^{-1}\|Ax_n - Ax_m\|$, so $\{x_n\}$ is a Cauchy sequence. Let $x = \lim x_n$; therefore $Ax = y$ and ran A is closed.

(b)$\Rightarrow$(a): Let $\mathscr{Y} = $ ran A; so $A: \mathscr{X} \to \mathscr{Y}$ is a continuous bijection. By the Inverse Mapping Theorem, there is a bounded operator $B: \mathscr{Y} \to \mathscr{X}$ such that $BAx = x$ for all x in $\mathscr{X}$. Thus if $\|x\| = 1$, $1 = \|BAx\| \leqslant \|B\| \|Ax\|$. That is, $\|Ax\| \geqslant \|B\|^{-1}$ whenever $\|x\| = 1$. Hence $0 \notin \sigma_{ap}(A)$. ■

It may be that $\sigma_p(A)$ is empty, but it will be shown that $\sigma_{ap}(A)$ is never empty. The first statement follows from the next result (or from other examples that have been presented); the second statement will be proved later.

6.5. Proposition. *If $1 \leqslant p \leqslant \infty$, define $S: l^p \to l^p$ by $S(x_1, x_2, \ldots) = (0, x_1, x_2, \ldots)$. Then $\sigma(S) = $ cl $\mathbb{D}$, $\sigma_p(S) = \square$, and $\sigma_{ap}(S) = \partial\mathbb{D}$. Moreover, for $|\lambda| < 1$, ran$(S - \lambda)$ is closed and $\dim[l^p/\text{ran}(S - \lambda)] = 1$.*

PROOF. Let S_p be the shift on l^p. For $1 \leqslant p \leqslant \infty$, define $T_p: l^p \to l^p$ by $T_p(x_1, x_2, \ldots) = (x_2, x_3, \ldots)$. It is easy to check that for $1 \leqslant p < \infty$ and $1/p + 1/q = 1$, $S_p^* = T_q$. Since $\|S_p\| = 1$, $\sigma(S_p) \subseteq$ cl $\mathbb{D}$.

Suppose $x = (x_1, x_2, \ldots) \in l^p$, $\lambda \neq 0$. If $S_p x = \lambda x$, $0 = \lambda x_1$, $x_1 = \lambda x_2, \ldots$. Hence $0 = x_1 = x_2 = \cdots$. Since S_p is an isometry, $\ker S_p = (0)$. Thus $\sigma_p(S_p) = \square$.

Let $1 \leqslant p \leqslant \infty$ and $|\lambda| < 1$. Put $x_\lambda = (1, \lambda, \lambda^2, \ldots)$. Then $\|x_\lambda\|_p^p = \sum_{n=0}^\infty |\lambda^p|^n < \infty$. Also, $T_p x_\lambda = (\lambda, \lambda^2, \ldots) = \lambda x_\lambda$. Hence $\lambda \in \sigma_p(T_p)$ and $x_\lambda \in \ker(T_p - \lambda)$. If $1 \leqslant p < \infty$ and $1/p + 1/q = 1$, $T_q = S_p^*$; so $\mathbb{D} \subseteq \sigma(T_q) = \sigma(S_p)$. Also, $S_\infty = T_1^*$, so $\mathbb{D} \subseteq \sigma(S_\infty)$. Thus for all p, $\mathbb{D} \subseteq \sigma(S_p) \subseteq$ cl $\mathbb{D}$. Since $\sigma(S_p)$ is necessarily closed, $\sigma(S_p) = $ cl $\mathbb{D}$.

If $|\lambda| \neq 1$ and $x \in l^p$, $\|(S_p - \lambda)x\|_p = \|S_p x - \lambda x\|_p \geqslant |\|S_p x\|_p - |\lambda| \|x\|_p| = |\|x\|_p - |\lambda| \|x\|_p| = |1 - |\lambda|| \|x\|_p$. By (6.4), $\lambda \notin \sigma_{ap}(S_p)$. Hence $\sigma_{ap}(S) \subseteq \partial\mathbb{D}$. The fact that $\sigma_{ap}(S_p) = \partial\mathbb{D}$ follows from the next proposition (6.7).

Fix $|\lambda| < 1$; we will show that $\dim \ker(T_p - \lambda) = 1$ for $1 \leqslant p \leqslant \infty$. Indeed, if $x \in l^p$ and $T_p x = \lambda x$, then $(x_2, x_3, \ldots) = (\lambda x_1, \lambda x_2, \ldots)$. So $x_{n+1} = \lambda x_n$ for all n. Thus $x_{n+1} = \lambda^n x_1$ for $n \geqslant 1$. That is, if $x_\lambda = (1, \lambda, \lambda^2, \ldots)$, then $x = x_1 x_\lambda$. Since it has already been shown that $x_\lambda \in \ker(T_p - \lambda)$, we have that the dimension of this kernel is 1. Therefore, if $1 \leqslant p < \infty$, $1 = \dim \ker(T_q - \lambda) = \dim \ker(S_p^* - \lambda) = \dim[\text{ran}(S_p - \lambda)^\perp]$ (VI.1.8) $= \dim[l^p/\text{ran}(S_p - \lambda)]^*$ (Why?). But this implies that $\dim[l^p/\text{ran}(S_p - \lambda)] = 1$, completing the proof for the case where p is finite. The proof for $p = \infty$ is similar and is left to the reader. ■

6.6. Corollary. *If $1 \leqslant p \leqslant \infty$ and $T: l^p \to l^p$ is defined by $T(x_1, x_2, \ldots) = (x_2, x_3, \ldots)$,* then $\sigma(T) = \mathrm{cl}\,\mathbb{D}$ and for $|\lambda| < 1$, $\ker(T - \lambda)$ is the one-dimensional space spanned by the vector $(1, \lambda, \lambda^2, \ldots)$.

The next result shows that if S is as in (6.5), then $\partial \mathbb{D} \subseteq \sigma_{ap}(S)$.

6.7. Proposition. If $A \in \mathcal{B}(\mathcal{X})$, then $\partial \sigma(A) \subseteq \sigma_{ap}(A)$.

PROOF. Let $\lambda \in \partial\sigma(A)$ and let $\{\lambda_n\} \subseteq \mathbb{C} \backslash \sigma(A)$ such that $\lambda_n \to \lambda$.

6.8. Claim. $\|(A - \lambda_n)^{-1}\| \to \infty$ as $n \to \infty$.

In fact, if the claim were false, then by passing to a subsequence if necessary, it follows that there is a constant M such that $\|(A - \lambda_n)^{-1}\| \leqslant M$ for all n. Choose n sufficiently large that $|\lambda_n - \lambda| < M^{-1}$. Then $\|(A - \lambda) - (A - \lambda_n)\| < \|(A - \lambda_n)^{-1}\|^{-1}$. By (2.3b), this implies that $(A - \lambda)$ is invertible, a contradiction. This establishes (6.8).

Let $\|x_n\| = 1$ such that $\alpha_n \equiv \|(A - \lambda_n)^{-1} x_n\| > \|(A - \lambda_n)^{-1}\| - n^{-1}$, so $\alpha_n \to \infty$. Put $y_n = \alpha_n^{-1}(A - \lambda_n)^{-1} x_n$; hence $\|y_n\| = 1$. Now

$$(A - \lambda)y_n = (A - \lambda_n)y_n + (\lambda - \lambda_n)y_n$$
$$= \alpha_n^{-1} x_n + (\lambda - \lambda_n)y_n.$$

Thus $\|(A - \lambda)y_n\| \leqslant \alpha_n^{-1} + |\lambda - \lambda_n|$, so that $\|(A - \lambda)y_n\| \to 0$ as $n \to \infty$. That is, $\lambda \in \sigma_{ap}(A)$. ∎

Let $A \in \mathcal{B}(\mathcal{X})$ and suppose Δ is a *clopen* subset of $\sigma(A)$; that is, Δ is a subset of $\sigma(A)$ that is both closed and relatively open. So $\sigma(A) = \Delta \cup (\sigma(A) \backslash \Delta)$. As in Proposition 4.11 (and Exercise 4.9),

6.9 $$E(\Delta) = E(\Delta; A) = \frac{1}{2\pi i} \int_\Gamma (z - A)^{-1}\, dz,$$

where Γ is a positively oriented Jordan system such that $\Delta \subseteq \mathrm{ins}\,\Gamma$ and $\sigma(A) \backslash \Delta \subseteq \mathrm{out}\,\Gamma$, is an idempotent. Moreover, $E(\Delta)B = BE(\Delta)$ whenever $AB = BA$ and if $\mathcal{X}_\Delta = E(\Delta)\mathcal{X}$, $\sigma(A|\mathcal{X}_\Delta) = \Delta$. Call $E(\Delta)$ the *Riesz idempotent* corresponding to Δ. If $\Delta = $ a singleton set $\{\lambda\}$, let $E(\lambda) = E(\{\lambda\})$ and $\mathcal{X}_\lambda = \mathcal{X}_{\{\lambda\}}$. Note that if λ is an isolated point of $\sigma(A)$, then $\{\lambda\}$ is a clopen subset of $\sigma(A)$.

6.10. Example. Let $\{\alpha_n\} \in l^\infty$, $1 \leqslant p \leqslant \infty$, and define A: $l^p \to l^p$ by $(Ax)(n) = \alpha_n x(n)$. Then $\sigma(A) = \mathrm{cl}\{\alpha_n\}$ and $\sigma_p(A) = \{\alpha_n\}$. For each k, define $N_k = \{n \in \mathbb{N}; \alpha_n = \alpha_k\}$ and define P_k: $l^p \to l^p$ by $P_k x = \chi_{N_k} x$. If α_k is an isolated point of $\sigma(A)$, then $\{\alpha_k\}$ is a clopen subset of $\sigma(A)$ and $E(\{\alpha_k\}; A) = P_k$.

Suppose $A \in \mathcal{B}(\mathcal{X})$ and λ_0 is an isolated point in $\sigma(A)$. Hence $E(\lambda_0) = E(\lambda_0; A)$ is a well-defined idempotent. Also, λ_0 is an isolated singularity of the analytic function $z \mapsto (z - A)^{-1}$ on $\mathbb{C} \backslash \sigma(A)$. Perhaps the nature of this singularity (pole or essential singularity) will reveal something of the nature of λ_0 as an element of $\sigma(A)$. First it is helpful to get the precise form of the Laurent expansion of $(z - A)^{-1}$ about λ_0.

6.11. Lemma. *If λ_0 is an isolated point of $\sigma(A)$, then*

$$(z - A)^{-1} = \sum_{n = -\infty}^{\infty} (z - \lambda_0)^n A_n$$

for $0 < |z - \lambda_0| < r_0 = \text{dist}(\lambda_0, \sigma(A) \backslash \{\lambda\})$, where

$$A_n = \frac{1}{2\pi i} \int_\gamma (z - \lambda_0)^{-n-1}(z - A)^{-1} dz$$

for $\gamma = $ any circle centered at λ_0 with radius $< r_0$.

The proof follows the lines of the usual Laurent series development (Conway [1978]).

6.12. Proposition. *If λ_0 is an isolated point of $\sigma(A)$, then λ_0 is a pole of $(z - A)^{-1}$ of order n if and only if $(\lambda_0 - A)^n E(\lambda_0) = 0$ and $(\lambda_0 - A)^{n-1} E(\lambda_0) \neq 0$.*

PROOF. Let $(z - A)^{-1} = \sum_{n=-\infty}^{\infty} (z - \lambda_0)^n A_n$ as is (6.11). Now λ_0 is a pole of order n if and only if $A_{-n} \neq 0$ and $A_{-k} = 0$ for $k > n$. Let Γ be a positively oriented system of curves such that $\sigma(A) \backslash \{\lambda_0\} \subseteq \text{ins} \, \Gamma$ and $\lambda_0 \in \text{out} \, \Gamma$. Let γ be a circle centered at λ_0 and contained in out Γ. Let $e(z) \equiv 1$ in a neighborhood of $\gamma \cup \text{ins} \, \gamma$ and $e(z) \equiv 0$ in a neighborhood of $\Gamma \cup \text{ins} \, \Gamma$. So $e \in \text{Hol}(A)$ and $e(A) = E(\lambda_0)$. If $k \geq 1$,

$$A_{-k} = \frac{1}{2\pi i} \int_\gamma (z - \lambda_0)^{k-1}(z - A)^{-1} dz$$

$$= \frac{1}{2\pi i} \int_{\gamma + \Gamma} e(z)(z - \lambda_0)^{k-1}(z - A)^{-1} dz$$

$$= E(\lambda_0)(A - \lambda_0)^{k-1}$$

since $\sigma(A) \subseteq \text{ins}(\gamma + \Gamma) = \text{ins} \, \gamma \cup \text{ins} \, \Gamma$. The proposition now follows. ∎

6.13. Corollary. *If λ_0 is an isolated point of $\sigma(A)$ and is a pole of $(z - A)^{-1}$, then $\lambda_0 \in \sigma_p(A)$.*

In fact, the preceding result implies that if n is the order of the pole, then $(0) \neq (\lambda_0 - A)^{n-1} E(\lambda_0) \mathcal{X} \subseteq \ker(A - \lambda_0)$.

6.14. Example. A measurable function $k: [0, 1] \times [0, 1] \to \mathbb{C}$ is called a *Volterra kernel* if k is bounded and $k(x, y) = 0$ when $x < y$. If $1 \leq p \leq \infty$ and

k is a Volterra kernel, define $V_k\colon L^p(0,1) \to L^p(0,1)$ by

$$V_k f(x) = \int_0^1 k(x,y) f(y)\, dy = \int_0^x k(x,y) f(y)\, dy.$$

Then $V_k \in \mathscr{B}(L^p)$ and $\|V_k\| \leqslant \|k\|_\infty$ (III.2.3).

If k, h are Volterra kernels and

$$(hk)(x,y) = \int_0^1 h(x,t) k(t,y)\, dt,$$

then hk is a Volterra kernel, $\|hk\|_\infty \leqslant \|h\|_\infty \|k\|_\infty$, and $V_{hk} = V_h V_k$. Note that if $k(x,y)$ is the characteristic function of $\{(x,y) \in [0,1] \times [0,1]\colon y < x\}$, then V_k is the Volterra operator (II.1.7).

If k is a Volterra kernel, then

$$\sigma(V_k) = \{0\}.$$

Indeed, from the preceding paragraph it is known that $V_k^n = V_{k^n}$. This will be used to show that the spectral radius of V_k is 0.

6.15. Claim. $|k^n(x,y)| \leqslant (\|k\|_\infty^n/(n-1)!)(x-y)^{n-1}$ for $y < x$.

This is proved by induction. Clearly it holds for $n = 1$. Suppose (6.15) is true for some $n \geqslant 1$. Then

$$|k^{n+1}(x,y)| = \left| \int_y^x k(x,t) k^n(t,y)\, dt \right|$$

$$\leqslant \int_y^x |k(x,t)|\, |k^n(t,y)|\, dt$$

$$\leqslant \|k\|_\infty \frac{\|k\|_\infty^n}{(n-1)!} \int_y^x (t-y)^{n-1}\, dt$$

$$\leqslant \frac{\|k\|_\infty^{n+1}}{n!} (x-y)^n.$$

This establishes the claim.

From (6.15) it follows that

$$\|V_k^n\| \leqslant \|k^n\|_\infty \leqslant \frac{\|k\|_\infty^n}{(n-1)!}.$$

Therefore

$$\|V_k^n\|^{1/n} \leqslant \|k\|_\infty [(n-1)!]^{-1/n}.$$

Since $[(n-1)!]^{-1/n} \to 0$ as $n \to \infty$, $r(V_k) = 0$. Thus $\square \neq \sigma(V_k) \subseteq \{\lambda \in \mathbb{C}\colon |\lambda| \leqslant 0\}$; that is, $\sigma(V_k) = \{0\}$.

It is possible for $\ker V_k$ to be nontrivial. For example, if $k(x,y) = \chi_{(0,1/2)}(y)$

when $y < x$ and 0 otherwise, then

$$V_k f(x) = \begin{cases} \displaystyle\int_0^x f(y)dy & \text{if } x \leqslant \tfrac{1}{2}, \\ \displaystyle\int_0^{1/2} f(y)dy & \text{if } x \geqslant \tfrac{1}{2}. \end{cases}$$

So if $f(y) = 0$ for $0 \leqslant y \leqslant \tfrac{1}{2}$, $V_k f = 0$.

On the other hand, the Volterra operator V $[= V_k$ for $k(x, y) =$ the characteristic function of $\{(x, y): y < x\}]$ has $\ker V = (0)$. In fact, if $0 = Vf$, then for all x, $0 = \int_0^x f(y)dy$. Differentiating gives that $f = 0$.

Is there an analogy between V_k for a Volterra kernel k and a lower triangular matrix?

EXERCISES

1. Prove Proposition 6.1.

2. Show that for $\mathscr{X}$ a Banach space and A in $\mathscr{B}(\mathscr{X})$, $\sigma_l(A) = \sigma_r(A^*)$. What happens in a Hilbert space?

3. If $\mathscr{H}$ is an infinite dimension Hilbert space and K is a non-empty compact subset of $\mathbb{C}$, show that there is an A in $\mathscr{B}(\mathscr{H})$ such that $\sigma(A) = K$. Can A be found such that $\sigma(A) = \sigma_{ap}(A) = K$?

4. Let K be a compact subset of $\mathbb{C}$. Does there exist an operator A in $\mathscr{B}(C[0,1])$ such that $\sigma(A) = K$?

5. If $\mathscr{X}$ is a Banach space and $A \in \mathscr{B}(\mathscr{X})$, show that A is left invertible if and only if $\ker A = (0)$ and $\operatorname{ran} A$ is a closed complemented subspace of $\mathscr{X}$.

6. If $\mathscr{X}$ is a Banach space and $A \in \mathscr{B}(\mathscr{X})$, show that A is right invertible if and only if $\operatorname{ran} A = \mathscr{X}$ and $\ker A$ is a complemented subspace of $\mathscr{X}$.

7. If $\mathscr{X}$ is a Banach space and $T: \mathscr{X} \to \mathscr{X}$ is an isometry, then either $\sigma(T) \subseteq \partial \mathbb{D}$ or $\sigma(T) = \operatorname{cl} \mathbb{D}$.

8. Verify the statements made in Example 6.10.

9. Let $1 \leqslant p \leqslant \infty$ and suppose $0 < \alpha_1 \leqslant \alpha_2 \cdots$ such that $r = \lim \alpha_n < \infty$. Define A: $l^p \to l^p$ by $A(x_1, x_2, \ldots) = (0, \alpha_1 x_1, \alpha_2 x_2, \ldots)$. Show that $\sigma(A) = \{z \in \mathbb{C}: |z| \leqslant r\}$ and $\sigma_{ap}(A) = \partial \sigma(A)$. If $|\lambda| < r$, then $\operatorname{ran}(A - \lambda)$ is closed and has codimension 1. Also, $\sigma_p(A) = \square$.

10. Verify the statements made in Example 6.14.

11. Let $1 \leqslant p \leqslant \infty$ and let (X, Ω, μ) be a σ-finite measure space. For ϕ in $L^\infty(\mu)$, define M_ϕ on $L^p(\mu)$ as in Example III.2.2. Find $\sigma(M_\phi)$, $\sigma_{ap}(M_\phi)$, and $\sigma_p(M_\phi)$.

12. If $A \in \mathscr{B}(\mathscr{X})$, $f \in \operatorname{Hol}(A)$, and $\lambda \in \sigma_p(A)$, is $f(\lambda) \in \sigma_p(f(A))$? If $\lambda \in \sigma_{ap}(A)$, is $f(\lambda) \in \sigma_{ap}(f(A))$? Is there a relation between $f(\sigma_{ap}(A))$ and $\sigma_{ap}(f(A))$?

13. If $A \in \mathscr{B}(\mathscr{X})$ say that a complex number λ has *finite index* if there is a positive integer k such that $\ker(A - \lambda)^k = \ker(A - \lambda)^{k+1}$; the *index* of λ, denoted by $\nu(\lambda)$

or $v_A(\lambda)$, is the smallest such integer k. (a) Show that if λ is an isolated point of $\sigma(A)$ and a pole of order n of $(z-A)^{-1}$, then $v(\lambda)=n$. (b) If $v(\lambda)<\infty$, show that

$$\ker(A-\lambda)^{v(\lambda)}=\ker(A-\lambda)^{v(\lambda)+k}\text{ for all }k\geqslant0.\text{ (c) If }\mathscr{X}=\mathbb{C}^n\text{ and }A=\begin{bmatrix}0&&&\\1&0&&\\&1&\ddots&\\&&\ddots&\ddots\\&&&1&0\end{bmatrix},$$

then $\sigma(A)=\{0\}$ and $v(0)=n$.

14. If V is the Volterra operator, show that 0 is an essential singularity of $(z-V)^{-1}$.

15. Let $\mathscr{A}$ be a Banach algebra with identity. If $a\in\mathscr{A}$, define $L_a, R_a\in\mathscr{B}(\mathscr{A})$ by $L_a(x)=ax$ and $R_a(x)=xa$. Show that $\sigma(L_a)=\sigma(R_a)=\sigma(a)$.

16. If E is a projection on a Hilbert space and E is neither 0 nor 1, then $\sigma(E)=\{0,1\}$.

17. (McCabe [1984]) If $\mathscr{X}$ is a complex Banach space and $T\in\mathscr{B}(\mathscr{X})$, show that the following statements are equivalent. (a) $r(T)<1$. (b) $\|T^m\|<1$ for some positive integer m. (c) $\sum_n\|T^n(x)\|<\infty$ for every x in $\mathscr{X}$.

§7. The Spectral Theory of a Compact Operator

Recall that for a Banach space $\mathscr{X}$, $\mathscr{B}_0(\mathscr{X})$ is the algebra of all compact operators. This Banach algebra has no identity, so if $A\in\mathscr{B}_0(\mathscr{X})$, then $\sigma(A)$ refers to the spectrum of A as an element of $\mathscr{B}(\mathscr{X})$. Of course, if $\mathscr{A}=\mathscr{B}_0(\mathscr{X})+\mathbb{C}$, then $\mathscr{A}$ is a Banach algebra with identity (Why?) and we could consider $\sigma_{\mathscr{A}}(A)$ for A in $\mathscr{B}_0(\mathscr{X})$. By Theorem 5.4, $\sigma(A)\subseteq\sigma_{\mathscr{A}}(A)$, $\partial\sigma_{\mathscr{A}}(A)\subseteq\sigma(A)$, and $\sigma(A)=\sigma_{\mathscr{A}}(A)$. Below, in Theorem 7.1, it will be shown that $\sigma(A)$ is a countable set and hence $\sigma(A)=\partial\sigma(A)=\sigma(A)\hat{}$. Thus $\sigma(A)=\sigma_{\mathscr{A}}(A)$.

7.1. Theorem. (F. Riesz) *If* $\dim\mathscr{X}=\infty$ *and* $A\in\mathscr{B}_0(\mathscr{X})$, *then one and only one of the following possibilities occurs.*

(a) $\sigma(A)=\{0\}$.
(b) $\sigma(A)=\{0,\lambda_1,\ldots,\lambda_n\}$, *where for* $1\leqslant k\leqslant n$, $\lambda_k\neq0$, *each* λ_k *is an eigenvalue of* A, *and* $\dim\ker(A-\lambda_k)<\infty$.
(c) $\sigma(A)=\{0,\lambda_1,\lambda_2,\ldots\}$, *where for each* $k\geqslant1$, λ_k *is an eigenvalue of* A, $\dim\ker(A-\lambda_k)<\infty$, *and, moreover,* $\lim\lambda_k=0$.

The proof will use several lemmas. The first lemma was given in the case that $\mathscr{X}$ is a Hilbert space in Proposition II.4.14. The proof is identical and will not be repeated here.

7.2. Lemma. *If* $A\in\mathscr{B}_0(\mathscr{X})$, $\lambda\neq0$, *and* $\ker(A-\lambda)=(0)$, *then* $\operatorname{ran}(A-\lambda)$ *is closed.*

The proof of the next lemma is like that of Corollary II.4.15.

7.3. Lemma. *If $A \in \mathcal{B}_0(\mathcal{X})$, $\lambda \neq 0$, and $\lambda \in \sigma(A)$, then either $\lambda \in \sigma_p(A)$ or $\lambda \in \sigma_p(A^*)$.*

7.4. Lemma. *If $\mathcal{M} \leq \mathcal{N}$, $\mathcal{M} \neq \mathcal{N}$, and $\varepsilon > 0$, then there is a y in $\mathcal{N}$ such that $\|y\| = 1$ and $\mathrm{dist}(y, \mathcal{M}) \geq 1 - \varepsilon$.*

PROOF. Let $\delta(y) = \mathrm{dist}(y, \mathcal{M})$ for every y in $\mathcal{N}$. Now if $y_1 \in \mathcal{N} \backslash \mathcal{M}$, there is an x_0 in $\mathcal{M}$ such that $\delta(y_1) \leq \|x_0 - y_1\| \leq (1 + \varepsilon)\delta(y_1)$. Let $y_2 = y_1 - x_0$. Then $(1 + \varepsilon)\delta(y_2) = (1 + \varepsilon)\inf\{\|y_2 - x\|: x \in \mathcal{M}\} = (1 + \varepsilon)\inf\{\|y_1 - x_0 - x\|: x \in \mathcal{M}\} = (1 + \varepsilon)\delta(y_1)$ since $x_0 \in \mathcal{M}$. Thus $(1 + \varepsilon)\delta(y_2) > \|x_0 - y_1\| = \|y_2\|$. Let $y = \|y_2\|^{-1}y_2$. So $\|y\| = 1$, $y \in \mathcal{N}$, and if $x \in \mathcal{M}$, then

$$\|y - x\| = \|\,\|y_2\|^{-1}y_2 - x\,\|$$
$$= \|y_2\|^{-1}\|y_2 - \|y_2\|x\| > [(1 + \varepsilon)\delta(y_2)]^{-1}\|y_2 - \|y_2\|x\|$$
$$\geq (1 + \varepsilon)^{-1} > 1 - \varepsilon. \qquad \blacksquare$$

If $\mathcal{M}$ and $\mathcal{N}$ are finite dimensional in the preceding lemma, then y can be chosen in $\mathcal{N}$ such that $\|y\| = 1$ and $\mathrm{dist}(y, \mathcal{M}) = 1$ (see Exercise 1).

7.5. Lemma. *If $A \in \mathcal{B}_0(\mathcal{X})$ and $\{\lambda_n\}$ is a sequence of distinct elements in $\sigma_p(A)$, then $\lim \lambda_n = 0$.*

PROOF. For each n let $x_n \in \ker(A - \lambda_n)$ such that $x_n \neq 0$. It follows that if $\mathcal{M}_n = \bigvee\{x_1, \ldots, x_n\}$, then $\dim \mathcal{M}_n = n$ (Exercise). Hence $\mathcal{M}_n \leq \mathcal{M}_{n+1}$ and $\mathcal{M}_n \neq \mathcal{M}_{n+1}$. By the preceding lemma there is a vector y_n in $\mathcal{M}_n$ such that $\|y_n\| = 1$ and $\mathrm{dist}(y_n, \mathcal{M}_{n-1}) > \frac{1}{2}$. Let $y_n = \alpha_1 x_1 + \cdots + \alpha_n x_n$. Hence

$$(A - \lambda_n)y_n = \alpha_1(\lambda_1 - \lambda_n)x_1 + \cdots + \alpha_{n-1}(\lambda_{n-1} - \lambda_n)x_{n-1} \in \mathcal{M}_{n-1}.$$

So if $n > m$,

$$A(\lambda_n^{-1}y_n) - A(\lambda_m^{-1}y_m) = \lambda_n^{-1}(A - \lambda_n)y_n - \lambda_m^{-1}(A - \lambda_m)y_m + y_n - y_m$$
$$= y_n - [y_m + \lambda_m^{-1}(A - \lambda_m)y_m - \lambda_n^{-1}(A - \lambda_n)y_n].$$

But the bracketed expression belongs to $\mathcal{M}_{n-1}$. Hence $\|A(\lambda_n^{-1}y_n) - A(\lambda_m^{-1}y_m)\| \geq \mathrm{dist}(y_n, \mathcal{M}_{n-1}) > \frac{1}{2}$. Therefore $A(\lambda_n^{-1}y_n)$ can have no convergent subsequence. But A is a compact operator so that if S is any bounded subset of $\mathcal{X}$, $\mathrm{cl}\,A(S)$ is compact. Thus it must be that $\{\lambda_n^{-1}y_n\}$ has no bounded subsequence. Since $\|y_n\| = 1$ for all n, it must be that $\|\lambda_n^{-1}y_n\| = |\lambda_n|^{-1} \to \infty$. That is, $0 = \lim \lambda_n$. $\qquad \blacksquare$

PROOF OF THEOREM 7.1. The first step is to establish the following.

7.6. Claim. *If $\lambda \in \sigma(A)$ and $\lambda \neq 0$, then λ is an isolated point of $\sigma(A)$.*

In fact, if $\{\lambda_n\} \subseteq \sigma(A)$ and $\lambda_n \to \lambda$, then each λ_n belongs to either $\sigma_p(A)$ or $\sigma_p(A^*)$ (7.3). So either there is a subsequence $\{\lambda_{n_k}\}$ that is contained in $\sigma_p(A)$

or there is a subsequence contained in $\sigma_p(A^*)$. If $\{\lambda_{n_k}\} \subseteq \sigma_p(A)$, then Lemma 7.5 implies $\lambda_{n_k} \to 0$, a contradiction. If $\{\lambda_{n_k}\} \subseteq \sigma_p(A^*)$, then the fact that A^* is compact gives the same contradiction. Thus λ must be isolated if $\lambda \neq 0$.

7.7. Claim. If $\lambda \in \sigma(A)$ and $\lambda \neq 0$, then $\lambda \in \sigma_p(A)$ and $\dim \ker(A - \lambda) < \infty$.

By (7.6), λ is an isolated point of $\sigma(A)$ so that $E(\lambda)$ can be defined as in (6.9). Let $\mathscr{X}_\lambda = E(\lambda)\mathscr{X}$ and $A_\lambda = A|\mathscr{X}_\lambda$. By Exercise 4.9 [also see (4.11)], $\sigma(A_\lambda) = \{\lambda\}$. Thus A_λ is an invertible compact operator. By Exercise VI.3.5, $\dim \mathscr{X}_\lambda < \infty$. If $n = \dim \mathscr{X}_\lambda$, then $A_\lambda - \lambda$ is a nilpotent operator on an n-dimensional space. Thus $(A_\lambda - \lambda)^n = 0$. Let $\nu =$ the positive integer such that $(A_\lambda - \lambda)^\nu = 0$ but $(A_\lambda - \lambda)^{\nu-1} \neq 0$. Let $x \in \mathscr{X}_\lambda$ such that $0 \neq (A_\lambda - \lambda)^{\nu-1}x = y$; then $(A - \lambda)y = 0$. Thus $\lambda \in \sigma_p(A)$.

Also, $\ker(A - \lambda) \in \operatorname{Lat} A$ and $A|\ker(A - \lambda)$ is compact. But $Ax = \lambda x$ for all x in $\ker(A - \lambda)$, so $\dim \ker(A - \lambda) < \infty$.

Now for the *dénouement*. If $\dim \mathscr{X} = \infty$ and $A \in \mathscr{B}_0(\mathscr{X})$, then A cannot be invertible (Exercise VI.3.5). Thus $0 \in \sigma(A)$. If $\lambda \in \sigma(A)$ and $\lambda \neq 0$, then Claim 7.7 says that $\lambda \in \sigma_p(A)$ and $\dim \ker(A - \lambda) < \infty$. So if $\sigma(A)$ is finite, either (a) or (b) of (7.1) hold. If $\sigma(A)$ is infinite, then Claim 7.6 implies that $\sigma(A)$ is countable. So let $\sigma(A) = \{0, \lambda_1, \lambda_2, \ldots\}$. By Lemma 7.5 and Claim 7.7, (c) holds. ∎

Part of the following surfaced in the proof of the theorem.

7.8. Corollary. *If $A \in \mathscr{B}_0(\mathscr{X})$ and $\lambda \in \sigma(A)$ with $\lambda \neq 0$, then λ is a pole of $(z - A)^{-1}$, $\ker(A - \lambda) \subseteq E(\lambda)\mathscr{X}$, and $\dim E(\lambda)\mathscr{X} < \infty$.*

PROOF. The only part of this corollary that did not appear in the preceding proof is the fact that $\ker(A - \lambda) \subseteq E(\lambda)\mathscr{X}$.

Let $\Delta = \sigma(A)\backslash\{\lambda\}$, $\mathscr{X}_\Delta = E(\Delta)\mathscr{X}$, $A_\Delta = A|\mathscr{X}_\Delta$. By Exercise 4.9, $\sigma(A_\Delta) = \Delta$; so $A_\Delta - \lambda$ is invertible on $\mathscr{X}_\Delta$. If $x \in \ker(A - \lambda)$, then $x = E(\lambda)x + E(\Delta)x$. Hence $0 = (A - \lambda)x = (A - \lambda)E(\lambda)x + (A - \lambda)E(\Delta)x = (A_\lambda - \lambda)E(\lambda)x + (A_\Delta - \lambda)E(\Delta)x$. But $\mathscr{X}_\lambda$ and $\mathscr{X}_\Delta \in \operatorname{Lat} A$, so $(A_\lambda - \lambda)E(\lambda)x \in \mathscr{X}_\lambda$ and $(A_\Delta - \lambda)E(\Delta)x \in \mathscr{X}_\Delta$; since $\mathscr{X}_\lambda \cap \mathscr{X}_\Delta = (0)$, $0 = (A_\lambda - \lambda)E(\lambda)x = (A_\Delta - \lambda)E(\Delta)x$. But $A_\Delta - \lambda$ is invertible so $E(\Delta)x = 0$; that is, $x = E(\lambda)x \in \mathscr{X}_\lambda$. Hence $\ker(A - \lambda) \subseteq \mathscr{X}_\lambda$. ∎

If k is a Volterra kernel (6.14), then V_k is a compact operator (Exercise VI.3.6) and $\sigma(V_k) = \{0\}$. So the first possibility of Theorem 7.1 can occur. If V is the Volterra operator, then $\sigma_p(V) = \square$.

Let V be the Volterra operator on $L^p(0, 1)$, $1 < p < \infty$. If $\lambda_1, \ldots, \lambda_n \in \mathbb{C}$, let $D: \mathbb{C}^n \to \mathbb{C}^n$ be defined by $D(z_1, \ldots, z_n) = (\lambda_1 z_1, \ldots, \lambda_n z_n)$. Then $A = V \oplus D$ on $L^p(0, 1) \oplus \mathbb{C}^n$ is compact and $\sigma(A) = \{0, \lambda_1, \ldots, \lambda_n\}$. So the second possibility of (7.1) occurs. If $\{\lambda_n\} \subseteq \mathbb{C}$ and $\lim \lambda_n = 0$, then define $D: l^p \to l^p$ ($1 \leq p \leq \infty$) by $(Dx)(n) = \lambda_n x(n)$. If $A = V \oplus D$ on $L^p(0, 1) \oplus l^p$, A is compact and $\sigma(A) = \{0, \lambda_1, \lambda_2, \ldots\}$ (see Exercise 3).

The next result has a number of applications in the theory of integral equations.

7.9. The Fredholm Alternative. *If* $A \in \mathscr{B}_0(\mathscr{X})$, $\lambda \in \mathbb{C}$, *and* $\lambda \neq 0$, *then* $\operatorname{ran}(A - \lambda)$ *is closed and* $\dim \ker(A - \lambda) = \dim \ker(A - \lambda)^* < \infty$.

PROOF. It suffices to assume that $\lambda \in \sigma(A)$. Put $\Delta = \sigma(A) \backslash \{\lambda\}$, $\mathscr{X}_\lambda = E(\lambda)\mathscr{X}$, $\mathscr{X}_\Delta = E(\Delta)\mathscr{X}$, $A_\lambda = A|\mathscr{X}_\lambda$, and $A_\Delta = A|\mathscr{X}_\Delta$. Now $\lambda \notin \Delta = \sigma(A_\Delta)$, so $A_\Delta - \lambda$ is invertible. Thus $\operatorname{ran}(A_\Delta - \lambda) = \mathscr{X}_\Delta$. Hence $\operatorname{ran}(A - \lambda) = (A - \lambda)\mathscr{X}_\lambda + (A - \lambda)\mathscr{X}_\Delta = \operatorname{ran}(A_\lambda - \lambda) + \mathscr{X}_\Delta$. Since $\dim \mathscr{X}_\lambda < \infty$, $\operatorname{ran}(A - \lambda)$ is closed (III.4.3).

Also note that

$$\mathscr{X}/\operatorname{ran}(A - \lambda) = (\mathscr{X}_\Delta + \mathscr{X}_\lambda)/[\operatorname{ran}(A_\lambda - \lambda) + \mathscr{X}_\Delta]$$
$$\approx \mathscr{X}_\lambda/\operatorname{ran}(A_\lambda - \lambda).$$

Since $\dim \mathscr{X}_\lambda < \infty$, $\dim[\mathscr{X}/\operatorname{ran}(A - \lambda)] = \dim \mathscr{X}_\lambda - \dim \operatorname{ran}(A_\lambda - \lambda) = \dim \ker(A_\lambda - \lambda) = \dim \ker(A - \lambda) < \infty$ since $\ker(A - \lambda) \subseteq \mathscr{X}_\lambda$ (7.8). But $[\mathscr{X}/\operatorname{ran}(A - \lambda)]^* = [\operatorname{ran}(A - \lambda)]^\perp$ (III.10.2) $= \ker(A - \lambda)^*$. Hence $\dim \ker(A - \lambda) = \dim \ker(A - \lambda)^*$. ∎

7.10. Corollary. *If* $A \in \mathscr{B}_0(\mathscr{X})$, $\lambda \in \mathbb{C}$, *and* $\lambda \neq 0$, *then for every* y *in* $\mathscr{X}$ *there is an* x *in* $\mathscr{X}$ *such that*

7.11 $(A - \lambda)x = y$

if and only if the only vector x *such that* $(A - \lambda)x = 0$ *is* $x = 0$. *If this condition is satisfied, then the solution to* (7.11) *is unique.*

This corollary is a rephrasing of part of the Fredholm Alternative together with the fact that an operator has dense range if and only if its adjoint has a trivial kernel.

The applications of the Fredholm Alternative occur by taking the compact operator to be an integral operator.

EXERCISES

1. If $\mathscr{M}$, $\mathscr{N}$ are finite dimensional spaces and $\mathscr{M} \leqslant \mathscr{N}$, $\mathscr{M} \neq \mathscr{N}$, then there is a y in $\mathscr{N}$ such that $\|y\| = 1$ and $\operatorname{dist}(y, \mathscr{M}) = 1$.

2. Let $A \in \mathscr{B}(\mathscr{X})$ and let $\lambda_1, \dots, \lambda_n$ be distinct points in $\sigma_p(A)$. If $x_k \in \ker(A - \lambda_k)$, $1 \leqslant k \leqslant n$, and $x_k \neq 0$, show that $\{x_1, \dots, x_n\}$ is a linearly independent set.

3. Let $\mathscr{X}_1, \mathscr{X}_2 \dots$ be Banach spaces and put $\mathscr{X} = \oplus_p \mathscr{X}_n$. Let $A_n \in \mathscr{B}(\mathscr{X}_n)$ such that $\sup_n \|A_n\| < \infty$ and define $A \colon \mathscr{X} \to \mathscr{X}$ by $A\{x_n\} = \{A_n x_n\}$. Show that $A \in \mathscr{B}(\mathscr{X})$ and $\|A\| = \sup_n \|A_n\|$. Show that $A \in \mathscr{B}_0(\mathscr{X})$ if and only if each $A_n \in \mathscr{B}_0(\mathscr{X})$ and $\lim \|A_n\| = 0$.

4. Suppose $A \in \mathscr{B}(\mathscr{X})$ and there is a polynomial p such that $p(A) \in \mathscr{B}_0(\mathscr{X})$. What can be said about $\sigma(A)$?

5. Suppose $A \in \mathcal{B}(\mathcal{X})$ and there is an entire function f such that $f(A) \in \mathcal{B}_0(\mathcal{X})$. What can be said about $\sigma(A)$?

6. With the terminology of Exercise 6.13, if $A \in \mathcal{B}_0(\mathcal{X})$, $\lambda \in \sigma(A)$, and $\lambda \neq 0$, what can be said about the index of λ?

§8. Abelian Banach Algebras

Recall that it is assumed that every Banach algebra is over $\mathbb{C}$. Also assume that all Banach algebras contain an identity.

A *division algebra* is an algebra such that every nonzero element has a multiplicative inverse. It may seem incongruous that the first theorem in this section allows the algebra to be nonabelian. However, the conclusion is that the algebra is abelian—and much more.

8.1. The Gelfand–Mazur Theorem. *If $\mathcal{A}$ is a Banach algebra that is also a division ring, then $\mathcal{A} = \mathbb{C}$ ($\equiv \{\lambda 1 : \lambda \in \mathbb{C}\}$).*

PROOF. If $a \in \mathcal{A}$, then $\sigma(a) \neq \square$. If $\lambda \in \sigma(a)$, then $a - \lambda$ has no inverse. But $\mathcal{A}$ is a division ring, so $a - \lambda = 0$. That is, $a = \lambda$. ■

As a corollary of the preceding theorem, the algebra of quaternions, $\mathbb{H}$, is not a Banach algebra. That is, it is impossible to put a norm on $\mathbb{H}$ that makes it into a Banach algebra over $\mathbb{C}$. Can you show this directly?

8.2. Proposition. *If $\mathcal{A}$ is an abelian Banach algebra and $\mathcal{M}$ is a maximal ideal, then there is a homomorphism $h: \mathcal{A} \to \mathbb{C}$ such that $\mathcal{M} = \ker h$. Conversely, if $h: \mathcal{A} \to \mathbb{C}$ is a nonzero homomorphism, then $\ker h$ is a maximal ideal. Moreover, this correspondence $h \mapsto \ker h$ between homomorphisms and maximal ideals is bijective.*

PROOF. If $\mathcal{M}$ is a maximal ideal, then $\mathcal{M}$ is closed (2.4b). Hence $\mathcal{A}/\mathcal{M}$ is a Banach algebra with identity. Let $\pi: \mathcal{A} \to \mathcal{A}/\mathcal{M}$ be the natural map. If $a \in \mathcal{A}$ and $\pi(a)$ is not invertible in $\mathcal{A}/\mathcal{M}$, then $\pi(\mathcal{A}a) = \pi(a)[\mathcal{A}/\mathcal{M}]$ is an ideal in $\mathcal{A}/\mathcal{M}$ that is proper. Let $I = \{b \in \mathcal{A} : \pi(b) \in \pi(\mathcal{A}a)\} = \pi^{-1}(\pi(\mathcal{A}a))$. Then I is a proper ideal of $\mathcal{A}$ and $\mathcal{M} \subseteq I$. Since $\mathcal{M}$ is maximal, $\mathcal{M} = I$. Thus $\pi(a\mathcal{A}) \subseteq \pi(I) = \pi(\mathcal{M}) = (0)$. That is, $\pi(a) = 0$. This says that $\mathcal{A}/\mathcal{M}$ is a field. By the Gelfand–Mazur Theorem $\mathcal{A}/\mathcal{M} = \mathbb{C} = \{\lambda + \mathcal{M} : \lambda \in \mathbb{C}\}$. Define $\tilde{h}$: $\mathcal{A}/\mathcal{M} \to \mathbb{C}$ by $\tilde{h}(\lambda + \mathcal{M}) = \lambda$ and define $h: \mathcal{A} \to \mathbb{C}$ by $h = \tilde{h} \circ \pi$. Then h is a homomorphism and $\ker h = \mathcal{M}$.

Conversely, suppose $h: \mathcal{A} \to \mathbb{C}$ is a nonzero homomorphism. Then $\ker h = \mathcal{M}$ is a nontrivial ideal and $\mathcal{A}/\mathcal{M} \approx \mathbb{C}$. (Why?) So $\mathcal{M}$ is maximal.

If h, h' are two nonzero homomorphisms and $\ker h = \ker h'$, then there is an α in $\mathbb{C}$ such that $h = \alpha h'$ (A.1.4). But $1 = h(1) = \alpha h'(1) = \alpha$, so $h = h'$. ■

8.3. Corollary. *If $\mathscr{A}$ is an abelian Banach algebra and h: $\mathscr{A} \to \mathbb{C}$ is a homomorphism, then h is continuous.*

PROOF. Maximal ideals are closed (2.4b). ∎

The next result improves the preceding corollary a little. Remember that by (8.3) if h: $\mathscr{A} \to \mathbb{C}$ is a homomorphism, then $h \in \mathscr{A}^*$ (the Banach space dual of $\mathscr{A}$).

8.4. Proposition. *If $\mathscr{A}$ is abelian and h: $\mathscr{A} \to \mathbb{C}$ is a non-zero homomorphism, then $\|h\| = 1$.*

PROOF. Let $a \in \mathscr{A}$ and put $\lambda = h(a)$. If $|\lambda| > \|a\|$, then $\|a/\lambda\| < 1$. Hence $1 - a/\lambda$ is invertible. Let $b = (1 - a/\lambda)^{-1}$, so $1 = b(1 - a/\lambda) = b - ba/\lambda$. Since $h(1) = 1$, $1 = h(b - ba/\lambda) = h(b) - h(b)h(a)/\lambda = h(b) - h(b) = 0$, a contradiction. Hence $\|a\| \geqslant |\lambda| = |h(a)|$; so $\|h\| \leqslant 1$. Since $h(1) = 1$, $\|h\| = 1$. ∎

8.5. Definition. If $\mathscr{A}$ is an abelian Banach algebra, let $\Sigma = $ the collection of all nonzero homomorphisms of $\mathscr{A} \to \mathbb{C}$. Give Σ the relative weak* topology that it has as a subset of $\mathscr{A}^*$. Σ with this topology is called the *maximal ideal space* of $\mathscr{A}$.

8.6. Theorem. *If $\mathscr{A}$ is an abelian Banach algebra, then its maximal ideal space Σ is a compact Hausdorff space. Moreover, if $a \in \mathscr{A}$, then $\sigma(a) = \Sigma(a) \equiv \{h(a): h \in \Sigma\}$.*

PROOF. Since $\Sigma \subseteq$ ball $\mathscr{A}^*$, it suffices for the proof of the first part of the theorem to show that Σ is weak* closed. Let $\{h_i\}$ be a net in Σ and suppose $h \in$ ball $\mathscr{A}^*$ such that $h_i \to h$ weak*. If $a, b \in \mathscr{A}$, then $h(ab) = \lim_i h_i(ab) = \lim_i h_i(a)h_i(b) = h(a)h(b)$. So h is a homomorphism. Since $h(1) = \lim_i h_i(1) = 1$, $h \in \Sigma$. Thus Σ is compact.

If $h \in \Sigma$ and $\lambda = h(a)$, then $a - \lambda \in \ker h$. So $a - \lambda$ is not invertible and $\lambda \in \sigma(a)$; that is, $\Sigma(a) \subseteq \sigma(a)$. Now assume that $\lambda \in \sigma(a)$; so $a - \lambda$ is not invertible and, hence, $(a - \lambda)\mathscr{A}$ is a proper ideal. Let $\mathscr{M}$ be a maximal ideal in $\mathscr{A}$ such that $(a - \lambda)\mathscr{A} \subseteq \mathscr{M}$. If $h \in \Sigma$ such that $\mathscr{M} = \ker h$, then $0 = h(a - \lambda) = h(a) - \lambda$; hence $\sigma(a) \subseteq \Sigma(a)$. ∎

Now it is time for an example. Here is one that is a little more than an example. If X is compact and $x \in X$, let δ_x: $C(X) \to \mathbb{C}$ be defined by $\delta_x(f) = f(x)$. It is easy to see that δ_x is a homomorphism on the algebra $C(X)$.

8.7. Theorem. *If X is compact and Σ is the maximal ideal space of $C(X)$, then the map $x \mapsto \delta_x$ is a homeomorphism of X onto Σ.*

PROOF. Let Δ: $X \to \Sigma$ be defined by $\Delta(x) = \delta_x$. As was pointed out before, $\Delta(X) \subseteq \Sigma$. It was shown in Proposition V.6.1 that Δ: $X \to (\Delta(X), \text{weak*})$ is a homeomorphism. Thus it only remains to show that $\Delta(X) = \Sigma$. If $h \in \Sigma$, then there is a measure μ in $M(X)$ such that $h(f) = \int f \, d\mu$ for all f in $C(X)$. Also, $\|\mu\| = \|h\| = 1$ and $\mu(X) = \int 1 \, d\mu = h(1) = 1$. Hence $\mu \geq 0$ (Exercises III.7.2). Let $x \in$ support (μ). It will be shown that $h = \delta_x$.

Let $\mathcal{M} = \{f \in C(X) : f(x) = 0\}$. So $\mathcal{M}$ is a maximal ideal of $C(X)$. Note that if it can be shown that $\ker h \subseteq \mathcal{M}$, then it must be that $\ker h = \mathcal{M}$ and so $h = \delta_x$. So let $f \in \ker h$. Because $\ker h$ is an ideal, $|f|^2 = f\bar{f} \in \ker h$. Hence $0 = h(|f|^2) = \int |f|^2 \, d\mu$. Since $\mu \geq 0$ and $|f|^2 \geq 0$, it must be that $f = 0$ a.e. $[\mu]$. Since f is continuous, $f \equiv 0$ on support (μ). In particular, $f(x) = 0$ and so $f \in \mathcal{M}$. ∎

It follows from the preceding theorem that the maximal ideals of $C(X)$ are all of the form $\{f \in C(X) : f(x) = 0\}$ for some x in X.

8.8. Definition. Let $\mathcal{A}$ be an abelian Banach algebra with maximal ideal space Σ. If $a \in \mathcal{A}$, then the *Gelfand transform* of a is the function $\hat{a} : \Sigma \to \mathbb{C}$ defined by $\hat{a}(h) = h(a)$.

8.9. Theorem. *If $\mathcal{A}$ is an abelian Banach algebra with maximal ideal space Σ and $a \in \mathcal{A}$, then the Gelfand transform of a, $\hat{a}$, belongs to $C(\Sigma)$. The map $a \mapsto \hat{a}$ of $\mathcal{A}$ into $C(\Sigma)$ is a continuous homomorphism of $\mathcal{A}$ into $C(\Sigma)$ of norm 1 and its kernel is*

$$\bigcap \{\mathcal{M} : \mathcal{M} \text{ is a maximal ideal of } \mathcal{A}\}.$$

Moreover, for each a in $\mathcal{A}$,

$$\|\hat{a}\|_\infty = \lim_{n \to \infty} \|a^n\|^{1/n}.$$

PROOF. If $h_i \to h$ in Σ, then $h_i \to h$ weak* in $\mathcal{A}^*$. So if $a \in \mathcal{A}$, $\hat{a}(h_i) = h_i(a) \to h(a) = \hat{a}(h)$. Thus $\hat{a} \in C(\Sigma)$.

Define $\gamma : \mathcal{A} \to C(\Sigma)$ by $\gamma(a) = \hat{a}$. If $a, b \in \mathcal{A}$, then $\gamma(ab)(h) = \widehat{ab}(h) = h(ab) = h(a)h(b) = \hat{a}(h)\hat{b}(h)$. Therefore $\gamma(ab) = \gamma(a)\gamma(b)$. It is easy to see that γ is linear, so γ is a homomorphism. Also, by (8.4), if $a \in \mathcal{A}$, $|\hat{a}(h)| = |h(a)| \leq \|a\|$; thus $\|\gamma(a)\|_\infty = \|\hat{a}\|_\infty \leq \|a\|$. So γ is continuous and $\|\gamma\| \leq 1$. Since $\gamma(1) = 1$, $\|\gamma\| = 1$.

Note that $a \in \ker \gamma$ if and only if $\hat{a} \equiv 0$; that is, $a \in \ker \gamma$ if and only if $h(a) = 0$ for each h in Σ. Thus $a \in \ker \gamma$ if and only if a belongs to every maximal ideal of $\mathcal{A}$.

Finally, by Theorem 8.6, if $a \in \mathcal{A}$, then $\|\hat{a}\|_\infty = \sup\{|\lambda| : \lambda \in \sigma(a)\}$. The last part of this theorem is thus a consequence of this observation and Proposition 3.8. ∎

The homomorphism $a \mapsto \hat{a}$ of $\mathcal{A}$ into $C(\Sigma)$ is called the *Gelfand transform* of $\mathcal{A}$. The kernel of the *Gelfand* transform is called the *radical* of $\mathcal{A}$, rad $\mathcal{A}$. So

$$\operatorname{rad} \mathcal{A} = \cap \{\mathcal{M} : \mathcal{M} \text{ is a maximal ideal of } \mathcal{A}\}.$$

If X is compact and Σ, the maximal ideal space of $C(X)$, is identified with X as in Theorem 8.7, then the Gelfand transform $C(X) \to C(\Sigma)$ becomes the identity map.

If $\mathscr{A}$ is an abelian algebra, say that a in $\mathscr{A}$ is a *generator* of $\mathscr{A}$ if $\{p(a): p$ is a polynomial$\}$ is dense in $\mathscr{A}$.

Recall that if $\tau: X \to Y$ is a homeomorphism, then $A: C(Y) \to C(X)$ defined by $Af = f \circ \tau$ is an isometric isomorphism (VI.2.1). Denote the relationship between A and τ by $A = \tau^{\#}$.

8.10. Proposition. *If $\mathscr{A}$ is an abelian Banach algebra with identity and a is a generator of $\mathscr{A}$, then there is a homeomorphism $\tau: \Sigma \to \sigma(a)$ such that if $\gamma: \mathscr{A} \to C(\Sigma)$ is the Gelfand transform and p is a polynomial, then $\gamma(p(a)) = \tau^{\#}(p)$.*

PROOF. Define $\tau: \Sigma \to \sigma(a)$ by $\tau(h) = h(a)$. By Theorem 8.6 τ is surjective. It is easy to see that τ is continuous. To see that τ is injective, suppose $\tau(h_1) = \tau(h_2)$, so $h_1(a) = h_2(a)$. Hence $h_1(a^n) = h_2(a^n)$ for all $n \geqslant 0$. By linearity, $h_1(p(a)) = h_2(p(a))$ for every polynomial p. Since a is a generator for $\mathscr{A}$ and h_1 and h_2 are continuous on $\mathscr{A}$, $h_1 = h_2$, and τ is injective. Since Σ is compact, τ is a homeomorphism.

The remainder of the proposition follows from the fact that γ and $\tau^{\#}$ are homomorphisms. Hence $\gamma(p(a))(h) = p(\gamma(a))(h) = p(\hat{a})(h) = p(\hat{a}(h)) = p(\tau(h)) = \tau^{\#}(p)(h)$. ∎

8.11. Corollary. *If $\mathscr{A}$ has two elements a_1 and a_2 each of which is a generator, then $\sigma(a_1)$ and $\sigma(a_2)$ are homeomorphic.*

The converse to (8.11) is not true. If $\mathscr{A} = C[-1, 1]$, then $f(x) = x$ defines a generator f for $\mathscr{A}$. If $g(x) = x^2$, then $\sigma(g) = g([-1, 1]) = [0, 1]$. So $\sigma(f)$ and $\sigma(g)$ are homeomorphic. However, g is not a generator for $\mathscr{A}$. In fact, the Banach algebra generated by g consists of the even functions in $C[-1, 1]$.

8.12. Example. If $V: L^2(0, 1) \to L^2(0, 1)$ is the Volterra operator and $\mathscr{A}$ is the closure in $\mathscr{B}(L^2(0, 1))$ of $\{p(V): p$ is a polynomial in $z\}$, then $\mathscr{A}$ is an abelian Banach algebra and $\operatorname{rad} \mathscr{A} = \operatorname{cl}\{p(V): p$ is a polynomial in z and $p(0) = 0\}$. In other words, $\mathscr{A}$ has a unique maximal ideal, $\operatorname{rad} \mathscr{A}$. In fact, if $\mathscr{B} = \mathscr{B}(L^2(0, 1))$, Theorem 5.4 implies that $\partial \sigma_{\mathscr{A}}(V) \subseteq \sigma_{\mathscr{B}}(V) \subseteq \sigma_{\mathscr{A}}(V)$. Since $\sigma_{\mathscr{B}}(V) = \{0\}$ (6.14), $\sigma_{\mathscr{A}}(V) = \{0\}$. The statement above now follows by Proposition 8.10.

8.13. Example. Let $\mathscr{A}$ be the closure in $C(\partial \mathbb{D})$ of the polynomials in z. If Σ is the maximal ideal space of $\mathscr{A}$, then Σ is homeomorphic to $\sigma_{\mathscr{A}}(z)$. (Here z is the function whose value at λ in $\partial \mathbb{D}$ is λ.) Now $\sigma_{\mathscr{A}}(z) = \operatorname{cl} \mathbb{D}$ as was shown in Example 5.1. If $f \in \mathscr{A}$, then the Maximum Modulus Theorem shows that f has a continuous extension to $\operatorname{cl} \mathbb{D}$ that is analytic in $\mathbb{D}$ [see (5.1)]. Also denote this extension by f. The proof of (8.10) shows that the continuous homomorphisms on $\mathscr{A}$ are of the form $f \mapsto f(\lambda)$ for some λ in $\operatorname{cl} \mathbb{D}$.

In the next section the Banach algebra $L^1(G)$ is examined for a locally compact abelian group and its maximal ideals are characterized.

EXERCISES

1. Let $\mathscr{A}$ be a Banach algebra with identity and let J be the smallest closed two-sided ideal of $\mathscr{A}$ containing $\{xy - yx: x, y \in \mathscr{A}\}$. J is called the *commutator ideal* of $\mathscr{A}$. (a) Show that $\mathscr{A}/J$ is an abelian Banach algebra. (b) If I is a closed ideal of $\mathscr{A}$ such that $\mathscr{A}/I$ is abelian, then $I \supseteq J$. (c) If $h: \mathscr{A} \to \mathbb{C}$ is a homomorphism, then $J \subseteq \ker h$ and h induces a homomorphism $\tilde{h}: \mathscr{A}/J \to \mathbb{C}$ such that $\tilde{h} \circ \pi = h$, where $\pi: \mathscr{A} \to \mathscr{A}/J$ is the natural map. Hence $\|\tilde{h}\| = 1$. (d) Let Σ be the set of homomorphisms of $\mathscr{A} \to \mathbb{C}$ and let $\tilde{\Sigma}$ be the set of homomorphisms of $\mathscr{A}/J$. Show that the map $h \mapsto \tilde{h}$ defined in (c) is a homeomorphism of Σ onto $\tilde{\Sigma}$.

2. Using the terminology of Exercises 2.6 and 2.7, let $\mathscr{A}$ be an abelian Banach algebra without identity and show that if $\mathscr{M}$ is a maximal modular ideal, then there is a homomorphism $h: \mathscr{A} \to \mathbb{C}$ such that $\mathscr{M} = \ker h$. Conversely, if $h: \mathscr{A} \to \mathbb{C}$ is a nonzero homomorphism, then $\ker h$ is a maximal modular ideal. Moreover, the correspondence $h \to \ker h$ is a bijection between homomorphisms and maximal modular ideals.

3. If $\mathscr{A}$ is an abelian Banach algebra and $h: \mathscr{A} \to \mathbb{C}$ is a homomorphism, then h is continuous and $\|h\| \leqslant 1$. If $\mathscr{A}$ has an approximate identity $\{e_i\}$ such that $\|e_i\| \leqslant 1$ for all i, then $\|h\| = 1$ (see Exercise 2.8).

4. Let $\mathscr{A}$ be an abelian Banach algebra and let Σ be the set of nonzero homomorphisms of $\mathscr{A} \to \mathbb{C}$. Show that Σ is locally compact if it has the relative weak* topology from $\mathscr{A}^*$ (Exercise 3).

5. With the notation of Exercise 4, assume that $\mathscr{A}$ has no identity and let $\mathscr{A}_1$ be the algebra obtained by adjoining an identity. For a in $\mathscr{A}$, let $\sigma(a)$ be the spectrum of a as an element of $\mathscr{A}_1$ and show that $\sigma(a) = \{h(a): h \in \Sigma\} \cup \{0\}$. Also, show that the maximal ideal space of $\mathscr{A}_1$, Σ_1, is the one-point compactification of Σ.

6. With the notation of Exercise 4, for each a in $\mathscr{A}$ define $\hat{a}: \Sigma \to \mathbb{C}$ by $\hat{a}(h) = h(a)$. Show that $\hat{a} \in C_0(\Sigma)$ and the map $a \mapsto \hat{a}$ of $\mathscr{A}$ into $C_0(\Sigma)$ is a contractive homomorphism with kernel $= \cap \{\mathscr{M}: \mathscr{M}$ is a maximal modular ideal of $\mathscr{A}\}$.

7. If X is locally compact, show that $x \mapsto \delta_x$ is a homeomorphism of X onto the maximal ideal space of $C_0(X)$.

8. Let X be locally compact and for each open subset U of X let $C_0(U) = \{f \in C_0(X): f(x) = 0$ for x in $X \backslash U\}$. Show that $C_0(U)$ is a closed ideal of $C_0(X)$ and that every closed ideal of $C_0(X)$ has this form. Moreover, the map $U \mapsto C_0(U)$ is a lattice isomorphism from the lattice of open subsets of X onto the lattice of ideals of $C_0(X)$.

9. With the notation of the preceding exercise, show that $C_0(U)$ is a modular ideal if and only if $X \backslash U$ is compact.

10. If $\mathscr{A}$ is an abelian Banach algebra and $a \in \mathscr{A}$, say that a is a *rational generator* of $\mathscr{A}$ if $\{f(a): f$ is a rational function with poles off $\sigma(a)\}$ is dense in $\mathscr{A}$. Show that if a is a rational generator of $\mathscr{A}$, then Σ is homeomorphic to $\sigma(a)$.

11. Verify the statements made in Example 8.12.

12. Say that $a_1, \ldots, a_n$ are generators of $\mathscr{A}$ if $\mathscr{A}$ is the smallest Banach algebra with identity that contains $\{a_1, \ldots, a_n\}$. Show that $a_1, \ldots, a_n$ are generators of $\mathscr{A}$ if and only if $\mathscr{A} = \text{cl}\{p(a_1, \ldots, a_n): p$ is a polynomial in n complex variables $z_1, \ldots, z_n\}$, and if Σ is the maximal ideal space, then there is a homeomorphism τ of Σ onto a compact subset K of $\mathbb{C}^n$ such that if p is a polynomial in n variables, then $\gamma(p(a_1, \ldots, a_n)) = \tau^{\#}(p)$.

13. Verify the statements made in Example 8.13.

14. (Zelazko [1968].) Let $\mathscr{A}$ be an algebra and suppose $\phi: \mathscr{A} \to \mathbb{C}$ is a linear functional such that $\phi(a^2) = \phi(a)^2$ for all a in $\mathscr{A}$. Show that ϕ is a homomorphism.

15. Let $\mathscr{A}$ be an abelian Banach algebra with identity that is semisimple [that is, rad $\mathscr{A} = (0)$]. If $\|\cdot\|$ is the norm on $\mathscr{A}$ and $\|\cdot\|_1$ is another norm on $\mathscr{A}$ that also makes $\mathscr{A}$ into a Banach algebra, then these two norms are equivalent. (Hint: use the Closed Graph Theorem to show that the identity map $i: (\mathscr{A}, \|\cdot\|) \to (\mathscr{A}, \|\cdot\|_1)$ is continuous.)

16. Let $\mathscr{A}$ be as in Example 8.13 and let $K = \{\phi \in \mathscr{A}^*: \phi(1) = \|\phi\| = 1\}$. Show that ext $K = \{\delta_z: |z| = 1\}$. (See (V.7).)

17. Show that $f(x) = \exp(\pi i x)$ is a generator of $C([0, 1])$ but $g(x) = \exp(2\pi i x)$ is not.

18. Show that $C(\partial \mathbb{D})$ does not have a single generator though it does have a single rational generator (that is, an element a such that $\{r(a): r$ is a rational function with poles off $\sigma(a)\}$ is dense.)

§9*. The Group Algebra of a Locally Compact Abelian Group

If G is a locally compact abelian group and m is Haar measure on G, then $L^1(G) \equiv L^1(m)$ is a Banach algebra (Example 1.11), where for f, g in $L^1(G)$ the product $f * g$ is the convolution of f and g:

$$f * g(x) = \int_G f(xy^{-1})g(y)\,dy.$$

Note that dy is used to designate integration with respect to m rather than $dm(y)$. Because G is abelian, $L^1(G)$ is abelian. In fact, $g * f(x) = \int g(xy^{-1})f(y)\,dy$. If $y^{-1}x$ is substituted for y in this integral, the value of the integral does not change because Haar measure is translation invariant. Hence $g * f(x) = \int g(y)f(y^{-1}x)\,dy = \int g(y)f(xy^{-1})\,dy = f * g(x)$.

Let e denote the identity of G. If G is discrete, then $\delta_e \in L^1(G)$ and δ_e is an identity for $L^1(G)$. If G is not discrete, then $L^1(G)$ does not have an identity (Exercise 1).

Some examples of nondiscrete locally compact abelian groups are $\mathbb{R}^n$ and $\mathbb{T}^n$, where $\mathbb{T} = $ the unit circle $\partial \mathbb{D}$ in $\mathbb{C}$ with the usual multiplication. Note that $\mathbb{T}^\infty$ is also a compact abelian group while $\mathbb{R}^\infty$ fails to be locally compact. The Cantor set can be identified with the product of a countable number of copies of $\mathbb{Z}_2$ and is thus a compact abelian group. Indeed, the product of a countable number of finite sets (with the discrete topology) is homeomorphic to the Cantor set, so that the Cantor set has infinitely many nonisomorphic group structures.

For a topological group G, $L^1(G)$ is called the *group algebra* for G. If G is discrete, the algebraists talk of the *group algebra* over a field K as the set of all $f = \sum_{g \in G} a_g g$, where $a_g \in K$ and $a_g \neq 0$ for at most a finite number of g in G. If $K = \mathbb{C}$, this is the set of functions $f: G \to \mathbb{C}$ with finite support. Thus in the discrete case the group algebra of the algebraists can be identified with a dense manifold in $L^1(G) = l^1(G)$.

Unlike §V.11, if $f: G \to \mathbb{C}$ and $x \in G$, define $f_x: G \to \mathbb{C}$ by $f_x(y) = f(yx^{-1})$; so $f_x(y) = f(x^{-1}y)$ for G abelian. We want to examine the function $x \mapsto f_x$ of $G \to L^p(G)$, $1 \leq p < \infty$. To do this we first prove the following (see Exercise V.11.10).

9.1. Proposition. *If G is a topological group and $f: G \to \mathbb{C}$ is a continuous function with compact support, then for any $\varepsilon > 0$ there is a neighborhood U of e such that $|f(x) - f(y)| < \varepsilon$ whenever $x^{-1}y \in U$.*

PROOF. Let $\mathscr{U}$ be the collection of open neighborhoods U of e such that $U = U^{-1}$. Note that if V is any neighborhood of e, then $U = V \cap V^{-1} \in \mathscr{U}$ and $U \subseteq V$. Order $\mathscr{U}$ by reverse inclusion.

Suppose the result is false. Then there is an $\varepsilon > 0$ such that for every U in $\mathscr{U}$ there are points x_U, y_U in G with $x_U^{-1}y_U$ in U and $|f(x_U) - f(y_U)| \geq \varepsilon$. Note that either x_U or $y_U \in K \equiv$ support f. Since $U = U^{-1}$, we may assume that $x_U \in K$ for every U in $\mathscr{U}$. Now $\{x_U : U \in \mathscr{U}\}$ is a net in K. Since K is compact, there is a point x in K such that $x_U \xrightarrow{\text{cl}} x$. But $x_U^{-1}y_U \to e$. Since multiplication is continuous, $y_U = x_U(x_U^{-1}y_U) \xrightarrow{\text{cl}} x$. Therefore if W is any neighborhood of x, there is a U in $\mathscr{U}$ with $x_U, y_U \in W$. But f is continuous at x so W can be chosen such that $|f(x) - f(w)| < \varepsilon/2$ whenever $w \in W$. With this choice of W, $|f(x_U) - f(y_U)| < \varepsilon$, a contradiction. ■

One can rephrase (9.1) by saying that continuous functions on a topological group that have compact support are uniformly continuous.

In the next result it is the case $p = 1$ which is of principal interest for us at this time. The proof of the general theorem is, however, no more difficult than this special case.

9.2. Proposition. *If G is a locally compact group, $1 \leq p < \infty$, and $f \in L^p(G)$, then the map $x \mapsto f_x$ is a continuous function from G into $L^p(G)$.*

PROOF. Fix f in $L^p(G)$, x in G, and $\varepsilon > 0$; it must be shown that there is a neighborhood V of x such that for y in V, $\|f_y - f_x\|_p < \varepsilon$. First note that there is a continuous function $\phi: G \to \mathbb{C}$ having compact support such that

$\|f - \phi\|_p < \varepsilon/3$. Let $K = \text{spt } \phi$. Note that because Haar measure is translation invariant, for any y in G, $\|f_y - \phi_y\|_p = \|f - \phi\|_p < \varepsilon/3$. Now by Proposition 9.1, there is a neighborhood U of e such that $|\phi(y) - \phi(w)| < \frac{1}{3}\varepsilon[2m(K)]^{-1/p}$ whenever $y^{-1}w \in U$. Put $V = Ux$. If $y \in V$, then

$$\|\phi_y - \phi_x\|_p^p = \int |\phi(zy^{-1}) - \phi(zx^{-1})|^p dz.$$

But $y = ux$ for some u in U, so $(zy^{-1})^{-1}(zx^{-1}) = yx^{-1} = u \in U$. Thus

$$\|\phi_y - \phi_x\|_p^p = \int_{Ky \cup Kx} |\phi(zy^{-1}) - \phi(zx^{-1})|^p dz$$

$$\leqslant \left(\frac{\varepsilon}{3}\right)^p [2m(K)]^{-1} m(Ky \cup Kx)$$

$$\leqslant \left(\frac{\varepsilon}{3}\right)^p.$$

Therefore if $y \in V$, $\|f_x - f_y\|_p \leqslant \|f_x - \phi_x\|_p + \|\phi_x - \phi_y\|_p + \|\phi_y - f_y\|_p < \varepsilon.$ ∎

The aim of this section is to discuss the homomorphisms on $L^1(G)$ when G is abelian and to examine the Gelfand transform. There is a bit of a difficulty here since $L^1(G)$ does not have an identity when G is not discrete. If δ_e is the unit point mass at e, then δ_e is the identity for $M(G)$ and hence acts as an identity for $L^1(G)$. Nevertheless $\delta_e \notin L^1(G)$ if G is not discrete. All is not lost as $L^1(G)$ has an approximate identity (Exercise 2.8) of a nice type.

9.3. Proposition. *If $f \in L^1(G)$ and $\varepsilon > 0$, then there is a neighborhood U of e such that if g is a non-negative Borel function on G that vanishes off U and has $\int g(x)dx = 1$, then $\|f - f*g\|_1 < \varepsilon$.*

PROOF. By the preceding proposition, there is a neighborhood U of e such that $\|f - f_y\|_1 < \varepsilon$ whenever $y \in U$. If g satisfies the conditions, then $f(x) - f*g(x) = \int [f(x) - f(xy^{-1})]g(y)dy$ for all x. Thus,

$$\|f - f*g\|_1 = \int \left| \int_U [f(x) - f(xy^{-1})]g(y)dy \right| dx$$

$$\leqslant \int_U g(y) \int |f(x) - f(xy^{-1})| dx\, dy$$

$$= \int_U g(y) \|f - f_y\|_1 dy$$

$$\leqslant \varepsilon.$$ ∎

9.4. Corollary. *There is a net $\{e_i\}$ of non-negative functions in $L^1(G)$ such that $\int e_i dm = 1$ for all i and $\|e_i * f - f\|_1 \to 0$ for all f in $L^1(G)$.*

PROOF. Let $\mathcal{U}$ be the collection of all neighborhoods of e and order $\mathcal{U}$ by reverse inclusion. Let $\mathcal{U} = \{U_i: i \in I\}$ where $i \leqslant j$ if and only if $U_j \subseteq U_i$. For each i in I put $e_i = m(U_i)^{-1}\chi_{U_i}$, so $e_i \geqslant 0$ and $\int e_i dm = 1$. If $f \in L^1(G)$ and $\varepsilon > 0$, let U_i be as in the preceding proposition. So if $j \geqslant i$, e_j satisfies the conditions on g in (9.3) and hence $\| f - f * e_j \|_1 < \varepsilon$. ∎

9.5. Corollary. *If h: $L^1(G) \to \mathbb{C}$ is a nonzero homomorphism, then h is bounded and $\| h \| = 1$.*

PROOF. The fact that h is bounded and $\| h \| \leqslant 1$ is Exercise 8.3. In light of the preceding corollary if $h(f) \neq 0$, $h(f) = \lim h(f * e_i) = h(f)\lim h(e_i)$. Hence $h(e_i) \to 1$. Since $\| e_i \| = 1$ for all i, $\| h \| = 1$. ∎

Even though Haar measure on most of the popular examples is σ-finite, this is not true in general. For example, if D is an uncountable discrete group, the Haar measure on D is counting measure and, hence, not σ-finite. Similarly, Haar measure on $D \times \mathbb{R}$ is not σ-finite. Nevertheless, it is true that $L^1(G)^* = L^\infty(G)$ for any locally compact group because (G, m) is an example of a decomposable measure space, though $L^\infty(G)$ must be redefined to be the equivalence classes of bounded Borel functions that are equal a.e. on every set of finite Haar measure. This fact will be assumed here. The interested reader can consult Hewitt and Ross [1963].

9.6. Theorem. *If G is a locally compact abelian group and γ: $G \to \mathbb{T}$ is a continuous homomorphism, define $\hat{f}(\gamma)$ by*

$$9.7 \qquad \hat{f}(\gamma) = \int f(x)\gamma(x^{-1})dx$$

for every f in $L^1(G)$. Then $f \mapsto \hat{f}(\gamma)$ is a nonzero homomorphism on $L^1(G)$. Conversely, if h: $L^1(G) \to \mathbb{C}$ is a nonzero homomorphism, there is a continuous homomorphism γ: $G \to \mathbb{T}$ such that $h(f) = \hat{f}(\gamma)$.

PROOF. First note that if γ: $G \to \mathbb{T}$ is a homomorphism, $\gamma(xy) = \gamma(x)\gamma(y)$ and $\gamma(x^{-1}) = \gamma(x)^{-1} = \overline{\gamma(x)}$, the complex conjugate of $\gamma(x)$. If $f, g \in L^1(G)$, then

$$\widehat{f * g}(\gamma) = \int (f * g)(x)\gamma(x^{-1})dx$$

$$= \int \gamma(x^{-1}) \int f(xy^{-1})g(y)dy\,dx$$

$$= \int g(y)\gamma(y^{-1})\left[\int f(xy^{-1})\gamma((xy^{-1})^{-1})dx \right] dy.$$

But the invariance of the Haar integral gives that $\int f(xy^{-1})\gamma((xy^{-1})^{-1})dx = \int f(x)\gamma(x^{-1})dx$. Hence

$$\widehat{f*g}(\gamma) = \int g(y)\gamma(y^{-1})\left[\int f(x)\gamma(x^{-1})dx\right]dy = \hat{f}(\gamma)\hat{g}(\gamma).$$

So $f \mapsto \hat{f}(\gamma)$ is a homomorphism. Since γ is continuous and $\gamma(G) \subseteq \mathbb{T}$, $\gamma \in L^\infty(G)$ and $\|\gamma\|_\infty = 1$. Thus $f \mapsto \hat{f}(\gamma)$ is not identically zero.

Now assume that $h: L^1(G) \to \mathbb{C}$ is a nonzero homomorphism. Since h is a bounded linear functional, there is a ϕ in $L^\infty(G)$ such that $h(f) = \int f(x)\phi(x)dx$ and $\|\phi\|_\infty = \|h\| = 1$. If f, $g \in L^1(G)$, then $h(f*g) = \int(f*g)(x)\phi(x)dx = \int g(y)[\int f(xy^{-1})\phi(x)dx]dy = \int g(y)h(f_y)dy$. [Note that $y \mapsto h(f_y)$ is a continuous scalar-valued function by Proposition 9.2.] But $h(f*g) = h(f)h(g) = \int g(y)h(f)\phi(y)dy$. So

$$0 = \int g(y)[h(f_y) - h(f)\phi(y)]dy$$

for every g in $L^1(G)$. But $y \mapsto h(f_y) - h(f)\phi(y)$ belongs to $L^\infty(G)$, so for any f in $L^1(G)$,

9.8
$$h(f_y) = h(f)\phi(y)$$

for locally almost all y in G. Pick f in $L^1(G)$ such that $h(f) \neq 0$. By (9.8), $\phi(y) = h(f_y)/h(f)$ a.e. But the right-hand side of this equation is continuous. Hence we may assume that ϕ is a continuous function. Thus for every f in $L^1(G)$, (9.8) holds everywhere.

In (9.8), replace y by xy and we obtain $h(f)\phi(xy) = h(f_{xy}) = h((f_x)_y)$. Now replace f in (9.8) by f_x to get $h(f_x)\phi(y) = h(f_{xy})$. Thus $h(f)\phi(xy) = h(f_x)\phi(y) = [h(f)\phi(x)]\phi(y)$. If $h(f) \neq 0$, this implies $\phi(xy) = \phi(x)\phi(y)$ for all x, y in G. Thus $\phi: G \to \mathbb{C}$ is a homomorphism and $|\phi(x)| \leq 1$ for all x. But $1 = \phi(e) = \phi(x)\phi(x^{-1}) = \phi(x)\phi(x)^{-1}$ and $|\phi(x)|, |\phi(x)^{-1}| \leq 1$. Hence $|\phi(x)| = 1$ for all x in G. If $\gamma(x) = \phi(x^{-1})$, then $\gamma: G \to \mathbb{T}$ is a continuous homomorphism and $h(f) = \hat{f}(\gamma)$ for all f in $L^1(G)$. ∎

Let Σ be the set of nonzero homomorphisms on $L^1(G)$, where G is assumed to be abelian (both here and throughout the rest of the chapter). So $\Sigma \subseteq \text{ball } L^1(G)^*$. If $h \in \text{ball } L^1(G)^*$ and $\{h_i\}$ is a net in Σ such that $h_i \to h$ weak*, then it is easy to see that h is multiplicative. Thus the weak* closure of $\Sigma \subseteq \Sigma \cup \{0\}$. Hence the relative weak* topology on Σ makes Σ into a locally compact Hausdorff space (see Exercise 8.4).

Let $\Gamma = $ all the continuous homomorphisms $\gamma: G \to \mathbb{T}$. By Theorem 9.6, Σ and Γ can be identified using formula (9.7). In fact, the map defined in (9.7) is the Gelfand transform when this identification is made. (Just look at the definitions.) Since Σ and Γ are identified and Σ has a topology, Γ can be given a topology. Thus Γ becomes a locally compact space with this topology. (For another description of the topology, see Exercise 6.) The functions in Γ are called *characters* and are sometimes denoted by $\Gamma = \hat{G}$ and called the *dual group*.

228 VII. Banach Algebras and Spectral Theory

Also notice that in a natural way Γ is a group. If $\gamma_1, \gamma_2 \in \Gamma$, then $(\gamma_1 \gamma_2)(x) \equiv \gamma_1(x)\gamma_2(x)$ and $\gamma_1 \gamma_2 \in \Gamma$.

9.9. Proposition. Γ *is a locally compact abelian group.*

Clearly Γ is an abelian group and we know that Γ is a locally compact space. It must be shown that Γ is a topological group. To do this we first prove a lemma.

9.10. Lemma.

(a) *The map* $(x, \gamma) \mapsto \gamma(x)$ *of* $G \times \Gamma \to \mathbb{T}$ *is continuous.*
(b) *If* $\{\gamma_i\}$ *is a net in* Γ *and* $\gamma_i \to \gamma$ *in* Γ, *then* $\gamma_i(x) \to \gamma(x)$ *uniformly for* x *belonging to any compact subset of* G.

PROOF. First note that if $x \in G$ and $f \in L^1(G)$, then for every γ in Γ,

$$\hat{f}_x(\gamma) = \int f_x(y)\gamma(y^{-1})\,dy$$

$$= \int f(yx^{-1})\gamma(y^{-1})\,dy$$

$$= \int f(z)\gamma(z^{-1}x^{-1})\,dz$$

$$= \gamma(x^{-1})\hat{f}(\gamma).$$

So if $\gamma_i \to \gamma$ in Γ and $x_i \to x$ in G,

$$|\hat{f}(\gamma_i)\gamma_i(x_i) - \hat{f}(\gamma)\gamma(x)| = |\hat{f}_{x_i^{-1}}(\gamma_i) - \hat{f}_{x^{-1}}(\gamma)|$$

$$\leqslant |\hat{f}_{x_i^{-1}}(\gamma_i) - \hat{f}_{x^{-1}}(\gamma_i)| + |\hat{f}_{x^{-1}}(\gamma_i) - \hat{f}_{x^{-1}}(\gamma)|.$$

But $|\hat{f}_{x_i^{-1}}(\gamma_i) - \hat{f}_{x^{-1}}(\gamma_i)| \leqslant \|f_{x_i^{-1}} - f_{x^{-1}}\|_1 \to 0$ by (9.2). Because $f_{x^{-1}} \in L^1(G)$, $\hat{f}_{x^{-1}}(\gamma_i) \to \hat{f}_{x^{-1}}(\gamma)$ since $\gamma_i \to \gamma$. Thus $\hat{f}(\gamma_i)\gamma_i(x_i) \to \hat{f}(\gamma)\gamma(x)$. If f is chosen so that $\hat{f}(\gamma) \neq 0$, then because $\hat{f}(\gamma_i) \to \hat{f}(\gamma)$, there is an i_0 such that $\hat{f}(\gamma_i) \neq 0$ for $i \geqslant i_0$. Therefore $\gamma_i(x_i) \to \gamma(x)$ and (a) is proven.

Now let K be a compact subset of G and let $\{\gamma_i\}$ be a net in Γ such that $\gamma_i \to \gamma_0$. Suppose $\{\gamma_i(x)\}$ does not converge uniformly on K to $\gamma_0(x)$. Then there is an $\varepsilon > 0$ such that for every i, there is a $j_i \geqslant i$ and an x_i in K such that $|\gamma_{j_i}(x_i) - \gamma_0(x_i)| \geqslant \varepsilon$. Now $\{\gamma_{j_i}\}$ is a net and $\gamma_{j_i} \to \gamma_0$ (Exercise). Since K is compact, there is an x_0 in K such that $x_i \xrightarrow{\text{cl}} x_0$. Now part (a) implies that the map $(x, \gamma) \mapsto (\gamma(x), \gamma_0(x))$ of $G \times \Gamma$ into $\mathbb{T} \times \mathbb{T}$ is continuous. Since $(x_i, \gamma_{j_i}) \xrightarrow{\text{cl}} (x_0, \gamma_0)$ in $G \times \Gamma$, $(\gamma_{j_i}(x_i), \gamma_0(x_i)) \xrightarrow{\text{cl}} (\gamma_0(x_0), \gamma_0(x_0))$. So for any i_0, there is an $i \geqslant i_0$ such that $|\gamma_{j_i}(x_i) - \gamma_0(x_0)| < \varepsilon/2$ and $|\gamma_0(x_i) - \gamma_0(x_0)| < \varepsilon/2$. Hence $|\gamma_{j_i}(x_i) - \gamma_0(x_i)| < \varepsilon$, a contradiction. ∎

PROOF OF PROPOSITION 9.9. Let $\{\gamma_i\}, \{\lambda_i\}$ be nets in Γ such that $\gamma_i \to \gamma$ and $\lambda_i \to \lambda$. It must be shown that $\gamma_i \lambda_i^{-1} \to \gamma\lambda^{-1}$. Let $\phi \in C_c(G)$ and put $K = \text{spt } \phi$.

Then $\hat{\phi}(\gamma_i\lambda_i^{-1}) = \int_K \phi(x)\gamma_i(x^{-1})\lambda_i(x)\,dx$. By the preceding lemma, $\gamma_i(x^{-1}) \to \gamma(x^{-1})$ and $\lambda_i(x) \to \lambda(x)$ uniformly for x in K. Thus $\hat{\phi}(\gamma_i\lambda_i^{-1}) \to \hat{\phi}(\gamma\lambda^{-1})$. If $f \in L^1(G)$ and $\varepsilon > 0$, let $\phi \in C_c(G)$ such that $\|f - \phi\|_1 < \varepsilon/3$. Then

$$|\hat{f}(\gamma_i\lambda_i^{-1}) - \hat{f}(\gamma\lambda^{-1})| < \frac{2\varepsilon}{3} + |\hat{\phi}(\gamma_i\lambda_i^{-1}) - \hat{\phi}(\gamma\lambda^{-1})|.$$

It follows that $\hat{f}(\gamma_i\lambda_i^{-1}) \to \hat{f}(\gamma\lambda^{-1})$ for every f in $L^1(G)$. Hence $\gamma_i\lambda_i^{-1} \to \gamma\lambda^{-1}$ in Γ. ∎

Since Γ is a locally compact abelian group, it too has a dual group. Let $\hat{\Gamma}$ be this dual group. If $x \in G$, define $\rho(x): \Gamma \to \mathbb{T}$ by $\rho(x)(\gamma) = \gamma(x)$. It is easy to see that ρ is a homomorphism. It is a rather deep fact, entitled the Pontryagin Duality Theorem, that $\rho: G \to \hat{\Gamma}$ is a homeomorphism and an isomorphism. That is, G "is" the dual group of its dual group. The interested reader may consult Rudin [1962]. We turn now to some examples.

9.11. Theorem. *If $y \in \mathbb{R}$, then $\gamma_y(x) = e^{ixy}$ defines a character on $\mathbb{R}$ and every character on $\mathbb{R}$ has this form. The map $y \mapsto \gamma_y$ is a homeomorphism and an isomorphism of $\mathbb{R}$ onto $\hat{\mathbb{R}}$. If $y \in \mathbb{R}$ and $f \in L^1(\mathbb{R})$, then*

9.12
$$\hat{f}(\gamma_y) = \hat{f}(y) = \int_{-\infty}^{\infty} f(x)e^{-ixy}\,dx,$$

the Fourier transform of f.

PROOF. If $y \in \mathbb{R}$, then $|\gamma_y(x)| = 1$ for all x and $\gamma_y(x_1 + x_2) = \gamma_y(x_1)\gamma_y(x_2)$. So $\gamma_y \in \hat{\mathbb{R}}$. Also, $\gamma_{y_1 + y_2}(x) = \gamma_{y_1}(x)\gamma_{y_2}(x)$. Hence $y \mapsto \gamma_y$ is a homomorphism of $\mathbb{R}$ into $\hat{\mathbb{R}}$.

Now let $\gamma \in \hat{\mathbb{R}}$. $\gamma(0) = 1$ so that there is a $\delta > 0$ such that $\int_0^\delta \gamma(x)\,dx = a \neq 0$. Thus

$$a\gamma(x) = \gamma(x)\int_0^\delta \gamma(t)\,dt$$

$$= \int_0^\delta \gamma(x + t)\,dt$$

$$= \int_x^{x+\delta} \gamma(t)\,dt.$$

Hence $\gamma(x) = a^{-1}\int_x^{x+\delta}\gamma(t)\,dt$. Because γ is continuous, the Fundamental Theorem of Calculus implies that γ is differentiable. Also,

$$\frac{\gamma(x + h) - \gamma(x)}{h} = \gamma(x)\left[\frac{\gamma(h) - 1}{h}\right].$$

So $\gamma'(x) = \gamma'(0)\gamma(x)$. Since $\gamma(0) = 1$ and $|\gamma(x)| = 1$ for all x, the elementary theory of differential equations implies that $\gamma = \gamma_y$ for some y in $\mathbb{R}$. This implies that $y \mapsto \gamma_y$ is an isomorphism of $\mathbb{R}$ onto $\mathbb{R}$.

It is clear from (9.7) that (9.12) holds. From here it is easy to see that $y \mapsto \gamma_y$ is a homeomorphism of $\mathbb{R}$ onto $\hat{\mathbb{R}}$.   ∎

So the preceding result says that $\mathbb{R}$ is its own dual group. Because of (9.12), the function $\hat{f}$ as defined in (9.7) is called the *Fourier transform* of f. The next result lends more weight to the use of this terminology.

9.13. Theorem. *If $n \in \mathbb{Z}$, define $\gamma_n: \mathbb{T} \to \mathbb{T}$ by $\gamma_n(z) = z^n$. Then $\gamma_n \in \hat{\mathbb{T}}$ and the map $n \mapsto \gamma_n$ is a homeomorphism and an isomorphism of $\mathbb{Z}$ onto $\hat{\mathbb{T}}$. If $n \in \mathbb{Z}$ and $f \in L^1(\mathbb{T})$, then*

$$9.14 \qquad \hat{f}(\gamma_n) = \hat{f}(n) \equiv \frac{1}{2\pi} \int_0^{2\pi} f(e^{i\theta}) e^{-in\theta} \, d\theta.$$

PROOF. It is left to the reader to check that $\gamma_n \in \hat{\mathbb{T}}$ and $n \mapsto \gamma_n$ is an injective homomorphism of $\mathbb{Z}$ into $\hat{\mathbb{T}}$. If $\gamma \in \hat{\mathbb{T}}$, define $\sigma: \mathbb{R} \to \mathbb{T}$ by $\sigma(t) = \gamma(e^{it})$; it follows that $\sigma \in \hat{\mathbb{R}}$. By (9.11), $\sigma(t) = e^{iyt}$ for some y in $\mathbb{R}$. But $\sigma(t + 2\pi) = \sigma(t)$, so $e^{2\pi i y} = 1$. Hence $y = n \in \mathbb{Z}$. Thus $\gamma(e^{i\theta}) = \sigma(\theta) = e^{in\theta}$, $\gamma = \gamma_n$, and $n \mapsto \gamma_n$ is an isomorphism of $\mathbb{Z}$ onto $\hat{\mathbb{T}}$. Formula (9.14) is immediate from (9.7). The fact that $n \mapsto \gamma_n$ is a homeomorphism is left as an exercise.   ∎

So $\hat{\mathbb{T}} = \mathbb{Z}$, a discrete group. This can be generalized.

9.15. Theorem. *If G is compact, $\hat{G}$ is discrete; if G is discrete, $\hat{G}$ is compact.*

PROOF. Put $\Gamma = \hat{G}$. If G is discrete, then $L^1(G)$ has an identity. Hence its maximal ideal space is compact. That is, Γ is compact.

Now assume that G is compact. Hence $\Gamma \subseteq L^1(G)$ since $m(G) = 1$. Suppose $\gamma \in \Gamma$ and $\gamma \neq$ the identity for Γ, then there is a point x_0 in G such that $\gamma(x_0) \neq 1$. Thus

$$\int \gamma(x)\,dx = \int \gamma(xx_0^{-1}x_0)\,dx$$

$$= \gamma(x_0) \int \gamma(xx_0^{-1})\,dx$$

$$= \gamma(x_0) \int \gamma(x)\,dx,$$

since Haar measure is translation invariant. Since $\gamma(x_0) \neq 1$, this implies that

$$\int_G \gamma(x)\,dx = 0 \qquad \text{if } \gamma \neq 1.$$

Of course if $\gamma = 1$, $\int 1 \, dx = m(G) = 1$. So if $f = 1$ on G, $f \in L^1(G)$ and

$\hat{f}(\gamma) = \int \gamma(x^{-1})dx = \chi_{\{1\}}(\gamma)$. Since $\hat{f}$ is continuous on Γ, $\{1\}$ is an open set. By translation, every singleton set in Γ is open and hence Γ is discrete. ∎

9.16. Theorem. *If $a \in \mathbb{T}$, define $\gamma_a: \mathbb{Z} \to \mathbb{T}$ by $\gamma_a(n) = a^n$. Then $\gamma_a \in \hat{\mathbb{Z}}$ and the map $a \mapsto \gamma_a$ is a homeomorphism and an isomorphism of $\mathbb{T}$ onto $\hat{\mathbb{Z}}$. If $a \in \mathbb{T}$ and $f \in L^1(\mathbb{Z}) = l^1(\mathbb{Z})$, then*

9.17
$$\hat{f}(\gamma_a) = \hat{f}(a) = \sum_{n=-\infty}^{\infty} f(n)a^{-n}.$$

PROOF. Again the proof that $a \mapsto \gamma_a$ is a monomorphism of $\mathbb{T} \to \hat{\mathbb{Z}}$ is left to the reader. If $\gamma \in \hat{\mathbb{Z}}$, let $\gamma(1) = a \in \mathbb{T}$. Also, $\gamma(n) = \gamma(1)^n = a^n$, so $\gamma = \gamma_a$. Hence $a \mapsto \gamma_a$ is an isomorphism. It is easy to show that this map is continuous and hence, by compactness, a homeomorphism. ∎

For additional reading, consult Rudin [1962].

EXERCISES
1. Prove that if $L^1(G)$ has an identity, then G is discrete.

2. If $f \in L^\infty(G)$, show that $x \mapsto f_x$ is a continuous function from G into $(L^\infty(G), \text{wk}^*)$.

3. Is there a measure μ on $\mathbb{R}$ different from Lebesgue measure such that for f in $L^1(\mu)$, $x \mapsto f_x$ is continuous? Is there a measure for which this map is discontinuous?

4. If $f \in C_0(G)$, show that $x \mapsto f_x$ is a continuous map from $G \to C_0(G)$.

5. If $f \in L^\infty(G)$ and f is uniformly continuous on G, show that $x \mapsto f_x$ is a continuous function from $G \to L^\infty(G)$. Is the converse true? See Edwards [1961].

6. If K is a compact subset of G, $\gamma_0 \in \Gamma$, and $\varepsilon > 0$, let $U(K, \gamma_0, \varepsilon) = \{\gamma \in \Gamma: |\gamma(x) - \gamma_0(x)| < \varepsilon \text{ for all } x \text{ in } K\}$. Show that the collection of all such sets is a base for the topology of Γ. (This says that the topology on Γ is the *compact-open topology*.)

7. Show that there is a discontinuous homomorphism $\gamma: \mathbb{R} \to \mathbb{T}$. If $\gamma: \mathbb{R} \to \mathbb{T}$ is a homomorphism that is a Borel function, show that γ is continuous.

8. If G is a compact abelian group, show that the linear span of Γ is dense in $C(G)$.

9. If G is a compact abelian group, show that Γ forms an orthonormal basis in $L^2(G)$.

10. If G is a compact abelian group, show that G is metrizable if and only if Γ is countable.

11. Let $\{G_\alpha\}$ be a family of compact abelian groups and $G = \Pi_\alpha G_\alpha$. If $\Gamma_\alpha = \hat{G}_\alpha$, show that the character group of G is $\{\{\gamma_\alpha\} \in \Pi_\alpha \Gamma_\alpha: \gamma_\alpha = e \text{ except for at most a finite number of } \alpha\}$.

CHAPTER VIII

C^*-Algebras

A C^*-algebra is a particular type of Banach algebra that is intimately connected with the theory of operators on a Hilbert space. If $\mathcal{H}$ is a Hilbert space, then $\mathcal{B}(\mathcal{H})$ is an example of a C^*-algebra. Moreover, if $\mathcal{A}$ is any C^*-algebra, then it is isomorphic to a subalgebra of $\mathcal{B}(\mathcal{H})$ (see Section 5). Some of the general theory developed in this chapter will be used in the next chapter to prove the Spectral Theorem, which reveals the structure of normal operators.

A more thorough treatment of C^*-algebras is available in Arveson [1976] or Sakai [1971].

§1. Elementary Properties and Examples

If $\mathcal{A}$ is a Banach algebra, an *involution* is a map $a \mapsto a^*$ of $\mathcal{A}$ into $\mathcal{A}$ such that the following properties hold for a and b in $\mathcal{A}$ and α in $\mathbb{C}$:

(i) $(a^*)^* = a$;
(ii) $(ab)^* = b^*a^*$;
(iii) $(\alpha a + b)^* = \bar{\alpha}a^* + b^*$.

Note that if $\mathcal{A}$ has involution and an identity, then $1^* \cdot a = (1^* \cdot a)^{**} = (a^* \cdot 1)^* = (a^*)^* = a$; similarly, $a \cdot 1^* = a$. Since the identity is unique, $1^* = 1$. Also, for any α in $\mathbb{C}$, $\alpha^* = \bar{\alpha}$.

1.1. Definition. A C^*-*algebra* is a Banach algebra $\mathcal{A}$ with an involution such that for every a in $\mathcal{A}$,

$$\|a^*a\| = \|a\|^2.$$

1.2. Example. If $\mathcal{H}$ is a Hilbert space, $\mathcal{A} = \mathcal{B}(\mathcal{H})$ is a C^*-algebra where for each A in $\mathcal{B}(\mathcal{H})$, $A^* =$ the adjoint of A. (See Proposition II.2.7.)

1.3. Example. If $\mathcal{H}$ is a Hilbert space, $\mathcal{B}_0(\mathcal{H})$ is a C^*-subalgebra of $\mathcal{B}(\mathcal{H})$, though $\mathcal{B}_0(\mathcal{H})$ does not have an identity if $\mathcal{H}$ is infinite dimensional.

1.4. Example. If X is a compact space, $C(X)$ is a C^*-algebra where $f^*(x) = \overline{f(x)}$ for f in $C(X)$ and x in X.

1.5. Example. If (X, Ω, μ) is a σ-finite measure space, $L^\infty(X, \Omega, \mu)$ is a C^*-algebra where the involution is defined as in (1.4).

1.6. Example. If X is locally compact but not compact, $C_0(X)$ is a C^*-algebra without identity.

1.7. Proposition. *If $\mathcal{A}$ is a C^*-algebra and $a \in \mathcal{A}$, then $\|a^*\| = \|a\|$.*

PROOF. Note that $\|a\|^2 = \|a^*a\| \leqslant \|a^*\| \|a\|$; so $\|a\| \leqslant \|a^*\|$. Since $a = a^{**}$, substituting a^* for a in this inequality gives $\|a^*\| \leqslant \|a\|$. ∎

1.8. Proposition. *If $\mathcal{A}$ is a C^*-algebra and $a \in \mathcal{A}$, then*

$$\|a\| = \sup\{\|ax\|: x \in \mathcal{A}, \|x\| \leqslant 1\}$$
$$= \sup\{\|xa\|: x \in \mathcal{A}, \|x\| \leqslant 1\}.$$

PROOF. Let $\alpha = \sup\{\|ax\|: x \in \mathcal{A}, \|x\| \leqslant 1\}$. Then $\|ax\| \leqslant \|a\| \|x\|$ for any x in $\mathcal{A}$; hence $\alpha \leqslant \|a\|$. If $x = a^*/\|a\|$, then $\|x\| = 1$ by the preceding proposition. For this x, $\|ax\| = \|a\|$, so $\alpha = \|a\|$. The proof of the other equality is similar. ∎

This last proposition has an alternative formulation that is useful. If $a \in \mathcal{A}$, define $L_a: \mathcal{A} \to \mathcal{A}$ by $L_a(x) = ax$. By (1.8), $L_a \in \mathcal{B}(\mathcal{A})$ and $\|L_a\| = \|a\|$. If $\rho: \mathcal{A} \to \mathcal{B}(\mathcal{A})$ is defined by $\rho(a) = L_a$, then ρ is a homomorphism and an isometry. That is, $\mathcal{A}$ is isometrically isomorphic to a subalgebra of $\mathcal{B}(\mathcal{A})$. The map ρ is called the *left regular representation* of $\mathcal{A}$.

The left regular representation can be used to discuss the process of adjoining an identity to $\mathcal{A}$. Since $\mathcal{A}$ is isomorphic to a subalgebra $\mathcal{B}(\mathcal{A})$ and $\mathcal{B}(\mathcal{A})$ has an identity, why not just look at the subalgebra of $\mathcal{B}(\mathcal{A})$ generated by $\mathcal{A}$ and the identity operator? Why not, indeed. This is just what is done below.

If $\mathcal{A}$ and $\mathcal{C}$ are Banach algebras and $v: \mathcal{A} \to \mathcal{C}$, then v is *-homomorphism if v is an algebra homomorphism such that $v(a^*) = v(a)^*$ for all a in $\mathcal{A}$.

1.9. Proposition. *If $\mathcal{A}$ is a C^*-algebra, then there is a C^*-algebra $\mathcal{A}_1$ with an identity such that $\mathcal{A}_1$ contains $\mathcal{A}$ as an ideal. If $\mathcal{A}$ does not have an identity, then $\mathcal{A}_1/\mathcal{A}$ is one dimensional. If $\mathcal{C}$ is a C^*-algebra with identity, and $v: \mathcal{A} \to \mathcal{C}$*

is a *-homomorphism, then $v_1: \mathscr{A}_1 \to \mathscr{C}$, defined by $v_1(a + \alpha) = v(a) + \alpha$ for a in $\mathscr{A}$ and α in $\mathbb{C}$, is a *-homomorphism.

PROOF. It may be assumed that $\mathscr{A}$ does not have an identity. Let $\mathscr{A}_1 = \{a + \alpha: a \in \mathscr{A}, \alpha \in \mathbb{C}\}$ $(a + \alpha$ is just a formal sum). Define multiplication and addition in the obvious way. Let $(a + \alpha)^* = a^* + \bar{\alpha}$ and define the norm on $\mathscr{A}_1$ by

$$\|a + \alpha\| = \sup\{\|ax + \alpha x\|: x \in \mathscr{A}, \|x\| \leqslant 1\}.$$

Clearly, this is a norm on $\mathscr{A}_1$. It must be shown that $\|y^*y\| = \|y\|^2$ for every y in $\mathscr{A}_1$.

Fix a in $\mathscr{A}$ and α in $\mathbb{C}$. If $\varepsilon > 0$, then there is an x in $\mathscr{A}$ such that $\|x\| \leqslant 1$ and

$$\|a + \alpha\|^2 - \varepsilon < \|ax + \alpha x\|^2 = \|(x^*a^* + \bar{\alpha}x^*)(ax + \alpha x)\|$$
$$= \|x^*(a + \alpha)^*(a + \alpha)x\| \leqslant \|(a + \alpha)^*(a + \alpha)\|.$$

Thus $\|a + \alpha\|^2 \leqslant \|(a + \alpha)^*(a + \alpha)\|$.

It is left to the reader to prove that the norm on $\mathscr{A}_1$ makes $\mathscr{A}_1$ a Banach algebra. For the other inequality, note that $\|(a + \alpha)^*(a + \alpha)\| \leqslant \|(a + \alpha)^*\| \|a + \alpha\|$. So the proof will be complete if it can be shown that $\|(a + \alpha)^*\| \leqslant \|a + \alpha\|$. Now if $x, y \in \mathscr{A}$ and $\|x\|, \|y\| \leqslant 1$, then $\|y(a + \alpha)^*x\| = \|ya^*x + \bar{\alpha}yx\| = \|x^*ay^* + \alpha x^*y^*\| = \|x^*(a + \alpha)y^*\| \leqslant \|a + \alpha\|$. Taking the supremum over all such x, y gives the desired inequality.

It remains to prove the statement concerning the *-homomorphism v, that $v(a^*) = v(a)^*$. The details are left to the reader. ∎

If $\mathscr{A}$ is a C*-algebra with identity and $a \in \mathscr{A}$, then $\sigma(a)$, the spectrum of a, is well defined. If $\mathscr{A}$ does not have an identity, $\sigma(a)$ is defined as the spectrum of a as an element of the C*-algebra $\mathscr{A}_1$ obtained in Proposition 1.9.

1.10. Definition. If $\mathscr{A}$ is a C*-algebra and $a \in \mathscr{A}$, then (a) a is hermitian if $a = a^*$; (b) a is normal if $a^*a = aa^*$; (c) a is unitary if $a^*a = aa^* = 1$ (this only makes sense if $\mathscr{A}$ has an identity).

1.11. Proposition. Let $\mathscr{A}$ be a C*-algebra and let $a \in \mathscr{A}$.

(a) If a is invertible, then a^* is invertible and $(a^*)^{-1} = (a^{-1})^*$.
(b) $a = x + iy$ where x and y are hermitian elements of $\mathscr{A}$.
(c) If u is a unitary element of $\mathscr{A}$, $\|u\| = 1$.
(d) If $\mathscr{B}$ is a C*-algebra and $\rho: \mathscr{A} \to \mathscr{B}$ is a *-homomorphism, then $\|\rho(a)\| \leqslant \|a\|$.
(e) If $a = a^*$, then $\|a\| = r(a)$.

PROOF. The proofs of (a), (b), and (c) are left as exercises.

(e) Since $a^* = a$, $\|a^2\| = \|a^*a\| = \|a\|^2$; by induction, $\|a^{2^n}\| = \|a\|^{2^n}$ for $n \geqslant 1$. That is, $\|a^{2^n}\|^{1/2^n} = \|a\|$ for $n \geqslant 1$. Hence $r(a) = \lim \|a^{2^n}\|^{1/2^n} = \|a\|$.

(d) If $\mathscr{A}$ has an identity, it is not assumed that $\rho(1) =$ the identity of $\mathscr{B}$. However, it is easy to see that $\rho(1)$ is the identity for cl $\rho(\mathscr{A})$. If $\mathscr{A}$ does not have an identity, then ρ can be extended to a *-homomorphism $\rho_1: \mathscr{A}_1 \to \mathscr{B}_1$

such that $\rho_1(1) = 1$ (1.9). Thus it suffices to prove the proposition under the additional assumption that $\mathscr{A}$ and $\mathscr{B}$ have identities and $\rho(1) = 1$.

If $x \in \mathscr{A}$, then it follows that $\sigma(\rho(x)) \subseteq \sigma(x)$ (Verify!) and hence $r(\rho(x)) \leqslant r(x)$. So, using part (e) and the fact that a^*a is hermitian, $\|\rho(a)\|^2 = \|\rho(a^*a)\| = r(\rho(a^*a)) \leqslant r(a^*a) = \|a^*a\| = \|a\|^2$. ∎

1.12. Proposition. *If $\mathscr{A}$ is a C^*-algebra and $h\colon \mathscr{A} \to \mathbb{C}$ is a non-zero homomorphism, then:*

(a) $h(a) \in \mathbb{R}$ *whenever* $a = a^*$;
(b) $h(a^*) = \overline{h(a)}$ *for all a in $\mathscr{A}$;*
(c) $h(a^*a) \geqslant 0$ *for all a in $\mathscr{A}$;*
(d) *if $1 \in \mathscr{A}$ and u is unitary, then $|h(u)| = 1$.*

PROOF. If $\mathscr{A}$ has no identity, extend h to $\mathscr{A}_1$ by letting $h(1) = 1$. Thus, we may assume that $\mathscr{A}$ has an identity. By Exercise VII.8.1, $\|h\| = 1$. If $a = a^*$ and $t \in \mathbb{R}$,

$$|h(a + it)|^2 \leqslant \|a + it\|^2 = \|(a + it)^*(a + it)\|$$
$$= \|(a - it)(a + it)\|$$
$$= \|a^2 + t^2\| \leqslant \|a\|^2 + t^2.$$

If $h(a) = \alpha + i\beta, \alpha, \beta$ in $\mathbb{R}$, then this yields

$$\|a\|^2 + t^2 \geqslant |\alpha + i(\beta + t)|^2$$
$$= \alpha^2 + (\beta + t)^2$$
$$= \alpha^2 + \beta^2 + 2\beta t + t^2;$$

hence $\|a\|^2 \geqslant \alpha^2 + \beta^2 + 2\beta t$ for all t in $\mathbb{R}$. If $\beta \neq 0$, then letting $t \to \pm \infty$, depending on the sign of β, gives a contradiction. Therefore $\beta = 0$ or $h(a) \in \mathbb{R}$. This proves (a).

Let $a = x + iy$, where x and y are hermitian. Since $h(x), h(y) \in \mathbb{R}$ by (a) and $a^* = x - iy$, (b) follows. Also, $h(a^*a) = h(a^*)h(a) = |h(a)|^2 \geqslant 0$, so (c) holds. Finally, if u is unitary, $|h(u)|^2 = h(u^*)h(u) = h(u^*u) = h(1) = 1$. ∎

Note that part (b) of the preceding proposition implies that any homomorphism $h\colon \mathscr{A} \to \mathbb{C}$ is a *-homomorphism. This, coupled with (VII.8.6), gives the following corollary.

1.13. Corollary. *If $\mathscr{A}$ is an abelian C^*-algebra and a is a hermitian element of $\mathscr{A}$, then $\sigma(a) \subseteq \mathbb{R}$.*

This corollary is short-lived as the conclusion remains valid even if $\mathscr{A}$ is not abelian.

1.14. Proposition. *Let $\mathscr{A}$ and $\mathscr{B}$ be C^*-algebras with a common identity and norm such that $\mathscr{A} \subseteq \mathscr{B}$. If $a \in \mathscr{A}$, then $\sigma_{\mathscr{A}}(a) = \sigma_{\mathscr{B}}(a)$.*

PROOF. First assume that a is hermitian and let $\mathscr{C} = C^*(a)$, the C^*-algebra generated by a and 1. So $\mathscr{C}$ is abelian. By Corollary 1.13 $\sigma_{\mathscr{C}}(a) \subseteq \mathbb{R}$. By Theorem VII.5.4, $\sigma_{\mathscr{A}}(a) \subseteq \sigma_{\mathscr{C}}(a) = \partial\sigma_{\mathscr{C}}(a) \subseteq \sigma_{\mathscr{A}}(a)$; so $\sigma_{\mathscr{A}}(a) = \sigma_{\mathscr{C}}(a) \subseteq \mathbb{R}$. By similar reasoning, $\sigma_{\mathscr{B}}(a) = \sigma_{\mathscr{C}}(a)$, and hence $\sigma_{\mathscr{A}}(a) = \sigma_{\mathscr{B}}(a)$.

Now let a be arbitrary. It suffices to show that if a is invertible in $\mathscr{B}$, a is invertible in $\mathscr{A}$. So suppose there is a b in $\mathscr{B}$ such that $ab = ba = 1$. Thus, $(a^*a)(bb^*) = (bb^*)(a^*a) = 1$. Since a^*a is hermitian, the first part of the proof implies a^*a is invertible in $\mathscr{A}$. But inverses are unique, so $bb^* = (a^*a)^{-1} \in \mathscr{A}$. Hence $b = b(b^*a^*) = (bb^*)a^* \in \mathscr{A}$. ∎

This result must, of course, be contrasted with Theorem VII.5.4.

EXERCISES

1. Verify the statements made in Examples 1.2 through 1.6.

2. Let $\mathscr{A} = \{f \in C(\mathrm{cl}\,\mathbb{D}): f \text{ is analytic in } \mathbb{D}\}$ and for f in $\mathscr{A}$ define f^* by $f^*(z) = \overline{f(\bar{z})}$. Show that $\mathscr{A}$ is a Banach algebra, $f^* \in \mathscr{A}$ when $f \in \mathscr{A}$, and $\|f^*\| = \|f\|$, but $\mathscr{A}$ is not a C^*-algebra.

3. If $\{\mathscr{A}_i: i \in I\}$ is a collection of C^*-algebras, show that $\oplus_\infty \mathscr{A}_i$ and $\oplus_0 \mathscr{A}_i$ are C^*-algebras.

4. Let X be a locally compact space and let $\mathscr{A}$ be a C^*-algebra. If $C_b(X, \mathscr{A}) = $ the collection of bounded continuous functions from $X \to \mathscr{A}$, show that $C_b(X, \mathscr{A})$ is a C^*-algebra. Let $C_0(X, \mathscr{A}) = $ all of the continuous functions $f: X \to \mathscr{A}$ such that for every $\varepsilon > 0, \{x \in X: \|f(x)\| \geq \varepsilon\}$ is compact. Show that $C_0(X, \mathscr{A})$ is a C^*-algebra.

§2. Abelian C^*-Algebras and the Functional Calculus in C^*-Algebras

The next theorem is the basic result of this section. It will be used to develop a functional calculus for normal elements that extends the Riesz Functional Calculus.

2.1. Theorem. *If $\mathscr{A}$ is an abelian C^*-algebra with identity and Σ is its maximal ideal space, then the Gelfand transform $\gamma: \mathscr{A} \to C(\Sigma)$ is an isometric $*$-isomorphism of $\mathscr{A}$ onto $C(\Sigma)$.*

PROOF. By Theorem VII.8.9, $\|\hat{x}\|_\infty \leq \|x\|$ for every x in $\mathscr{A}$. But $\|\hat{x}\|_\infty$ is the spectral radius of x, so by (1.11e), $\|x\| = \|\hat{x}\|_\infty$ for every hermitian element x of $\mathscr{A}$. In particular, $\|x^*x\| = \|x^*x\|_\infty$ for every x in $\mathscr{A}$.

If $a \in \mathscr{A}$ and $h \in \Sigma$, then $a^*(h) = h(a^*) = \overline{h(a)} = \overline{\hat{a}(h)}$. That is, $\widehat{a^*} = \overline{\hat{a}}$. Equivalently, $\gamma(a^*) = \gamma(a)^*$ since the involution on $C(\Sigma)$ is defined by complex conjugation. Thus, γ is a $*$-homomorphism. Also, $\|a\|^2 = \|a^*a\| = \|\widehat{a^*a}\|_\infty = \| |\hat{a}|^2 \|_\infty = \|\hat{a}\|_\infty^2$; therefore $\|a\| = \|\hat{a}\|_\infty$ and γ is an isometry.

Because γ is an isometry, it has closed range. To show that γ is surjective, therefore, it suffices to show that it has dense range. This is accomplished by applying the Stone–Weierstrass Theorem. Note that $\hat{1} = 1$, so $\gamma(\mathscr{A})$ is a subalgebra of $C(\Sigma)$ containing the constants. Because γ preserves the involution, $\gamma(\mathscr{A})$ is closed under complex conjugation. It remains to show that $\gamma(\mathscr{A})$ separates the points of Σ. But if h_1 and h_2 are distinct homomorphisms in Σ, they are distinct because there is an a in $\mathscr{A}$ such that $h_1(a) \neq h_2(a)$. Hence $\hat{a}(h_1) \neq \hat{a}(h_2)$. ∎

By combining the preceding theorem with Proposition 1.9 and Exercise VII.8.6, the following is obtained.

2.2. Corollary. *If $\mathscr{A}$ is an abelian C*-algebra without identity and Σ is its maximal ideal space, then the Gelfand transform $\gamma: \mathscr{A} \to C_0(\Sigma)$ is an isometric *-isomorphism of $\mathscr{A}$ onto $C_0(\Sigma)$.*

In order to focus our attention on the key concepts and not be distracted by peripheral considerations, we now make the following.

Assumption. *All C*-algebras that are considered have an identity.*

Let $\mathscr{B}$ be an arbitrary C*-algebra and let a be a normal element of $\mathscr{B}$. So if $\mathscr{A} = C^*(a)$, the C*-algebra generated by a (and 1), $\mathscr{A}$ is abelian. Hence $\mathscr{A} \cong C(\Sigma)$, where Σ is the maximal ideal space of $\mathscr{A}$. So by Theorem 2.1 if $f \in C(\Sigma)$, there is a unique element x of $\mathscr{A}$ such that $\hat{x} = f$. We want to think of x as $f(a)$ and thus define a functional calculus for normal elements of a C*-algebra. To be useful, however, we should have a ready way of identifying Σ. Moreover, since $\mathscr{A} = C^*(a)$ and thus depends on a, it should be that Σ depends on a in a clear way. The idea embodied in Proposition VII.8.10 that Σ and $\sigma(a)$ are homeomorphic via a natural map is the key here, although (VII.8.10) is not directly applicable here since a is not a generator of $C^*(a)$ as a Banach algebra but only as a C*-algebra. [If $a = a^*$, then a is a generator of $C^*(a)$ as a Banach algebra.] Nevertheless the result is true.

2.3. Proposition. *If $\mathscr{A}$ is an abelian C*-algebra with maximal ideal space Σ and $a \in \mathscr{A}$ such that $\mathscr{A} = C^*(a)$, then the map $\tau: \Sigma \to \sigma(a)$ defined by $\tau(h) = h(a)$ is a homeomorphism. If $p(z, \bar{z})$ is a polynomial in z and $\bar{z}$ and $\gamma: \mathscr{A} \to C(\Sigma)$ is the Gelfand transform, then $\gamma(p(a, a^*)) = p \circ \tau$.*

The proof of this result follows, with a few variations, along the lines of the proof of Proposition VII.8.10 and is left to the reader.

If $\tau: \Sigma \to \sigma(a)$ is defined as in the preceding proposition, then $\tau^{\#}: C(\sigma(a)) \to C(\Sigma)$ is defined by $\tau^{\#}(f) = f \circ \tau$. Note that $\tau^{\#}$ is a *-isomorphism and an isometry, because τ is a homeomorphism. Note that $\mathscr{A} = C^*(a)$ is the closure of $\{p(a, a^*): p(z, \bar{z})$ is a polynomial in z and $\bar{z}\}$. Now such a polynomial $p(z, \bar{z})$

is, of course, a function on $\sigma(a)$. [Just evaluate the polynomial at any z in $\sigma(a)$.] The last part of (2.3) says that $\gamma(p(a, a^*)) = \tau^\#(p)$. We define a map ρ: $C(\sigma(a)) \to C^*(a)$ so that the following diagram commutes:

2.4
$$C^*(a) \xrightarrow{\gamma} C(\Sigma)$$
$$\underset{\rho}{\nwarrow} \qquad \underset{\tau^\#}{\nearrow}$$
$$C(\sigma(a))$$

Note that if $\mathscr{B}$ is any C^*-algebra and a is a normal element of $\mathscr{B}$, then $\mathscr{A} = C^*(a)$ is an abelian C^*-algebra contained in $\mathscr{B}$ and so (2.4) applies. Moreover, in light of Proposition 1.14, the spectrum of a does not depend on whether a is considered as an element of $\mathscr{A}$ or $\mathscr{B}$. The following definition is, therefore, unambiguous.

2.5. Definition. If $\mathscr{B}$ is a C^*-algebra with identity and a is a normal element of $\mathscr{B}$, let ρ: $C(\sigma(a)) \to C^*(a) \subseteq \mathscr{B}$ be as in (2.4). If $f \in C(\sigma(a))$ define

$$f(a) \equiv \rho(f).$$

The map $f \mapsto f(a)$ of $C(\sigma(a)) \to \mathscr{B}$ is called the *functional calculus for a.*

Note that if $p(z, \bar{z})$ is a polynomial in z and $\bar{z}$, then $\rho(p(z, \bar{z})) = p(a, a^*)$. In particular, $\rho(z^n \bar{z}^m) = a^n a^{*m}$ so that $\rho(z) = a$ and $\rho(\bar{z}) = a^*$. Also, $\rho(1) = 1$.

The properties of this functional calculus can be obtained from the fact that ρ is an isometric $*$-isomorphism of $C(\sigma(a))$ into $\mathscr{B}$—with one exception. How does this functional calculus compare with the Riesz Functional Calculus? If $f \in \text{Hol}(a)$, $f|\sigma(a) \in C(\sigma(a))$; so $f(a)$ has two possible interpretations. Or does it?

2.6. Theorem. *If $\mathscr{B}$ is a C^*-algebra and a is a normal element of $\mathscr{B}$, then the functional calculus has the following properties.*

(a) $f \mapsto f(a)$ *is a $*$-monomorphism.*
(b) $\|f(a)\| = \|f\|_\infty$.
(c) $f \mapsto f(a)$ *is an extension of the Riesz Functional Calculus.*

Moreover, the functional calculus is unique in the sense that if τ: $C(\sigma(a)) \to C^*(a)$ is a $*$-homomorphism that extends the Riesz Functional Calculus, then $\tau(f) = f(a)$ for every f in $C(\sigma(a))$.

PROOF. Let ρ: $C(\sigma(a)) \to C^*(a)$ be the map defined by $\rho(f) = f(a)$. From (2.4), (a) and (b) are immediate.

Let π: $\text{Hol}(a) \to \mathscr{A} \subseteq \mathscr{B}$ denote the map defined by the Riesz Functional Calculus. Since $\rho(z) = \pi(z) = a$, an algebraic manipulation gives that $\rho(f) = \pi(f)$ for every rational function f with poles off $\sigma(a)$. If $f \in \text{Hol}(a)$, then by Runge's Theorem there is a sequence $\{f_n\}$ of such rational functions such that $f_n(z) \to f(z)$ uniformly in a neighborhood of $\sigma(a)$. Thus $\pi(f_n) \to \pi(f)$. By (b), $\rho(f_n) \to \rho(f)$. Thus $\rho(f) = \pi(f)$.

To prove uniqueness, let τ: $C(\sigma(a)) \to \mathscr{B}$ be a $*$-homomorphism that extends the Riesz Functional Calculus. If $f \in C(\sigma(a))$, then there is a sequence $\{p_n\}$ of polynomials in z and $\bar{z}$ such that $p_n(z, \bar{z}) \to f(z)$ uniformly on $\sigma(a)$. But $\tau(p_n) = p_n(a, a^*)$, $\tau(p_n) \to \tau(f)$ (1.11d), and $p_n(a, a^*) \to f(a)$. Hence $\tau(f) = f(a)$. ∎

Because of the uniqueness statement in the preceding theorem, it is not necessary to remember the form of the functional calculus $f \mapsto f(a)$, but only the fact that it is an isometric *-monomorphism that extends the Riesz Functional Calculus. Indeed, by the uniqueness of the Riesz Functional Calculus, it suffices to have that $f \mapsto f(a)$ is an isometric *-monomorphism such that if $f(z) \equiv 1$, then $f(a) = 1$, and if $f(z) = z$, then $f(a) = a$. Any properties or applications of the functional calculus can be derived or justified using only these properties. There may, however, be an occasion when the precise form of the functional calculus [viz., (2.4)] facilitates a proof. There are also situations in which the definition of the functional calculus gets in the way of a proof and the properties in (2.6) give the clean way of applying this powerful tool.

2.7. Spectral Mapping Theorem. *If $\mathscr{A}$ is a C*-algebra and a is a normal element of $\mathscr{A}$, then for every f in $C(\sigma(a))$,*

$$\sigma(f(a)) = f(\sigma(a)).$$

PROOF. Let $\rho: C(\sigma(a)) \to C^*(a)$ be defined by $\rho(f) = f(a)$. So ρ is a *-isomorphism. Hence $\sigma(f(a)) = \sigma(\rho(f)) = \sigma(f)$. But $\sigma(f) = f(\sigma(a))$ (VII.3.2). ∎

Once again (1.14) was used implicitly in the preceding proof.

EXERCISES

1. Prove a converse to Proposition 2.3. If K is a compact subset of $\mathbb{C}$, $C(K)$ is a singly generated C*-algebra.

2. If $\mathscr{A}$ is an abelian C*-algebra with a finite number of C*-generators $a_1, \ldots, a_n$, then there is a compact subset X of $\mathbb{C}^n$ and an isometric *-isomorphism $\rho: \mathscr{A} \to C(X)$ such that $\rho(a_k) = z_k$, $1 \leq k \leq n$, where $z_k(\lambda_1, \ldots, \lambda_n) = \lambda_k$ (see Exercise VII.8.12).

3. If X is a compact Hausdorff space, show that X is totally disconnected if and only if $C(X)$ is the closed linear span of its projections ($\equiv$ hermitian idempotents).

4. Using the terminology of Exercise 3, show that if (X, Ω, μ) is a σ-finite measure space, the maximal ideal space of $L^\infty(X, \Omega, \mu)$ is totally disconnected.

5. If $\mathscr{A}$ is a C*-algebra with identity and $a = a^*$, show that $\exp(ia) = u$ is unitary. Is the converse true?

6. Let X be compact and fix a point x_0 in X. Let $\mathscr{A} = \{\{f_n\}: f_n \in C(X), \sup_n \|f_n\| < \infty,$ and $\{f_n(x_0)\}$ is a convergent sequence$\}$. Show that $\mathscr{A}$ is an abelian C*-algebra with identity and find its maximal ideal space.

7. If X is completely regular, then $C_b(X)$ is a C^*-algebra and its maximal ideal space is the Stone–Čech compactification of X.

§3. The Positive Elements in a C^*-Algebra

This section is an application of the functional calculus developed in the preceding section. The results here are very useful in the study of operators on Hilbert space and they demonstrate the power of the functional calculus.

If $\mathscr{A}$ is a C^*-algebra, let $\mathrm{Re}\,\mathscr{A}$ denote the hermitian elements of $\mathscr{A}$.

3.1. Definition. If $\mathscr{A}$ is a C^*-algebra and $a \in \mathscr{A}$, then a is *positive* if $a \in \mathrm{Re}\,\mathscr{A}$ and $\sigma(a) \subseteq [0, \infty)$. If a is positive, this is denoted by $a \geqslant 0$. Let $\mathscr{A}_+$ be the set of all positive elements of $\mathscr{A}$.

3.2. Example. If $\mathscr{A} = C(X)$, then f is positive in $\mathscr{A}$ if and only if $f(x) \geqslant 0$ for all x in X.

3.3. Example. If $\mathscr{A} = L^\infty(\mu)$ and $f \in L^\infty(\mu)$, then $f \geqslant 0$ if and only if $f(x) \geqslant 0$ a.e. $[\mu]$.

3.4. Proposition. *If $a \in \mathrm{Re}\,\mathscr{A}$, then there are unique positive elements u, v in $\mathscr{A}$ such that $a = u - v$ and $uv = vu = 0$.*

PROOF. Let $f(t) = \max(t, 0)$, $g(t) = -\min(t, 0)$. Then $f, g \in C(\mathbb{R})$ and $f(t) - g(t) = t$. Using the functional calculus, let $u = f(a)$ and $v = g(a)$. So u and v are hermitian and by the Spectral Mapping Theorem $u, v \geqslant 0$. Also, $u - v = f(a) - g(a) = a$ and $uv = vu = (gf)(a) = 0$ since $fg \equiv 0$.

To show uniqueness, let $u_1, v_1 \in \mathscr{A}_+$ such that $u_1 - v_1 = a$ and $u_1 v_1 = v_1 u_1 = 0$. Let $\{p_n\}$ be a sequence of polynomials such that $p_n(0) = 0$ for all n and $p_n(t) \to f(t)$ uniformly on $\sigma(a)$. Hence $p_n(a) \to u$ in $\mathscr{A}$. But $u_1 a = au_1$. So $u_1 p_n(a) = p_n(a)u_1$ for all n; hence $u_1 u = uu_1$. Similarly, it follows that a, u, v, u_1, and v_1 are pairwise commuting hermitian elements of $\mathscr{A}$. Let $\mathscr{B} = $ the C^*-algebra generated by a, u, v, u_1, and v_1; so $\mathscr{B}$ is abelian. Hence $\mathscr{B} \cong C(\Sigma)$ where Σ is the maximal ideal space of $\mathscr{B}$. The uniqueness now follows from the uniqueness statement for $C(\Sigma)$ (Exercise 1). ∎

The next result follows in a similar way.

3.5. Proposition. *If $a \in \mathscr{A}_+$ and $n \geqslant 1$, there is a unique element b in $\mathscr{A}_+$ such that $a = b^n$.*

The decomposition $a = u - v$ of a hermitian element a is sometimes called the orthogonal decomposition of a. The elements u and v are called the *positive* and *negative parts* of a and are denoted by $u = a_+$ and $v = a_-$. Note that $a_- \geqslant 0$.

If $a \in \mathscr{A}_+$, then the unique b obtained in (3.5) is called the *nth root* of a and is denoted by $b = a^{1/n}$. Note that if b is not assumed to be positive, it is not necessarily unique (see Exercise 5).

If X is compact and $f \in C(X)_+$, then notice that $|f(x) - t| \leqslant t$ for every real number $t \geqslant \|f\|$. Conversely, if $|f(x) - t| \leqslant t$ for some $t \geqslant \|f\|$, then $f(x) \geqslant 0$ for all x and so $f \geqslant 0$. These observations are behind some of the statements in the next result.

3.6. Theorem. *If $\mathscr{A}$ is a C*-algebra and $a \in \mathscr{A}$, then the following statements are equivalent.*

(a) $a \geqslant 0$.
(b) $a = b^2$ *for some b in* $\operatorname{Re}\mathscr{A}$.
(c) $a = x^*x$ *for some x in* $\mathscr{A}$.
(d) $a = a^*$ *and* $\|t - a\| \leqslant t$ *for all* $t \geqslant \|a\|$.
(e) $a = a^*$ *and* $\|t - a\| \leqslant t$ *for some* $t \geqslant \|a\|$.

PROOF. It is clear that (b) implies (c) and (d) implies (e). By (3.5), (a) implies (b).

(e) $\Rightarrow$ (a): Since $a = a^*$, $C^*(a)$ is abelian. If $X = \sigma(a)$, $X \subseteq \mathbb{R}$ and $f \mapsto f(a)$ is a *-isomorphism of $C(X)$ onto $C^*(a)$. Using this isomorphism and (e), $|t - x| \leqslant t$ for some $t \geqslant \|a\| = \sup\{|s| : s \in X\}$ and all x in X. From the discussions preceding this theorem (with $f(x) = x$), $x \geqslant 0$ for all x in X. That is, $X = \sigma(a) \subseteq [0, \infty)$. Hence $a \geqslant 0$.

(a) $\Rightarrow$ (d): This proof follows the lines of the preceding paragraph and is left to the reader.

(c) $\Rightarrow$ (a): If $a = x^*x$ for some x in $\mathscr{A}$, then it is clear that $a = a^*$. Let $a = u - v$, where $u, v \geqslant 0$ and $uv = vu = 0$. It must be shown that $v = 0$.

If $xv^{1/2} = b + ic$, where $b, c \in \operatorname{Re}\mathscr{A}$, then $(xv^{1/2})^*(xv^{1/2}) = (b - ic)(b + ic) = b^2 + c^2 + i(bc - cb)$. But also $(xv^{1/2})^*(xv^{1/2}) = v^{1/2}x^*xv^{1/2} = v^{1/2}(u - v)v^{1/2} = -v^2$. Hence $i(bc - cb) = -v^2 - b^2 - c^2$. By Proposition 3.7 below, $i(bc - cb) \leqslant 0$. Also $(xv^{1/2})^*(xv^{1/2}) = -v^2 \leqslant 0$ because (a) and (b) are equivalent. By Exercise VII.3.7, $(xv^{1/2})(xv^{1/2})^* \leqslant 0$. Put $(xv^{1/2})(xv^{1/2})^* = -y$, where $y \in \mathscr{A}_+$. So $-y = (b + ic)(b - ic) = b^2 + c^2 - i(bc - cb)$. Hence $i(bc - cb) = b^2 + c^2 + y \in \mathscr{A}_+$ by (3.7). Therefore $i(bc - cb) \in (-\mathscr{A}_+) \cap \mathscr{A}_+ = (0)$. But this implies that $-v^2 = (xv^{1/2})^*(xv^{1/2}) = b^2 + c^2 \in (-\mathscr{A}_+) \cap \mathscr{A}_+$, so that $v^2 = 0$. But $v \leqslant 0$ so $v = 0$. That is, $a = u \geqslant 0$. ∎

The next result will be proved only using the equivalence of (a), (d), and (e) from the preceding theorem.

3.7. Proposition. *If $\mathscr{A}$ is a C*-algebra, then $\mathscr{A}_+$ is a closed cone.*

PROOF. Let $\{a_n\} \subseteq \mathscr{A}_+$ and suppose $a_n \to a$. Clearly $a \in \operatorname{Re}\mathscr{A}$. By (3.6d), $\|a_n - \|a_n\|\| \leqslant \|a_n\|$. Hence $\|a - \|a\|\| \leqslant \|a\|$, so by (3.6e), $a \geqslant 0$.

Clearly, $\lambda a \geqslant 0$ if $a \geqslant 0$ and $\lambda \geqslant 0$. Let $a, b \in \mathscr{A}_+$; it must be shown that $a + b \geqslant 0$. It suffices to assume that $\|a\|, \|b\| \leqslant 1$. But $\|1 - \frac{1}{2}(a + b)\| = \frac{1}{2}\|(1 - a) + (1 - b)\| \leqslant 1$ by (3.6d). So by (3.6e), $\frac{1}{2}(a + b) \geqslant 0$.

If $a \in \mathscr{A}_+ \cap (-\mathscr{A}_+)$, then $a = a^*$ and $\sigma(a) = \{0\}$. But $\|a\| = r(a)$ (1.11e). ∎

Now to look at one more example—a very important one.

3.8. Theorem. *If $\mathscr{H}$ is a Hilbert space and $A \in \mathscr{B}(\mathscr{H})$, then $A \geqslant 0$ if and only if $\langle Ah, h \rangle \geqslant 0$ for all h in $\mathscr{H}$.*

PROOF. If $A \geqslant 0$, then (3.6c) $A = T^*T$ for some T in $\mathscr{B}(\mathscr{H})$. Hence $\langle Ah, h \rangle = \|Th\|^2 \geqslant 0$. Conversely, suppose $\langle Ah, h \rangle \geqslant 0$ for all h in $\mathscr{H}$. By (II.2.12), $A = A^*$. It remains to show that $\sigma(A) \subseteq [0, \infty)$. If $h \in \mathscr{H}$ and $\lambda < 0$, then

$$\|(A - \lambda)h\|^2 = \|Ah\|^2 - 2\lambda \langle Ah, h \rangle + \lambda^2 \|h\|^2$$
$$\geqslant -2\lambda \langle Ah, h \rangle + \lambda^2 \|h\|^2 \geqslant \lambda^2 \|h\|^2$$

since $\lambda < 0$ and $\langle Ah, h \rangle \geqslant 0$. By (VII.6.4), $\lambda \notin \sigma_{ap}(A)$. But this implies that $A - \lambda$ is left invertible (Exercise VII.6.5). Since $A - \lambda$ is self-adjoint, $A - \lambda$ is also right invertible. Thus $\lambda \notin \sigma(A)$ and $A \geqslant 0$. ∎

3.9. Definition. If $\mathscr{A}$ is a C*-algebra and $a, b \in \operatorname{Re} \mathscr{A}$, then $a \leqslant b$ if $b - a \in \mathscr{A}_+$.

This ordering makes a C*-algebra as well as $\operatorname{Re} a$ into ordered vector spaces.

Note that if A and B are hermitian operators on the Hilbert space $\mathscr{H}$, then $A \leqslant B$ if and only if $\langle Ah, h \rangle \leqslant \langle Bh, h \rangle$ for all h in $\mathscr{H}$.

This section closes with an application of positivity to obtain the polar decomposition of an operator. If $\lambda \in \mathbb{C}$, then $\lambda = |\lambda|e^{i\theta}$ for some θ; this is the polar decomposition of λ. Can an analogue be found for operators? To answer this question we might first ask what is the analogue of $|\lambda|$ and $e^{i\theta}$ among operators. If $A \in \mathscr{B}(\mathscr{H})$, then the proper definition for $|A|$ would seem to be $|A| \equiv (A^*A)^{1/2}$ [see (3.5)]. How about an analogue of $e^{i\theta}$? Should it be a unitary operator? An isometry? For an arbitrary operator neither of these is appropriate. The following new class of operators is needed.

3.10. Definition. A *partial isometry* is an operator W such that for h in $(\ker W)^{\perp}$, $\|Wh\| = \|h\|$. The space $(\ker W)^{\perp}$ is called the *initial space* of W and the space $\operatorname{ran} W$ is called the *final space* of W. See Exercises 15–20 for more on partial isometries.

3.11. Polar Decomposition. *If $A \in \mathscr{B}(\mathscr{H})$, then there is a partial isometry W with $(\ker A)^{\perp}$ as its initial space and $\operatorname{cl}(\operatorname{ran} A)$ as its final space such that $A = W|A|$. Moreover, if $A = UP$ where $P \geqslant 0$ and U is a partial isometry with $\ker U = \ker P$, then $P = |A|$ and $U = W$.*

PROOF. If $h \in \mathcal{H}$, then $\|Ah\|^2 = \langle Ah, Ah \rangle = \langle A^*Ah, h \rangle = \langle |A|h, |A|h \rangle$. Thus

3.12 $$\|Ah\|^2 = \||A|h\|^2.$$

Since $(\operatorname{ran} A^*)^\perp = \ker A$, $\operatorname{ran} A^*$ is dense in $(\ker A)^\perp$. If $f \in \operatorname{ran} A^*$, $f = A^*g$ for g in $(\ker A^*)^\perp = \operatorname{cl} \operatorname{ran} A$. Therefore, $\{A^*Ak \colon k \in \mathcal{H}\}$ is dense in $\operatorname{cl}[\operatorname{ran} A^*] = (\ker A)^\perp$. But $A^*Ak = |A|^2 k = |A|h$, where $h = |A|k$. That is, $\{|A|h \colon h \in \mathcal{H}\}$ is dense in $(\ker A)^\perp$. If $W \colon \operatorname{ran} |A| \to \operatorname{ran} A$ is defined by

3.13 $$W(|A|h) = Ah,$$

then (3.12) implies that W is a well-defined isometry. Thus W extends to an isometry $W \colon (\ker A)^\perp \to \operatorname{cl}(\operatorname{ran} A)$. If Wh is defined to be 0 for h in $\ker A$, W is a partial isometry. By (3.13), $W|A| = A$.

For the uniqueness, note that $A^*A = PU^*UP$. Now $U^*U = E \equiv$ the projection onto the initial space of U (Exercise 16), $(\ker U)^\perp = (\ker P)^\perp = \operatorname{cl}(\operatorname{ran} P)$. Thus $A^*A = PEP = P^2$. By the uniqueness of the positive square root, $P = |A|$. Since $A = U|A|$, $U|A|h = Ah = W|A|h$. That is, U and W agree on a dense subset of their common initial space. Hence $U = W$. ∎

EXERCISES

1. Prove the uniqueness statement in Proposition 3.4 for the case that $\mathcal{A}$ is abelian.

2. Prove Proposition 3.5.

3. Let $A \in \mathcal{B}(L^2(0,1))$ be defined by $(Af)(t) = tf(t)$. Show that $A \geq 0$ and find $A^{1/n}$.

4. Let (X, Ω, μ) be a σ-finite measure space, let $\phi \in L^\infty(X, \Omega, \mu)$, and define M_ϕ as in Theorem II.1.5. Show that $M_\phi \geq 0$ if and only if $\phi(x) \geq 0$ a.e. $[\mu]$. What is $M_\phi^{1/n}$? If $M_\phi \in \operatorname{Re} \mathcal{B}(\mathcal{H})$, find the positive and negative parts of M_ϕ.

5. Find an example of a positive operator on a Hilbert space that has a nonhermitian square root.

6. If $a \in \operatorname{Re} \mathcal{A}$, show that $|a| \equiv (a^2)^{1/2} = a_+ + a_-$.

7. If $a \in \mathcal{A}_+$, show that $x^*ax \in \mathcal{A}_+$ for every x in $\mathcal{A}$.

8. If $a, b \in \mathcal{A}$, $0 \leq a \leq b$, and a is invertible, then b is invertible and $b^{-1} \leq a^{-1}$.

9. If $a, b \in \operatorname{Re} \mathcal{A}$, $a \leq b$, and $ab = ba$, then $f(a) \leq f(b)$ for every increasing continuous function f on $\mathbb{R}$.

10. If $a \in \operatorname{Re} \mathcal{A}$ and $\|a\| \leq 1$, show that a is the sum of two unitaries. (Hint: First solve this for $\mathcal{A} = \mathbb{C}$.)

11. If $\alpha > 0$, define $f_\alpha \colon (-\alpha^{-1}, \infty) \to \mathbb{R}$ by $f_\alpha(t) = t/(1 + \alpha t) = \alpha^{-1}[1 - (1 + \alpha t)^{-1}]$. Show:

   (a) If $0 \leq a \leq b$ in $\mathcal{A}$, $f_\alpha(a) \leq f_\alpha(b)$ for all $\alpha > 0$;
   (b) $f_\alpha(t) < \min\{t, \alpha^{-1}\}$ for $t > 0$;
   (c) $\lim_{\alpha \to 0} f_\alpha(t) = t$ uniformly on bounded intervals in $[0, \infty)$;
   (d) if $0 \leq \alpha \leq \beta$, $f_\alpha \leq f_\beta$ on $[0, \infty)$;
   (e) $f_\alpha \circ f_\beta = f_{\alpha + \beta}$;
   (f) $\lim_{\alpha \to \infty} \alpha f_\alpha(t) = 1$ uniformly on bounded intervals in $[0, \infty)$.

12. If $a, b \in \mathscr{A}_+$ and $a \leqslant b$, show that $a^\beta \leqslant b^\beta$ for $0 \leqslant \beta \leqslant 1$. ($a^\beta = f(a)$ where $f(t) = t^\beta$.) (Hint: Let f_α be as in Exercise 11 and show that $\int_0^\infty f_\alpha(t)\alpha^{-\beta} d\alpha = \gamma t^\beta$ where $\gamma > 0$. Use the definition of the improper integral and the functional calculus.)

13. Give an example of a C^*-algebra $\mathscr{A}$ and positive elements a, b in $\mathscr{A}$ such that $a \leqslant b$ but $b^2 - a^2 \notin \mathscr{A}_+$.

14. Let $\mathscr{A} = \mathscr{B}(l^2)$, let $a =$ the unilateral shift on l^2, and let $b = a^*$. Show that $\sigma(ab) \neq \sigma(ba)$.

15. Let $W \in \mathscr{B}(\mathscr{H})$ and show that the following statements are equivalent: (a) W is a partial isometry; (b) W^* is a partial isometry; (c) W^*W is a projection; (d) WW^* is a projection; (e) $WW^*W = W$; (f) $W^*WW^* = W^*$.

16. If W is a partial isometry, show that W^*W is the projection onto the initial space of W and WW^* is the projection onto the final space of W.

17. If W_1, W_2 are partial isometries, define $W_1 \lesssim W_2$ to mean that $W_1^*W_1 \leqslant W_2^*W_2$, $W_1 W_1^* \leqslant W_2 W_2^*$, and $W_2 h = W_1 h$ whenever h is in the initial space of W_1. Show that $\lesssim$ is a partial ordering on the set of partial isometries and that a partial isometry W is a maximal element in this ordering if and only if either W or W^* is an isometry.

18. Using the terminology of Exercise 17, show that the extreme points of ball $\mathscr{B}(\mathscr{H})$ are the maximal partial isometries. (See Exercise V.7.10.)

19. Find the polar decomposition of each of the following operators: (a) M_ϕ as defined in (II.1.5); (b) the unilateral shift; (c) the weighted unilateral shift $[A(x_1, x_2, \ldots) = (0, \alpha_1 x_1, \alpha_2 x_2, \ldots)$ for x in l^2 and $\sup_n |\alpha_n| < \infty]$ with nonzero weights; (d) $A \oplus \alpha$ (in terms of the polar decomposition of A).

20. Let $A \in \mathscr{B}(\mathscr{H})$ such that $\ker A = (0)$ and $A \geqslant 0$ and define S on $\mathscr{K} = \mathscr{H} \oplus \mathscr{H} \oplus \cdots$ by $S(h_1, h_2, \ldots) = (0, Ah_1, Ah_2, \ldots)$. Find the polar decomposition of $S, S = W|S|$, and show that $S = |S|W$.

21. Show that the parts of the polar decomposition of a normal operator commute.

22. If $A \in \mathscr{B}(\mathscr{H})$, show that there is a positive operator P and a partial isometry W such that $A = PW$. Discuss the uniqueness of P and W.

23. If A is normal and invertible, show that the parts of the polar decomposition of A belong to $C^*(A)$.

24. Give an example of a normal operator A such that the partial isometry in the polar decomposition of A does not belong to $C^*(A)$.

25. Show that for an arbitrary C^*-algebra $\mathscr{A}$ it is not necessarily true that $ab \in \mathscr{A}_+$ whenever a and $b \in \mathscr{A}_+$. If, however, $a, b \in \mathscr{A}_+$ and $ab = ba$, then $ab \in \mathscr{A}_+$.

§4*. Ideals and Quotients of C*-Algebras

We begin with a basic result.

4.1. Proposition. *If I is a closed left or right ideal in the C*-algebra $\mathscr{A}$, $a \in I$ with $a = a^*$, and if $f \in C(\sigma(a))$ with $f(0) = 0$, then $f(a) \in I$.*

PROOF. Note that if I is proper, then $0 \in \sigma(a)$ since a cannot be invertible. Since $\sigma(a) \subseteq \mathbb{R}$, the Weierstrass Theorem implies there is a sequence $\{p_n\}$ of polynomials such that $p_n(t) \to f(t)$ uniformly for t in $\sigma(a)$. Hence $p_n(0) \to f(0) = 0$. Thus $q_n(t) = p_n(t) - p_n(0) \to f(t)$ uniformly on $\sigma(a)$ and $q_n(0) = 0$ for all n. Thus $q_n(a) \in I$ and by the functional calculus, $\| q_n(a) - f(a) \| \to 0$. Hence $f(a) \in I$. ∎

4.2. Corollary. *If I is a closed left or right ideal, $a \in I$ with $a = a^*$, then a_+, a_-, $|a|$, and $|a|^{1/2} \in I$.*

Note that if I is a left ideal of $\mathscr{A}$, then $\{a^*: a \in I\}$ is a right ideal. Therefore a left ideal I is an ideal if $a^* \in I$ whenever $a \in I$.

4.3. Theorem. *If I is a closed ideal in the C*-algebra $\mathscr{A}$, then $a^* \in I$ whenever $a \in I$.*

PROOF. Fix a in I. Thus $a^*a \in I$ since I is an ideal. The idea is to construct a sequence $\{u_n\}$ of continuous functions defined on $[0, \infty)$ such that

4.4
$$(i) \quad u_n(0) = 0 \text{ and } u_n(t) \geq 0 \text{ for all } t;$$
$$(ii) \quad \| a u_n(a^*a) - a \| \to 0 \text{ as } n \to \infty.$$

Note that if such a sequence $\{u_n\}$ can be constructed, then $u_n(a^*a) \geq 0$ and $u_n(a^*a) \in I$ by Proposition 4.1. Also, $u_n(a^*a)a^* \in I$ since I is an ideal and $\| u_n(a^*a)a^* - a^* \| = \| a u_n(a^*a) - a \| \to 0$ by (ii). Thus $a^* \in I$ whenever $a \in I$. It remains to construct the sequence $\{u_n\}$.

Note that

$$\| a u_n(a^*a) - a \|^2$$
$$= \| [a u_n(a^*a) - a]^* [a u_n(a^*a) - a] \|$$
$$= \| u_n(a^*a)a^*a u_n(a^*a) - a^* a u_n(a^*a) - u_n(a^*a)a^*a + a^*a \|.$$

If $b = a^*a$, then the fact that $b u_n(b) = u_n(b)b$ implies that $\| a u_n(a^*a) - a \|^2 = \| f_n(b) \| \leq \sup\{|f_n(t)|: t \geq 0\}$, where $f_n(t) = t u_n(t)^2 - 2t u_n(t) + t = t[u_n(t) - 1]^2$. If $u_n(t) = nt$ for $0 \leq t \leq n^{-1}$ and $u(t) = 1$ for $t \geq n^{-1}$, then it is seen that $\sup\{|f_n(t)|: t \geq 0\} = 4/27n \to 0$ as $n \to \infty$; so (4.4) is satisfied. ∎

Notice that the construction of the sequence $\{u_n\}$ satisfying (4.4) actually proves more. It shows that there is a "local" approximate identity. That is, the proof of the preceding theorem shows that the following holds.

4.5. Proposition. *If $\mathscr{A}$ is a C*-algebra and I is an ideal of $\mathscr{A}$, then for every a in I there is a sequence $\{e_n\}$ of positive elements in I such that:*

(a) $e_1 \leqslant e_2 \leqslant \cdots$ *and* $\|e_n\| \leqslant 1$ *for all* n;
(b) $\|ae_n - a\| \to 0$ *as* $n \to \infty$.

In the preceding proposition the sequence $\{e_n\}$ depends on the element a. It is also true that there is a positive increasing net $\{e_i\}$ in I such that $\|e_i a - a\| \to 0$ and $\|ae_i - a\| \to 0$ for every a in I (see p. 36 of Arveson [1976]).

We turn now to an important consequence of Theorem 4.3.

4.6. Theorem. *If $\mathscr{A}$ is a C*-algebra and I is a closed ideal of $\mathscr{A}$, then for each $a + I$ in $\mathscr{A}/I$ define $(a + I)^* = a^* + I$. Then $\mathscr{A}/I$ with its quotient norm is a C*-algebra.*

To prove (4.6), a lemma is needed.

4.7. Lemma. *If I is an ideal in a C*-algebra $\mathscr{A}$ and $a \in \mathscr{A}$, then $\|a + I\| = \inf\{\|a - ax\|: x \in I, x \geqslant 0, \text{ and } \|x\| \leqslant 1\}$.*

PROOF. If $(\text{ball } I)_+ = \{x \in \text{ball } I: x \geqslant 0\}$, then clearly $\|a + I\| \leqslant \inf\{\|a - ax\|: x \in (\text{ball } I)_+\}$ since $aI \subseteq I$. Let $y \in I$ and let $\{e_n\}$ be a sequence in $(\text{ball } I)_+$ such that $\|y - ye_n\| \to 0$ as $n \to \infty$. Now $0 \leqslant 1 - e_n \leqslant 1$, so $\|(a + y)(1 - e_n)\| \leqslant \|a + y\|$. Hence

$$\|a + y\| \geqslant \liminf \|(a + y)(1 - e_n)\|$$
$$= \liminf \|(a - ae_n) + (y - ye_n)\|$$
$$= \liminf \|a - ae_n\|$$

since $\|y - ye_n\| \to 0$. Thus $\|a + y\| \geqslant \inf_n \|a - ae_n\| \geqslant \inf\{\|a - ax\|: x \in (\text{ball } I)_+\}$. Taking the infimum over all y in I gives the desired remaining inequality. ∎

PROOF OF THEOREM 4.6. The only difficult part of this proof is to show that $\|a + I\|^2 = \|a^*a + I\|$ for every a in $\mathscr{A}$. Since $x^* \in I$ whenever $x \in I$ (4.3), $\|a^* + I\| = \|a + I\|$ for all a in $\mathscr{A}$. Thus the submultiplicativity of the norm in $\mathscr{A}/I$ (VII.2.6) implies

$$\|a^*a + I\| = \|(a^* + I)(a + I)\|$$
$$\leqslant \|a^* + I\| \|a + I\|$$
$$= \|a + I\|^2.$$

On the other hand, the preceding lemma gives that

$$\|a + I\|^2 = \inf\{\|a - ax\|^2: x \in (\text{ball } I)_+\}$$
$$= \inf\{\|a(1 - x)\|^2: x \in (\text{ball } I)_+\}$$
$$= \inf\{\|(1 - x)a^*a(1 - x)\|: x \in (\text{ball } I)_+\}$$

$$\leqslant \inf\{\|a^*a(1-x)\|: x \in (\text{ball } I)_+\}$$
$$= \inf\{\|a^*a - a^*ax\|: x \in (\text{ball } I)_+\}$$
$$= \|a^*a + I\|. \quad \blacksquare$$

If $\mathscr{A}, \mathscr{B}$ are C^*-algebras with ideals I, J, respectively, and $\rho: \mathscr{A} \to \mathscr{B}$ is a *-homomorphism such that $\rho(I) \subseteq J$, then ρ induces a *-homomorphism $\tilde{\rho}$: $\mathscr{A}/I \to \mathscr{B}/J$ defined by $\tilde{\rho}(a + I) = \rho(a) + J$. In particular, if $I = \ker \rho$ and $J = (0)$, then $\tilde{\rho}: \mathscr{A}/\ker \rho \to \mathscr{B}$ is a *-homomorphism and $\tilde{\rho} \circ \pi = \rho$, where π: $\mathscr{A} \to \mathscr{A}/\ker \rho$ is the natural map. Keep these facts in mind when reading the proof of the next result.

4.8. Theorem. *If $\mathscr{A}, \mathscr{B}$ are C^*-algebras and $\rho: \mathscr{A} \to \mathscr{B}$ is a *-homomorphism, then $\|\rho(a)\| \leqslant \|a\|$ for all a and $\mathrm{ran}\, \rho$ is closed in $\mathscr{B}$. If ρ is a *-monomorphism, then ρ is an isometry.*

PROOF. The fact that $\|\rho(a)\| \leqslant \|a\|$ is a restatement of (1.11d). Now assume that ρ is a *-monomorphism. As in the proof of (1.11d), it suffices to assume that $\mathscr{A}$ and $\mathscr{B}$ have identities and $\rho(1) = 1$. (Why?)

If $a \in \mathscr{A}$ and $a = a^*$, then it is easy to see that $\rho(a) = \rho(a)^*$ and $\sigma(\rho(a)) \subseteq \sigma(a)$. If $\sigma(\rho(a)) \neq \sigma(a)$, there is a continuous function f on $\sigma(a)$ such that $f(t) = 0$ for all t in $\sigma(\rho(a))$ but f is not identically zero on $\sigma(a)$. Thus $f(\rho(a)) = 0$, but $f(a) \neq 0$. Let $\{p_n\}$ be polynomials such that $p_n(t) \to f(t)$ uniformly on $\sigma(a)$. Thus $p_n(a) \to f(a)$ and $p_n(\rho(a)) \to f(\rho(a)) = 0$. But $p_n(\rho(a)) = \rho(p_n(a)) \to \rho(f(a))$. Thus $\rho(f(a)) = f(\rho(a)) = 0$. Since ρ was assumed injective, $f(a) = 0$, a contradiction. Hence $\sigma(a) = \sigma(\rho(a))$ if $a = a^*$. Thus by (1.11e), $\|a\| = r(a) = r(\rho(a)) = \|\rho(a)\|$ if $a = a^*$. But then for arbitrary a, $\|a\|^2 = \|a^*a\| = \|\rho(a^*a)\| = \|\rho(a)^*\rho(a)\| = \|\rho(a)\|^2$ and ρ is an isometry.

To complete the proof let $\rho: \mathscr{A} \to \mathscr{B}$ be a *-homomorphism and let $\tilde{\rho}$: $\mathscr{A}/\ker \rho \to \mathscr{B}$ be the induced *-monomorphism. So $\tilde{\rho}$ is an isometry and hence $\mathrm{ran}\, \tilde{\rho}$ is closed. But $\mathrm{ran}\, \tilde{\rho} = \mathrm{ran}\, \rho$. $\quad \blacksquare$

We turn now to some specific examples of C^*-algebras and their ideals.

4.9. Proposition. *If X is compact and I is a closed ideal of $C(X)$, then there is a closed subset F of X such that $I = \{f \in C(X): f(x) = 0 \text{ for all } x \text{ in } F\}$. Moreover, $C(X)/I$ is isometrically isomorphic to $C(F)$.*

PROOF. Let $F = \{x \in X: f(x) = 0 \text{ for all } f \text{ in } I\}$, so F is a closed subset of X. If $\mu \in M(X)$ and $\mu \perp I$, then $\int |f|^2 d\mu = 0$ for every f in I since $|f|^2 = \bar{f}f \in I$ whenever $f \in I$. Thus each f must vanish on the support of μ; hence $|\mu|(X \backslash F) = 0$. Conversely, if $\mu \in M(X)$ and the support of μ is contained in F, $\int f d\mu = 0$ for every f in I. Thus $I^{\perp} = \{\mu \in M(X): |\mu|(X \backslash F) = 0\}$. Since I is closed, $I = {}^{\perp}(I^{\perp}) = \{f \in C(X): f(x) = 0 \text{ for all } x \text{ in } F\}$. The remainder of the proof is left to the reader. $\quad \blacksquare$

4.10. Proposition. *If I is a closed ideal of $\mathscr{B}(\mathscr{H})$, then $I \supseteq \mathscr{B}_0(\mathscr{H})$ or $I = (0)$.*

PROOF. Suppose $I \neq (0)$ and let T be a nonzero operator in I. Thus there are vectors f_0, f_1 in $\mathscr{H}$ such that $Tf_0 = f_1 \neq 0$. Let g_0, g_1 be arbitrary nonzero vectors $\mathscr{H}$. Define $A: \mathscr{H} \to \mathscr{H}$ by letting $Ah = \|g_0\|^{-2}\langle h, g_0 \rangle f_0$. Then $Ag_0 = f_0$ and $Ah = 0$ if $h \perp g_0$. Define $B: \mathscr{H} \to \mathscr{H}$ by letting $Bh = \|f_1\|^{-2}\langle h, f_1 \rangle g_1$. So $Bf_1 = g_1$ and $Bh = 0$ if $h \perp f_1$. Thus $BTAh = 0$ if $h \perp g_0$ and $BTAg_0 = g_1$. Hence for any pair of nonzero vectors g_0, g_1 in $\mathscr{H}$ the rank-one operator that takes g_0 to g_1 and is zero on $[g_0]^{\perp}$ belongs to I. From here it easily follows that I contains all finite-rank operators. Since I is closed, $I \supseteq \mathscr{B}_0(\mathscr{H})$. ∎

It will be shown in (IX.4.2), after we have spectral theorem, that if I is a closed ideal in $\mathscr{B}(\mathscr{H})$ and $\mathscr{H}$ is separable, then $I = (0)$, $\mathscr{B}_0(\mathscr{H})$, or $\mathscr{B}(\mathscr{H})$.

EXERCISES

1. Complete the proof of Proposition 4.9.

2. Show that $M_n(\mathbb{C})$ has no nontrivial ideals. Find all of the left ideals.

3. If α is an infinite cardinal number, let $I_\alpha = \{A \in \mathscr{B}(\mathscr{H}): \dim \text{cl}(\text{ran } A) \leq \alpha\}$. Show that I_α is a closed ideal in $\mathscr{B}(\mathscr{H})$.

4. Let S be the unilateral shift on l^2. Show that $C^*(S) \supseteq \mathscr{B}_0(l^2)$ and $C^*(S)/\mathscr{B}_0(l^2)$ is abelian. Show that the maximal ideal space of $C^*(S)/\mathscr{B}_0(l^2)$ is homeomorphic to $\partial\mathbb{D}$.

5. If V is the Volterra operator on $L^2(0,1)$, show that $C^*(V) = \mathbb{C} + \mathscr{B}_0(L^2(0,1))$.

6. If $\mathscr{A}$ is a C*-algebra, I is a closed ideal of $\mathscr{A}$, and $\mathscr{B}$ is a C*-subalgebra of $\mathscr{A}$, show that the C*-algebra generated by $I \cup \mathscr{B}$ is $I + \mathscr{B}$.

7. If $\mathscr{A}$ is a C*-algebra and I and J are closed ideals in $\mathscr{A}$, show that $I + J$ is a closed ideal of $\mathscr{A}$.

§5*. Representations of C*-Algebras and the Gelfand–Naimark–Segal Construction

5.1. Definition. A *representation* of a C*-algebera is a pair $(\pi, \mathscr{H})$, where $\mathscr{H}$ is a Hilbert space and $\pi: \mathscr{A} \to \mathscr{B}(\mathscr{H})$ is a *-homomorphism. If $\mathscr{A}$ has an identity, it is assumed that $\pi(1) = 1$. (The algebras considered in this book are assumed to have an identity. The reader should be aware that not all authors make this assumption and so he should be cautious when consulting the literature.) Often mention of $\mathscr{H}$ is suppressed and we say that π is a representation.

5.2. Example. If $\mathscr{H}$ is a Hilbert space and $\mathscr{A}$ is a C*-subalgebra of $\mathscr{B}(\mathscr{H})$, then the inclusion map $\mathscr{A} \hookrightarrow \mathscr{B}(\mathscr{H})$ is a representation.

5.3. Example. If n is any cardinal number and $\mathscr{H}$ is a Hilbert space, let $\mathscr{H}^{(n)}$ denote the direct sum of $\mathscr{H}$ with itself n times. If $A \in \mathscr{B}(\mathscr{H})$, then $A^{(n)}$ is the direct sum of A with itself n times; so $A^{(n)} \in \mathscr{B}(\mathscr{H}^{(n)})$ and $\|A^{(n)}\| = \|A\|$. The operator $A^{(n)}$ is called the *inflation* of A. If $\pi\colon \mathscr{A} \to \mathscr{B}(\mathscr{H})$ is a representation, the *inflation* of π is the map $\pi^{(n)}\colon \mathscr{A} \to \mathscr{B}(\mathscr{H}^{(n)})$ defined by $\pi^{(n)}(a) = \pi(a)^{(n)}$ for all a in $\mathscr{A}$.

5.4. Example. If (X, Ω, μ) is a σ-finite measure space and $\mathscr{H} = L^2(\mu)$, then $\pi\colon L^\infty(\mu) \to \mathscr{B}(\mathscr{H})$ defined by $\pi(\phi) = M_\phi$ is a representation.

5.5. Example. If X is a compact space and μ is a positive Borel measure on X, then $\pi\colon C(X) \to \mathscr{B}(L^2(\mu))$ defined by $\pi(f) = M_f$ is a representation.

5.6. Definition. A representation π of a C^*-algebra $\mathscr{A}$ is *cyclic* if there is a vector e in $\mathscr{H}$ such that $\mathrm{cl}[\pi(\mathscr{A})e] = \mathscr{H}$; e is said to be a *cyclic vector* for the representation π.

Note that the representations in Examples 5.4 and 5.5 are cyclic (Exercises 2 and 3). Also, the identity representation $i\colon \mathscr{B}(\mathscr{H}) \to \mathscr{B}(\mathscr{H})$ is cyclic and every nonzero vector is a cyclic vector for this representation. If $\mathscr{A} = \mathbb{C} + \mathscr{B}_0(\mathscr{H})$, then the identity representation is cyclic. On the other hand, if $n \geqslant 2$, then the inflation $\pi^{(n)}$ of a representation of $C(X)$ is never cyclic (Exercise 4).

There is another way to obtain representations.

5.7. Definition. If $\{(\pi_i, \mathscr{H}_i)\colon i \in I\}$ is a family of representations of $\mathscr{A}$, then the *direct sum* of this family is the representation $(\pi, \mathscr{H})$, where $\mathscr{H} = \oplus_i \mathscr{H}_i$ and $\pi(a) = \{\pi_i(a)\}$ for every a in $\mathscr{A}$.

Note that since $\|\pi_i(a)\| \leqslant \|a\|$ for every i (4.8), $\pi(a)$ is a bounded operator on $\mathscr{H}$. It is easy to check that π is a representation.

5.8. Example. Let X be a compact space and let $\{\mu_n\}$ be a sequence of measures on X. For each n let $\pi_n\colon C(X) \to \mathscr{B}(L^2(\mu_n))$ be defined by $\pi_n(f) = M_f$ on $L^2(\mu_n)$. Then $\pi = \oplus_n \pi_n$ is a representation. If the measures $\{\mu_n\}$ are pairwise mutually singular, then π is equivalent (below) to the representation $f \to M_f$ of $C(X) \to \mathscr{B}(L^2(\mu))$, where $\mu = \sum_{n=1}^\infty \mu_n / 2^n \|\mu_n\|$ (Exercise 5).

The concept of equivalence for representations is that of unitary equivalence. That is, two representations of a C^*-algebra $\mathscr{A}$, $(\pi_1, \mathscr{H}_1)$ and $(\pi_2, \mathscr{H}_2)$, are *equivalent* if there is an isomorphism $U\colon \mathscr{H}_1 \to \mathscr{H}_2$ such that $U\pi_1(a)U^{-1} = \pi_2(a)$ for every a in $\mathscr{A}$. The importance of cyclic representations arises from the fact, given in the next result, that every representation is equivalent to the direct sum of cyclic representations.

5.9. Theorem. *If π is a representation of the C^*-algebra $\mathscr{A}$, then there is a family of cyclic representations $\{\pi_i\}$ of $\mathscr{A}$ such that π and $\oplus_i \pi_i$ are equivalent.*

PROOF. let $\mathscr{E}$ = the collection of all subsets E of nonzero vectors in $\mathscr{H}$ such that $\pi(\mathscr{A})e \perp \pi(\mathscr{A})f$ for e, f in E with $e \neq f$. Order $\mathscr{E}$ by inclusion. An application of Zorn's Lemma implies that $\mathscr{E}$ has a maximal element E_0. Let $\mathscr{H}_0 = \oplus \{\mathrm{cl}[\pi(\mathscr{A})e]: e \in E_0\}$. If $h \in \mathscr{H} \ominus \mathscr{H}_0$, then $0 = \langle \pi(a)e, h \rangle$ for every a in $\mathscr{A}$ and e in E_0. So if $a, b \in \mathscr{A}$ and $e \in E_0, 0 = \langle \pi(b^*a)e, h \rangle = \langle \pi(b)^*\pi(a)e, h \rangle = \langle \pi(a)e, \pi(b)h \rangle$. That is, $\pi(\mathscr{A})e \perp \pi(\mathscr{A})h$ for all e in E_0. Hence $E_0 \cup \{h\} \in \mathscr{E}$; by the maximality of E_0 it must be that $h = 0$. Therefore $\mathscr{H} = \mathscr{H}_0$.

For e in E_0 let $\mathscr{H}_e = \mathrm{cl}[\pi(\mathscr{A})e]$. If $a \in \mathscr{A}$, clearly $\pi(a)\mathscr{H}_e \subseteq \mathscr{H}_e$. Since $a^* \in \mathscr{A}$ and $\pi(a)^* = \pi(a^*)$, $\mathscr{H}_e$ reduces $\pi(a)$. So if $\pi_e: \mathscr{A} \to \mathscr{B}(\mathscr{H}_e)$ is defined by $\pi_e(a) = \pi(a)|\mathscr{H}_e$, π_e is a representation of a. Clearly $\pi = \oplus \{\pi_e: e \in E_0\}$. ∎

In light of the preceding theorem, it becomes important to understand cyclic representations. To do this, let $\pi: \mathscr{A} \to \mathscr{B}(\mathscr{H})$ be a cyclic representation with cyclic vector e. Define $f: \mathscr{A} \to \mathbb{C}$ by $f(a) = \langle \pi(a)e, e \rangle$. Note that f is a bounded linear functional on $\mathscr{A}$ with $\|f\| \leqslant \|e\|^2$. Since $f(1) = \|e\|^2$, $\|f\| = \|e\|^2$. Moreover, $f(a^*a) = \langle \pi(a^*a)e, e \rangle = \langle \pi(a)^*\pi(a)e, e \rangle = \|\pi(a)e\|^2 \geqslant 0$.

5.10. Definition. If $\mathscr{A}$ is a C^*-algebra, a linear functional $f: \mathscr{A} \to \mathbb{C}$ is *positive* if $f(a) \geqslant 0$ whenever $a \in \mathscr{A}_+$. A *state* on $\mathscr{A}$ is a positive linear functional on $\mathscr{A}$ of norm 1.

5.11. Proposition. *If f is a positive linear functional on a C^*-algebra $\mathscr{A}$, then*

$$|f(y^*x)|^2 \leqslant f(y^*y)f(x^*x)$$

for every x, y in $\mathscr{A}$.

PROOF. If $[x, y] = f(y^*x)$ for x, y in $\mathscr{A}$, then $[\cdot, \cdot]$ is a semi-inner product on $\mathscr{A}$. The proposition now follows by the CBS inequality (I.1.4). ∎

5.12. Corollary. *If f is a non-zero positive linear functional on the C^*-algebra $\mathscr{A}$, then f is bounded and $\|f\| = f(1)$.*

5.13. Example. If X is a compact space, then the positive linear functionals on $C(X)$ correspond to the positive measures on X. The states correspond to the probability measures on X.

As was shown above, each cyclic representation gives rise to a positive linear functional. It turns out that each positive linear functional gives rise to a cyclic representation.

5.14. Gelfand–Naimark–Segal Construction. *Let $\mathscr{A}$ be a C^*-algebra with identity.*

(a) *If f is a positive linear functional on $\mathscr{A}$, then there is a cyclic representation $(\pi_f, \mathscr{H}_f)$ of $\mathscr{A}$ with cyclic vector e such that $f(a) = \langle \pi_f(a)e, e \rangle$ for all a in $\mathscr{A}$.*

(b) *If $(\pi, \mathscr{H})$ is a cyclic representation of $\mathscr{A}$ with cyclic vector e and*

$f(a) \equiv \langle \pi(a)e, e \rangle$ and if $(\pi_f, \mathcal{H}_f)$ is constructed as in (a), then π and π_f are equivalent.

Before beginning the proof, it will be helpful if the theorem is examined when $\mathcal{A}$ is abelian. So let $\mathcal{A} = C(X)$ where X is compact. If f is a positive linear functional on $\mathcal{A}$, then there is a positive measure μ on X such that $f(\phi) = \int \phi \, d\mu$ for all ϕ in $\mathcal{A}$. The representation $(\pi_f, \mathcal{H}_f)$ is the one obtained by letting $\mathcal{H}_f = L^2(\mu)$ and $\pi_f(\phi) = M_\phi$, but let us look a little closer. One way to obtain $L^2(\mu)$ from $C(X)$ and μ is to let $\mathcal{L} = \{\phi \in C(X): \int |\phi|^2 \, d\mu = 0\}$. Note that $\mathcal{L}$ is an ideal in $C(X)$. Define an inner product on $C(X)/\mathcal{L}$ by $\langle \phi + \mathcal{L}, \psi + \mathcal{L} \rangle = \int \phi \bar\psi \, d\mu$. The completion of $C(X)/\mathcal{L}$ with respect to this inner product is $L^2(\mu)$.

To see part (b) in the abelian case, let $\pi: C(X) \to \mathcal{B}(\mathcal{H})$ be a cyclic representation with cyclic vector e. Let μ be the positive measure on X such that $\int \phi \, d\mu = \langle \pi(\phi)e, e \rangle = f(\phi)$. Now define $U_1: C(X) \to \mathcal{H}$ by $U_1(\phi) = \pi(\phi)e$. Note that U_1 is linear and has dense range. If $\mathcal{L}$ is in the preceding paragraph and $\phi \in \mathcal{L}$, then $\| U_1(\phi) \|^2 = \langle \pi(\phi)e, \pi(\phi)e \rangle = \langle \pi(\phi^*\phi)e, e \rangle = \int |\phi|^2 \, d\mu = 0$. So $U_1 \mathcal{L} = 0$. Thus U_1 induces a linear map $U: C(X)/\mathcal{L} \to \mathcal{H}$ where $U(\phi + \mathcal{L}) = \pi(\phi)e$. If $\langle \phi + \mathcal{L}, \psi + \mathcal{L} \rangle \equiv \int \phi \bar\psi \, d\mu$, then $\langle U(\phi + \mathcal{L}), U(\psi + \mathcal{L}) \rangle = \langle \pi(\phi)e, \pi(\psi)e \rangle = \langle \pi(\phi\psi^*)e, e \rangle = \int \phi \bar\psi \, d\mu = \langle \phi + \mathcal{L}, \psi + \mathcal{L} \rangle$. Thus U extends to an isomorphism U from the completion of $\mathcal{A}/\mathcal{L} = L^2(\mu)$ onto $\mathcal{H}$. So $U: L^2(\mu) \to \mathcal{H}$ and if $\phi \in C(X)$ and we think of $C(X)$ as a (dense) subset of $L^2(\mu)$, $U\phi = \pi(\phi)e$. If $\phi, \psi \in C(X)$, then $UM_\phi \psi = U(\phi\psi) = \pi(\phi\psi)e = \pi(\phi)\pi(\psi)e = \pi(\phi)U(\psi)$; that is, $UM_\phi = \pi(\phi)U$ on a dense subset of $L^2(\mu)$ and, hence, $UM_\phi = \pi(\phi)U$ for every ϕ in $C(X)$. In other words, π is equivalent to the representation $\phi \mapsto M_\phi$.

PROOF OF THEOREM 5.14. Let f be a positive linear functional on $\mathcal{A}$ and put $\mathcal{L} = \{x \in \mathcal{A}: f(x^*x) = 0\}$. It is easy to see that $\mathcal{L}$ is closed in $\mathcal{A}$. Also if $a \in \mathcal{A}$ and $x \in \mathcal{L}$, then (5.11) implies that

$$
\begin{aligned}
f((ax)^*(ax))^2 &= f(x^*(a^*ax))^2 \\
&\leq f(x^*x)f(x^*a^*aa^*ax) \\
&= 0.
\end{aligned}
$$

That is, $\mathcal{L}$ is a closed left ideal in $\mathcal{A}$. Now consider $\mathcal{A}/\mathcal{L}$ as a vector space. (Since $\mathcal{L}$ is only a left ideal, $\mathcal{A}/\mathcal{L}$ is not an algebra.) For x, y in $\mathcal{A}$, define

$$ \langle x + \mathcal{L}, y + \mathcal{L} \rangle = f(y^*x). $$

It is left as an exercise for the reader to show that $\langle \cdot, \cdot \rangle$ is a well-defined inner product on $\mathcal{A}/\mathcal{L}$. Let $\mathcal{H}_f$ be the completion of $\mathcal{A}/\mathcal{L}$ with respect to the norm defined on $\mathcal{A}/\mathcal{L}$ by this inner product.

Because $\mathcal{L}$ is a left ideal of $\mathcal{A}, x + \mathcal{L} \mapsto ax + \mathcal{L}$ is a well-defined linear transformation on $\mathcal{A}/\mathcal{L}$. Also, $\| ax + \mathcal{L} \|^2 = \langle ax + \mathcal{L}, ax + \mathcal{L} \rangle = f(x^*a^*ax)$. Now if $\| aa^* \|$ is considered as an element of $\mathcal{A}$ (it is a multiple of the identity), then an appeal to the functional calculus for a^*a shows that $\| a^*a \| - a^*a \geqslant 0$.

Hence (Exercise 3.7) $0 \leqslant x^*(\|a^*a\| - a^*a)x = \|a\|^2 x^*x - x^*a^*ax$; that is, $x^*a^*ax \leqslant \|a\|^2 x^*x$. Therefore $\|ax + \mathscr{L}\|^2 \leqslant \|a\|^2 f(x^*x) = \|a\|^2 \|x + \mathscr{L}\|^2$. Thus if $\pi_f(a)\colon \mathscr{A}/\mathscr{L} \to \mathscr{A}/\mathscr{L}$ is defined by $\pi_f(a)(x + \mathscr{L}) = ax + \mathscr{L}$, $\pi_f(a)$ is a bounded linear operator with $\|\pi_f(a)\| \leqslant \|a\|$. Hence $\pi_f(a)$ extends to an element of $\mathscr{B}(\mathscr{H}_f)$. It is left to the reader to verify that $\pi_f\colon \mathscr{A} \to \mathscr{B}(\mathscr{H}_f)$ is a representation.

Put $e = 1 + \mathscr{L}$ in $\mathscr{H}_f$. Then $\pi_f(\mathscr{A})e = \{a + \mathscr{L}\colon a \in \mathscr{A}\} = \mathscr{A}/\mathscr{L}$ which, by definition, is dense in $\mathscr{H}_f$. Thus e is a cyclic vector for π_f. [Also note that $\langle \pi_f(a)e, e \rangle = f(a)$.] This proves (a).

Now let $(\pi, \mathscr{H})$, e, and f be as in (b) and let $(\pi_f, \mathscr{H}_f)$ be the representation constructed. Let e_f be the cyclic vector for π_f so that $f(a) = \langle \pi_f(a)e_f, e_f \rangle$ for all a in $\mathscr{A}$. Hence $\langle \pi_f(a)e_f, e_f \rangle = \langle \pi(a)e, e \rangle$ for all a in $\mathscr{A}$. Define U on the dense manifold $\pi_f(\mathscr{A})e_f$ in $\mathscr{H}_f$ by $U\pi_f(a)e_f = \pi(a)e$. Note that $\|\pi(a)e\|^2 = \langle \pi(a)e, \pi(a)e \rangle = \langle \pi(a^*a)e, e \rangle = \langle \pi_f(a^*a)e_f, e_f \rangle = \|\pi_f(a)e_f\|^2$. This implies that U is well defined and an isometry. Thus U extends to an isomorphism of $\mathscr{H}_f$ onto $\mathscr{H}$. If $x, a \in \mathscr{A}$, then $U\pi_f(a)\pi_f(x)e_f = U\pi_f(ax)e_f = \pi(a)\pi(x)e = \pi(a)U\pi_f(x)e_f$. Thus $\pi(a)U = U\pi_f(a)$ so that π and π_f are equivalent. $\blacksquare$

The Gelfand–Naimark–Segal construction is often called the GNS construction.

It is not difficult to show that if f is a positive linear functional on $\mathscr{A}$ and $\alpha > 0$, then the representations π_f and $\pi_{\alpha f}$ are equivalent (Exercise 8). So it is appropriate to only consider the cyclic representations corresponding to states. If $\mathscr{A}$ is a C^*-algebra, let $S_{\mathscr{A}} =$ the collection of all states on $\mathscr{A}$. Note that $S_{\mathscr{A}} \subseteq$ ball $\mathscr{A}^*$. $S_{\mathscr{A}}$ is called the *state space* of $\mathscr{A}$.

5.15. Proposition. *If $\mathscr{A}$ is a C^*-algebra with identity, then $S_{\mathscr{A}}$ is a weak* compact convex subset of $\mathscr{A}^*$ and if $a \in \mathscr{A}_+$, then $\|a\| = \sup\{f(a)\colon f \in S_{\mathscr{A}}\}$ and this supremum is attained.*

PROOF. Since $S_{\mathscr{A}} \subseteq$ ball $\mathscr{A}^*$, to show that $S_{\mathscr{A}}$ is weak* compact, it suffices to show that $S_{\mathscr{A}}$ is weak* closed. The reader can supply this proof using nets. Clearly $S_{\mathscr{A}}$ is convex.

If $\mathscr{A} = C(X)$ with X compact and $f \in C(X)_+$, then there is a point x in X such that $f(x) = \|f\|$. Thus $\|f\| = \int f d\delta_x = \sup\{\int f d\mu\colon \mu \in (\text{ball } M(X))_+\}$. If $\mathscr{A}$ is arbitrary and $a \in \mathscr{A}_+$, then $\|a\| \geqslant \sup\{f(a)\colon f \in S_{\mathscr{A}}\}$. Also, from the argument in the abelian case, there is a state f_1 on $C^*(a)$ such that $f_1(a) = \|a\|$. If we can show that f_1 extends to a state f on $\mathscr{A}$, the proof is complete. That this can be done is a consequence of the next result. $\blacksquare$

5.16. Proposition. *Let $\mathscr{A}, \mathscr{B}$ be C^*-algebras with $\mathscr{B} \subseteq \mathscr{A}$. If f_1 is a state on $\mathscr{B}$, then there is a state f on $\mathscr{A}$ such that $f|\mathscr{B} = f_1$.*

PROOF. Consider the real linear spaces Re $\mathscr{A}$ and Re $\mathscr{B}$. If $a \in \mathscr{A}_+$, then $a \leqslant \|a\|$ in $\mathscr{A}$. Since $1 \in \text{Re }\mathscr{B}$, Re $\mathscr{B}$ has an order unit. By Corollary III.9.12, if $f_1 \in S_{\mathscr{B}}$ there is a positive linear functional f on Re $\mathscr{A}$ such that $f | \text{Re } \mathscr{B} = f_1$. Since $1 \in \mathscr{B}$, $f(1) = f_1(1) = 1$. Now let $f(a) = f((a + a^*)/2) + if((a - a^*)/2i)$ for an arbitrary a in $\mathscr{A}$. It follows that $f \in S_{\mathscr{A}}$ and $f | \mathscr{B} = f_1$. ∎

The next result says that every C^*-algebra is isomorphic to a C^*-algebra contained in $\mathscr{B}(\mathscr{H})$ for some $\mathscr{H}$. Thus each C^*-algebra "is" an algebra of operators.

5.17. Theorem. *If $\mathscr{A}$ is a C^*-algebra, then there is a representation $(\pi, \mathscr{H})$ of $\mathscr{A}$ such that π is an isometry. If $\mathscr{A}$ is separable, then $\mathscr{H}$ can be chosen separable.*

PROOF. Let F be a weak* dense subset of $S_{\mathscr{A}}$ and let $\pi = \oplus \{\pi_f : f \in F\}$, $\mathscr{H} = \oplus \{\mathscr{H}_f : f \in F\}$. Thus $\|a\|^2 \geq \|\pi(a)\|^2 = \sup_f \|\pi_f(a)\|^2$. If e_f is the cyclic vector for π_f, then $\|e_f\|^2 = \langle e_f, e_f \rangle = \langle \pi_f(1)e_f, e_f \rangle = f(1) = 1$. Hence $\|\pi_f(a)\|^2 \geq \|\pi_f(a)e_f\|^2 = \langle \pi_f(a^*a)e_f, e_f \rangle = f(a^*a)$, and $\|a\|^2 \geq \|\pi(a)\|^2 \geq \sup\{f(a^*a): f \in F\}$. Since F is weak* dense in $S_{\mathscr{A}}$, Proposition 5.15 implies $\sup\{f(a^*a): f \in F\} = \|a^*a\| = \|a\|^2$. Hence π is an isometry.

If $\mathscr{A}$ is separable, (ball $\mathscr{A}^*$, wk*) is a compact metric space (V.5.1). Hence $S_{\mathscr{A}}$ is weak* separable so that the set F of the preceding paragraph can be chosen to be countable. Now if $f \in F$, $\pi(\mathscr{A})e_f$ is a separable dense submanifold in $\mathscr{H}_f$ since $\mathscr{A}$ is separable. Thus $\mathscr{H}_f$ is separable. It follows that $\mathscr{H}$ is separable. ∎

Actually, more can be said if $\mathscr{A}$ is separable. In fact, every separable C^*-algebra has a cyclic representation that is isometric (Exercise 11).

EXERCISES

1. Let $\mathscr{A}$ be a C^*-algebra with identity and let $\pi : \mathscr{A} \to \mathscr{B}(\mathscr{H})$ be a *-homomorphism [but don't assume that $\pi(1) = 1$]. Let $P_1 = \pi(1)$. Show that P_1 is a projection and $\mathscr{H}_1 \equiv P_1 \mathscr{H}$ reduces $\pi(\mathscr{A})$. If $\pi_1(a) = \pi(a) | \mathscr{H}_1$, show that $\pi_1 : \mathscr{A} \to \mathscr{B}(\mathscr{H}_1)$ is a representation.

2. Show that the representation in Example 5.4 is a cyclic representation and find all of the cyclic vectors.

3. Show that the representation in Example 5.5 is a cyclic representation and find all the cyclic vectors.

4. If X is compact and μ is a positive measure on X, let $\pi_\mu : C(X) \to \mathscr{B}(L^2(\mu))$ be the representation defined in Example 5.5. If μ, ν are positive measures on X show that $\pi_\mu \oplus \pi_\nu$ is cyclic if and only if $\mu \perp \nu$. If $\mu \perp \nu$, then $\pi_\mu \oplus \pi_\nu$ is equivalent to $\pi_{\mu + \nu}$. Also, $\pi_\mu^{(n)}$ is not cyclic if $n \geq 2$.

5. Verify the statements in Example 5.8.

6. If $\mathscr{A} = \mathbb{C} + \mathscr{B}_0(\mathscr{H})$ and $\pi : \mathscr{A} \to \mathscr{B}(\mathscr{H})$ is the identity representation, show that $\pi^{(\infty)}$ is a cyclic representation.

7. Fix a Banach limit LIM on $l^\infty(\mathbb{N})$ and let $\mathscr{H}$ be a separable Hilbert space with an orthonormal basis $\{e_n\}$. Define $f : \mathscr{B}(\mathscr{H}) \to \mathbb{C}$ by $f(T) = \text{LIM}\{\langle Te_n, e_n \rangle\}$. Show

that f is a state on $\mathscr{B}(\mathscr{H})$. If π_f is the corresponding cyclic representation, show that $\ker \pi_f = \mathscr{B}_0(\mathscr{H})$. Hence π_f induces a cyclic representation of $\mathscr{B}(\mathscr{H})/\mathscr{B}_0(\mathscr{H})$ that is isometric. Is $\mathscr{H}_f$ separable?

8. If f is a positive linear functional on $\mathscr{A}$ and $\alpha \in (0, \infty)$, show that π_f and $\pi_{\alpha f}$ are equivalent representations.

9. If $a \in \mathscr{A}$, then $a \geqslant 0$ if and only if $f(a) \geqslant 0$ for every state f.

10. If $a \in \mathscr{A}$ and $a \neq 0$, then there is a state f on $\mathscr{A}$ such that $f(a) \neq 0$.

11. If $\mathscr{A}$ is a separable C^*-algebra and $\{f_n\}$ is a countable weak* dense subset of $S_{\mathscr{A}}$, let $f = \sum_n 2^{-n} f_n$. Show that π_f is an isometry.

CHAPTER IX

Normal Operators on Hilbert Space

In this chapter the Spectral Theorem for normal operators on a Hilbert space is proved. This theorem is then used to answer a number of questions concerning normal operators. In fact, the Spectral Theorem can be used to answer essentially every question about normal operators.

§1. Spectral Measures and Representations of Abelian C^*-Algebras

Before beginning this section the reader should familiarize himself with the definitions and examples in (VIII.5.1) through (VIII.5.8).

In this secton we want to focus our attention on representations of abelian C^*-algebras. The reason for this is that the Spectral Theorem and its generalizations can be obtained as a special case of such a theory. The idea is the following. Let N be a normal operator on $\mathcal{H}$. Then $C^*(N)$ is an abelian C^*-algebra and the functional calculus $f \mapsto f(N)$ is a $*$-isomorphism of $C(\sigma(N))$ onto $C^*(N)$ (VIII.2.6). Thus $f \mapsto f(N)$ is a representation $C(\sigma(N)) \to \mathcal{B}(\mathcal{H})$ of the abelian C^*-algebra $C(\sigma(N))$. A diagnosis of such representations yields the Spectral Theorem.

A representation $\rho: C(X) \to \mathcal{B}(\mathcal{H})$ is a $*$-homomorphism with $\rho(1) = 1$. Also, $\|\rho\| = 1$ (VIII.1.11d). If $f \in C(X)_+$, then $f = g^2$ where $g \in C(X)_+$; hence $\rho(f) = \rho(g)^2 = \rho(g)^* \rho(g) \geq 0$. So ρ is a positive map. One might expect, by analogy with the Riesz Representation Theorem, that $\rho(f) = \int f \, dE$ for some type of measure E whose values are operators rather than scalars. This is indeed the case. We begin by introducing these measures and defining the integral of a scalar-valued function with respect to one of them.

1.1. Definition. If X is a set, Ω is a σ-algebra of subsets of X, and $\mathscr{H}$ is a Hilbert space, a *spectral measure* for $(X, \Omega, \mathscr{H})$ is a function $E: \Omega \to \mathscr{B}(\mathscr{H})$ such that:

(a) for each Δ in Ω, $E(\Delta)$ is a projection;
(b) $E(\square) = 0$ and $E(X) = 1$;
(c) $E(\Delta_1 \cap \Delta_2) = E(\Delta_1) E(\Delta_2)$ for Δ_1 and Δ_2 in Ω;
(d) if $\{\Delta_n\}_{n=1}^{\infty}$ are pairwise disjoint sets from Ω, then

$$E\left(\bigcup_{n=1}^{\infty} \Delta_n\right) = \sum_{n=1}^{\infty} E(\Delta_n).$$

A word or two concerning condition (d) in the preceding definition. If $\{E_n\}$ is a sequence of pairwise orthogonal projections on $\mathscr{H}$, then it was shown in Exercise II.3.5 that for each h in $\mathscr{H}$, $\sum_{n=1}^{\infty} E_n(h)$ converges in $\mathscr{H}$ to $E(h)$, where E is the orthogonal projection of $\mathscr{H}$ onto $\vee \{E_n(\mathscr{H}): n \geq 1\}$. Thus it is legitimate to write $E = \sum_{n=1}^{\infty} E_n$. Now if $\Delta_1 \cap \Delta_2 = \square$, then (b) and (c) above imply that $0 = E(\Delta_1)E(\Delta_2) = E(\Delta_2)E(\Delta_1)$; that is, $E(\Delta_1)$ and $E(\Delta_2)$ have orthogonal ranges. So if $\{\Delta_n\}_1^{\infty}$ is a sequence of pairwise disjoint sets in Ω, the ranges of $\{E(\Delta_n)\}$ are pairwise orthogonal. Thus the equation $E(\bigcup_1^{\infty} \Delta_n) = \Sigma_1^{\infty} E(\Delta_n)$ in (d) has the precise meaning just discussed.
    Another way to discuss this is by the introduction of two topologies that will also be of value later.

1.2. Definition. If $\mathscr{H}$ is a Hilbert space, the *weak operator topology* (WOT) on $\mathscr{B}(\mathscr{H})$ is the locally convex topology defined by the seminorms $\{p_{h,k}: h, k \in \mathscr{H}\}$ where $p_{h,k}(A) = |\langle Ah, k \rangle|$. The *strong operator topology* (SOT) is the topology defined on $\mathscr{B}(\mathscr{H})$ by the family of seminorms $\{p_h: h \in \mathscr{H}\}$, where $p_h(A) = \|Ah\|$.

1.3. Proposition. *Let $\mathscr{H}$ be a Hilbert space and let $\{A_i\}$ be a net in $\mathscr{B}(\mathscr{H})$.*

(a) *$A_i \to A$ (WOT) if and only if $\langle A_i h, k \rangle \to \langle Ah, k \rangle$ for all h, k in $\mathscr{H}$.*
(b) *If $\sup_i \|A_i\| < \infty$ and $\mathscr{T}$ is a total subset of $\mathscr{H}$, then $A_i \to A$ (WOT) if and only if $\langle A_i h, k \rangle \to \langle Ah, k \rangle$ for all h, k in $\mathscr{T}$.*
(c) *$A_i \to A$ (SOT) if and only if $\|A_i h - Ah\| \to 0$ for all h in $\mathscr{H}$.*
(d) *If $\sup_i \|A_i\| < \infty$ and $\mathscr{T}$ is a total subset of $\mathscr{H}$, then $A_i \to A$ (SOT) if and only if $\|A_i h - Ah\| \to 0$ for all h in $\mathscr{T}$.*
(e) *If $\mathscr{H}$ is separable, then the WOT and SOT are metrizable on bounded subsets of $\mathscr{B}(\mathscr{H})$.*

PROOF. The proofs of (a) through (d) are left as exercises. For (e), let $\{h_n\}$ be any countable total subset of ball $\mathscr{H}$. If $A, B \in \mathscr{B}(\mathscr{H})$, let

$$d_s(A, B) = \sum_{n=1}^{\infty} 2^{-n} \|(A - B)h_n\|,$$

$$d_w(A, B) = \sum_{m,n=1}^{\infty} 2^{-n-m} |\langle (A - B)h_n, h_m \rangle|.$$

Then d_s and d_w are metrics on $\mathcal{B}(\mathcal{H})$. It is left as an exercise to show that d_s and d_w define the SOT and WOT on bounded subsets of $\mathcal{B}(\mathcal{H})$. ∎

1.4. Example. Let (X, Ω, μ) be a σ-finite measure space. If $\phi \in L^\infty(\mu)$, let M_ϕ be the multiplication operator on $L^2(\mu)$. Then a net $\{\phi_i\}$ in $L^\infty(\mu)$ converges weak* to ϕ if and only if $M_{\phi_i} \to M_\phi$ (WOT). In fact, if $f, g \in L^2(\mu)$ and $\phi_i \to \phi$ weak* in $L^\infty(\mu)$, then $\langle M_{\phi_i} f, g \rangle = \int \phi_i f \bar{g} \, d\mu \to \int \phi f \bar{g} \, d\mu = \langle M_\phi f, g \rangle$ since $f \bar{g} \in L^1(\mu)$. Conversely, if $M_{\phi_i} \to M_\phi$ (WOT) and $f \in L^1(\mu)$, then $f = g_1 \bar{g}_2$, where $g_1, g_2 \in L^2(\mu)$. (Why?) So $\int \phi_i f \, d\mu = \langle M_{\phi_i} g_1, g_2 \rangle \to \langle M_\phi g_1, g_2 \rangle = \int \phi f \, d\mu$.

1.5. Example. If $\{E_n\}$ is a sequence of pairwise orthogonal projections on $\mathcal{H}$, then $\sum_1^\infty E_n$ converges (SOT) to the projection of $\mathcal{H}$ onto $\vee \{E_n(\mathcal{H}): n \geq 1\}$.

In light of (1.5), a spectral measure for $(X, \Omega, \mathcal{H})$ could be defined as a SOT-countably additive projection-valued measure.

1.6. Example. Let X be a compact set. $\Omega =$ the Borel subsets of X, $\mu = $ a measure on Ω, and $\mathcal{H} = L^2(\mu)$. For Δ in Ω, let $E(\Delta) =$ multiplication by χ_Δ, the characteristic function of Δ. E is a spectral measure for $(X, \Omega, \mathcal{H})$.

1.7. Example. If E is a spectral measure for $(X, \Omega, \mathcal{H})$, the *inflation*, $E^{(n)}$, of E, defined by $E^{(n)}(\Delta) = E(\Delta)^{(n)}$, is a spectral measure for $(X, \Omega, \mathcal{H}^{(n)})$.

1.8. Example. Let X be any set, $\Omega =$ all the subsets of X, $\mathcal{H} =$ any separable Hilbert space, and fix a sequence $\{x_n\}$ in X. If $\{e_1, e_2, \ldots\}$ is some orthonormal basis for $\mathcal{H}$, define $E(\Delta) =$ the projection onto $\vee \{e_n: x_n \in \Delta\}$. E is a spectral measure for $(X, \Omega, \mathcal{H})$.

The next lemma is useful in studying spectral measures as it allows us to prove things about spectral measures from known facts about complex-valued measures.

1.9. Lemma. *If E is a spectral measure for $(X, \Omega, \mathcal{H})$ and $g, h \in \mathcal{H}$, then*

$$E_{g,h}(\Delta) \equiv \langle E(\Delta)g, h \rangle$$

defines a countably additive measure on Ω with total variation $\leq \|g\| \, \|h\|$.

PROOF. That $\mu = E_{g,h}$ as defined above, is a countably additive measure is left for the reader to verify. If $\Delta_1, \ldots, \Delta_n$ are pairwise disjoint sets in Ω, let $\alpha_j \in \mathbb{C}$ such that $|\alpha_j| = 1$ and $|\langle E(\Delta_j)g, h \rangle| = \alpha_j \langle E(\Delta_j)g, h \rangle$. So $\sum_j |\mu(\Delta_j)| = \sum_j \alpha_j \langle E(\Delta_j)g, h \rangle = \langle \sum_j E(\Delta_j)\alpha_j g, h \rangle \leq \|\sum_j E(\Delta_j)\alpha_j g\| \, \|h\|$. Now $\{E(\Delta_j)\alpha_j g: 1 \leq j \leq n\}$ is a finite sequence of pairwise orthogonal vectors so that $\|\sum_j E(\Delta_j)\alpha_j g\|^2 = \sum_j \|E(\Delta_j)g\|^2 = \|E(\bigcup_{j=1}^n \Delta_j)g\|^2 \leq \|g\|^2$; hence $\sum_j |\mu(\Delta_j)| \leq \|g\| \, \|h\|$. Thus $\|\mu\| \leq \|g\| \, \|h\|$. ∎

It is possible to use spectral measures to define representations. The next result is crucial for this purpose. It tells us how to integrate with respect to a spectral measure.

1.10. Proposition. *If E is a spectral measure for $(X, \Omega, \mathcal{H})$ and $\phi: X \to \mathbb{C}$ is a bounded Ω-measurable function, then there is a unique operator A in $\mathcal{B}(\mathcal{H})$ such that if $\varepsilon > 0$ and $\{\Delta_1, \dots, \Delta_n\}$ is an Ω-partition of X with $\sup\{|\phi(x) - \phi(x')|: x, x' \in \Delta_k\} < \varepsilon$ for $1 \leqslant k \leqslant n$, then for any x_k in Δ_k,*

$$\left\| A - \sum_{k=1}^{n} \phi(x_k) E(\Delta_k) \right\| < \varepsilon.$$

PROOF. Define $B(g, h) \equiv \int \phi \, dE_{g,h}$ for g, h in $\mathcal{H}$. By the preceding lemma it is easy to see that B is a sesquilinear form with $|B(g, h)| \leqslant \|\phi\|_\infty \|g\| \|h\|$. Hence there is a unique operator A such that $B(g, h) = \langle Ag, h \rangle$ for all g and h in $\mathcal{H}$.

Let $\{\Delta_1, \dots, \Delta_n\}$ be an Ω-partition satisfying the condition in the statement of the proposition. If g and h are arbitrary vectors in $\mathcal{H}$ and $x_k \in \Delta_k$ for $1 \leqslant k \leqslant n$, then

$$\left| \langle Ag, h \rangle - \sum_{k=1}^{n} \phi(x_k) \langle E(\Delta_k) g, h \rangle \right| = \left| \sum_{k=1}^{n} \int_{\Delta_k} [\phi(x) - \phi(x_k)] d\langle E(x)g, h \rangle \right|$$

$$\leqslant \sum_{k=1}^{n} \int_{\Delta_k} |\phi(x) - \phi(x_k)| \, d|\langle E(x)g, h \rangle|$$

$$\leqslant \varepsilon \int d|\langle E(x)g, h \rangle| \leqslant \varepsilon \|g\| \|h\|. \quad \blacksquare$$

The operator A obtained in the preceding proposition is the *integral of ϕ with respect to E* and is denoted by

$$\int \phi \, dE.$$

Therefore if $g, h \in \mathcal{H}$ and ϕ is a bounded Ω-measurable function on X, the preceding proof implies that

1.11
$$\left\langle \left(\int \phi \, dE \right) g, h \right\rangle = \int \phi \, dE_{g,h}.$$

Let $B(X, \Omega)$ denote the set of bounded Ω-measurable functions $\phi: X \to \mathbb{C}$ and let $\|\phi\| = \sup\{|\phi(x)|: x \in X\}$. It is easy to see that $B(X, \Omega)$ is a Banach algebra with identity. In fact, if $\phi^*(x) \equiv \overline{\phi(x)}$, then $B(X, \Omega)$ is an abelian C^*-algebra. The properties of the integral $\int \phi \, dE$ are summarized by the following result.

1.12. Proposition. *If E is a spectral measure for $(X, \Omega, \mathcal{H})$ and $\rho: B(X, \Omega) \to \mathcal{B}(\mathcal{H})$ is defined by $\rho(\phi) = \int \phi \, dE$, then ρ is representation of $B(X, \Omega)$ and $\rho(\phi)$ is a normal operator for every ϕ in $B(X, \Omega)$.*

PROOF. It will only be shown that ρ is multiplicative; the remainder is an exercise. Let ϕ and $\psi \in B(X,\Omega)$. Let $\varepsilon > 0$ and choose a Borel partition $\{\Delta_1, \ldots, \Delta_n\}$ of X such that $\sup\{|\omega(x) - \omega(x')|: x, x' \in \Delta_k\} < \varepsilon$ for $\omega = \phi, \psi$ or $\phi\psi$ and for $1 \leqslant k \leqslant n$. Hence, if $x_k \in \Delta_k$ $(1 \leqslant k \leqslant n)$,

$$\left\| \int \omega \, dE - \sum_{k=1}^{n} \omega(x_k) E(\Delta_k) \right\| < \varepsilon$$

for $\omega = \phi, \psi$, or $\phi\psi$. Thus, using the triangle inequality,

$$\left\| \int \phi\psi \, dE - \left(\int \phi \, dE \right)\left(\int \psi \, dE \right) \right\|$$

$$\leqslant \varepsilon + \left\| \sum_{k=1}^{n} \phi(x_k)\psi(x_k) E(\Delta_k) - \left[\sum_{i=1}^{n} \phi(x_i) E(\Delta_i) \right]\left[\sum_{j=1}^{n} \psi(x_j) E(\Delta_j) \right] \right\|$$

$$+ \left\| \left[\sum_{i=1}^{n} \phi(x_i) E(\Delta_i) \right]\left[\sum_{j=1}^{n} \psi(x_j) E(\Delta_j) \right] - \left(\int \phi \, dE \right)\left(\int \psi \, dE \right) \right\|.$$

But $E(\Delta_i)E(\Delta_j) = E(\Delta_i \cap \Delta_j)$ and $\{\Delta_1, \ldots, \Delta_n\}$ is a partition. So the middle term in this sum is zero. Hence

$$\left\| \int \phi\psi \, dE - \left(\int \phi \, dE \right)\left(\int \phi \, dE \right) \right\|$$

$$\leqslant \varepsilon + \left\| \left[\sum_{i=1}^{n} \phi(x_i) E(\Delta_i) \right]\left[\sum_{j=1}^{n} \psi(x_j) E(\Delta_j) - \int \psi \, dE \right] \right\|$$

$$+ \left\| \left[\sum_{i=1}^{n} \phi(x_i) E(\Delta_i) - \int \phi \, dE \right]\left[\int \psi \, dE \right] \right\| \leqslant \varepsilon[1 + \|\phi\| + \|\psi\|].$$

Since ε was arbitrary, $\int \phi\psi \, dE = (\int \phi \, dE)(\int \psi \, dE)$. ∎

1.13. Corollary. *If X is a compact Hausdorff space and E is a spectral measure defined on the Borel subsets of X, then $\rho: C(X) \to \mathscr{B}(\mathscr{H})$ defined by $\rho(u) = \int u \, dE$ is a representation of $C(X)$.*

The next result is the main result of this section and it states that the converse to the preceding corollary holds.

1.14. Theorem. *If $\rho: C(X) \to \mathscr{B}(\mathscr{H})$ is a representation, there is a unique spectral measure E defined on the Borel subsets of X such that for all g and h in $\mathscr{H}$, $E_{g,h}$ is a regular measure and*

$$\rho(u) = \int u \, dE$$

for every u in $C(X)$.

PROOF. The idea of the proof is similar to the idea of the proof of the Riesz

Representation Theorem for linear functionals on $C(X)$. We wish to extend ρ to a representation $\tilde{\rho}: B(X) \to \mathcal{B}(\mathcal{H})$, where $B(X)$ is the C^*-algebra of bounded Borel functions. The measure E of a Borel set Δ is then defined by letting $E(\Delta) = \tilde{\rho}(\chi_\Delta)$. In fact, it is possible to give a proof of the theorem patterned on the proof of the Riesz Representation Theorem. Here, however, the proof will use the Riesz Representation Theorem to simplify the technical details.

If $g, h \in \mathcal{H}$, then $u \mapsto \langle \rho(u)g, h \rangle$ is a linear functional on $C(X)$ with norm $\leq \|g\| \|h\|$. Hence there is a unique measure, $\mu_{g,h}$ in $M(X)$ such that

$$\textbf{1.15} \qquad\qquad \langle \rho(u)g, h \rangle = \int u \, d\mu_{g,h}$$

for all u in $C(X)$. It is easy to verify that the map $(g, h) \mapsto \mu_{g,h}$ is sesquilinear (use uniqueness) and $\|\mu_{g,h}\| \leq \|g\| \|h\|$. Now fix ϕ in $B(X)$ and define $[g, h] = \int \phi \, d\mu_{g,h}$. Then $[\cdot, \cdot]$ is a sesquilinear form and $|[g, h]| \leq \|\phi\| \|g\| \|h\|$. Hence there is a unique bounded operator A such that $[g, h] = \langle Ag, h \rangle$ and $\|A\| \leq \|\phi\|$ (II.2.2). Denote the operator A by $\tilde{\rho}(\phi)$. So $\tilde{\rho}: B(X) \to \mathcal{B}(\mathcal{H})$ is a well-defined function, $\|\tilde{\rho}(\phi)\| \leq \|\phi\|$, and for all g, h in $\mathcal{H}$,

$$\textbf{1.16} \qquad\qquad \langle \tilde{\rho}(\phi)g, h \rangle = \int \phi \, d\mu_{g,h}.$$

1.17. Claim. $\tilde{\rho}: B(X) \to \mathcal{B}(\mathcal{H})$ is a representation and $\tilde{\rho}|C(X) = \rho$.

The fact that $\tilde{\rho}(u) = \rho(u)$ whenever $u \in C(X)$ follows immediately from (1.15) and (1.16). If $\phi \in B(X)$, consider ϕ as an element of $M(X)^*$ $(= C(X)^{**})$; that is, ϕ corresponds to the linear functional $\mu \mapsto \int \phi \, d\mu$. By Proposition V.4.1, $\{u \in C(X): \|u\| \leq \|\phi\|\}$ is $\sigma(M(X)^*, M(X))$ dense in $\{L \in M(X)^*: \|L\| \leq \|\phi\|\}$. Thus there is a net $\{u_i\}$ in $C(X)$ such that $\|u_i\| \leq \|\phi\|$ for all u_i and $\int u_i \, d\mu \to \int \phi \, d\mu$ for every μ in $M(X)$. If $\psi \in B(X)$, then $\psi \mu \in M(X)$ whenever $\mu \in M(X)$. Hence $\int u_i \psi \, d\mu \to \int \phi \psi \, d\mu$ for every ψ in $B(X)$ and μ in $M(X)$. By (1.16), $\tilde{\rho}(u_i \psi) \to \tilde{\rho}(\phi \psi)$ (WOT) for all ψ in $B(X)$. In particular, if $\psi \in C(X)$, then $\tilde{\rho}(\phi \psi) = \text{WOT} - \lim \tilde{\rho}(u_i \psi) = \text{WOT} - \lim \rho(u_i)\rho(\psi) = \tilde{\rho}(\phi)\rho(\psi)$. That is,

$$\tilde{\rho}(\phi\psi) = \tilde{\rho}(\phi)\rho(\psi)$$

whenever $\phi \in B(X)$ and $\psi \in C(X)$. Hence $\tilde{\rho}(u_i \psi) = \rho(u_i)\tilde{\rho}(\psi)$ for any ψ in $B(X)$ and for all u_i. Since $\tilde{\rho}(u_i) \to \tilde{\rho}(\phi)$ (WOT) and $\tilde{\rho}(u_i\psi) \to \tilde{\rho}(\phi\psi)$ (WOT), this implies that

$$\tilde{\rho}(\phi\psi) = \tilde{\rho}(\phi)\tilde{\rho}(\psi)$$

whenever $\phi, \psi \in B(X)$.

The proof that $\tilde{\rho}$ is linear is immediate by (1.16). To see that $\tilde{\rho}(\phi)^* = \tilde{\rho}(\bar{\phi})$. Let $\{u_i\}$ be the net obtained in the preceding paragraph. If $\mu \in M(X)$, let $\bar{\mu}$ be the measure defined by $\bar{\mu}(\Delta) = \overline{\mu(\Delta)}$. Then $\rho(u_i) \to \tilde{\rho}(\phi)$ (WOT) and so $\rho(u_i)^* \to \tilde{\rho}(\phi)^*$ (WOT). But $\int \bar{u}_i \, d\mu = \overline{\int u_i \, d\bar{\mu}} \to \overline{\int \phi \, d\bar{\mu}} = \int \bar{\phi} \, d\mu$ for every measure

μ. Hence $\rho(\bar{u}_i) \to \tilde{\rho}(\bar{\phi})$. But $\rho(u_i)^* = \rho(\bar{u}_i)$ since ρ is a $*$-homomorphism. Thus $\tilde{\rho}(\phi)^* = \tilde{\rho}(\bar{\phi})$ and $\tilde{\rho}$ is a representation.

For any Borel subset Δ of X let $E(\Delta) \equiv \tilde{\rho}(\chi_\Delta)$. We want to show that E is a spectral measure. Since χ_Δ is a hermitian idempotent in $B(X)$, $E(\Delta)$ is a projection by (1.17). Since $\chi_\square = 0$ and $\chi_X = 1$, $E(\square) = 0$ and $E(X) = 1$. Also, $E(\Delta_1 \cap \Delta_2) = \tilde{\rho}(\chi_{\Delta_1 \cap \Delta_2}) = \tilde{\rho}(\chi_{\Delta_1} \chi_{\Delta_2}) = E(\Delta_1)E(\Delta_2)$. Now let $\{\Delta_n\}$ be a pairwise disjoint sequence of Borel sets and put $\Lambda_n = \bigcup_{k=n+1}^\infty \Delta_k$. It is easy to see that E is finitely additive so if $h \in \mathcal{H}$, then

$$\left\| E\left(\bigcup_{k=1}^\infty \Delta_k \right)h - \sum_{k=1}^n E(\Delta_k)h \right\|^2 = \langle E(\Lambda_n)h, E(\Lambda_n)h \rangle$$
$$= \langle E(\Lambda_n)h, h \rangle$$
$$= \langle \tilde{\rho}(\chi_{\Lambda_n})h, h \rangle$$
$$= \int \chi_{\Lambda_n} d\mu_{h,h}$$
$$= \sum_{k=n+1}^\infty \mu_{h,h}(\Delta_k) \to 0$$

as $n \to \infty$. Therefore E is a spectral measure.

It remains to show that $\rho(u) = \int u \, dE$. It will be shown that $\tilde{\rho}(\phi) = \int \phi \, dE$ for every ϕ in $B(X)$. Fix ϕ in $B(X)$ and $\varepsilon > 0$. If $\{\Delta_1, \ldots, \Delta_n\}$ is any Borel partition of X such that $\sup\{|\phi(x) - \phi(x')| : x, x' \in \Delta_k\} < \varepsilon$ for $1 \leqslant k \leqslant n$, then $\| \phi - \sum_{k=1}^n \phi(x_k)\chi_{\Delta_k} \|_\infty < \varepsilon$ for any choice of x_k in Δ_k. Since $\|\tilde{\rho}\| = 1$, $\varepsilon > \| \tilde{\rho}(\phi) - \sum_{k=1}^n \phi(x_k)E(\Delta_k) \|$. This implies that $\tilde{\rho}(\phi) = \int \phi \, dE$ for any ϕ in $B(X)$.

The proof of the uniqueness of E is left to the reader.   ∎

EXERCISES

1. Prove Proposition 1.3.

2. Show that ball $\mathcal{B}(\mathcal{H})$ is WOT compact.

3. Show that Re $\mathcal{B}(\mathcal{H})$ and $\mathcal{B}(\mathcal{H})$, are WOT and SOT closed.

4. If $L: \mathcal{B}(\mathcal{H}) \to \mathbb{C}$ is a linear functional, show that the following statements are equivalent: (a) L is SOT-continuous; (b) L is WOT-continuous; (c) there are vectors $h_1, \ldots, h_n, g_1, \ldots, g_n$ in $\mathcal{H}$ such that $L(A) = \sum_{j=1}^n \langle Ah_j, g_j \rangle$.

5. Show that a convex subset of $\mathcal{B}(\mathcal{H})$ is WOT closed if and only if it is SOT closed.

6. Verify the statement in Example 1.5.

7. Verify the statements made in Examples 1.6, 1.7, and 1.8.

8. For the spectral measures in (1.6), (1.7), and (1.8), give the corresponding representations.

9. If $\{E_i\}$ is a net of projections and E is a projection, show that $E_i \to E$ (WOT) if and only if $E_i \to E$ (SOT).

10. For the representation in (VIII.5.5), find the corresponding spectral measure.

11. In Example VIII.5.4, the representation is not quite covered by Theorem 1.14 since it is a representation of $L^\infty(\mu)$ and not $C(X)$. Nevertheless, this representation is given by a spectral measure defined on Ω. Find it.

12. Let X be a compact Hausdorff space and let $\{x_n\}$ be a sequence in X. Let $\{e_n\}$ be an orthonormal basis for $\mathscr{H}$ and for each u in $C(X)$ define $\rho(u)$ in $\mathscr{B}(\mathscr{H})$ by $\rho(u)e_n = u(x_n)e_n$. Show that ρ is a representation and find the corresponding spectral measure.

13. A representation $\rho: \mathscr{A} \to \mathscr{B}(\mathscr{H})$ is *irreducible* if the only projections in $\mathscr{B}(\mathscr{H})$ that commute with every $\rho(a)$, a in $\mathscr{A}$, are 0 and 1. Prove that if $\mathscr{A}$ is abelian and ρ is an irreducible representation of $\mathscr{A}$, then dim $\mathscr{H} = 1$. Find the corresponding spectral measure.

14. Show that a representation $\rho: C(X) \to \mathscr{B}(\mathscr{H})$ is injective if and only if $E(G) \neq 0$ for every non-empty open set G, where E is the corresponding spectral measure.

15. Let $\{A_i\}$ be a net of hermitian operators on $\mathscr{H}$ and suppose that there is a hermitian operator T such that $A_i \leqslant T$ for all i. If $\{\langle A_i h, h \rangle\}$ is an increasing net in $\mathbb{R}$ for every h in $\mathscr{H}$, then there is a hermitian operator A such that $A_i \to A$(WOT).

16. Show that there is a contraction $\tau: \mathscr{B}(\mathscr{H})^{**} \to \mathscr{B}(\mathscr{H})$ such that $\tau(T) = T$ for T in $\mathscr{B}(\mathscr{H})$. If $\rho: C(X) \to \mathscr{B}(\mathscr{H})$ is a representation, show that the map $\tilde{\rho}$ in the proof of Theorem 1.14 is given by $\tilde{\rho}(\phi) = \tau \circ \rho^{**}(\phi)$.

§2. The Spectral Theorem

The Spectral Theorem is a landmark in the theory of operators on a Hilbert space. It provides a complete statement about the nature and structure of normal operators. This accolade will be seen to be deserved when in Section 10 the Spectral Theorem is used to give a complete set of unitary invariants. Two operators A and B are *unitarily equivalent* if there is a unitary operator U such that $UAU^* = B$; in symbols, $A \cong B$. Using the Spectral Theorem, a (countable) set of objects is attached to a normal operator N on a (separable) Hilbert space. It is then shown that two normal operators are unitarily equivalent if and only if these objects are equal.

The Spectral Theorem for a normal operator N on a Hilbert space with dim $\mathscr{H} = d < \infty$ says that N can be diagonalized. That is, if $\alpha_1, \ldots, \alpha_d$ are the eigenvalues of N (repeated as often as their multiplicities), then the corresponding eigenvectors $e_1, e_2, \ldots, e_d$ from an orthonormal basis for $\mathscr{H}$. In infinite dimensional spaces a normal operator need not have eigenvalues. For example, let N = multiplication by the independent variable on $L^2(0, 1)$. So an alternative formulation that can be generalized is desired.

Let N be normal on $\mathscr{H}$, dim $\mathscr{H} = d < \infty$. Let $\lambda_1, \ldots, \lambda_n$ be the distinct eigenvalues of N and let E_k be the orthogonal projection of $\mathscr{H}$ onto

$\ker(N - \lambda_k)$, $1 \leqslant k \leqslant n$. Then the Spectral Theorem says that

2.1
$$N = \sum_{k=1}^{n} \lambda_k E_k.$$

In this form a generalization is possible. Rather than discuss orthogonal projections on eigenspaces (which may not exist), the concept of a spectral measure is used; rather than the sum that appears in (2.1), an integral is used. It is worth mentioning that the finite dimensional version is a corollary of the general theorem (see Exercise 4).

2.2. The Spectral Theorem. *If N is a normal operator, there is a unique spectral measure E on the Borel subsets of $\sigma(N)$ such that:*

(a) $N = \int z\, dE(z)$;
(b) *if G is a nonempty relatively open subset of $\sigma(N)$, $E(G) \neq 0$;*
(c) *if $A \in \mathscr{B}(\mathscr{H})$, then $AN = NA$ and $AN^* = N^*A$ if and only if $AE(\Delta) = E(\Delta)A$ for every Δ.*

PROOF. Let $\mathscr{A} = C^*(N)$, the C^*-algebra generated by N. So $\mathscr{A}$ is the closure of all polynomials in N and N^*. By Theorem VIII.2.6, there is an isometric isomorphism $\rho\colon C(\sigma(N)) \to \mathscr{A} \subseteq \mathscr{B}(\mathscr{H})$ given by $\rho(u) = u(N)$ (the functional calculus). By Theorem 1.14 there is a unique spectral measure E defined on the Borel subsets of $\sigma(N)$ such that $\rho(u) = \int u\, dE$ for all u in $C(\sigma(N))$. In particular, (a) holds since $N = \rho(z)$.

If G is a nonempty relatively open subset of $\sigma(N)$, there is a nonzero continuous function u on $\sigma(N)$ such that $0 \leqslant u \leqslant \chi_G$. Using Claim 1.17, one obtains that $E(G) = \tilde{\rho}(\chi_G) \geqslant \rho(u) \neq 0$; so (b) holds.

Now let $A \in \mathscr{B}(\mathscr{H})$ such that $AN = NA$ and $AN^* = N^*A$. It is not hard to see that this implies, by the Stone–Weierstrass Theorem, that $A\rho(u) = \rho(u)A$ for every u in $C(\sigma(N))$; that is, $Au(N) = u(N)A$ for all u in $C(\sigma(N))$. Let $\Omega = \{\Delta\colon \Delta$ is a Borel set and $AE(\Delta) = E(\Delta)A\}$. It is left to the reader to show that Ω is a σ-algebra. If G is an open set in $\sigma(N)$, there is a sequence $\{u_n\}$ of positive continuous functions on $\sigma(N)$ such that $u_n(z) \uparrow \chi_G(z)$ for all z. Thus

$$\langle AE(G)g, h \rangle = \langle E(G)g, A^*h \rangle$$
$$= E_{g, A^*h}(G)$$
$$= \lim \int u_n\, dE_{g, A^*h}$$
$$= \lim \langle u_n(N)g, A^*h \rangle$$
$$= \lim \langle Au_n(N)g, h \rangle$$
$$= \lim \langle u_n(N)Ag, h \rangle$$
$$= \langle E(G)Ag, h \rangle.$$

So Ω contains every open set and, hence, it must be the collection of Borel sets. The converse is left to the reader.   ∎

The unique spectral measure E obtained in the Spectral Theorem is called the *spectral measure* for N. An abbreviation for the Spectral Theorem is to say, "Let $N = \int \lambda\, dE(\lambda)$ be the *spectral decomposition* of N." If ϕ is a bounded Borel function on $\sigma(N)$, define $\phi(N)$ by

$$\phi(N) \equiv \int \phi\, dE,$$

where E is the spectral measure for N.

2.3. Theorem. *If N is a normal operator on $\mathscr{H}$ with spectral measure E and $B(\sigma(N))$ is the C^*-algebra of bounded Borel functions on $\sigma(N)$, then the map*

$$\phi \mapsto \phi(N)$$

is a representation of the C^*-algebra $B(\sigma(N))$. If $\{\phi_i\}$ is a net in $B(\sigma(N))$ such that $\int \phi_i\, d\mu \to 0$ for every μ in $M(\sigma(N))$, then $\phi_i(N) \to 0$ (WOT). This map is unique in the sense that if $\tau \colon B(\sigma(N)) \to \mathscr{B}(\mathscr{H})$ is a representation such that $\tau(z) = N$ and $\tau(\phi_i) \to 0$ (WOT) whenever $\{\phi_i\}$ is a bounded net in $B(\sigma(N))$ such that $\int \phi_i\, d\mu \to 0$ for every μ in $M(\sigma(N))$, then $\tau(\phi) = \phi(N)$ for all ϕ in $B(\sigma(N))$.

PROOF. The fact that $\phi \mapsto \phi(N)$ is a representation is a consequence of Proposition 1.12. If $\{\phi_i\}$ is as in the statement, then the fact that $E_{g,h} \in M(\sigma(N))$ implies that $\phi_i(N) \to 0$ (WOT).

To prove uniqueness, let $\tau \colon B(\sigma(N)) \to \mathscr{B}(\mathscr{H})$ be a representation with the appropriate properties. Then $\tau(u) = u(N)$ if $u \in C(\sigma(N))$ by the uniqueness of the functional calculus for normal elements of a C^*-algebra (VIII.2.6). If $\phi \in B(\sigma(N))$, then Proposition V.4.1 implies that there is a net $\{u_i\}$ in $C(\sigma(N))$ such that $\|u_i\| \leq \|\phi\|$ for all u_i and $\int u_i\, d\mu \to \int \phi\, d\mu$ for every μ in $M(\sigma(N))$. Thus $u_i(N) \to \phi(N)$ (WOT). But $\tau(\phi) = \text{WOT} - \lim \tau(u_i) = \text{WOT} - \lim u_i(N)$; therefore $\tau(\phi) = \phi(N)$.   ∎

It is worthwhile to rewrite (1.11) as

2.4 $$\langle \phi(N)g, h \rangle = \int \phi\, dE_{g,h}$$

for ϕ in $B(\sigma(N))$ and g, h in $\mathscr{H}$. If $\phi \in B(\mathbb{C})$, then the restriction of ϕ to $\sigma(N)$ belongs to $B(\sigma(N))$. Since the support of each measure $E_{g,h}$ is contained in $\sigma(N)$, (2.4) holds for every bounded Borel function ϕ on $\mathbb{C}$. This has certain technical advantages that will become apparent when we begin to apply (2.4).

Proposition 2.3 thus extends the functional calculus for normal operators. This functional calculus or, equivalently, the Spectral Theorem, will be exploited in this chapter. But right now we look at some examples.

2.5. Example. If μ is a regular Borel measure on $\mathbb{C}$ with compact support K, define N_μ on $L^2(\mu)$ by $N_\mu f = zf$ for each f in $L^2(\mu)$. It is easy to check that $N_\mu^* f = \bar{z}f$, and, hence, N_μ is normal.

(a) $\sigma(N_\mu) = K = $ support of μ. (Exercise.)
(b) If, for a bounded Borel function ϕ, we define M_ϕ on $L^2(\mu)$ by $M_\phi f = \phi f$, then $\phi(N_\mu) = M_\phi$.

Indeed, this is an easy application of the uniqueness part of (2.3).

(c) If E is the spectral measure for N_μ, then $E(\Delta) = M_{\chi_\Delta}$.

Just note that $E(\Delta) = \chi_\Delta(N_\mu)$.

2.6. Example. Let (X, Ω, μ) be any σ-finite measure space and put $\mathscr{H} = L^2(X, \Omega, \mu)$. For ϕ in $L^\infty(\mu) \equiv L^\infty(X, \Omega, \mu)$, define M_ϕ on $\mathscr{H}$ by $M_\phi f = \phi f$.

(a) M_ϕ is normal and $M_\phi^* = M_{\bar\phi}$ (II.2.8).
(b) $\phi \mapsto M_\phi$ is a representation of $L^\infty(\mu)$ (VIII.5.4).
(c) If $\phi \in L^\infty(\mu)$, $\|\phi\|_\infty = \|M_\phi\|$ (II.1.5).
(d) Define the *essential range* of ϕ by

$$\text{ess-ran}(\phi) \equiv \cap \{\text{cl}(\phi(\Delta)): \Delta \in \Omega \text{ and } \mu(X \backslash \Delta) = 0\}.$$

Then $\sigma(M_\phi) = \text{ess-ran}(\phi)$. (This appears as Exercise VII.3.3, but a proof is given here.)

First assume that $\lambda \notin \text{ess-ran}(\phi)$. So there is a set Δ in Ω with $\mu(X \backslash \Delta) = 0$ and λ not in $\text{cl}(\phi(\Delta))$; thus there is a $\delta > 0$ with $|\phi(x) - \lambda| \geq \delta$ for all x in Δ. If $\psi = (\phi - \lambda)^{-1}$, $\psi \in L^\infty(\mu)$ and $M_\psi = (M_\phi - \lambda)^{-1}$.

Conversely, assume $\lambda \in \text{ess-ran}(\phi)$. It follows that for every integer n there is a set Δ_n in Ω such that $0 < \mu(\Delta_n) < \infty$ and $|\phi(x) - \lambda| < 1/n$ for all x in Δ_n. Put $f_n = (\mu(\Delta_n))^{-1/2} \chi_{\Delta_n}$; so $f_n \in L^2(\mu)$ and $\|f_n\|_2 = 1$. However, $\|(M_\phi - \lambda)f_n\|_2^2 = (\mu(\Delta_n))^{-1} \int_{\Delta_n} |\phi - \lambda|^2 \, d\mu \leq 1/n^2$, showing that $\lambda \in \sigma_{ap}(M_\phi)$.

(e) If E is the spectral measure for M_ϕ [so E is defined on the Borel subsets of $\sigma(M_\phi) = \text{ess-ran}(\phi) \subseteq \mathbb{C}$], then for every Borel subset Δ of $\sigma(M_\phi)$, $E(\Delta) = M_{\chi_{\phi^{-1}(\Delta)}}$.

2.7. Proposition. *If for $k \geq 1$, N_k is a normal operator on $\mathscr{H}_k$ with $\sup_k \|N_k\| < \infty$, E_k is the spectral measure for N_k, and if $N = \oplus_{k=1}^\infty N_k$ on $\mathscr{H} = \oplus_{k=1}^\infty \mathscr{H}_k$, then:*

(a) $\sigma(N) = \text{cl}[\bigcup_{k=1}^\infty \sigma(N_k)]$;
(b) *if E is the spectral measure for N, $E(\Delta) = \oplus_{k=1}^\infty E_k(\Delta \cap \sigma(N_k))$ for every Borel subset Δ of $\sigma(N)$.*

PROOF. Exercise.

A historical account of the spectral theorem is an enormous undertaking by itself. One such account in Steen [1973]. You might also consult the notes in Dunford and Schwartz [1963] and Halmos [1951].

EXERCISES

Throughout these exercises, N is a normal operator on $\mathscr{H}$ with spectral measure E.

1. Show that $\lambda \in \sigma_p(N)$ if and only if $E(\{\lambda\}) \neq 0$. Moreover, if $\lambda \in \sigma_p(N)$, $E(\{\lambda\})$ is the orthogonal projection onto $\ker(N - \lambda)$.

2. If Δ is a clopen subset of $\sigma(N)$, show that $E(\Delta)$ is the Riesz idempotent associated with Δ.

3. Prove Theorem II.5.1 and its corollaries by using the Spectral Theorem.

4. Prove Theorem II.7.6 and its corollaries by using the Spectral Theorem.

5. Obtain Theorem II.7.11 as a consequence of (2.3).

6. Verify the statements in Example 2.5.

7. Verify (2.6e).

8. Show that if $\mathscr{H}$ is separable, there are at most a countable number of points $\{z_n\}$ in $\sigma(N)$ such that $E(z_n) \neq 0$. By Exercise 1, these are the eigenvalues of N.

9. Show that a normal operator N is (a) hermitian if and only if $\sigma(N) \subseteq \mathbb{R}$; (b) positive if and only if $\sigma(N) \subseteq [0, \infty)$; (c) unitary if and only if $\sigma(N) \subseteq \partial \mathbb{D}$.

10. Let A be a hermitian operator with spectral measure E on a separable space. For each real number t define a projection $P(t) = E(-\infty, t)$. Show:
    (a) $P(s) \leqslant P(t)$ for $s \leqslant t$;
    (b) if $t_n \leqslant t_{n+1}$ and $t_n \to t$, $P(t_n) \to P(t)$ (SOT);
    (c) for all but a countable number of points t, $P(t_n) \to P(t)$ (SOT) if $t_n \to t$;
    (d) for f in $C(\sigma(A))$, $f(A) = \int_{-\infty}^{\infty} f(t)\, dP(t)$, where this integral is to be defined (by the reader) in the Riemann–Stieltjes sense.

11. If $\sigma_p(N)$ is a Borel set, show that $E(\sigma(N)\backslash\sigma_p(N)) = 0$ if and only if N is diagonalizable; that is, there is a basis for $\mathscr{H}$ consisting of eigenvectors for N. If $\sigma_p(N)$ is not assumed to be a Borel set, is it still possible to characterize diagonalizable normal operators in a similar way? Give an example of a normal operator N such that $\sigma_p(N)$ is not a Borel set.

12. Show that if $N = U|N|$ ($|N| = (N^*N)^{1/2}$) is the polar decomposition of N, $U = \phi(N)$ for some Borel function ϕ on $\sigma(N)$. Hence $U|N| = |N|U$ (see Exercise VIII.3.21).

13. Show that $N = W|N|$ for some unitary W that is a function of N.

14. Prove that if A is hermitian, $\exp(iA)$ is unitary. Is the converse true?

15. Show that there is a normal operator M such that $M^2 = N$ and $M = \phi(N)$ for some Borel function ϕ. How many such normal operators M are there?

16. Define $N: L^2(\mathbb{R}) \to L^2(\mathbb{R})$ by $(Nf)(t) = f(t + 1)$. Show that N is normal and find its spectral decomposition. ((X.6.17) is useful here.)

17. Suppose that $N_1, \ldots, N_d$ are normal operators such that $N_j N_k^* = N_k^* N_j$ for

$1 \leqslant j, k \leqslant d$. Show that there is a subset X of $\mathbb{C}^d$ and a spectral measure E defined on the Borel subsets of X such that $N_k = \int z_k \, dE(z)$ for $1 \leqslant k \leqslant d$ (z_k = the kth coordinate function) (see Exercise VIII.2.2).

18. If $N_1, \ldots, N_d$ are as in Exercise 17 and each is compact, show that there is a basis for $\mathcal{H}$ consisting of eigenvectors for each N_k. (This is the *simultaneous diagonalization of* $N_1, \ldots, N_d$.)

19. This exercise gives the properties of Hilbert–Schmidt operators (defined below). (a) If $\{e_i\}$ and $\{f_j\}$ are two orthonormal bases for $\mathcal{H}$ and $A \in \mathcal{B}(\mathcal{H})$, then

$$\sum_i \| Ae_i \|^2 = \sum_j \| Af_j \|^2 = \sum_i \sum_j |\langle Ae_i, f_j \rangle|^2.$$

(b) If $A \in \mathcal{B}(\mathcal{H})$ and $\{e_i\}$ is a basis for $\mathcal{H}$, define

$$\| A \|_2 = \left[\sum_i \| Ae_i \|^2 \right]^{1/2}.$$

By (a) $\| A \|_2$ is independent of the basis chosen and hence is well defined. If $\| A \|_2 < \infty$, A is called a *Hilbert–Schmidt operator*. $\mathcal{B}_2 = \mathcal{B}_2(\mathcal{H})$ denotes the set of all Hilbert–Schmidt operators. (c) $\| A \| \leqslant \| A \|_2$ for every A in $\mathcal{B}(\mathcal{H})$ and $\| \cdot \|_2$ is a norm on $\mathcal{B}_2$. (d) If $T \in \mathcal{B} = \mathcal{B}(\mathcal{H})$ and $A \in \mathcal{B}_2$, then $\| TA \|_2 \leqslant \| T \| \, \| A \|_2$, $\| A^* \|_2 = \| A \|_2$, and $\| AT \|_2 \leqslant \| A \|_2 \| T \|$. (e) $\mathcal{B}_2$ is an ideal of $\mathcal{B}$ that contains $\mathcal{B}_{00}$, the finite-rank operators. (f) $A \in \mathcal{B}_2$ if and only if $|A| \equiv (A^*A)^{1/2} \in \mathcal{B}_2$; in this case $\| A \|_2 = \| \, |A| \, \|_2$. (g) $\mathcal{B}_2 \subseteq \mathcal{B}_0$; moreover, if A is a compact operator and $\lambda_1, \lambda_2, \ldots$ are the eigenvalues of $|A|$, each repeated as often as its multiplicity, then $A \in \mathcal{B}_2(\mathcal{H})$ iff $\sum_{n=1}^{\infty} \lambda_n^2 < \infty$. In this case, $\| A \|_2 = (\sum \lambda_n^2)^{1/2}$. (h) If (X, Ω, μ) is a measure space and $k \in L^2(\mu \times \mu)$, let $K : L^2(\mu) \to L^2(\mu)$ be the integral operator with kernel k. Then $K \in \mathcal{B}_2(L^2(\mu))$ and $\| K \|_2 = \| k \|_2$ (see Proposition II.4.7 and Lemma II.4.8). (i) Interpret part (h) for a purely atomic measure space. More information on $\mathcal{B}_2$ is contained in the next exercise.

20. This exercise discusses trace-class operators (defined below) and assumes a knowledge of Exercise 19. $\mathcal{B}_1(\mathcal{H}) = \{AB : A \text{ and } B \in \mathcal{B}_2(\mathcal{H})\}$. Operators belonging to $\mathcal{B}_1(\mathcal{H})$ are called *trace-class operators* and $\mathcal{B}_1(\mathcal{H}) = \mathcal{B}_1$ is called the *trace class*. (a) If $A \in \mathcal{B}_1(\mathcal{H})$ and $\{e_i\}$ is a basis, then $\sum |\langle Ae_i, e_i \rangle| < \infty$. Moreover, the sum $\sum \langle Ae_i, e_i \rangle$ is independent of the choice of basis. (Hint: If $A = C^*B$, B, C in $\mathcal{B}_2$, show that $|\langle Ae_i, e_i \rangle| \equiv \frac{1}{2}(\| Be_i \|^2 + \| Ce_i \|^2)$.) (b) If $\{e_i\}$ is a basis for $\mathcal{H}$, define $\mathrm{tr} : \mathcal{B}_1 \to \mathbb{C}$ by

$$\mathrm{tr}(A) = \sum_i \langle Ae_i, e_i \rangle.$$

By (a) the definition of $\mathrm{tr}(A)$ does not depend on the choice of a basis; $\mathrm{tr}(A)$ is called the *trace* of A. If $\dim \mathcal{H} < \infty$, then $\mathrm{tr}(A)$ is precisely the sum of the diagonal terms of any matrix representation of A. (c) If $A \in \mathcal{B}(\mathcal{H})$, then the following are equivalent: (1) $A \in \mathcal{B}_1$; (2) $|A| = (A^*A)^{1/2} \in \mathcal{B}_1$; (3) $|A|^{1/2} \in \mathcal{B}_2$; (4) $\mathrm{tr}(|A|) < \infty$. (d) If $A \in \mathcal{B}_1$ and $T \in \mathcal{B}$, then AT and TA are in $\mathcal{B}_1$ and $\mathrm{tr}(AT) = \mathrm{tr}(TA)$. Moreover, $\mathrm{tr} : \mathcal{B}_1 \to \mathbb{C}$ is a positive linear functional such that if $A \in \mathcal{B}_1$, $A \geqslant 0$, and $\mathrm{tr}(A) = 0$, then $A = 0$. (e) If $A \in \mathcal{B}_1$, define $\| A \|_1 \equiv \mathrm{tr}(|A|)$. If $A \in \mathcal{B}_1$ and $T \in \mathcal{B}$, show that $|\mathrm{tr}(TA)| \leqslant \| T \| \, \| A \|_1$. (f) $\| A \|_1 = \| A^* \|_1$ if $A \in \mathcal{B}_1$. (g) If $T \in \mathcal{B}$ and $A \in \mathcal{B}_1$, then $\| TA \|_1 \leqslant \| T \| \, \| A \|_1$ and $\| AT \|_1 \leqslant \| T \| \, \| A \|_1$. (h) $\| \cdot \|_1$ is a norm on $\mathcal{B}_1$. It is called the *trace norm*. (i) $\mathcal{B}_1$ is an ideal in $\mathcal{B}(\mathcal{H})$ that contains $\mathcal{B}_{00}$. (j) If $A \in \mathcal{B}_1$

and $\{e_i\}$ and $\{f_i\}$ are two bases for $\mathscr{H}$, then $\sum_i |\langle Ae_i, f_i \rangle| \leqslant \|A\|_1$. (d) $\mathscr{B}_1 \subseteq \mathscr{B}_0$. Also, if $A \in \mathscr{B}_0$ and $\lambda_1, \lambda_2, \ldots$ are the eigenvalues of $|A|$, each repeated as often as its multiplicity, then $A \in \mathscr{B}_1$ if and only if $\sum_{n=1}^{\infty} \lambda_n < \infty$. In this case, $\|A\|_1 = \sum_{n=1}^{\infty} \lambda_n$. (1) If A and $B \in \mathscr{B}_2$, define $(A, B) = \operatorname{tr}(B^*A)$. Then $(\cdot, \cdot)$ is an inner product on $\mathscr{B}_2$, $\|\cdot\|_2$ is the norm defined by this inner product, and $\mathscr{B}_2$ is $\|\cdot\|_2$ complete. In other words, $\mathscr{B}_2$ is a Hilbert space. (m) $(\mathscr{B}_1, \|\cdot\|_1)$ is a Banach space. (n) $\mathscr{B}_{00}$ is dense in both $\mathscr{B}_1$ and $\mathscr{B}_2$. (For more on these matters, see Ringrose [1971] and Schatten [1960].)

21. This exercise assumes a knowledge of Exercise 20. If $g, h \in \mathscr{H}$, let $g \otimes h$ denote the rank-one operator defined by $(g \otimes h)(f) = \langle f, h \rangle g$. (a) If $g, h \in \mathscr{H}$ and $A \in \mathscr{B}(\mathscr{H})$, $\operatorname{tr}(A(g \otimes h)) = \langle Ag, h \rangle$. (b) If $T \in \mathscr{B}_1$, then $\|T\|_1 = \sup\{|\operatorname{tr}(CT)|: C \in \mathscr{B}_0, \|C\| \leqslant 1\}$. (c) If $T \in \mathscr{B}_1$, define $L_T: \mathscr{B}_0 \to \mathbb{C}$ by $L_T(C) = \operatorname{tr}(TC) \, (= \operatorname{tr}(CT))$. Show that the map $T \mapsto L_T$ is an isometric isomorphism of $\mathscr{B}_1$ onto $\mathscr{B}_0^*$. (d) If $B \in \mathscr{B}$, define $F_B: \mathscr{B}_1 \to \mathbb{C}$ by $F_B(T) = \operatorname{tr}(BT)$. Show that $B \mapsto F_B$ is an isometric isomorphism of $\mathscr{B}$ onto $\mathscr{B}_1^*$. (e) If $L \in \mathscr{B}^*$ show that $L = L_0 + L_1$ where $L_0, L_1 \in \mathscr{B}^*$, $L_1(B) = \operatorname{tr}(BT)$ for some T in $\mathscr{B}_1$, and $L_0(C) = 0$ for every compact operator C. Show that $\|L\| = \|L_0\| + \|L_1\|$ and that L_0 and L_1 are unique. Give necessary and sufficient conditions on $\mathscr{H}$ for $\mathscr{B}(\mathscr{H})$ to be a reflexive Banach space.

22. Prove that if U is any unitary operator on $\mathscr{H}$, then there is a continuous function u: $[0, 1] \to \mathscr{B}(\mathscr{H})$ such that $u(t)$ is unitary for all t, $u(0) = U$, and $u(1) = 1$.

23. If N is normal, show that there is a sequence of invertible normal operators that converges to N.

§3. Star-Cyclic Normal Operators

Recall the definition of a reducing subspace and some of its equivalent formulations (Section II.3).

3.1. Definition. A vector e_0 in $\mathscr{H}$ is a *star-cyclic vector* for A if $\mathscr{H}$ is the smallest reducing subspace for A that contains e_0. The operator A is *star cyclic* if it has a star-cyclic vector. A vector e_0 is *cyclic* for A if $\mathscr{H}$ is the smallest invariant subspace for A that contains e_0; A is *cyclic* if it has a cyclic vector.

3.2. Proposition. (a) *A vector e_0 is a star-cyclic vector for A if and only if $\mathscr{H} = \operatorname{cl}\{Te_0: T \in C^*(A)\}$, where $C^*(A) = $ the C^*-algebra generated by A. (b) A vector e_0 is a cyclic vector for A if and only if $\mathscr{H} = \operatorname{cl}\{p(A)e_0: p = a\ polynomial\}$.*

PROOF. Exercise.

Note that if e_0 is a star-cyclic vector for A, then it is a cyclic vector for the algebra $C^*(A)$.

3.3. Proposition. *If A has either a cyclic or a star-cyclic vector, then $\mathscr{H}$ is separable.*

PROOF. It is easy to see that $C^*(A)$ and $\{p(A): p = \text{a polynomial}\}$ are separable subalgebras of $\mathcal{B}(\mathcal{H})$. Now use (3.2). ∎

Let μ be a compactly supported measure on $\mathbb{C}$ and let N_μ be defined on $L^2(\mu)$ as in Example 2.5. If $K = \text{support } \mu$, then $C^*(N_\mu) = \{M_u: u \in C(K)\}$. Since $C(K)$ is dense in $L^2(\mu)$, it follows that 1 is a star-cyclic vector for N_μ. The converse of this is also true.

3.4. Theorem. *A normal operator N is star-cyclic if and only if N is unitarily equivalent to N_μ for some compactly supported measure μ on $\mathbb{C}$. If e_0 is a star-cyclic vector for N, then μ can be chosen such that there is an isomorphism $V: \mathcal{H} \to L^2(\mu)$ with $Ve_0 = 1$ and $VNV^{-1} = N_\mu$. Under these conditions, V is unique.*

PROOF. If $N \cong N_\mu$, then we have already seen that N is star-cyclic. So suppose that N has a star-cyclic vector e_0. If E is the spectral measure for N, put $\mu(\Delta) = \| E(\Delta)e_0 \|^2 = \langle E(\Delta)e_0, e_0 \rangle$ for every Borel subset Δ of $\mathbb{C}$ (see Lemma 1.9). Let $K = \text{support } \mu$.

If $\phi \in B(K)$, then (2.4) implies

$$\begin{aligned}
\| \phi(N)e_0 \|^2 &= \langle \phi(N)e_0, \phi(N)e_0 \rangle \\
&= \langle |\phi|^2(N)e_0, e_0 \rangle \\
&= \int |\phi(z)|^2 d\langle E(z)e_0, e_0 \rangle \\
&= \int |\phi|^2 d\mu.
\end{aligned}$$

So if $B(K)$ is considered as a submanifold of $L^2(\mu)$, $U\phi = \phi(N)e_0$ defines an isometry from $B(K)$ onto $\{\phi(N)e_0: \phi \in B(K)\}$. But e_0 is a star-cyclic vector, so the range of U is dense in $\mathcal{H}$. Hence U extends to an isomorphism $U: L^2(\mu) \to \mathcal{H}$.

If $\phi \in B(K)$, then $UN_\mu U^{-1}(\phi(N)e_0) = UN_\mu(\phi) = U(z\phi) = N\phi(N)e_0$. Hence $UN_\mu U^{-1} = N$ on $\{\phi(N)e_0: \phi \in B(K)\}$, which is dense in $\mathcal{H}$. So $UN_\mu U^{-1} = N$. Let $V = U^{-1}$.

The proof of the uniqueness statement is an exercise. ∎

Any theorem about the operators N_μ is a theorem about star-cyclic normal operators. With this in mind, the next theorem gives a complete unitary invariant for star-cyclic normal operators. But first, a definition.

3.5. Definition. Two measures, μ_1 and μ_2, are *mutually absolutely continuous* if they have the same sets of measure zero; that is, $\mu_1(\Delta) = 0$ if and ony if $\mu_2(\Delta) = 0$. This will be denoted by $[\mu_1] = [\mu_2]$. (The more standard notation in the literature is $\mu_1 \equiv \mu_2$, but this seems insufficient.) If $[\mu_1] = [\mu_2]$, then the Radon–Nikodym derivatives $d\mu_1/d\mu_2$ and $d\mu_2/d\mu_1$ are well defined. Say that μ_1 and μ_2 are *boundedly mutually absolutely continuous* if $[\mu_1] = [\mu_2]$ and the Radon–Nikodym derivatives are essentially bounded functions.

3.6. Theorem. $N_{\mu_1} \cong N_{\mu_2}$ *if and only if* $[\mu_1] = [\mu_2]$.

PROOF. Suppose $[\mu_1] = [\mu_2]$ and put $\phi = d\mu_1/d\mu_2$. So if $g \in L^1(\mu_1)$, $g\phi \in L^1(\mu_2)$ and $\int g\phi \, d\mu_2 = \int g \, d\mu_1$. Hence, if $f \in L^2(\mu_1)$, $\sqrt{\phi} f \in L^2(\mu_2)$ and $\|\sqrt{\phi} f\|_2 = \|f\|_2$; that is, $U: L^2(\mu_1) \to L^2(\mu_2)$ defined by $Uf = \sqrt{\phi} f$ is an isometry. If $g \in L^2(\mu_2)$, then $f = \phi^{-1/2} g \in L^2(\mu_1)$ and $Uf = g$; hence U is surjective and $U^{-1} g = \phi^{-1/2} g$ for g in $L^2(\mu_2)$. If $g \in L^2(\mu_2)$, then $U N_{\mu_1} U^{-1} g = U N_{\mu_1} \phi^{-1/2} g = U z \phi^{-1/2} g = zg$, and so $U N_{\mu_1} U^{-1} = N_{\mu_2}$.

Now assume that $V: L^2(\mu_1) \to L^2(\mu_2)$ is an isomorphism such that $V N_{\mu_1} V^{-1} = N_{\mu_2}$. Put $\psi = V(1)$; so $\psi \in L^2(\mu_2)$. For convenience, put $N_j = N_{\mu_j}, j = 1, 2$. It is easy to see that $V N_1^k V^{-1} = N_2^k$ and $V N_1^{*k} V^{-1} = N_2^{*k}$. Hence $V p(N_1, N_1^*) V^{-1} = p(N_2, N_2^*)$ for any polynomial p in z and $\bar{z}$. Since $N_1 \cong N_2$, $\sigma(N_1) = \sigma(N_2)$; hence support $\mu_1 = $ support $\mu_2 = K$. By taking uniform limits of polynomials in z and $\bar{z}$, $V u(N_1) V^{-1} = u(N_2)$ for u in $C(K)$. Hence for u in $C(K)$, $V(u) = V u(N_1) 1 = u(N_2) V 1 = u\psi$. Because V is an isometry, this implies that $\int |u|^2 d\mu_1 = \int |u|^2 |\psi|^2 d\mu_2$ for every u in $C(K)$. Hence $\int v \, d\mu_1 = \int v |\psi|^2 d\mu_2$ for v in $C(K)$, $v \geq 0$. By the uniqueness part of the Riesz Representation Theorem, $\mu_1 = |\psi|^2 \mu_2$, so $\mu_1 \ll \mu_2$.

By using V^{-1} instead of V and reversing the roles of N_1 and N_2 in the preceding argument, it follows that $\mu_2 \ll \mu_1$. Hence $[\mu_1] = [\mu_2]$. ∎

EXERCISES

1. If μ is a compactly supported measure on $\mathbb{C}$ and $f \in L^2(\mu)$, f is a star-cyclic vector for N_μ if and only if $\mu(\{x: f(x) = 0\}) = 0$.

2. Prove Proposition 3.2.

3. If μ_1 and μ_2 are compactly supported measures on $\mathbb{C}$, show that the following statements are equivalent: (a) μ_1 and μ_2 are boundedly mutually absolutely continuous; (b) there is an isomorphism $V: L^2(\mu_1) \to L^2(\mu_2)$ such that $V N_{\mu_1} V^{-1} = N_{\mu_2}$ and $V L^\infty(\mu_1) = L^\infty(\mu_2)$; (c) there is a bounded bijection $R: L^2(\mu_1) \to L^2(\mu_2)$ such that $R p(z, \bar{z}) = p(z, \bar{z})$ for every polynomial in z and $\bar{z}$.

4. Show that if N is a star-cyclic normal operator and $\lambda \in \sigma_p(N)$, then $\dim \ker(N - \lambda) = 1$.

5. If N is diagonalizable and star-cyclic and if $\sigma_p(N) = \{\lambda_1, \lambda_2, \ldots\}$, show that N is unitarily equivalent to N_μ, where $\mu = \sum_{n=1}^\infty 2^{-n} \delta_{\lambda_n}$ (see Exercise 2.11).

6. Let N be a diagonalizable normal operator. Show that $N \cong M$ if and only if M is a diagonalizable normal operator, $\sigma_p(N) = \sigma_p(M)$, and $\dim \ker(N - \lambda) = \dim \ker(M - \lambda)$ for all λ. (Compare this with Theorem II.8.3.)

7. Let U be the bilateral shift on $l^2(\mathbb{Z})$. If e_0 is the vector in $l^2(\mathbb{Z})$ that has 1 in the zeroth place and zeros elsewhere, then e_0 is a star-cyclic vector for U. If μ is the compactly supported measure on $\mathbb{C}$ and $V: l^2(\mathbb{Z}) \to L^2(\mu)$ is the isomorphism such that $V e_0 = 1$ and $V U V^{-1} = N_\mu$, then

(a) $\mu = m$ = normalized arc length on $\partial\mathbb{D}$;
(b) V^{-1} = the Fourier transform on $L^2(m) = L^2(\partial\mathbb{D})$.

8. Suppose $N_1, \ldots, N_d$ are normal operators such that $N_j N_k^* = N_k^* N_j$ for $1 \leqslant j, k \leqslant d$ and suppose there is a vector e_0 in $\mathscr{H}$ such that $\mathscr{H}$ is the only subspace of $\mathscr{H}$ containing e_0 that reduces each of the operators $N_1, \ldots, N_d$. Show that there is a compactly supported measure μ on $\mathbb{C}^d$ and an isomorphism $V: \mathscr{H} \to L^2(\mu)$ such that $V N_k V^{-1} f = z_k f$ for f in $L^2(\mu)$ and $1 \leqslant k \leqslant d$ (z_k = the kth coordinate function) (see Exercise 2.17).

§4. Some Applications of the Spectral Theorem

In this section a few diverse applications of the Spectral Theorem are presented. These will show the power and finesse of the Spectral Theorem as well as demonstrate some of the methods used to apply it. One result in this section (Theorem 4.6) is more than an application. Indeed, many regard this as the optimal statement of the Spectral Theorem.

If N is a normal operator and $N = \int z \, dE(z)$ is its spectral representation, then $\phi \mapsto \phi(N) \equiv \int \phi \, dE$ is a *-homomorphism of $B(\mathbb{C})$ into $\mathscr{B}(\mathscr{H})$. Thus, if ϕ, $\psi \in B(\mathbb{C})$, $(\int \phi \, dE)(\int \psi \, dE) = \int \phi\psi \, dE$ and $\| \int \phi \, dE \| \leqslant \sup\{|\phi(z)|: z \in \sigma(N)\}$.

4.1. Proposition. *If N is a normal operator and $N = \int z \, dE(z)$, then N is compact if and only if for every $\varepsilon > 0$, $E(\{z: |z| > \varepsilon\})$ has finite rank.*

PROOF. If $\varepsilon > 0$, let $\Delta_\varepsilon = \{z: |z| > \varepsilon\}$ and $E_\varepsilon = E(\Delta_\varepsilon)$. Then

$$N - NE_\varepsilon = \int z \, dE(z) - \int z\chi_{\Delta_\varepsilon}(z) \, dE(z)$$

$$= \int z\chi_{\mathbb{C}\backslash\Delta_\varepsilon}(z) \, dE(z) = \phi(N)$$

where $\phi(z) = z\chi_{\mathbb{C}\backslash\Delta_\varepsilon}(z)$. Thus $\| N - NE_\varepsilon \| \leqslant \sup\{|z|: z \in \mathbb{C}\backslash\Delta_\varepsilon\} \leqslant \varepsilon$. If E_ε has finite rank for every $\varepsilon > 0$, then so does NE_ε. Thus $N \in \mathscr{B}_0(\mathscr{H})$.

Now assume that N is compact and let $\varepsilon > 0$. Put $\phi(z) = z^{-1}\chi_{\Delta_\varepsilon}(z)$; so $\phi \in B(\mathbb{C})$. Since N is compact, so is $N\phi(N)$. But $N\phi(N) = \int zz^{-1}\chi_{\Delta_\varepsilon}(z) \, dE(z) = E_\varepsilon$. Since E_ε is a compact projection, it must have finite rank. (Why?) ∎

The preceding result could have been proved by using the fact that compact normal operators are diagonalizable and the eigenvalues must converge to 0.

4.2. Theorem. *If $\mathscr{H}$ is separable and I is an ideal of $\mathscr{B}(\mathscr{H})$ that contains a noncompact operator, then $I = \mathscr{B}(\mathscr{H})$.*

PROOF. If $A \in I$ and $A \notin \mathscr{B}_0(\mathscr{H})$, consider A^*A; let $A^*A = \int t \, dE(t)$ ($\sigma(A^*A) \subseteq [0, \infty)$). By the preceding proposition, there is an $\varepsilon > 0$ such that

$P = E(\varepsilon, \infty)$ has infinite rank. But $P = (\int t^{-1}\chi_{(\varepsilon, \infty)}(t)dE(t))A^*A \in I$. Since $\mathscr{H}$ is separable, dim $P\mathscr{H} = \dim \mathscr{H} = \aleph_0$. Let $U: \mathscr{H} \to P\mathscr{H}$ be a surjective isometry. It is easy to check that $1 = U^*PU$. But $P \in I$, so $1 \in I$. Hence $I = \mathscr{B}(\mathscr{H})$. ∎

In Proposition VIII.4.10, it was shown that every nonzero ideal of $\mathscr{B}(\mathscr{H})$ contains the finite-rank operators. When combined with the preceding result, this yields the following.

4.3. Corollary. *If $\mathscr{H}$ is separable, then the only nontrivial closed ideal of $\mathscr{B}(\mathscr{H})$ is the ideal of compact operators.*

The next proposition is related to Theorem VIII.5.9. Indeed, it is a consequence of it so that the proof will only be sketched.

Let N be a normal operator on $\mathscr{H}$ and for every vector e in $\mathscr{H}$ let $\mathscr{H}_e \equiv \bigvee \{N^{*k}N^j e: k, j \geqslant 0\}$. So $\mathscr{H}_e$ is the smallest subspace of $\mathscr{H}$ that contains e and reduces N. Also, $N|\mathscr{H}_e$ is a star-cyclic normal operator.

4.4. Proposition. *If N is a normal operator on $\mathscr{H}$, then there are reducing subspaces $\{\mathscr{H}_i: i \in I\}$ for N such that $\mathscr{H} = \oplus_i \mathscr{H}_i$ and $N|\mathscr{H}_i$ is star cyclic.*

PROOF. Using Zorn's Lemma find a maximal set of vectors $\mathscr{E}$ in $\mathscr{H}$ such that if $e, f \in \mathscr{E}$ and $e \neq f$, then $\mathscr{H}_e \perp \mathscr{H}_f$. It follows that $\mathscr{H} = \oplus_e \mathscr{H}_e$. ∎

4.5. Corollary. *Every normal operator is unitarily equivalent to the direct sum of star-cyclic normal operators.*

By combining the preceding proposition with Theorem 3.4 on the representation of star-cyclic normal operators we can obtain the following theorem.

4.6. Theorem. *If N is a normal operator on $\mathscr{H}$, then there is a measure space (X, Ω, μ) and a function ϕ in $L^\infty(X, \Omega, \mu)$ such that N is unitarily equivalent to M_ϕ on $L^2(X, \Omega, \mu)$.*

PROOF. If $\mathscr{M}$ is a reducing subspace for N, then $N \cong N|\mathscr{M} \oplus N|\mathscr{M}^\perp$; thus $\sigma(N|\mathscr{M}) \subseteq \sigma(N)$. So if $\{N_i\}$ is a collection of star-cyclic normal operators such that $N \cong \oplus_i N_i$ (4.5), then $\sigma(N_i) \subseteq \sigma(N)$ for every N_i. By Theorem 3.4 there is a measure μ_i supported on $\sigma(N)$ such that $N_i \cong N_{\mu_i}$. Let X_i = the support of μ_i and let Ω_i = the Borel subsets of X_i. Let X = the disjoint union of $\{X_i\}$. Define Ω to be the collection of all subsets Δ of X such that $\Delta \cap X_i \in \Omega_i$ for all i. It is easy to check that Ω is a σ-algebra. If $\Delta \in \Omega$ let $\mu(\Delta) \equiv \sum_i \mu_i(\Delta \cap X_i)$; then (X, Ω, μ) is a measure space. If $f \in L^2(X, \Omega, \mu)$ then $f_i = f|X_i \in L^2(\mu_i)$. Moreover, the map $U: L^2(\mu) \to \oplus_i L^2(\mu_i)$ defined by $Uf = \oplus_i (f|X_i)$ is easily seen to be an isomorphism. Define $\phi: X \to \mathbb{C}$ by letting $\phi(z) = z$ if $z \in X_i (\subseteq \mathbb{C})$; since $X_i \subseteq \sigma(N)$ for every i, ϕ is a bounded function. If G is an open subset of $\mathbb{C}$,

$\phi^{-1}(G) \cap X_i = G \cap X_i \in \Omega_i$; hence ϕ is Ω-measurable. Therefore $\phi \in L^\infty(X, \Omega, \mu)$. It is left to the reader to check that $UM_\phi U^{-1} = \oplus_i N_{\mu_i} \cong N$. ∎

4.7. Proposition. *If $\mathscr{H}$ is separable, then the measure space in Theorem 4.6 is σ-finite.*

PROOF. First note that the measure space (X, Ω, μ) constructed in the preceding theorem has no infinite atoms. Now let $\mathscr{E}$ be a collection of pairwise disjoint sets from Ω having non-zero finite measure. A computation shows that $\{(\mu(\Delta))^{-1/2}\chi_\Delta : \Delta \in \mathscr{E}\}$ are pairwise orthogonal vectors in $L^2(\mu)$. If $L^2(\mu)$ is separable, then $\mathscr{E}$ must be countable. Therefore (X, Ω, μ) is σ-finite. ∎

Of course if (X, Ω, μ) is finite it is not necessarily true that $L^2(\mu)$ is separable.

The next result will be useful later in this book and it also provides a different type of application of the Spectral Theorem.

4.8. Proposition. *If $\mathscr{A}$ is an SOT closed C*-subalgebra of $\mathscr{B}(\mathscr{H})$, then $\mathscr{A}$ is the norm closed linear span of the projections in $\mathscr{A}$.*

PROOF. If $A \in \mathscr{A}$, $A + A^*$ and $A - A^* \in \mathscr{A}$; hence $\mathscr{A}$ is the linear span of Re $\mathscr{A}$. Suppose $A \in \text{Re }\mathscr{A}$ and $A = \int t \, dE(t)$. If $[a, b] \subseteq \mathbb{R}$, then there is a sequence $\{u_n\}$ in $C(\mathbb{R})$ such that $0 \leqslant u_n \leqslant 1$, $u_n(t) = 1$ for $a \leqslant t \leqslant b - n^{-1}$, $u_n(t) = 0$ for $t \leqslant a - n^{-1}$ and $t \geqslant b$. Hence $u_n(t) \to \chi_{[a,b)}(t)$ as $n \to \infty$. If $h \in \mathscr{H}$, then

$$\| \{u_n(A) - E[a, b)\}h \|^2 = \int |u_n(t) - \chi_{[a,b)}(t)|^2 dE_{h,h}(t) \to 0$$

by the Lebesgue Dominated Convergence Theorem. That is, $u_n(A) \to E[a, b)$ (SOT). Since $\mathscr{A}$ is SOT-closed, $E[a, b) \in \mathscr{A}$. Now let (α, β) be an open interval containing $\sigma(A)$. If $\varepsilon > 0$, then there is a partition $\{\alpha = t_0 < \cdots < t_n = \beta\}$ such that $|t - \sum_{k=1}^n t_k \chi_{[t_{k-1}, t_k)}(t)| < \varepsilon$ for t in $\sigma(A)$; hence $\| A - \sum_{k=1}^n t_k E[t_{k-1}, t_k) \| < \varepsilon$. Thus every self-adjoint operator in $\mathscr{A}$ belongs to the closed linear span of the projections in $\mathscr{A}$. ∎

EXERCISES

1. If N is a normal operator show that ran N is closed if and only if 0 is not a limit point of $\sigma(N)$.

2. Give an example of a non-normal operator A such that 0 is an isolated point of $\sigma(A)$ and ran A is closed. Give an example of a non-normal operator B such that ran B is closed and 0 is not an isolated point of $\sigma(B)$.

3. If $\mathscr{H}$ is a nonseparable Hilbert space find an example of a nontrivial closed ideal of $\mathscr{B}(\mathscr{H})$ that is different from $\mathscr{B}_0(\mathscr{H})$.

4. Let (X, Ω, μ) be the measure space obtained in the proof of Theorem 4.6 and show that $L^1(X, \Omega, \mu)^*$ is isometrically isomorphic to $L^\infty(X, \Omega, \mu)$.

5. Show that $\mathscr{H}$ is separable if and only if every collection of pairwise orthogonal projections in $\mathscr{B}(\mathscr{H})$ is countable.

6. If (X, Ω, μ) is a measure space, then (X, Ω, μ) is σ-finite or $L^2(\mu)$ is finite dimensional if and only if every collection of pairwise orthogonal projections in $\{M_\phi \in L^\infty(\mu)\}$ is countable.

7. If $N = \int z \, dE(z)$ and $\varepsilon > 0$, show that ran $E(\{z: |z| > \varepsilon\}) \subseteq$ ran N.

8. (Calkin [1939]) Let $\mathcal{M}$ be a linear manifold in $\mathcal{H}$ and show that $\mathcal{M}$ has the property that $\mathcal{M}$ contains no closed infinite dimensional subspaces if and only if whenever $A \in \mathcal{B}(\mathcal{H})$ and ran $A \subseteq \mathcal{M}$, then A is compact.

9. Show that the extreme points of $\{A \in \mathcal{B}(\mathcal{H}): 0 \leqslant A \leqslant 1\}$ are the projections.

10. (Halmos [1972]) If N is a normal operator, show that there is a hermitian operator A and a continuous function f such that $N = f(A)$. (Hint: Use Theorem 4.6.)

§5. Topologies on $\mathcal{B}(\mathcal{H})$

In this section some results on the SOT and WOT on $\mathcal{B}(\mathcal{H})$ are presented. These results are necessary for understanding some of the results that are to follow in later sections and also for a proper comprehension of a number of other subjects in mathematics.

The first result appeared as Exercise 1.4.

5.1. Proposition. *If $L: \mathcal{B}(\mathcal{H}) \to \mathbb{C}$ is a linear functional, then the following statements are equivalent.*

(a) *L is SOT continuous.*
(b) *L is WOT continuous.*
(c) *There are vectors $g_1, \ldots, g_n, h_1, \ldots, h_n$ in $\mathcal{H}$ such that $L(A) = \sum_{k=1}^{n} \langle Ag_k, h_k \rangle$ for every A in $\mathcal{B}(\mathcal{H})$.*

PROOF. Clearly (c) implies (b) and (b) implies (a). So assume (a). By (IV.3.1f) there are vectors $g_1, \ldots, g_n$ in $\mathcal{H}$ such that

$$|L(A)| \leqslant \sum_{k=1}^{n} \| Ag_k \| \leqslant \sqrt{n} \left[\sum_{k=1}^{n} \| Ag_k \|^2 \right]^{1/2}$$

for every A in $\mathcal{B}(\mathcal{H})$. Replacing g_k by $\sqrt{n} g_k$, it may be assumed that

$$|L(A)| \leqslant \left[\sum_{k=1}^{n} \| Ag_k \|^2 \right]^{1/2} \equiv p(A).$$

Now p is a seminorm and $p(A) = 0$ implies $L(A) = 0$. Let $\mathcal{K} = \text{cl}\, \{Ag_1 \oplus Ag_2 \oplus \cdots \oplus Ag_n: A \in \mathcal{B}(\mathcal{H})\}$; so $\mathcal{K} \subseteq \mathcal{H} \oplus \cdots \oplus \mathcal{H}$ (n times). Note that if $Ag_1 \oplus \cdots \oplus Ag_n = 0$, $p(A) = 0$, and hence, $L(A) = 0$. Thus $F(Ag_1 \oplus \cdots \oplus Ag_n) = L(A)$ is a well-defined linear functional on a dense manifold in $\mathcal{K}$. But

$$|F(Ag_1 \oplus \cdots \oplus Ag_n)| \leqslant p(A) = \| Ag_1 \oplus \cdots \oplus Ag_n \|.$$

So F can be extended to a bounded linear functional F_1 on $\mathscr{H}^{(n)}$. Hence there are vectors $h_1, \ldots, h_n$ in $\mathscr{H}$ such that

$$F_1(f_1 \oplus \cdots \oplus f_n) = \langle f_1 \oplus \cdots \oplus f_n, h_1 \oplus \cdots \oplus h_n \rangle$$

$$= \sum_{k=1}^{n} \langle f_k, h_k \rangle.$$

In particular, $L(A) = F(Ag_1 \oplus \cdots \oplus Ag_n) = \sum_{k=1}^{n} \langle Ag_k, h_k \rangle$. $\blacksquare$

5.2. Corollary. *If $\mathscr{C}$ is a convex subset of $\mathscr{B}(\mathscr{H})$, the WOT closure of $\mathscr{C}$ equals the SOT closure of $\mathscr{C}$.*

PROOF. Combine the preceding proposition with Corollary V.1.4. $\blacksquare$

When discussing the closure (WOT or SOT) of a convex set it is usually better to discuss the SOT. Shortly an "algebraic" characterization of the SOT closure of a subalgebra of $\mathscr{B}(\mathscr{H})$ will be given. But first recall (VIII.5.3) that if $1 \leqslant n \leqslant \infty$, $\mathscr{H}^{(n)}$ denotes the direct sum of $\mathscr{H}$ with itself n times ($\aleph_0$ times if $n = \infty$). If $A \in \mathscr{B}(\mathscr{H})$, $A^{(n)}$ is the operator on $\mathscr{H}^{(n)}$ defined by $A^{(n)}(h_1, \ldots, h_n) = (Ah_1, \ldots, Ah_n)$. If $\mathscr{S} \subset \mathscr{B}(\mathscr{H})$, $\mathscr{S}^{(n)} \equiv \{A^{(n)}: A \in \mathscr{S}\}$. It is rather interesting that the SOT closure of an algebra can be characterized using its lattice of invariant subspaces.

5.3. Proposition. *If $\mathscr{A}$ is a subalgebra of $\mathscr{B}(\mathscr{H})$ containing 1, then the SOT closure of $\mathscr{A}$ is*

5.4 $\{B \in \mathscr{B}(\mathscr{H}): \text{for every finite } n, \text{ Lat } \mathscr{A}^{(n)} \subseteq \text{Lat } B^{(n)}\}$.

PROOF. It is left as an exercise for the reader to show that if $B \in \text{SOT} - \text{cl } \mathscr{A}$, B belongs to the set (5.4). Now assume that B belongs to the set (5.4). Fix $f_1, f_2, \ldots, f_n$ in $\mathscr{H}$ and $\varepsilon > 0$. It must be shown that there is an A in $\mathscr{A}$ such that $\|(A - B)f_k\| < \varepsilon$ for $1 \leqslant k \leqslant n$.

Let $\mathscr{M} = \vee \{(Af_1, \ldots, Af_n): A \in \mathscr{A}\}$. Because $\mathscr{A}$ is an algebra, $\mathscr{M} \in \text{Lat } \mathscr{A}^{(n)}$; hence $\mathscr{M} \in \text{Lat } B^{(n)}$. Because $1 \in \mathscr{A}$, $(f_1, \ldots, f_n) \in \mathscr{M}$. Since $\{(Af_1, \ldots, Af_n); A \in \mathscr{A}\}$ is a dense manifold and $(Bf_1, \ldots, Bf_n) \in \mathscr{M}$, there is an A in $\mathscr{A}$ with $\varepsilon^2 > \sum_{k=1}^{n} \|(A - B)f_k\|^2$; hence $B \in \text{SOT} - \text{cl } \mathscr{A}$. $\blacksquare$

5.5. Proposition. *The closed unit ball of $\mathscr{B}(\mathscr{H})$ is WOT compact.*

PROOF. The proof of this proposition follows along the lines of the proof of Alaoglu's Theorem. For each h in ball $\mathscr{H}$ let $X_h = $ a copy of ball $\mathscr{H}$ with the weak topology. Put $X = \Pi\{X_h: \|h\| \leqslant 1\}$. If $A \in \text{ball } \mathscr{B}(\mathscr{H})$ let $\tau(A) \in X$ defined by $\tau(A)_h = Ah$. Give X the product topology. Then $\tau: (\text{ball } \mathscr{B}(\mathscr{H}),$ WOT$) \to X$ is a continuous function and a homeomorphism onto its image (verify). Now show that $\tau(\text{ball } \mathscr{B}(\mathscr{H}))$ is closed in X. From here it follows that ball $\mathscr{B}(\mathscr{H})$ is WOT compact. $\blacksquare$

EXERCISES

1. Show that if $B \in \text{SOT} - \text{cl } \mathscr{A}$, then B belongs to the set defined in (5.4).

2. Show that $\mathscr{B}_{00}$ is SOT dense in $\mathscr{B}$.

3. If $\{A_k\}$ and $\{B_k\}$ are sequences in $\mathscr{B}(\mathscr{H})$ such that $A_k \to A(\text{WOT})$ and $B_k \to B(\text{SOT})$, then $A_k B_k \to AB(\text{WOT})$.

4. With the notation of Exercise 3, show that if $A_k \to A(\text{SOT})$, then $A_k B_k \to AB(\text{SOT})$.

5. Let S be the unilateral shift on $l^2(\mathbb{N})$ (II.2.10). Examine the sequences $\{S^k\}$ and $\{S^{*k}\}$ and their relation to Exercises 3 and 4.

6. (Halmos.) Fix an orthonormal basis $\{e_n: n \geq 1\}$ for $\mathscr{H}$. (a) Show that $0 \in \text{weak}$ closure of $\{\sqrt{n}e_n: n \geq 1\}$ (Halmos [1982], Solution 28). (b) Let $\{n_i\}$ be a net of integers such that $\sqrt{n_i}e_{n_i} \to 0$ weakly. Define $A_i f = \sqrt{n_i}\langle f, e_{n_i}\rangle e_{n_i}$ for f in $\mathscr{H}$. Show that $A_i \to 0$ (SOT) but $\{A_i^2\}$ does not converge to 0 (SOT).

§6. Commuting Operators

If $\mathscr{S} \subseteq \mathscr{B}(\mathscr{H})$, let $\mathscr{S}' \equiv \{A \in \mathscr{B}(\mathscr{H}): AS = SA$ for every S in $\mathscr{S}\}$. $\mathscr{S}'$ is called the *commutant* of $\mathscr{S}$. It is not difficult to see that $\mathscr{S}'$ is always an algebra. Similarly, $\mathscr{S}'' \equiv (\mathscr{S}')'$ is called the *double commutant* of $\mathscr{S}$. This process can continue, but (happily) $\mathscr{S}''' = \mathscr{S}'$ (Exercise 1). In some circumstances, $\mathscr{S} = \mathscr{S}''$.

The problem of determining the commutant or double commutant of a single operator or a collection of operators leads to some exciting and interesting mathematics. The commutant is an algebraic object and the idea is to bring the force of analysis to bear in the characterization of this algebra.

We begin by examining the commutant of a direct sum of operators. Recall that if $\mathscr{H} = \mathscr{H}_1 \oplus \mathscr{H}_2 \oplus \cdots$ and $A_n \in \mathscr{B}(\mathscr{H}_n)$ for $n \geq 1$, then $A = A_1 \oplus A_2 \oplus \cdots$ defines a bounded operator on $\mathscr{H}$ if and only if $\sup_n \|A_n\| < \infty$; in this case $\|A\| = \sup_n \|A_n\|$. Also, each operator B on $\mathscr{H}$ has a matrix representation $[B_{ij}]$ where $B_{ij} \in \mathscr{B}(\mathscr{H}_j, \mathscr{H}_i)$.

6.1. Proposition. (a) *If $A = A_1 \oplus A_2 \oplus \cdots$ is a bounded operator on $\mathscr{H} = \mathscr{H}_1 \oplus \mathscr{H}_2 \oplus \cdots$ and $B = [B_{ij}] \in \mathscr{B}(\mathscr{H})$, then $AB = BA$ if and only if $B_{ij}A_j = A_iB_{ij}$ for all i, j.*

(b) *If $B = [B_{ij}] \in \mathscr{B}(\mathscr{H}^{(n)})$, $BA^{(n)} = A^{(n)}B$ if and only if $B_{ij}A = AB_{ij}$ for all i, j.*

The proof of this proposition is an easy exercise in matrix manipulation and is left to the reader.

6.2. Proposition. *If $A \in \mathscr{B}(\mathscr{H})$ and $1 \leq n \leq \infty$, then $\{A^{(n)}\}'' = \{B^{(n)}: B \in \{A\}''\} = \{\{A\}''\}^{(n)}$.*

PROOF. The second equality in the statement is a tautology and it is the first equality that forms the substance of the proposition. If $B \in \{A\}''$, then the preceding proposition implies that $B^{(n)} \in \{A^{(n)}\}''$. Now let $B \in \{A^{(n)}\}''$. To simplify the notation, assume $n = 2$. So $B \in \{A \oplus A\}''$; let $B = [B_{ij}]$, $B_{ij} \in \mathscr{B}(\mathscr{H})$.

Since $\begin{bmatrix} 0 & 1 \\ 0 & 0 \end{bmatrix} \in \{A \oplus A\}'$, matrix multiplication shows that $B_{11} = B_{22}$ and

$B_{21} = 0$. Similarly, the fact that $\begin{bmatrix} 0 & 0 \\ 1 & 0 \end{bmatrix}$ commutes with $A \oplus A$ implies that

$B_{12} = 0$. If $C = B_{11} \ (= B_{22})$, $B = C \oplus C$. If $T \in \{A\}'$, then $T \oplus T \in \{A \oplus A\}'$, so $B(T \oplus T) = (T \oplus T)B$. This shows that $C \in \{A\}''$. ∎

The next result is a corollary of the preceding proof.

6.3. Corollary. *If $\mathscr{S} \subseteq \mathscr{B}(\mathscr{H})$, $\{\mathscr{S}^{(n)}\}'' = \{\mathscr{S}''\}^{(n)}$.*

Say that a subspace $\mathscr{M}$ of $\mathscr{H}$ reduces a collection $\mathscr{S}$ of operators if it reduces each operator in $\mathscr{S}$. By Proposition II.3.7, $\mathscr{M}$ reduces $\mathscr{S}$ if and only if the projection of $\mathscr{H}$ onto $\mathscr{M}$ belongs to $\mathscr{S}'$. This is important in the next theorem, due to von Neumann [1929].

6.4. The Double Commutant Theorem. *If $\mathscr{A}$ is a C^*-subalgebra of $\mathscr{B}(\mathscr{H})$ containing 1, then $\mathrm{SOT} - \mathrm{cl}\, \mathscr{A} = \mathrm{WOT} - \mathrm{cl}\, \mathscr{A} = \mathscr{A}''$.*

PROOF. By Corollary 5.2, $\mathrm{WOT} - \mathrm{cl}\, \mathscr{A} = \mathrm{SOT} - \mathrm{cl}\, \mathscr{A}$. Also, since $\mathscr{A}''$ is SOT closed (Exercise 2) and $\mathscr{A} \subseteq \mathscr{A}''$, $\mathrm{SOT} - \mathrm{cl}\, \mathscr{A} \subseteq \mathscr{A}''$.

It remains to show that $\mathscr{A}'' \subseteq \mathrm{SOT} - \mathrm{cl}\, \mathscr{A}$. To do this Proposition 5.3 will be used.

Let $B \in \mathscr{A}''$, $n \geq 1$, and let $\mathscr{M} \in \mathrm{Lat}\, \mathscr{A}^{(n)}$. It must be shown that $B^{(n)}\mathscr{M} \subseteq \mathscr{M}$. Because $\mathscr{A}$ is a C^*-algebra, so is $\mathscr{A}^{(n)}$. So the fact that $\mathscr{M} \in \mathrm{Lat}\, \mathscr{A}^{(n)}$ and $A^{*(n)} \in \mathscr{A}^{(n)}$ whenever $A^{(n)} \in \mathscr{A}^{(n)}$ implies that $\mathscr{M}$ reduces $A^{(n)}$ for each A in $\mathscr{A}$. Si if P is the projection of $\mathscr{H}^{(n)}$ onto $\mathscr{M}$, $P \in \{\mathscr{A}^{(n)}\}'$. But $B \in \mathscr{A}''$; so by Corollary 6.3, $B^{(n)} \in \{\mathscr{A}^{(n)}\}''$. Hence $B^{(n)}P = PB^{(n)}$ and $\mathscr{M} \in \mathrm{Lat}\, B^{(n)}$. ∎

6.5. Corollary. *If $\mathscr{A}$ is a SOT closed C^*-subalgebra of $\mathscr{B}(\mathscr{H})$ containing 1 and $A \in \mathscr{B}(\mathscr{H})$ such that $A(P\mathscr{H}) \subseteq P\mathscr{H}$ for every projection P in $\mathscr{A}'$, then $A \in \mathscr{A}$.*

PROOF. This uses, in addition to the Double Commutant Theorem, Proposition 4.8 as applied to $\mathscr{A}'$. Indeed, $\mathscr{A}'$ is a SOT closed C^*-algebra and hence it is the norm-closed linear span of its projections. So if $A \in \mathscr{B}(\mathscr{H})$ and $AP\mathscr{H} \subseteq P\mathscr{H}$ for every projection P in $\mathscr{A}'$, then $A(1 - P)\mathscr{H} \subseteq (1 - P)\mathscr{H}$ for every projection P in $\mathscr{A}'$. Thus $P\mathscr{H}$ reduces A and, hence, $AP = PA$. By (4.8), $A \in \mathscr{A}'' = \mathscr{A}$. ∎

6.6. Theorem. *If (X, Ω, μ) is a σ-finite measure space and $\phi \in L^\infty(\mu)$, define M_ϕ on $L^2(\mu)$ by $M_\phi f = \phi f$. If $\mathscr{A}_\mu \equiv \{M_\phi : \phi \in L^\infty(\mu)\}$, then $\mathscr{A}_\mu' = \mathscr{A}_\mu = \mathscr{A}_\mu''$.*

PROOF. It is easy to see that if $\mathscr{A} = \mathscr{A}'$, then $\mathscr{A} = \mathscr{A}''$. Since $\mathscr{A}_\mu \subseteq \mathscr{A}_\mu'$, it suffices to show that $\mathscr{A}_\mu' \subseteq \mathscr{A}_\mu$. So fix A in $\mathscr{A}_\mu'$; it must be shown that $A = M_\phi$ for some ϕ in $L^\infty(\mu)$.

Case 1: $\mu(X) < \infty$. Here $1 \in L^2(\mu)$; put $\phi = A(1)$. Thus $\phi \in L^2(\mu)$. If $\psi \in L^\infty(\mu)$, then $\psi \in L^2(\mu)$ and $A(\psi) = AM_\psi 1 = M_\psi A1 = M_\psi \phi = \phi\psi$. Also, $\|\phi\psi\|_2 = \|A\psi\|_2 \leqslant \|A\| \|\psi\|_2$.

Let $\Delta_n = \{x \in X: |\phi(x)| \geqslant n\}$. Putting $\psi = \chi_{\Delta_n}$ in the preceding argument gives

$$\|A\|^2 \mu(\Delta_n) = \|A\|^2 \|\psi\|^2 \geqslant \|\phi\psi\|^2 = \int_{\Delta_n} |\phi|^2 d\mu \geqslant n^2 \mu(\Delta_n).$$

So if $\mu(\Delta_n) \neq 0$, $\|A\| \geqslant n$. Since A is bounded, $\mu(\Delta_n) = 0$ for some n; equivalently, $\phi \in L^\infty(\mu)$. But $A = M_\phi$ on $L^\infty(\mu)$ and $L^\infty(\mu)$ is dense in $L^2(\mu)$, so $A = M_\phi$ on $L^2(\mu)$.

Case 2: $\mu(X) = \infty$. If $\mu(\Delta) < \infty$, let $L^2(\mu|\Delta) = \{f \in L^2(\mu): f = 0 \text{ off } \Delta\}$. For f in $L^2(\mu|\Delta)$, $Af = A\chi_\Delta f = \chi_\Delta Af \in L^2(\mu|\Delta)$. Let $A_\Delta =$ the restriction of A to $L^2(\mu|\Delta)$. By Case 1, there is a ϕ_Δ in $L^\infty(\mu|\Delta)$ such that $A_\Delta = M_{\phi_\Delta}$. Now if $\mu(\Delta_1) < \infty$ and $\mu(\Delta_2) < \infty$, $\phi_{\Delta_1}|\Delta_1 \cap \Delta_2 = \phi_{\Delta_2}|\Delta_1 \cap \Delta_2$ (Exercise).

Write $X = \bigcup_{n=1}^\infty \Delta_n$, where $\Delta_n \in \Omega$ and $\mu(\Delta_n) < \infty$. From the argument above, if $\phi(x) = \phi_{\Delta_n}(x)$ when $x \in \Delta_n$, ϕ is a well-defined measurable function on X. Now $\|\phi_\Delta\|_\infty = \|M_{\phi_\Delta}\|$ (II.1.5) $= \|A_\Delta\| \leqslant \|A\|$; hence $\|\phi\| \leqslant A\|$. It is easy to check that $A = M_\phi$. ∎

The next result will enable us to solve a number of problems concerning normal operators. It can be considered as a result that removes a technicality, but it is much more than that.

6.7. The Fuglede–Putnam Theorem. *If N and M are normal operators on $\mathcal{H}$ and $\mathcal{K}$, and $B: \mathcal{K} \to \mathcal{H}$ is an operator such that $NB = BM$, then $N^*B = BM^*$.*

PROOF. Note that it follows from the hypothesis that $N^k B = BM^k$ for all $k \geqslant 0$. So if $p(z)$ is a polynomial, $p(N)B = Bp(M)$. Since for a fixed z in $\mathbb{C}$, $\exp(i\bar{z}N)$ and $\exp(i\bar{z}M)$ are limits of polynomials in N and M, respectively, it follows that $\exp(i\bar{z}N)B = B\exp(i\bar{z}M)$ for all z in $\mathbb{C}$. Equivalently, $B = e^{-i\bar{z}N}Be^{i\bar{z}M}$. Because $\exp(X + Y) = (\exp X)(\exp Y)$ when X and Y commute, the fact that N and M are normal implies that

$$f(z) \equiv e^{-izN^*}Be^{izM^*}$$

$$= e^{-izN^*}e^{-i\bar{z}N}Be^{i\bar{z}M}e^{izM^*}$$

$$= e^{-i(zN^* + \bar{z}N)}Be^{i(\bar{z}M + zM^*)}.$$

But for every z in $\mathbb{C}$, $zN^* + \bar{z}N$ and $zM^* + \bar{z}M$ are hermitian operators. Hence $\exp[-i(zN^* + \bar{z}N)]$ and $\exp[i(zM^* + \bar{z}M)]$ are unitary (Exercise 2.14). Therefore $\|f(z)\| \leqslant \|B\|$. But $f: \mathbb{C} \to \mathcal{B}(\mathcal{K}, \mathcal{H})$ is an entire function. By Liouville's Theorem, f is constant.

Thus, $0 = f'(z) = -iN^*e^{-izN^*}Be^{izM^*} + ie^{-izN^*}BM^*e^{izM^*}$. Putting $z = 0$ gives $0 = -iN^*B + iBM^*$, whence the theorem. ∎

This theorem was originally proved in Fuglede [1950] under the assumption that $N = M$. As stated, the theorem was proved in Putnam [1951]. The proof given here is due to Rosenblum [1958]. Another proof is in Radjavi and Rosenthal [1973]. Berberian [1959] observed that Putnam's version can be derived from Fuglede's original theorem by the following matrix trick. If

$$L = \begin{bmatrix} N & 0 \\ 0 & M \end{bmatrix} \quad \text{and} \quad A = \begin{bmatrix} 0 & B \\ 0 & 0 \end{bmatrix}$$

then L is normal on $\mathscr{H} \oplus \mathscr{H}$ and $LA = AL$. Hence $L^*A = AL^*$, and this gives Putnam's version.

6.8. Corollary. *If $N = \int z \, dE(z)$ and $BN = NB$, then $BE(\Delta) = E(\Delta)B$ for every Borel set Δ.*

PROOF. If $BN = NB$, then $BN^* = N^*B$; the conclusion now follows by The Spectral Theorem. ∎

The Fuglede–Putnam Theorem can be combined with some other results we have obtained to yield the following.

6.9. Corollary. *If μ is a compactly supported measure on $\mathbb{C}$, then*

$$\{N_\mu\}' = \mathscr{A}_\mu \equiv \{M_\phi \colon \phi \in L^\infty(\mu)\}.$$

PROOF. Clearly $\mathscr{A}_\mu \subseteq \{N_\mu\}'$. If $A \in \{N_\mu\}'$, then Theorem 6.7 implies $AN_\mu^* = N_\mu^* A$. By an easy algebraic argument, $AM_\phi = M_\phi A$ whenever ϕ is a polynomial in z and $\bar{z}$. By taking weak* limits of such polynomials, it follows that $A \in \mathscr{A}_\mu'$. By Theorem 6.6 $A \in \mathscr{A}_\mu$. ∎

Putnam applied his generalization of Fuglede's Theorem to show that similar normal operators must be unitarily equivalent. This has a formal generalization which is useful.

6.10. Proposition. *Let N_1 and N_2 be normal operators on $\mathscr{H}_1$ and $\mathscr{H}_2$. If $X \colon \mathscr{H}_1 \to \mathscr{H}_2$ is an operator such that $XN_1 = N_2 X$, then:*

(a) *$\mathrm{cl}(\mathrm{ran}\, X)$ reduces N_2;*
(b) *$\ker X$ reduces N_1;*
(c) *If $M_1 = N_1 | (\ker X)^\perp$ and $M_2 = N_2 | \mathrm{cl}(\mathrm{ran}\, X)$, then $M_1 \cong M_2$.*

PROOF. (a) If $f_1 \in \mathscr{H}_1$, $N_2 X f_1 = X N_1 f_1 \in \mathrm{ran}\, X$; so $\mathrm{cl}(\mathrm{ran}\, X)$ is invariant for N_2. By the Fuglede–Putnam Theorem, $X N_1^* = N_2^* X$, so $\mathrm{cl}(\mathrm{ran}\, X)$ is invariant for N_2^*.
  (b) Exercise.
  (c) Since $X(\ker X)^\perp \subseteq \mathrm{cl}(\mathrm{ran}\, X)$, part (c) will be proved if it can be shown that $N_1 \cong N_2$ when $\ker X = (0)$ and $\mathrm{ran}\, X$ is dense. So make these assumptions and consider the polar decomposition of X, $X = UA$ (see Exercise 11).

Because $\ker X = (0)$ and ran X is dense, A is a positive operator on $\mathcal{H}_1$ and $U: \mathcal{H}_1 \to \mathcal{H}_2$ is an isomorphism. Now $X^* N_2^* = N_1^* X^*$, so $X^* N_2 = N_1 X^*$. A calculation shows that $A^2 = X^* X \in \{N_1\}'$, so $A \in \{N_1\}'$. (Why?) Hence $N_2 UA = N_2 X = UAN_1 = UN_1 A$; that is, $N_2 U = UN_1$ on the range of A. But $\ker A = (0)$, so ran A is dense in $\mathcal{H}_1$. Therefore $N_2 U = UN_1$, or $N_2 = UN_1 U^{-1}$. ■

6.11. Corollary. *Two similar normal operators are unitarily equivalent.*

The corollary appears in Putnam [1951], while Proposition 6.10 first appeared in Douglas [1969].

EXERCISES

1. If $\mathcal{S} \subseteq \mathcal{B}(\mathcal{H})$, show that $\mathcal{S}' = \mathcal{S}'''$.

2. If $\mathcal{S} \subseteq \mathcal{B}(\mathcal{H})$, show that $\mathcal{S}'$ is always a SOT closed subalgebra of $\mathcal{B}(\mathcal{H})$.

3. Prove Proposition 6.1.

4. Let $\mathcal{H}$ be a Hilbert space of dimension α and define $S: \mathcal{H}^{(\infty)} \to \mathcal{H}^{(\infty)}$ by $S(h_1, h_2, \ldots) = (0, h_1, h_2, \ldots)$. S is called the *unilateral shift of multiplicity* α. (a) Show that $A = [A_{ij}] \in \{S\}'$ if and only if $A_{ij} = 0$ for $j > i$ and $A_{ij} = A_{i+1,j+1}$ for $i \geq j$. (b) Show that $A = [A_{ij}] \in \{S\}''$ if and only if $A_{ij} = 0$ for $j > i$ and $A_{ij} = A_{i+1,j+1} = $ a multiple of the identity for $i \geq j$.

5. What is $\{N_\mu \oplus N_\mu\}'$? $\{N_\mu \oplus N_\mu\}''$?

6. If $\mathscr{A}$ is a subalgebra of $\mathcal{B}(\mathcal{H})$, show that $\mathscr{A}$ is a maximal abelian subalgebra of $\mathcal{B}(\mathcal{H})$ if and only if $\mathscr{A} = \mathscr{A}'$.

7. Find a non-normal operator that is similar to a normal operator. (Hint: Try $\dim \mathcal{H} = 2$.)

8. Let μ be a compactly supported measure on $\mathbb{C}$ and let $\mathcal{X}$ be a separable Hilbert space. A function $f: \mathbb{C} \to \mathcal{X}$ is a Borel function if $f^{-1}(G)$ is a Borel set when G is weakly open in $\mathcal{X}$. Define $L^2(\mu, \mathcal{X})$ to be the equivalence classes of Borel functions $f: \mathbb{C} \to \mathcal{X}$ such that $\int \|f(x)\|^2 \, d\mu(x) < \infty$. Define $\langle f, g \rangle = \int \langle f(x), g(x) \rangle \, d\mu(x)$ for f and g in $L^2(\mu, \mathcal{X})$. (a) Show that $L^2(\mu, \mathcal{X})$ is a Hilbert space. Define N on $L^2(\mu, \mathcal{X})$ by $(Nf)(z) = zf(z)$. (b) Show that N is a normal operator and $\sigma(N) = \text{support } \mu$. Calculate N^*. (c) Show that $N \cong N_\mu^{(\alpha)}$, where $\alpha = \dim \mathcal{X}$. (d) Find $\{N\}'$. (Hint: Use 6.1.) (e) Find $\{N\}''$.

9. Let $\mathcal{H}$ be separable with basis $\{e_n\}$. Let A be the diagonal operator on $\mathcal{H}$ given by $Ae_n = \lambda_n e_n$, where $\sup_n |\lambda_n| < \infty$. Determine $\{A\}'$ and $\{A\}''$. Give necessary and sufficient conditions on $\{\lambda_n\}$ such that $\{A\}' = \{A\}''$.

10. Let $\mathscr{A}$ be a C^*-subalgebra of $\mathcal{B}(\mathcal{H})$ but do not assume that $\mathscr{A}$ contains the identity operator. Let $\mathcal{M} = \vee \{\text{ran } A: A \in \mathscr{A}\}$ and let $P = $ the projection of $\mathcal{H}$ onto $\mathcal{M}$. Show that $\text{SOT} - \text{cl } \mathscr{A} = \mathscr{A}'' P = P\mathscr{A}''$.

11. Formulate and prove a polar decomposition for operators between different Hilbert spaces.

12. Let (X, Ω, μ) be an arbitrary measure space and let $L \in L^1(\mu)^*$. (a) Show that for every f in $L^2(\mu)$, there is an h in $L^2(\mu)$ such that $L(g\bar{f}) = \int g\bar{h}\,d\mu$ for all g in $L^2(\mu)$. (b) If $f \in L^2(\mu)$, let Tf be the function h in $L^2(\mu)$ obtained in part (a). Show that $T: L^2(\mu) \to L^2(\mu)$ defines a bounded linear operator and T commutes with M_ϕ for every ϕ in $L^\infty(\mu)$. (c) In light of parts (a) and (b), compare Theorem 6.6 and Example 20.17 in Hewitt and Stromberg [1975].

§7. Abelian von Neumann Algebras

7.1. Definition. A *von Neumann algebra* $\mathscr{A}$ is a C^*-subalgebra of $\mathscr{B}(\mathscr{H})$ such that $\mathscr{A} = \mathscr{A}''$.

Note that if $\mathscr{A}$ is a von Neumann algebra, then $1 \in \mathscr{A}$ and $\mathscr{A}$ is SOT closed. Conversely, if $1 \in \mathscr{A}$ and $\mathscr{A}$ is a SOT closed C^*-subalgebra of $\mathscr{B}(\mathscr{H})$, then $\mathscr{A}$ is a von Neumann algebra by the Double Commutant Theorem.

It is a result of S. Sakai that a C^*-algebra is isomorphic to a von Neumann algebra if it is the dual of a Banach space. The converse to this is an easy consequence of the fact that $\mathscr{B}(\mathscr{H})$ is a dual space (Exercise 2.21). For an account of the history of this result and its predecessors, as well as a number of proofs, see Kadison [1985].

7.2. Examples. (a) $\mathscr{B}(\mathscr{H})$ and $\mathbb{C}$ are von Neumann algebras.

(b) If (X, Ω, μ) is a σ-finite measure space, then $\mathscr{A}_\mu \equiv \{M_\phi : \phi \in L^\infty(\mu)\} \subseteq \mathscr{B}(L^2(\mu))$ is an abelian von Neumann algebra by Theorem 6.6. In fact, it is a maximal abelian von Neumann algebra.

It will be shown in this section that $\mathscr{A}_\mu$ is the only abelian von Neumann algebra up to a $*$-isomorphism. However, there are many others that are not unitarily equivalent to $\mathscr{A}_\mu$.

For $\mathscr{A}_j \subseteq \mathscr{B}(\mathscr{H}_j), j \geq 1, \mathscr{A}_1 \oplus \mathscr{A}_2 \oplus \cdots$ is used to denote the l^∞ direct sum of $\mathscr{A}_1, \mathscr{A}_2, \dots$. That is, $\mathscr{A}_1 \oplus \mathscr{A}_2 \oplus \cdots = \{A_1 \oplus A_2 \oplus \cdots : A_j \in \mathscr{A}_j \text{ for } j \geq 1 \text{ and } \sup_j \|A_j\| < \infty\}$. Note that $\mathscr{A}_1 \oplus \mathscr{A}_2 \oplus \cdots \subseteq \mathscr{B}(\mathscr{H}_1 \oplus \mathscr{H}_2 \oplus \cdots)$ and $\|A_1 \oplus A_2 \oplus \cdots\| = \sup_j \|A_j\|$.

7.3. Proposition. (a) *If* $\mathscr{A}_1, \mathscr{A}_2, \dots$ *are von Neumann, algebras, then so is* $\mathscr{A}_1 \oplus \mathscr{A}_2 \oplus \cdots$. (b) *If* $\mathscr{A}$ *is a von Neumann algebra and* $1 \leq n \leq \infty$, *then* $\mathscr{A}^{(n)}$ *is a von Neumann algebra.*

PROOF. Exercise.

The proof of the next result is also an exercise.

7.4. Proposition. *Let* $\mathscr{A}_j$ *be a von Neumann algebra on* $\mathscr{H}_j, j = 1, 2$. *If* $U: \mathscr{H}_1 \to \mathscr{H}_2$ *is an isomorphism such that* $U\mathscr{A}_1 U^{-1} = \mathscr{A}_2$, *then* $U\mathscr{A}_1' U^{-1} = \mathscr{A}_2'$.

Now let (X, Ω, μ) be a σ-finite measure space and define $\rho: \mathscr{A}_\mu \to \mathscr{A}_\mu^{(2)}$ by

$\rho(T) = T \oplus T$. Then ρ is a *-isomorphism. However, $\mathscr{A}_\mu$ and $\mathscr{A}_\mu^{(2)}$ are not
spatially isomorphic. That is, there is no Hilbert space isomorphism U:
$L^2(\mu) \to L^2(\mu) \oplus L^2(\mu)$ such that $U\mathscr{A}_\mu U^{-1} = \mathscr{A}_\mu^{(2)}$. Why? One way to see that no
such U exists is to note that $\mathscr{A}_\mu$ has a cyclic vector (give an example).
However, $\mathscr{A}_\mu^{(2)}$ does not have a cyclic vector as shall be seen presently
(Theorem 7.8).

7.5. Definition. If $\mathscr{A} \subseteq \mathscr{B}(\mathscr{H})$ and $e_0 \in \mathscr{H}$, then e_0 is a *separating vector* for
$\mathscr{A}$ if the only operator A in $\mathscr{A}$ such that $Ae_0 = 0$ is the operator $A = 0$.

If (X, Ω, μ) is a σ-finite measure space and $f \in L^2(\mu)$ such that $\mu(\{x \in X: f(x) = 0\}) = 0$ (Why does such an f exist?), then f is a separating vector for
$\mathscr{A}_\mu$ as well as a cyclic vector. If $\mathscr{A} = \mathscr{B}(\mathscr{H})$, then no vector in $\mathscr{H}$ is separating
for $\mathscr{A}$ while every nonzero vector is a cyclic vector. If $\mathscr{A} = \mathbb{C}$ and $\dim \mathscr{H} > 1$,
then $\mathscr{A}$ has no cyclic vectors but every nonzero vector is separating for $\mathscr{A}$.

7.6. Proposition. *If e_0 is a cyclic vector for $\mathscr{A}$, then e_0 is a separating vector
for $\mathscr{A}'$.*

PROOF. If $T \in \mathscr{A}'$ and $Te_0 = 0$, then for every A in $\mathscr{A}$, $TAe_0 = ATe_0 = 0$. Since
$\bigvee \mathscr{A}e_0 = \mathscr{H}$, $T = 0$. ∎

7.7. Corollary. *If $\mathscr{A}$ is an abelian subalgebra of $\mathscr{B}(\mathscr{H})$, then every cyclic vector
for $\mathscr{A}$ is a separating vector for $\mathscr{A}$.*

PROOF. Because $\mathscr{A}$ is abelian, $\mathscr{A} \subseteq \mathscr{A}'$. ∎

Since $\mathscr{B}(\mathscr{H})' = \mathbb{C}$, Proposition 7.6 explains some of the duality exhibited
prior to (7.6). Also note that if (X, Ω, μ) is a finite measure space, $1 \oplus 0$, $0 \oplus 1$,
and $1 \oplus 1$ are all separating vectors for $\mathscr{A}_\mu^{(2)}$. Because $\mathscr{A}_\mu^{(2)} \neq (\mathscr{A}_\mu^{(2)})'$, the next
theorem says that $\mathscr{A}_\mu^{(2)}$ has no cyclic vector.

Although it is easy to see that conditions (a) and (b) in the next result are
equivalent, irrespective of any assumption on $\mathscr{H}$, the equivalence of the
remaining parts to (a) and (b) is not true unless some additional assumption
is made on $\mathscr{H}$ or $\mathscr{A}$ (see Exercise 5). We are content to assume that $\mathscr{H}$ is
separable.

7.8. Theorem. *Assume that $\mathscr{H}$ is separable and $\mathscr{A}$ is an abelian C^*-subalgebra
of $\mathscr{B}(\mathscr{H})$. The following statements are equivalent.*

(a) *$\mathscr{A}$ is a maximal abelian von Neumann algebra.*
(b) *$\mathscr{A} = \mathscr{A}'$.*
(c) *$\mathscr{A}$ has a cyclic vector, contains 1, and is SOT closed.*
(d) *There is a compact metric space X, a positive Borel measure μ with support
X, and an isomorphism $U: L^2(\mu) \to \mathscr{H}$ such that $U\mathscr{A}_\mu U^{-1} = \mathscr{A}$.*

PROOF. The proof that (a) and (b) are equivalent is left as an exercise.

(b)⇒(c): By Zorn's Lemma and the separability of $\mathcal{H}$, there is a maximal sequence of unit vectors $\{e_n\}$ such that for $n \neq m$, $\mathrm{cl}[\mathcal{A}e_n] \perp \mathrm{cl}[\mathcal{A}e_m]$. It follows from the maximality of $\{e_n\}$ that $\mathcal{H} = \oplus_{n=1}^{\infty} \mathrm{cl}[\mathcal{A}e_n]$.

Let $e_0 = \sum_{n=1}^{\infty} e_n/\sqrt{2^n}$. Since $e_n \perp e_m$ for $n \neq m$, $\|e_0\|^2 = \sum 2^{-n} = 1$. Let P_n = the projection of $\mathcal{H}$ onto $\mathcal{H}_n = \mathrm{cl}[\mathcal{A}e_n]$. Clearly $\mathcal{A}$ leaves $\mathcal{H}_n$ invariant and so, since $\mathcal{A}$ is a *-algebra, $\mathcal{H}_n$ reduces $\mathcal{A}$. Thus $P_n \in \mathcal{A}' = \mathcal{A}$ and $\mathrm{cl}[\mathcal{A}e_0] \supseteq \mathrm{cl}[\mathcal{A}P_ne_0] = \mathrm{cl}[\mathcal{A}e_n] = \mathcal{H}_n$. Therefore $\mathrm{cl}[\mathcal{A}e_0] = \mathcal{H}$ and e_0 is a cyclic vector for $\mathcal{A}$.

(c)⇒(d): Since $\mathcal{H}$ is separable, ball $\mathcal{A}$ is WOT metrizable and compact (1.3 and 5.5). By picking a countable WOT dense subset of ball $\mathcal{A}$ and letting $\mathcal{A}_1$ be the C*-algebra generated by this countable dense subset, it follows that $\mathcal{A}_1$ is a separable C*-algebra whose SOT closure is $\mathcal{A}$. Let X be the maximal ideal space of $\mathcal{A}_1$ and let $\rho: C(X) \to \mathcal{A}_1 \subseteq \mathcal{A} \subseteq \mathcal{B}(\mathcal{H})$ be the inverse of the Gelfand map. By Theorem 1.14 there is a spectral measure E defined on the Borel subsets of X such that $\rho(u) = \int u\,dE$ for u in $C(X)$. If $\phi \in B(X)$ and $\{u_i\}$ is a net in $C(X)$ such that $\int u_i\,dv \to \int \phi\,dv$ for every v in $M(X)$, then $\rho(u_i) = \int u_i\,dE \to \int \phi\,dE$ (WOT). Thus $\{\int \phi\,dE: \phi \in B(X)\} \subseteq \mathcal{A}$ since $\mathcal{A}$ is SOT closed.

Let e_0 be a cyclic vector for $\mathcal{A}$ and put $\mu(\Delta) = \|E(\Delta)e_0\|^2 = \langle E(\Delta)e_0, e_0 \rangle$. Thus $\langle (\int \phi\,dE)e_0, e_0 \rangle = \int \phi\,d\mu$ for every ϕ in $B(X)$. Consider $B(X)$ as a linear manifold in $L^2(\mu)$ by identifying functions that agree a.e. $[\mu]$. If $\phi \in B(X)$, then

$$\left\|\left(\int \phi\,dE\right)e_0\right\|^2 = \left\langle \left(\int \phi\,dE\right)^* \left(\int \phi\,dE\right)e_0, e_0 \right\rangle$$

$$= \int |\phi|^2\,d\mu.$$

This says two things. First, if $\phi = 0$ a.e. $[\mu]$, then $(\int \phi\,dE)e_0 = 0$. Hence U: $B(X) \to \mathcal{H}$ defined by $U\phi = (\int \phi\,dE)e_0$ is a well-defined map from the dense manifold $B(X)$ in $L^2(\mu)$ into $\mathcal{H}$. Second, U is an isometry. Since the domain and range of U are dense (Why?), U extends to an isomorphism $U: L^2(\mu) \to \mathcal{H}$.

If $\phi \in B(X)$ and $\psi \in L^{\infty}(\mu)$, then $UM_\psi\phi = U(\psi\phi) = (\int \psi\phi\,dE)e_0 = (\int \psi\,dE)(\int \phi\,dE)e_0 = (\int \psi\,dE)U\phi$. Hence $UM_\psi U^{-1} = \int \psi\,dE$ and $U\mathcal{A}_\mu U^{-1} \subseteq \mathcal{A}$. On the other hand, $U\mathcal{A}_\mu U^{-1}$ is a SOT closed C*-subalgebra of $\mathcal{B}(\mathcal{H})$ that contains $UC(X)U^{-1} = \mathcal{A}_1$. (Why?) So $U\mathcal{A}_\mu U^{-1} = \mathcal{A}$.

Because $\mathcal{A}_1$ is separable, X is metrizable.

(d)⇒(b): This is a consequence of Theorem 6.6 and Proposition 7.4. ∎

Mercer (1986) shows that for a maximal abelian von Neumann algebra $\mathcal{A}$, there is an orthonormal basis for the underlying Hilbert space consisting of vectors that are cyclic and separating for $\mathcal{A}$.

7.9. Corollary. *If $\mathcal{A}$ is an abelian C*-subalgebra of $\mathcal{B}(\mathcal{H})$ and $\mathcal{H}$ is separable, then $\mathcal{A}$ has a separating vector.*

PROOF. By Zorn's Lemma, $\mathscr{A}$ is contained in a maximal abelian C^*-algbera, $\mathscr{A}_m$. It is easy to see that $\mathscr{A}_m$ must be SOT closed, so $\mathscr{A}_m$ is a maximal abelian von Neumann algebra. By the preceding theorem, there is a cyclic vector e_0 for $\mathscr{A}_m$. But (7.7) e_0 is separating for $\mathscr{A}_m$, and hence for any subset of $\mathscr{A}_m$.   ∎

The preceding corollary may seem innocent, but it is, in fact, the basis for the next section.

EXERCISES

1. Prove Proposition 7.3.

2. Prove Proposition 7.4.

3. Why are $\mathscr{A}_\mu$ and $\mathscr{A}_\mu^{(2)}$ not spatially isomorphic?

4. Show that if X is any compact metric space, there is a separable Hilbert space $\mathscr{H}$ and a ∗-monomorphism $\tau: C(X) \to \mathscr{B}(\mathscr{H})$. Find the spectral measure for τ.

5. Let $\mathscr{A}$ be an abelian C^*-subalgebra of $\mathscr{B}(\mathscr{H})$ such that $\mathscr{A}'$ contains no uncountable collection of pairwise orthogonal projections. Show that the following statements are equivalent: (a) $\mathscr{A}$ is a maximal abelian von Neumann algebra; (b) $\mathscr{A} = \mathscr{A}'$; (c) $\mathscr{A}$ has a cyclic vector and is SOT closed; (d) there is a finite measure space (X, Ω, μ) and an isomorphism $U: L^2(\mu) \to \mathscr{H}$ such that $U \mathscr{A}_\mu U^{-1} = \mathscr{A}$.

6. Let $\{P_n\}$ be a sequence of commuting projections in $\mathscr{B}(\mathscr{H})$ and put $A = \sum_{n=1}^{\infty} 3^{-n}(2P_n - 1)$. Show that $C^*(A)$ is the C^*-algebra generated by $\{P_n\}$. (Do you see a connection between A and the Cantor–Lebesgue function?)

7. If $\mathscr{A}$ is an abelian von Neumann algebra on a separable Hilbert space $\mathscr{H}$, show that there is a hermitian operator A such that $\mathscr{A}$ equals the smallest von Neumann algebra containing A. (Hint: Let $\{P_n\}$ be a countable WOT dense subset of the set projections in $\mathscr{A}$ and use Exercise 6. This proof is due to Rickart [1960], pp. 293–294. Also see Jenkins [1972].)

8. If X is a compact space, show that $C(X)$ is generated as a C^*-algebra by its characteristic functions if and only if X is totally disconnected. If A is as in Exercise 6, show that $\sigma(A)$ is totally disconnected.

9. If X and Y are compact spaces and $\tau: C(X) \to C(Y)$ is a homomorphism with $\tau(1) = 1$, show that there is a continuous function $\phi: Y \to X$ such that $\tau(u) = u \circ \phi$ for every u in $C(X)$. Show that τ is injective if and only if ϕ is surjective, and, in this case, τ is an isometry. Show that τ is surjective if and only if ϕ is injective.

10. Let X and Z be compact spaces, $Y = X \times Z$, and let $\phi: Y \to X$ be the projection onto the first coordinate. Define $\tau: C(X) \to C(Y)$ by $\tau(u) = u \circ \phi$. Describe the range of τ.

11. Adopt the notation of Exercise 9. Define an equilvalence relation $\sim$ on Y by saying $y_1 \sim y_2$ if and only if $\phi(y_1) = \phi(y_2)$. Let $q: Y \to Y/\sim$ be the natural map and $q^*: C(Y/\sim) \to C(Y)$ the induced homomorphism. Show that there is a ∗-epimorphism $\rho: C(X) \to C(Y/\sim)$ such that the diagram

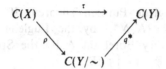

commutes. Find the corresponding injection $Y/\sim \to X$.

12. If X is any compact metric space, show that there is a totally disconnected compact metric space Y and a continuous surjection $\phi\colon Y \to X$. (Hint: Start by embedding $C(X)$ into $\mathscr{B}(\mathscr{H})$ and use Exercises 7 and 8.)

13. Show that every totally disconnected compact metric space is the continuous image of the Cantor ternary set. (Do this directly; do not try to use C^*-algebras.) Combine this with Exercise 12 to get that every compact metric space is the continuous image of the Cantor set.

14. If $\mathscr{A}$ is an abelian von Neumann algebra and X is its maximal ideal space, then X is a *Stonean space*; that is, if U is open in X, then cl U is open in X.

15. This exercise assumes Exercise 2.21 where it was proved that $\mathscr{B}_1^* = \mathscr{B}$. When referring to the weak* topology on $\mathscr{B} = \mathscr{B}(\mathscr{H})$, we mean the topology $\mathscr{B}$ has as the Banach dual of $\mathscr{B}_1$. (a) Show that on bounded subsets of $\mathscr{B}$ the weak* topology = WOT. (b) Show that a C^*-subalgebra of $\mathscr{B}(\mathscr{H})$ is von Neumann algebra if and only if $\mathscr{A}$ is weak* closed. (c) Show that WOT and the weak* topology agree on abelian von Neumann algebras (See Takeda [1951] and Pallu de la Barrière [1954].) (d) Give an example of a weak* closed subspace of $\mathscr{B}$ that is not WOT closed. (See von Neumann [1936].)

16. Prove the converse of Proposition 7.6.

§8. The Functional Calculus for Normal Operators: The Conclusion of the Saga

In this section it will always be assumed that

all Hilbert spaces are separable.

Indeed, this assumption will remain in force for the rest of the chapter. This assumption is necessary for the validity of some of the results and minimizes the technical details in others.

If N is a normal operator on $\mathscr{H}$, let $W^*(N)$ be the von Neumann algebra generated by N. That is, $W^*(N)$ is the intersection of all of the von Neumann algebras containing N. Hence $W^*(N)$ is the WOT closure of $\{p(N, N^*)\colon p(z, \bar{z})$ is a polynomial in z and $\bar{z}\}$.

8.1. Proposition. *If N is a normal operator, then $W^*(N) = \{N\}'' \supseteq \{\phi(N)\colon \phi \in B(\sigma(N))\}$.*

PROOF. The equality results from combining the Double Commutant

Theorem and the Fuglede–Putnam Theorem. If $\phi \in B(\sigma(N))$, $N = \int z\, dE(z)$, and $T \in \{N\}'$, then $T \in \{N, N^*\}'$ by the Fuglede–Putnam Theorem and $TE(\Delta) = E(\Delta)T$ for every Borel set Δ by the Spectral Theorem. Hence $T\phi(N) = \phi(N)T$ since $\phi(N) = \int \phi\, dE$.   ∎

The purpose of this section is to prove that the containment in the preceding proposition is an equality. In fact, more will be proved. A measure μ whose support is $\sigma(N)$ will be found such that $\phi(N)$ is well defined if $\phi \in L^\infty(\mu)$ and the map $\phi \mapsto \phi(N)$ is a *-isomorphism of $L^\infty(\mu)$ onto $W^*(N)$. To find μ, Corollary 7.9 (which requires the separability of $\mathcal{H}$) is used.

By Corollary 7.9, $W^*(N)$, being an abelian von Neumann algebra, has a separating vector e_0. Define a measure μ on $\sigma(N)$ by

8.2 $\mu(\Delta) = \langle E(\Delta)e_0, e_0 \rangle = \| E(\Delta)e_0 \|^2.$

8.3. Proposition. $\mu(\Delta) = 0$ *if and only if* $E(\Delta) = 0$.

PROOF. If $\mu(\Delta) = 0$, then $E(\Delta)e_0 = 0$. But $E(\Delta) = \chi_\Delta(N) \in W^*(N)$. Since e_0 is a separating vector, $E(\Delta) = 0$. The reverse implication is clear.   ∎

8.4. Definition. A *scalar-valued spectral measure for N* is a positive Borel measure μ on $\sigma(N)$ such that $\mu(\Delta) = 0$ if and only if $E(\Delta) = 0$; that is, μ and E are mutually absolutely continuous.

So Proposition 8.3 says that scalar-valued spectral measures exist. It will be shown (8.9) that every scalar-valued spectral measure is defined by (8.2) where e_0 is a separating vector for $W^*(N)$. In the process additional information is obtained about a normal operator and its functional calculus.

If $h \in \mathcal{H}$, let $\mu_h \equiv E_{h,h}$ and let $\mathcal{H}_h \equiv \mathrm{cl}[W^*(N)h]$. Note that $\mathcal{H}_h$ is the smallest reducing subspace for N that contains h. Let $N_h \equiv N | \mathcal{H}_h$. Thus N_h is a *-cyclic normal operator with *-cyclic vector h. The uniqueness of the spectral measure for a normal operator implies that the spectral measure for N_h is $E(\Delta) | \mathcal{H}_h$: that is, $\chi_\Delta(N_h) = \chi_\Delta(N) | \mathcal{H}_h = E(\Delta) | \mathcal{H}_h$. Thus Theorem 3.4 implies there is a unique isomorphism $U_h : \mathcal{H}_h \to L^2(\mu_h)$ such that $U_h h = 1$ and $U_h N_h U_h^{-1} f = zf$ for all f in $L^2(\mu_h)$. The notation of this paragraph is used repeatedly in this section.

The way to understand what is going on is to consider each N_h as a localization of N. Since N_h is unitarily equivalent to M_z on $L^2(\mu_h)$ we can agree that we thoroughly understand the local behavior of N. Can we put together this local behavior of N to understand the global behavior of N? This is precisely what is done in §10.

In the present section the objective is to show that if h is a separating vector for $W^*(N)$, then the functional calculus for N is completely determined by the functional calculus for N_h. The sense in which this "determination" is made is the following. If $A \in W^*(N)$, then the definition of $\mathcal{H}_h$ shows that $A\mathcal{H}_h \subseteq \mathcal{H}_h$. Since $A^* \in W^*(N)$, $\mathcal{H}_h$ reduces each operator in $W^*(N)$; thus

$A|\mathcal{H}_h$ is meaningful. It will be shown that the map $A \to A|\mathcal{H}_h$ is a *-isomorphism of $W^*(N)$ onto $W^*(N_h)$ if h is a separating vector for $W^*(N)$. Since N_h is *-cyclic, Theorem 6.6 and Corollary 6.9 show how to determine $W^*(N_h)$.

We begin with a modest lemma.

8.5. Lemma. *If $h \in \mathcal{H}$ and ρ_h: $W^*(N) \to W^*(N_h)$ is defined by $\rho_h(A) = A|\mathcal{H}_h$, then ρ_h is a *-epimorphism that is WOT-continuous. Moreover, If $\psi \in B(\sigma(N))$, then $\rho_h(\psi(N)) = \psi(N_h)$ and if $A \in W^*(N)$, then there is a ϕ in $B(\sigma(N_h))$ such that $\rho_h(A) = \phi(N_h)$.*

PROOF. First let us see that ρ_h maps $W^*(N)$ into $W^*(N_h)$. If $p(z, \bar{z})$ is a polynomial in z and $\bar{z}$ then $\rho_h[p(N, N^*)] = p(N_h, N_h^*)$ as an algebraic manipulation shows. If $\{p_i\}$ is a net of such polynomials such that $p_i(N, N^*) \to A$(WOT), then for f, g in $\mathcal{H}_h$, $\langle p_i(N, N^*)f, g \rangle \to \langle Af, g \rangle$; thus $p_i(N_h, N_h^*) \to \rho_h(A)$(WOT) and so $\rho_h(A) \in W^*(N_h)$. It is left as an exercise for the reader to show that ρ_h is a *-homomorphism. Also, the preceding argument can be used to show that ρ_h is WOT continuous.

If $\psi \in B(\sigma(N))$, there is a net $\{p_i(z, \bar{z})\}$ of polynomials in z and $\bar{z}$ such that $\int p_i d\nu \to \int \psi \, d\nu$ for every ν in $M(\sigma(N))$. (Why?) Since $\sigma(N_h) \subseteq \sigma(N)$ (Why?), $\int p_i d\eta \to \int \psi \, d\eta$ for every η in $M(\sigma(N_h))$. Therefore $p_i(N, N^*) \to \psi(N)$(WOT) and $p_i(N_h, N_h^*) \to \psi(N_h)$(WOT). But $\rho_h(p_i(N, N^*)) = p_i(N_h, N_h^*)$ and $\rho_h(p_i(N, N^*)) \to \rho_h(\psi(N))$; hence $\rho_h(\psi(N)) = \psi(N_h)$.

Let U_h: $\mathcal{H}_h \to L^2(\mu_h)$ be the isomorphism such that $U_h h = 1$ and $U_h N_h U_h^{-1} = N_{\mu_h}$. If $A \in W^*(N)$ and $A_h = \rho_h(A)$, then $A_h N_h = N_h A_h$; thus $U_h A_h U_h^{-1} \in \{N_{\mu_h}\}'$. By Corollary 6.9, there is a ϕ in $B(\sigma(N_h))$ such that $U_h A_h U_h^{-1} = M_\phi$. It follows (How?) that $A_h = \phi(N_h)$.

Finally, to show that ρ_h is surjective note that if $B \in W^*(N_h)$, then (use the argument in the preceding paragraph) $B = \psi(N_h)$ for some ψ in $B(\sigma(N_h))$. Extend ψ to $\sigma(N)$ by letting $\psi = 0$ on $\sigma(N) \setminus \sigma(N_h)$. Then $\psi(N) \in W^*(N)$ and $\rho_h(\psi(N)) = \psi(N_h) = B$. ∎

8.6. Lemma. *If $e \in \mathcal{H}$ such that μ_e is a scalar-valued spectral measure for N and if ν is a positive measure on $\sigma(N)$ such that $\nu \ll \mu_e$, then there is an h in $\mathcal{H}_e$ such that $\nu = \mu_h$.*

PROOF. This proof is just an application of the Radon–Nikodym Theorem once certain identifications are made; namely, $f = [d\nu/d\mu_e]^{1/2} \in L^2(\mu_e)$, so put $h = U_e^{-1} f$. Hence $h \in \mathcal{H}_e$. For any Borel set Δ, $\nu(\Delta) = \int \chi_\Delta \, d\nu = \int \chi_\Delta \, ff d\mu_e = \langle M_{\chi_\Delta} f, f \rangle = \langle U_e^{-1} M_{\chi_\Delta} f, U_e^{-1} f \rangle = \langle E(\Delta)h, h \rangle = \mu_h(\Delta)$. ∎

8.7. Lemma. $W^*(N) = \{\phi(N): \phi \in B(\sigma(N))\}$.

PROOF. Let $\mathcal{A} = \{\phi(N): \phi \in B(\sigma(N))\}$. Hence $\mathcal{A}$ is a *-algebra and $\mathcal{A} \subseteq W^*(N)$ by Proposition 8.1. Since $N \in \mathcal{A}$ it suffices to prove that $\mathcal{A}$ is WOT closed. Let $\{\phi_i\}$ be a net in $B(\sigma(N))$ such that $\phi_i(N) \to A$ (WOT); so $A \in W^*(N)$. By

(8.5) $\phi_i(N_h) \to A|\mathcal{H}_h(\text{WOT})$ for any h in $\mathcal{H}$. Also, by Lemma 8.5. for every h in $\mathcal{H}$ there is a ϕ_h in $B(\mathbb{C})$ such that $A|\mathcal{H}_h = \phi_h(N_h)$. Fix a separating vector e for $W^*(N)$; hence μ_e is a scalar-valued spectral measure for N.

If $h \in \mathcal{H}$, then the fact that $\phi_i(N_h) \to \phi_h(N_h)(\text{WOT})$ implies $\phi_i \to \phi_h$ weak* in $L^\infty(\mu_h)$. Also, $\phi_i \to \phi_e$ weak* in $L^\infty(\mu_e)$. But $\mu_h \ll \mu_e$ so that $d\mu_h/d\mu_e \in L^1(\mu_e)$; hence for any Borel set Δ

$$\int_\Delta \phi_i \, d\mu_h = \int_\Delta \phi_i \frac{d\mu_h}{d\mu_e} \, d\mu_e \to \int_\Delta \phi_e \, d\mu_h.$$

But also

$$\int_\Delta \phi_i \, d\mu_h \to \int_\Delta \phi_h \, d\mu_h.$$

So $0 = \int_\Delta (\phi_e - \phi_h) \, d\mu_h$ for every Borel set Δ. Therefore $\phi_h = \phi_e$ a.e. $[\mu_h]$. But if $g \in \mathcal{H}_h$, then $\langle \phi_h(N_h)g, g \rangle = \langle \phi_h(N)g, g \rangle = \int \phi_h \, d\mu_g = \int \phi_e \, d\mu_g$ since $\mu_g \ll \mu_h$. Thus $\langle \phi_h(N_h)g, g \rangle = \langle \phi_e(N_h)g, g \rangle$; that is, $\phi_h(N_h) = \phi_e(N_h)$. In particular, $Ah = \phi_h(N_h)h = \phi_e(N_h)h = \phi_e(N)h$. Since h was arbitrary, $A = \phi_e(N)$. ∎

8.8. Corollary. *If ρ_h: $W^*(N) \to W^*(N_h)$ is the *-epimorphism of Lemma 8.5, then* $\ker \rho_h = \{\phi(N): \phi = 0 \text{ a.e. } [\mu_h]\}$.

8.9. Theorem. *If N is a normal operator and $e \in \mathcal{H}$, the following statements are equivalent.*

(a) *e is a separating vector for $W^*(N)$.*
(b) *μ_e is a scalar-valued spectral measure for N.*
(c) *The map ρ_e: $W^*(N) \to W^*(N_e)$ defined in (8.5) is a *-isomorphism.*
(d) *$\{\phi \in B(\sigma(N)): \phi(N) = 0\} = \{\phi \in B(\sigma(N)): \phi = 0 \text{ a.e. } [\mu_e]\}$.*

PROOF. (a)$\Rightarrow$(b): Proposition 8.3.

(b)$\Rightarrow$(c): By Lemma 8.5, ρ_e is a *-epimorphism. By Corollary 8.8, $\ker \rho_e = \{\phi(N): \phi = 0 \text{ a.e. } [\mu_e]\}$. But if $\phi = 0$ a.e. $[\mu_e]$, (b) implies that $\phi = 0$ off a set Δ such that $E(\Delta) = 0$. Thus $\phi(N) = \int_\Delta \phi \, dE = 0$.

(c)$\Rightarrow$(d): Combine (c) with Corollary 8.8.

(d)$\Rightarrow$(a): Suppose $A \in W^*(N)$ and $Ae = 0$. By Lemma 8.7, there is a ϕ in $B(\sigma(N))$ such that $\phi(N) = A$. Thus, $0 = \|Ae\|^2 = \langle A^*Ae, e \rangle = \int |\phi|^2 \, d\mu_e$. So $\phi = 0$ a.e. $[\mu_e]$. By (d), $A = 0$. ∎

These results can now be combined to yield the final statement of the functional calculus for normal operators.

8.10. The Functional Calculus for a Normal Operator. *If N is a normal operator on the separable Hilbert space $\mathcal{H}$ and μ is a scalar-valued spectral measure for N, then there is a well-defined map $\rho: L^\infty(\mu) \to W^*(N)$ given by the formula $\rho(\phi) = \phi(N)$ such that*

(a) ρ is a *-isomorphism and an isometry;

(b) $\rho: (L^\infty(\mu), \text{weak}^*) \to (W^*(N), \text{WOT})$ is a homeomorphism.

PROOF. Let e be a separating vector such that $\mu = \mu_e$ [by (8.6) and (8.9)]. If $\phi \in B(\sigma(N))$ and $\phi = 0$ a.e. $[\mu]$, then $\phi(N) = 0$ by (8.9d); so $\rho(\phi) = \phi(N)$ is a well-defined map. It is left to the reader to show that ρ is a *-homomorphism. By Lemma 8.7, ρ is surjective. Also, if $\rho(\phi) = \phi(N) = 0$, then $\phi = 0$ a.e. $[\mu]$ by (8.9d). Thus ρ is a *-isomorphism. By (VIII.4.8) ρ is an isometry. (A proof avoiding (VIII.4.8) is possible—it is left as an exercise.) This proves (a).

Let $\{\phi_i\}$ be a net in $L^\infty(\mu)$ and suppose that $\phi_i(N) \to 0 (\text{WOT})$. If $f \in L^1(\mu)$ and $f \geq 0$, $f\mu \ll \mu = \mu_e$. By Lemma 8.6 there is a vector h such that $f\mu = \mu_h$. Thus $\int \phi_i f \, d\mu = \int \phi_i \, d\mu_h = \langle \phi_i(N)h, h \rangle \to 0$. Thus $\phi_i \to 0$ (weak*) in $L^\infty(\mu)$. This proves half of (b); the other half is left as an exercise. ∎

8.11. The Spectral Mapping Theorem. *If N is a normal operator on a separable space and μ is a scalar-valued spectral measure for N and if $\phi \in L^\infty(\mu)$, then $\sigma(\phi(N)) = $ the μ-essential range of ϕ.*

PROOF. Use (8.10) and the fact (2.6) that the μ-essential range of ϕ is the spectrum of ϕ as an element of $L^\infty(\mu)$. ∎

8.12. Proposition. *Let N, μ, ϕ be as in (8.11). If $N = \int z \, dE$, then $\mu \circ \phi^{-1}$ is a scalar-valued spectral measure for $\phi(N)$ and $E \circ \phi^{-1}$ is its spectral measure.*

EXERCISES

1. What is a scalar-valued spectral measure for a diagonalizable normal operator?

2. Let N_1 and N_2 be normal operators with scalar-valued spectral measures μ_1 and μ_2. What is a scalar spectral measure for $N_1 \oplus N_2$?

3. Let $\{e_n\}$ be an orthonormal basis for $\mathcal{H}$ and put $\mu(\Delta) = \sum_{n=1}^\infty 2^{-n} \| E(\Delta)e_n \|^2$. Show that μ is a scalar-valued spectral measure for N.

4. Give an example of a normal operator on a nonseparable space which has no scalar-valued spectral measure.

5. Prove that the map ρ in (8.10) is an isometry without using (VIII.4.8).

6. Prove Proposition 8.12.

7. Show that if μ and v are compactly supported measures on $\mathbb{C}$, the following statements are equivalent: (a) $N_\mu \oplus N_v$ is *-cyclic; (b) $W^*(N_\mu \oplus N_v) = W^*(N_\mu) \oplus W^*(N_v)$; (c) $\mu \perp v$.

8. If M and N are normal operators with scalar-valued spectral measures μ and v, respectively, show that the following are equivalent: (a) $W^*(M \oplus N) = W^*(M) \oplus W^*(N)$; (b) $\{M \oplus N\}' = \{M\}' \oplus \{N\}'$; (c) there is no operator A such that $MA = AN$ other than $A = 0$; (d) $\mu \perp v$.

9. If M and N are normal operators, show that $C^*(M \oplus N) = C^*(M) \oplus C^*(N)$ if and only if $\sigma(M) \cap \sigma(N) = \square$.

10. Give an example of two normal operators M and N such that $W^*(M \oplus N) = W^*(M) \oplus W^*(N)$ but $C^*(M \oplus N) \neq C^*(M) \oplus C^*(N)$. In fact, find M and N such that $W^*(M \oplus N)$ splits, but $\sigma(M) = \sigma(N)$.

11. If U is the bilateral shift and V is any unitary operator, show that $W^*(U \oplus V) = W^*(U) \oplus W^*(V)$ if and only if V has a spectral measure that is singular to arc length on $\partial\mathbb{D}$. (See Exercise 3.7.)

12. If $\mathscr{A}$ is an abelian von Neumann algebra on a separable space, show that there is a compactly supported measure μ on $\mathbb{R}$ such that $\mathscr{A}$ is $*$-isomorphic to $L^\infty(\mu)$. (Hint: Use Exercise 7.7.)

13. (This exercise assumes a knowledge of Exercise 2.21.) Let $N = \int z\, dE(z)$ be a normal operator with scalar-valued spectral measure μ and define $\alpha: \mathscr{B}_1(\mathscr{H}) \to L^1(\mu)$ by $\alpha(T)(\Delta) = \operatorname{tr}(T E(\Delta))$. Show that α is a surjective contraction. What is α^*? [$L^1(\mu)$ is identified, via the Radon–Nikodym Theorem, with the set of complex-valued measures that are absolutely continuous with respect to μ.]

14. Let $\pi: \mathscr{B}(\mathscr{H}) \to \mathscr{B}(\mathscr{H})/\mathscr{B}_0(\mathscr{H})$ be the natural map and let $A \in \mathscr{B}(\mathscr{H})$. (a) If $\pi(A)$ is hermitian, show that there is a hermitian operator B such that $A - B$ is compact. (b) If $\pi(A)$ is positive, show that there is a positive operator B such that $A - B$ is compact. (See Exercise XI.3.14.)

15. (L.G. Brown) If $\pi: \mathscr{B}(\mathscr{H}) \to \mathscr{B}(\mathscr{H})/\mathscr{B}_0(\mathscr{H})$ is the natural map and A and B are hermitian operators such that $\pi(A) \leqslant 0 \leqslant \pi(B)$, then there is a hermitian compact operator K such that $A \leqslant K \leqslant B$.

§9. Invariant Subspaces for Normal Operators

Remember that we continue to assume that all Hilbert spaces are separable.

Every normal operator on a Hilbert space of dimension at least 2 has a nontrivial invariant subspace. This is an easy consequence of the Spectral Theorem. Indeed, if $N = \int z\, dE(z)$, $E(\Delta)\mathscr{H}$ is a reducing subspace for every Borel set Δ.

If $A \in \mathscr{B}(\mathscr{H})$, $\mathscr{M}$ is a linear subspace of $\mathscr{H}$, and P is the orthogonal projection of $\mathscr{H}$ onto $\mathscr{M}$, then $\mathscr{M}$ reduces A if and only if $P \in \{A\}'$. Also, $\mathscr{M} \in \operatorname{Lat} A$ (= the lattice of invariant subspaces for A) if and only if $AP = PAP$. Since the spectral projections of a normal operator belong to $W^*(N)$, they are even more than reducing.

9.1. Definition. An operator A is *reductive* if every invariant subspace for A reduces A. Equivalently, A is reductive if and only if $\operatorname{Lat} A \subseteq \operatorname{Lat} A^*$.

Thus, every self-adjoint operator is reductive. Every normal operator on a finite dimensional space is reductive. More generally, every normal compact operator is reductive (Andô [1963]). However, the bilateral shift is not reductive. Indeed, if U is the bilateral shift on $l^2(\mathbb{Z})$, $\mathscr{H} = \{f \in l^2(\mathbb{Z}): f(n) = 0$

if $n < 0\} \in \text{Lat } U$, but $\mathcal{H} \notin \text{Lat } U^*$. Wermer [1952] first studied reductive normal operators and characterized the reductive unitary operators. A first step towards characterizing the reductive normal operators will be taken here. The final step has been taken but it will not be viewed in this book. The result is due to Sarason [1972]. Also see Conway [1981], §VII.5.

9.2. Definition. If μ is a compactly supported measure on $\mathbb{C}$, $P^{\infty}(\mu)$ denotes the weak* closure of the polynomials in $L^{\infty}(\mu)$.

Because the support of μ is compact, every polynomial, when restricted to that support, belongs to $L^{\infty}(\mu)$.

For any operator A, let $W(A)$ denote the WOT closed subalgebra of $\mathcal{B}(\mathcal{H})$ generated by A; that is, $W(A)$ is the WOT closure in $\mathcal{B}(\mathcal{H})$ of the polynomials in A. The next result is an immediate consequence of The Functional Calculus for Normal Operators.

9.3. Theorem. *If N is a normal operator and μ is a scalar-valued spectral measure for N, then the functional calculus, when restricted to $P^{\infty}(\mu)$, is an isometric isomorphism $\rho: P^{\infty}(\mu) \to W(N)$ and a weak*-WOT homeomorphism. Also, $\rho(z) = N$.*

9.4. Definition. An operator A is *reflexive* if whenever $B \in \mathcal{B}(\mathcal{H})$ and $\text{Lat } A \subseteq \text{Lat } B$, then $B \in W(A)$.

It is easy to see that if $B \in W(A)$, then $\text{Lat } A \subseteq \text{Lat } B$ (Exercise). An operator is reflexive precisely when it has sufficiently many invariant subspaces to characterize $W(A)$. For a survey of reflexive operators and some related topics, see Radjavi and Rosenthal [1973].

9.5. Theorem. *(Sarason [1966].). Every normal operator is reflexive.*

PROOF. Suppose N is normal and $\text{Lat } N \subseteq \text{Lat } A$. If P is a projection in $\{N\}'$, then $P\mathcal{H}$ and $(P\mathcal{H})^{\perp} \in \text{Lat } N \subseteq \text{Lat } A$, so $AP = PA$. By Corollary 6.5, $A \in W^*(N)$. Let μ be a scalar-valued spectral measure for N. By Theorem 8.10, there is a ϕ in $L^{\infty}(\mu)$ such that $A = \phi(N)$. By Theorem 9.3, it must be shown that $\phi \in P^{\infty}(\mu)$.

Now let's focus our attention on a special case. Assume that N is *-cyclic; thus $N = N_{\mu}$. Suppose $f \in L^1(\mu)$ and $\int f\psi\, d\mu = 0$ for every ψ in $P^{\infty}(\mu)$. If it can be shown that $\int f\phi\, d\mu = 0$, then the Hahn–Banach Theorem implies that $\phi \in P^{\infty}(\mu)$. This is the strategy we follow. Let $f = g\bar{h}$ for some g, h in $L^2(\mu)$. Put $\mathcal{M} = \bigvee\{z^k g: k \geq 0\}$. Clearly $\mathcal{M} \in \text{Lat } N$, so $\mathcal{M} \in \text{Lat } A = \text{Lat } \phi(N) = \text{Lat } M_{\phi}$. Hence $\phi g \in \mathcal{M}$. But $0 = \int z^k f\, d\mu = \int z^k g\bar{h}\, d\mu = \langle N^k g, h \rangle$ for all $k \geq 0$; hence $h \perp \mathcal{M}$. Thus $0 = \langle \phi g, h \rangle = \int \phi g\bar{h}\, d\mu = \int \phi f\, d\mu$, and $\phi \in P^{\infty}(\mu)$.

Now we return to the general case. By (7.9) and (8.9) there is a separating vector e for $W^*(N)$ such that $\mu(\Delta) = \|E(\Delta)e\|^2$, where $N = \int z\, dE(z)$. Let

$\mathscr{K} = \vee\{N^{*k}N^j e: k, j \geq 0\}$. Clearly $\mathscr{K}$ reduces N and $N|\mathscr{K}$ is *-cyclic. Hence $\mathscr{K}$ reduces A and $A|\mathscr{K} = \phi(N|\mathscr{K})$. By the preceding paragraph $\phi \in P^\infty(\mu)$.   ∎

An immediate consequence of the preceding theorem is the first step in the characterization of reductive normal operators.

9.6. Corollary. *If N is a normal operator and μ is a scalar-valued spectral measure for N, then N is reductive if and only if $P^\infty(\mu) = L^\infty(\mu)$.*

PROOF. If $\mathscr{M} \in \operatorname{Lat} N$, then $\mathscr{M} \in \operatorname{Lat} \phi(N)$ for every ϕ in $P^\infty(\mu)$. So if $P^\infty(\mu) = L^\infty(\mu)$, $\bar{z} \in P^\infty(\mu)$ and, hence, $\mathscr{M} \in \operatorname{Lat} N^*$ whenever $\mathscr{M} \in \operatorname{Lat} N$.

Now suppose N is reductive. This means that $\operatorname{Lat} N \subseteq \operatorname{Lat} N^*$. By the preceding theorem, this implies $N^* \in W(N)$; equivalently, $\bar{z} \in P^\infty(\mu)$. Since $P^\infty(\mu)$ is an algebra, every polynomial in z and $\bar{z}$ belongs to $P^\infty(\mu)$. By taking weak* limits this implies that $P^\infty(\mu) = L^\infty(\mu)$.   ∎

The preceding corollary fails to be a good characterization of reductive normal operators since it only says that one difficult problem is equivalent to another. A way is needed to determine when $P^\infty(\mu) = L^\infty(\mu)$. This is what was done in Sarason [1972].

Are there any reductive operators that are not normal? This natural and seemingly innocent question has much more to it than meets the eye. Dyer, Pedersen, and Porcelli [1972] have shown that this question has an affirmative answer if and only if every operator on a Hilbert space has a nontrivial invariant subspace.

EXERCISES

1. (Andô [1963].) Use Corollary 9.6 to show that every compact normal operator is reductive.

2. Determine all of the invariant subspaces of a compact normal operator.

3. (Andô [1963].) Show that every reductive compact operator is normal. (Also see Rosenthal [1968].)

4. Show that an operator A is reductive if and only if $\operatorname{Lat} A = \operatorname{Lat} A^*$.

5. Let N be a normal operator on a Hilbert space $\mathscr{H}$, let $\mathscr{R}$ be an invariant subspace for N, and let $S = N|\mathscr{R}$. Show that S is normal if and only if $\mathscr{R}$ is a reducing subspace for N.

6. Let μ be a compactly supported regular Borel measure on $\mathbb{C}$ and let $P^2(\mu)$ denote the closure in $L^2(\mu)$ of the analytic polynomials; that is, $P^2(\mu)$ is the closed linear span of $\{z^n: n \geq 0\}$. Show that $P^2(\mu)$ is invariant for N_μ and that $N_\mu|P^2(\mu)$ is normal if and only if $P^2(\mu) = L^2(\mu)$.

7. With the notation of Exercise 6, show that if $\mathscr{R}$ is a reducing subspace for N_μ and $P^2(\mu) \subseteq \mathscr{R}$, then $\mathscr{R} = L^2(\mu)$.

§10. Multiplicity Theory for Normal Operators: A Complete Set of Unitary Invariants

Throughout this section only separable Hilbert spaces are considered.

When are two normal operators unitarily equivalent? The answer to this question must be given in the following way: to each normal operator we must attach a collection of objects such that two normal operators are unitarily equivalent if and only if the two collections are equal (or equivalent). Furthermore, it should be easier to verify that these collections are equivalent than to verify that the normal operators are equivalent. This is contained in the following result due to Hellinger [1907]. Note that it generalizes Theorem 3.6.

10.1. Theorem. (a) *If N is a normal operator, then there is a sequence (possibly finite) of measures $\{\mu_n\}$ on $\mathbb{C}$ such that $\mu_{n+1} \ll \mu_n$ for all n and*

10.2
$$N \cong N_{\mu_1} \oplus N_{\mu_2} \oplus \cdots.$$

(b) *If N and $\{\mu_n\}$ are as in (a) and $M \cong N_{\nu_1} \oplus N_{\nu_2} \oplus \cdots$, where $\nu_{n+1} \ll \nu_n$ for all n, then $N \cong M$ if and only if $[\mu_n] = [\nu_n]$ for all n.*

The proof of this theorem requires several lemmas. Before beginning, we will examine a couple of false starts for a proof. This will cause us to arrive at the correct strategy for a proof and show us the necessity for some of the lemmas.

Let $N = \int z \, dE(z)$. If $e \in \mathcal{H}$ and $\mathcal{H}_e = \mathrm{cl}[W^*(N)e]$, then $N|\mathcal{H}_e$ is a *-cyclic normal operator. An application of Zorn's Lemma and the separability of $\mathcal{H}$ produces a maximal sequence $\{e_n\}$ in $\mathcal{H}$ such that $\mathcal{H}_{e_n} \perp \mathcal{H}_{e_m}$. By the maximality of $\{e_n\}$, $\mathcal{H} = \oplus_n \mathcal{H}_{e_n}$. If $N_n = N|\mathcal{H}_{e_n}$, $N_n = N_{\mu_n}$, where $\mu_n(\Delta) = \|E(\Delta)e_n\|^2$; thus $N \cong \oplus_n N_{\mu_n}$. The trouble here is that μ_{n+1} is not necessarily absolutely continuous with respect to μ_n. Just using Zorn's Lemma to produce the sequence $\{\mu_n\}$ eliminates any possibility of having $\{\mu_n\}$ canonical and producing the unitary invariant desired for normal operators. Let's try again.

Note that if $\mu_{n+1} \ll \mu_n$ for all n in (10.2), then $\mu_n \ll \mu_1$ for all n. This in turn implies that μ_1 is a scalar-valued spectral measure for N. Using Lemma 8.6 we are thus led to choose μ_1 as follows. Let e_1 be a separating vector for $W^*(N)$; this exists by Corollary 7.9 and the separability of $\mathcal{H}$. Put $\mu_1(\Delta) = \|E(\Delta)e_1\|^2$. If $\mathcal{H}_1 = \mathrm{cl}[W^*(N)e_1]$. then $N|\mathcal{H}_1 \cong N_{\mu_1}$. Let $N_2 \equiv N|\mathcal{H}_1^\perp$; so N_2 is normal. A pair of easy exercises shows that the spectral measure E_2 for N_2 is given by $E_2(\Delta) = E(\Delta)|\mathcal{H}_1^\perp$ and $W^*(N_2) = W^*(N)|\mathcal{H}_1^\perp$ ($\equiv \{A \in \mathcal{B}(\mathcal{H}_1^\perp): A = T|\mathcal{H}_1^\perp$ for some T in $W^*(N)\}$). Let e_2 be a separating vector for $W^*(N_2)$ and put $\mu_2(\Delta) = \|E_2(\Delta)e_2\|^2$. By the easy exercises above, $\mu_2(\Delta) = \|E(\Delta)e_2\|^2$, so that $\mu_2 \ll \mu_1$, and $\mathcal{H}_2 \equiv \mathrm{cl}[W^*(N)e_2] = \mathrm{cl}[W^*(N_2)e_2] \leqslant \mathcal{H}_1^\perp$. Also, $N|\mathcal{H}_2 \cong N_{\mu_2}$.

Continuing in this way produces a sequence of vectors $\{e_n\}$ such that if $\mathcal{H}_n = \mathrm{cl}[W^*(N)e_n]$ and $\mu_n(\Delta) = \|E(\Delta)e_n\|^2$, then $\mathcal{H}_n \perp \mathcal{H}_m$ for $n \neq m$, $\mu_{n+1} \ll \mu_n$, and $N|\mathcal{H}_n \cong N_{\mu_n}$. The difficulty here is that $\mathcal{H}$ is not necessarily equal to $\oplus_n \mathcal{H}_n$ so that N and $\oplus_n N_{\mu_n}$ cannot be proved to be unitarily equivalent. (Actually, N and $\oplus_n N_{\mu_n}$ are unitarily equivalent, but to show this we need the force of Theorem 10.1. See Exercise 2.) The following provides us with a look at an example to see what can go wrong.

10.3. Example. For $n \geqslant 1$ let $\mu_n =$ Lebesgue measure on $[0, 1 + 2^{-n}]$ and let $\mu_\infty =$ Lebesgue measure on $[0, 1]$. Put $N = \oplus_{n=1}^\infty N_{\mu_n} \oplus N_{\mu_\infty}$. If the process of the preceding paragraph is followed, it might be that vectors $\{e_n\}$ that are chosen are the vectors with a 1 in the $L^2(\mu_n)$ coordinate and zeros elsewhere. Thus the spaces $\{\mathcal{H}_n\}$ are precisely the spaces $\{L^2(\mu_n)\}$ and $[\oplus_1^\infty \mathcal{H}_n]^\perp = L^2(\mu_\infty)$.

(Nevertheless, $N \cong \oplus_1^\infty N_{\mu_n}$. Indeed, let $v_n = \mu_n|[1, 1 + 2^{-n}]$; so $\mu_n = \mu_\infty + v_n$ and $\mu_\infty \perp v_n$. Thus $N_{\mu_n} \cong N_{v_n} \oplus N_{\mu_\infty}$. Therefore

$$\begin{aligned} N &= \oplus_1^\infty N_{\mu_n} \oplus N_{\mu_\infty} \\ &\cong \oplus_1^\infty N_{v_n} \oplus N_{\mu_\infty}^{(\infty)} \oplus N_{\mu_\infty} \\ &\cong \oplus_1^\infty N_{v_n} \oplus N_{\mu_\infty}^{(\infty)} \\ &\cong \oplus_1^\infty N_{\mu_n}.) \end{aligned}$$

After an examination of the statement of Theorem 10.1, it becomes clear that some procedure like the one outlined in the paragraph preceding Example 10.3 should be used. It only becomes necessary to modify this procedure so that the vectors $\{e_n\}$ can be chosen in such a way that $\mathcal{H} = \oplus_1^\infty \mathcal{H}_n$. For example, let, as above, e_1 be a separating vector for $W^*(N)$ and let $\{f_n\}$ be an orthonormal basis for $\mathcal{H}$ such that $f_1 = e_1$. We now want to choose the vectors $\{e_n\}$ such that $\{f_1, \ldots, f_n\} \subseteq \mathcal{H}_1 \oplus \cdots \oplus \mathcal{H}_n$. In this way we will meet success. The vital link here is the next result.

10.4. Proposition. *If N is a normal operator on $\mathcal{H}$ and $e \in \mathcal{H}$, then there is a separating vector e_0 for $W^*(N)$ such that $e \in \mathrm{cl}[W^*(N)e_0]$.*

PROOF. Let f_0 be any separating vector for $W^*(N)$, let E be the spectral measure for N, let $\mu(\Delta) = \|E(\Delta)f_0\|^2$, and put $\mathcal{G} = \mathrm{cl}[W^*(N)f_0]$. Write $e = g_1 + h_1$, where $g_1 \in \mathcal{G}$ and $h_1 \in \mathcal{G}^\perp$.

Let $\eta(\Delta) = \|E(\Delta)h_1\|^2$ and let $\mathcal{L} = \mathrm{cl}[W^*(N)h_1]$. Hence $\eta \ll \mu$, N is reduced by both $\mathcal{L}$ and $\mathcal{G}$, and $\mathcal{L} \perp \mathcal{G}$. Moreover, $N|\mathcal{G} \cong N_\mu$ and $N|\mathcal{L} \cong N_\eta$. Now the fact that $\eta \ll \mu$ implies that there is a Borel set Δ such that $[\eta] = [\mu|\Delta]$. (Why?) Hence $N|\mathcal{L} \cong N_v$ if $v = \mu|\Delta$ (Theorem 3.6). Let $U: \mathcal{G} \oplus \mathcal{L} \to L^2(\mu) \oplus L^2(v)$ be an isomorphism such that $U((0) \oplus \mathcal{L}) \subseteq (0) \oplus L^2(v)$ and $U(N|\mathcal{G} \oplus \mathcal{L}) \cdot U^{-1} = N_\mu \oplus N_v$. Since $e = g_1 + h_1 \in \mathcal{G} \oplus \mathcal{L}$, let $Ue = g \oplus h$. Because h_1 is a *-cyclic vector for $N|\mathcal{L}$, $h(z) \neq 0$ a.e. $[v]$.

This reduces the proof of this proposition to proving the next lemma. ∎

10.5. Lemma. *Let μ be a compactly supported measure on $\mathbb{C}$, Δ a Borel subset of the support of μ, and put $\nu = \mu | \Delta$. If $N = N_\mu \oplus N_\nu$ on $L^2(\mu) \oplus L^2(\nu)$ and $g \oplus h \in L^2(\mu) \oplus L^2(\nu)$ such that $h(z) \neq 0$ a.e. $[\nu]$, then there is an f in $L^2(\mu)$ such that $f \oplus h$ is a separating vector for $W^*(N)$ and $g \oplus h \in \mathrm{cl}[W^*(N)(f \oplus h)]$.*

PROOF. Define $f(z) = g(z)$ for z in Δ and $f(z) = 1$ for z not in Δ. Put $\mathcal{H} = \mathrm{cl}[W^*(N)(f \oplus h)] = \mathrm{cl}\{\phi f \oplus \phi h : \phi \in L^\infty(\mu)\}$ since μ is a scalar-valued spectral measure for N. If $\Delta' =$ the complement of Δ, then note that $\phi \chi_{\Delta'} \oplus 0 = \phi \chi_{\Delta'}(f \oplus h) \in \mathcal{H}$ for all ϕ in $L^\infty(\mu)$. Hence $L^2(\mu | \Delta') \oplus 0 \subseteq \mathcal{H}$. This implies that $(1 - g)\chi_{\Delta'} \oplus 0 \in \mathcal{H}$, so $g \oplus h = f \oplus h - (1 - g)\chi_{\Delta'} \oplus 0 \in \mathcal{H}$.

On the other hand, if $\phi \in L^\infty(\mu)$ and $0 = \phi f \oplus \phi h$, then $\phi f = \phi h = 0$ a.e. $[\mu]$. Since $h(z) \neq 0$ a.e. $[\nu]$, $\phi(z) = 0$ a.e. $[\mu]$ on Δ. But for z in Δ', $f(z) = 1$; hence $\phi(z) = 0$ a.e. $[\mu]$ on Δ'. Thus, $f \oplus h$ is a separating vector for $W^*(N)$. ∎

PROOF OF THEOREM 10.1 (a): Let e_1 be a separating vector for $W^*(N)$ and let $\{f_n\}$ be an orthonormal basis for $\mathcal{H}$ such that $f_1 = e_1$. Put $\mathcal{H}_1 = \mathrm{cl}[W^*(N)e_1]$, $\mu_1(\Delta) = \| E(\Delta)e_1 \|^2$, and $N_2 = N | \mathcal{H}_1^\perp$. Let $f_2' =$ the orthogonal projection of f_2 onto $\mathcal{H}_1^\perp$. By Proposition 10.4 there is a separating vector e_2 for $W^*(N_2)$ such that $f_2' \in \mathrm{cl}[W^*(N_2)e_2] \equiv \mathcal{H}_2$. Note that $\mathcal{H}_2 = \mathrm{cl}[W^*(N)e_2]$ and $\{f_1, f_2\} \subseteq \mathcal{H}_1 \oplus \mathcal{H}_2$. Put $\mu_2(\Delta) = \| E(\Delta)e_2 \|^2$. Now continue by induction. ∎

Now for part (b) of Theorem 10.1. If $[\mu_n] = [\nu_n]$ (the notation is that of Theorem 10.1) for every n, then $N_{\mu_n} \cong N_{\nu_n}$ by Theorem 3.6. Therefore $N \cong M$. Thus it is the converse that causes difficulties. So assume that $N \cong M$. If $M \in \mathcal{B}(\mathcal{K})$, $U: \mathcal{H} \to \mathcal{K}$ is an isomorphism such that $UNU^{-1} = M$, and e_1 is a separating vector for $W^*(N)$, then $Ue_1 = f_1$ is easily seen to be a separating vector for $W^*(M)$. Since μ_1 and ν_1 are scalar-valued spectral measures for N and M, respectively, it follows that $[\mu_1] = [\nu_1]$; thus $N_{\mu_1} \cong N_{\nu_1}$ by Theorem 3.6. However, here is the difficulty—the isomorphism that shows that $N_{\mu_1} \cong N_{\nu_1}$ may not be related to U; that is, if $\mathcal{H} = \oplus_n \mathcal{H}_n$, $\mathcal{K} = \oplus_n \mathcal{K}_n$, where $N | \mathcal{H}_n \cong N_{\mu_n}$ and $M | \mathcal{K}_n \cong N_{\nu_n}$, then $N | \mathcal{H}_1 \cong M | \mathcal{K}_1$, but we do not know that $U\mathcal{H}_1 = \mathcal{K}_1$. Thus we want to argue that because $N \cong M$ and $N | \mathcal{H}_1 \cong M | \mathcal{K}_1$, then $N | \mathcal{H}_0^\perp \cong M | \mathcal{K}_1^\perp$. In this way we can prove (10.1.b) by induction. This step is justified by the following.

10.6. Proposition. *If N, A, and B are normal operators, N is $*$-cyclic, and $N \oplus A \cong N \oplus B$, then $A \cong B$.*

PROOF. Let $N \in \mathcal{B}(\mathcal{K})$, $A \in \mathcal{B}(\mathcal{H}_A)$, $B \in \mathcal{B}(\mathcal{H}_B)$, and let $U: \mathcal{K} \oplus \mathcal{H}_A \to \mathcal{K} \oplus \mathcal{H}_B$ be an isomorphism such that $U(N \oplus A)U^{-1} = N \oplus B$. Now U can be written as a 2×2 matrix,

$$U = \begin{bmatrix} U_{11} & U_{12} \\ U_{21} & U_{22} \end{bmatrix},$$

where $U_{11}: \mathcal{K} \to \mathcal{K}$, $U_{12}: \mathcal{H}_A \to \mathcal{K}$, $U_{21}: \mathcal{K} \to \mathcal{H}_B$, $U_{22}: \mathcal{H}_A \to \mathcal{H}_B$.

Expressing $N \oplus A$ and $N \oplus B$ as

$$\begin{bmatrix} N & 0 \\ 0 & A \end{bmatrix} \text{ and } \begin{bmatrix} N & 0 \\ 0 & B \end{bmatrix}$$

respectively, the equation $U(N \oplus A) = (N \oplus B)U$ becomes

10.7
$$\begin{bmatrix} U_{11}N & U_{12}A \\ U_{21}N & U_{22}A \end{bmatrix} = \begin{bmatrix} NU_{11} & NU_{12} \\ BU_{21} & BU_{22} \end{bmatrix}.$$

Similarly, $U(N \oplus A)^* = (N \oplus B)^*U$ becomes

10.7*
$$\begin{bmatrix} U_{11}N^* & U_{12}A^* \\ U_{21}N^* & U_{22}A^* \end{bmatrix} = \begin{bmatrix} N^*U_{11} & N^*U_{12} \\ B^*U_{21} & B^*U_{22} \end{bmatrix}.$$

Parts of the preceding equations will be referred to as $(10.7)_{ij}$ and $(10.7)^*_{ij}$, $i, j = 1, 2$.

An examination of the equation $U^*U = 1$ and $UU^* = 1$, written in matrix form, yields the equations

10.8
$$\left. \begin{array}{ll} \text{(a) } U^*_{11}U_{12} + U^*_{21}U_{22} = 0 & \text{(on } \mathscr{H}_A) \\ \text{(b) } U_{11}U^*_{11} + U_{12}U^*_{12} = 1 & \text{(on } \mathscr{H}) \\ \text{(c) } U_{21}U^*_{11} + U_{22}U^*_{12} = 0 & \text{(on } \mathscr{H}). \end{array} \right\}$$

Now equation $(10.7)_{22}$ and Proposition 6.10 imply that $(\ker U_{22})^\perp$ reduces A, $\mathrm{cl}(\mathrm{ran}\, U_{22}) = (\ker U^*_{22})^\perp$ reduces B, and

10.9
$$A|(\ker U_{22})^\perp \cong B|(\ker U^*_{22})^\perp.$$

What about $A|\ker U_{22}$ and $B|\ker U^*_{22}$? If they are unitarily equivalent, then $A \cong B$ and we are done. If $h \in \ker U_{22} \subseteq \mathscr{H}_A$, then

$$\begin{bmatrix} U_{11} & U_{12} \\ U_{21} & U_{22} \end{bmatrix} \begin{bmatrix} 0 \\ h \end{bmatrix} = \begin{bmatrix} U_{12}h \\ 0 \end{bmatrix}.$$

Since U is an isometry, it follows that U_{12} maps $\ker U_{22}$ isometrically onto a closed subspace of $\mathscr{H}$. Put $\mathscr{M}_1 = U_{12}(\ker U_{22})$. Equations $(10.7)_{12}$ and $(10.7)^*_{12}$ and the fact that $\ker U_{22}$ reduces A imply that $\mathscr{M}_1$ reduces N. Thus, the restriction of U_{12} to $\ker U_{22}$ is the required isomorphism to show that

10.10
$$A|\ker U_{22} \cong N|\mathscr{M}_1.$$

Similarly, U^*_{21} maps $\ker U^*_{22} = (\mathrm{ran}\, U_{22})^\perp$ isometrically into $\mathscr{M}_2 = U^*_{21}(\ker U^*_{22})$, $\mathscr{M}_2$ reduces N, and

10.11
$$B|\ker U^*_{22} \cong N|\mathscr{M}_2.$$

Note that if $\mathscr{M}_1 = \mathscr{M}_2$, then (10.9), (10.10), and (10.11) show that $A \cong B$. Could it be that $\mathscr{M}_1$ and $\mathscr{M}_2$ are equal?

If $h \in \ker U_{22}$, then (10.8.a) implies that $U^*_{11}U_{12}h = -U^*_{21}U_{22}h = 0$. Hence $\mathscr{M}_1 = U_{12}(\ker U_{22}) \subseteq \ker U^*_{11}$. On the other hand, if $f \in \ker U^*_{11}$, then (10.8.b) implies $f = (U_{11}U^*_{11} + U_{12}U^*_{12})f = U_{12}U^*_{12}f$. But by (10.8.c), $U_{22}U^*_{12}f = -$

$U_{21}U_{11}^*f = 0$, so $U_{12}^*f \in \ker U_{22}$. Hence $f \in U_{12}(\ker U_{22})$. Thus

$$\mathcal{M}_1 = \ker U_{11}^*.$$

Similarly,

$$\mathcal{M}_2 = \ker U_{11}.$$

Until this point we have not used the fact that N is $*$-cyclic. Equation $(10.7)_{11}$ implies that $U_{11} \in \{N'\}'$. By Theorem 3.4 and Corollary 6.9 this implies that U_{11} is normal. Hence, $\ker U_{11}^* = \ker U_{11}$, or $\mathcal{M}_1 = \mathcal{M}_2$. ∎

If the hypothesis in the preceding proposition that N is $*$-cyclic is deleted, the conclusion is no longer valid. For example, let N and A be the identities on separable infinite dimensional spaces and let B be the identity on a finite dimensional space. Then $N \oplus A \cong N \oplus B$, but A and B are not equivalent. However, the requirement that N be $*$-cyclic can be replaced by another, even when N, A, and B are not assumed to be normal. For the details see Kadison and Singer [1957].

The proof of Theorem 10.1(b) is now a straightforward argument as outlined before the statement of Proposition 10.6. The details are left to the reader.

If μ and ν are measures and $\nu \ll \mu$, then there is a Borel set Δ such that $[\nu] = [\mu|\Delta]$. Using this fact, Theorem 10.1 can be restated as follows.

10.12. Corollary. (a) *If N is a normal operator with scalar-valued spectral measure μ, then there is a decreasing sequence $\{\Delta_n\}$ of Borel subsets of $\sigma(N)$ such that $\Delta_1 = \sigma(N)$ and*

$$N \cong N_\mu \oplus N_{\mu|\Delta_2} \oplus N_{\mu|\Delta_3} \oplus \cdots.$$

(b) *If M is another normal operator with scalar-valued spectral measure ν and if $\{\Sigma_n\}$ is a decreasing sequence of Borel subsets of $\sigma(M)$ such that $M \cong N_\nu \oplus N_{\nu|\Sigma_2} \oplus N_{\nu|\Sigma_3} \oplus \cdots$, then $N \cong M$ if and only if (i) $[\mu] = [\nu]$ and (ii) $\mu(\Delta_n \setminus \Sigma_n) = 0 = \mu(\Sigma_n \setminus \Delta_n)$ for all n.*

10.13. Example. Let μ be Lebesgue measure on $[0,1]$ and let μ_n be Lebesgue measure on $[1/(n+1), 1/n]$ for $n \geq 1$. (So $\mu = \sum \mu_n$.) Let $N = N_{\mu_1} \oplus N_{\mu_2}^{(2)} \oplus N_{\mu_3}^{(3)} \oplus \cdots$. The direct sum decomposition of N that appears in Corollary 10.12 is obtained by letting $\Delta_n = [0, 1/n]$, $n \geq 1$. Then $N \cong N_\mu \oplus N_{\mu|\Delta_2} \oplus N_{\mu|\Delta_3} \oplus \cdots$.

What does Theorem 10.1 say for normal operators on a finite dimensional space? If $\dim \mathcal{H} < \infty$, there is an orthonormal basis $\{e_n\}$ for $\mathcal{H}$ consisting of eigenvectors for N. Observe that N is $*$-cyclic if and only if each eigenvalue has multiplicity 1. So each summand that appears in (10.2) must operate on a subspace of $\mathcal{H}$ that contains only one basis element e_n per eigenvalue. Moreover, since μ_1 is a scalar-valued spectral measure for N, it must be that the first summand in (10.2) contains one basis element for each eigenvalue for N. Thus, if $\sigma(N) = \{\lambda_1, \lambda_2, \ldots, \lambda_n\}$, where $\lambda_i \neq \lambda_j$ for $i \neq j$, then (10.2) becomes

10.14
$$N \cong D_1 \oplus D_2 \oplus \cdots \oplus D_m,$$

where $D_1 = \mathrm{diag}(\lambda_1, \lambda_2, \ldots, \lambda_n)$ and, for $k \geq 2$, D_k is a diagonalizable operator whose diagonal consists of one, and only one, of each of the eigenvalues of N having multiplicity at least k.

There is another decomposition for normal operators that furnishes a complete set of unitary invariants and has a connection with the concept of multiplicity. For normal operators on a finite dimensional space, this decomposition takes on the following form.

Let $\Lambda_k =$ the eigenvalues of N having multiplicity k. So for λ in Λ_k, $\dim \ker(N - \lambda) = k$. If $\Lambda_k = \{\lambda_j^{(k)}: 1 \leq j \leq m_k\}$, let N_k be the diagonalizable operator on a km_k dimensional space whose diagonal contains each $\lambda_j^{(k)}$ repeated k times. So $N \cong N_1 \oplus N_2 \oplus \cdots \oplus N_p$, if $\sigma(N) = \Lambda_1 \cup \cdots \cup \Lambda_p$. Now $\sigma(N_k) = \Lambda_k$ and each eigenvalue of N_k has multiplicity k. Thus $N_k \cong A_k^{(k)}$, where A_k is a diagonalizable operator on an m_k dimensional space with $\sigma(A_k) = \Lambda_k$. Thus

10.15
$$N \cong A_1 \oplus A_2^{(2)} \oplus \cdots \oplus A_p^{(p)},$$

and $\sigma(A_i) \cap \sigma(A_j) = \square$ for $i \neq j$.

Now the big advantage of the decomposition (10.15) is that it permits a discussion of $\{N\}'$. Because the spectra of the operators A_k are disjoint,

$$\{N\}' = \{N_1\}' \oplus \{N_2\}' \oplus \cdots \oplus \{N_p\}'.$$

(Why?) If $\mathscr{H}_j^{(k)} = \ker(N - \lambda_j^{(k)})$, then $\dim \mathscr{H}_j^{(k)} = k$ and $\oplus_{j=1}^{m_k} \mathscr{H}_j^{(k)} =$ the domain of N_k. Since $\lambda_i^{(k)} \neq \lambda_j^{(k)}$ for $i \neq j$,

$$\{N_k\}' = \mathscr{B}(\mathscr{H}_1^{(k)}) \oplus \cdots \oplus \mathscr{B}(\mathscr{H}_{m_k}^{(k)}),$$

and each $\mathscr{B}(\mathscr{H}_j^{(k)})$ is isomorphic to the $k \times k$ matrices.

The decomposition of an arbitrary normal operator that is analogous to decomposition (10.15) for finite dimensional normal operators is contained in the next result. The corresponding discussion of the commutant will follow this theorem.

10.16. Theorem. *If N is a normal operator, then there are mutually singular measures $\mu_\infty, \mu_1, \mu_2, \ldots$ (some of which may be zero) such that*

$$N \cong N_{\mu_\infty}^{(\infty)} \oplus N_{\mu_1} \oplus N_{\mu_2}^{(2)} \oplus \cdots .$$

If M is another normal operator with corresponding measures $v_\infty, v_1, v_2, \ldots$, then $N \cong M$ if and only if $[\mu_n] = [v_n]$ for $1 \leq n \leq \infty$.

PROOF. Let μ be a scalar-valued spectral measure for N and let $\{\Delta_n\}$ be the sequence of Borel subsets of $\sigma(N)$ obtained in Corollary 10.12. Put $\Sigma_\infty = \bigcap_{n=1}^{\infty} \Delta_n$ and $\Sigma_n = \Delta_n \backslash \Delta_{n+1}$ for $1 \leq n < \infty$; let $\mu_n = \mu | \Sigma_n$, $1 \leq n \leq \infty$. Put

$v_n = \mu | \Delta_n,\ 1 \leqslant n < \infty.$ Now $\Delta_n = \Sigma_\infty \cup (\Delta_n \backslash \Delta_{n+1}) \cup (\Delta_{n+1} \backslash \Delta_{n+2}) \cup \cdots = \Sigma_\infty \cup \Sigma_n \cup \Sigma_{n+1} \cup \cdots$. Hence $v_n = \mu_\infty + \mu_n + \mu_{n+1} + \cdots$ and the measures μ_∞, $\mu_n,\ \mu_{n+1}, \ldots$ are pairwise singular. Hence $N_{v_n} \cong N_{\mu_x} \oplus N_{\mu_n} \oplus N_{\mu_{n+1}} \cdots$. Combining this with Corollary 10.12 gives

$$N \cong N_{v_1} \oplus N_{v_2} \oplus N_{v_3} \oplus \cdots$$
$$\cong (N_{\mu_x} \oplus N_{\mu_1} \oplus N_{\mu_2} \oplus \cdots) \oplus (N_{\mu_x} \oplus N_{\mu_2} \oplus N_{\mu_3} \oplus \cdots)$$
$$\oplus (N_{\mu_x} \oplus N_{\mu_3} \oplus N_{\mu_4} \oplus \cdots) \oplus \cdots$$
$$\cong N_{\mu_x}^{(\infty)} \oplus N_{\mu_1} \oplus N_{\mu_2}^{(2)} \oplus N_{\mu_3}^{(3)} \oplus \cdots.$$

The proof of the uniqueness part of the theorem is left to the reader. ∎

Note that the form of the normal operator presented in Example 10.13 is the form of the operator given in the conclusion of the preceding theorem.

Now to discuss $\{N\}'$. Fix a compactly supported positive Borel measure μ on $\mathbb{C}$ and let $\mathcal{H}_n$ be an n-dimensional Hilbert space, $1 \leqslant n \leqslant \infty$. Define a function $f: \mathbb{C} \to \mathcal{H}_n$ to be a Borel function if $z \mapsto \langle f(z), g \rangle$ is a Borel function for each g in $\mathcal{H}_n$. If $f: \mathbb{C} \to \mathcal{H}_n$ is a Borel function and $\{e_j\}$ is an orthonormal basis for $\mathcal{H}_n$, then $\| f(z) \|^2 = \sum_j |\langle f(z), e_j \rangle|^2$, so $z \to \| f(z) \|^2$ is a Borel function. Let $L^2(\mu; \mathcal{H}_n)$ be the space of all Borel functions $f: \mathbb{C} \to \mathcal{H}_n$ such that $\| f \|^2 \equiv \int \| f(z) \|^2 d\mu(z) < \infty$, where two functions agreeing a.e. $[\mu]$ are identified. If f and $g \in L^2(\mu; \mathcal{H}_n)$, $\langle f, g \rangle \equiv \int \langle f(z), g(z) \rangle d\mu(z)$ defines an inner product on $L^2(\mu; \mathcal{H}_n)$. It is not difficult to show that $L^2(\mu; \mathcal{H}_n)$ is a Hilbert space.

10.17. Proposition. *If N is multiplication by z on $L^2(\mu; \mathcal{H}_n)$, then $N \cong N_\mu^{(n)}$.*

PROOF. Let $\{e_j: 1 \leqslant j \leqslant n\}$ be an orthonormal basis for $\mathcal{H}_n$ and define $U: L^2(\mu; \mathcal{H}_n) \to L^2(\mu)^{(n)}$ by $Uf = (\langle f(\cdot), e_1 \rangle, \langle f(\cdot), e_2 \rangle, \ldots)$. Then U is an isomorphism and $UNU^{-1} = N_\mu^{(n)}$. The details are left to the reader. ∎

Combining the preceding proposition with Proposition 6.1(b), we can find $\{N\}'$; namely, $\{N_\mu^{(n)}\}' =$ all matrices (T_{ij}) on $\mathcal{B}(L^2(\mu)^{(n)})$ such that $T_{ij} \in \{N_\mu\}'$ for all i, j. By Corollary 6.9, $\{N_\mu^{(n)}\}' =$ matrices $(M_{\phi_{ij}})$ that belong to $\mathcal{B}(L^2(\mu)^{(n)})$, such that $\phi_{ij} \in L^\infty(\mu)$. Now the idea is to use Proposition 10.17 to bring this back to $\mathcal{B}(L^2(\mu; \mathcal{H}_n))$ and describe $\{N\}'$.

A function $\phi: \mathbb{C} \to \mathcal{B}(\mathcal{H}_n)$ is defined to be a Borel function if for each f and g in $\mathcal{H}_n$, $z \mapsto \langle \phi(z)f, g \rangle$ is a Borel function. If $\{f_j\}$ is a countable dense subset of the unit ball of $\mathcal{H}_n$, $\| \phi(z) \| = \sup\{|\langle \phi(z)f_i, f_j \rangle|: 1 \leqslant i, j < \infty\}$, so $z \mapsto \| \phi(z) \|$ is a Borel function. Let $L^\infty(\mu; \mathcal{B}(\mathcal{H}_n))$ be the equivalence classes of bounded Borel functions from $\mathbb{C}$ into $\mathcal{B}(\mathcal{H}_n)$ furnished with the μ-essential supremum norm.

If $\phi \in L^\infty(\mu; \mathcal{B}(\mathcal{H}_n))$ and $f \in L^2(\mu; \mathcal{H}_n)$, let $f(z) = \sum f_j(z)e_j$, where $\{e_j\}$ is an orthonormal basis for $\mathcal{H}_n$ and $f_j(z) = \langle f(z), e_j \rangle$, so $\sum |f_j(z)|^2 = \| f(z) \|^2$. Thus $\phi(z)f(z) = \sum_j f_j(z)\phi(z)e_j$. So for any e in $\mathcal{H}_n$, $\langle \phi(z)f(z), e \rangle = \sum f_j(z)\langle \phi(z)e_j, e \rangle$ is a Borel function. It is easy to check that $\phi f \in L^2(\mu; \mathcal{H}_n)$ and

$\|\phi f\| \leqslant \|\phi\|_\infty \|f\|$. Let $M_\phi\colon L^2(\mu; \mathcal{H}_n) \to L^2(\mu; \mathcal{H}_n)$ be defined by $M_\phi f = \phi f$. Combined with the preceding remarks, the following result can be shown to hold. (The proof is left to the reader.)

10.18. Proposition. *If N is multiplication by z on $L^2(\mu; \mathcal{H}_n)$, then*

$$\{N\}' = \{M_\phi\colon \phi \in L^\infty(\mu; \mathcal{B}(\mathcal{H}_n))\}.$$

Also, $\|M_\phi\| = \|\phi\|_\infty$ for every ϕ in $L^\infty(\mu; \mathcal{B}(\mathcal{H}_n))$.

The next lemma is a consequence of Proposition 6.10 and the fact that unitarily equivalent normal operators have mutually absolutely continuous scalar-valued spectral measures.

10.19. Lemma. *If N_1 and N_2 are normal operators with mutually singular scalar spectral measures and $XN_1 = N_2X$, then $X = 0$.*

Using the observation made prior to Corollary 6.8, the preceding lemma implies that $\{N_1 \oplus N_2\}' = \{N_1\}' \oplus \{N_2\}'$ whenever N_1 and N_2 are as in the lemma.

The next theorem of this section can be proved by piecing together Theorem 10.16 and the remaining results of this section. The details are left to the reader.

10.20. Theorem. *If N is a normal operator on $\mathcal{H}$, there are mutually singular measures $\mu_\infty, \mu_1, \mu_2, \ldots$ and an isomorphism*

$$U\colon \mathcal{H} \to L^2(\mu_\infty; \mathcal{H}_\infty) \oplus L^2(\mu_1) \oplus L^2(\mu_2; \mathcal{H}_2) \oplus \cdots$$

such that

$$UNU^{-1} = N_\infty \oplus N_1 \oplus N_2 \oplus \cdots$$

where $N_n = $ multiplication by z on $L^2(\mu_n; \mathcal{H}_n)$. Also,

$$\{N_\infty \oplus N_1 \oplus N_2 \oplus \cdots\}' = L^\infty(\mu_\infty; \mathcal{B}(\mathcal{H}_\infty)) \oplus L^\infty(\mu_1)$$
$$\oplus L^\infty(\mu_2; \mathcal{B}(\mathcal{H}_2)) \oplus \cdots.$$

Using the notation of the preceding theorem, if μ is a scalar-valued spectral measure for N, then there are pairwise disjoint Borel sets $\Delta_\infty, \Delta_1, \ldots$ such that $[\mu_n] = [\mu|\Delta_n]$. Define a function $m_N\colon \mathbb{C} \to \{0, 1, \ldots \infty\}$ by letting $m_N = \infty\chi_{\Delta_x} + \chi_{\Delta_1} + 2\chi_{\Delta_2} + \cdots$. As it stands the definition of m_N depends on the choice of the sets $\{\Delta_n\}$ as well as N. However, any two choices of the sets $\{\Delta_n\}$ differ from one another by sets of μ-measure zero. The function m_N is called the *multiplicity function* for N. Note that m_N is a Borel function.

If $m\colon \mathbb{C} \to \{\infty, 0, 1, 2, \ldots\}$ is a Borel function and μ is a compactly supported measure such that $\mu(\{z\colon m(z) = 0\}) = 0$, let $\Delta_n = \{z\colon m(z) = n\}$, $n = \infty, 1, 2, \ldots$. If $N_n = N_{\mu|\Delta_n}$, then $N = N_\infty^{(\infty)} \oplus N_1 \oplus N_2^{(2)} \oplus \cdots$ is a normal operator whose spectral measure is μ and whose multiplicity function agrees with m a.e. $[\mu]$.

10.21. Theorem. *Two normal operators are unitarily equivalent if and only if they have the same scalar-valued spectral measure μ and their multiplicity functions are equal* a.e. $[\mu]$.

There is some notation that is used by many and we should mention its connection with what we have just finished. Suppose $m: \mathbb{C} \to \{\infty, 1, 2, \ldots\}$ is a Borel function and μ is a compactly supported measure on $\mathbb{C}$ such that $\mu(\{z: m(z) = 0\}) = 0$. If $z \in \mathbb{C}$ let $\mathscr{H}(z)$ be a Hilbert space of dimension $m(z)$. The *direct integral* of the spaces $\mathscr{H}(z)$, denoted by $\int \mathscr{H}(z) d\mu(z)$, is precisely the space

$$L^2(\mu \mid \Delta_\infty; \mathscr{H}_\infty) \oplus L^2(\mu \mid \Delta_1) \oplus L^2(\mu \mid \Delta_2; \mathscr{H}_2) \oplus \cdots,$$

where $\Delta_n = \{z: m(z) = n\}$ and $\dim \mathscr{H}_n = n$. If $\phi: \mathbb{C} \to \mathscr{B}(\mathscr{H}_\infty) \cup \mathscr{B}(\mathbb{C}) \cup \mathscr{B}(\mathscr{H}_2) \cup \cdots$ such that $\phi(z) \in \mathscr{B}(\mathscr{H}_n)$ when $z \in \Delta_n$, $\phi: \Delta_n \to \mathscr{B}(\mathscr{H}_n)$ is a Borel function, and there is a constant M such that $\| \phi(z) \| \leqslant M$ a.e. $[\mu]$, then $\int \phi(z) d\mu(z)$ denotes the operator $M_{\phi \mid \Delta_x} \oplus \cdots$ as in (10.20). Although the direct integral notation is quite suggestive, one must revert to the notation of (10.20) to produce proofs.

Remarks. There are several sources for multiplicity theory. Most begin by proving Theorem 10.16. This is done for nonseparable spaces in Halmos [1951] and Brown [1974]. Another source is Arveson [1976], where the theory is set in the context of C^*-algebras which is its proper milieu. Also, Arveson shows that the theory can be applied to some non-normal operators. The details of this more general multiplicity theory are carried out in Ernest [1976] as part of a more general classification scheme. Another source for multiplicity theory is Dunford and Schwartz [1963].

By Theorem 4.6, every normal operator is unitarily equivalent to a multiplication operator M_ϕ on $L^2(X, \Omega, \mu)$ for some measure space (X, Ω, μ). The scalar-valued spectral measure for M_ϕ is $\mu \circ \phi^{-1}$. What is the multiplicity function for M_ϕ? One is tempted to say that $m_{M_\phi}(z) =$ the number of points in $\phi^{-1}(z)$. This is not quite correct. The answer can be found in Abrahamse and Kriete [1973]. Also, Abrahamse [1978] contains a survey of spectral multiplicity for normal operators treated from this point of view. An especially accessible and readable account of this can be found in Kriete [1986].

EXERCISES

1. Let A and B be operators on $\mathscr{H}$ and $\mathscr{K}$, respectively. Let $\mathscr{H}_0$ and $\mathscr{K}_0$ be reducing subspaces for A and B and suppose that $A \cong B \mid \mathscr{K}_0$ and $B \cong A \mid \mathscr{H}_0$. Show that $A \cong B$.

2. Let $\mu_1, \mu_2, \ldots$ be compactly supported measures on $\mathbb{C}$ such that $\mu_{n+1} \ll \mu_n$ for all n. Show that if M is any normal operator whose spectral measure is absolutely continuous with respect to each μ_n, then $N_{\mu_1} \oplus N_{\mu_2} \oplus \cdots \cong (N_{\mu_1} \oplus N_{\mu_2} \oplus \cdots) \oplus M$.

3. If μ = Lebesgue measure on $[0,1]$, show that $N_\mu \cong N_\mu^p$ for $0 < p < \infty$.

4. Let μ = Lebesgue measure on $[0,1]$ and characterize the functions ϕ in $L^\infty(\mu)$ such that $N_\mu \cong \phi(N_\mu)$.

5. Let μ = area measure on $\mathbb{D}$ and show that N_μ and N_μ^2 are not unitarily equivalent.

6. Let μ = Lebesgue measure on $[0,1]$ and let ν = Lebesgue measure on $[-1,1]$. Show that $N_\nu^2 \cong N_\mu \oplus N_\mu$. How about N_ν^3?

7. Let μ be Lebesgue measure on $\mathbb{R}$ and N = multiplication by $\sin x$ on $L^2(\mu)$. Find the decompositions of N obtained in Theorem 10.1 and 10.16.

8. If μ is Lebesgue measure on $\mathbb{R}$ and N = multiplication by e^{ix} on $L^2(\mu)$, show that $N \cong N_m^{(\infty)}$ where m = arc length measure on $\partial \mathbb{D}$.

9. Define $U: L^2(\mathbb{R}) \to L^2(\mathbb{R})$ by $(Uf)(t) = f(t-1)$. Show that U is unitary and find its scalar-valued spectral measure and multiplicity function.

10. Represent N as in Theorem 10.1 and find the corresponding representation for $N \oplus N = N^{(2)}$; for $N^{(3)}$, for $N^{(\infty)}$. (Are you surprised by the result for $N^{(\infty)}$?)

11. Prove the results and solve the exercises from §II.8.

12. Let N be a normal operator and show that $N \cong N^{(2)}$ if and only if there is a $*$-cyclic normal operator M such that $N \cong M^{(\infty)}$. What does this say about the multiplicity function for N?

13. Let (X, Ω, μ) be a measure space such that $L^2(\mu)$ is separable, let $\phi \in L^\infty(\mu)$, and let $N = M_\phi$ on $L^2(\mu)$. Find the decompositions of N obtained in Theorems 10.1 and 10.16.

14. Let μ be a compactly supported measure on $\mathbb{C}$, ϕ a bounded Borel function on $\mathbb{C}$, and suppose $\{\Delta_n\}$ are pairwise disjoint Borel sets such that ϕ is one-to-one on each Δ_n and $\mu(\mathbb{C} \setminus \bigcup_{n=1}^\infty \Delta_n) = 0$. Let $\phi_n = \phi \chi_{\Delta_n}$ and $\mu_n = \mu \circ \phi_n^{-1}$ for $n \geqslant 1$. Prove that M_ϕ on $L^2(\mu)$ is unitarily equivalent to $\bigoplus_{n=1}^\infty N_{\mu_n}$.

CHAPTER X
Unbounded Operators

It is unfortunate for the world we live in that all of the operators that arise naturally are not bounded. But that is indeed the case. Thus it is important to study such operators.

The idea here is not to study an arbitrary linear transformation on a Hilbert space. In fact, such a study is the province of linear algebra rather than analysis. The operators that are to be studied do possess certain properties that connect them to the underlying Hilbert space. The properties that will be isolated are inspired by natural examples.

All Hilbert spaces in this chapter are assumed separable.

§1. Basic Properties and Examples

The first relaxation in the concept of operator is not to assume that the operators are defined everywhere on the Hilbert space.

1.1. Definition. If $\mathcal{H}, \mathcal{K}$ are Hilbert spaces, a *linear operator* $A: \mathcal{H} \to \mathcal{K}$ is a function whose domain of definition is a linear manifold, dom A, in $\mathcal{H}$ and such that $A(\alpha f + \beta g) = \alpha Af + \beta Ag$ for f, g in dom A and α, β in $\mathbb{C}$. A is *bounded* if there is a constant $c > 0$ such that $\|Af\| \leqslant c\|f\|$ for all f in dom A.

Note that if A is bounded, then A can be extended to a bounded linear operator on cl[dom A] and then extended to $\mathcal{H}$ by letting A be 0 on (dom $A)^{\perp}$. So unless it is specified to the contrary, a bounded operator will always be assumed to be defined on all of $\mathcal{H}$.

If A is a linear operator from $\mathcal{H}$ into $\mathcal{K}$, then A is also a linear operator from cl[dom A] into $\mathscr{K}$. So we will often only consider those A such that dom A is dense in $\mathscr{H}$; such an operator A is said to be *densely defined*. $\mathscr{B}(\mathscr{H})$ still denotes the bounded operators defined on $\mathscr{H}$.

If A, B are linear operators from $\mathscr{H}$ into $\mathscr{K}$, then $A + B$ is defined with dom$(A + B) = $ dom $A \cap$ dom B. If $B: \mathscr{H} \to \mathscr{K}$ and $A: \mathscr{K} \to \mathscr{L}$, then AB is a linear operator from $\mathscr{H}$ into $\mathscr{L}$ with dom$(AB) = B^{-1}($dom $A)$.

1.2. Definition. If A, B are operators from $\mathscr{H}$ into $\mathscr{K}$, then A is an *extension* of B if dom $B \subseteq$ dom A and $Ah = Bh$ whenever $h \in$ dom B. In symbols this is denoted by $B \subseteq A$.

Note that if $A \in \mathscr{B}(\mathscr{H})$, then the only extension of A is itself. So this concept is only of value for unbounded operators.

If $A: \mathscr{H} \to \mathscr{K}$, the *graph* of A is the set

$$\text{gra } A \equiv \{h \oplus Ah \in \mathscr{H} \oplus \mathscr{K}: h \in \text{dom } A\}.$$

It is easy to see that $B \subseteq A$ if and only if gra $B \subseteq$ gra A.

1.3. Definition. An operator $A: \mathscr{H} \to \mathscr{K}$ is *closed* if its graph is closed in $\mathscr{H} \oplus \mathscr{K}$. An operator is *closable* if it has a closed extension. Let $\mathscr{C}(\mathscr{H}, \mathscr{K}) = $ the collection of all closed densely defined operators from $\mathscr{H}$ into $\mathscr{K}$. Let $\mathscr{C}(\mathscr{H}) = \mathscr{C}(\mathscr{H}, \mathscr{H})$. (It should be emphasized that the operators in $\mathscr{C}(\mathscr{H}, \mathscr{K})$ are densely defined.)

When is a subset of $\mathscr{H} \oplus \mathscr{K}$ a graph of an operator from $\mathscr{H}$ into $\mathscr{K}$? If $\mathscr{G} = $ gra A for some $A: \mathscr{H} \to \mathscr{K}$, then $\mathscr{G}$ is a submanifold of $\mathscr{H} \oplus \mathscr{K}$ such that if $k \in \mathscr{K}$ and $0 \oplus k \in \mathscr{G}$, then $k = 0$. The converse is also true. That is, suppose that $\mathscr{G}$ is a submanifold of $\mathscr{H} \oplus \mathscr{K}$ such that if $k \in \mathscr{K}$ and $0 \oplus k \in \mathscr{G}$, then $k = 0$. Let $\mathscr{D} = \{h \in \mathscr{H}:$ there exists a k in $\mathscr{K}$ with $h \oplus k$ in $\mathscr{G}\}$. If $h \in \mathscr{D}$ and $k_1, k_2 \in \mathscr{K}$ such that $h \oplus k_1, h \oplus k_2 \in \mathscr{G}$, then $0 \oplus (k_1 - k_2) = h \oplus k_1 - h \oplus k_2 \in \mathscr{G}$. Hence $k_1 = k_2$. That is, for every h in $\mathscr{D}$ there is a unique k in $\mathscr{K}$ such that $h \oplus k \in \mathscr{G}$; denote k by $k = Ah$. It is easy to check that A is a linear map and $\mathscr{G} = $ gra A. This gives an internal characterization of graphs that will be useful in the next proposition.

1.4. Proposition. *An operator* $A: \mathscr{H} \to \mathscr{K}$ *is closable if and only if* cl[gra A] *is a graph.*

PROOF. Let cl[gra A] be a graph. That is, there is an operator $B: \mathscr{H} \to \mathscr{K}$ such that gra $B = $ cl[gra A]. Clearly gra $A \subseteq$ gra B, so A is closable.

Now assume that A is closable; that is, there is a closed operator $B: \mathscr{H} \to \mathscr{K}$ with $A \subseteq B$. If $0 \oplus k \in$ cl[gra A], $0 \oplus k \in$ gra B and hence $k = 0$. By the remarks preceding this proposition, cl[gra A] is a graph. ∎

If A is closable, call the operator whose graph is cl[gra A] the *closure* of A.

1.5. Definition. If $A: \mathscr{H} \to \mathscr{K}$ is densely defined, let dom $A^* = \{k \in \mathcal{K}: h \mapsto \langle Ah, k \rangle$ is a bounded linear functional on dom $A\}$.

Because dom A is dense in $\mathcal{H}$, if $k \in$ dom A^*, then there is a unique vector f in $\mathcal{H}$ such that $\langle Ah, k \rangle = \langle h, f \rangle$ for all h in dom A. Denote this unique vector f by $f = A^*k$. Thus

$$\langle Ah, k \rangle = \langle h, A^*k \rangle$$

for h in dom A and k in dom A^*.

1.6. Proposition. *If $A: \mathcal{H} \to \mathcal{K}$ is a densely defined operator, then:*

(a) A^* *is a closed operator;*
(b) A^* *is densely defined if and only if A is closable;*
(c) *if A is closable, then its closure is $(A^*)^* \equiv A^{**}$.*

Before proving this, a lemma is needed which will also be useful later.

1.7. Lemma. *If $A: \mathcal{H} \to \mathcal{K}$ is densely defined and $J: \mathcal{H} \oplus \mathcal{K} \to \mathcal{K} \oplus \mathcal{H}$ is defined by $J(h \oplus k) = (-k) \oplus h$, then J is an isomorphism and*

$$\text{gra } A^* = [J \text{ gra } A]^{\perp}.$$

PROOF. It is clear that J is an isomorphism. To prove the formula for gra A^*, note that gra $A^* = \{k \oplus A^*k \in \mathcal{K} \oplus \mathcal{H}: k \in$ dom $A^*\}$. So if $k \in$ dom A^* and $h \in$ dom A,

$$\langle k \oplus A^*k, J(h \oplus Ah) \rangle = \langle k \oplus A^*k, -Ah \oplus h \rangle$$
$$= -\langle k, Ah \rangle + \langle A^*k, h \rangle = 0.$$

Thus gra $A^* \subseteq [J \text{ gra } A]^{\perp}$. Conversely, if $k \oplus f \in [J \text{ gra } A]^{\perp}$, then for every h is dom A, $0 = \langle k \oplus f, -Ah \oplus h \rangle = -\langle k, Ah \rangle + \langle f, h \rangle$, so $\langle Ah, k \rangle = \langle h, f \rangle$. By definition $k \in$ dom A^* and $A^*k = f$. ∎

PROOF OF PROPOSITION 1.6. The proof of (a) is clear from Lemma 1.7. For the remainder of the proof notice that because the map J in (1.7) is an isomorphism, $J^* = J^{-1}$ and so $J^*(k \oplus h) = h \oplus (-k)$.

(b) Assume A is closable and let $k_0 \in (\text{dom } A^*)^{\perp}$. We want to show that $k_0 = 0$. Thus $k_0 \oplus 0 \in [\text{gra } A^*]^{\perp} = [J \text{ gra } A]^{\perp\perp} = \text{cl}[J \text{ gra } A] = J[\text{cl}(\text{gra } A)]$. So $0 \oplus -k_0 = J^*(k_0 \oplus 0) \in J^*J[(\text{gra } A)] = \text{cl}(\text{gra } A)$. But because A is closable, cl(gra A) is a graph; hence $k_0 = 0$. For the converse, assume dom A^* is dense in $\mathcal{K}$. Thus $A^{**} \equiv (A^*)^*$ is defined. By (a), A^{**} is a closed operator. It is easy to see that $A \subseteq A^{**}$, so A has a closed extension.

(c) Note that by Lemma 1.7 gra $A^{**} = [J^* \text{ gra } A^*]^{\perp} = [J^*[J \text{ gra } A]^{\perp}]^{\perp}$. But for any linear manifold $\mathcal{M}$ and any isomorphism $J, (J\mathcal{M})^{\perp} = J(\mathcal{M}^{\perp})$. Hence $J^*[(J\mathcal{M})^{\perp}] = \mathcal{M}^{\perp}$ and, thus, $[J^*[J\mathcal{M}]^{\perp}]^{\perp} = \mathcal{M}^{\perp\perp} = \text{cl}\,\mathcal{M}$. Putting $\mathcal{M} = \text{gra } A$ gives that gra $A^{**} = \text{cl gra } A$. ∎

1.8. Corollary. *If $A \in \mathcal{C}(\mathcal{H}, \mathcal{K})$, then $A^* \in \mathcal{C}(\mathcal{K}, \mathcal{H})$ and $A^{**} = A$.*

1.9. Example. Let $e_0, e_1, \ldots$ be an orthonormal basis for $\mathcal{H}$ and let $\alpha_0, \alpha_1, \ldots$ be complex numbers. Define $\mathcal{D} = \{h \in \mathcal{H}: \sum_0^\infty |\alpha_n \langle h, e_n \rangle|^2 < \infty\}$ and let $Ah = \sum_0^\infty \alpha_n \langle h, e_n \rangle e_n$ for h in $\mathcal{D}$. Then $A \in \mathscr{C}(\mathcal{H})$ with dom $A = \mathcal{D}$. Also, dom $A^* = \mathcal{D}$ and $A^*h = \sum_0^\infty \bar{\alpha}_n \langle h, e_n \rangle e_n$ for all h in $\mathcal{D}$.

1.10. Example. Let (X, Ω, μ) be a σ-finite measure space and let $\phi: X \to \mathbb{C}$ be an Ω-measurable function. Let $\mathcal{D} = \{f \in L^2(\mu): \phi f \in L^2(\mu)\}$ and define $Af = \phi f$ for all f in $\mathcal{D}$. Then $A \in \mathscr{C}(L^2(\mu))$, dom $A^* = \mathcal{D}$, and $A^*f = \bar{\phi}f$ for f in $\mathcal{D}$.

1.11. Example. Let $\mathcal{D} = $ all functions $f: [0,1] \to \mathbb{C}$ that are absolutely continuous with $f' \in L^2(0,1)$ and such that $f(0) = f(1) = 0$. $\mathcal{D}$ includes all polynomials p with $p(0) = p(1) = 0$. So the uniform closure of $\mathcal{D}$ is $\{f \in C[0,1]: f(0) = f(1) = 0\}$. Thus $\mathcal{D}$ is dense in $L^2(0,1)$. Define $A: L^2(0,1) \to L^2(0,1)$ by $Af = if'$ for f in $\mathcal{D}$. To see that A is closed, suppose $\{f_n\} \subseteq \mathcal{D}$ and $f_n \oplus if'_n \to f \oplus g$ in $L^2 \oplus L^2$. Let $h(x) = -i\int_0^x g(t)dt$; so h is absolutely continuous. Now using the Cauchy-Schwarz inequality we get that $|f_n(x) - h(x)| = |\int_0^x [f'_n(t) + ig(t)]dt| \leq \|f'_n + ig\|_2 = \|if'_n - g\|_2$. Thus $f_n(x) \to h(x)$ uniformly on $[0,1]$. Since $f_n \to f$ in $L^2(0,1)$, $f(x) = h(x)$ a.e. So we may assume that $f(x) = -i\int_0^x g(t)dt$ for all x. Therefore f is absolutely continuous and $f_n(x) \to f(x)$ uniformly on $[0,1]$; thus $f(0) = f(1) = 0$ and $f' = -ig \in L^2(0,1)$. So $f \in \mathcal{D}$ and $f \oplus g = f \oplus g = f \oplus if' \in $ gra A; that is, $A \in \mathscr{C}(L^2(0,1))$. Note that $\{f': f \in \mathcal{D}\} = \{h \in L^2(0,1): \int_0^1 h(x)dx = 0\} = [1]^\perp$.

Claim. dom $A^* = \{g: g \text{ is absolutely continuous on } [0,1], g' \in L^2(0,1)\}$ and for g in dom A^*, $A^*g = ig'$.

In fact, suppose $g \in $ dom A^* and let $h = A^*g$. Put $H(x) = \int_0^x h(t)\,dt$. Using integration by parts, for every f in $\mathcal{D}$, $i\int_0^1 f'\bar{g} = \langle Af, g \rangle = \langle f, h \rangle = \int_0^1 f\bar{h} = \int_0^1 f(x)d\bar{H}(x) = -\int_0^1 f'(x)\overline{H(x)}dx$; that is, $\langle f', -ig \rangle = \langle f', -H \rangle$ for all f in $\mathcal{D}$. Thus $H - ig \in \{f': f \in \mathcal{D}\}^\perp = [1]^{\perp\perp}$; hence $H - ig = c$, a constant function. Thus $g = ic - iH$ so that g is absolutely continuous and $g' = -ih \in L^2$. Also note that $A^*g = h = ig'$. The proof of the other inclusion is left to the reader.

1.12. Example. Let $\mathscr{E} = \{f \in L^2(0,1): f \text{ is absolutely continuous}, f' \in L^2, \text{ and } f(0) = f(1)\}$. Define $Bf = if'$ for f in $\mathscr{E}$. As in (1.11), $B \in \mathscr{C}(L^2(0,1))$ and ran $B = [1]^\perp$.

Claim. dom $B^* = \mathscr{E}$ and $B^*g = ig'$ for g in $\mathscr{E}$.

Let $g \in $ dom B^*. Put $h = B^*g$ and $H(x) = \int_0^x h(t)dt$. As in (1.11), $H(0) = H(1) = 0$ and for every f in $\mathscr{E}$, $i\int_0^1 f'\bar{g} = -\int_0^1 f'\bar{H}$. Hence $0 = \int_0^1 (if'\bar{g} + f'\bar{H}) = \int_0^1 if'\overline{(g + iH)}$. Thus $g + iH \perp $ ran B and so $g + iH = c$, a constant function. Thus $g = c - iH$ is absolutely continuous, $g' = -ih \in L^2$, and $g(0) = g(1) = c$. Thus $g \in \mathscr{E}$ and $B^*g = h = ig'$. The proof of the other inclusion is left to the reader.

The preceding two examples illustrate the fact that the calculation of the adjoint depends on the domain of the operator, not just the formal definition of the operator. Note the fact that the next result generalizes (II.2.19).

1.13. Proposition. *If* $A: \mathcal{H} \to \mathcal{K}$ *is densely defined, then*

$$(\operatorname{ran} A)^{\perp} = \ker A^*.$$

If A *is also closed, then*

$$(\operatorname{ran} A^*)^{\perp} = \ker A.$$

PROOF. If $h \perp \operatorname{ran} A$, then for every f in dom $A, 0 = \langle Af, h \rangle$. Hence $h \in$ dom A^* and $A^* h = 0$. The other inclusion is clear. By Corollary 1.8, if $A \in \mathscr{C}(\mathcal{H}, \mathcal{K})$, $A^{**} = A$. So the second equality follows from the first. ∎

1.14. Definition. If $A: \mathcal{H} \to \mathcal{K}$ is a linear operator, A is *boundedly invertible* if there is a bounded linear operator $B: \mathcal{K} \to \mathcal{H}$ such that $AB = 1$ and $BA \subseteq 1$.

Note that if $BA \subseteq 1$, then BA is bounded on its domain. Call B a *(bounded) inverse* of A.

1.15. Proposition. *Let* $A: \mathcal{H} \to \mathcal{K}$ *be a linear operator.*

(a) A *is boundedly invertible if and only if* $\ker A = (0)$, $\operatorname{ran} A = \mathcal{K}$, *and the graph of* A *is closed.*

(b) *If* A *is boundedly invertible, its inverse is unique and denoted by* A^{-1}.

PROOF. (a) Let B be a bounded inverse of A. So dom $B = \mathcal{K}$. Since $BA \subseteq 1$, $\ker A = (0)$; since $AB = 1$, $\operatorname{ran} A = \mathcal{K}$. Also, gra $A = \{ h \oplus Ah : h \in$ dom $A \} = \{ Bk \oplus k : k \in \mathcal{K} \}$. Since B is bounded, gra A is closed. Conversely, if A has the stated properties, $Bk = A^{-1}k$ for k in $\mathcal{K}$ is a well-defined operator on $\mathcal{K}$. Because gra A is closed, gra B is closed. By the Closed Graph Theorem, $B \in \mathscr{B}(\mathcal{K}, \mathcal{H})$.

(b) This is an exercise. ∎

1.16. Definition. If $A: \mathcal{H} \to \mathcal{H}$ is a linear operator, $\rho(A)$, the *resolvent set* for A, is defined by $\rho(A) = \{ \lambda \in \mathbb{C} : \lambda - A$ is boundedly invertible$\}$. The *spectrum* of A is the set $\sigma(A) = \mathbb{C} \setminus \rho(A)$:

It is easy to see that if $A: \mathcal{H} \to \mathcal{K}$ is a linear operator and $\lambda \in \mathbb{C}$, gra A is closed if and only if gra$(A - \lambda)$ is closed. So if A does not have closed graph, $\sigma(A) = \mathbb{C}$. Even if A has closed graph, it is possible that $\sigma(A)$ is empty (see Exercise 10). The spectrum of an unbounded operator, however, does enjoy some of the properties possessed by the spectrum of an element of a Banach algebra. The proof of the next result is left to the reader.

1.17. Proposition. *If* $A: \mathcal{H} \to \mathcal{H}$ *is a linear operator, then* $\sigma(A)$ *is closed and* $z \mapsto (z - A)^{-1}$ *is an analytic function on* $\rho(A)$.

Note that if A is defined as in Example 1.9, then $\sigma(A) = \mathrm{cl}\{\alpha_n\}$. Hence it is possible for $\sigma(A)$ to equal any closed subset of $\mathbb{C}$.

1.18. Proposition. *Let $A \in \mathscr{C}(\mathscr{H})$.*

(a) $\lambda \in \rho(A)$ *if and only if* $\ker(A - \lambda) = (0)$ *and* $\mathrm{ran}(A - \lambda) = \mathscr{H}$.

(b) $\sigma(A^*) = \{\bar{\lambda}: \lambda \in \sigma(A)\}$ *and for λ in* $\rho(A)$, $(A - \lambda)^{*-1} = [(A - \lambda)^{-1}]^*$.

PROOF. Exercise.

EXERCISES

1. If A, B, and AB are densely defined linear operators, show that $(AB)^* \supseteq B^*A^*$.

2. Verify the statements in Example 1.9.

3. Verify the statements in Example 1.10.

4. Define an unbounded weighted shift and determine its adjoint.

5. Verify the statements in Example 1.11.

6. If $\mathscr{H}$ is infinite dimensional, show that there is a linear operator $A: \mathscr{H} \to \mathscr{H}$ such that gra A is dense in $\mathscr{H} \oplus \mathscr{H}$. What does this say about dom A^*? (See Lindsay [1984].)

7. Let $\mathscr{D}$ be the set of absolutely continuous functions f such that $f' \in L^2(0, 1)$. Let $Df = f'$ for f in $\mathscr{D}$ and let $(Af)(x) = xf(x)$ for f in $L^2(0, 1)$. Show that $DA - AD \subseteq 1$.

8. If $\mathscr{A}$ is a Banach algebra with identity, show that there are no elements a, b in $\mathscr{A}$ such that $ab - ba = 1$. (Hint: compute $a^n b - ba^n$.)

9. Prove Proposition 1.18.

10. Define $A: L^2(\mathbb{R}) \to L^2(\mathbb{R})$ by $(Af)(x) = \exp(-x^2)f(x-1)$ for all f in $L^2(\mathbb{R})$. (a) Show that $A \in \mathscr{B}(L^2(\mathbb{R}))$. (b) Find $\|A^n\|$ and show that $r(A) = 0$ so that $\sigma(A) = \{0\}$. (c) Show that A is injective. (d) Find A^* and show that ran A is dense. (e) Define $B = A^{-1}$ with dom $B = $ ran A and show that $B \in \mathscr{C}(L^2(\mathbb{R}))$ with $\sigma(B) = \square$.

11. If $A \in \mathscr{C}(\mathscr{H})$, show that $A^*A \in \mathscr{C}(\mathscr{H})$. Show that $-1 \notin \sigma(A^*A)$ and that if $B = (1 + A^*A)^{-1}$, $\|B\| \leq 1$.

12. If B is the bounded operator obtained in Exercise 11, show that $C = AB$ is also bounded and $\|C\| \leq 1$.

13. If A is a self-adjoint operator, then $\lambda \in \rho(A)$ if and only if $A - \lambda$ is surjective.

§2. Symmetric and Self-Adjoint Operators

An appropriate introduction to this section consists in a careful examination of Examples 1.11 and 1.12 in the preceding section. In (1.11) we saw that the operator A seemed to be inclined to be self-adjoint, but dom A^* was different from dom A so we could not truly say that $A = A^*$. In (1.12), $B = B^*$ in any sense of the concept of equality. This points out the distinction between symmetric and self-adjoint operators that it is necessary to make in the theory of unbounded operators.

2.1. Definition. An operator $A: \mathcal{H} \to \mathcal{H}$ is *symmetric* if A is densely defined and $\langle Af, g \rangle = \langle f, Ag \rangle$ for all f, g in dom A.

The proof of the next proposition is left to the reader.

2.2. Proposition. *If A is densely defined, the following statements are equivalent.*

(a) *A is symmetric.*
(b) *$\langle Af, f \rangle \in \mathbb{R}$ for all f in dom A.*
(c) *$A \subseteq A^*$.*

If A is symmetric, then the fact that $A \subseteq A^*$ implies dom A^* is dense. Hence A is closable by Proposition 1.6. Symmetric operators can behave cantankerously. For example, there is an example of a closed symmetric operator T such that dom $(T^2) = (0)$. See Chernoff [1983].

It is easy to check that the operators in Examples 1.11 and 1.12 are symmetric.

2.3. Definition. A densely defined operator $A: \mathcal{H} \to \mathcal{H}$ is *self-adjoint* if $A = A^*$.

Let us emphasize that the condition that $A = A^*$ in the preceding definition carries with it the requirement that dom $A =$ dom A^*. Now clearly every self-adjoint operator is symmetric, but the operator A in Example 1.11 shows that there are symmetric operators that are not self-adjoint. If, however, an operator is bounded, then it is self-adjoint if and only if it is symmetric. The operator B in Example 1.12 is an unbounded self-adjoint operator and Examples 1.9 and 1.10 can be used to furnish additional examples of unbounded self-adjoint operators.

Note that Proposition 1.6 implies that a self-adjoint operator is necessarily closed.

2.4. Proposition. *Suppose A is a symmetric operator on $\mathcal{H}$.*

(a) *If ran A is dense, then A is injective.*
(b) *If $A = A^*$ and A is injective, then ran A is dense and A^{-1} is self-adjoint.*
(c) *If dom $A = \mathcal{H}$, then $A = A^*$ and A is bounded.*
(d) *If ran $A = \mathcal{H}$, then $A = A^*$ and $A^{-1} \in \mathcal{B}(\mathcal{H})$.*

PROOF. The proof of (a) is trivial and (b) is an easy consequence of (1.13) and some manipulation.

(c) We have $A \subseteq A^*$. If dom $A = \mathcal{H}$, then $A = A^*$ and so A is closed. By the Closed Graph Theorem $A \in \mathcal{B}(\mathcal{H})$.

(d) If ran $A = \mathcal{H}$, then A is injective by (a). Let $B = A^{-1}$ with dom $B = $ ran $A = \mathcal{H}$. If $f = Ag$ and $h = Ak$, with g, k in dom A, then $\langle Bf, h \rangle = \langle g, Ak \rangle = \langle Ag, k \rangle = \langle f, k \rangle = \langle f, Bh \rangle$. Hence B is symmetric. By (c), $B = B^* \in \mathcal{B}(\mathcal{H})$. By (b), $A = B^{-1}$ is self-adjoint. ∎

We now will turn our attention to the spectral properties of symmetric and self-adjoint operators. In particular, it will be seen that symmetric operators can have nonreal numbers in their spectra, though the nature of the spectrum can be completely diagnosed (2.8). Self-adjoint operators, however, must have real spectra. The next result begins this spectral discussion.

2.5. Proposition. *Let A be a symmetric operator and let $\lambda = \alpha + i\beta, \alpha$ and β real numbers.*

(a) *For each f in dom A, $\|(A - \lambda)f\|^2 = \|(A - \alpha)f\|^2 + \beta^2\|f\|^2$.*

(b) *If $\beta \neq 0$, $\ker(A - \lambda) = (0)$.*

(c) *If A is closed and $\beta \neq 0$, $\operatorname{ran}(A - \lambda)$ is closed.*

PROOF. Note that

$$\|(A - \lambda)f\|^2 = \|(A - \alpha)f - i\beta f\|^2$$
$$= \|(A - \alpha)f\|^2 + 2\operatorname{Re} i\langle (A - \alpha)f, \beta f\rangle + \beta^2\|f\|^2.$$

But

$$\langle (A - \alpha)f, \beta f \rangle = \beta\langle Af, f\rangle - \alpha\beta\|f\|^2 \in \mathbb{R},$$

so (a) follows. Part (b) is immediate from (a). To prove (c), note that $\|(A - \lambda)f\|^2 \geq \beta^2\|f\|^2$. Let $\{f_n\} \subseteq \operatorname{dom} A$ such that $(A - \lambda)f_n \to g$. The preceding inequality implies that $\{f_n\}$ is a Cauchy sequence in $\mathcal{H}$; let $f = \lim f_n$. But $f_n \oplus (A - \lambda)f_n \in \operatorname{gra}(A - \lambda)$ and $f_n \oplus (A - \lambda)f_n \to f \oplus g$. Hence $f \oplus g \in \operatorname{gra}(A - \lambda)$ and so $g = (A - \lambda)f \in \operatorname{ran}(A - \lambda)$. This proves (c). ∎

2.6. Lemma. *If $\mathcal{M}, \mathcal{N}$ are closed subspaces of $\mathcal{H}$ and $\mathcal{M} \cap \mathcal{N}^\perp = (0)$, then $\dim \mathcal{M} \leq \dim \mathcal{N}$.*

PROOF. Let P be the orthogonal projection of $\mathcal{H}$ onto $\mathcal{N}$ and define $T: \mathcal{M} \to \mathcal{N}$ by $Tf = Pf$ for f in $\mathcal{M}$. Since $\mathcal{M} \cap \mathcal{N}^\perp = (0)$, T is injective. If $\mathcal{L}$ is a finite dimensional subspace of $\mathcal{M}$, $\dim \mathcal{L} = \dim T\mathcal{L} \leq \dim \mathcal{N}$. Since $\mathcal{L}$ was arbitrary, $\dim \mathcal{M} \leq \dim \mathcal{N}$. ∎

2.7. Theorem. *If A is a closed symmetric operator, then $\dim \ker(A^* - \lambda)$ is constant for $\operatorname{Im} \lambda > 0$ and constant for $\operatorname{Im} \lambda < 0$.*

PROOF. Let $\lambda = \alpha + i\beta$, α and β real numbers and $\beta \neq 0$.

Claim. If $|\lambda - \mu| < |\beta|$, $\ker(A^* - \mu) \cap [\ker(A^* - \lambda)]^\perp = (0)$.

Suppose this is not so. Then there is an f in $\ker(A^* - \mu) \cap [\ker(A^* - \lambda)]^\perp$ with $\|f\| = 1$. By (2.5c), $\operatorname{ran}(A - \bar{\lambda})$ is closed. Hence $f \in [\ker(A^* - \lambda)]^\perp = \operatorname{ran}(A - \bar{\lambda})$. Let $g \in \operatorname{dom} A$ such that $f = (A - \bar{\lambda})g$. Since $f \in \ker(A^* - \mu)$,

$$
\begin{aligned}
0 = \langle (A^* - \mu)f, g \rangle &= \langle f, (A - \bar{\mu})g \rangle \\
&= \langle f, (A - \bar{\lambda} + \bar{\lambda} - \bar{\mu})g \rangle \\
&= \|f\|^2 + (\lambda - \mu)\langle f, g \rangle.
\end{aligned}
$$

Hence $1 = \|f\|^2 = |\lambda - \mu| |\langle f, g \rangle| \leqslant |\lambda - \mu| \|g\|$. But (2.5a) implies that $1 = \|f\| = \|(A - \bar{\lambda})g\| \geqslant |\beta| \|g\|$; so $\|g\| \leqslant |\beta|^{-1}$. Hence $1 \leqslant |\lambda - \mu| \|g\| \leqslant |\lambda - \mu| |\beta|^{-1} < 1$ if $|\lambda - \mu| < |\beta|$. This contradiction establishes the claim.

Combining the claim with Lemma 2.6 gives that $\dim \ker(A^* - \mu) \leqslant \dim \ker(A^* - \lambda)$ if $|\lambda - \mu| < |\beta| = |\operatorname{Im} \lambda|$. Note that if $|\lambda - \mu| < \frac{1}{2}|\beta|$, then $|\lambda - \mu| < |\operatorname{Im} \mu|$, so that the other inequality also holds. This shows that the function $\lambda \mapsto \dim \ker(A^* - \lambda)$ is locally constant on $\mathbb{C} \backslash \mathbb{R}$. A simple topological argument demonstrates the theorem. ∎

2.8. Theorem. *If A is a closed symmetric operator, then one and only one of the following possibilities occurs*:

(a) $\sigma(A) = \mathbb{C}$;
(b) $\sigma(A) = \{\lambda \in \mathbb{C}: \operatorname{Im} \lambda \geqslant 0\}$;
(c) $\sigma(A) = \{\lambda \in \mathbb{C}: \operatorname{Im} \lambda \leqslant 0\}$;
(d) $\sigma(A) \subseteq \mathbb{R}$.

PROOF. Let $H_\pm = \{\lambda \in \mathbb{C}: \pm \operatorname{Im} \lambda > 0\}$. By (2.5) for λ in $H_\pm$, $A - \lambda$ is injective and has closed range. So if $A - \lambda$ is surjective, $\lambda \in \rho(A)$. But $[\operatorname{ran}(A - \lambda)]^\perp = \ker(A^* - \bar{\lambda})$. So the preceding theorem implies that either $H_\pm \subset \sigma(A)$ or $H_\pm \cap \sigma(A) = \square$. Since $\sigma(A)$ is closed, if $H_\pm \subseteq \sigma(A)$, then either $\sigma(A) = \mathbb{C}$ or $\sigma(A) = \operatorname{cl} H_\pm$. If $H_\pm \cap \sigma(A) = \square$, $\sigma(A) \subseteq \mathbb{R}$. ∎

2.9. Corollary. *If A is a closed symmetric operator, the following statements are equivalent.*

(a) *A is self-adjoint.*
(b) $\sigma(A) \subseteq \mathbb{R}$.
(c) $\ker(A^* - i) = \ker(A^* + i) = (0)$.

PROOF. If A is symmetric, every eigenvalue of A is real (Exercise 1). So if $A = A^*$ and $\operatorname{Im} \lambda \neq 0$, $\ker(A^* - \lambda) = \ker(A - \lambda) = (0)$. Thus $A - \lambda$ is injective and has dense range. By (2.5), $A - \lambda$ has closed range and so $A - \lambda$ has a bounded inverse (1.15) whenever $\operatorname{Im} \lambda \neq 0$. That is, $\sigma(A) \subseteq \mathbb{R}$ and so (a) implies (b).

If $\sigma(A) \subseteq \mathbb{R}$, $\ker(A^* \pm i) = [\operatorname{ran}(A \mp i)]^\perp = \mathcal{H}^\perp = (0)$. Hence (b) implies (c).

If (c) holds, then this, combined with (2.5c) and (1.13), implies $A + i$ is surjective. Let $h \in \operatorname{dom} A^*$. Then there is an f in dom A such that $(A + i)f = (A^* + i)h$. But $A^* + i \supseteq A + i$, so $(A^* + i)f = (A^* + i)h$. But $A^* + i$ is injective, so $h = f \in \operatorname{dom} A$. Thus $A = A^*$. ∎

2.10. Corollary. *If A is a closed symmetric operator and $\sigma(A)$ does not contain* $\mathbb{R}$, *then $A = A^*$.*

It may have occurred to the reader that a symmetric operator A fails to be self-adjoint because its domain is too small and that this can be rectified by merely increasing the size of the domain. Indeed, if A is the symmetric operator in Example 1.11, then the operator B of Example 1.12 is a self-adjoint extension of A. However, the general situation is not always so cooperative.

Fix a symmetric operator A and suppose B is a symmetric extension of A: $A \subseteq B$. It is easy to verify that $B^* \subseteq A^*$. Since $B \subseteq B^*$, we get $A \subseteq B \subseteq B^* \subseteq A^*$. Thus every symmetric extension of A is a restriction of A^*.

2.11. Proposition. (a) *A symmetric operator has a maximal symmetric extension.* (b) *Maximal symmetric extensions are closed.* (c) *A self-adjoint operator is a maximal symmetric operator.*

PROOF. Part (a) is an easy application of Zorn's Lemma. If A is symmetric, $A \subseteq A^*$ and so A is closable. The closure of a symmetric operator is symmetric (Exercise 3), so part (b) is immediate. Part (c) is a consequence of the comments preceding this proposition. ∎

2.12. Definition. Let A be a closed symmetric operator. The *deficiency subspaces* of A are the spaces

$$\mathscr{L}_+ = \ker(A^* - i) = [\operatorname{ran}(A + i)]^\perp,$$
$$\mathscr{L}_- = \ker(A^* + i) = [\operatorname{ran}(A - i)]^\perp.$$

The *deficiency indices* of A are the numbers $n_\pm = \dim \mathscr{L}_\pm$.

It is possible for any pair of deficiency indices to occur (see Exercise 6).
In order to study the closed symmetric extensions of a symmetric operator we also introduce the spaces

$$\mathscr{K}_+ = \{f \oplus if : f \in \mathscr{L}_+\},$$
$$\mathscr{K}_- = \{g \oplus (-ig) : g \in \mathscr{L}_-\}.$$

So $\mathscr{K}_\pm \leq \mathscr{H} \oplus \mathscr{H}$. Notice that $\mathscr{K}_\pm$ are contained in gra A^* and are the portions of graph of A^* that lie above $\mathscr{L}_\pm$. The next lemma will indicate why the deficiency subspaces are so named.

2.13. Lemma. *If A is a closed symmetric operator,*

$$\operatorname{gra} A^* = \operatorname{gra} A \oplus \mathscr{K}_+ \oplus \mathscr{K}_-.$$

PROOF. Let $f \in \mathscr{L}_+$ and $h \in \text{dom } A$. Then

$$\langle h \oplus Ah, f \oplus if \rangle = \langle h, f \rangle - i\langle Ah, f \rangle$$
$$= -i\langle (A+i)h, f \rangle$$
$$= 0$$

since $\mathscr{L}_+ = [\text{ran}(A+i)]^\perp$. The remainder of the proof that gra A, $\mathscr{K}_+$, and $\mathscr{K}_-$ are pairwise orthogonal is left to the reader. Since it is clear that gra $A \oplus \mathscr{K}_+ \oplus \mathscr{K}_- \subseteq$ gra A^*, it remains to show that this direct sum is dense in gra A^*.

Let $h \in \text{dom } A^*$ and assume $h \oplus A^*h \perp \text{gra } A \oplus \mathscr{K}_+ \oplus \mathscr{K}_-$. Since $h \oplus A^*h \perp \text{gra } A$, for every f in dom A, $0 = \langle h \oplus A^*h, f \oplus Af \rangle = \langle h, f \rangle + \langle A^*h, Af \rangle$. So $\langle A^*h, Af \rangle = -\langle h, f \rangle$ for every f in dom A. This implies that $A^*h \in \text{dom } A^*$ and $A^*A^*h = -h$. Therefore $(A^*-i)(A^*+i)h = (A^*A^* + 1)h = 0$. Thus $(A^*+i)h \in \mathscr{L}_+$. Reversing the order of these factors also shows that $(A^*-i)h \in \mathscr{L}_-$. But if $g \in \mathscr{L}_+$, $0 = \langle h \oplus A^*h, g \oplus ig \rangle = \langle h, g \rangle - i\langle A^*h, g \rangle = -i\langle (A^*+i)h, g \rangle$. Since g can be taken equal to $(A^*+i)h$, we get that $(A^*+i)h = 0$, or $h \in \mathscr{L}_-$. Similarly, $h \in \mathscr{L}_+$. So $h \in \mathscr{L}_+ \cap \mathscr{L}_- = (0)$. ∎

2.14. Definition. If A is a closed symmetric operator and $\mathscr{M}$ is a linear manifold in dom A^*, then $\mathscr{M}$ is A-symmetric if $\langle A^*f, g \rangle = \langle f, A^*g \rangle$ for all f, g in $\mathscr{M}$. Call such a manifold A-closed if $\{f \oplus A^*f : f \in \mathscr{M}\}$ is closed in $\mathscr{H} \oplus \mathscr{H}$.

So $\mathscr{M}$ is both A-symmetric and A-closed precisely when $A^*|\mathscr{M}$, the restriction of A^* to $\mathscr{M}$, is a closed symmetric operator; if $\mathscr{M} \supseteq \text{dom } A$, then $A^*|\mathscr{M}$ is a closed symmetric extension of A.

2.15. Lemma. *If A is a closed symmetric operator on $\mathscr{H}$ and B is a closed symmetric extension of A, then there is an A-closed, A-symmetric submanifold $\mathscr{M}$ of $\mathscr{L}_+ + \mathscr{L}_-$ such that*

2.16 $$\text{gra } B = \text{gra } A + \text{gra}(A^*|\mathscr{M}).$$

Conversely, if $\mathscr{M}$ is an A-closed, A-symmetric manifold in $\mathscr{L}_+ + \mathscr{L}_-$, then there is a closed symmetric extension B of A such that (2.16) holds.

PROOF. If the A-symmetric manifold $\mathscr{M}$ in $\mathscr{L}_+ + \mathscr{L}_-$ is given, let $\mathscr{D} = \text{dom } A + \mathscr{M}$. Since $\mathscr{D} \subseteq \text{dom } A^*$, $B = A^*|\mathscr{D}$ is well defined. Let $f = f_0 + f_1, g = g_0 + g_1, f_0, g_0$ in dom A and f_1, g_1 in $\mathscr{M}$. Then

$$\langle A^*f, g \rangle = \langle A^*f_0 + A^*f_1, g_0 + g_1 \rangle$$
$$= \langle Af_0, g_0 \rangle + \langle Af_0, g_1 \rangle + \langle A^*f_1, g_0 \rangle + \langle A^*f_1, g_1 \rangle.$$

Using the A-symmetry of $\mathscr{M}$, the symmetry of A, and the definition of A^* we get

$$\langle A^*f, g \rangle = \langle f_0, Ag_0 \rangle + \langle f_0, A^*g_1 \rangle + \langle f_1, Ag_0 \rangle + \langle f_1, A^*g_1 \rangle$$
$$= \langle f, A^*g \rangle.$$

So $B = A^*|\mathcal{D}$ is symmetric. Note that gra $A \perp \text{gra}(A^*|\mathcal{M})$ in $\mathcal{H} \oplus \mathcal{H}$. Since both of these spaces are closed, gra B, given by (2.16), is closed.

Now let B be any closed symmetric extension of A. As discussed before, $A \subseteq B \subseteq A^*$; so gra $A \subseteq \text{gra } B \subseteq \text{gra } A^* = \text{gra } A \oplus \mathcal{K}_+ \oplus \mathcal{K}_-$. Let $\mathcal{G} = \text{gra } B \cap (\mathcal{K}_+ \oplus \mathcal{K}_-)$ and let $\mathcal{M} =$ the set of first coordinates of elements in $\mathcal{G}$. Clearly, $\mathcal{M}$ is a manifold in $\mathcal{L}_+ + \mathcal{L}_-$ and $\mathcal{M} \subseteq \text{dom } B$. Hence for f, g in $\mathcal{M}, \langle A^*f, g \rangle = \langle Bf, g \rangle = \langle f, Bg \rangle = \langle f, A^*g \rangle$. So $\mathcal{M}$ is A-symmetric. Clearly, $\text{gra}(A^*|\mathcal{M}) = \mathcal{G}$, so $\mathcal{M}$ is A-closed. If $h \oplus Bh \in \text{gra } B$. let $h \oplus Bh = (f \oplus Af) + k$ where $f \in \text{dom } A$ and $k \in \mathcal{K}_+ \oplus \mathcal{K}_-$. Since $A \subseteq B, k \in \text{gra } B$; so $k \in \mathcal{G}$. This shows that (2.16) holds. ∎

2.17. Theorem. *Let A be a closed symmetric operator. If W is a partial isometry with initial space in $\mathcal{L}_+$ and final subspace in $\mathcal{L}_-$, let*

2.18 $$\mathcal{D}_W = \{f + g + Wg : f \in \text{dom } A, g \in \text{initial } W\}$$

and define A_W on $\mathcal{D}_W$ by

2.19 $$A_W(f + g + Wg) = Af + ig - iWg.$$

Then A_W is a closed symmetric extension of A. Conversely, if B is any closed symmetric extension of A, then there is a unique partial isometry W such that $B = A_W$ as in (2.19).

If W is such a partial isometry and W has finite rank, then

$$n_\pm(A_W) = n_\pm(A) - \dim(\text{ran } W).$$

PROOF. Let W be a partial isometry with initial space I_+ in $\mathcal{L}_+$ and final space I_- in $\mathcal{L}_-$. Define $\mathcal{D}_W$ and A_W as in (2.18) and (2.19). Let $\mathcal{M} = \{g + Wg : g \in I_+\}$; so $\mathcal{M}$ is a manifold in $\mathcal{L}_+ + \mathcal{L}_-$. If $g, h \in I_+$, then $\langle Wg, Wh \rangle = \langle g, h \rangle$. Hence $\langle A^*(g + Wg), h + Wh \rangle = \langle A^*g, h \rangle + \langle A^*g, Wh \rangle + \langle A^*Wg, h \rangle + \langle A^*Wg, Wh \rangle$. Since $g \in \text{ker}(A^* - i)$ and $Wg \in \text{ker}(A^* + i)$,

$$\langle A^*(g + Wg), h + Wh \rangle = i\langle g, h \rangle + i\langle g, Wh \rangle - i\langle Wg, h \rangle - i\langle Wg, Wh \rangle$$
$$= i\langle g, Wh \rangle - i\langle Wg, h \rangle.$$

Similarly, $\langle g + Wg, A^*(h + Wh) \rangle = i\langle g, Wh \rangle - i\langle Wg, h \rangle$, so that $\mathcal{M}$ is A-symmetric. If $\{g_n\} \subseteq I_+$ and $(g_n + Wg_n) \oplus (ig_n - iWg_n) \to f \oplus h$ in $\mathcal{H} \oplus \mathcal{H}$, then $2ig_n = i(g_n + Wg_n) + (ig_n - iWg_n) \to if + h$ and $2iWg_n = i(g_n + Wg_n) - (ig_n - iWg_n) \to if - h$. If $g = (2i)^{-1}(if + h)$, then $f = g + Wg$ and $h = ig - iWg$. Hence $\mathcal{M}$ is A-closed. By Lemma 2.15, A_W is a closed symmetric extension of A.

To prove that $n_+(A_W) = n_+(A) - \dim I_+$, let $f \in \text{dom } A$, $g \in I_+$. Then

$$(A_W + i)(f + g + Wg) = (A + i)f + ig - iWg + ig + iWg$$
$$= (A + i)f + 2ig.$$

Thus $\text{ran}(A_W + i) = \text{ran}(A + i) \oplus I_+$, and so $n_+(A_W) = \dim[\text{ran}(A_W + i)]^\perp = \dim(\mathcal{L}_+ \ominus I_+) = n_+(A) - \dim I_+$. Similarly, $n_-(A_W) = n_-(A) - \dim I_- = n_-(A) - \dim I_+$.

Now let B be a closed symmetric extension of A. By Lemma 2.15 there is an A-symmetric, A-closed manifold $\mathcal{M}$ in $\mathcal{L}_+ + \mathcal{L}_-$ such that $\operatorname{gra} B = \operatorname{gra} A + \operatorname{gra}(A^* | \mathcal{M})$. If $f \in \mathcal{M}$, let $f = f^+ + f^-$, where $f^{\pm} \in \mathcal{L}_{\pm}$; put $I_+ = \{f^+ : f \in \mathcal{M}\}$. Since $\mathcal{M}$ is A-symmetric, $0 = \langle A^*f, f \rangle - \langle f, A^*f \rangle = 2i\langle f^+, f^+ \rangle - 2i\langle f^-, f^- \rangle$; hence $\|f^+\| = \|f^-\|$ for all f in $\mathcal{M}$. So if $Wf^+ = f^-$ whenever $f = f^+ + f^- \in \mathcal{M}$ and if I_+ is closed, W is a partial isometry and (2.18) and (2.19) are easily seen to hold. It remains to show that I_+ is closed. Suppose $\{f_n\} \subseteq \mathcal{M}$ and $f_n^+ \to g^+$ in $\mathcal{L}_+$. Since $\|f_n^+ - f_m^+\| = \|f_n^- - f_m^-\|$, there is a g^- in $\mathcal{L}_-$ such that $f_n^- \to g^-$. Clearly $f_n \to g^+ + g^- = g$. Also, $A^*f_n^{\pm} = \pm if_n^{\pm} \to \pm ig^{\pm}$. It follows that $g \oplus A^*g \in \operatorname{cl} \operatorname{gra}(A^* | \mathcal{M}) = \operatorname{gra}(A^* | \mathcal{M})$; thus $g^+ \in I_+$. ∎

2.20. Theorem. *Let A be a closed symmetric operator with deficiency indices $n_{\pm}$.*

(a) *A is self-adjoint if and only if $n_+ = n_- = 0$.*

(b) *A has a self-adjoint extension if and only if $n_+ = n_-$. In this case the set of self-adjoint extensions is in natural correspondence with the set of isomorphisms of $\mathcal{L}_+$ onto $\mathcal{L}_-$.*

(c) *A is a maximal symmetric operator that is not self-adjoint if and only if either $n_+ = 0$ and $n_- > 0$ or $n_+ > 0$ and $n_- = 0$.*

PROOF. Part (a) is a rephrasing of Corollary 2.9. For (b), $n_+ = n_-$ if and only if $\mathcal{L}_+$ and $\mathcal{L}_-$ are isomorphic. But this is equivalent to stating that there is a partial isometry on $\mathcal{H}$ with initial and final spaces $\mathcal{L}_+$ and $\mathcal{L}_-$, respectively. Part (c) follows easily from the preceding theorem. ∎

2.21. Example. Let A and $\mathcal{D}$ be as in Example 1.11; so A is symmetric. The operator B of Example 1.12 is a self-adjoint extension of A. Let us determine all self-adjoint extensions of A. To do this it is necessary to determine $\mathcal{L}_{\pm}$. Now $f \in \mathcal{L}_{\pm}$ if and only if $f \in \operatorname{dom} A^*$ and $\pm if = A^*f = if'$, so $\mathcal{L}_{\pm} = \{\alpha e^{\pm x} : \alpha \in \mathbb{C}\}$. Hence $n_{\pm} = 1$. Also, the isomorphisms of $\mathcal{L}_+$ onto $\mathcal{L}_-$ are all of the form $W_{\lambda} e^x = \lambda e^{-x}$ where $|\lambda| = e$. If $|\lambda| = e$, let

$$\mathcal{D}_{\lambda} \equiv \{f + \alpha e^x + \lambda\alpha e^{-x} : \alpha \in \mathbb{C}, f \in \mathcal{D}\}.$$
$$A_{\lambda}(f + \alpha e^x + \lambda\alpha e^{-x}) = if' + \alpha i e^x - i\lambda\alpha e^{-x},$$

if $f \in \mathcal{D}$, $\alpha \in \mathbb{C}$.

According to Theorem 2.17, $\{(A_{\lambda}, \mathcal{D}_{\lambda}) : |\lambda| = e\}$ are all of the self-adjoint extensions of A. The operator B of Example 1.12 is the extension A_e.

For more information on symmetric operators and the relation of the problem of finding self-adjoint extensions to physical problems, see Reed and Simon [1975] from which much of the present development is taken.

EXERCISES

1. If A is symmetric, show that all of the eigenvalues of A are real.

2. If A is symmetric and λ, μ are distinct eigenvalues, show that $\ker(A - \lambda) \perp \ker(A - \mu)$.

316 X. Unbounded Operators

3. Show that the closure of a symmetric operator is symmetric.

4. Let $\mathscr{D} = \{f \in L^2(0, \infty): \text{for every } c > 0, f \text{ is absolutely continuous on } [0, c], f(0) = 0, \text{and } f' \in L^2(0, \infty)\}$. Define $Af = if'$ for f in $\mathscr{D}$. Show that A is a densely defined closed operator and find dom A^*. Show that A is symmetric with deficiency indices $n_+ = 0$ and $n_- = 1$.

5. Let $\mathscr{E} = \{f \in L^2(-\infty, 0): \text{for every } c < 0, f \text{ is absolutely continuous on } [c, 0], f(0) = 0, \text{and } f' \in L^2(-\infty, 0)\}$. Define $Af = if'$ for f in $\mathscr{E}$. Show that A is a densely defined closed operator and find dom A^*. Show that A is symmetric with deficiency indices $n_+ = 1, n_- = 0$.

6. If k, l are any nonnegative integers or ∞, show that there is a closed symmetric operator A with $n_+ = k$ and $n_- = l$. (Hint: Use Exercises 4 and 5.)

7. Let $C_c^2(0, 1)$ be all twice continuously differentiable functions on $(0, 1)$ with compact support and let $Af = -f''$ for f in $C_c^2(0, 1)$. Show that the closure of A is a densely defined symmetric operator and determine all of its self-adjoint extensions.

8. If $A \in \mathscr{C}(\mathscr{H})$, show that A^*A is self-adjoint (see Exercise 1.11).

9. Say that an operator A is positive if $\langle Ah, h \rangle \geq 0$ for all h in dom A. Prove that if A is positive and self-adjoint, then $\sigma(A) \subseteq [0, \infty)$. If A is only assumed to be closed and positive, show that this conclusion may fail. (Hint: Look at the operator in Exercise 7.)

10. (Lasser [1972]) Let $\mathscr{M}$ be a dense linear manifold in $\mathscr{H}$ and let $\mathscr{A}$ consist of all linear transformations A such that dom $A = \mathscr{M}$, $A\mathscr{M} \subseteq \mathscr{M}$, the adjoint of A exists, $\mathscr{M} \subseteq$ dom A^*, and $A^*\mathscr{M} \subseteq \mathscr{M}$. Prove that $A \to A^*|\mathscr{M}$ defines an involution on $\mathscr{A}$.

§3. The Cayley Transform

Consider the Möbius transformation

$$M(z) = \frac{z - i}{z + i}.$$

It is immediate that $M(0) = -1$, $M(1) = -i$, and $M(\infty) = 1$. Thus M maps the upper half plane onto $\mathbb{D}$ and $M(\mathbb{R} \cup \infty) = \partial\mathbb{D}$. So if A is self-adjoint, $M(A)$ should be unitary. Suppose A is symmetric; does $M(A)$ make sense? What is $M(A)$?

To answer these questions, we should first investigate the meaning of $M(A)$ if A is symmetric. We want to define $M(A)$ as $(A - i)(A + i)^{-1}$. As was seen in the last section, however, ran$(A + i)$ is not necessarily all of $\mathscr{H}$ if A is not self-adjoint. In fact, $(\text{ran}(A + i))^\perp = \mathscr{L}_+$ and $(\text{ran}(A - i))^\perp = \mathscr{L}_-$, the deficiency spaces for A. However (2.5), if A is closed and symmetric, ran$(A \pm i)$ is closed. Also, realize that if $w = M(z)$, then $z = M^{-1}(w) = i(1 + w)/(1 - w)$.

3.1. Theorem. (a) *If A is a closed densely defined symmetric operator with deficiency subspaces $\mathscr{L}_{\pm}$, and if $U: \mathscr{H} \to \mathscr{H}$ is defined by letting $U = 0$ on $\mathscr{L}_{+}$ and*

3.2
$$U = (A - i)(A + i)^{-1}$$

on $\mathscr{L}_{+}^{\perp}$, then U is a partial isometry with initial space $\mathscr{L}_{+}^{\perp}$, final space $\mathscr{L}_{-}^{\perp}$, and such that $(1 - U)(\mathscr{L}_{+}^{\perp})$ is dense in $\mathscr{H}$.

(b) *If U is a partial isometry with initial and final spaces $\mathscr{M}$ and $\mathscr{N}$, respectively, and such that $(1 - U)\mathscr{M}$ is dense in $\mathscr{H}$, then*

3.3
$$A = i(1 + U)(1 - U)^{-1}$$

is a densely defined closed symmetric operator with deficiency subspaces $\mathscr{L}_{+} = \mathscr{M}^{\perp}$ and $\mathscr{L}_{-} = \mathscr{N}^{\perp}$.

(c) *If A is given as in (a) and U is defined by (3.2), then A and U satisfy (3.3). If U is given as in (b) and A is defined by (3.3), then A and U satisfy (3.2).*

PROOF. (a) By (2.5c), ran$(A \pm i)$ is closed and so $\mathscr{L}_{\pm}^{\perp} = \text{ran}(A \pm i)$. By (2.5b), ker$(A + i) = (0)$, so $(A + i)^{-1}$ is well defined on $\mathscr{L}_{+}^{\perp}$. Moreover, $(A + i)^{-1}\mathscr{L}_{+}^{\perp} \subseteq$ dom A so that U defined by (3.2) makes sense and gives a well-defined operator. If $h \in \mathscr{L}_{+}^{\perp}$, then $h = (A + i)f$ for a unique f in dom A. Hence $\| Uh \|^2 = \|(A - i)f\|^2 = (2.5a) \| Af \|^2 + \| f \|^2 = \|(A + i)f\|^2 = \| h \|^2$. Hence U is a partial isometry, $(\ker U)^{\perp} = \mathscr{L}_{+}^{\perp}$, and ran $U = \mathscr{L}_{-}^{\perp}$. Once again, if $f \in$ dom A and $h = (A + i)f$, then $(1 - U)h = h - (A - i)f = (A + i)f - (A - i)f = 2if$. So $(1 - U)\mathscr{L}_{+}^{\perp} =$ dom A and is dense in $\mathscr{H}$.

(b) Now assume that U is a partial isometry as in (b). It follows that ker$(1 - U) = (0)$. In fact, if $f \in \ker(1 - U)$, then $Uf = f$; so $\| f \| = \| Uf \|$ and hence $f \in$ initial U. Since U^*U is the projection onto initial U, $f = U^*Uf = U^*f$; so $f \in \ker(1 - U^*) = \text{ran}(1 - U)^{\perp} \subseteq [(1 - U)\mathscr{M}]^{\perp} = (0)$ by hypothesis. Thus $f = 0$ and $1 - U$ is injective.

Let $\mathscr{D} = (1 - U)\mathscr{M}$ and define $(1 - U)^{-1}$ on $\mathscr{D}$. Because $1 - U$ is bounded, gra$(1 - U)^{-1}$ is closed. If A is defined as in (3.3), it follows that A is a closed densely defined operator. If $f, g \in \mathscr{D}$, let $f = (1 - U)h$ and $g = (1 - U)k, h, k \in \mathscr{M}$. Hence

$$\langle Af, g \rangle = i\langle (1 + U)h, (1 - U)k \rangle$$
$$= i[\langle h, k \rangle + \langle Uh, k \rangle - \langle h, Uk \rangle - \langle Uh, Uk \rangle].$$

Since $h, k \in \mathscr{M}$, $\langle Uh, Uk \rangle = \langle h, k \rangle$; hence $\langle Af, g \rangle = i[\langle Uh, k \rangle - \langle h, Uk \rangle]$. Similarly, $\langle f, Ag \rangle = -i\langle (1 - U)h, (1 + U)k \rangle = -i[\langle h, Uk \rangle - \langle Uh, k \rangle] = \langle Af, g \rangle$. Hence A is symmetric.

Finally, if $h \in \mathscr{M}$ and $f = (1 - U)h$, then $(A + i)f = Af + if = i(1 + U)h + i(1 - U)h = 2ih$. Thus ran$(A + i) = \mathscr{M}$. Similarly, $(A - i)f = i(1 + U)h - i(1 - U)h = 2Uh$, so that ran$(A - i) = \text{ran } U = \mathscr{N}$.

(c) Suppose A is as in (a) and U is defined as in (3.2). If $g \in (1 - U)\mathscr{L}_{+}^{\perp}$, put $g = (1 - U)h$, where $h \in \mathscr{L}_{+}^{\perp} = \text{ran}(A + i)$. Hence $h = (A + i)f$ for some f in dom A. Thus $g = h - Uh = (A + i)f - (A - i)f = 2if$; so $f = -\frac{1}{2}ig$.

Also,

$$
\begin{aligned}
i(1 + U)(1 - U)^{-1}g &= i(1 + U)h \\
&= i[h + Uh] \\
&= i[(A + i)f + (A - i)f] \\
&= 2iAf \\
&= Ag.
\end{aligned}
$$

Therefore (3.3) holds.

The proof of the remainder of (c) is left to the reader. ∎

3.4. Definition. If A is a densely defined closed symmetric operator, the partial isometry U defined by (3.2) is called the *Cayley transform* of A.

3.5. Corollary. *If A is a self-adjoint operator and U is its Cayley transform, then U is a unitary operator with $\ker(1 - U) = (0)$. Conversely, if U is a unitary with $1 \notin \sigma_p(U)$, then the operator A defined by (3.3) is self-adjoint.*

PROOF. If A is a densely defined symmetric operator, then A is self-adjoint if and only if $\mathscr{L}_{\pm} = (0)$. A partial isometry is a unitary operator if and only if its initial and final spaces are all of $\mathscr{H}$. This corollary is now seen to follow from Theorem 3.1. ∎

One use of the Cayley transform is to study self-adjoint operators by using the theory of unitary operators. Indeed, the preceding results say that there is a bijective correspondence between self-adjoint operators and the set of unitary operators without 1 as an eigenvalue.

EXERCISES

1. If U is a partial isometry, show that the following statements are equivalent: (a) $\ker(1 - U) = (0)$; (b) $\ker(1 - U^*) = (0)$; (c) $\operatorname{ran}(1 - U)$ is dense; (d) $\operatorname{ran}(1 - U^*)$ is dense.

2. Let U be a partial isometry with initial and final spaces $\mathscr{M}$ and $\mathscr{N}$, respectively. Show that the following statements are equivalent: (a) $(1 - U)\mathscr{M}$ is dense; (b) $(1 - U^*)\mathscr{N}$ is dense; (c) $\ker(U^* - U^*U) = (0)$; (d) $\ker(U - UU^*) = (0)$.

3. Find a partial isometry U such that $\ker(1 - U) = (0)$ but $(1 - U)(\ker U)^{\perp}$ is not dense.

4. If A is a densely defined closed symmetric operator and B and C are the operators defined in Exercises 1.11 and 1.12, then the Cayley transform of A is an extension of $(C - iB)(C + iB)^{-1}$.

5. Find the Cayley transform of the operator in Example 1.9 when each α_n is real.

6. Find the Cayley transform of the operator in Example 1.10 when ϕ is real valued.

7. Let S be the unilateral shift of multiplicity 1 (see Exercise IX.6.4) and find the symmetric operator A such that S is the Cayley transform of A.

8. Let $U = S^*$, where S is the unilateral shift of multiplicity 1. Is U the Cayley transform of a symmetric operator A? If so, find it.

§4. Unbounded Normal Operators and the Spectral Theorem

If A is self-adjoint, the classical way to obtain the spectral decomposition of A is to let U be the Cayley transform of A, obtain the spectral decomposition of U, and then use the inverse Cayley transform to translate this back to a decomposition for A. There is a spectral theorem for unbounded normal operators, however, and the Cayley transform is not applicable here.

In this section the approach is to prove the spectral theorem for normal operators by using that theorem for the bounded case. The spectral theorem for self-adjoint operators is then only a special case.

4.1. Definition. A linear operator N on $\mathcal{H}$ is *normal* if N is closed, densely defined, and $N^*N = NN^*$.

Note that the equation $N^*N = NN^*$ that appears in Definition 4.1 implicitly carries the condition that $\operatorname{dom} N^*N = \operatorname{dom} NN^*$. The operators in Examples 1.9 and 1.10 are normal and every self-adjoint operator is normal. Examining Example 1.9 it is easy to see that for a normal operator it is not necessarily the case that $\operatorname{dom} N^*N = \operatorname{dom} N$.

Parts of the next result have appeared in various exercises in this chapter, but a complete proof is given here.

4.2. Proposition. *If $A \in \mathscr{C}(\mathcal{H})$, then*

(a) $1 + A^*A$ *has a bounded inverse defined on all of $\mathcal{H}$.*
(b) *If $B = (1 + A^*A)^{-1}$, then $\|B\| \leqslant 1$ and $B \geqslant 0$.*
(c) *The operator $C = A(1 + A^*A)^{-1}$ is a contraction.*
(d) A^*A *is self-adjoint.*
(e) $\{h \oplus Ah : h \in \operatorname{dom} A^*A\}$ *is dense in* gra A.

PROOF. Define $J: \mathcal{H} \oplus \mathcal{H} \to \mathcal{H} \oplus \mathcal{H}$ by $J(h \oplus k) = (-k) \oplus h$. By Lemma 1.7, gra $A^* = [J \text{ gra } A]^{\perp}$. So if $h \in \mathcal{H}$, there are f in dom A and g in dom A^* such that $0 \oplus h = J(f \oplus Af) + g \oplus A^*g = (-Af) \oplus f + g \oplus A^*g$. Hence $0 = -Af + g$, or $g = Af$; also, $h = f + A^*g = f + A^*Af = (1 + A^*A)f$. Thus ran$(1 + A^*A) = \mathcal{H}$.

Also, for f in dom A^*A, $Af \in \operatorname{dom} A^*$ and $\|f + A^*Af\|^2 = \|f\|^2 + 2\|Af\|^2 + \|A^*Af\|^2 \geqslant \|f\|^2$. Hence ker$(1 + A^*A) = (0)$. Thus $(1 + A^*A)^{-1}$ exists and is defined on all of $\mathcal{H}$. In the next paragraph (the proof of (b)) it will be shown that $(1 + A^*A)^{-1}$ is a contraction, completing the proof of (a).

It was shown that $\|(1 + A^*A)f\| \geqslant \|f\|$ whenever $f \in \operatorname{dom} A^*A$. If $h = (1 + A^*A)f$ and $B = (1 + A^*A)^{-1}$, then this implies that $\|Bh\| \leqslant \|h\|$.

Hence $\| B \| \leqslant 1$. In addition, $\langle Bh, h \rangle = \langle f, (1 + A^*A)f \rangle = \| f \|^2 + \| Af \|^2 \geqslant 0$, so (b) holds.

Put $C = A(1 + A^*A)^{-1} = AB$; if $f \in \text{dom } A^*A$ and $(1 + A^*A)f = h$, then $\| Ch \|^2 = \| Af \|^2 \leqslant \| (1 + A^*A)f \|^2 = \| h \|^2$ by the argument used to prove (a). Hence $\| C \| \leqslant 1$, so (c) is proved.

Now to prove (e). Since A is closed, it suffices to show that no nonzero vector in gra A is orthogonal to $\{h \oplus Ah : h \in \text{dom } A^*A\}$. So let $g \in \text{dom } A$ and suppose that for every h in dom A^*A,

$$
\begin{aligned}
0 &= \langle g \oplus Ag, h \oplus Ah \rangle \\
&= \langle g, h \rangle + \langle Ag, Ah \rangle \\
&= \langle g, h \rangle + \langle g, A^*Ah \rangle \\
&= \langle g, (1 + A^*A)h \rangle.
\end{aligned}
$$

So $g \perp \text{ran}(1 + A^*A) = \mathscr{H}$; hence $g = 0$.

To prove (d), note that (e) implies that dom A^*A is dense. Now let $f, g \in \text{dom } A^*A$; so $f, g \in \text{dom } A$ and $Af, Ag \in \text{dom } A^*$. Hence $\langle A^*Af, g \rangle = \langle Af, Ag \rangle = \langle f, A^*Ag \rangle$. Thus A^*A is symmetric. Also, $1 + A^*A$ has a bounded inverse. This implies two things. First, $1 + A^*A$ is closed, and so A^*A is closed. Also, $-1 \notin \sigma(A^*A)$ so that by Corollary 2.10, A^*A is self-adjoint. ∎

4.3. Proposition. *If N is a normal operator, then* dom $N =$ dom N^* *and* $\| Nf \| = \| N^*f \|$ *for every f in* dom N.

PROOF. First observe that if $h \in \text{dom } N^*N = \text{dom } NN^*$, then $Nh \in \text{dom } N^*$ and $N^*h \in \text{dom } N$. Hence $\| Nh \|^2 = \langle N^*Nh, h \rangle = \langle NN^*h, h \rangle = \| N^*h \|^2$. Now if $f \in \text{dom } N$, (4.2e) implies that there is a sequence $\{h_n\}$ in dom N^*N such that $h_n \oplus Nh_n \to f \oplus Nf$; so $\| Nh_n - Nf \| \to 0$. But from the first part of this proof, $\| N^*h_n - N^*h_m \| = \| Nh_n - Nh_m \|$. So there is a g in $\mathscr{H}$ such that $N^*h_n \to g$. Thus $h_n \oplus N^*h_n \to f \oplus g$. But N^* is closed; thus $f \in \text{dom } N^*$ and $g = N^*f$. So dom $N \subseteq \text{dom } N^*$ and $\| Nf \| = \lim \| Nh_n \| = \lim \| N^*h_n \| = \| N^*f \|$.

On the other hand, N^* is normal (Why?), and so dom $N^* \subseteq \text{dom } N^{**} = \text{dom } N$. ∎

4.4. Lemma. *Let $\mathscr{H}_1, \mathscr{H}_2, \ldots$ be Hilbert spaces and let $A_n \in \mathscr{B}(\mathscr{H}_n)$ for all $n \geqslant 1$. If $\mathscr{D} = \{(h_n) \in \oplus_n \mathscr{H}_n : \sum_{n=1}^{\infty} \| A_n h_n \|^2 < \infty\}$ and A is defined on $\mathscr{H} = \oplus_n \mathscr{H}_n$ by $A(h_n) = (A_n h_n)$ whenever $(h_n) \in \mathscr{D}$, then $A \in \mathscr{C}(\mathscr{H})$. A is a normal operator if and only if each A_n is normal.*

PROOF. Since $\mathscr{H}_n \subseteq \mathscr{D}$ for each $n, \mathscr{D}$ is dense in $\mathscr{H}$. Clearly A is linear. If $\{h^{(j)}\} \subseteq \text{dom } A$ and $h^{(j)} \oplus Ah^{(j)} \to h \oplus g$ in $\mathscr{H} \oplus \mathscr{H}$, then for each n, $h_n^{(j)} \oplus A_n h_n^{(j)} \to h_n \oplus g_n$. Since A_n is bounded, $A_n h_n = g_n$. Hence $\sum_n \| A_n h_n \|^2 = \sum \| g_n \|^2 = \| g \|^2 < \infty$; so $h \in \text{dom } A$. Clearly $Ah = g$, so $A \in \mathscr{C}(\mathscr{H})$.

It is left to the reader to show that dom $A^* = \{(h_n) \in \mathscr{H} : \sum_{n=1}^{\infty} \| A_n^* h_n \|^2 < \infty\}$ and $A^*(h_n) = (A_n^* h_n)$ when $(h_n) \in \text{dom } A^*$. From this the rest of the lemma easily follows. ∎

If (X, Ω) is a measurable space and $\mathcal{H}$ is a Hilbert space, recall the definition of a spectral measure E for $(X, \Omega, \mathcal{H})$ (IX.1.1). If $h, k \in \mathcal{H}$, let $E_{h,k}$ be the complex-valued measure given by $E_{h,k}(\Delta) = \langle E(\Delta)h, k \rangle$ for each Δ in Ω.

Let $\phi \colon X \to \mathbb{C}$ be an Ω-measurable function and for each n let $\Delta_n = \{x \in X : n - 1 \leqslant |\phi(x)| < n\}$. So $\chi_{\Delta_n} \phi$ is a bounded Ω-measurable function. Put $\mathcal{H}_n = E(\Delta_n)\mathcal{H}$. Since $\bigcup_{n=1}^{\infty} \Delta_n = X$ and the sets $\{\Delta_n\}$ are pairwise disjoint, $\bigoplus_{n=1}^{\infty} \mathcal{H}_n = \mathcal{H}$. If $E_n(\Delta) = E(\Delta \cap \Delta_n)$, E_n is a spectral measure for $(X, \Omega, \mathcal{H}_n)$. Also, $\int \phi \, dE_n$ is a normal operator on $\mathcal{H}_n$. Define

4.5
$$\mathcal{D}_\phi \equiv \left\{ h \in \mathcal{H} : \sum_{n=1}^{\infty} \left\| \left(\int \phi \, dE_n \right) E(\Delta_n) h \right\|^2 < \infty \right\}.$$

By Lemma 4.4, $N_\phi \colon \mathcal{H} \to \mathcal{H}$ given by

4.6
$$N_\phi h = \sum_{n=1}^{\infty} \left(\int \phi \, dE_n \right) E(\Delta_n) h$$

for h in $\mathcal{D}_\phi$ is a normal operator. The operator N_ϕ is also denoted by

$$N_\phi = \int \phi \, dE.$$

4.7. Theorem. *If E is a spectral measure for $(X, \Omega, \mathcal{H})$, $\phi \colon X \to \mathbb{C}$ is an Ω-measurable function, and $\mathcal{D}_\phi$ and N_ϕ are defined as in (4.5) and (4.6), then:*

(a) $\mathcal{D}_\phi = \{ h \in \mathcal{H} : \int |\phi|^2 \, dE_{h,h} < \infty \}$;
(b) *for h in $\mathcal{D}_\phi$ and f in $\mathcal{H}$, $\phi \in L^1(|E_{h,f}|)$ with*

4.8
$$\int |\phi| \, d|E_{h,f}| \leqslant \|f\| \left(\int |\phi|^2 \, dE_{h,h} \right)^{1/2},$$

4.9
$$\left\langle \left(\int \phi \, dE \right) h, f \right\rangle = \int \phi \, dE_{h,f},$$

and

$$\left\| \left(\int \phi \, dE \right) h \right\|^2 = \int |\phi|^2 \, dE_{h,h}.$$

PROOF. Using the $*$-homomorphic properties associated with a spectral measure (IX.1.12), one obtains

$$\left\| \left(\int \phi \, dE_n \right) E(\Delta_n) h \right\|^2 = \left\langle \left(\int \chi_{\Delta_n} \phi \, dE \right)^* \left(\int \chi_{\Delta_n} \phi \, dE \right) h, h \right\rangle$$

$$= \int_{\Delta_n} |\phi|^2 \, dE_{h,h}.$$

From here, (a) is immediate.

Now let $h \in \mathcal{D}_\phi$, $f \in \mathcal{H}$. By the Radon–Nikodym Theorem, there is an Ω-measurable function u such that $|u| \equiv 1$ and $|E_{h,f}| = uE_{h,f}$, where $|E_{h,f}|$ is the variation for $E_{h,f}$. Let $\phi_n = \sum_{k=1}^n \chi_{\Delta_k}\phi$; so ϕ_n is bounded, as is $u\phi_n$. Thus

$$
\int |\phi_n| d|E_{h,f}| = \int |\phi_n| u dE_{h,f}
$$

$$
= \left\langle \left(\int |\phi_n| u dE \right) h, f \right\rangle
$$

$$
\leq \| f \| \left\| \left(\int |\phi_n| u dE \right) h \right\|.
$$

But

$$
\left\| \left(\int |\phi_n| u dE \right) h \right\|^2 = \left\langle \left(\int |\phi_n| u dE \right) h, \left(\int |\phi_n| u dE \right) h \right\rangle
$$

$$
= \left\langle \left(\int |\phi_n|^2 dE \right) h, h \right\rangle
$$

$$
= \int |\phi_n|^2 dE_{h,h}
$$

$$
\leq \int |\phi|^2 dE_{h,h}.
$$

Combining this with the preceding inequality gives that $\int |\phi_n| d|E_{h,f}| \leq \| f \| (\int |\phi|^2 dE_{h,h})^{1/2}$ for all n. Letting $n \to \infty$ gives (4.8). Since ϕ_n is bounded, $\langle (\int \phi_n dE) h, f \rangle = \int \phi_n dE_{h,f}$. If $h \in \mathscr{D}_\phi$ and $f \in \mathscr{H}$, then (4.8) and the Lebesgue Dominated Convergence Theorem imply that $\int \phi_n dE_{h,f} \to \int \phi dE_{h,f}$ as $n \to \infty$. But

$$
\left(\int \phi_n dE \right) h = \left(\int \phi dE \right) E \left(\bigcup_{j=1}^n \Delta_j \right) h
$$

$$
= E \left(\bigcup_{j=1}^n \Delta_j \right) \left(\int \phi dE \right) h.
$$

Since $E(\bigcup_{j=1}^n \Delta_j) \to E(X) = 1$ (SOT) as $n \to \infty$, $\langle (\int \phi_n dE) h, f \rangle \to \langle (\int \phi dE) h, f \rangle$ as $n \to \infty$. This proves (4.9).   ■

Note that as a consequence of (4.7) dom N_ϕ and the definition of N_ϕ do not depend on the choice of the sets $\{\Delta_n\}$, as would seem to be the case from (4.5) and (4.6). Also, by (4.7.a), $E(\Delta)h \in \text{dom } N_\phi$ if $h \in \text{dom } N_\phi$.

4.10. Theorem. *If (X, Ω) is a measurable space, $\mathscr{H}$ is a Hilbert space, and E is a spectral measure for $(X, \Omega, \mathscr{H})$, let $\Phi(X, \Omega)$ be the algebra of all Ω-measurable functions $\phi \colon X \to \mathbb{C}$ and define $\rho \colon \Phi(X, \Omega) \to \mathscr{C}(\mathscr{H})$ by $\rho(\phi) = \int \phi dE$. Then for ϕ, ψ in $\Phi(X, \Omega)$:*

(a) $\rho(\phi)^* = \rho(\bar{\phi})$;
(b) $\rho(\phi\psi) \supseteq \rho(\phi)\rho(\psi)$ and $\text{dom}(\rho(\phi)\rho(\psi)) = \mathscr{D}_\psi \cap \mathscr{D}_{\phi\psi}$;

(c) *If ψ is bounded, $\rho(\phi)\rho(\psi) = \rho(\psi)\rho(\phi) = \rho(\phi\psi)$;*
(d) $\rho(\phi)^*\rho(\phi) = \rho(|\phi|^2)$.

The proof of this theorem is left as an exercise.

4.11. The Spectral Theorem. *If N is a normal operator on $\mathcal{H}$, then there is a unique spectral measure E defined on the Borel subsets of $\mathbb{C}$ such that:*

(a) $N = \int z\, dE(z)$;
(b) $E(\Delta) = 0$ *if* $\Delta \cap \sigma(N) = \square$;
(c) *if U is an open subset of $\mathbb{C}$ and $U \cap \sigma(N) \neq \square$, then $E(U) \neq 0$;*
(d) *if $A \in \mathcal{B}(\mathcal{H})$ such that $AN \subseteq NA$ and $AN^* \subseteq N^*A$, then $A(\int \phi\, dE) \subseteq (\int \phi\, dE)A$ for every Borel function ϕ on $\mathbb{C}$.*

Before launching into the proof, a few words motivating the proof are appropriate. Suppose a spectral measure E defined on the Borel subsets of $\mathbb{C}$ is given and let $N = \int z\, dE(z)$. It is not difficult to see that if $0 \leqslant a \leqslant b < \infty$ and Δ is the annulus $\{z: a \leqslant |z| \leqslant b\}$, then $\mathcal{H}_\Delta = E(\Delta)\mathcal{H} = \{h \in \mathrm{dom}\, N: h \in \mathrm{dom}\, N^n$ for all n and $a^n \|h\| \leqslant \|N^n h\| \leqslant b^n \|h\|\}$. $\mathcal{H}_\Delta$ is a closed subspace of $\mathcal{H}$ that reduces N and $N|\mathcal{H}_\Delta$ is bounded. The idea behind the proof is to write $\mathbb{C}$ as the disjoint union of annuli $\{\Delta_j\}$ such that for each Δ_j there is a reducing subspace $\mathcal{H}_{\Delta_j}$ for N with $N_j \equiv N|\mathcal{H}_{\Delta_j}$ bounded, and, moreover, such that $\mathcal{H} = \bigoplus_j \mathcal{H}_{\Delta_j}$. Once this is done the Spectral Theorem for bounded normal operators can be applied to each N_j and direct sums of these can be formed to obtain the spectral measure for N.

So we would like to show that for the annulus $\{z: a \leqslant |z| \leqslant b\}$, $\{h \in \mathrm{dom}\, N: h \in \mathrm{dom}\, N^n$ for all n and $a^n \|h\| \leqslant \|N^n h\| \leqslant b^n \|h\|\}$ is a reducing subspace for N. To facilitate this, we will use the operator $B = (1 + N^*N)^{-1}$ which is a positive contraction (4.2). To understand what is done below note that $z \mapsto (1 + |z|^2)^{-1}$ maps $\mathbb{C}$ onto $(0, 1]$ and $a \leqslant |z| \leqslant b$ if and only if $(1 + a^2)^{-1} \geqslant (1 + |z|^2)^{-1} \geqslant (1 + b^2)^{-1}$.

4.12. Lemma. *If N is a normal operator, $B = (1 + N^*N)^{-1}$, and $C = N(1 + N^*N)^{-1}$, then $BC = CB$ and $(1 + N^*N)^{-1}N \subseteq C$.*

PROOF. From (4.2), B and C are contractions and $B \geqslant 0$. It will first be shown that $(1 + N^*N)^{-1}N \subseteq C$; that is, $BN \subseteq NB$. If $f \in \mathrm{dom}\, BN$, then $f \in \mathrm{dom}\, N$. Let $g \in \mathrm{dom}\, N^*N$ such that $f = (1 + N^*N)g$. Then $N^*Ng \in \mathrm{dom}\, N$; hence $Ng \in \mathrm{dom}\, NN^* = \mathrm{dom}\, N^*N$. Thus $Nf = Ng + NN^*Ng = (1 + N^*N)Ng$. Therefore $BNf = B(1 + N^*N)Ng = Ng$. But $NBf = Ng$, so $BN = NB$ on $\mathrm{dom}\, N$. Thus $BN \subseteq NB$.

If $h \in \mathcal{H}$, let $f \in \mathrm{dom}\, N^*N$ such that $h = (1 + N^*N)f$. So $BCh = BNBh = BNf = NBf = NBBh = CBh$. Hence $BC = CB$. $\blacksquare$

4.13. Lemma. *With the same notation as in Lemma 4.12, if $B = \int t\, dP(t)$ is its spectral representation, $1 > \delta > 0$, and Δ is a Borel subset of $[\delta, 1]$, then*

$\mathscr{H}_\Delta = P(\Delta)\mathscr{H} \subseteq \operatorname{dom} N$, $\mathscr{H}_\Delta$ is invariant for both N and N^*, and $N|\mathscr{H}_\Delta$ is a bounded normal operator with $\|N|\mathscr{H}_\Delta\| \leqslant [(1-\delta)/\delta]^{1/2}$.

PROOF. If $h \in \mathscr{H}_\Delta$, then because $P(\Delta) = \chi_\Delta(B)$, $\|Bh\|^2 = \langle B^2 P(\Delta)h, h \rangle = \int_\Delta t^2 dP_{h,h} \geqslant \delta^2 \|h\|^2$. So $B|\mathscr{H}_\Delta$ is invertible and there is a g in $\mathscr{H}_\Delta$ such that $h = Bg$. But $\operatorname{ran} B = \operatorname{dom}(1 + N^*N) \subseteq \operatorname{dom} N$. Hence $h \in \operatorname{dom} N$; that is, $\mathscr{H}_\Delta \subseteq \operatorname{dom} N$.

Let $h \in \mathscr{H}_\Delta$ and again let $g \in \mathscr{H}_\Delta$ such that $h = Bg$. Hence $Nh = NBg = Cg$. By Lemma 4.12, $BC = CB$; so by (IX.2.2), $P(\Delta)C = CP(\Delta)$. Since $g \in \mathscr{H}_\Delta$, $Nh = Cg \in \mathscr{H}_\Delta$. Note that if $M = N^*$ and $B_1 = (1 + M^*M)^{-1}$, then $B_1 = B$. From the preceding argument $N^*\mathscr{H}_\Delta = M\mathscr{H}_\Delta \subseteq \mathscr{H}_\Delta$. It easily follows that $N|\mathscr{H}_\Delta$ is normal.

Finally, if $h \in \mathscr{H}_\Delta$, then

$$
\begin{aligned}
\|Nh\|^2 &= \langle N^*Nh, h \rangle \\
&= \langle [(N^*N + 1) - 1]h, h \rangle \\
&= \int_\delta^1 (t^{-1} - 1) dP_{h,h}(t) \leqslant \|h\|^2 (1 - \delta)/\delta.
\end{aligned}
$$

Hence $\|N|\mathscr{H}_\Delta\| \leqslant [(1-\delta)/\delta]^{1/2}$.   ∎

PROOF OF THE SPECTRAL THEOREM. Let $B = (1 + N^*N)^{-1}$ and $C = N(1 + N^*N)^{-1}$ as in Lemma 4.12. Let $B = \int_0^1 t dP(t)$ be the spectral decomposition of B and put $P_n = P(1/(n+1), 1/n]$ for $n \geqslant 1$. Since $\ker B = (0) = P(\{0\})\mathscr{H}$, $\sum_{n=1}^\infty P_n = 1$. Let $\mathscr{H}_n = P_n \mathscr{H}$. By Lemma 4.13, $\mathscr{H}_n \subseteq \operatorname{dom} N$, $\mathscr{H}_n$ reduces N, and $N_n \equiv N|\mathscr{H}_n$ is bounded normal operator with $\|N_n\| \leqslant n^{1/2}$. Also, if $h \in \mathscr{H}_n$, $(1 + N_n^*N_n)Bh = B(1 + N_n^*N_n)h = h$; that is,

$$
B|\mathscr{H}_n = (1 + N_n^*N_n)^{-1}.
$$

Thus if $\lambda \in \sigma(N_n)$, $(1 + |\lambda|^2)^{-1} \in \sigma(B|\mathscr{H}_n) \subseteq [1/(n+1), 1/n]$. Thus $\sigma(N_n) \subseteq \{z \in \mathbb{C} : (n-1)^{1/2} \leqslant |z| \leqslant n^{1/2}\} \equiv \Delta_n$. Let $N_n = \int z dE_n(z)$ be the spectral decomposition of N_n. For any Borel subset Δ of $\mathbb{C}$, let $E(\Delta)$ be defined by

4.14
$$
E(\Delta) = \sum_{n=1}^\infty E_n(\Delta \cap \Delta_n).
$$

Note that $E_n(\Delta \cap \Delta_n)$ is a projection with range in $\mathscr{H}_n$. Since $\mathscr{H}_n \perp \mathscr{H}_m$ for $n \neq m$, (4.14) defines a projection in $\mathscr{B}(\mathscr{H})$. (Technically $E(\Delta)$ should be defined by $E(\Delta) = \sum_{n=1}^\infty E_n(\Delta \cap \Delta_n)P_n$. But this technicality does not add anything to understanding.)

Now to show that E is a spectral measure. Clearly $E(\mathbb{C}) = 1$ and $E(\square) = 0$. If Λ_1 and Λ_2 are Borel subsets of $\mathbb{C}$, then

$$
\begin{aligned}
E(\Lambda_1 \cap \Lambda_2) &= \sum_{n=1}^\infty E_n(\Lambda_1 \cap \Lambda_2 \cap \Delta_n) \\
&= \sum_{n=1}^\infty E_n(\Lambda_1 \cap \Delta_n)E_n(\Lambda_2 \cap \Delta_n).
\end{aligned}
$$

Again, the fact that the spaces $\{\mathscr{H}_n\}$ are pairwise orthogonal implies

$$E(\Lambda_1 \cap \Lambda_2) = \left(\sum_{n=1}^{\infty} E_n(\Lambda_1 \cap \Delta_n) \right) \left(\sum_{n=1}^{\infty} E_n(\Lambda_2 \cap \Delta_n) \right)$$

$$= E(\Lambda_1)E(\Lambda_2).$$

If $h \in \mathscr{H}$, then $\langle E(\Delta)h, h \rangle = \sum_{n=1}^{\infty} \langle E_n(\Delta \cap \Delta_n)h, h \rangle$. So if $\{\Lambda_j\}_{j=1}^{\infty}$ are pairwise disjoint Borel sets,

$$\left\langle E\left(\bigcup_{j=1}^{\infty} \Lambda_j \right)h, h \right\rangle = \sum_{n=1}^{\infty} \left\langle E_n \left(\left(\bigcup_{j=1}^{\infty} \Lambda_j \right) \cap \Delta_n \right)h, h \right\rangle$$

$$= \sum_{n=1}^{\infty} \sum_{j=1}^{\infty} \langle E_n(\Lambda_j \cap \Delta_n)h, h \rangle.$$

Since each term in this double summation is non-negative, the order of summation can be reversed. Thus

$$\left\langle E\left(\bigcup_{j=1}^{\infty} \Lambda_j \right)h, h \right\rangle = \sum_{j=1}^{\infty} \sum_{n=1}^{\infty} \langle E_n(\Lambda_j \cap \Delta_n)h, h \rangle$$

$$= \sum_{j=1}^{\infty} \langle E(\Lambda_j)h, h \rangle.$$

So $E(\bigcup_{j=1}^{\infty}\Lambda_j) = \sum_{j=1}^{\infty} E(\Lambda_j)$; therefore E is a spectral measure.

Let $M = \int z\, dE(z)$ be defined as in Theorem 4.7. Thus $\mathscr{H}_n \subseteq \operatorname{dom} M$ and by the Spectral Theorem for bounded operators, $Mh = N_n h = Nh$ if $h \in \mathscr{H}_n$. If h is any vector in dom M, $h = \sum_1^{\infty} h_n$, $h_n \in \mathscr{H}_n$, and $\sum_1^{\infty} \|Nh_n\|^2 < \infty$. Because N is closed, $h \in \operatorname{dom} N$ and $Nh = Mh$. Thus $M \subseteq N$. To prove the other inclusion, note that M is a closed operator by Lemma 4.4. Thus, by (4.2.e), it suffices to show that $\{h \oplus Nh\colon h \in \operatorname{dom} N^*N\} \subseteq \operatorname{gra} M$. If $h \in \operatorname{dom} N^*N$, there is a vector g such that $h = Bg$. Then $P_n Nh = P_n NBg = P_n Cg = CP_n g$ (Why?) $= NP_n h$. If $h_n = P_n h$, then $\sum \|Nh_n\|^2 = \sum \|P_n Nh\|^2 = \|Nh\|^2 < \infty$. Therefore $h \in \operatorname{dom} M$ and so, by the preceding argument, $Nh = Mh$. That is, $h \oplus Nh \in \operatorname{gra} M$. This proves (a).

4.15. Claim.
$$\sigma(N) = \operatorname{cl}\left[\bigcup_{n=1}^{\infty} \sigma(N_n) \right].$$

It is left to the reader to show that $\bigcup_{n=1}^{\infty} \sigma(N_n) \subseteq \sigma(N)$. Since $\sigma(N)$ is closed, this proves half of (4.15). If $\lambda \notin \operatorname{cl}[\bigcup_{n=1}^{\infty} \sigma(N_n)]$, then there is a $\delta > 0$ such that $|\lambda - z| \geq \delta$ for all z in $\bigcup_{n=1}^{\infty} \sigma(N_n)$. Thus $(N_n - \lambda)^{-1}$ exists and $\|(N_n - \lambda)^{-1}\| \leq \delta^{-1}$ for all n. Thus $A = \bigoplus_{n=1}^{\infty}(N_n - \lambda)^{-1}$ is a bounded operator. It follows that $A = (N - \lambda)^{-1}$, so $\lambda \notin \sigma(N)$.

By (4.15) if $\Delta \cap \sigma(N) = \square$, $\Delta \cap \sigma(N_n) = \square$ for all n. Thus $E_n(\Delta) = 0$ for all n. Hence $E(\Delta) = 0$ and (b) holds.

If U is open and $U \cap \sigma(N) \neq \square$, then (4.15) implies $U \cap \sigma(N_n) \neq \square$ for some n. Since $E_n(U) \neq 0$, $E(U) \neq 0$ and (c) is true.

Now let $A \in \mathcal{B}(\mathcal{H})$ such that $AN \subseteq NA$ and $AN^* \subseteq N^*A$. Thus $A(1 + N^*N) \subseteq (1 + N^*N)A$. It follows that $AB = BA$. By the Spectral Theorem for bounded operators, A commutes with the spectral projections of B. In particular, each $\mathcal{H}_n$ reduces A and if $A_n \equiv A | \mathcal{H}_n$, then $A_n N_n = N_n A_n$. Hence $A_n E_n(\Delta) = E_n(\Delta) A_n$ for every Borel set Δ contained in Δ_n. It follows that $AE(\Delta) = E(\Delta)A$ for every Borel set Δ. The remaining details of the proof of (d) are left to the reader. ∎

The Fuglede–Putnam Theorem holds for unbounded normal operators (Exercise 8), so that the hypothesis in part (d) of the Spectral Theorem can be weakened to $AN \subseteq NA$.

4.16. Definition. If N is a normal operator on $\mathcal{H}$, then a vector e_0 is a *star-cyclic* vector for N if for all non-negative integers k and l, $e_0 \in \mathrm{dom}(N^{*k} N^l)$ and $\mathcal{H} = \bigvee \{ N^{*k} N^l e_0 : k, l \geq 0 \}$.

4.17. Example. Let μ be a finite measure on $\mathbb{C}$ such that every polynomial in z and $\bar{z}$ belongs to $L^2(\mu)$ and the collection of such polynomials is dense in $L^2(\mu)$. Let $\mathcal{D}_\mu = \{ f \in L^2(\mu): zf \in L^2(\mu) \}$ and define $N_\mu f = zf$ for f in $\mathcal{D}_\mu$. Then N_μ is a normal operator and 1 is a star-cyclic vector for N_μ.

Note that $d\mu(z) = e^{-|z|} d\,\mathrm{Area}(z)$ is a measure satisfying the conditions of (4.17).

4.18. Theorem. *If N is a normal operator on $\mathcal{H}$ with a star-cyclic vector e_0, then there is a finite measure μ on $\mathbb{C}$ such that every polynomial in z and $\bar{z}$ belongs to $L^2(\mu)$ and there is an isomorphism $W: \mathcal{H} \to L^2(\mu)$ such that $We_0 = 1$ and $WNW^{-1} = N_\mu$.*

The proof of Theorem 4.18 can be accomplished by using the Spectral Theorem to write N as the direct sum (in the sense of Lemma 4.4) of bounded normal operators N_n on $\mathcal{H}_n$ with spectral measures that are pairwise mutually singular and such that each N_n has e_n, the projection of e_0 onto $\mathcal{H}_n$, as a *-cyclic vector. If $\mu_n = E_{e_n,e_n}$, then (IX.3.4) implies that there is an isomorphism $W_n: \mathcal{H}_n \to L^2(\mu_n)$ such that $W_n N_n W_n^{-1} = N_{\mu_n}$. If $W = \bigoplus_1^\infty W_n$, then W is an isomorphism of $\mathcal{H}$ onto $\bigoplus_1^\infty L^2(\mu_n)$. But the fact that the measures μ_n are pairwise mutually singular implies that $\bigoplus_1^\infty L^2(\mu_n) = L^2(\mu)$, where $\mu = \sum_{n=1}^\infty \mu_n = E_{e_0,e_0}$. Clearly $WNW^{-1} = N_\mu$.

4.19. Theorem. *If N is a normal operator on the separable Hilbert space $\mathcal{H}$, then there is a σ-finite measure space (X, Ω, μ) and an Ω-measurable function ϕ such that N is unitarily equivalent to M_ϕ on $L^2(\mu)$.*

The proof of Theorem 4.19 is only sketched. Write N as the (unbounded) direct sum of bounded normal operators $\{N_n\}$. By Theorem IX.4.6, there is a σ-finite measure space (X_n, Ω_n, μ_n) and a bounded Ω_n-measurable function

ϕ_n such that $N_n \cong M_{\phi_n}$. Let $X = $ the disjoint union of $\{X_n\}$ and let $\Omega = \{\Delta \subseteq X: \Delta \cap X_n \in \Omega_n$ for every $n\}$. If $\Delta \in \Omega$, let $\mu(\Delta) = \sum_1^\infty \mu_n(\Delta \cap X_n)$. Let $\phi: X \to \mathbb{C}$ be defined by $\phi(x) = \phi_n(x)$ if $x \in X_n$. Then ϕ is Ω-measurable and $N \cong M_\phi$ on $L^2(X, \Omega, \mu)$.

EXERCISES

1. Prove Theorem 4.10.

2. Show that if A is a symmetric operator that is normal, then A is self-adjoint.

3. With the notation of Theorem 4.7, show that for h in $\mathscr{D}_\phi$, $\| (\int \phi \, dE)h \|^2 = \int |\phi|^2 \, dE_{h,h}$.

4. Using the notation of Theorem 4.10, what is $\sigma(\int \phi \, dE)$?

5. If Δ_n and E_n are as in the proof of the Spectral Theorem, show that $E_n(\Delta_{n+1}) = E_n(\Delta_{n-1}) = 0$.

6. Use the Spectral Theorem to show that if $0 < a \leqslant b < \infty$, $\Delta = \{z \in \mathbb{C}: a \leqslant |z| \leqslant b\}$, and $N = \int z \, dE(z)$ is the spectral decomposition of the normal operator N, then $E(\Delta)\mathscr{H} = \{h \in \text{dom } N: a^n \|h\| \leqslant \|N^n h\| \leqslant b^n \|h\|$ for all $n \geqslant 1\}$.

7. State and prove a polar decomposition for operators in $\mathscr{C}(\mathscr{H}, \mathscr{K})$.

8. If A is self-adjoint, prove that $\exp(iA)$ is unitary.

9. (Fuglede–Putnam Theorem.) If N, M are normal operators and A is a bounded operator such that $AN \subseteq MA$, then $AN^* \subseteq M^*A$.

10. Prove Theorem 4.18.

11. If μ_1, μ_2 are finite measures on $\mathbb{C}$ and N_{μ_1}, N_{μ_2} are defined as in Example 4.17, show that $N_{\mu_1} \cong N_{\mu_2}$ iff $[\mu_1] = [\mu_2]$.

12. Fill in the details of the proof of Theorem 4.19.

§5. Stone's Theorem

If A is a self-adjoint operator on $\mathscr{H}$, then $\exp(iA)$ is a unitary operator (Exercise 4.7). Hence $U(t) = \exp(itA)$ is unitary for all t in $\mathbb{R}$. The purpose of this section is not to investigate the individual operators $\exp(itA)$, but rather the entire collection of operators $\{\exp(itA): t \in \mathbb{R}\}$. In fact, as the first theorem shows, $U: \mathbb{R} \to$ unitaries on $\mathscr{H}$ is a group homomorphism with certain properties. Stone's Theorem provides a converse to this; every such homomorphism arises in this way.

5.1. Theorem. *If A is self-adjoint and $U(t) = \exp(itA)$ for t in $\mathbb{R}$, then*

(a) *$U(t)$ is unitary;*
(b) *$U(s + t) = U(s)U(t)$ for all s in $\mathbb{R}$;*
(c) *if $h \in \mathscr{H}$, then $\lim_{s \to t} U(s)h = U(t)h$;*

(d) *if* $h \in \operatorname{dom} A$, *then*

5.2
$$\lim_{t \to 0} \frac{1}{t} [U(t)h - h] = iAh;$$

(e) *if* $h \in \mathcal{H}$ *and* $\lim_{t \to 0} t^{-1}[U(t)h - h]$ *exists, then* $h \in \operatorname{dom} A$. *Consequently,* dom A *is invariant under each* $U(t)$.

PROOF. As was mentioned, part (a) is an exercise. Since $\exp(itx)\exp(isx) = \exp(i(s+t)x)$ for all x in $\mathbb{R}$, (b) is a consequence of the functional calculus for normal operators [(4.10) and (4.11)]. Also note that $U(0)U(t) = U(t)$, so that $U(0) = 1$.

(c) If $h \in \mathcal{H}$, then $\| U(t)h - U(s)h \| = \| U(t - s + s)h - U(s)h \| = $ [by (b)] $\| U(s)[U(t - s)h - h] \| = \| U(t - s)h - h \|$ since $U(s)$ is unitary. Thus (c) will be shown if it is proved that $\| U(t)h - h \| \to 0$ as $t \to 0$. If $A = \int_{-\infty}^{\infty} x \, dE(x)$ is the spectral decomposition of A, then

$$\| U(t)h - h \|^2 = \int_{-\infty}^{\infty} |e^{itx} - 1|^2 \, dE_{h,h}(x).$$

Now $E_{h,h}$ is a finite measure on $\mathbb{R}$; for each x in $\mathbb{R}$, $|e^{itx} - 1|^2 \to 0$ as $t \to 0$; and $|e^{itx} - 1|^2 \leq 4$. So the Lebesgue Dominated Convergence Theorem implies that $U(t)h \to h$ as $t \to 0$.

(d) Note that $t^{-1}[U(t) - 1] - iA = f_t(A)$, where $f_t(x) = t^{-1}[\exp(itx) - 1] - ix$. So if $h \in \operatorname{dom} A$,

$$\left\| \frac{1}{t}[U(t)h - h] - iAh \right\|^2 = \| f_t(A)h \|^2$$

$$= \int_{-\infty}^{\infty} \left| \frac{e^{itx} - 1}{t} - ix \right|^2 \, dE_{h,h}(x).$$

As $t \to 0$, $t^{-1}[e^{itx} - 1] - ix \to 0$ for all x in $\mathbb{R}$. Also, $|e^{is} - 1| \leq |s|$ for all real numbers s (Why?), hence $|f_t(x)| \leq |t|^{-1}|e^{itx} - 1| + |x| \leq 2|x|$. But $|x| \in L^2(E_{h,h})$ by Theorem 4.7(a). So again the Lebesgue Dominated Convergence Theorem implies that (5.2) is true.

(e) Let $\mathcal{D} = \{h \in \mathcal{H} : \lim_{t \to 0} t^{-1}[U(t)h - h] \text{ exists in } \mathcal{H}\}$. For h in $\mathcal{D}$, let Bh be defined by

$$Bh = -i \lim_{t \to 0} \frac{U(t)h - h}{t}.$$

It is easy to see that $\mathcal{D}$ is a linear manifold in $\mathcal{H}$ and B is linear on $\mathcal{D}$. Also, by (d), $B \supseteq A$ so that B is densely defined. Moreover, if $h, g \in \mathcal{D}$, then

$$\langle Bh, g \rangle = -i \lim_{t \to 0} \left\langle \frac{U(t)h - h}{t}, g \right\rangle.$$

By (b) and the fact that each $U(t)$ is unitary, it follows that $U(t)^* = U(t)^{-1} =$

$U(-t)$. Hence

$$\langle Bh, g \rangle = -i \lim_{t \to 0} \left\langle h, \frac{U(-t)g - g}{t} \right\rangle$$

$$= \lim_{t \to 0} \left\langle h, -i \left[\frac{U(-t)g - g}{-t} \right] \right\rangle$$

$$= \langle h, Bg \rangle.$$

Hence B is a symmetric extension of A. Since self-adjoint operators are maximal symmetric operators (2.11), $B = A$ and $\mathcal{D} = \text{dom } A$. ∎

The following definition is inspired by the preceding theorem.

5.3. Definition. A *strongly continuous one parameter unitary group* is a function $U: \mathbb{R} \to \mathcal{B}(\mathcal{H})$ such that for all s and t in $\mathbb{R}$: (a) $U(t)$ is a unitary operator; (b) $U(s + t) = U(s)U(t)$; (c) if $h \in \mathcal{H}$ and $t_0 \in \mathbb{R}$, then $U(t)h \to U(t_0)h$ as $t \to t_0$.

Note that by Theorem 5.1, if A is self-adjoint, then $U(t) = \exp(itA)$ defines a strongly continuous one parameter unitary group.

Also, $U(0) = 1$ and $U(-t) = U(t)^{-1}$, so that $\{U(t): t \in \mathbb{R}\}$ is indeed a group. Property (c) also implies that $U: \mathbb{R} \to (\mathcal{B}(\mathcal{H}), \text{SOT})$ is continuous. By Exercise 1, if U is only assumed to be WOT-continuous, then U is SOT-continuous. However, this condition can be relaxed even further as the following result of von Neumann [1932] shows.

5.4. Theorem. *If $\mathcal{H}$ is separable, $U: \mathbb{R} \to \mathcal{B}(\mathcal{H})$ satisfies conditions (a) and (b) of Definition 5.3, and if for all h, g in $\mathcal{H}$ the function $t \mapsto \langle U(t)h, g \rangle$ is Lebesgue measurable, then U is a strongly continuous one-parameter unitary group.*

PROOF. If $0 < a < \infty$ and $h, g \in \mathcal{H}$, then $t \mapsto \langle U(t)h, g \rangle$ is a bounded measurable function on $[0, a]$ and hence

$$\int_0^a |\langle U(t)h, g \rangle| \, dt \leq a \|h\| \|g\|.$$

Thus

$$h \mapsto \int_0^a \langle U(t)h, g \rangle \, dt$$

is a bounded linear function on $\mathcal{H}$. Therefore there is a g_a in $\mathcal{H}$ such that

5.5 $$\langle h, g_a \rangle = \int_0^a \langle U(t)h, g \rangle \, dt$$

and $\|g_a\| \leq a \|g\|$.

Claim. $\{g_a: g \in \mathcal{H}, a > 0\}$ is total in $\mathcal{H}$.

In fact, suppose $h \in \mathcal{H}$ and $h \perp \{g_a : g \in \mathcal{H}, a > 0\}$. Then by (5.5), for every $a > 0$ and every g in $\mathcal{H}$,

$$0 = \int_0^a \langle U(t)h, g \rangle dt.$$

Thus for every g in $\mathcal{H}$, $\langle U(t)h, g \rangle = 0$ a.e. on $\mathbb{R}$. Because $\mathcal{H}$ is separable there is a subset Δ of $\mathbb{R}$ having measure zero such that if $t \notin \Delta$, $\langle U(t)h, g \rangle = 0$ whenever g belongs to a preselected countable dense subset of $\mathcal{H}$. Thus $U(t)h = 0$ if $t \notin \Delta$. But $\| h \| = \| U(t)h \|$, so $h = 0$ and the claim is established.

Now if $s \in \mathbb{R}$,

$$\langle h, U(s)g_a \rangle = \langle U(-s)h, g_a \rangle$$

$$= \int_0^a \langle U(t-s)h, g \rangle dt$$

$$= \int_{-s}^{a-s} \langle U(t)h, g \rangle dt.$$

Thus $\langle h, U(s)g_a \rangle \to \langle h, g_a \rangle$ as $s \to 0$. By the claim and the fact that the group is uniformly bounded, $U : \mathbb{R} \to (\mathcal{B}(\mathcal{H}), \text{WOT})$ is continuous at 0. By the group property, $U : \mathbb{R} \to (\mathcal{B}(\mathcal{H}), \text{WOT})$ is continuous. Hence U is SOT-continuous (Exercise 1). ∎

We now turn our attention to the principal result of this section, Stone's Theorem, which states that the converse of Theorem 5.1 is valid. Note that if $U(t) = \exp(itA)$ for a self-adjoint operator A, then part (d) of Theorem 5.1 instructs us how to recapture A. This is the route followed in the proof of Stone's Theorem, proved in Stone [1932].

5.6. Stone's Theorem. *If U is a strongly continuous one parameter unitary group, then there is a self-adjoint operator A such that $U(t) = \exp(itA)$.*

PROOF. Begin by defining $\mathcal{D}$ to be the set of all vectors h in $\mathcal{H}$ such that $\lim_{t \to 0} t^{-1}[U(t)h - h]$ exists; since $0 \in \mathcal{D}$, $\mathcal{D} \neq \square$. Clearly $\mathcal{D}$ is a linear manifold in $\mathcal{H}$.

5.7. Claim. $\mathcal{D}$ is dense in $\mathcal{H}$.

Let $\mathcal{L} = $ all continuous functions ϕ on $\mathbb{R}$ such that $\phi \in L^1(0, \infty)$. Hence for any h in $\mathcal{H}$, $t \mapsto \phi(t)U(t)h$ is a continuous function of $\mathbb{R}$ into $\mathcal{H}$. Because $\| U(t)h \| = \| h \|$ for all t, a Riemann integral, $\int_0^\infty \phi(t)U(t)h\,dt$, can be defined and is a vector in $\mathcal{H}$. Put

5.8
$$T_\phi h = \int_0^\infty \phi(t)U(t)h\,dt.$$

It is easy to see that $T_\phi : \mathcal{H} \to \mathcal{H}$ is linear and bounded with $\| T_\phi \| \leq \int_0^\infty |\phi(t)|\,dt$.

Similarly, for each ϕ in $\mathscr{L}$

5.9
$$S_\phi h = \int_0^\infty \phi(t)U(-t)h\,dt$$

defines a bounded operator on $\mathscr{H}$.

For any ϕ in $\mathscr{L}$ and t in $\mathbb{R}$,

$$U(t)T_\phi h = U(t)\int_0^\infty \phi(s)U(s)h\,ds$$

$$= \int_0^\infty \phi(s)U(t+s)h\,ds$$

$$= \int_t^\infty \phi(s-t)U(s)h\,ds.$$

Similarly,

$$U(t)S_\phi h = \int_{-t}^\infty \phi(s+t)U(-s)h\,ds.$$

Now let $\mathscr{L}^{(1)} = $ all ϕ in $\mathscr{L}$ that are continuously differentiable with ϕ' in $\mathscr{L}$. For ϕ in $\mathscr{L}^{(1)}$,

$$-\frac{i}{t}[U(t)-1]T_\phi h = -\frac{i}{t}\int_t^\infty \phi(s-t)U(s)h\,ds + \frac{i}{t}\int_0^\infty \phi(s)U(s)h\,ds$$

$$= -i\int_t^\infty \left[\frac{\phi(s-t)-\phi(s)}{t}\right]U(s)h\,ds + \frac{i}{t}\int_0^t \phi(s)U(s)h\,ds.$$

Now

$$\left\| \int_0^t \left[\frac{\phi(s-t)-\phi(s)}{t}\right]U(s)h\,ds \right\| \le \|h\|\sup\{|\phi(s-t)-\phi(s)|: 0 \le s \le 1\} \to 0$$

as $t \to 0$. Hence

$$\lim_{t\to 0} \int_t^\infty \left[\frac{\phi(s-t)-\phi(s)}{t}\right]U(s)h\,ds = -\int_0^\infty \phi'(s)U(s)h\,ds$$

$$= -T_{\phi'}h.$$

Since $s \mapsto \phi(s)U(s)h$ is continuous and $U(0) = 1$, the Fundamental Theorem of Calculus implies that

$$\lim_{t\to 0}\frac{1}{t}\int_0^t \phi(s)U(s)h\,ds = \phi(0)h.$$

Hence for ϕ in $\mathscr{L}^{(1)}$ and h in $\mathscr{H}$,

5.10
$$\lim_{t\to 0} -\frac{i}{t}[U(t)-1]T_\phi h = iT_{\phi'}h + i\phi(0)h.$$

Similarly, for ϕ in $\mathscr{L}^{(1)}$ and h in $\mathscr{H}$,

5.11 $\lim\limits_{t \to 0} -\dfrac{i}{t}[U(t) - 1]S_\phi h = -iS_{\phi'}h - i\phi(0)h.$

So (5.10) implies that

$$\mathscr{D} \supseteq \{T_\phi h\colon \phi \in \mathscr{L}^{(1)} \text{ and } h \in \mathscr{H}\}.$$

But for every positive integer n there is a ϕ_n in $\mathscr{L}^{(1)}$ such that $\phi_n \geq 0$, $\phi_n(t) = 0$ for $t \geq 1/n$, and $\int_0^\infty \phi_n(t)dt = 1$ (Exercise 2). Hence

$$T_{\phi_n}h - h = \int_0^{1/n} \phi_n(t)[U(t) - 1]h\,dt$$

and so $\|T_{\phi_n}h - h\| \leq \sup\{\|U(t)h - h\|\colon 0 \leq t \leq 1/n\}$. Therefore $\|T_{\phi_n}h - h\| \to 0$ as $n \to \infty$ since U is strongly continuous. This says that $\mathscr{D}$ is dense.

For h in $\mathscr{D}$, define

5.12 $Ah = -i\lim\limits_{t \to 0} \dfrac{1}{t}[U(t) - 1]h.$

5.13. Claim. A is symmetric.

The proof of this is left to the reader.

By (2.2c), A is closable; also denote the closure of A by A. According to Corollary 2.9, to prove that A is self-adjoint it suffices to prove that $\ker(A^* \pm i) = (0)$. Equivalently, it suffices to show that $\operatorname{ran}(A \pm i)$ is dense. It will be shown that there are operators $B_\pm$ such that $(A \pm i)B_\pm = 1$, so that $A \pm i$ is surjective.

Notice that according to (5.10),

$$(A + i)T_\phi = AT_\phi + iT_\phi = i(T_{\phi'} + T_\phi) + i\phi(0).$$

So taking $\phi(t) = -ie^{-1}$, $(A + i)T_\phi = 1$. According to (5.11),

$$(A - i)S_\psi = AS_\psi - iS_\psi = -i(S_{\psi'} + S_\psi) - i\psi(0).$$

Taking $\psi(t) = ie^{-1}$, $(A - i)S_\psi = 1$. Hence A is self-adjoint.

Put $V(t) = \exp(iAt)$. It remains to show that $V = U$. Let $h \in \mathscr{D}$. By Theorem 5.1(d),

$$s^{-1}[V(t + s) - V(t)]h = s^{-1}[V(s) - 1]V(t)h \to iAV(t)h;$$

that is, $V'(t)h = iAV(t)h$. Similarly,

$$s^{-1}[U(t + s) - U(t)]h = s^{-1}[U(s) - 1]U(t)h \to iAU(t)h.$$

So if $h(t) = U(t)h - V(t)h$, then $h\colon \mathbb{R} \to \mathscr{H}$ is differentiable and

$$h'(t) = iAU(t)h - iAV(t)h = iAh(t).$$

But

$$\frac{d}{dt}\|h(t)\|^2 = \langle h'(t), h(t)\rangle + \langle h(t), h'(t)\rangle$$

$$= \langle iAh(t), h(t)\rangle + \langle h(t), iAh(t)\rangle.$$

Thus $(d/dt)\|h(t)\|^2 = 0$ and so $\|h\|: \mathbb{R} \to \mathbb{R}$ is a constant function. But $h(0) = 0$, so $h(t) \equiv 0$. This says that $U(t)h = V(t)h$ for all h in $\mathscr{D}$ and all t in $\mathbb{R}$. Since $\mathscr{D}$ is dense, $U = V$. ∎

5.14. Definition. If U is a strongly continuous one parameter unitary group, then the self-adjoint operator A such that $U(t) = \exp(itA)$ is called the *infinitesimal generator* of U.

By virtue of Stone's Theorem and Theorem 5.1, there is a one-to-one correspondence between self-adjoint operators and strongly continuous one-parameter unitary groups. Thus, it should be possible to characterize certain properties of a group in terms of its infinitesimal generator and vice versa. For example, suppose the infinitesimal generator is bounded; what can be said about the group? (Also see Exercise 6.)

5.15. Proposition. *If U is a strongly continuous one parameter unitary group with infinitesimal generator A, then A is bounded if and only if $\lim_{t\to 0}\|U(t) - 1\| = 0$.*

PROOF. First assume that A is bounded. Hence $\|U(t) - 1\| = \|\exp(itA) - 1\| = \sup\{|e^{itx} - 1|: x\in\sigma(A)\} \to 0$ as $t \to 0$ since $\sigma(A)$ is compact.

Now assume that $\|U(t) - 1\| \to 0$ as $t \to 0$. Let $0 < \varepsilon < \pi/4$; then there is a $t_0 > 0$ such that $\|U(t) - 1\| < \varepsilon$ for $|t| < t_0$. Since $U(t) - 1 = \int_{\sigma(A)}(e^{itx} - 1)dE(t)$, $\sup\{|e^{itx} - 1|: x\in\sigma(A)\} = \|U(t) - 1\| < \varepsilon$ for $|t| < t_0$. Thus for a small δ, $tx\in\bigcup_{n=-\infty}^{\infty}(2\pi n - \delta, 2\pi n + \delta) \equiv G$ whenever $x\in\sigma(A)$ and $|t| < t_0$. In fact, if ε is chosen sufficiently small, then δ is small enough that the intervals $\{(2\pi n - \delta, 2\pi n + \delta)\}$ are the components of G. If $x\in\sigma(A)$, $\{tx: 0\le t < t_0\}$ is the interval from 0 to $t_0 x$ and is contained in G. Hence $tx\in(-\delta, \delta)$ for x in $\sigma(A)$ and $|t| < t_0$. In particular, $t_0\sigma(A) \subseteq [-\delta, \delta]$ so $\sigma(A)$ is compact and A is bounded. ∎

Let μ be a positive measure on $\mathbb{R}$ and let $A_\mu f = xf$ for f in $\mathscr{D}_\mu = \{f\in L^2(\mu): xf\in L^2(\mu)\}$. We have already seen that A_μ is self-adjoint. Clearly $\exp(itA_\mu) = M_{e_t}$ on $L^2(\mathbb{R})$, where e_t is the function $e_t(x) = \exp(itx)$. This can be generalized a bit.

5.16. Proposition. *Let (X, Ω, μ) be a σ-finite measure space and let ϕ be a real-valued Ω-measurable function on X. If $A = M_\phi$ on $L^2(\mu)$ and $U(t) = \exp(itA)$, then $U(t) = M_{e_t}$, where $e_t(x) = \exp(it\phi(x))$.*

Since each self-adjoint operator on a separable Hilbert space can be represented as a multiplication operator (Theorem 4.19), the preceding proposition gives a representation of all strongly continuous one parameter semigroups.

1. If $U: \mathbb{R} \to \mathscr{B}(\mathscr{H})$ is such that $U(t)$ is unitary for all t, $U(s+t) = U(s)U(t)$ for all s, t, and $U: \mathbb{R} \to (\mathscr{B}(\mathscr{H}), \text{WOT})$ is continuous, then U is SOT-continuous.

2. Show that for every integer n there is a continuously differentiable function ϕ_n such that both ϕ_n and $\phi_n' \in L^1(0, \infty)$, $\phi_n(t) = 0$ if $t \geq 1/n$, and $\int_0^\infty \phi_n(t)dt = 1$.

3. Prove Claim 5.13.

4. Adopt the notation from the proof of Stone's Theorem. Let $\phi, \psi \in \mathscr{L}$ and show:
   (a) $T_\phi^* = S_{\bar\phi}$; (b) $T_\phi T_\psi = T_{\phi * \psi}$ and $S_\phi S_\psi = S_{\phi * \psi}$; (c) $T_\phi A \subseteq A T_\phi$.

5. Let U be a strongly continuous one parameter unitary group with infinitesimal generator A. Suppose e is a nonzero vector in $\mathscr{H}$ such that $Ae = \lambda e$. What is $U(t)e$? Conversely, suppose there is a nonzero t such that $U(t)$ has an eigenvector. What can be said about A? $U(s)$?

6. (This exercise is designed to give another proof of Proposition 5.15 as well as give additional information. My thanks to R.B. Burckel for pointing this out to me.) Let U be a strongly continuous one- parameter unitary group with infinitesimal generator A. Show that if $\| U(t) - 1 \| \to 0$ as $t \to 0$, then as $t \to 0$, $t^{-1} \int_a^{a+t} U(s)ds \to U(a)$ in norm. From here show that as $t \to 0$, $t^{-1}[U(t) - 1]$ has a norm limit, and hence A is a bounded operator since it is the norm limit of bounded operators.

§6. The Fourier Transform and Differentiation

Perhaps the best way to begin this section is by examining an example.

6.1. Example. Let $\mathscr{D} = \{f \in L^2(\mathbb{R}): f$ is absolutely continuous on every bounded interval in $\mathbb{R}$ and $f' \in L^2(\mathbb{R})\}$. For f in $\mathscr{D}$, let $Af = if'$. Then A is self-adjoint.

First let's show that A is symmetric. If $f \in \mathscr{D}$, note that $f(x) \to 0$ as $x \to \pm \infty$ since f and $f' \in L^2(\mathbb{R})$. So if $f, g \in \mathscr{D}$, $0 < a < \infty$,

$$ i \int_{-a}^a f'(x)\overline{g(x)}dx = i[f(a)\overline{g(a)} - f(-a)\overline{g(-a)}] - i \int_{-a}^a f(x)\overline{g'(x)}dx. $$

Hence $\langle Af, g \rangle = \langle f, Ag \rangle$ and A is symmetric.

Now let $g \in \text{dom } A^*$ and for $0 < a < \infty$ let $\mathscr{D}_a = \{f \in \mathscr{D}: f(x) = 0$ for $|x| \geq a\}$. The proof that $g \in \text{dom } A$ follows the lines of the argument used in Example 1.11. In fact, let $h = A^* g$. So if $f \in \mathscr{D}$, then $\int f(x)\overline{h(x)}dx = i \int f'(x)\overline{g(x)}dx$. Let $H(x) = \int_0^x h(t) dt$. Then using integration by parts we get that for f in $\mathscr{D}_a$,

$$\int_{-a}^{a} f\bar{h} = \overline{H(a)}f(a) - \overline{H(-a)}f(-a) - \int_{-a}^{a} f'\bar{H}$$

$$= -\int_{-a}^{a} f'\bar{H}.$$

Therefore $\int_{-a}^{a} f'[\bar{H} - (i\bar{g})] = 0$ for every f in $\mathscr{D}_a$. As in (1.11), it follows that $H - ig$ is constant on $[-a, a]$ and g is absolutely continuous. Moreover, $0 = H' - ig' = h - ig'$; hence $A^*g = h = ig'$. Thus $g \in \mathscr{D}$ and A is self-adjoint.

If A is the differentiation operator in Example 6.1, what is the group $U(t) = \exp(itA)$? Since A is not represented as a multiplication operator, Proposition 5.16 cannot be applied. One could proceed to try and discover the spectral measure for A. Since $A = \int x \, dE(x)$, $U(t) = \int e^{itx} \, dE(x)$. Or one could be clever.

Later in this section it will be shown that if $\mathscr{F}: L^2(\mathbb{R}) \to L^2(\mathbb{R})$ is the Fourier–Plancherel transform, then $\mathscr{F}$ is a unitary operator (6.17) and $\mathscr{F}^{-1}A\mathscr{F} =$ the operator on $L^2(\mathbb{R})$ of multiplication by the independent variable (6.18). Thus $\mathscr{F}^{-1}U(t)\mathscr{F}$ is multiplication by e^{itx}. But it is possible to find $U(t)$ directly.

Recall that if $f \in \mathrm{dom}\, A$,

$$Af = -i\lim_{t \to 0} \frac{U(t)f - f}{t}.$$

So

$$f'(x) = \lim_{t \to 0} -\frac{(U(t)f)(x) - f(x)}{t}.$$

Being clever, one might guess that $(U(t)f)(x) = f(x - t)$.

6.2. Theorem. *If A and $\mathscr{D}$ are as in Example 6.1 and $U(t) = \exp(itA)$, then $(U(t)f)(x) = f(x - t)$ for all f in $L^2(\mathbb{R})$ and x, t in $\mathbb{R}$.*

PROOF. Let $(V(t)f)(x) = f(x - t)$. It is easy to see that V is a strongly continuous one parameter unitary group. Let B be the infinitesimal generator of V. It must be shown that $B = A$.

Note that $f \in \mathrm{dom}\, B$ if and only if $\lim_{t \to 0} t^{-1}(V(t)f - f)$ exists. Let $f \in C_c^{(1)}(\mathbb{R})$; that is, f is continuously differentiable and has compact support. Thus for $t > 0$,

$$\left[\frac{V(t)f - f}{t}\right](x) = \frac{f(x - t) - f(x)}{t} = -\frac{1}{t}\int_{x-t}^{x} f'(y)\,dy$$

and

$$\left|\frac{V(t)f(x) - f(x)}{t} + f'(x)\right| \leq \frac{1}{t}\int_{x-t}^{x} |f'(x) - f'(y)|\,dy$$

$$\leq \sup\{|f'(x) - f'(y)| : |x - y| \leq t\}.$$

Because f' is continuous with compact support, f' is uniformly continuous. Let $K = \{x: \operatorname{dist}(x, \operatorname{spt} f') \leqslant 2\}$. So K is compact. For $\varepsilon > 0$, let $\delta(\varepsilon) < 1$ be such that if $|x - y| < \delta(\varepsilon)$, then $|f'(x) - f'(y)| < \varepsilon$. Hence $\|t^{-1}[Vf - f] + f'\|_2 \leqslant \varepsilon^2 |K|$, where $|K| =$ the Lebesgue measure of K. Thus $C_c^{(1)}(\mathbb{R}) \subseteq \operatorname{dom} B$ and $Bf = Af$ for f in $C_c^{(1)}(\mathbb{R})$. But if $f \in \operatorname{dom} A$, there is a sequence $\{f_n\} \subseteq C_c^{(1)}(\mathbb{R})$ such that $f_n \oplus Af_n \to f \oplus Af$ in gra A (Exercise 1). But $f_n \oplus Af_n = f_n \oplus Bf_n \in$ gra B, so $f \oplus Af \in$ gra B; that is, $A \subseteq B$. Since self-adjoint operators are maximal symmetric operators (2.11), $A = B$. ∎

To show that the Fourier transform demonstrates that M_x and $i\, d/dx$ are unitarily equivalent, we introduce the Schwartz space of rapidly decreasing functions.

6.3. Definition. A function $\phi \colon \mathbb{R} \to \mathbb{R}$ is *rapidly decreasing* if ϕ is infinitely differentiable and for all integers $m, n \geqslant 0$,

6.4
$$\|\phi\|_{m,n} \equiv \sup\{|x^m \phi^{(n)}(x)|: x \in \mathbb{R}\} < \infty.$$

Let $\mathscr{S} = \mathscr{S}(\mathbb{R})$ be the set of all rapidly decreasing functions on $\mathbb{R}$.

Note that if $\phi \in \mathscr{S}$, then for all $m, n \geqslant 0$ there is a constant $C_{m,n}$ such that

$$|\phi^{(n)}(x)| \leqslant C_{m,n}|x|^{-m}.$$

Thus if p is any polynomial and $n \geqslant 0$, $|p(x)\phi^{(n)}(x)| \to 0$ as $|x| \to \infty$. In fact, this is equivalent to ϕ belonging to $\mathscr{S}$ (Exercise 3). Also note that if $\phi \in \mathscr{S}$, then $x^m \phi^{(n)} \in \mathscr{S}$ for all $m, n \geqslant 0$.

It is not difficult to see that $\|\cdot\|_{m,n}$ is a seminorm on $\mathscr{S}$. Also, $\mathscr{S}$ with all of these seminorms is a Fréchet space (Exercise 2). The space $\mathscr{S}$ is sometimes called the *Schwartz space* after Laurent Schwartz.

6.5. Proposition. *If* $1 \leqslant p \leqslant \infty$, $\mathscr{S} \subseteq L^p(\mathbb{R})$. *If* $1 \leqslant p < \infty$, $\mathscr{S}$ *is dense in* $L^p(\mathbb{R})$; $\mathscr{S}$ *is weak-star dense in* $L^\infty(\mathbb{R})$.

PROOF. We already have that $\mathscr{S} \subseteq L^\infty(\mathbb{R})$. If $1 \leqslant p < \infty$ and $\phi \in \mathscr{S}$, then

$$\int_{-\infty}^{\infty} |\phi|^p \, dx = \int_{-\infty}^{\infty} (1 + x^2)^{-p}(1 + x^2)^p |\phi|^p \, dx$$

$$\leqslant \|(1 + x^2)^p |\phi|^p\|_\infty \int_{-\infty}^{\infty} (1 + x^2)^{-p} \, dx.$$

Since $(1 + x^2)^p \geqslant 1 + x^2$,

$$\|\phi\|_p \leqslant \pi^{1/p} \|(1 + x^2)\phi\|_\infty$$
$$\leqslant \pi^{1/p}[\|\phi\|_{0,0} + \|\phi\|_{2,0}].$$

Since $C_c^{(\infty)}(\mathbb{R}) \subseteq \mathscr{S}$, the density statements are immediate. ∎

6.6. Definition. If $f \in L^1(\mathbb{R})$, the *Fourier transform* of f is the function $\hat{f}$ defined by

$$\hat{f}(x) = \frac{1}{\sqrt{2\pi}} \int_{\mathbb{R}} f(t)e^{-ixt} dt.$$

Because $f \in L^1(\mathbb{R})$, this integral is well defined.

The interested reader may want to peruse §VII.9, where the Fourier transform is presented in the more general context of locally compact abelian groups. That section will not be assumed here.

Recall that if $f, g \in L^1$, then the convolution of f and g,

$$f * g(x) = (2\pi)^{-1/2} \int_{\mathbb{R}} f(x - t)g(t)dt,$$

belongs to $L^1(\mathbb{R})$ and $\| f * g \|_1 \leqslant \| f \|_1 \| g \|_1$ if the norm of a function f in $L^1(\mathbb{R})$ is defined as $\| f \|_1 \equiv (2\pi)^{-1/2} \int |f(x)| dx$. It is also true that if $f \in L^p(\mathbb{R})$, $1 \leqslant p \leqslant \infty$, then $f * g \in L^p(\mathbb{R})$ and $\| f * g \|_p \leqslant \| f \|_p \| g \|_1$ (see Exercise 4).

6.7. Theorem. (a) *If $f \in L^1(\mathbb{R})$, then $\hat{f}$ is a continuous function on $\mathbb{R}$ that vanishes at $\pm \infty$. Also, $\| \hat{f} \|_\infty \leqslant \| f \|_1$.*
(b) *If $\phi \in \mathscr{S}$, $\hat{\phi} \in \mathscr{S}$. Also for $m, n \geqslant 0$,*

6.8 $$(ix)^m \left(\frac{d}{dx}\right) \hat{\phi} = \left[\left(\frac{d}{dx}\right)^m ((-ix)^n \phi)\right]^{\char94}.$$

(c) *If $f, g \in L^1(\mathbb{R})$, then $(f * g)^{\char94} = \hat{f}\hat{g}$. If f and $g \in \mathscr{S}$, then $f * g \in \mathscr{S}$.*
(*Note:* [ ]^ $=$ *the Fourier transform of the function defined in the brackets.*)

PROOF. (a) The fact that $\hat{f}$ is continuous is an easy consequence of Lebesgue's Dominated Convergence Theorem; it is clear that $\| \hat{f} \|_\infty \leqslant \| f \|_1$. For the other part of (a), let $f =$ the characteristic function of the interval (a, b). Then $\hat{f}(x) = i(2\pi)^{-1/2}x^{-1}[e^{-ixb} - e^{-ixa}] \to 0$ as $|x| \to \infty$. So $\hat{f}(x)$ vanishes at $\pm \infty$ if f is a linear combination of such characteristic functions. The result for a general f follows by approximation.

(b) It is convenient to introduce the notation $D\phi = \phi'$. Thus $D^n\phi = \phi^{(n)}$. Also in this proof, as in many others of this section, x will be used to denote the function whose value at t is t and it will also be used occasionally to denote the independent variable.

If $\phi \in \mathscr{S}$, then differentiation under the integral sign (Why is this justified?) gives

$$(D\hat{\phi})(y) = \frac{1}{\sqrt{2\pi}} \int_{-\infty}^{\infty} (-it)e^{-iyt}\phi(t)dt$$

$$= [(-ix)\phi]^{\char94}(y).$$

By induction we get that for $n \geqslant 0$,

6.9 $$D^n\hat{\phi} = [(-ix)^n\phi]^{\char94}.$$

Using integration by parts,

$$(D\phi)\hat{\ }(y) = \frac{1}{\sqrt{2\pi}} \int_{-\infty}^{\infty} e^{-iyt}\phi'(t)dt$$

$$= \frac{-1}{\sqrt{2\pi}} \int_{-\infty}^{\infty} \phi(t)\frac{d}{dt}[e^{-iyt}]dt$$

$$= \frac{iy}{\sqrt{2\pi}} \int_{-\infty}^{\infty} e^{-iyt}\phi(t)dt.$$

That is, $(D\phi)\hat{\ } = (ix)\hat{\phi}$. By induction,

6.10 $$(D^n\phi)\hat{\ } = (ix)^n\hat{\phi}$$

for all $n \geqslant 0$. Combining (6.9) and (6.10) gives (6.8).

By (6.8) if $m, n \geqslant 0$, then for ϕ in $\mathscr{S}$,

$$\|\hat{\phi}\|_{m,n} = \sup\{|x^m(D^n\hat{\phi})(x)|: x \in \mathbb{R}\}$$

$$= \sup\left\{\left|\frac{1}{\sqrt{2\pi}} \int_{-\infty}^{\infty} e^{-ixt}\left(\frac{d}{dt}\right)^m[(-it)^n\phi(t)]dt\right|: x \in \mathbb{R}\right\}$$

$$\leqslant \frac{1}{\sqrt{2\pi}} \int_{-\infty}^{\infty} \left|\left(\frac{d}{dt}\right)^m[t^n\phi(t)]\right|dt$$

$$< \infty$$

since $D^m(x^n\phi) \in L^1(\mathbb{R})$ (6.5).

(c) This is an easy exercise in integration theory and is left to the reader. ∎

The fact that $\hat{f}(x) \to 0$ as $|x| \to \infty$ is called the *Riemann–Lebesgue Lemma.*

The process now begins whereby it will be shown that the Fourier transform on $L^1 \cap L^2$ extends to a unitary operator on $L^2(\mathbb{R})$. Moreover, the adjoint of this unitary will be calculated and it will be shown that if id/dx is conjugated by this unitary, then the resulting self-adjoint operator is M_x.

Changing notation a little, let U_y [instead of $U(y)$] denote the translation operator. Moreover, think of U_y as operating on all of the L^p spaces, not just L^2, so $(U_yf)(x) = f(x-y)$ for f in $L^p(\mathbb{R})$. Also, let e_y be the function $e_y(x) = \exp(ixy)$.

6.11. Proposition. *If $f \in L^1(\mathbb{R})$ and $y \in \mathbb{R}$, then*

$$[U_yf]\hat{\ } = e_{-y}\hat{f},$$
$$[e_yf]\hat{\ } = U_y\hat{f}.$$

PROOF. If $f \in L^1(\mathbb{R})$,

$$[U_y f]\hat{}(x) = (2\pi)^{-1/2} \int [U_y f](t) e^{-ixt} dt$$

$$= (2\pi)^{-1/2} \int f(t-y) e^{-ixt} dt$$

$$= (2\pi)^{-1/2} \int f(s) e^{-ix(s+y)} ds$$

$$= e_{-y}(x) \hat{f}(x).$$

The proof of the other equation is left as an exercise. ∎

In the proof of the next lemma the fact that $\int_{-\infty}^{\infty} e^{-t^2} dt = \sqrt{\pi}$ is needed. Those who have never seen this can verify it by putting $I = \int_0^{\infty} e^{-x^2} dx$, noting that $I^2 = \int_0^{\infty} \int_0^{\infty} e^{-(x^2+y^2)} dx\,dy$, and using polar coordinates.

6.12. Lemma. *If $\varepsilon > 0$ and $\rho_\varepsilon(t) = e^{-\varepsilon^2 t^2}$, then*

$$\hat{\rho}_\varepsilon(x) = \frac{1}{\varepsilon\sqrt{2}} e^{-x^2/4\varepsilon^2}.$$

PROOF. Note that $\rho_\varepsilon \in \mathcal{S}$. By (6.8), $D\hat{\rho}_\varepsilon = (-ix\rho_\varepsilon)\hat{}$. Using integration by parts,

$$(D\hat{\rho}_\varepsilon)(x) = \frac{-i}{\sqrt{2\pi}} \int_{-\infty}^{\infty} e^{-\varepsilon^2 t^2} t e^{-ixt} dt$$

$$= \frac{-i}{\sqrt{2\pi}} \left(\frac{-1}{2\varepsilon^2} \right) \int_{-\infty}^{\infty} e^{-ixt} d(e^{-\varepsilon^2 t^2})$$

$$= \frac{-i}{2\varepsilon^2\sqrt{2\pi}} \int_{-\infty}^{\infty} e^{-\varepsilon^2 t^2} (-ix) e^{-ixt} dt$$

$$= \frac{-x}{2\varepsilon^2} \hat{\rho}_\varepsilon(x).$$

Let $\psi_\varepsilon(x) = e^{-x^2/4\varepsilon^2}$. Then both $\hat{\rho}_\varepsilon$ and ψ_ε satisfy the differential equation $u'(x) = -(x/2\varepsilon^2) u(x)$. Hence $\hat{\rho}_\varepsilon = c\psi_\varepsilon$ for some constant c. But $\psi_\varepsilon(0) = 1$, and

$$\hat{\rho}_\varepsilon(0) = \frac{1}{\sqrt{2\pi}} \int_{-\infty}^{\infty} e^{-\varepsilon^2 t^2} dt$$

$$= \frac{1}{\varepsilon\sqrt{2\pi}} \int_{-\infty}^{\infty} e^{-s^2} ds$$

$$= \frac{1}{\varepsilon\sqrt{2\pi}} \sqrt{\pi} = \frac{1}{\varepsilon\sqrt{2}}. \qquad \blacksquare$$

6.13. Proposition. *If* $\psi \in L^1(\mathbb{R})$ *such that* $(2\pi)^{-1/2} \int_{\mathbb{R}} \psi(x) dx = 1$ *and if for* $\varepsilon > 0$, $\psi_\varepsilon(x) = \varepsilon^{-1} \psi(x/\varepsilon)$, *then for every* f *in* $C_0(\mathbb{R})$, $\psi_\varepsilon * f(x) \to f(x)$ *uniformly on* $\mathbb{R}$.

PROOF. Note that $(2\pi)^{-1/2} \int \psi_\varepsilon(x) dx = 1$ for all $\varepsilon > 0$. Hence for any x in $\mathbb{R}$,

$$\psi_\varepsilon * f(x) - f(x) = (2\pi)^{-1/2} \int [f(x-t) - f(x)] \frac{1}{\varepsilon} \psi\left(\frac{t}{\varepsilon}\right) dt$$

$$= (2\pi)^{-1/2} \int [f(x - s\varepsilon) - f(x)] \psi(s) ds.$$

Put $\omega(y) = \sup\{|f(x-y) - f(x)|: y \in \mathbb{R}\}$. Now f is uniformly continuous (Why?), so if $\varepsilon > 0$, then there is a $\delta > 0$ such that $\omega(y) < \varepsilon$ if $|y| < \delta$. Thus $\omega(y) \to 0$ as $|y| \to 0$. Moreover, the inequality above implies

$$\|\psi_\varepsilon * f - f\|_\infty \leqslant (2\pi)^{-1/2} \int \omega(s\varepsilon) |\psi(s)| ds.$$

Since $\psi \in L^1(\mathbb{R})$, the Lebesgue Dominated Convergence Theorem implies that $\|\psi_{\varepsilon_k} * f - f\|_\infty \to 0$ whenever $\varepsilon_k \to 0$. This proves the proposition. ∎

The next result is often called the *Multiplication Formula*. Remember that if $f \in L^1(\mathbb{R})$, $\hat{f} \in C_0(\mathbb{R})$. Hence $\hat{f}g \in L^1(\mathbb{R})$ when both f and $g \in L^1(\mathbb{R})$.

6.14. Theorem. *If* $f, g \in L^1(\mathbb{R})$, *then*

$$\int_{\mathbb{R}} \hat{f}(x) g(x) dx = \int_{\mathbb{R}} f(x) \hat{g}(x) dx.$$

PROOF. The proof is an easy consequence of Fubini's Theorem. In fact, if $f, g \in L^1(\mathbb{R})$, then

$$\int \hat{f}(x) g(x) dx = \int \left[\frac{1}{\sqrt{2\pi}} \int f(t) e^{-ixt} dt \right] g(x) dx$$

$$= \int f(t) \left[\frac{1}{\sqrt{2\pi}} \int g(x) e^{-ixt} dx \right] dt$$

$$= \int f(t) \hat{g}(t) dt. \qquad ∎$$

6.15. Inversion Formula. *If* $\phi \in \mathcal{S}$, *then*

$$\phi(x) = \frac{1}{\sqrt{2\pi}} \int_{-\infty}^{\infty} \hat{\phi}(t) e^{ixt} dt.$$

PROOF. Let $\rho_\varepsilon(x) = e^{-\varepsilon^2 x^2}$ and put $\psi(x) = \hat{\rho}_1(x)$. Then by Lemma 6.12

$\psi_\varepsilon(x) = \varepsilon^{-1}\psi(x/\varepsilon) = \hat{\rho}_\varepsilon(x)$. Also,

$$(2\pi)^{-1/2} \int \psi(x)\,dx = (2\pi)^{-1/2} \int_{-\infty}^{\infty} 2^{-1/2} e^{-x^2/4}\,dx = 1.$$

So $\psi_\varepsilon * h(x) \to h(x)$ uniformly for any h in $C_0(\mathbb{R})$. If $\phi \in \mathscr{S}$, put $f = \phi$ and $g = e_x \rho_\varepsilon$ in (6.14). By Proposition 6.11 and Lemma 6.12, $\hat{g} = U_x \hat{\rho}_\varepsilon = U_x \psi_\varepsilon$. Thus

$$\frac{1}{\sqrt{2\pi}} \int_{-\infty}^{\infty} \hat{\phi}(t)e^{itx}e^{-\varepsilon^2 t^2}\,dt = \frac{1}{\sqrt{2\pi}} \int_{-\infty}^{\infty} \phi(t)\psi_\varepsilon(t-x)\,dt$$

$$= \phi * \psi_\varepsilon(x)$$

$$\to \phi(x)$$

as $\varepsilon \to 0$. The Lebesgue Dominated Convergence Theorem implies the left-hand side converges to $(2\pi)^{-1/2}\int \hat{\phi}(t)e^{ixt}\,dt$ and the theorem is proved. ∎

In many ways the next result is a rephrasing of the preceding theorem.

6.16. Theorem. *If $\mathscr{F}: \mathscr{S} \to \mathscr{S}$ is defined by $\mathscr{F}\phi = \hat{\phi}$, $\mathscr{F}$ is a bijection with*

$$(\mathscr{F}^{-1}\phi)(x) = \frac{1}{\sqrt{2\pi}} \int_{-\infty}^{\infty} \phi(t)e^{ixt}\,dt.$$

Moreover, if $\mathscr{S}$ is given the topology induced by the seminorms $\{\|\cdot\|_{m,n}: m, n \geq 0\}$ that were defined in (6.4), $\mathscr{F}$ is a homeomorphism.

PROOF. By (6.7b), $\mathscr{F}\mathscr{S} \subseteq \mathscr{S}$. The preceding theorem says that $\mathscr{F}$ is bijective and gives the formula for $\mathscr{F}^{-1}$. The proof of the topological statement is left to the reader. ∎

6.17. Plancherel's Theorem. *If $\phi \in \mathscr{S}$, then $\|\phi\|_2 = \|\hat{\phi}\|_2$ and the Fourier transform $\mathscr{F}$ extends to a unitary operator on $L^2(\mathbb{R})$.*

PROOF. Let $\phi \in \mathscr{S}$ and put $\psi(x) = \overline{\phi(-x)}$. So $\rho = \phi * \psi \in L^1(\mathbb{R})$ and $\hat{\rho} = \hat{\phi}\hat{\psi}$. An easy calculation shows that $\hat{\psi} = \overline{\hat{\phi}}$; hence $\hat{\rho} = |\hat{\phi}|^2$. Also, the Inversion Formula shows that $\rho(0) = (2\pi)^{-1/2}\int \hat{\rho}(x)\,dx = (2\pi)^{-1/2}\int |\hat{\phi}(x)|^2\,dx$. Thus

$$\int |\hat{\phi}(x)|^2\,dx = (2\pi)^{1/2}\rho(0)$$

$$= (2\pi)^{1/2}\phi * \psi(0)$$

$$= \int \phi(x)\psi(0-x)\,dx$$

$$= \int |\phi(x)|^2\,dx.$$

So if $\mathscr{S}$ is considered as a subspace of $L^2(\mathbb{R})$, $\mathscr{F}$, the Fourier transform, is an isometry on $\mathscr{S}$. By Proposition 6.5 and the preceding theorem, $\mathscr{F}$ extends to a unitary operator on $L^2(\mathbb{R})$.   ∎

Warning! The content of the Plancherel Theorem is that the Fourier transform extends to an isometry. The formula for this isometry is not given by the formula for the Fourier transform. Indeed, this formula does not make sense when f is not an L^1 function. However, the same symbol, $\mathscr{F}$, will be used to denote this unitary operator on $L^2(\mathbb{R})$. For emphasis it is called the Plancherel transform.

6.18. Theorem. *Let A be the operator on $L^2(\mathbb{R})$ given by $Af = if'$ and let M be the operator defined by $Mf = xf$. If $\mathscr{F}: L^2(\mathbb{R}) \to L^2(\mathbb{R})$ is the Plancherel Transform, then $\mathscr{F}$ dom M = dom A and*

$$\mathscr{F}^{-1} A \mathscr{F} = M.$$

PROOF. The fact that $A\mathscr{F} = \mathscr{F}M$ on $\mathscr{S}$ is an immediate consequence of Theorem 6.7(b). Since $\mathscr{S}$ is dense in both dom A and dom M, the rest of the result follows (with some work—give the details).   ∎

Fourier analysis is a subject unto itself. One source is Stein and Weiss [1971]; another is Reed and Simon [1975].

EXERCISES

1. If $\mathscr{D}$ is as in Example 6.1, show that for every f in $\mathscr{D}$ there is a sequence $\{f_n\}$ in $C_c^{(1)}(\mathbb{R})$ such that $f_n \to f$ and $f_n' \to f'$ in $L^2(\mathbb{R})$.

2. Show that the Schwartz space $\mathscr{S}$ with the seminorms $\{\|\cdot\|_{m,n}: m, n \geq 0\}$ is a Fréchet space.

3. If ϕ is infinitely differentiable on $\mathbb{R}$, show that $\phi \in \mathscr{S}$ if and only if for every integer $n \geq 0$ and every polynomial p, $\phi^{(n)}(x)p(x) \to 0$ as $|x| \to \infty$.

4. If $f \in L^p(\mathbb{R})$, $1 \leq p \leq \infty$, and $g \in L^1(\mathbb{R})$, show that $f*g \in L^p(\mathbb{R})$ and $\|f*g\|_p \leq \|f\|_p \|g\|_1$. (Hint: See Dunford and Schwartz [1958], p. 530, Exercise 13 for a generalization of Minkowski's Inequality.)

5. If ψ and ψ_ε are as Proposition 6.13 and $f \in L^p(\mathbb{R})$, $1 \leq p < \infty$, show that $\|f*\psi_\varepsilon - f\|_p \to 0$ as $\varepsilon \to 0$. If $f \in L^\infty(\mathbb{R})$, show that $f*\psi_\varepsilon \to f$ (weak*).

6. If $f \in L^1(\mathbb{R})$ and $\hat{f} \in L^1(\mathbb{R})$, show that $f(x) = (2\pi)^{-1/2} \int_{\mathbb{R}} \hat{f}(t) e^{ixt} \, dt$ a.e.

7. If $\mathscr{F}: L^2(\mathbb{R}) \to L^2(\mathbb{R})$ is the Plancherel Transform and $f \in L^2(\mathbb{R})$, show that $(\mathscr{F}^{-1}f)(x) = (\mathscr{F}f)(-x)$.

8. Show that $\mathscr{F}^4 = 1$ but $\mathscr{F}^2 \neq 1$. What does this say about $\sigma(\mathscr{F})$?

9. Find the Fourier transform of the Hermite polynomials. What do you think? (This exercise is broken up into a series of easier steps on pages 98–99 of Dym and McKean [1972].)

§7. Moments

To understand this section, the preceding two sections are unnecessary.

Let μ be a positive Borel measure on $\mathbb{R}$ such that $\int |t|^n \, d\mu(t) = m_n < \infty$ for every $n \geq 0$. The numbers $\{m_n\}$ are called the *moments* of μ in analogy with the corresponding concept from mechanics. The central problem here, called the *Hamburger moment problem*, is to characterize those sequences of numbers that are moment sequences. Just as self-adjoint operators are connected to measures, the theory of self-adjoint operators is connected to the solution of this moment problem.

7.1. Theorem. *If $\{m_n : n \geq 0\}$ is a sequence of real numbers, the following statements are equivalent.*

(a) *There is a positive regular Borel measure μ on $\mathbb{R}$ such that $\int |t|^n \, d\mu(t) < \infty$ for all $n \geq 0$ and $m_n = \int t^n \, d\mu(t)$.*

(b) *If $\alpha_0, \ldots, \alpha_n \in \mathbb{C}$, then $\sum_{j,k=0}^n m_{j+k} \alpha_j \bar{\alpha}_k \geq 0$.*

(c) *There is a self-adjoint operator A and a vector e such that $e \in \text{dom } A^n$ for all n and $m_n = \langle A^n e, e \rangle$ for all $n \geq 0$.*

Before proving this theorem, a preliminary result is needed. This result is useful in many other situations and is one of the standard ways to show that a symmetric operator has a self-adjoint extension.

7.2. Proposition. *Let T be a symmetric operator on $\mathcal{H}$ and suppose there is a function $J: \mathcal{H} \to \mathcal{H}$ having the following properties:*

(a) *J is conjugate linear (that is, $J(h + g) = Jh + Jg$ and $J(\alpha h) = \bar{\alpha} Jh$);*

(b) *$J^2 = 1$;*

(c) *J is continuous;*

(d) *$J \, \text{dom } T \subseteq \text{dom } T$ and $TJ \subseteq JT$.*

Then T has a self-adjoint extension.

PROOF. First note that if $h \in \text{dom } T$, then $Jh \in \text{dom } T$ and $h = J(Jh)$. Hence $J \, \text{dom } T = \text{dom } T$ and $JT = TJ$.

Let $h \in \mathcal{H}$ and define $L: \mathcal{H} \to \mathbb{C}$ by $L(f) = \langle h, Jf \rangle$. Since J is conjugate linear, L is a linear functional. By (c), L is continuous. Thus there is a unique vector h^* in $\mathcal{H}$ such that $L(f) = \langle f, h^* \rangle$. Let $J^* h = h^*$. Thus $J^*: \mathcal{H} \to \mathcal{H}$ and

7.3
$$\langle f, J^* h \rangle = \langle h, Jf \rangle.$$

It is clear that J^* is additive. If $\alpha \in \mathbb{C}$, then $\langle f, J^*(\alpha h) \rangle = \langle \alpha h, Jf \rangle = \alpha \langle f, J^* h \rangle = \langle f, \bar{\alpha} J^* h \rangle$. Thus J^* is conjugate linear. Since $J^2 = 1$, it follows that $J^{*2} = 1$.

Let $h \in \text{dom } T^*$ and $f \in \text{dom } T$. Then $\langle TJf, h \rangle = \langle Jf, T^* h \rangle = \langle J^* T^* h, f \rangle$ by (7.3). But also by (d), $\langle TJf, h \rangle = \langle JTf, h \rangle = \langle J^* h, Tf \rangle$. So $\langle J^* T^* h, f \rangle = \langle J^* h, Tf \rangle$ for all h in $\text{dom } T^*$ and f in $\text{dom } T$. But this says that $J^* h \in \text{dom } T^*$

whenever $h \in \operatorname{dom} T^*$ and, furthermore, $T^*J^*h = J^*T^*h$. Since $J^{*2} = 1$, it follows that $J^* \operatorname{dom} T^* = \operatorname{dom} T^*$ and $J^*T^* = T^*J^*$.

Now let $h \in \ker(T^* \pm i)$. Then $T^*J^*h = J^*T^*h = J^*(\pm ih) = \mp iJ^*h$. Thus $J^* \ker(T^* \pm i) \subseteq \ker(T^* \mp i)$. Since $J^{*2} = 1$, $J^* \ker(T^* \pm i) = \ker(T^* \mp i)$. But J^* is injective. Indeed, if $J^*h = 0$, then $h = J^*(J^*h) = 0$. Thus the deficiency indices of T are equal. By Theorem 2.20, T has a self-adjoint extension. ∎

PROOF OF THEOREM 7.1. (a) *implies* (b). If $\alpha_0, \ldots, \alpha_n \in \mathbb{C}$, then

$$\sum_{j,k=0}^{n} m_{j+k} \alpha_j \bar{\alpha}_k = \int \sum_{j,k=0}^{n} \alpha_j \bar{\alpha}_k t^{j+k} \, d\mu(t)$$

$$= \int \left(\sum_{j=0}^{n} \alpha_j t^j \right) \left(\sum_{k=0}^{n} \bar{\alpha}_k t^k \right) d\mu(t)$$

$$= \int \left| \sum_{k=0}^{n} \alpha_k t^k \right|^2 d\mu(t) \geq 0.$$

(b) *implies* (c). Let $\mathcal{H}_0 =$ the collection of all finitely nonzero sequences of complex numbers $\{\alpha_n : n \geq 0\}$. That is, $\{\alpha_0, \alpha_1, \ldots\} \in \mathcal{H}_0$ if $\alpha_n \in \mathbb{C}$ for all $n \geq 0$ and $\alpha_n = 0$ for all but a finite number of values of n. If $x = \{\alpha_n\}$, $y = \{\beta_n\} \in \mathcal{H}_0$ define $[x, y]$ by

7.4
$$[x, y] \equiv \sum_{j,k=0}^{\infty} m_{j+k} \alpha_j \bar{\beta}_k.$$

It is easy to see that $\mathcal{H}_0$ is a vector space and (7.4) defines a semi-inner product on $\mathcal{H}_0$. In fact, it is routine that $[\cdot, \cdot]$ is sesquilinear and condition (b) implies that $[x, x] \geq 0$ for all x in $\mathcal{H}_0$.

Let $\mathcal{K}_0 = \{x \in \mathcal{H}_0 : [x, x] = 0\}$ and let $\mathcal{H}_1$ be the quotient vector space $\mathcal{H}_0 / \mathcal{K}_0$. If $h = x + \mathcal{K}_0$ and $f = y + \mathcal{K}_0 \in \mathcal{H}_1$, then

7.5
$$\langle h, f \rangle \equiv [x, y]$$

can be verified to be a well-defined inner product on $\mathcal{H}_1$. Let $\mathcal{H}$ be the Hilbert space obtained by completing $\mathcal{H}_1$ with respect to the norm defined by the inner product (7.5).

Now to define some operators. If $x = \{\alpha_n\} \in \mathcal{H}_0$, let $T_0 x = \{0, \alpha_0, \alpha_1, \ldots\}$. It is easy to check that T_0 is a linear transformation on $\mathcal{H}_0$. Also, if $x = \{\alpha_n\}$, $y = \{\beta_n\} \in \mathcal{H}_0$, let $T_0 x = \{\gamma_n\}$. So $\gamma_0 = 0$ and $\gamma_n = \alpha_{n-1}$ if $n \geq 1$. Hence

$$[T_0 x, y] = \sum_{j,k=0}^{\infty} m_{j+k} \gamma_j \bar{\beta}_k$$

$$= \sum_{\substack{j=1 \\ k=0}}^{\infty} m_{j+k} \alpha_{j-1} \bar{\beta}_k$$

$$= \sum_{j,k=0}^{\infty} m_{j+k+1} \alpha_j \bar{\beta}_k$$

$$= \sum_{\substack{j=0 \\ k=1}}^{\infty} m_{j+k} \alpha_j \bar{\beta}_{k-1}$$

$$= [x, T_0 y].$$

In particular, if $x \in \mathcal{K}_0$, then the preceding equation and the CBS inequality imply that

$$|[T_0 x, T_0 x]| = |[T_0^2 x, x]| \leqslant [T_0^2 x, T_0^2 x][x, x]$$
$$= 0$$

Hence $T_0 \mathcal{K}_0 \subseteq \mathcal{K}_0$. Thus T_0 induces a linear transformation T on $\mathcal{H}_1$ defined by $T(x + \mathcal{K}_0) = T_0 x + \mathcal{K}_0$. It follows that $\langle Th, f \rangle = \langle h, Tf \rangle$ for all h, f in $\mathcal{H}_1$. Since $\mathcal{H}_1$ is, by definition, dense in $\mathcal{H}$, T is a densely defined symmetric operator on $\mathcal{H}$. Now to show that T has a self-adjoint extension.

Define $J_0: \mathcal{H}_0 \to \mathcal{H}_0$ by $J_0(\{\alpha_n\}) = \{\bar{\alpha}_n\}$. It is easy to see that J_0 is conjugate linear and $J_0^2 = 1$. Also, $J_0 T_0 = T_0 J_0$. An easy calculation shows that $[J_0 x, J_0 y] = [x, y]$ for all x, y in $\mathcal{H}_0$. So $J_0 \mathcal{K}_0 \subseteq \mathcal{K}_0$ and J_0 induces a conjugate linear function $J_1: \mathcal{H}_1 \to \mathcal{H}_1$ defined by $J_1(x + \mathcal{K}_0) = J_0 x + \mathcal{K}_0$. It follows that $J_1 T = T J_1$, $J_1^2 = 1$, and $\|J_1 h\| = \|h\|$ for all h in $\mathcal{H}_1$. Thus J_1 extends to a conjugate linear $J: \mathcal{H} \to \mathcal{H}$ such that $J^2 = 1$ and $\|Jh\| = \|h\|$ for all h in $\mathcal{H}$. Hence J is continuous. Also, $J \operatorname{dom} T = J_1 \mathcal{H}_1 \subseteq \mathcal{H}_1 = \operatorname{dom} T$ and $TJ \subseteq JT$. By Proposition 7.2, T has a self-adjoint extension A.

Let $e_0 = \{1, 0, 0, \dots\} \in \mathcal{H}_0$. Hence $T_0^n e_0$ has a 1 in the nth place and zeros elsewhere. If $e = e_0 + \mathcal{K}_0$, then $e \in \operatorname{dom} T^n \subseteq \operatorname{dom} A^n$ for all $n \geqslant 0$. Also,

$$\langle A^n e, e \rangle = [T_0^n e_0, e_0] = m_n$$

for $n \geqslant 0$.

(c) *implies* (a). By the Spectral Theorem there is a spectral measure E for A. Let $\mu = E_{e,e}$; by (4.11) μ is supported on $\mathbb{R}$ and, since $e \in \operatorname{dom} A$, μ is finite. Moreover, since $e \in \operatorname{dom} A^n$ for every $n \geqslant 0$ it follows (supply the details) that $\int t^n d\mu(t) < \infty$ for every $n \geqslant 0$. Finally, by (4.9), $m_n = \langle A^n e, e \rangle = \int t^n d\mu(t)$ for every $n \geqslant 0$. ∎

The measure obtained in Theorem 7.1 need not be unique since, in the proof that (b) implies (c) above, the self-adjoint extension of T may not be unique. See pages 201–202 of Berg, Christensen, and Ressel [1984] for an example as well as further discussions of moment problems.

EXERCISES

1. (Stieltjes.) Let $\{m_n : n \geqslant 0\}$ be a sequence of real numbers and show that the following statements are equivalent. (a) There is a positive regular Borel measure μ on $[0, \infty)$ such that $m_n = \int t^n d\mu(t)$ for all $n \geqslant 0$. (b) If $\alpha_0, \dots, \alpha_n \in \mathbb{C}$, then $\sum_{j,k=0}^n m_{j+k} \alpha_j \bar{\alpha}_k \geqslant 0$ and $\sum_{j,k=0}^n m_{j+k+1} \alpha_j \bar{\alpha}_k \geqslant 0$. (c) There is a self-adjoint operator A with $\sigma(A) \subseteq [0, \infty)$ and a vector e in $\operatorname{dom} A^n$ for all $n \geqslant 0$ such that $m_n = \langle A^n e, e \rangle$ for $n \geqslant 0$.

2. (Bochner.) Let $m: \mathbb{R} \to \mathbb{C}$ be a function and show that the following statements are

equivalent. (a) There is a finite positive measure μ on $\mathbb{R}$ such that $m(t) = \int e^{ixt} d\mu(x)$ for all t in $\mathbb{R}$. (b) m is continuous and if $\alpha_0, \ldots, \alpha_n \in \mathbb{C}$ and $t_0, \ldots, t_n \in \mathbb{R}$, then $\sum_{j,k=0}^n m(t_j - t_k) \alpha_j \bar{\alpha}_k \geq 0$. (c) There is a strongly continuous one-parameter unitary group $U(t)$ and a vector e such that $m(t) = \langle U(t)e, e \rangle$ for all t. (Hint: Let $\mathscr{H}_0 = $ all functions $f: \mathbb{R} \to \mathbb{C}$ that vanish off a finite set.)

3. Let $\{m_n : n \in \mathbb{Z}\} \subseteq \mathbb{C}$ and show that the following statements are equivalent. (a) There is a positive measure μ on $\partial \mathbb{D}$ such that $m_n = \int z^n d\mu(z)$ for all n in $\mathbb{Z}$. (b) If $\alpha_{-n}, \ldots, \alpha_{-1}, \alpha_0, \alpha_1, \ldots, \alpha_n \in \mathbb{C}$, then $\sum_{j,k=-n}^n m_{j-k} \alpha_j \bar{\alpha}_k \geq 0$. (c) There is a unitary operator U and a vector e such that $m_n = \langle U^n e, e \rangle$ for all n.

4. Show that the operator A that appears in the proof that (7.1b) implies (7.1c) is cyclic.

CHAPTER XI

Fredholm Theory

This chapter is entirely independent of the preceding one and only tangentially dependent on Chapters VIII and IX.

The purpose of this chapter is to study certain properties of operators on a Hilbert space that are invariant under compact perturbations. That is, we want to study properties of an operator A in $\mathcal{B}(\mathcal{H})$ that are also possessed by $A + K$ for every K in $\mathcal{B}_0(\mathcal{H})$. The correct view here is to consider this undertaking as a study of the quotient algebra $\mathcal{B}(\mathcal{H})/\mathcal{B}_0(\mathcal{H}) = \mathcal{B}/\mathcal{B}_0$—the *Calkin algebra*. Any property associated with an element of the Calkin algebra is a property associated with a coset of operators and vice versa. It is useful—indeed essential—to relate these properties to the way in which the operators act on the underlying Hilbert space.

§1. The Spectrum Revisited

In Section VII.6 we saw several properties of the spectrum of an operator on a Banach space. In particular, the concepts of point spectrum, $\sigma_p(A)$, and approximate point spectrum, $\sigma_{ap}(A)$, were explored. It was also shown (VII. 6.7) that $\partial\sigma(A) \subseteq \sigma_{ap}(A)$. Recall that $\sigma_l(A)$ is the left spectrum of A and $\sigma_r(A)$ is the right spectrum of A.

1.1. Proposition. *If* $A \in \mathcal{B}(\mathcal{H})$, *the following statements are equivalent.*

(a) $\lambda \notin \sigma_{ap}(A)$; *that is,* $\inf\{\|(A - \lambda)h\|: \|h\| = 1\} > 0.$
(b) $\text{ran}(A - \lambda)$ *is closed and* $\dim\ker(A - \lambda) = 0.$
(c) $\lambda \notin \sigma_l(A).$

(d) $\bar{\lambda} \notin \sigma_r(A^*)$.

(e) $\operatorname{ran}(A^* - \bar{\lambda}) = \mathcal{H}$.

PROOF. By Proposition VII.6.4, (a) and (b) are equivalent. Also, if $B \in \mathcal{B}(\mathcal{H})$, then $B(A - \lambda) = 1$ if and only if $(A^* - \bar{\lambda})B^* = 1$ so that (c) and (d) are easily seen to be equivalent.

(b) *implies* (c). Let $\mathcal{M} = \operatorname{ran}(A - \lambda)$ and define $T: \mathcal{H} \to \mathcal{M}$ by $Th = (A - \lambda)h$; then T is bijective. By the Open Mapping Theorem, $T^{-1}: \mathcal{M} \to \mathcal{H}$ is continuous. Define $B: \mathcal{H} \to \mathcal{H}$ by letting $B = T^{-1}$ on $\mathcal{M}$ and $B = 0$ on $\mathcal{M}^{\perp}$. Then $B \in \mathcal{B}(\mathcal{H})$ and $B(A - \lambda) = 1$. (Note that we used a property of Hilbert spaces here; see Exercise VII.6.5.)

(d) *implies* (e). Since $\bar{\lambda} \notin \sigma_r(A^*)$, there is an operator C in $\mathcal{B}(\mathcal{H})$ such that $(A^* - \bar{\lambda})C = 1$. Hence $\mathcal{H} = (A^* - \bar{\lambda})C\mathcal{H} \subseteq \operatorname{ran}(A^* - \bar{\lambda})$.

(e) *implies* (a). Let $\mathcal{N} = \ker(A^* - \bar{\lambda})^{\perp}$ and define $T: \mathcal{N} \to \mathcal{H}$ by $Th = (A^* - \bar{\lambda})h$. Then T is bijective and hence invertible. Let $C: \mathcal{H} \to \mathcal{H}$ be defined by $Ch = T^{-1}h$. Then $C\mathcal{H} = \mathcal{N}$ and $(A^* - \bar{\lambda})C = 1$. Thus $C^*(A - \lambda) = 1$ so that if $h \in \mathcal{H}$, $\|h\| = \|C^*(A - \lambda)h\| \leqslant \|C^*\| \|(A - \lambda)h\|$. Hence $\inf\{\|(A - \lambda)h\|: \|h\| = 1\} \geqslant \|C^*\|^{-1}$. ∎

If $\Delta \subseteq \mathbb{C}$, $\Delta^* \equiv \{\bar{\lambda}: \lambda \in \Delta\}$.

1.2. Corollary. *If $A \in \mathcal{B}(\mathcal{H})$, then $\partial\sigma(A) \subseteq \sigma_l(A) \cap \sigma_r(A) = \sigma_{ap}(A) \cap \sigma_{ap}(A^*)^*$.*

PROOF. The equality is immediate from the preceding theorem. In fact, $\sigma_l(A) = \sigma_{ap}(A)$ and $\sigma_r(A) = \sigma_l(A^*)^* = \sigma_{ap}(A^*)^*$. If $\lambda \in \partial\sigma(A)$, then (VII.6.7) $\lambda \in \sigma_{ap}(A)$. But $\bar{\lambda} \in \partial\sigma(A^*)$ so that $\bar{\lambda} \in \sigma_{ap}(A^*)$. ∎

For normal elements there is less variety. The pertinent result is proved here in a more general setting than that of operators.

1.3. Proposition. *Let $\mathscr{A}$ be a C^*-algebra with identity. If a is a normal element of $\mathscr{A}$, then the following statements are equivalent.*

(a) *a is invertible.*

(b) *a is left invertible.*

(c) *a is right invertible.*

PROOF. Assume that a is left invertible, so there is a b in $\mathscr{A}$ such that $ba = 1$. Thus for any x in $\mathscr{A}$, $\|x\| = \|bax\| \leqslant \|b\| \|ax\|$, and hence $\|ax\| \geqslant \|b\|^{-1}\|x\|$. In particular, this is true whenever $x \in C^*(a)$. Because a is normal, $C^*(a)$ is isomorphic to $C(K)$ where $K = \sigma(a)$ and where the isomorphism takes a into the function z ($z(w) = w$). The inequality above thus becomes: $\|zf\| \geqslant \|b\|^{-1}\|f\|$ for every f in $C(K)$. It must be shown that $0 \notin K$ ($= \sigma(a)$). If $0 \in K$, then for every integer n there is a function f_n in $C(K)$ such that $0 \leqslant f_n \leqslant 1$, $f_n(0) = 1$ and $f_n(z) = 0$ for z in K and $|z| \geqslant n^{-1}$. Since $0 \in K$, $\|f_n\| = 1$. But $\|zf_n\| \leqslant 1/n$. This contradicts the inequality and so $0 \notin \sigma(a)$; that is, a is invertible.

The argument above shows that (b) implies (a). If a is right invertible, then

a^* is left invertible. By the preceding argument a^* is invertible, and hence so is a. ∎

1.4. Proposition. *If N is a normal operator, then $\sigma(N) = \sigma_r(N) = \sigma_l(N)$. If λ is an isolated point of $\sigma(N)$, then $\lambda \in \sigma_p(N)$.*

PROOF. The first part of the proposition is immediate from the preceding result. If λ is an isolated point of $\sigma(N)$ and $N = \int z \, dE(z)$, then $0 \neq E(\{\lambda\})\mathscr{H} = \ker(N - \lambda)$ (Exercise IX.2.1). ∎

EXERCISES

1. Let S be the unilateral shift of multiplicity 1 on $l^2(\mathbb{N})$ and find $\sigma_l(S)$ and $\sigma_r(S)$.

2. The *compression spectrum* of A, $\sigma_c(A)$, is defined by $\sigma_c(A) = \{\lambda \in \mathbb{C}: \operatorname{ran}(A - \lambda)$ is not dense in $\mathscr{H}\}$. Show: (a) $\lambda \in \sigma_c(A)$ if and only if $\bar{\lambda} \in \sigma_p(A^*)$. (b) $\sigma_c(A) \subseteq \sigma_r(A)$, but this inclusion may be proper. (c) $\sigma_c(A)$ is not necessarily closed. (d) $\sigma(A) = \sigma_{ap}(A) \cup \sigma_c(A)$.

3. If $A \in \mathscr{B}(\mathscr{H})$ and $f \in \operatorname{Hol}(A)$, then $f(\sigma_p(A)) \subseteq \sigma_p(f(A))$. If f is not constant on any component of its domain, then $f(\sigma_p(A)) = \sigma_p(f(A))$.

4. If $A \in \mathscr{B}(\mathscr{H})$ and $f \in \operatorname{Hol}(A)$, then $f(\sigma_{ap}(A)) = \sigma_{ap}(f(A))$.

§2. Fredholm Operators

We begin with a definition.

2.1. Definition. If $\mathscr{H}$ and $\mathscr{H}'$ are Hilbert spaces and $A: \mathscr{H} \to \mathscr{H}'$ is a bounded operator, then A is said to be *left semi-Fredholm* if there is a bounded operator $B: \mathscr{H}' \to \mathscr{H}$ and a compact operator K on $\mathscr{H}$ such that $BA = 1 + K$. Analogously, A is *right semi-Fredholm* if there is a such a bounded operator B and a compact operator K' on $\mathscr{H}'$ such that $AB = 1 + K'$. A is a *semi-Fredholm* operator if it is either left or right semi-Fredholm and A is a *Fredholm operator* if it is both left and right semi-Fredholm.

Observe that A is left semi-Fredholm if and only if A^* is right semi-Fredholm. Thus results about semi-Fredholm operators will usually only be phrased in terms of left semi-Fredholm operators and the reader will be allowed to make the appropriate statement for right semi-Fredholm operators.

Note that a left invertible operator is left semi-Fredholm. However it is easy to get left semi-Fredholm operators that are not left invertible.

2.2. Example. Let $\mathscr{H} = \mathscr{H}_0 \oplus \mathscr{H}_1 \oplus \cdots$, where $\dim \mathscr{H}_j = \alpha$ for all $j \geqslant 0$, and let S be the unilateral shift of multiplicity α with respect to this decomposition. (So S maps $\mathscr{H}_j$ isometrically onto $\mathscr{H}_{j+1}$.) Recall that $S^*S = 1$ so S is left invertible and hence left semi-Fredholm. Also $SS^* = 1 - P_0$, where P_0 is the projection of $\mathscr{H}$ onto $\mathscr{H}_0$. So S is Fredholm if $\alpha < \infty$.

In the next result, the equivalence of the first three conditions is referred to as Atkinson's Theorem. The rest of this theorem is from Wolf [1959], Schechter [1968], and Fillmore, Stampfli and Williams [1972].

2.3. Theorem. *If A: $\mathscr{H} \to \mathscr{H}'$ is a bounded operator, the following statements are equivalent.*

(a) *A is left semi-Fredholm.*

(b) *ran A is closed and $\dim \ker A < \infty$.*

(c) *There is a bounded operator B: $\mathscr{H}' \to \mathscr{H}$ and a finite rank operator F on $\mathscr{H}$ such that $BA = 1 + F$.*

(d) *There is no sequence $\{h_n\}$ of unit vectors in $\mathscr{H}$ such that $h_n \to 0$ weakly and $\lim \|Ah_n\| = 0$.*

(e) *There is no orthonormal sequences $\{e_n\}$ in $\mathscr{H}$ such that $\lim \|Ae_n\| = 0$.*

(f) *There is a $\delta > 0$ such that $\{h \in \mathscr{H}: \|Ah\| \leqslant \delta \|h\|\}$ contains no infinite dimensional manifold.*

(g) *If the positive operator $(A^*A)^{1/2} = \int_0^\infty t\, dE(t)$, then there is a $\delta > 0$ such that $E[0, \delta]\mathscr{H}$ is finite dimensional.*

(h) *If $K \in \mathscr{B}_0(\mathscr{H})$, then $\dim \ker(A + K) < \infty$.*

PROOF. (a) *implies* (b). According to (a) there is a bounded operator B such that $\pi(B)\pi(A) = 1$; that is, $\pi(BA - 1) = 0$. Hence $BA = 1 + K$ for some compact operator K. But $\ker A \subseteq \ker BA = \ker(1 + K)$. Since the eigenspace corresponding to nonzero eigenvalues of compact operators are finite dimensional, $\dim \ker A < \infty$. Also, the Fredholm Alternative (VII.7.9) implies $\operatorname{ran} BA = \operatorname{ran}(K + 1)$ is closed. Hence there is a constant $c > 0$ such that for $h \perp \ker(BA)$, $\|BAh\| \geqslant c\|h\|$. Thus if $h \in [\ker BA]^\perp$, $c\|h\| \leqslant \|B\| \|Ah\|$, or $\|Ah\| \geqslant (c/\|B\|)\|h\|$. This implies that $A([\ker BA]^\perp)$ is closed. But $\operatorname{ran} A = A([\ker BA]^\perp) + A(\ker BA)$. Since $A(\ker BA)$ is finite dimensional, $\operatorname{ran} A$ is closed.

(b) *implies* (c). First define A_1: $(\ker A)^\perp \to \operatorname{ran} A$ by $A_1 = A|(\ker A)^\perp$ and note that A_1 is invertible by The Open Mapping Theorem. Let P be the projection of $\mathscr{H}'$ onto $\operatorname{ran} A$ and define B: $\mathscr{H}' \to \mathscr{H}$ by $B = A_1^{-1}P$. It is left to the reader to check that $BA = 1 - F$, where F is the orthogonal projection of $\mathscr{H}$ onto $\ker A$. Since $\ker A$ is finite dimensional, this establishes (c).

(c) *implies* (a). This is clear.

(a) *implies* (d). Suppose $\{h_n\}$ is a sequence of unit vectors in $\mathscr{H}$ that converges weakly to 0 and let B: $\mathscr{H}' \to \mathscr{H}$ and K be as in the definition of a left semi-Fredholm operator. Since $BA = 1 + K$, $|1 - \|BAh_n\| | = | \|h_n\| - \|BAh_n\| | \leqslant \|Kh_n\|$ and $\|Kh_n\| \to 0$ since K is compact. Thus $\|BAh_n\| \to 1$ and so it is impossible for $\{Ah_n\}$ to converge to 0 in norm.

(d) *implies* (e). Orthonormal sequences converge weakly to 0.

(e) *implies* (f). If (f) is false, then for every positive integer n there is an infinite dimensional manifold $\mathscr{M}_n$ such that $\|Ah\| \leqslant (1/n)\|h\|$ for all h in $\mathscr{M}_n$.

Let e_1 be a unit vector in $\mathcal{M}_1$. Suppose $e_1, \ldots, e_n$ are orthonormal vectors such that $e_k \in \mathcal{M}_k$, $1 \leqslant k \leqslant n$. Let E be the projection of $\mathcal{H}$ onto $\vee \{e_1, \ldots, e_n\}$. If $\mathcal{M}_{n+1} \cap [e_1, \ldots, e_n]^\perp = (0)$, then E is injective on $\mathcal{M}_{n+1}$. Since $\dim \mathcal{M}_{n+1} = \infty$ and $\dim \operatorname{ran} E < \infty$, this is impossible. Thus there is a unit vector e_{n+1} in $\mathcal{M}_{n+1}$ such that $e_{n+1} \perp \{e_1, \ldots, e_n\}$. The orthonormal sequence $\{e_n\}$ shows that (e) does not hold.

(f) *implies* (g). Let $|A| = \int t \, dE(t)$ and let $\delta > 0$. If $h \in E[0, \delta]\mathcal{H}$, then

$$\|Ah\|^2 = \langle A^*Ah, h \rangle$$
$$= \langle |A|^2 h, h \rangle$$
$$= \int_0^\delta t^2 \, dE_{h,h}(t) \leqslant \delta^2 E_{h,h}[0, \delta]$$
$$= \delta^2 \|h\|^2.$$

So $E[0, \delta]\mathcal{H} \subseteq \{h \colon \|Ah\| \leqslant \delta \|h\|\}$. By (f) there is a $\delta > 0$ such that $E[0, \delta)\mathcal{H}$ is finite dimensional.

(g) *implies* (c). Let $\mathcal{M}_\delta = \{E[0, \delta]\mathcal{H}\}^\perp$. Now $|A|$ maps $\mathcal{M}_\delta$ bijectively onto $\mathcal{M}_\delta$. In fact, the inverse of $|A| \colon \mathcal{M}_\delta \to \mathcal{M}_\delta$ is $(\int_\delta^\infty t^{-1} dE(t))|\mathcal{M}_\delta$. Let $A = U|A|$ be the polar decomposition of A. Since $\mathcal{M}_\delta \subseteq \operatorname{ran}|A| \subseteq$ initial U, U maps $\mathcal{M}_\delta$ isometrically onto some closed subspace $\mathcal{L}$ of $\operatorname{ran} A$. Let $V =$ the inverse of U on $\mathcal{L}$ and $V = 0$ on $\mathcal{L}^\perp$; that is, $V|\mathcal{L}^\perp = 0$ and $V|\mathcal{L} = (U|\mathcal{M}_\delta)^{-1}$. Hence V is a partial isometry. Let $B_1 = \int_\delta^\infty t^{-1} dE(t)$ and put $B = B_1 V$. If $h \in \mathcal{M}_\delta$, then $BAh = B_1 VU|A|h = h$. If $h \in \mathcal{M}_\delta^\perp = E[0, \delta]h$, $|A|h \in \mathcal{M}_\delta^\perp$ and so $U|A|h \perp \mathcal{L}$; thus $BAh = 0$. Hence $BA = E(\delta, \infty) = 1 - E[0, \delta]$, and $E[0, \delta]$ has finite rank.

(a) *implies* (h). Let $B \colon \mathcal{H}' \to \mathcal{H}$ be a bounded operator such that $BA = 1 + L$, where L is a compact operator on $\mathcal{H}$. If $K \colon \mathcal{H} \to \mathcal{H}'$ is any compact operator, then $B(A + K) = 1 + (L + BK)$ and $L + BK$ is compact. By definition $A + K$ is left semi-Fredholm. Since we have already shown that (a) implies (b), $\dim \ker(A + K) < \infty$.

(h) *implies* (e). Suppose (e) does not hold. So there is an orthonormal sequence, $\{e_n\}$ usch that $\|Ae_n\| \to 0$. By passing to a subsequence if necessary, it may be assumed that $\sum_{n=1}^\infty \|Ae_n\|^2 < \infty$. Thus for any h in $\mathcal{H}$,

$$\sum |\langle h, e_n \rangle| \|Ae_n\| \leqslant [\sum |\langle h, e_n \rangle|^2]^{1/2} [\sum \|Ae_n\|^2]^{1/2}$$
$$\leqslant C \|h\|,$$

where $C = [\sum \|Ae_n\|^2]^{1/2}$. Thus $Kh = \sum_{n=1}^\infty \langle h, e_n \rangle Ae_n$ defines a bounded operator. Moreover, if $K_n h = \sum_{j=1}^n \langle h, e_j \rangle Ae_j$, it is easy to see that $\|K_n - K\| \to 0$. Thus K is compact. But $(A - K)e_n = 0$ for every n, so $\dim \ker(A - K) = \infty$. ∎

As mentioned previously, the result for right semi-Fredholm operators that is analogous to the preceding theorem is left for the reader to state. We will, however, make explicit part of this result for Fredholm operators.

2.4. Corollary. *A bounded operator* $A: \mathcal{H} \to \mathcal{H}'$ *is Fredholm if and only if* ran A *is closed and both* ker A *and* ker A^* *are finite dimensional.*

In the case of a bounded operator A from a Hilbert space $\mathcal{H}$ into itself, the concept of a semi-Fredholm operator can be rephrased in terms of the corresponding concept in the Calkin algebra, $\mathcal{B}/\mathcal{B}_0$. In fact, this will be the primary situation in which the ideas of Fredholm theory are applied. The next result is immediate from the definition.

2.5. Proposition. *For a Hilbert space* $\mathcal{H}$, *let* $\pi\colon \mathcal{B} \to \mathcal{B}/\mathcal{B}_0$ *be the natural map and let* $A \in \mathcal{B} = \mathcal{B}(\mathcal{H})$. *The operator* A *is left (respectively, right) semi-Fredholm if and only if* $\pi(A)$ *is left (respectively, right) invertible in the Calkin algebra.*

For notation, let $\mathcal{F}_\ell = \mathcal{F}_\ell(\mathcal{H})$ and $\mathcal{F}_r = \mathcal{F}_r(\mathcal{H})$ be the set of left and right semi-Fredholm operators on the Hilbert space $\mathcal{H}$. So $\mathcal{F} = \mathcal{F}_\ell \cap \mathcal{F}_r$ and $\mathcal{S}\mathcal{F} = \mathcal{F}_\ell \cup \mathcal{F}_r$ are the sets of Fredholm and semi-Fredholm operators on $\mathcal{H}$. Since $\mathcal{F}_\ell =$ the inverse image under π of the left invertible elements of the Banach algebra $\mathcal{B}/\mathcal{B}_0$, the next proposition is immediate.

2.6. Proposition. *Each of the sets* $\mathcal{F}_\ell, \mathcal{F}_r, \mathcal{F}$, *and* $\mathcal{S}\mathcal{F}$ *are open subjects of* $\mathcal{B}$.

EXERCISES

1. If $A \in \mathcal{B}(\mathcal{H})$ and ran A is closed, prove than ran A^* is closed without using Theorem VI.1.10. [Hint: Show that there is a bounded operator B on $\mathcal{H}$ such that $BA =$ the projection of $\mathcal{H}$ onto (ker A)$^\perp$.]

2. Give a direct proof that (b) implies (a) in Theorem 2.3.

3. Let $A, B, C \in \mathcal{B}(\mathcal{H})$ and define $X: \mathcal{H}^{(2)} \to \mathcal{H}^{(2)}$ by the matrix $X = \begin{bmatrix} A & B \\ 0 & C \end{bmatrix}$. (a) Show that if $A \in \mathcal{F}$, then $X \in \mathcal{F}$ if and only if $C \in \mathcal{F}$. (b) If $A \in \mathcal{F}$, show that $X \in \mathcal{S}\mathcal{F}$ if and only if $C \in \mathcal{S}\mathcal{F}$. (c) Suppose $A, C \in \mathcal{S}\mathcal{F}$ with dim ker $A = \infty$ and dim ker $C^* = \infty$. Show that $X \notin \mathcal{S}\mathcal{F}$.

4. If $A \in \mathcal{B}(\mathcal{H})$, show that $A\mathcal{M}$ is closed for every closed subspace $\mathcal{M}$ of $\mathcal{H}$ if and only if A has finite rank or A is left Fredholm.

5. If $\mathcal{X}$ and $\mathcal{Y}$ are Banach spaces and $T \in \mathcal{B}(\mathcal{X}, \mathcal{Y})$ such that the Hamel basis dimension (that is, the algebraic dimension) of $\mathcal{Y}/(\text{ran } T)$ is finite, then ran T is closed.

6. Show that a normal operator is Fredholm if and only if 0 is not a limit point of $\sigma(N)$ and dim ker $N < \infty$. (See Proposition 4.5 below.)

§3. The Fredholm Index

I wish to acknowledge that the basis of this section is a development of the Fredholm index which I learned from my colleague Hari Bercovici.

If A is a semi-Fredholm operator, define the *(Fredholm) index* of A, ind A, by

3.1
$$\text{ind } A = \dim \ker A - \dim(\text{ran } A)^{\perp}$$
$$= \dim \ker A - \dim \ker A^{*}.$$

Note that ind $A \in \mathbb{Z} \cup \{ \pm \infty \}$ and it is necessary for either ker A or ker A^{*} to be finite dimensional in order for (3.1) to make sense. For ind A to be well defined, it is not necessary that ran A be closed (the other part of the definition of semi-Fredholm operators), but this property will be used in a critical way when the properties of the index are established.

See Dieudonné [1985] for some historical notes on the Fredholm index.

The main properties of the Fredholm index are contained in Theorems 3.7, 3.11, and 3.12 below. But we will begin with some elementary results.

3.2. Proposition. *If A: $\mathscr{H} \to \mathscr{H}'$ and $\mathscr{H}$ and $\mathscr{H}'$ are finite dimensional, then A is Fredholm and* ind $A = \dim \mathscr{H} - \dim \mathscr{H}'$.

PROOF. Clearly A is Fredholm. From linear algebra we know that $\dim \mathscr{H} = \dim(\text{ran } A) + \dim(\ker A) = [\dim \mathscr{H}' - \dim(\text{ran } A)^{\perp}] + \dim(\ker A)$. Thus ind $A = \dim(\ker A) - \dim(\text{ran } A)^{\perp} = \dim \mathscr{H} - \dim \mathscr{H}'$. ∎

The next result is actually just a restatement of the Fredholm Alternative (VII. 7.9).

3.3. Proposition. *If K: $\mathscr{H} \to \mathscr{H}$ is a compact operator and $\lambda \neq 0$, then $\lambda + K$ is Fredholm and* $\text{ind}(\lambda + K) = 0$.

We already observed that the adjoint of a semi-Fredholm operator is also a semi-Fredholm operator. This same reasoning produces the calculation of the index.

3.4. Proposition. (a) *If A is a semi-Fredholm operator, then A^{*} is also semi-Fredholm and* ind $A^{*} = -$ ind A.
(b) *If N is a normal operator on a Hilbert space $\mathscr{H}$, then N is semi-Fredholm if and only if N is Fredholm, in which case* ind $N = 0$.
(c) *If A and B are Fredholm operators, then $A \oplus B$ is Fredholm and* $\text{ind}(A \oplus B) = \text{ind } A + \text{ind } B$.

PROOF. As stated, the proof of (a) is easy. For part (b), recall that for a normal operator N, $\| Nh \| = \| N^{*}h \|$ for every vector h in $\mathscr{H}$. Thus ker $N = \ker N^{*}$. The result is now immediate. The proof of (c) is left to the reader. ∎

Also see Exercise 2.6 and Proposition 4.5 below.

The next theorem is one of the main properties of the Fredholm index. But first two lemmas are required.

3.5. Lemma. *Let $A: \mathscr{H} \to \mathscr{H}'$, $\mathscr{H} = \mathscr{M} \oplus \mathscr{N}$, $\mathscr{H}' = \mathscr{M}' \oplus \mathscr{N}'$, and suppose A has the matrix*

$$\begin{bmatrix} A_1 & X \\ 0 & A_2 \end{bmatrix}$$

relative to these two decompositions of $\mathscr{H}$ and $\mathscr{H}'$. If A_1 is invertible and $\mathscr{N}$ and $\mathscr{N}'$ are finite dimensional, then A is Fredholm and $\operatorname{ind} A = \dim \mathscr{N} - \dim \mathscr{N}'$.

PROOF. It is easy to see that ran A is closed since ran $A_1 = \mathscr{M}'$ and $\dim \mathscr{N} < \infty$. Let's show that $\ker A^* = \ker A_2^*$ and $\dim(\ker A) = \dim(\ker A_2)$. If this is done, then $\operatorname{ind} A = \dim(\ker A) - \dim(\ker A^*) = \dim(\ker A_2) - \dim(\ker A_2^*) = \dim \mathscr{N} - \dim \mathscr{N}'$ by Proposition 3.2.

For the first of the two desired equalities, let $f' \in \mathscr{M}'$ and $g' \in \mathscr{N}'$. Then $A^*(f' \oplus g') = A_1^* f' \oplus (X^* f' + A_2^* g')$. So $f' \oplus g' \in \ker A^*$ if and only if $A_1^* f' = 0$ and $A_2^* g' = - X^* f'$. But A_1 is invertible, so this happens exactly when $f' = 0$, and hence $A_2^* g' = 0$. From here it is clear that $\ker A^* = \ker A_2^*$. For the second equality, note that $g \to - A_1^{-1} X g \oplus g$ is a bijection between $\ker A_2$ and $\ker A$. ∎

The second lemma is elementary and its proof is left to the reader.

3.6. Lemma. *Let $\mathscr{M}$ and $\mathscr{N}$ be two closed subspaces of the Hilbert space $\mathscr{H}$.*

(a) *If $\mathscr{M} \cap \mathscr{N} = (0)$ and $\dim \mathscr{N} = \infty$, then $\dim \mathscr{M}^\perp = \infty$.*
(b) *If $\dim \mathscr{M}^\perp = \infty$ and $\dim \mathscr{N} < \infty$, then $\dim(\mathscr{M} + \mathscr{N})^\perp < \infty$.*

3.7. Theorem. *If $A: \mathscr{H} \to \mathscr{H}'$ and $B: \mathscr{H}' \to \mathscr{H}''$ are left semi-Fredholm operators, then BA is a left semi-Fredholm operator and $\operatorname{ind} BA = \operatorname{ind} A + \operatorname{ind} B$.*

PROOF. By definition, there are operators X and Y such that $XA = 1 + K$ and $YB = 1 + K'$, where K and K' are compact operators on the appropriate spaces. Hence $(XY)(BA) = X(1 + K')A = 1 + (K + XK'A)$, and $K + XK'A$ is compact. Therefore BA is left semi-Fredholm.

To prove the formula for the index, we consider 4 cases.

Case 1. Both A and B are Fredholm.

Let $\mathscr{M}' \equiv (\operatorname{ran} A) \cap (\ker B)^\perp$ and put $\mathscr{N}' = \mathscr{M}'^\perp$.

Claim 1: $\mathscr{N}'$ is finite dimensional.

To see this note that $\mathscr{N}' = \mathscr{M}'^\perp = (\operatorname{ran} A)^\perp \vee (\ker B)$. Since both of these spaces are finite dimensional, so is $\mathscr{N}'$. Let $\mathscr{M} = A^{-1}(\mathscr{M}') \cap (\ker A)^\perp$; $\mathscr{N} = \mathscr{M}^\perp$; $\mathscr{M}'' = B\mathscr{M}'$; $\mathscr{N}'' = \mathscr{M}''^\perp$. Note the following: $A(\mathscr{M}) = \mathscr{M}'$; $A|\mathscr{M}$ is invertible (because $\ker(A|\mathscr{M}) = \ker A \cap \mathscr{M} = (0)$); $B|\mathscr{M}'$ is invertible.

Claim 2: $\mathcal{N}$ is finite dimensional.

Indeed, let $A_1 \equiv A|(\ker A)^\perp$; so A_1: $(\ker A)^\perp \rightarrow \operatorname{ran} A$ is invertible. But $A_1^{-1}(\mathcal{M}') = \mathcal{M}$, so $\dim[(\ker A)^\perp \cap \mathcal{M}^\perp] = \dim[(\operatorname{ran} A) \cap \mathcal{M}'^\perp]$, and this last dimension is finite since $\mathcal{N}'$ is finite dimensional.

Claim 3: $\mathcal{N}''$ is finite dimensional.

In fact, $\dim[(\ker B)^\perp \cap \mathcal{M}^\perp] < \infty$ and so $\dim[(\operatorname{ran} B) \cap \mathcal{M}''^\perp] < \infty$. This implies that $\dim \mathcal{N}'' < \infty$. Now represent the operators as 2×2 matrices:

$$A = \begin{bmatrix} A_1 & X \\ 0 & A_2 \end{bmatrix}: \begin{matrix} \mathcal{M} \\ \oplus \\ \mathcal{N} \end{matrix} \rightarrow \begin{matrix} \mathcal{M}' \\ \oplus \\ \mathcal{N}' \end{matrix},$$

$$B = \begin{bmatrix} B_1 & Y \\ 0 & B_2 \end{bmatrix}: \begin{matrix} \mathcal{M}' \\ \oplus \\ \mathcal{N}' \end{matrix} \rightarrow \begin{matrix} \mathcal{M}'' \\ \oplus \\ \mathcal{N}'' \end{matrix},$$

$$BA = \begin{bmatrix} B_1 A_1 & Z \\ 0 & B_2 A_2 \end{bmatrix}: \begin{matrix} \mathcal{M} \\ \oplus \\ \mathcal{N} \end{matrix} \rightarrow \begin{matrix} \mathcal{M}'' \\ \oplus \\ \mathcal{N}'' \end{matrix}.$$

It follows that B_1 and A_1 are invertible. Since $\mathcal{N}, \mathcal{N}'$, and $\mathcal{N}''$ are finite dimensional, the preceding lemma implies that $\operatorname{ind} A = \dim \mathcal{N} - \dim \mathcal{N}'$, $\operatorname{ind} B = \dim \mathcal{N}' - \dim \mathcal{N}''$, and $\operatorname{ind} BA = \dim \mathcal{N} - \dim \mathcal{N}'' = \operatorname{ind} A + \operatorname{ind} B$.

Case 2. Assume $\operatorname{ind} B = -\infty$.

This is equivalent to the assumption that $\dim(\operatorname{ran} B)^\perp = \infty$. But $\operatorname{ran} B \supseteq \operatorname{ran} BA$ and so $\dim(\operatorname{ran} BA)^\perp = \infty$. Hence $\operatorname{ind} BA = -\infty = \operatorname{ind} A + \operatorname{ind} B$.

Case 3. Assume B is invertible.

Without loss of generality we may assume that $\operatorname{ind} A = -\infty$ since the alternative situation is covered in Case 1. If $\mathcal{M} = B(\operatorname{ran} A) = \operatorname{ran} BA$ and $\mathcal{N} = B([\operatorname{ran} A]^\perp)$, then the fact that B is invertible implies that $\mathcal{M} \cap \mathcal{N} = (0)$ and $\mathcal{N}$ is infinite dimensional. Thus Lemma 3.6(a) implies that $\infty = \dim \mathcal{M}^\perp = \dim(\operatorname{ran} BA)^\perp$ and so $\operatorname{ind} BA = -\infty = \operatorname{ind} A + \operatorname{ind} B$.

Case 4. B is a Fredholm operator.

Again, without loss of generality we may assume that $\operatorname{ind} A = -\infty$. It must be shown that $\operatorname{ind} BA = -\infty$.

Put $\mathcal{H}_1' = (\ker B)^\perp$ and $\mathcal{H}_1'' = \operatorname{ran} B$; define B_1: $\mathcal{H}_1' \rightarrow \mathcal{H}_1''$ as the restriction of B to $\mathcal{H}'_1$. Clearly B_1 is invertible. If P is the orthogonal projecton of $\mathcal{H}'$ onto $\mathcal{H}'_1$, let $A_1: \mathcal{H} \to \mathcal{H}'_1$ be defined by $A_1 = PA$. Now both P and A are left semi-Fredholm, so A_1 is left semi-Fredholm as was established at the opening of the proof.

Note that Lemma 3.6(b) implies that $\dim(\operatorname{ran} A + \ker B)^\perp = \infty$. But $(\operatorname{ran} A_1)^\perp = \ker A_1^* = \ker A^*P = \ker B + (\ker A^*) \cap (\ker B)^\perp = \ker B + (\operatorname{ran} A + \ker B)^\perp$ and so $\operatorname{ind} A_1 = -\infty$. According to Case 3, $\operatorname{ind} B_1 A_1 = -\infty$.

This is equivalent to the condition that $\infty = \dim(\operatorname{ran} B_1 A_1)^\perp = \dim(\mathcal{H}''_1 \cap [B(\operatorname{ran} A_1)]^\perp)$. But $B(\operatorname{ran} A_1) = BP(\operatorname{ran} A) = B(\operatorname{ran} A) = \operatorname{ran} BA$ and so $\mathcal{H}''_1 \cap [B(\operatorname{ran} A_1)]^\perp \subseteq (\operatorname{ran} BA)^\perp$. Thus $\operatorname{ind} BA = -\infty$. ∎

3.8. Corollary. *If* $A \in \mathcal{F}_\ell$ *and* R *is an invertible operator, then* $RAR^{-1} \in \mathcal{F}_\ell$ *and* $\operatorname{ind} RAR^{-1} = \operatorname{ind} A$.

Before going further, let's look at some examples.

3.9. Example. Let S be the unilateral shift of multiplicity α. We saw in Example 2.2 that S is left semi-Fredholm. It is easy to calculate that $\operatorname{ind} S = -\alpha$. According to the preceding theorem, $\operatorname{ind} S^2 = -2\alpha$. But, of course, S^2 is the unilateral shift of multiplicity 2α.

3.10. Example. Let S be the unilateral shift of multiplicity α on the Hilbert space $\mathcal{H}$ and put $A = S \oplus S^*$. Note that $\ker A = (0) \oplus \ker S^*$ and $\operatorname{ran} A = (\operatorname{ran} S) \oplus \mathcal{H}$. Thus A has closed range. The operator A, however, is semi-Fredholm if and only if $\alpha < \infty$, in which case A is a Fredholm operator. Also when $\alpha < \infty$, $\operatorname{ind} A = 0$.

3.11. Theorem. *If* $A: \mathcal{H} \to \mathcal{H}'$ *is a left (respectively, right) semi-Fredholm operator and* $K: \mathcal{H} \to \mathcal{H}'$ *is a compact operator, then* $A + K$ *is left (respectively, right) semi-Fredholm and* $\operatorname{ind} A = \operatorname{ind}(A + K)$.

PROOF. Assume that A is a left semi-Fredholm operator. By definition there is an operator $X: \mathcal{H}' \to \mathcal{H}$ and a compact operator $K_0: \mathcal{H} \to \mathcal{H}$ such that $XA = 1 + K_0$. Thus $X(A + K) = 1 + (K_0 + XK)$ and so $A + K$ is left semi-Fredholm.

To verify that $\operatorname{ind} A = \operatorname{ind}(A + K)$, first assume that A is Fredholm. So there is a Fredholm operator X and a compact operator L such that $XA = 1 + L$. But Theorem 3.7 implies that XA is Fredholm and, using Proposition 3.3, $0 = \operatorname{ind}(1 + L) = \operatorname{ind} A + \operatorname{ind} X$; so $\operatorname{ind} A = -\operatorname{ind} X$. But $X(A + K) = 1 + (L + XK)$ and so the same type of reasoning implies that $\operatorname{ind}(A + K) = -\operatorname{ind} X = \operatorname{ind} A$.

Now assume that A is left semi-Fredholm; so $A + K$ is also left semi-Fredholm. If $\operatorname{ind} A$ is finite, then A is Fredholm and we are done by the preceding paragraph. If $\operatorname{ind} A$ is $-\infty$, then $\operatorname{ind}(A + K)$ must also be $-\infty$ for otherwise $A + K$ is Fredholm and it follows that $A = (A + K) - K$ is also Fredholm, a contradiction. ∎

The preceding theorem says that the value of the index is impervious to compact perturbations. The next result, the third in the list of important properties of the Fredholm index, says that the value of the index is unchanged for all perturbations of the operator, provided that the size of the perturbation is sufficiently small.

3.12. Theorem. *If $A: \mathcal{H} \to \mathcal{H}'$ is a Fredholm operator, then there is an $\varepsilon > 0$ such that if $Y \in \mathcal{B}(\mathcal{H}, \mathcal{H}')$ and $\|Y\| < \varepsilon$, then $A + Y$ is Fredholm and $\operatorname{ind} A = \operatorname{ind}(A + Y)$.*

PROOF. With respect to the decompositions $\mathcal{H} = (\ker A)^{\perp} \oplus \ker A$ and $\mathcal{H}' = \operatorname{ran} A \oplus \ker A^*$, the operator A has the matrix

$$\begin{bmatrix} A_1 & 0 \\ 0 & 0 \end{bmatrix}$$

and $A_1: (\ker A)^{\perp} \to \operatorname{ran} A$ is invertible since A is Fredholm. Thus there is an $\varepsilon > 0$ such that if $\|Y_1\| < \varepsilon$, then $A_1 + Y_1$ is invertible. If $Y: \mathcal{H} \to \mathcal{H}'$ and $\|Y\| < \varepsilon$, then, with respect to the same decomposition of $\mathcal{H}$,

$$Y = \begin{bmatrix} Y_1 & Y_2 \\ Y_3 & Y_4 \end{bmatrix}$$

and so

$$A + Y = \begin{bmatrix} A_1 + Y_1 & Y_2 \\ Y_3 & Y_4 \end{bmatrix} = \begin{bmatrix} A_1 + Y_1 & 0 \\ 0 & 0 \end{bmatrix} + \begin{bmatrix} 0 & Y_2 \\ Y_3 & Y_4 \end{bmatrix},$$

where the first matrix represents a Fredholm operator and the second represents a finite rank operator. Therefore $\operatorname{ind}(A + Y)$ is the index of the first matrix. But since $A_1 + Y_1$ is invertible, Lemma 3.5 implies that this index is equal to $\dim(\ker A) - \dim(\ker A^*) = \operatorname{ind} A$. ∎

3.13. Corollary. *If $\mathcal{SF}$ is given the norm topology and $\mathbb{Z} \cup \{\pm \infty\}$ is given the discrete topology, then the Fredholm index is a continuous function from $\mathcal{SF}$ into $\mathbb{Z} \cup \{\pm \infty\}$.*

This continuity statement has an equivalent formulation. Because $\mathcal{SF}$ is an open subset of $\mathcal{B}(\mathcal{H})$, its components are open sets. Thus the continuity of the index is equivalent to the statement that it is constant on the components of $\mathcal{SF}$.

For a treatment of the Fredholm index applicable to unbounded operators on a Banach space, see Kato [1966], pp. 229–244. For other approaches to Fredholm theory, with variations and generalizations of the material in this book, see Caradus, Pfaffenberger, and Yood [1974] and Harte [1982].

EXERCISES

1. Prove Lemma 3.6.

2. If $A \in \mathcal{B}(\mathcal{H})$ and $\operatorname{ran} A$ is closed, show that $\operatorname{ran} A^{(\infty)}$ is closed. If $A \in \mathcal{SF}$ and $\ker A = (0)$, show that $A^{(\infty)} \in \mathcal{SF}$ and $\operatorname{ind} A^{(\infty)} = -\infty$ or 0.

3. Does the unilateral shift of multiplicity 1 have a square root?

4. Show that for every n in $\mathbf{Z} \cup \{\pm\infty\}$ there is an operator A in $\mathscr{S}\mathscr{F}$ such that ind $A = n$.

5. If $A \in \mathscr{S}\mathscr{F}$, then for every $n \geqslant 1$, $A^n \in \mathscr{S}\mathscr{F}$ and ind $A^n = n(\text{ind } A)$.

6. If $A: \mathscr{H} \to \mathscr{H}'$ is a left semi-Fredholm operator, then there is a finite rank operator $F: \mathscr{H} \to \mathscr{H}'$ such that $\ker(A + F) = (0)$ and $\text{ind}(A + F) = \text{ind } A$.

7. If A is a Fredholm operator in $\mathscr{B}(\mathscr{H})$, prove that the following statements are equivalent. (a) ind $A = 0$. (b) There is a compact operator K such that $A + K$ is invertible. (c) There is a finite rank operator F such that $A + F$ is invertible.

8. If U is the unilateral shift of multiplicity 1 and $\pi: \mathscr{B}(\mathscr{H}) \to \mathscr{B}(\mathscr{H})/\mathscr{B}_0(\mathscr{H})$ is the natural map, show that $\pi(U)$ is normal in $\mathscr{B}(\mathscr{H})/\mathscr{B}_0(\mathscr{H})$ but there is no normal operator N such that $U - N$ is compact. (See Exercise IX.8.14.)

§4. The Essential Spectrum

Now concentrate on operators acting on a single Hilbert space $\mathscr{H}$ and let $\pi: \mathscr{B} \to \mathscr{B}/\mathscr{B}_0$ be the natural map from $\mathscr{B}(\mathscr{H})$ into the Calkin algebra. Since the Calkin algebra is a Banach algebra with identity, the next definition makes sense.

4.1. Definition. If $A \in \mathscr{B}(\mathscr{H})$, the *essential spectrum* of A, $\sigma_e(A)$, is the spectrum of $\pi(A)$ in $\mathscr{B}/\mathscr{B}_0$; that is, $\sigma_e(A) = \sigma(\pi(A))$. Similarly the *left* and *right essential spectrum* of A are defined by $\sigma_{\ell e}(A) = \sigma_\ell(\pi(A))$ and $\sigma_{re}(A) = \sigma_r(\pi(A))$, respectively.

The proof of the next proposition is a straightforward application of the general properties of the various spectra in an arbitrary Banach algebra.

4.2. Proposition. *Let* $A \in \mathscr{B}(\mathscr{H})$.
(a) $\sigma_e(A) = \sigma_{\ell e}(A) \cup \sigma_{re}(A)$.
(b) $\sigma_{\ell e}(A) = \sigma_{re}(A^*)^*$.
(c) $\sigma_{\ell e}(A) \subseteq \sigma_\ell(A)$, $\sigma_{re}(A) \subseteq \sigma_r(A)$, and $\sigma_e(A) \subseteq \sigma(A)$.
(d) $\sigma_{\ell e}(A)$, $\sigma_{re}(A)$, and $\sigma_e(A)$ are compact sets.
(e) If K is a compact operator, $\sigma_{\ell e}(A + K) = \sigma_{\ell e}(A)$, $\sigma_{re}(A + K) = \sigma_{re}(A)$, and $\sigma_e(A + K) = \sigma_e(A)$.

Our understanding of semi-Fredholm operators gained in the preceding sections can now be applied to better understand the essential spectrum. Indeed, $\sigma_{\ell e}(A) = \{\lambda \in \mathbf{C}: A - \lambda \notin \mathscr{F}_\ell\}$. Thus an application of Theorem 2.3 gives us the following.

4.3. Proposition. *Let* $A \in \mathcal{B}(\mathcal{H})$.

(a) $\lambda \in \sigma_{le}(A)$ *if and only if* $\dim \ker(A - \lambda) = \infty$ *or* $\operatorname{ran}(A - \lambda)$ *is not closed.*
(b) $\lambda \in \sigma_{re}(A)$ *if and only if* $\dim [\operatorname{ran}(A - \lambda)]^{\perp} = \infty$ *or* $\operatorname{ran}(A - \lambda)$ *is not closed.*

The reader should compare Proposition 4.3 and Proposition 1.1.

4.4. Proposition. *If* $A \in \mathcal{B}(\mathcal{H})$, *then* $\sigma_{ap}(A) = \sigma_{le}(A) \cup \{\lambda \in \sigma_p(A): \dim \ker(A - \lambda) < \infty\}$.

PROOF. If $\lambda \in \sigma_{ap}(A)$, then (1.1) either $\operatorname{ran}(A - \lambda)$ is not closed or $\ker(A - \lambda) \neq 0$. If $\operatorname{ran}(A - \lambda)$ is not closed or if $\dim \ker(A - \lambda) = \infty$, then $\lambda \in \sigma_{le}(A)$ by (4.3). The proof of other inclusion is left to the reader. ∎

4.5. Proposition. *If* N *is a normal operator and* $\lambda \in \sigma(N)$, *then* $\operatorname{ran}(N - \lambda)$ *is closed if and only if* λ *is not a limit point of* $\sigma(N)$.

PROOF. Assume λ is an isolated point of $\sigma(N)$; thus $X = \sigma(N) \backslash \{\lambda\}$ is a closed subset of $\sigma(N)$. If $N = \int z \, dE(z)$ and $\mathcal{H}_1 = E(X)\mathcal{H}$, then $\mathcal{H}_1$ reduces N and $\sigma(N | \mathcal{H}_1) = X$. Hence $(N - \lambda)\mathcal{H}_1$ is closed. Since $\mathcal{H}_1^{\perp} = \ker(N - \lambda)$, $\operatorname{ran}(N - \lambda) = (N - \lambda)\mathcal{H}_1$; hence $N - \lambda$ has closed range.

Now assume that $\lambda \in \sigma(N)$ but λ is not an isolated point. Then there is a strictly decreasing sequence $\{r_n\}$ of positive real numbers such that $r_n \to 0$ and such that each open annulus $A_n = \{z: r_{n+1} < |z - \lambda| < r_n\}$ has non-empty intersection with $\sigma(N)$. Thus $E(A_n)\mathcal{H} \neq (0)$; let e_n be a unit vector in $E(A_n)\mathcal{H}$. Then $e_n \perp \ker(N - \lambda) \, (= E(\{\lambda\})\mathcal{H})$ and

$$\|(N - \lambda)e_n\|^2 = \int_{A_n} |z - \lambda|^2 \, dE_{e_n, e_n}(z) \leqslant r_n^2 \to 0.$$

That is, $\inf \{ \|(N - \lambda)h\|: \|h\| = 1, \, h \perp \ker(N - \lambda) \} = 0$ and so, by the Open Mapping Theorem, $N - \lambda$ does not have closed range. ∎

4.6. Proposition. *If* N *is a normal operator, then* $\sigma_e(N) = \sigma_{le}(N) = \sigma_{re}(N)$ *and* $\sigma(N) \backslash \sigma_e(N) = \{\lambda \in \sigma(N): \lambda$ *is an isolated point of* $\sigma(N)$ *that is an eigenvalue of finite multiplicity*\}.

PROOF. The first part follows by applying Proposition 1.3 to the Calkin algebra. If λ is an isolated point of $\sigma(N)$, then $\operatorname{ran}(N - \lambda)$ is closed by the preceding proposition. So if $\dim \ker(N - \lambda) < \infty$, $\lambda \notin \sigma_{le}(N) = \sigma_e(N)$ by Proposition 4.3. Conversely, if $\lambda \in \sigma(N) \backslash \sigma_e(N)$, then $\operatorname{ran}(N - \lambda)$ is closed and $\dim \ker(N - \lambda) < \infty$. By the preceding proposition, λ is an isolated point of $\sigma(N)$. ∎

4.7. Example. Let G be a bounded region in $\mathbb{C}$ and, to avoid pathologies, assume $\partial G = \partial [\operatorname{cl} G]$. Let $\mathcal{H} = L_a^2(G)$ (I.1.10) and define $S: \mathcal{H} \to \mathcal{H}$ by $(Sf)(z) = zf(z)$. Then $\sigma(S) = \operatorname{cl} G$, $\sigma_e(S) = \sigma_{le}(S) = \sigma_{re}(S) = \partial G = \sigma_{ap}(S)$,

$\sigma_p(S) = \square$, and for λ in G, $\mathrm{ran}(S - \lambda)$ is closed and $\dim[\mathrm{ran}(S - \lambda)]^{\perp} = 1$. Thus $\mathrm{ind}(S - \lambda) = -1$ for λ in G.

To show that these statements are true, begin by proving:

4.8   If $\lambda \in G$, $\mathrm{ran}(S - \lambda) = \{f \in L_a^2(G): f(\lambda) = 0\}$.

In fact, if $h \in L_a^2(G)$, then $[(S - \lambda)h](z) = (z - \lambda)h(z)$ so that $f = (z - \lambda)h$ vanishes at λ. Conversely, suppose $f \in L_a^2(G)$ and $f(\lambda) = 0$; then $f(z) = (z - \lambda)h(z)$ for some analytic function h on G. It must be shown that $h \in L_a^2(G)$. Let $r > 0$ such that $D = \{z: |z - \lambda| \leqslant r\} \subseteq G$. Then

$$\iint |h|^2 = \iint_D |h|^2 + \iint_{G \setminus D} |h|^2.$$

Now $\iint_D |h|^2 < \infty$ since h is bounded on D. For z in $G \setminus D$, $|h(z)| = |f(z)|/|z - \lambda| \leqslant r^{-1}|f(z)|$. Hence

$$\iint_{G \setminus D} |h|^2 \leqslant r^{-2} \iint_G |f|^2 < \infty.$$

Thus $h \in L_a^2(G)$ and $f = (S - \lambda)h$. This proves (4.8).

Using Corollary I.1.12, $f \mapsto f(\lambda)$ is a bounded linear functional on $L_a^2(G)$ whenever $\lambda \in G$. By (4.8), $\mathrm{ran}(S - \lambda)$ is the kernel of this linear functional and hence is closed.

Because G is bounded, the constant functions belong to $L_a^2(G)$. So if $f \in L_a^2(G)$, $f = [f - f(\lambda)] + f(\lambda)$ and $f - f(\lambda) \in \mathrm{ran}(S - \lambda)$. Thus $L_a^2(G) = \mathrm{ran}(S - \lambda) + \mathbb{C}$. Therefore $\dim[\mathrm{ran}(S - \lambda)]^{\perp} = \dim[L_a^2(G)/\mathrm{ran}(S - \lambda)] = 1$ when $\lambda \in G$.

If $\lambda \in G$, then $S - \lambda$ is not surjective; hence $G \subseteq \sigma(S)$. If $\lambda \notin \mathrm{cl}\, G$, then $(z - \lambda)^{-1}$ is a bounded analytic function on G. If $Af = (z - \lambda)^{-1}f$, then A is a bounded operator on $L_a^2(G)$ and it is easy to check that $A(S - \lambda) = (S - \lambda)A = 1$. Thus $\sigma(S) \subseteq \mathrm{cl}\, G$. Combining these two containments, we get $\sigma(S) = \mathrm{cl}\, G$.

From Corollary 2.4 we have that $S - \lambda$ is a Fredholm operator whenever $\lambda \in G$; thus $G \cap \sigma_e(S) = \square$. So $\sigma_e(S) \subseteq \partial G = \partial[\mathrm{cl}\, G]$. If $\lambda \in \partial G$, then $\lambda \in \partial \sigma(S)$; thus $\lambda \in \sigma_{ap}(S)$ (1.2). Since $\ker(S - \lambda) = (0)$, $\mathrm{ran}(S - \lambda)$ is not closed. Thus $\partial G \subseteq \sigma_{le}(S) \cap \sigma_{re}(S)$. This proves that $\sigma_e(S) = \sigma_{le}(S) = \sigma_{re}(S) = \partial G = \sigma_{ap}(S)$.

One of the primary uses of Fredholm theory is the examination of the values of $\mathrm{ind}(A - \lambda)$ for all λ for which this makes sense. Such an examination often leads to structural information about the operator A. Note that $\mathrm{ind}(A - \lambda)$ is defined when $\lambda \notin \sigma_{le}(A) \cap \sigma_{re}(A)$.

4.9. Proposition. *If $A \in \mathcal{B}(\mathcal{H})$, then $\mathrm{ind}(A - \lambda)$ is constant on the components of $\mathbb{C} \setminus \sigma_{le}(A) \cap \sigma_{re}(A)$. If λ is a boundary point of $\sigma(A)$ and $\lambda \notin \sigma_{le}(A) \cap \sigma_{re}(A)$, then $\mathrm{ind}(A - \lambda) = 0$.*

PROOF. The map $\lambda \mapsto A - \lambda$ is a continuous map of $\mathbb{C} \setminus \sigma_{le}(A) \cap \sigma_{re}(A)$ into $\mathcal{S}\mathcal{F}$. So the first part of the proposition follows from Corollary 3.13. If λ is a boundary point of $\sigma(A)$ and $\lambda \notin \sigma_{le}(A) \cap \sigma_{re}(A)$, then there is a sequence $\{\lambda_n\}$ in $\mathbb{C} \backslash \sigma(A)$ such that $\lambda_n \to \lambda$. Thus $\text{ind}(A - \lambda_n) \to \text{ind}(A - \lambda)$. Since $\text{ind}(A - \lambda_n) = 0$ for all n, the result follows. ∎

Here is the remaining spectral information about an old friend.

4.10. Example. Let S be the unilateral shift on l^2. Then $\sigma_{le}(S) = \sigma_{re}(S) = \partial \mathbb{D}$ and $\text{ind}(S - \lambda) = -1$ for $|\lambda| < 1$.

In Proposition VII.6.5 it was shown that $\sigma(S) = \text{cl } \mathbb{D}$, $\sigma_p(S) = \square$, and $\sigma_{ap}(S) = \partial \mathbb{D}$. Thus for $|\lambda| = 1$, $\text{ran}(S - \lambda)$ is not closed and hence $\partial \mathbb{D} \subseteq \sigma_{le}(S) \cap \sigma_{re}(S)$. Also, if $|\lambda| < 1$, it was shown that $\text{ran}(S - \lambda)$ is closed and $\dim[\text{ran}(S - \lambda)]^\perp = 1$. This implies that $\partial \mathbb{D} = \sigma_{le}(S) = \sigma_{re}(S)$ and $\text{ind}(S - \lambda) = -1$ for λ in $\mathbb{D}$.

Using this information about the shift and Proposition 3.4(c) we can get complete information about another operator. (Also see Example 3.10.)

4.11. Example. Let S be the unilateral shift of multiplicity 1 and put $A = S \oplus S^*$. It follows that $\sigma_{le}(A) = \sigma_{re}(A) = \partial \mathbb{D}$, $\sigma(A) = \text{cl } \mathbb{D}$, and $\text{ind}(A - \lambda) = 0$ for $|\lambda| < 1$.

EXERCISES

1. Show that the material of this section is only significant for infinite dimensional Hilbert spaces by showing that the essential spectrum of every operator on $\mathscr{H}$ is non-empty if and only if $\mathscr{H}$ is infinite dimensional.

2. Let G be a bounded region in $\mathbb{C}$ such that $\partial G = \partial[\text{cl } G]$ and let ϕ be a function that is analytic in a neighborhood of cl G. Define $A: L_a^2(G) \to L_a^2(G)$ by $Af = \phi f$. Find all of the parts of the spectrum of A.

3. (Fillmore, Stampfli, and Williams [1972].) If $\lambda \in \sigma_{le}(A)$, then there is a projection P, having infinite rank, such that $\pi(A - \lambda)\pi(P) = 0$.

4. (Fillmore, Stampfli, and Williams [1972].) Let $A \in \mathscr{B}(\mathscr{H})$. (a) If A has a cyclic vector e, show that $\dim\{Ae, A^2e, \ldots\}^\perp \leqslant 1$. (b) Let $\lambda \in \sigma_{le}(A^*)$. If $\varepsilon > 0$, let f_1, f_2 be orthonormal vectors such that $\|(A^* - \lambda)f_j\| < \varepsilon$ for $j = 1, 2$ and let $P =$ the projection onto $\vee\{f_1, f_2\}$. Put $B = \bar{\lambda}P + (1 - P)A$. Show that $\|B - A\| < 2\varepsilon$. (c) Show that the noncyclic operators are dense in $\mathscr{B}(\mathscr{H})$ if $\dim \mathscr{H} > 1$.

5. Let $A \in \mathscr{F}$ and suppose f is analytic in a neighborhood of $\sigma(A)$ and does not vanish on $\sigma_e(A)$. Show that $f(A) \in \mathscr{F}$ and find $\text{ind } f(A)$.

6. Let S be the unilateral shift and let f be an analytic function in a neighborhood of cl $\mathbb{D}$ such that $f(z) \neq 0$ if $|z| = 1$. Let $\gamma(t) = f(\exp(2\pi it))$, $0 \leqslant t \leqslant 1$. Show that $\sigma_e(f(S)) = f(\partial \mathbb{D}) = \{\gamma(t): 0 \leqslant t \leqslant 1\}$ and that if $\lambda \notin f(\partial \mathbb{D})$, $\text{ind}(f(S) - \lambda) = -n(\gamma; \lambda)$, where $n(\gamma; \lambda) =$ the winding number of γ about λ. Moreover, show that if $\text{ind}(f(S) - \lambda) = 0$, then $\lambda \notin \sigma(f(S))$.

7. Let S be the operator defined in Example 2.11 where $G = \mathbb{D}$. Show that there is a compact operator K such that $S + K$ is unitarily equivalent to the unilateral shift.

8. If S is the unilateral shift, show that for every $\varepsilon > 0$ there is a rank one operator F with $\| F \| < \varepsilon$ such that $\sigma(S^* \oplus S + F) = \partial \mathbb{D}$.

9. Let G be an open connected subset of $\sigma(A) \backslash \sigma_{le}(A) \cup \sigma_{re}(A)$ and suppose $\lambda_0 \in G$ such that $\text{ind}(A - \lambda_0) = 0$. Show that there is a finite rank operator F such that $A + F - \lambda_0$ is invertible. Show that $A + F - \lambda$ is invertible for every λ in G.

§5. The Components of $\mathcal{S}\mathcal{F}$

Since the index is continuous on $\mathcal{S}\mathcal{F}$ and assumes every possible value (Exercise 3.4), $\mathcal{S}\mathcal{F}$ cannot be connected. What are its components?

Note that because $\mathcal{S}\mathcal{F}$ is an open subset of a Banach space, its components are arcwise connected (Exercise IV.1.24).

5.1. Theorem. *If $A, B \in \mathcal{S}\mathcal{F}$, then A and B belong to the same component of $\mathcal{S}\mathcal{F}$ if and only if $\text{ind}\, A = \text{ind}\, B$.*

Half of this theorem is easy. For the other half we first prove a lemma.

5.2. Lemma. *If $A \in \mathcal{F}$ and $\text{ind}\, A = 0$, then there is a path $\gamma: [0,1] \to \mathcal{F}$ such that $\gamma(0) = 1$ and $\gamma(1) = A$.*

PROOF. By Exercise 3.7 there is a finite-rank operator F such that $A + F$ is invertible. If $\gamma(t) = A + tF$, $\gamma(0) = A$, $\gamma(1) = A + F$, and $\gamma(t) \in \mathcal{F}$ for all t. Thus we may assume that A is invertible.

Let $A = U|A|$ be the polar decomposition of A. Because A is invertible, U is a unitary operator and $|A|$ is invertible. Using the Spectral Theorem, $U = \exp(iB)$ where B is hermitian. Also, since $0 \notin \sigma(|A|)$, $|A| = \int_{[\delta,r]} x\, dE(x)$, where $0 < \delta < r = \| A \|$. Define $\gamma: [0,1] \to \mathcal{B}(\mathcal{H})$ by

$$\gamma(t) = e^{itB} \int_{[\delta,r]} x^t\, dE(x) = e^{itB} |A|^t.$$

It is easy to check that γ is continuous, $\gamma(0) = 1$, and $\gamma(1) = A$. Also, each $\gamma(t)$ is invertible so $\gamma(t) \in \mathcal{F}$. ∎

PROOF OF THEOREM 5.1. First assume that $A, B \in \mathcal{F}$ and $\text{ind}\, A = \text{ind}\, B$. So there is an operator C such that $CB = 1 + K$ for some compact operator K. Thus $C \in \mathcal{F}_r$ and $\text{ind}\, C = -\text{ind}\, B = -\text{ind}\, A$. Hence $AC \in \mathcal{F}$ and $\text{ind}\, AC = 0$. By the preceding lemma there is a path $\gamma: [0,1] \to \mathcal{F}$ such that $\gamma(0) = 1$ and $\gamma(1) = AC$. Put $\rho(t) = \gamma(t)B - tAK$. Because $AK \in \mathcal{B}_0$, $\rho(t) \in \mathcal{F}$ for all t in $[0,1]$. Also, $\rho(0) = B$ and $\rho(1) = ACB - AK = A(1 + K) - AK = A$.

Now assume that $\text{ind}\, A = -\infty$; so $\dim(\text{ran}\, A)^\perp = \infty$ and $\dim \ker A < \infty$. Let F be a finite-rank operator such that $\ker(A + tF) = 0$ for $t \neq 0$. (Why does F exist?) This path shows that we may assume that $\ker A = (0)$. Let V be any isometry such that $\dim(\text{ran}\, V)^\perp = \infty$ and consider the polar decomposition $A = U|A|$ of A. Since $A \in \mathscr{S}\mathscr{F}$ and $\ker A = (0)$, $|A|$ is invertible and U is an isometry. Also, $\operatorname{ran} U = \operatorname{ran} A$, so (Exercise 4) there is a path $\rho\colon [0, 1] \to \mathscr{B}(\mathscr{H})$ such that $\rho(t)$ is an isometry for every t and $\rho(0) = U$ and $\rho(1) = V$. Let $\gamma\colon [0, 1] \to \mathscr{F}$ be a path such that $\gamma(0) = |A|$ and $\gamma(1) = 1$. Then $\sigma(t) = \rho(t)\gamma(t)$ for $0 \leqslant t \leqslant 1$ defines a path $\sigma\colon [0, 1] \to \mathscr{S}\mathscr{F}$ (Why?) such that $\sigma(0) = A$ and $\sigma(1) = V$. Similarly, if $\operatorname{ind} B = -\infty$, there is a path connecting B to V; so A and B belong to the same component of $\mathscr{S}\mathscr{F}$.

If $\operatorname{ind} A = \operatorname{ind} B = +\infty$, apply the preceding paragraph to A^* and B^*. ∎

5.3. Corollary. *The component of the identity in $\mathscr{F}$, $\mathscr{F}_0$, is a normal sub-group of $\mathscr{F}$ and $\mathscr{F}/\mathscr{F}_0$ is an infinite cyclic group.*

PROOF. By Theorem 3.7, $\operatorname{ind} \mathscr{F} \to \mathbb{Z}$ is a group homomorphism and it is surjective (Exercise 3.4). By Theorem 5.1, $\ker(\operatorname{ind}) = \mathscr{F}_0$. ∎

EXERCISES

1. Let G be any topological group and let G_0 be the component of the identity. Show that G_0 is a normal subgroup of G.

2. What are the components of the set of invertible elements in $C(\partial \mathbb{D})$?

3. If S = the unilateral shift, what are the components of the set of invertible elements of $C^*(S)$?

4. If U and V are isometries, show that there is a path consisting entirely of isometries connecting U to V if and only if $\dim(\operatorname{ran} U)^\perp = \dim(\operatorname{ran} V)^\perp$.

5. Find the components of the set of partial isometries.

§6. A Finer Analysis of the Spectrum

In this section we will examine the spectrum and index more closely. For example, one question that arises: If A is semi-Fredholm and $\delta > 0$ is chosen so that B is semi-Fredholm and $\operatorname{ind} B = \operatorname{ind} A$ whenever $\|B - A\| < \delta$ (Corollary 3.13), how does $\dim \ker B$ differ from $\dim \ker A$?

Begin this investigation by associating with every bounded operator A the following number:

$$\gamma(A) = \inf\{\|Ah\|\colon \|h\| = 1 \text{ and } h \perp \ker A\}.$$

The first proposition contains some elementary properties of γ and its proof is left to the reader. (Actually part (a) has appeared several times in this book under different guises.)

6.1. Proposition. *Let A be a bounded operator on $\mathscr{H}$.*

(a) $\gamma(A) > 0$ *if and only if A has closed range.*

(b) $\gamma(A) = \sup\{\gamma : \|Ah\| \geqslant \gamma\|h\|$ *for all* h *in* $(\ker A)^{\perp}\} = \inf\{\|Ah\|/\|h\|:$ $h \in (\ker A)^{\perp}\backslash\{0\}\}$.

6.2. Proposition. *If* $A \in \mathcal{B}(\mathcal{H})$, *then* $\gamma(A) = \gamma(A^*)$.

PROOF. Let $h \perp \ker A$. Then $\|A^*Ah\| = \||A|A|h\| = \||A|A|h\|$. But $|A|h \in$ cl ran A^* (Why?) $= \ker A^{\perp}$. Hence the definition of $\gamma(A)$ implies that $\|A^*Ah\| \geqslant \gamma(A)\||A|h\| = \gamma(A)\|Ah\|$; that is, $\|A^*f\| \geqslant \gamma(A)\|f\|$ for every f in ran A. Since ran A is dense in $(\ker A^*)^{\perp}$, $\gamma(A^*) \geqslant \gamma(A)$. But $A = A^{**}$, so $\gamma(A) \geqslant \gamma(A^*)$. ∎

Note that the preceding two propositions give a proof of the fact that an operator on a Hilbert space has closed range if and only if its adjoint does. See Theorem VI.1.10 and Exercise 2.1.

6.3. Lemma. *If* $\mathcal{M}, \mathcal{N} \leqslant \mathcal{H}$, $\mathcal{N}$ *is finite dimensional, and* $\dim \mathcal{M} > \dim \mathcal{N}$, *then there is a non-zero vector* m *in* $\mathcal{M}$ *such that* $\|m\| = \mathrm{dist}(m, \mathcal{N})$.

PROOF. Let P be the projection of $\mathcal{H}$ onto $\mathcal{M}$, so $\dim P(\mathcal{N}) \leqslant \dim \mathcal{N} < \dim \mathcal{M}$. Thus $P(\mathcal{N})$ is a proper subspace of $\mathcal{M}$; let $m \in \mathcal{M} \cap P(\mathcal{N})^{\perp}$. If $n \in \mathcal{N}$, then $0 = \langle Pn, m \rangle = \langle n, Pm \rangle = \langle n, m \rangle$, so $m \perp \mathcal{N}$. Hence $\|m\| = \mathrm{dist}(m, \mathcal{N})$. ∎

6.4. Lemma. *If* $h \in \mathcal{H}$, *then* $\gamma(A)\mathrm{dist}(h, \ker A) \leqslant \|Ah\|$.

PROOF. Let P be the projection of $\mathcal{H}$ onto $\ker A^{\perp}$; then $\|Ph\| = \mathrm{dist}(h, \ker A)$. Hence $\|Ah\| = \|APh\| \geqslant \gamma(A)\|Ph\| = \gamma(A)\mathrm{dist}(h, \ker A)$. ∎

If the role of $\gamma(A)$ in the next and subsequent propositions impresses the reader as somewhat mysterious, reflect that if A is invertible, then $\gamma(A) = \|A^{-1}\|^{-1}$ (Exercise 1). Now in Corollary VII.2.3, it was shown that if $\mathcal{A}$ is a Banach algebra, $a_0 \in \mathcal{A}$, and $b_0 a_0 = 1$, then $a + b$ is left invertible whenever $\|b\| < \|b_0\|^{-1}$. Of course, a similar result holds for right invertible elements. The number $\gamma(A)$ is trying to play the role of the reciprocal of the norm of a one-sided inverse.

For example, if A is left invertible, then ran A is closed and $\ker A = (0)$; hence $A \in \mathcal{SF}$. The next result implies that if $\|B\| < \gamma(A)$, then $A + B$ is left invertible.

6.5. Proposition. *If* $A \in \mathcal{SF}$ *and* $B \in \mathcal{B}(\mathcal{H})$ *such that* $\|B\| < \gamma(A)$, *then* $A + B \in \mathcal{SF}$ *and:*

(a) $\dim \ker(A + B) \leqslant \dim \ker A$;
(b) $\dim \mathrm{ran}(A + B)^{\perp} \leqslant \dim \mathrm{ran}\, A^{\perp}$.

PROOF. First note that because $A \in \mathcal{SF}$, $\gamma(A) > 0$.

If $h \in \ker(A + B)$ and $h \neq 0$, then $Ah = -Bh$. By Lemma 6.4, $\gamma(A)\mathrm{dist}(h, \ker A) \leqslant \|Bh\| \leqslant \|B\|\|h\| < \gamma(A)\|h\|$. Thus $\mathrm{dist}(h, \ker A) < \|h\|$ for every nonzero vector h in $\ker(A + B)$. By Lemma 6.3, (a) holds.

Since $\| B \| = \| B^* \|$ and $\gamma(A) = \gamma(A^*)$, (a) implies that $\dim \ker(A^* + B^*) \leqslant \dim \ker A^*$. But this inequality is equivalent to (b).

It remains to prove that $\operatorname{ran}(A + B)$ is closed. Since $A \in \mathscr{SF}$, either $\dim \ker A < \infty$ or $\dim \ker A^* < \infty$. Suppose $\dim \ker A < \infty$. It will be shown that $A + B \in \mathscr{F}_l$ by using Theorem 2.3(f) and showing that if $\delta < \gamma(A) - \| B \|$, then $\{ h: \| (A + B)h \| \leqslant \delta \| h \| \}$ contains no infinite dimensional manifold. Indeed, if it did, it would contain a finite dimensional subspace $\mathscr{M}$ with $\dim \mathscr{M} > \dim \ker A$. By Lemma 6.3 there is a non-zero vector h in $\mathscr{M}$ with $\| h \| = \operatorname{dist}(h, \ker A)$. Now $\| (A + B)h \| \leqslant \delta \| h \|$, so Lemma 6.4 implies that $\gamma(A) \| h \| = \gamma(A)\operatorname{dist}(h, \ker A) \leqslant \| Ah \| \leqslant \| (A + B)h \| + \| Bh \| \leqslant (\delta + \| B \|) \| h \| < \gamma(A) \| h \|$, a contradiction. Thus $A + B \in \mathscr{F}_l$ and so $\operatorname{ran}(A + B)$ is closed.

If $\dim \ker A = \infty$, then $\dim \ker A^* < \infty$. The argument of the preceding paragraph gives that $\operatorname{ran}(A^* + B^*)$ is closed. By (VI.1.10), $\operatorname{ran}(A + B)$ is closed. ∎

6.6. Proposition. *If $A \in \mathscr{SF}$ and either $\ker A = (0)$ or $\operatorname{ran} A = \mathscr{H}$, then there is a $\delta > 0$ such that if $\| B - A \| < \delta$, then $\dim \ker B = \dim \ker A$ and $\dim \operatorname{ran} B = \dim \operatorname{ran} A$.*

PROOF. By Proposition 6.5 and Theorem 3.12 there is a $\delta > 0$ such that if $\| B - A \| < \delta$, then $\operatorname{ind} A = \operatorname{ind} B$, $\dim \ker B \leqslant \dim \ker A$, and $\dim \operatorname{ran} B^{\perp} \leqslant \dim \operatorname{ran} A^{\perp}$. Since one of these dimensions for A is 0, the proposition is proved. ∎

If both $\ker A$ and $\operatorname{ran} A$ are nonzero, then there are semi-Fredholm operators B that are arbitrarily close to A such that $\dim \ker B < \dim \ker A$ (see Exercise 2). In fact, just about anything that can go wrong here does go wrong. However, $\dim \ker(A - \lambda)$ does behave rather nicely as a function of λ.

6.7. Theorem. *If $\lambda \notin \sigma_{le}(A) \cap \sigma_{re}(A)$, then there is a $\delta > 0$ such that $\dim \ker(A - \mu)$ and $\dim \operatorname{ran}(A - \mu)^{\perp}$ are constant for $0 < | \mu - \lambda | < \delta$.*

PROOF. We may assume that $\lambda = 0$. We may also assume that $\ker A$ is finite dimensional. Indeed, if this is not the case, then $\ker A^*$ must be finite dimensional and the proof that follows applies to A^*. But observe that if the conclusion of the theorem is demonstrated for A^*, then it also holds for A. It follows that for each $n \geqslant 1$, $A^n \in \mathscr{SF}$ (Theorem 3.7). Thus $\mathscr{M}_n = \operatorname{ran} A^n$ is closed. Note that $\mathscr{M}_{n+1} \subseteq \mathscr{M}_n$ and $A\mathscr{M}_n = \mathscr{M}_{n+1}$. Let $\mathscr{M} = \bigcap_{n=1}^{\infty} \mathscr{M}_n$ and put $B = A | \mathscr{M}$.

Claim. $B\mathscr{M} = \mathscr{M}$.

Since $\ker A$ is finite dimensional and the spaces $\mathscr{M}_n$ are decreasing, there is an integer m such that $\mathscr{M}_n \cap \ker A = \mathscr{M}_m \cap \ker A$ for all $n \geqslant m$. If $h \in \mathscr{M}$ and $n \geqslant m$, there is an f_n in $\mathscr{M}_n$ such that $h = Af_n$. But $f_m - f_n \in (\ker A) \cap \mathscr{M}_m = (\ker A) \cap \mathscr{M}_n$. Therefore $f_m \in \mathscr{M}_n$ for every $n \geqslant m$. That is, $f_m \in \mathscr{M}$ and so $h = Af_m = Bf_m \in B\mathscr{M}$.

Thus $B \in \mathscr{S}\mathscr{F}$ and $\operatorname{ind} B = \dim \ker B$. By (3.12) and (6.5) there is a $\delta > 0$ such that if $|\mu| < \delta$, then $\dim \ker(B - \mu) \leqslant \dim \ker B$, $\dim \operatorname{ran}(B - \mu)^\perp = 0$, and $\operatorname{ind}(B - \mu) = \operatorname{ind} B$. Thus $\dim \ker(B - \mu) = \dim \ker B$ for $|\mu| < \delta$. Also, choose δ such that $\operatorname{ind}(A - \mu) = \operatorname{ind} A$ for $|\mu| < \delta$.

On the other hand, if $\mu \neq 0$, then $\ker(A - \mu) \subseteq \mathscr{M}$. In fact, if $h \in \ker(A - \mu)$, then $A^n h = \mu^n h$, so that $h = A^n(\mu^{-n} h) \in \mathscr{M}_n$ for every n. Thus for $0 < |\mu| < \delta$, $\dim \ker(A - \mu) = \dim \ker(B - \mu) = \dim \ker B$; that is, $\dim \ker(A - \mu)$ is constant for $0 < |\mu| < \delta$. Since $\operatorname{ind}(A - \mu)$ is constant for these values of μ, $\dim \operatorname{ran}(A - \mu)^\perp$ is also constant. ∎

The next result is from Putnam [1968].

6.8. Theorem. *If* $\lambda \in \partial\sigma(A)$, *then either* λ *is an isolated point of* $\sigma(A)$ *or* $\lambda \in \sigma_{le}(A) \cap \sigma_{re}(A)$.

PROOF. Suppose $\lambda \in \partial\sigma(A)$ but λ is not a point of $\sigma_{le}(A) \cap \sigma_{re}(A)$. Thus $A - \lambda \in \mathscr{S}\mathscr{F}$. By the preceding theorem, there is a $\delta > 0$ such that $A - \mu \in \mathscr{S}\mathscr{F}$ and $\dim \ker(A - \mu)$ and $\dim \operatorname{ran}(A - \mu)^\perp$ are constant for $0 < |\mu - \lambda| < \delta$. Since $\lambda \in \partial\sigma(A)$, there is a ν with $0 < |\lambda - \nu| < \delta$ such that $A - \nu$ is invertible. Therefore $\ker(A - \mu) = (0) = \operatorname{ran}(A - \mu)^\perp$ for $0 < |\lambda - \mu| < \delta$. But $A - \mu \in \mathscr{S}\mathscr{F}$ for these values of μ and hence has closed range. It follows that $A - \mu$ is invertible whenever $0 < |\lambda - \mu| < \delta$. This says that λ is an isolated point of $\sigma(A)$. ∎

What happens if λ is an isolated point of $\sigma(A)$?

6.9. Proposition. *If* λ *is an isolated point of* $\sigma(A)$, *the following statements are equivalent.*

(a) $\lambda \notin \sigma_{le}(A) \cap \sigma_{re}(A)$.
(b) *The Riesz idempotent* $E(\lambda)$ *has finite rank.*
(c) $A - \lambda \in \mathscr{F}$ *and* $\operatorname{ind}(A - \lambda) = 0$.

PROOF. Exercise 4.

If $n \in \mathbb{Z} \cup \{\pm \infty\}$ and $A \in \mathscr{B}(\mathscr{H})$, define

$$P_n(A) \equiv \{\lambda \in \sigma(A): A - \lambda \in \mathscr{S}\mathscr{F} \text{ and } \operatorname{ind}(A - \lambda) = n\}.$$

So for $n \neq 0$, $P_n(A)$ is an open subset of the plane; the set $P_0(A)$ consists of an open set together with some isolated points of $\sigma(A)$. In fact, Proposition 6.9 can be used to show that $P_0(A)$ contains precisely the isolated points of $\sigma(A)$ for which the Riesz idempotent has finite rank. The proof of the next result is easy.

6.10. Proposition. *If* $A \in \mathscr{B}(\mathscr{H})$, *then* $\sigma_e(A) = [\sigma_{le}(A) \cap \sigma_{re}(A)] \cup P_{+\infty}(A) \cup P_{-\infty}(A)$.

6.11. Definition. If $A \in \mathcal{B}(\mathcal{H})$, then the *Weyl spectrum* of A, $\sigma_w(A)$, is defined by

$$\sigma_w(A) = \cap \{\sigma(A + K): K \in \mathcal{B}_0\}.$$

Note that since $\sigma_e(A + K) = \sigma_e(A)$ for every compact operator, $\sigma_w(A)$ is nonempty and $\sigma_e(A) \subseteq \sigma_w(A)$. The way to think of the Weyl spectrum is that it is the largest part of the spectrum of A that remains unchanged under compact perturbations. It is clear that $\sigma_w(A) = \sigma_w(A + K)$ for every K in $\mathcal{B}_0$ and $\sigma_w(A) \subseteq \sigma(A)$. The following is a result of Schechter [1965].

6.12. Theorem. *If* $A \in \mathcal{B}(\mathcal{H})$, *then* $\sigma_w(A) = \sigma_e(A) \cup \bigcup_{n \neq 0} P_n(A)$.

PROOF. Clearly $X \equiv \sigma_e(A) \cup \bigcup_{n \neq 0} P_n(A) \subseteq \sigma_w(A)$. Now suppose $\lambda \notin X$. Then $A - \lambda \in \mathcal{F}$ and $\text{ind}(A - \lambda) = 0$. By Exercise 3.7 there is a finite rank operator F such that $A + F - \lambda$ is invertible. Hence $\lambda \notin \sigma(A + F)$ so that $\lambda \notin \sigma_w(A)$. ∎

So for every operator A in $\mathcal{B}(\mathcal{H})$ there is a *spectral picture* for A (a term coined in Pearcy [1978]). There are the open sets $\{P_n(A): 0 < |n| \leqslant \infty\}$, the set $P_0(A) = G_0 \cup D$ where D consists of isolated points λ for which $\dim E(\lambda) = n_\lambda < \infty$, and there is the remainder of $\sigma(A)$, which is the set $\sigma_{le}(A) \cap \sigma_{re}(A)$. The next result is due to Conway [1985].

6.13. Proposition. *Let* K *be a compact subset of* $\mathbb{C}$, *let* $\{G_n: -\infty \leqslant n \leqslant \infty\}$ *be disjoint open subsets of* K *(some possibly empty), let* D *be a subset of the set of isolated points of* K, *and for each* λ *in* D *let* $n_\lambda \in \{1, 2, \ldots\}$. *Then there is an operator* A *on* $\mathcal{H}$ *such that* $\sigma(A) = K$, $P_n(A) = G_n$ *for* $0 < |n| \leqslant \infty$, $P_0(A) = G_0 \cup D$, *and* $\dim E(\lambda) = n_\lambda$ *for every* λ *in* D.

We prove only a special case of this result; the general case is left to the reader. Let K be any compact subset of $\mathbb{C}$ and let G be an open subset of K. Put $H = \text{int}[\text{cl } G]$; so $G \subseteq H$, but it may be that $H \neq G$. However, $\partial H = \partial[\text{cl } H]$. Let $Tf = zf$ for f in $L_a^2(H)$, so $H = P_{-1}(T)$, $\sigma(T) = \text{cl } H$, and $\sigma_{le}(T) \cap \sigma_{re}(T) = \partial H = \partial[\text{cl } G]$. Let $\{\lambda_k\}$ be a countable dense subset of $K \backslash G$ and let N be the diagonalizable normal operator with $\sigma_p(N) = \{\lambda_k\}$ and such that $\dim \ker(N - \lambda_k) = \infty$ for each λ_k. If $0 < n \leqslant \infty$ and $A = N \oplus T^{(n)}$, then $\sigma(A) = K$, $P_{-n}(A) = G$, and $K \backslash G = \sigma_{le}(A) \cap \sigma_{re}(A)$.

EXERCISES

1. If A is an invertible operator, show that $\gamma(A) = \|A^{-1}\|^{-1}$.

2. Let $A \in \mathcal{F}$ and suppose that $\ker A \neq (0)$ and $\text{ran } A^\perp \neq (0)$. Show that for every $\delta > 0$ there is an operator B in $\mathcal{F}$ such that $\|B - A\| < \delta$, $\dim \ker B < \dim \ker A$, and $\dim \text{ran } B^\perp < \dim \text{ran } A^\perp$.

3. If $A \in \mathcal{S}\mathcal{F}$, show that there is a $\delta > 0$ such that $\dim \ker(A - \mu) = \dim \ker A$ and $\dim \text{ran}(A - \mu)^\perp = \dim \text{ran } A^\perp$ for $|\mu| < \delta$ if $\ker A \subseteq \text{ran } A^n$ for every $n \geqslant 1$.

4. Prove Proposition 6.9.

5. Prove Proposition 6.10

6. If $\lambda \in \partial P_n(A)$ and $n \neq 0$, show that $\mathrm{ran}(A - \lambda)$ is not closed. What happens if $n = 0$?

7. (Stampfli [1974].) If $A \in \mathcal{B}(\mathcal{H})$, then there is a K in $\mathcal{B}_0(\mathcal{H})$ such that $\sigma(A + K) = \sigma_w(A)$.

8. Prove Proposition 6.13.

9. (Conway [1985].) If L and R are nonempty compact subsets of $\mathbb{C}$, then there is a bounded operator A on $\mathcal{H}$ such that $\sigma_l(A) = L$ and $\sigma_r(A) = R$ if and only if $\partial L \subseteq R$ and $\partial R \subseteq L$.

APPENDIX A
Preliminaries

As was stated in the Preface, the prerequisites for understanding this book are a good course in measure and integration theory and, as a corequisite, analytic function theory. In this and the succeeding appendices an attempt is made to fill in some of the gaps and standardize some notation. These sections are not meant to be a substitute for serious study of these topics.

In Section 1 of this appendix some results from infinite dimensional linear algebra are set forth. Most of this is meant as review. Proposition 1.4, however, seems to be a fact that is not stressed or covered in courses but that is used often in functional analysis. Section 2 on topology is presented mainly to discuss nets. This topic is often not covered in the basic courses and it is especially useful in discussing various ideas and proving results in functional analysis.

§1. Linear Algebra

Let $\mathcal{X}$ be a vector space over $\mathbb{F} = \mathbb{R}$ or $\mathbb{C}$. A subset E of $\mathcal{X}$ is *linearly independent* if for any finite subset $\{e_1, \ldots, e_n\}$ of E and for any finite set of scalars $\{\alpha_1, \ldots, \alpha_n\}$, if $\sum_{k=1}^{n} \alpha_k e_k = 0$, then $\alpha_1 = \cdots = \alpha_n = 0$. A *Hamel basis* is a maximal linearly independent subset of $\mathcal{X}$.

1.1. Proposition. *If E is a linearly independent subset of $\mathcal{X}$, then E is a Hamel basis if and only if every vector x in $\mathcal{X}$ can be written as $x = \sum_{k=1}^{n} \alpha_k e_k$ for scalars $\alpha_1, \ldots, \alpha_n$ and $\{e_1, \ldots, e_n\} \subseteq E$.*

PROOF. Suppose E is a basis and $x \in \mathcal{X}$, $x \notin E$. Then $E \cup \{x\}$ is not linearly independent. Thus there are $\alpha_0, \alpha_1, \ldots, \alpha_n$ in $\mathbb{F}$ and $e_1, \ldots, e_n$ in E such that $0 = \alpha_0 x + \alpha_1 e_1 + \cdots + \alpha_n e_n$, with $\alpha_0 \neq 0$. (Why?) Thus $x = \sum_{k=1}^{n} (-\alpha_k/\alpha_0) e_k$.

Conversely, if $\mathcal{X}$ is the linear span of E, then for every x in $\mathcal{X}\backslash E$, $E\cup\{x\}$ is not linearly independent. Thus E is a basis. ∎

1.2. Proposition. *If E_0 is a linearly independent subset of $\mathcal{X}$, then there is a basis E that contains E_0.*

PROOF. Use Zorn's Lemma.

A *linear functional* on $\mathcal{X}$ is a function $f\colon \mathcal{X}\to\mathbb{F}$ such that $f(\alpha x+\beta y)=\alpha f(x)+\beta f(y)$ for x,y in $\mathcal{X}$ and α,β in $\mathbb{F}$. If $\mathcal{X}$ and $\mathcal{Y}$ are vector spaces over $\mathbb{F}$, a *linear transformation* from $\mathcal{X}$ into $\mathcal{Y}$ is a function $T\colon \mathcal{X}\to\mathcal{Y}$ such that $T(\alpha_1 x_1+\alpha_2 x_2)=\alpha_1 T(x_1)+\alpha_2 T(x_2)$ for x_1,x_2 in $\mathcal{X}$ and α_1,α_2 in $\mathbb{F}$.

If $A,B\subseteq\mathcal{X}$, then $A+B\equiv\{a+b\colon a\in A,\ b\in B\}$; $A-B\equiv\{a-b\colon a\in A,\ b\in B\}$. For α in $\mathbb{F}$ and $A\subseteq\mathcal{X}$, $\alpha A\equiv\{\alpha a\colon a\in A\}$. If $\mathcal{M}$ is a *linear manifold* in $\mathcal{X}$ (that is, $\mathcal{M}\subseteq\mathcal{X}$ and $\mathcal{M}$ is also a vector space with the same operations defined on $\mathcal{X}$), then define $\mathcal{X}/\mathcal{M}$ to be the collection of all the subsets of $\mathcal{X}$ of the form $x+\mathcal{M}$. A set of the form $x+\mathcal{M}$ is called a *coset* of $\mathcal{M}$. Note that $(x+\mathcal{M})+(y+\mathcal{M})=(x+y)+\mathcal{M}$ and $\alpha(x+\mathcal{M})=\alpha x+\mathcal{M}$ since $\mathcal{M}$ is a linear manifold. Hence $\mathcal{X}/\mathcal{M}$ becomes a vector space over $\mathbb{F}$. It is called the *quotient space* of $\mathcal{X}$ mod $\mathcal{M}$.

Define $Q\colon \mathcal{X}\to\mathcal{X}/\mathcal{M}$ by $Q(x)=x+\mathcal{M}$. It is easy to see that Q is a linear transformation. It is called the *quotient map*.

If $T\colon \mathcal{X}\to\mathcal{Y}$ is a linear transformation,

$$\ker T\equiv\{x\in\mathcal{X}\colon Tx=0\},$$
$$\operatorname{ran} T\equiv\{Tx\colon x\in\mathcal{X}\};$$

$\ker T$ is the *kernel* of T and $\operatorname{ran} T$ is the *range* of T. If $\operatorname{ran} T=\mathcal{Y}$, T is *surjective*; if $\ker T=(0)$, T is *injective*. If T is both injective and surjective, then T is *bijective*. It is easy to see that the natural map $Q\colon \mathcal{X}\to\mathcal{X}/\mathcal{M}$ is surjective and $\ker Q=\mathcal{M}$.

Suppose now that $T\colon \mathcal{X}\to\mathcal{Y}$ is a linear transformation and $\mathcal{M}$ is a linear manifold in $\mathcal{X}$. We want to define a map $\hat{T}\colon \mathcal{X}/\mathcal{M}\to\mathcal{Y}$ by $\hat{T}(x+\mathcal{M})=Tx$. But $\hat{T}$ may not be well defined. To ensure that it is we must have $Tx_1=Tx_2$ if $x_1+\mathcal{M}=x_2+\mathcal{M}$. But $x_1+\mathcal{M}=x_2+\mathcal{M}$ if and only if $x_1-x_2\in\mathcal{M}$, and $Tx_1=Tx_2$ if and only if $x_1-x_2\in\ker T$. So $\hat{T}$ is well defined if $\mathcal{M}\subseteq\ker T$. It is easy to check that if $\hat{T}$ is well defined, $\hat{T}$ is linear.

1.3. Proposition. *If $T\colon \mathcal{X}\to\mathcal{Y}$ is a linear transformation and $\mathcal{M}$ is a linear manifold in $\mathcal{X}$ contained in $\ker T$, then there is a linear transformation $\hat{T}\colon \mathcal{X}/\mathcal{M}\to\mathcal{Y}$ such that the diagram*

$$\begin{array}{ccc} \mathcal{X} & \xrightarrow{\ T\ } & \mathcal{Y} \\ & {\scriptstyle Q}\searrow\quad\nearrow{\scriptstyle \hat{T}} & \\ & \mathcal{X}/\mathcal{M} & \end{array}$$

commutes.

The preceding proposition is especially useful if $\mathcal{M} = \ker T$. In that case $\hat{T}$ is injective.

The last proposition of this section will be quite helpful in the book.

1.4. Proposition. *Let* $f, f_1, \ldots, f_n$ *be linear functionals in* $\mathcal{X}$. *If* $\ker f \supseteq \bigcap_{k=1}^n \ker f_k$, *then there are scalars* $\alpha_1, \ldots, \alpha_n$ *such that* $f = \sum_{k=1}^n \alpha_k f_k$ *(that is,* $f(x) = \sum_{k=1}^n \alpha_k f_k(x)$ *for every* x *in* $\mathcal{X}$).

PROOF. It may be assumed without loss of generality that for $1 \leq k \leq n$,

$$\bigcap_{j \neq k} \ker f_j \neq \bigcap_{j=1}^n \ker f_j.$$

(Why?). So for $1 \leq k \leq n$, there is a y_k in $\bigcap_{j \neq k} \ker f_j$ such that $y_k \notin \bigcap_{j=1}^n \ker f_j$. So $f_j(y_k) = 0$ for $j \neq k$, but $f_k(y_k) \neq 0$. Let $x_k = [f_k(y_k)]^{-1} y_k$. Hence $f_k(x_k) = 1$ and $f_j(x_k) = 0$ for $j \neq k$.

Now let f be as in the statement of the proposition and put $\alpha_k = f(x_k)$. If $x \in \mathcal{X}$, let $y = x - \sum_{k=1}^n f_k(x) x_k$. Then $f_j(y) = f_j(x) - \sum_{k=1}^n f_k(x) f_j(x_k) = 0$. By hypothesis, $f(y) = 0$. Thus

$$0 = f(x) - \sum_{k=1}^n f_k(x) f(x_k)$$

$$= f(x) - \sum_{k=1}^n \alpha_k f_k(x);$$

equivalently, $f = \sum_{k=1}^n \alpha_k f_k$. ∎

§2. Topology

In this book all topological spaces are assumed to be Hausdorff.

This section will review some of the concepts and results using *nets*, as this idea is frequently used in the text.

A *directed set* is a partially ordered set $(I, \leq)$ such that if $i_1, i_2 \in I$, then there is an i_3 in I such that $i_3 \geq i_1$ and $i_3 \geq i_2$. A good example of a directed set is to let $(X, \mathscr{S})$ be a topological space and for a fixed x_0 in X let $\mathscr{U} = \{U \text{ in } \mathscr{S}: x_0 \in U\}$. If $U, V \in \mathscr{U}$, define $U \geq V$ if $U \subseteq V$ (so bigger is smaller). $\mathscr{U}$ is said to be *ordered by reverse inclusion*. Another example is found if S is any set and $\mathscr{F}$ is the collection of all finite subsets of S. Define $F_1 \geq F_2$ in $\mathscr{F}$ if $F_1 \supseteq F_2$ (bigger means bigger). Here $\mathscr{F}$ is said to be *ordered by inclusion*. Both of these examples are used frequently in the text.

A *net* in X is a pair $((I, \leq), x)$ where $(I, \leq)$ is a directed set and x is a function from I into X. Usually we will write x_i instead of $x(i)$ and will use the phrase "let $\{x_i\}$ be a net in X."

Note that $\mathbb{N}$, the natural numbers, is a directed set, so every sequence is a net. If $(X, \mathcal{T})$ is a topological space, $x_0 \in X$, and $\mathcal{U} = \{U \text{ in } \mathcal{T}: x_0 \in U\}$, then let $x_U \in U$ for every U in $\mathcal{U}$. So $\{x_U: U \in \mathcal{U}\}$ is a net in X.

2.1. Definition. If $\{x_i\}$ is a net in a topological space X, then $\{x_i\}$ *converges* to x_0 (in symbols, $x_i \to x_0$ or $x_0 = \lim x_i$) if for every open subset U of X such that $x_0 \in U$, there is an $i_0 = i_0(U)$ such that $x_i \in U$ for $i \geq i_0$. The net *clusters* at x_0 (in symbols, $x_i \xrightarrow[\text{cl}]{} x_0$) if for every i_0 and for every open neighborhood U of x_0, there exists an $i \geq i_0$ such that $x_i \in U$.

These notions generalize the corresponding concepts for sequences. Also, if $x_i \to x_0$, then $x_i \xrightarrow[\text{cl}]{} x_0$. Note that the net $\{x_U: U \in \mathcal{U}\}$ defined just prior to the definition converges to x_0. This is a very important example of a convergent net.

2.2. Proposition. *If X is a topological space and $A \subseteq X$, then $x \in \text{cl } A$ (closure of A) if and only if there is a net $\{a_i\}$ in A such that $a_i \to x$.*

PROOF. Let $\mathcal{U} = \{U: U \text{ is open and } x \in U\}$. If $x \in \text{cl } A$, then for each U in $\mathcal{U}$ there is a point a_U in $A \cap U$. If $U_0 \in \mathcal{U}$, then $a_U \in U_0$ for every $U \geq U_0$; therefore $x = \lim a_U$. Conversely, if $\{a_i\}$ is a net in A and $a_i \to x$, then each U in $\mathcal{U}$ contains a point a_i and $a_i \in A \cap U$. Thus $x \in \text{cl } A$. ∎

2.3. Proposition. *If $A \subseteq X$, $\{a_i\}$ is a net in A, and $a_i \xrightarrow[\text{cl}]{} x$, then $x \in \text{cl } A$.*

PROOF. Exercise.

There is a concept of a subnet of a net and with this concept it is possible to prove that if a net clusters at a point x, then there is a subnet that converges to x. The concept of a subnet is, however, somewhat technical and is not what you might at first think it should be. Since this concept is not used in this book, the interested reader is referred to Kelley [1955]. It might also be appropriate to mention that a topological space is Hausdorff if and only if each convergent net has a unique limit point.

2.4. Proposition. *If X and Y are topological spaces and $f: X \to Y$, then f is continuous at x_0 if and only if $f(x_i) \to f(x_0)$ whenever $x_i \to x_0$.*

PROOF. First assume that f is continuous at x_0 and let $\{x_i\}$ be a net in X such that $x_i \to x_0$ in X. If V is open in Y and $f(x_0) \in V$, then there is an open set U in X such that $x_0 \in U$ and $f(U) \subseteq V$. Let i_0 be such that $x_i \in U$ for $i \geq i_0$. Hence $f(x_i) \in V$ for $i \geq i_0$. This says that $f(x_i) \to f(x_0)$.

Let $\mathcal{U} = \{U: U \text{ is open in } X \text{ and } x_0 \in U\}$. Suppose f is not continuous at x_0. Then there is an open subset V of Y such that $f(x_0) \in V$ and $f(U) \backslash V \neq \square$ for every U in $\mathcal{U}$. Thus for each U in $\mathcal{U}$ there is a point x_U in U with $f(x_U) \notin V$. But $\{x_U\}$ is a net in X with $x_U \to x_0$ and clearly $\{f(x_U)\}$ cannot converge to $f(x_0)$. ∎

2.5. Proposition. *If $f: X \to Y$, f is continuous at x_0, and $\{x_i\}$ is a net in X that clusters at x_0, then $\{f(x_i)\}$ clusters at $f(x_0)$.*

PROOF. Exercise.

2.6. Proposition. *Let $K \subseteq X$. Then K is compact if and only if each net in K has a cluster point in K.*

PPROOF. Suppose that K is compact and let $\{x_i : i \in I\}$ be a net in K. For each i let $F_i = \operatorname{cl}\{x_j : j \geqslant i\}$, so each F_i is a closed subset of K. It will be shown that $\{F_i : i \in I\}$ has the finite intersection property. In fact, since I is directed, if $i_1, \ldots, i_n \in I$, then there is an $i \geqslant i_1, \ldots, i_n$. Thus $F_i \subseteq \bigcap_{k=1}^{n} F_{i_k}$ and $\{F_i\}$ has the finite intersection property. Because K is compact, there is an x_0 in $\bigcap_i F_i$. But if U is open with x_0 in U and $i_0 \in I$, the fact that $x_0 \in \operatorname{cl}\{x_i : i \geqslant i_0\}$, implies there is an $i \geqslant i_0$ with x_i in U. Thus $x_i \xrightarrow[\text{cl}]{} x_0$.

Now assume that each net in K has a cluster point in K. Let $\{K_\alpha : \alpha \in A\}$ be a collection of relatively closed subsets of K having the finite intersection property. If $\mathscr{F} = $ the collection of all finite subsets of A, order $\mathscr{F}$ by inclusion. By hypothesis, if $F \in \mathscr{F}$, there is a point x_F in $\bigcap \{K_\alpha : \alpha \in F\}$. Thus $\{x_F\}$ is a net in K. By hypothesis, $\{x_F\}$ has a cluster point x_0 in K. Let $\alpha \in A$, so $\{\alpha\} \in \mathscr{F}$. Thus if U is any open set containing x_0 there is an F in $\mathscr{F}$ such that $\alpha \in F$ and $x_F \in U$. Thus $x_F \in U \cap K_\alpha$; that is, for each α in A and for every open set U containing x_0, $U \cap K_\alpha \neq \square$. Since K_α is relatively closed, $x_0 \in K_\alpha$ for each α in A. Thus $x_0 \in \bigcap_\alpha K_\alpha$ and K must be compact. ∎

The next result is used repeatedly in this book.

2.7. Proposition. *If X is compact, $\{x_i\}$ is a net in X, and x_0 is the only cluster point of $\{x_i\}$, then the net $\{x_i\}$ converges to x_0.*

PROOF. Let U be an open neighborhood of x_0 and let $J = \{j \in I : x_j \notin U\}$. If $\{x_i\}$ does not converge to x_0, then for every i in I there is a j in J such that $j \geqslant i$. In particular, J is also a directed set. Hence $\{x_j : j \in J\}$ is a net in the compact set $X \backslash U$. Thus it has a cluster point y_0. But the property of J mentioned before implies that y_0 is also a cluster point of $\{x_i : j \in I\}$, contradicting the assumption. Thus $x_i \to x_0$. ∎

The next result is rather easy, but it will be used so often that it should be explicitly stated and proved.

2.8. Proposition. *If $f: X \to Y$ is bijective and continuous and X is compact, then f is a homeomorphism.*

PROOF. If F is a closed subset of X, then F is compact. Thus $f(F)$ is compact in Y and hence closed. Since f maps closed sets to closed sets, f^{-1} is continuous. ∎

Note that the Hausdorff property was used in the preceding proof when we said that a compact subset of Y is closed.

In the study of functional analysis it is often the case that the mathematician is presented with a set that has two topologies. It is useful to know how properties of one topology relate to the other and when the two topologies are, in fact, one.

If X is a set and $\mathcal{T}_1, \mathcal{T}_2$ are two topologies on X, say that $\mathcal{T}_2$ is *larger* or *stronger* than $\mathcal{T}_1$ if $\mathcal{T}_2 \supseteq \mathcal{T}_1$; in this case you may also say that $\mathcal{T}_1$ is *smaller* or *weaker*. In the literature there is also an unfortunate nomenclature for these concepts; the words "finer" and "coarser" are used.

The following result is easy to prove (it is an exercise) but it is enormously useful in discussing a set with two topologies.

2.9. Lemma. *If $\mathcal{T}_1, \mathcal{T}_2$ are topologies on X, then $\mathcal{T}_2$ is larger than $\mathcal{T}_1$ if and only if the identity map $i: (X, \mathcal{T}_2) \to (X, \mathcal{T}_1)$ is continuous.*

2.10. Proposition. *Let $\mathcal{T}_1, \mathcal{T}_2$ be topologies on X and assume that $\mathcal{T}_2$ is larger than $\mathcal{T}_1$.*

(a) *If F is $\mathcal{T}_1$-closed, F is $\mathcal{T}_2$-closed.*
(b) *If $f: Y \to (X, \mathcal{T}_2)$ is continuous, then $f: Y \to (X, \mathcal{T}_1)$ is continuous.*
(c) *If $f(X, \mathcal{T}_1) \to Y$ is continuous, then $f: (X, \mathcal{T}_2) \to Y$ is continuous.*
(d) *If K is $\mathcal{T}_2$-compact, then K is $\mathcal{T}_1$-compact.*
(e) *If X is $\mathcal{T}_2$-compact, then $\mathcal{T}_1 = \mathcal{T}_2$.*

PROOF. (b) Note that $f: Y \to (X, \mathcal{T}_1)$ is the composition of $f: Y \to (X, \mathcal{T}_2)$ and $i: (X, \mathcal{T}_2) \to (X, \mathcal{T}_1)$ and use Lemma 2.9.
(d) Use Lemma 2.9.
(e) Use Lemma 2.9 and Proposition 2.8.
    The remainder of the proof is an exercise.   ■

APPENDIX B

The Dual of $L^p(\mu)$

In this section we will prove the following which appears as III.5.5 and III.5.6 in the text.

Theorem. *Let (X, Ω, μ) be a measure space, let $1 \leqslant p < \infty$, and let $1/p + 1/q = 1$. If $g \in L^q(\mu)$, define $F_g: L^p(\mu) \to \mathbb{F}$ by*

$$F_g(f) = \int f g \, d\mu.$$

If $1 < p < \infty$, the map $g \mapsto F_g$ defines an isometric isomorphism of $L^q(\mu)$ onto $L^p(\mu)^*$. If $p = 1$ and (X, Ω, μ) is σ-finite, $g \mapsto F_g$ is an isometric isomorphism of $L^\infty(\mu)$ onto $L^1(\mu)^*$.

PROOF. If $g \in L^q(\mu)$, then Hölder's Inequality implies that $|F_g(f)| \leqslant \|f\|_p \|g\|_q$ for all f in $L^p(\mu)$. Hence $F_g \in L^p(\mu)^*$ and $\|F_g\| \leqslant \|g\|_q$. Therefore $g \mapsto F_g$ is a linear contraction. It must be shown that this map is surjective and an isometry. Assume $F \in L^p(\mu)^*$.

Case 1: $\mu(X) < \infty$. Here $\chi_\Delta \in L^p(\mu)$ for every Δ in Ω. Define $\nu(\Delta) = F(\chi_\Delta)$. It is easy to see that ν is finitely additive. If $\{\Delta_n\} \subseteq \Omega$ with $\Delta_1 \supseteq \Delta_2 \supseteq \cdots$ and $\bigcap_{n=1}^{\infty} \Delta_n = \square$, then

$$\|\chi_{\Delta_n}\|_p = \left[\int |\chi_{\Delta_n}|^p d\mu \right]^{1/p}$$

$$= \mu(\Delta_n)^{1/p} \to 0.$$

Hence $\nu(\Delta_n) \to 0$ since F is bounded. It follows by standard measure theory that ν is a countably additive measure. Moreover, if $\mu(\Delta) = 0$, $\chi_\Delta = 0$ in $L^p(\mu)$; hence $\nu(\Delta) = 0$. that is, $\nu \ll \mu$. By the Radon–Nikodym Theorem there is an Ω-measurable function g such $\nu(\Delta) = \int_\Delta g \, d\mu$ for every Δ in Ω; that is, $F(\chi_\Delta) =$

$\int \chi_\Delta g \, d\mu$ for every Δ in Ω. It follows that

B.1
$$F(f) = \int fg \, d\mu$$

for every simple function f.

B.2. Claim. $g \in L^q(\mu)$ and $\|g\|_q \leqslant \|F\|$.

Note that once this claim is proven, the proof of Case 1 is complete. Indeed, (B.2) says that $F_g \in L^p(\mu)^*$ and since F and F_g agree on a dense subset of $L^p(\mu)$ (B.1), $F = F_g$. Also, $\|g\|_q \leqslant \|F\| = \|F_g\| \leqslant \|g\|_q$.

To prove (B.2), let $t > 0$ and put $E_t = \{x \in X : |g(x)| \leqslant t\}$. If $f \in L^p(\mu)$ such that $f = 0$ off E_t, then there is a sequence $\{f_n\}$ of simple functions such that for every n, $f_n = 0$ off E_t, $|f_n| \leqslant |f|$, and $f_n(x) \to f(x)$ a.e. $[\mu]$. (Why?) Thus $|(f_n - f)g| \leqslant 2t|f|$ and $\int |f| \, d\mu = \int |f| \cdot 1 \, d\mu \leqslant \|f\|_p \mu(X)^{1/q} < \infty$. By the Lebesgue Dominated Convergence Theorem, $F(f_n) = \int f_n g \, d\mu \to \int fg \, d\mu$. Also, $|f_n - f|^p \leqslant 2^p |f|^p$, so $\|f_n - f\|_p \to 0$; thus $F(f_n) \to F(f)$. Combining these results we get that for any $t > 0$ and any f in $L^p(\mu)$ that vanishes off E_t, (B.1) holds.

Case 1a: $1 < p < \infty$. So $1 < q < \infty$. Let $f = \chi_{E_t} |g|^q / g$, where $g(x) \neq 0$, and put $f(x) = 0$ when $g(x) = 0$. If $A = \{x : g(x) \neq 0\}$, then

$$\int |f|^p \, d\mu = \int_{E_t \cap A} \frac{|g|^{pq}}{|g|^p} \, d\mu = \int_{E_t} |g|^q \, d\mu$$

since $pq - p = q$. Therefore

$$\int_{E_t} |g|^q \, d\mu = \int fg \, d\mu = F(f) \leqslant \|F\| \, \|f\|_p = \|F\| \left[\int_{E_t} |g|^q \, d\mu \right]^{1/p}.$$

Thus

$$\|F\| \geqslant \left[\int_{E_t} |g|^q \, d\mu \right]^{1 - 1/p} \geqslant \left[\int_{E_t} |g|^q \, d\mu \right]^{1/q}.$$

Letting $t \to \infty$ gives that $\|g\|_q \leqslant \|F\|$.

Case 1b: $p = 1$. So $q = \infty$. For $\varepsilon > 0$ let $A = \{x : |g(x)| > \|F\| + \varepsilon\}$. For $t > 0$ let $f = \chi_{E_t \cap A} \bar{g} / |g|$. Then $\|f\|_1 = \mu(A \cap E_t)$, and so

$$\|F\| \mu(A \cap E_t) \geqslant \int fg \, d\mu = \int_{A \cap E_t} |g| \, d\mu \geqslant (\|F\| + \varepsilon) \mu(A \cap E_t).$$

Letting $t \to \infty$ we get that $\|F\| \mu(A) \geqslant (\|F\| + \varepsilon) \mu(A)$, which can only be if $\mu(A) = 0$. Thus $\|g\|_\infty \leqslant \|F\|$.

Case 2: (X, Ω, μ) is arbitrary. Let $\mathscr{E} = $ all of the sets E in Ω such that $\mu(E) < \infty$. For E in Ω let $\Omega_E = \{\Delta \in \Omega : \Delta \subseteq E\}$ and define $(\mu|E)(\Delta) = \mu(\Delta)$ for Δ in Ω_E. Put $L^p(\mu|E) = L^p(E, \Omega_E, \mu|E)$ and notice that $L^p(\mu|E)$ can be identified in a natural way with the functions in $L^p(X, \Omega, \mu)$ that vanish off E. Make this identification and consider the restriction of $F: L^p(\mu) \to \mathbb{F}$ to $L^p(\mu|E)$;

denote the restriction by F_E: $L^p(\mu|E) \to \mathbb{F}$. Clearly F_E is bounded and $\|F_E\| \leqslant \|F\|$ for every E in $\mathscr{E}$.

By Case 1, for every E in $\mathscr{E}$ there is a g_E in $L^q(\mu|E)$ such that for f in $L^p(\mu|E)$,

B.3 $$F(f) = \int_E f g_E \, d\mu \text{ and } \|g_E\|_q \leqslant \|F\|.$$

If $D, E \in \mathscr{E}$, then $L^p(\mu|D \cap E)$ is contained in both $L^p(\mu|D)$ and $L^p(\mu|E)$. Moreover, $F_D|L^p(\mu|D \cap E) = F_E|L^p(\mu|D \cap E) = F_{D \cap E}$. Hence $g_D = g_E = g_{D \cap E}$ a.e. $[\mu]$ on $D \cap E$. Thus, a function g can be defined on $\bigcup \{E \colon E \in \mathscr{E}\}$ by letting $g = g_E$ on E; put $g = 0$ off $\bigcup \{E \colon E \in \mathscr{E}\}$. A difficulty arises here in trying to show that g is measurable.

Case 2a: $1 < p < \infty$. Put $\sigma = \sup\{\|g_E\|_q \colon E \in \mathscr{E}\}$; so $\sigma \leqslant \|F\| < \infty$. Since $\|g_D\|_q \leqslant \|g_E\|_q$ if $D \subseteq E$, there is a sequence $\{E_n\}$ in $\mathscr{E}$ such that $E_n \subseteq E_{n+1}$ for all n and $\|g_{E_n}\|_q \to \sigma$. Let $G = \bigcup_{n=1}^{\infty} E_n$. If $E \in \mathscr{E}$ and $E \cap G = \square$, then $\|g_{E \cup E_n}\|_q^q = \|g_E\|_q^q + \|g_{E_n}\|_q^q \to \|g_E\|_q^q + \sigma^q$; thus $g_E = 0$. Therefore $g = 0$ off G and clearly g is measurable. Moreover, $g \in L^q(\mu)$ with $\|g\|_q = \sigma$.

If $f \in L^p(\mu)$, then $\{x \colon f(x) \neq 0\} = \bigcup_{n=1}^{\infty} D_n$ where $D_n \in \mathscr{E}$ and $D_n \subseteq D_{n+1}$ for all n. Thus $\chi_{D_n} f \to f$ in $L^p(\mu)$ and so $F(f) = \lim F(\chi_{D_n} f) = (\text{B.3}) \lim \int_{D_n} g f \, d\mu = \int g f \, d\mu$. Thus $F = F_g$ and $\|F\| = \|F_g\| \leqslant \|g\|_q \leqslant \sigma \leqslant \|F\|$.

Case 2b: $p = \infty$ and (X, Ω, μ) is σ-finite. This is left to the reader. ∎

EXERCISE

Look at the proof of the theorem and see if you can represent $L^1(X, \Omega, \mu)^*$ for an arbitrary measure space.

APPENDIX C

The Dual of $C_0(X)$

The purpose of this section is to show that the dual of $C_0(X)$ is the space of regular Borel measures on X and to put this result, and the accompanying definitions, in the context of complex-valued measures and functions.

Let X be any set and let Ω be a σ-algebra of subsets of X; so (X,Ω) is a *measurable space*. If μ is a countably additive function defined on Ω such that $\mu(\square) = 0$ and $0 \leqslant \mu(\Delta) \leqslant \infty$ for all Δ in Ω, call μ a *positive measure* on (X,Ω); (X,Ω,μ) is called a *measure space*.

If (X,Ω) is a measurable space, a *signed measure* is a countably additive function μ defined on Ω such that $\mu(\square) = 0$ and μ takes its values in $\mathbb{R} \cup \{\pm\infty\}$. (Note: μ can assume only one of the values $\pm\infty$.) It is assumed that the reader is familiar with the following result.

C.1. Hahn–Jordan Decomposition. *If μ is a signed measure on (X, Ω), then $\mu = \mu_1 - \mu_2$, where μ_1 and μ_2 are positive measures, and $X = E_1 \cup E_2$, where $E_1, E_2 \in \Omega$, $E_1 \cap E_2 = \square$, $\mu_1(E_2) = 0 = \mu_2(E_1)$. The measures μ_1 and μ_2 are unique and the sets E_1 and E_2 are unique up to sets of $\mu_1 + \mu_2$ measure zero.*

A *measure* (or *complex-valued measure*) is a complex-valued function μ defined on Ω that is countably additive and such that $\mu(\square) = 0$. Note that μ does not assume any infinite values. If μ is a measure, then $(\operatorname{Re}\mu)(\Delta) \equiv \operatorname{Re}(\mu(\Delta))$ is a signed measure, as is $(\operatorname{Im}\mu)(\Delta) \equiv \operatorname{Im}(\mu(\Delta))$; hence $\mu = \operatorname{Re}\mu + i\operatorname{Im}\mu$. Applying (C.1) to $\operatorname{Re}\mu$ and $\operatorname{Im}\mu$ we get

C.2 $$\mu = (\mu_1 - \mu_2) + i(\mu_3 - \mu_4)$$

where μ_j ($1 \leqslant j \leqslant -4$) are positive measures, $\mu_1 \perp \mu_2$ (μ_1 and μ_2 are *mutually singular*) and $\mu_3 \perp \mu_4$. (C.2) will also be called the Hahn–Jordan decomposition of μ.

C.3. Definition. If μ is a measure on (X, Ω) and $\Delta \in \Omega$, define the *variation* of μ, $|\mu|$, by

$$|\mu|(\Delta) = \sup \left\{ \sum_{j=1}^{m} |\mu(E_j)| : \{E_j\}_1^m \text{ is a measurable partition of } \Delta \right\}.$$

C.4. Proposition. *If μ is a measure on (X, Ω), then $|\mu|$ is a positive finite measure on (X, Ω). If μ is a signed measure, $|\mu|$ is a positive measure. If (C.2) is satisfied, then $|\mu|(\Delta) \leqslant \sum_{k=1}^{4} \mu_k(\Delta)$; if μ is a signed measure, then $|\mu| = \mu_1 + \mu_2$.*

PROOF. Clearly $|\mu|(\Delta) \geqslant 0$. Let $\{\Delta_n\}$ be pairwise disjoint measurable sets and let $\Delta = \bigcup_{n=1}^{\infty} \Delta_n$. If $\varepsilon > 0$, then there is a measurable partition $\{E_j\}_{j=1}^{m}$ of Δ such that $|\mu|(\Delta) - \varepsilon < \sum_{j=1}^{m} |\mu(E_j)|$. Hence

$$|\mu|(\Delta) - \varepsilon \leqslant \sum_{j=1}^{m} \left| \sum_{n=1}^{\infty} \mu(E_j \cap \Delta_n) \right|$$

$$\leqslant \sum_{n=1}^{\infty} \sum_{j=1}^{m} |\mu(E_j \cap \Delta_n)|.$$

But $\{E_j \cap \Delta_n\}_{j=1}^{m}$ is a partition of Δ_n, so $|\mu|(\Delta) - \varepsilon \leqslant \sum_{n=1}^{\infty} |\mu|(\Delta_n)$. Therefore $|\mu|(\Delta) \leqslant \sum_{n=1}^{\infty} |\mu|(\Delta_n)$. For the reverse inequality we may assume that $|\mu|(\Delta) < \infty$. It follows that $|\mu|(\Delta_n) < \infty$ for every n. (Why?) Let $\varepsilon > 0$ and for each $n \geqslant 1$ let $\{E_1^{(n)}, \ldots, E_{m_n}^{(n)}\}$ be a partition of Δ_n such that $\sum_j |\mu(E_j^{(n)})| > |\mu|(\Delta_n) - \varepsilon/2^n$. Then

$$\sum_{n=1}^{N} |\mu|(\Delta_n) < \sum_{n=1}^{N} \left[\frac{\varepsilon}{2^n} + \sum_j |\mu(E_j^{(n)})| \right]$$

$$\leqslant \varepsilon + \sum_{n=1}^{N} \sum_j |\mu(E_j^{(n)})|$$

$$\leqslant \varepsilon + |\mu|(\Delta).$$

Letting $N \to \infty$ and $\varepsilon \to 0$ gives that $\sum_1^{\infty} |\mu|(\Delta_n) \leqslant |\mu|(\Delta)$.

Clearly $|\mu(\Delta)| \leqslant \sum_{k=1}^{4} \mu_k(\Delta)$, so $|\mu| \leqslant \sum_{k=1}^{4} \mu_k$. It is left to the reader to show that $|\mu| = \mu_1 + \mu_2$ if μ is a signed measure. Since $\mu_1, \mu_2, \mu_3, \mu_4$ are all finite, $|\mu|$ is finite if μ is complex-valued. ∎

C.5. Definition. If μ is a measure on (X, Ω) and ν is a positive measure on (X, Ω), say that μ is *absolutely continuous* with respect to ν ($\mu \ll \nu$) if $\mu(\Delta) = 0$ whenever $\nu(\Delta) = 0$. If ν is complex-valued, $\mu \ll \nu$ means $\mu \ll |\nu|$.

C.6. Proposition. *Let μ be a measure and ν a positive measure on (X, Ω). The following statements are equivalent.*

(a) $\mu \ll \nu$.
(b) $|\mu| \ll \nu$.
(c) *If (C.2) holds, $\mu_k \ll \nu$ for $1 \leqslant k \leqslant 4$.*

PROOF. Exercise.

The Radon–Nikodym Theorem can now be proved for complex-valued measures μ by using (C.6) and applying the usual theorem to the real and imaginary parts of μ. The details are left to the reader.

C.7. Radon–Nikodym Theorem. *If (X, Ω, v) is a σ-finite measure space and μ is a complex-valued measure on (X, Ω) such that $\mu \ll v$, then there is a unique complex-valued function f in $L^1(X, \Omega, v)$ such that $\mu(\Delta) = \int_\Delta f\, dv$ for every Δ in Ω.*

The function f obtained in (C.7) is called the *Radon–Nikodym derivative* of μ with respect to v and is denoted by $f = d\mu/dv$.

C.8. Theorem. *Let (X, Ω, v) be a σ-finite measure space and let μ be a complex-valued measure on (X, Ω) such that $\mu \ll v$ and let $f = d\mu/dv$.*

(a) *If $g \in L(X, \Omega, |\mu|)$, then $gf \in L^1(X, \Omega, v)$ and $\int g\, d\mu = \int gf\, dv$.*
(b) *For Δ in Ω, $|\mu|(\Delta) = \int_\Delta |f|\, dv$.*

PROOF. Part (a) follows from the corresponding result for signed measures by using (C.2) and a similar decomposition for f.
To prove (b), let $\{E_j\}$ be a measurable partition of Δ. Then

$$\sum_j |\mu(E_j)| \leqslant \sum_j \int_{E_j} |f|\, dv = \int_\Delta |f|\, dv.$$

For the reverse inequality, let $g(x) = \overline{f(x)}/|f(x)|$ if $x \in \Delta$ and $f(x) \neq 0$; let $g(x) = 0$ otherwise. Let $\{g_n\}$ be a sequence of Ω-measurable simple functions such that $g_n(x) = 0$ off Δ, $|g_n| \leqslant |g| \leqslant 1$, and $g_n(x) \to g(x)$ a.e. $[v]$. Thus $fg_n \to |f|\chi_\Delta$ a.e. $[v]$. Also, $|fg_n| \leqslant |f|\chi_\Delta$ and $f\chi_\Delta \in L^1(v)$ [see (C.2)]. By the Lebesgue Dominated Convergence Theorem, $\int fg_n\, dv \to \int_\Delta |f|\, dv$. If $g_n = \sum_j \alpha_j \chi_{E_j}$, where $\{E_j\}$ is a partition of Δ and $|\alpha_j| \leqslant 1$, then $|\int fg_n\, dv| = |\int g_n\, d\mu| = |\sum_j \alpha_j \mu(E_j)| \leqslant |\mu|(\Delta)$. Thus $\int_\Delta |f|\, dv \leqslant |\mu|(\Delta)$. ∎

One way of phrasing (C.8b) is that $|d\mu/dv| = d|\mu|/dv$. The next result is left to the reader.

C.9. Corollary. *If μ is a complex-valued measure on (X, Ω), then there is an Ω-measurable function f on X such that $|f| = 1$ a.e. $[|\mu|]$ and $\mu(\Delta) = \int_\Delta f\, d|\mu|$ for each Δ in Ω.*

C.10. Definition. Let X be a locally compact space and let Ω be the smallest σ-algebra of subsets of X that contains the open sets. Sets in Ω are called *Borel sets*. A positive measure μ on (X, Ω) is a *regular Borel measure* if (a) $\mu(K) < \infty$ for every compact subset K of X; (b) for any E in Ω, $\mu(E) = \sup\{\mu(K): K \subseteq E \text{ and } K \text{ is compact}\}$; (c) for any E in Ω, $\mu(E) = \inf\{\mu(U): U \supseteq E \text{ and } U$ is open$\}$. If μ is a complex-valued measure on (X, Ω), μ is a regular Borel measure if $|\mu|$ is. Let $M(X) =$ all of the complex-valued regular Borel measures on X. Note that $M(X)$ is a vector space over $\mathbb{C}$. For μ in $M(X)$, let

C.11
$$\|\mu\| \equiv |\mu|(X).$$

C.12. Proposition. (C.11) *defines a norm on* $M(X)$.

PROOF. Exercise.

C.13. Lemma. *If* $\mu \in M(X)$, *define* $F_\mu: C_0(X) \to \mathbb{C}$ *by* $F_\mu(f) = \int f\, d\mu$. *Then* $F_\mu \in C_0(X)^*$ *and* $\|F_\mu\| = \|\mu\|$.

PROOF. If $f \in C_0(X)$, then $|F_\mu(f)| \leqslant \int |f|\, d|\mu| \leqslant \|f\|\,\|\mu\|$. Hence $F_\mu \in C_0(X)^*$ and $\|F_\mu\| \leqslant \|\mu\|$.

To show equality, let f_0 be a Borel function such that $|f_0| = 1$ a.e. $[|\mu|]$ and $\mu(\Delta) = \int_\Delta f_0\, d|\mu|$. By Lusin's Theorem, if $\varepsilon > 0$, there is a continuous function ϕ on X with compact support such that $\int |\phi - \bar{f}_0|\, d|\mu| < \varepsilon$ and $\|\phi\| \leqslant \sup|f_0(x)| = 1$. Thus $\|\mu\| = \int f_0 \bar{f}_0\, d|\mu|$ (C.8a) $= \int \bar{f}_0\, d\mu = |\int \bar{f}_0\, d\mu| \leqslant |\int (\bar{f}_0 - \phi)\, d\mu| + |\int \phi\, d\mu| \leqslant \varepsilon + |F_\mu(\phi)| \leqslant \varepsilon + \|F_\mu\|$. Hence $\|\mu\| \leqslant \|F_\mu\|$. ∎

C.14. Corollary. (a) *If* U *is an open subset of* X *and* $\mu \in M(X)$, *then* $|\mu|(U) = \sup\{|\int \phi\, d\mu|: \phi \in C_c(X),\ \mathrm{spt}\, \phi \subseteq U,\ \text{and}\ \|\phi\| \leqslant 1\}$. (b) *If* $\mu \geqslant 0$, $\mu(K) = \inf\{\int \phi\, d\mu: \phi \in C_0(X)\ \text{and}\ \phi \geqslant \chi_K\}$.

PROOF. (a) If U is given the relative topology from X, U is locally compact. Let v be the restriction of μ to U. Then (a) becomes a restatement of (C.13) for the space U together with the fact that $C_c(U)$ is norm dense in $C_0(U)$.

(b) If $\phi \geqslant \chi_K$, then because μ is positive, $\int \phi\, d\mu \geqslant \mu(K)$. Thus $\mu(K) \leqslant \alpha \equiv \inf\{\int \phi\, d\mu: \phi \in C_0(X)\ \text{and}\ \phi \geqslant \chi_K\}$. Using the regularity of μ, for every integer n there is an open set U_n such that $K \subseteq U_n$ and $\mu(U_n \setminus K) < n^{-1}$. Let $\psi_n \in C_c(X)$ such that $0 \leqslant \psi_n \leqslant 1$, $\psi_n = 1$ on K, and $\psi_n = 0$ off U_n. Thus $\psi_n \geqslant \chi_K$ and so $\alpha \leqslant \int \psi_n\, d\mu \leqslant \mu(U_n) < \mu(K) + n^{-1}$. ∎

The next step in the process of representing bounded linear functionals on $C_0(X)$ by measures is to associate with each such functional a positive functional. If $\mu \in M(X)$, then the next lemma would associate with the functional F_μ the positive functional $I = F_{|\mu|}$.

C.15. Lemma. *If* $F: C_0(X) \to \mathbb{C}$ *is a bounded linear functional, then there is a unique linear functional* $I: C_0(X) \to \mathbb{C}$ *such that if* $f \in C_0(X)$ *and* $f \geqslant 0$, *then*

C.16
$$I(f) = \sup\{|F(g)|: g \in C_0(X)\ \text{and}\ |g| \leqslant f\}.$$

Moreover $\|I\| = \|f\|$.

PROOF. Let $C_0(X)_+$ be the positive functions in $C_0(X)$ and for f in $C_0(X)_+$ define $I(f)$ as in (C.16). If $\alpha > 0$, then clearly $I(\alpha f) = \alpha I(f)$ if $f \in C_0(X)_+$. Also, if $g \in C_0(X)$ and $|g| \leqslant f$, then $|F(g)| \leqslant \|F\|\,\|g\| \leqslant \|F\|\,\|f\|$. Hence $I(f) \leqslant \|F\|\,\|f\| < \infty$.

Now we will show that $I(f_1 + f_2) = I(f_1) + I(f_2)$ whenever $f_1, f_2 \in C(X)_+$. If $\varepsilon > 0$, let $g_1, g_2 \in C_0(X)$ such that $|g_j| \leqslant f_j$ and $|F(g_j)| > I(f_j) - \frac{1}{2}\varepsilon$ for $j = 1, 2$. There are complex numbers β_j, $j = 1, 2$, with $|\beta_j| = 1$ and $F(g_j) = \beta_j |F(g_j)|$. Thus

$$I(f_1) + I(f_2) < \varepsilon + |F(g_1)| + |F(g_2)|$$
$$= \varepsilon + \bar{\beta}_1 F(g_1) + \bar{\beta}_2 F(g_2)$$
$$= \varepsilon + |F(\bar{\beta}_1 g_1 + \bar{\beta}_2 g_2)|.$$

But $|\bar{\beta}_1 g_1 + \bar{\beta}_2 g_2| \leqslant |g_1| + |g_2| \leqslant f_1 + f_2$. Hence $I(f_1) + I(f_2) \leqslant \varepsilon + I(f_1 + f_2)$. Since ε was arbitrary, we have half of the desired equality.

For the other half of the equality, let $g \in C_0(X)$ such that $|g| \leqslant f_1 + f_2$ and $I(f_1 + f_2) < |F(g)| + \varepsilon$. Let $h_1 = \min(|g|, f_1)$ and $h_2 = |g| - h_1$. Clearly $h_1, h_2 \in C_0(X)_+$, $h_1 \leqslant f_1$, $h_2 \leqslant f_2$, and $h_1 + h_2 = |g|$. Define $g_j \colon X \to \mathbb{C}$ by

$$g_j(x) = \begin{cases} 0 & \text{if } g(x) = 0, \\ \dfrac{h_j(x)g(x)}{|g(x)|} & \text{if } g(x) \neq 0. \end{cases}$$

It is left to the reader to verify that $g_j \in C_0(X)$ and $g_1 + g_2 = g$. Hence

$$I(f_1 + f_2) < |F(g_1) + F(g_2)| + \varepsilon$$
$$\leqslant |F(g_1)| + |F(g_2)| + \varepsilon$$
$$\leqslant I(f_1) + I(f_2) + \varepsilon.$$

Now let $\varepsilon \to 0$.

If f is a real-valued function in $C_0(X)$, then $f = f_1 - f_2$ where $f_1, f_2 \in C_0(X)_+$. If also $f = g_1 - g_2$ for some g_1, g_2 in $C_0(X)_+$, then $g_1 + f_2 = f_1 + g_2$. By the preceding argument $I(g_1) + I(f_2) = I(f_1) + I(g_2)$. Hence if we define $I \colon \operatorname{Re} C_0(X) \to \mathbb{R}$ by $I(f) = I(f_1) - I(f_2)$ where $f = f_1 - f_2$ with f_1, f_2 in $C_0(X)_+$, I is well defined. It is left to the reader to verify that I is $\mathbb{R}$-linear.

If $f \in C_0(X)$, then $f = f_1 + if_2$, where $f_1, f_2 \in \operatorname{Re} C_0(X)$. Let $I(f) = I(f_1) + iI(f_2)$. It is left to the reader to show that $I \colon C_0(X) \to \mathbb{C}$ is a linear functional.

To prove that $\|I\| = \|F\|$, first let $f \in C_0(X)$ and put $I(f) = \alpha |I(f)|$ where $|\alpha| = 1$. Hence $\bar{\alpha}f = f_1 + if_2$, where $f_1, f_2 \in \operatorname{Re} C_0(X)$. Thus $|I(f)| = \bar{\alpha}I(f) = I(f_1) + iI(f_2)$. Since $|I(f)|$ is a positive real number, $I(f_2) = 0$ and $I(f_1) = |I(f)|$. But $f_1 = \operatorname{Re}(\bar{\alpha}f) \leqslant |f|$. Hence

$$|I(f)| \leqslant I(|f|).$$

From here we get, as in the beginning of this proof, that $\|I\| \leqslant \|F\|$. For the other half, if $\varepsilon > 0$, let $f \in C_0(X)$ such that $\|f\| \leqslant 1$ and $\|F\| < |F(f)| + \varepsilon$. Thus $\|F\| < I(|f|) + \varepsilon \leqslant \|I\| + \varepsilon$.   ∎

C.17. Theorem. *If $I \colon C_0(X) \to \mathbb{C}$ is a bounded linear functional such that $I(f) \geqslant 0$ whenever $f \in C_0(X)_+$, then there is a positive measure ν in $M(X)$ such that $I(f) = \int f \, d\nu$ for every f in $C_0(X)$ and $\|I\| = \nu(X)$.*

The proof of this is an involved construction. Inspired by Corollary C.14, one defines $v(U)$ for an open set U by

$$v(U) = \sup\{I(\phi): \phi \in C_c(X)_+, \ \phi \leqslant 1, \ \text{spt } \phi \subseteq U\}.$$

Then for any Borel set E, let

$$v(E) = \inf\{v(U): E \subseteq U \text{ and } U \text{ is open}\}.$$

It must now be shown that v is a positive measure and $I(f) = \int f \, dv$. For the details see (12.36) in Hewitt and Stromberg [1975] or §56 in Halmos [1974]. Indeed, Theorem C.17 is often called the Riesz Representation Theorem.

C.18. Riesz Representation Theorem. *If X is a locally compact space and $\mu \in M(X)$, define $F_\mu: C_0(X) \to \mathbb{C}$ by*

$$F_\mu(f) = \int f \, d\mu.$$

Then $F_\mu \in C_0(X)^*$ and the map $\mu \mapsto F_\mu$ is an isometric isomorphism of $M(X)$ onto $C_0(X)^*$.

PROOF. The fact that $\mu \mapsto F_\mu$ is an isometry is the content of Lemma C.13. It remains to show that $\mu \mapsto F_\mu$ is surjective. Let $F \in C_0(x)^*$ and define I: $C_0(X) \to \mathbb{C}$ as in Lemma C.15. By Theorem C.17, there is a positive measure v in $M(X)$ such that $I(f) = \int f \, dv$ for all f in $C_0(X)$. If $f \in C_0(X)$, then the definition of I implies that $|F(f)| \leqslant I(|f|) = \int |f| \, dv$. Thus, $f \mapsto F(f)$ defines a bounded linear functional on $C_0(X)$ considered as a linear manifold in $L^1(v)$. Now $C_0(X)$ is dense in $L^1(v)$ (Why?), so F has a unique extension to a bounded linear functional on $L^1(v)$. By Theorem B.1 there is a function ϕ in $L^\infty(v)$ such that $F(f) = \int f \phi \, dv$ for every f in $C_0(X)$ and $\|\phi\|_\infty \leqslant 1$. Let $\mu(E) = \int_E \phi \, dv$ for every Borel set E. Then $\mu \in M(X)$ and by Theorem C.8(a), $F(f) = \int f \, d\mu$; that is, $F = F_\mu$. ■

Bibliography

M.B. Abrahamse [1978]. Multiplication operators. In *Hilbert Space Operators*. New York: Springer-Verlag Notes, Vol. 693, pp. 17–36.

M.B. Abrahamse and T.L. Kriete [1973]. The spectral multiplicity function of a multiplication operator. *Indiana J. Math.*, **22**, 845–857.

T. Andô [1963]. Note on invariant subspaces of a compact normal operator. *Archiv. Math.*, **14**, 337–340.

N. Aronszajn and K.T. Smith [1954]. Invariant subspaces of completely continuous operators. *Ann. Math.*, **60**, 345–350.

W. Arveson [1956]. *An Invitation to C*-algebras*. New York: Springer-Verlag.

S. Banach [1955]. *Theorie des opérations linéaires*. New York: Chelsea Publ. Co.

N.K. Bari [1951]. Biorthogonal systems and bases in Hilbert space. *Moskov Gos Univ Ucenye Zapinski* **148**, *Matematika* **4**, 69–107. (Russian)

B. Beauzamy [1985]. Un operateur sans sous-espace invariant: simplification de l'exemple de P. Enflo. *Integral Eqs. and Operator Theory*, **8**, 314–384.

S.K. Berberian [1959]. Note on a theorem of Fuglede and Putnam. *Proc. Amer. Math. Soc.*, **10**, 175–182.

C. Berg, J.P.R. Christensen, and P. Ressel [1984]. *Harmonic Analysis on Semigroups*. New York: Springer-Verlag.

A.R. Bernstein and A. Robinson [1966]. Solution of an invariant subspace problem of K.T. Smith and P.R. Halmos. *Pacific J. Math.*, **16**, 421–431.

A. Beurling [1949]. On two problems concerning linear transformations in Hilbert space. *Acta. Math.*, **81**, 239–255.

F.F. Bonsall [1986]. Decompositions of functions as sums of elementary functions. *Quart J. Math.* **37**, 129–136.

F.F. Bonsall and J. Duncan [1973]. *Complete Normed Algebras*. Berlin: Springer-Verlag.

N. Bourbaki [1967]. *Espaces Vectoriels Topologiques*. Paris: Hermann.

L. de Branges [1959]. The Stone-Weierstrass Theorem. *Proc. Amer. Math. Soc.*, **10**, 822–824.

A Brown [1974]. A version of multiplicity theory. *Topics in Operator Theory*. Math. Surveys A.M.S., Vol. 13, 129–160.

R.C. Buck [1958]. Bounded continuous functions on a locally compact space. *Michigan Math. J.*, **5**, 95–104.

J.W. Calkin [1939]. Abstract symmetric boundary conditions. *Trans. Amer. Math. Soc.*, **45**, 369-442.

S.R. Caradus, W.E. Pfaffenberger, and B. Yood [1974]. *Calkin Algebras and Algebras of Operators of Banach Spaces.* New York: Marcel Dekker.

L. Carleson [1966]. On convergence and growth of partial sums of Fourier series. *Acta. Math.*, **116**, 135-157.

P. Chernoff [1983]. A semibounded closed symmetric operator whose square has trivial domain. *Proc. Amer. Math. Soc.*, **89**, 289-290.

M.D. Choi [1983]. Tricks or treats with the Hilbert matrix. *Amer. Math. Monthly*, **90**, 301-312.

J.A. Clarkson [1936]. Uniformly convex spaces. *Trans. Amer. Math. Soc.*, **40**, 396-414.

J.B. Conway [1978]. *Functions of One Complex Variable.* New York: Springer-Verlag.

J.B. Conway [1981]. *Subnormal Operators.* Boston: Pitman.

J.B. Conway [1985]. Arranging the disposition of the spectrum. *Proc. Royal Irish Acad.* **85A**, 139-142.

R. Courant and D. Hilbert [1953]. *Methods of Mathematical Physics.* New York: Interscience.

A.M. Davie [1973]. The approximation problem for Banach spaces. *Bull. London Math. Soc.*, **5**, 261-266.

A.M. Davie [1975]. The Banach approximation problem. *J. Approx. Theory*, **13**, 392-394.

W.J. Davis, T. Figiel, W.B. Johnson, and A. Pelczynski [1974]. Factoring weakly compact operators. *J. Functional Anal.*, **17**, 311-327.

J. Diestel [1984]. *Sequences and series in Banach spaces.* New York: Springer-Verlag.

J. Dieudonné [1985]. The index of operators in Banach spaces. *Integral Equations and Operator Theory* **8**, 580-589.

W.F. Donoghue [1957]. The lattice of invariant subspaces of a completely continuous quasi-nilpotent transformation. *Pacific J. Math.*, **7**, 1031-1035.

R.G. Douglas [1969]. On the operator equation $S^* X T = X$ and related topics. *Acta. Sci. Math. (Szeged)*, **30**, 19-32.

J. Dugundji [1966]. *Topology.* Boston: Allyn and Bacon.

N. Dunford and J. Schwartz [1958]. *Linear Operators. I.* New York: Interscience.

N. Dunford and J. Schwartz [1963]. *Linear Operators. II.* New York: Interscience.

J. Dyer, E. Pedersen, and P. Porcelli [1972]. An equivalent formulation of the invariant subspace conjecture. *Bull. Amer. Math. Soc.*, **78**, 1020-1023.

H. Dym and H.P. Mckean [1972]. *Fourier Series and Integrals.* New York and London: Academic Press.

D.A. Edwards [1961]. On translates of L^∞-functions. *J. London Math. Soc.* **36**, 431-432.

D.A. Edwards [1986]. A short proof of a theorem of Machado. *Math. Proc. Camb. Phil. Soc.* **99**, 111-114.

P. Enflo [1973]. A counterexample to the approximation problem in Banach spaces. *Acta. Math.*, **130**, 309-317.

P. Enflo [1987]. On the invariant subspace problem for Banach spaces. *Acta. Math.* **158**, 213-313.

J. Ernest [1976]. Charting the operator terrain. *Memoirs Amer. Math. Soc.*, Vol. 71.

P.A. Fillmore, J.G. Stampfli, and J.P. Williams [1972]. On the essential numerical range, the essential spectrum, and a problem of Halmos. *Acta Sci. Math. (Szeged)*, **33**, 179-192.

B. Fuglede [1950]. A commutativity theorem for normal operators. *Proc. Nat. Acad. Sci.*, **36**, 35-40.

T.W. Gamelin [1969]. *Uniform Algebras.* Englewood Cliffs: Prentice-Hall.

L. Gillman and M. Jerison [1960]. *Rings of Continuous Functions.* Princeton: Van Nostrand. Reprinted by Springer-Verlag, New York.

D.B. Goodner [1964]. The closed convex hull of certain extreme points. *Proc. Amer. Math. Soc.*, **15**, 256–258.

S. Grabiner [1986]. The Tietze extension theorem and the open mapping theorem. *Amer. Math. Monthly*, **93**, 190–191.

F. Greenleaf [1969]. *Invariant Means on Topological Groups.* New York: Van Nostrand.

A. Grothendieck [1953]. Sur les applications linéaires faiblement compactes d'espaces du type $C(K)$. *Canadian J. Math.*, **5**, 129–173.

P.R. Halmos [1951]. *Introduction to Hilbert Space and the Theory of Spectral Multiplicity.* New York: Chelsea Publ. Co.

P.R. Halmos [1961]. Shifts on Hilbert spaces. *J. Reine Angew. Math.*, **208**, 102–112.

P.R. Halmos [1963]. What does the spectral theorem say? *Amer. Math. Monthly*, **70**, 241–247.

P.R. Halmos [1966]. Invariant subspaces of polynomially compact operators. *Pacific J. Math.*, **16**, 433–437.

P.R. Halmos [1972]. Continuous functions of Hermitian operators. *Proc. Amer. Math. Soc.*, **31**, 130–132.

P.R. Halmos [1974]. *Measure Theory*, New York: Springer-Verlag.

P.R. Halmos [1982]. *A Hilbert Space Problem Book, Second ed.* New York: Springer-Verlag.

P.R. Halmos and V. Sunder [1978]. *Bounded Integral Operators on L^2 Spaces.* New York: Springer-Verlag.

G. Hansel and J.P. Troallic [1976]. Demonstration du théorèm de point fixe de Ryll–Nardzewskii par extension de la méthode de F. Hahn. *C.R. Acad. Sci. Paris Sér. A-B*, **282**, A857–A859.

R. Harte [1982]. Fredholm theory relative to a Banach algebra homomorphism. *Math. Zeit.*, **179**, 431–436.

E. Hellinger [1907]. Die Orthogonal invarianten quadratischer Formen. Inaugural Dissertation, Göttingen.

H. Helson [1964]. *Invariant Subspaces.* New York: Academic Press.

H. Helson [1986]. *The Spectral Theorem.* Lecture Notes in Mathematics, vol. 1227. Heidelberg: Springer-Verlag.

J. Hennefeld [1980]. A nontopological proof of the uniform boundedness theorem. *Amer. Math. Monthly*, **87**, 217.

E. Hewitt and K.A. Ross [1963]. *Abstract Harmonic Analysis, I.* New York: Springer-Verlag.

E. Hewitt and K.A. Ross [1970]. *Abstract Harmonic Analysis, II.* New York: Springer-Verlag.

E. Hewitt and K. Stromberg [1975]. *Real and Abstract Analysis.* New York: Springer-Verlag.

R.A. Hunt [1967]. On the convergence of Fourier series, orthogonal expansions and their continuous analogies. *Proc. of Conference at Edwardsville, Ill.* Southern Illinois Univ. Press, pp. 235–255.

R.C. James [1951]. A non-reflexive Banach space isometric with its second conjugate space. *Proc. Nat. Acad. Sci. USA*, **37**, 174–177.

R.C. James [1964a]. Weakly compact sets. *Trans. Amer. Math. Soc.*, **113**, 129–140.

R.C. James [1964b]. Weak compactness and reflexivity. *Israel J. Math.*, **2**, 101–119.

R.C. James [1971]. A counterexample for a sup theorem in normed spaces. *Israel J. Math.*, **9**, 511–512.

J.W. Jenkins [1972]. On the characterization of abelian W*-algebras. *Proc. Amer. Math. Soc.*, **35**, 436–438.

P.T. Johnstone [1982]. *Stone Spaces.* Cambridge: Cambridge University Press.

P. Jordan and J. von Neumann [1935]. On inner products in linear, metric spaces. *Ann. Math.*, (2)**36**, 719–723.

R.V. Kadison [1985]. The von Neumann algebra characterization theorems. *Exp. Math.*, 3, 193–227.

R.V. Kadison and I.M. Singer [1957]. Three test problems in operator theory. *Pacific J. Math.*, 7, 1101–1106.

T. Kato [1966]. *Perturbation Theory for Linear Operators.* New York: Springer-Verlag.

J.L. Kelley [1955]. *General Topology.* New York: Springer-Verlag.

J.L. Kelley [1966]. Decomposition and representation theorems in measure theory. *Math. Ann.*, 163, 89–94.

G. Köthe [1969]. *Topological Vector Spaces, I.* New York: Springer-Verlag.

S. Kremp [1986]. An elementary proof of the Eberlein–Smulian Theorem and the double limit criterion. *Arch. der Math.*, 47, 66–69.

G. Lasser [1972]. Topological algebras of operators. *Rep. Math. Phys.*, 3, 279–293.

C. Léger [1968]. Convexes compacts et leurs points extrêmaux. *Comptes Rend. Acad. Sci. Paris*, 267, 92–93.

J. Lindenstrauss [1967]. On complemented subspaces of *m. Israel J. Math.*, 5, 153–156.

J. Lindenstrauss and L. Tzafriri [1971]. On complemented subspaces problem. *Israel J. Math.*, 9, 263–269.

J.M. Lindsay [1984]. A family of operators with everywhere dense graph. *Expos. Math.*, 2, 375–378.

V. Lomonosov [1973]. On invariant subspaces of families of operators commuting with a completely continuous operator. *Funkcional. Anal. i Prilozen*, 7, 55–56.

T.F. McCabe [1984]. A note on iterates that are contractions. *J. Math. Anal. Appl.*, 104, 64–66.

I.J. Maddox [1980]. *Infinite Matrices of Operators.* Lecture Notes in Mathematics, vol. 768. New York: Springer-Verlag.

R. Mercer [1986]. Dense G_δ's contain orthonormal bases. *Proc. Amer. Math. Soc.*, 97, 449–52.

A.J. Michaels [1977]. Hilden's simple proof of Lomonosov's invariant subspace theorem. *Adv. Math.*, 25, 56–58.

F.J. Murray [1937]. On complementary manifolds and projections in spaces L_p and l_p. *Trans. Amer. Math. Soc.*, 41, 138–152.

L. Nachbin [1965]. *The Haar Integral.* Princeton: D. Van Nostrand.

I. Namioka and E. Asplund [1967]. A geometric proof of Ryll-Nardzewski's fixed point theorem. *Bull. Amer. Math. Soc.*, 73, 443–445.

J. von Neumann [1929]. Zur algebra der funktional-operatoren und theorie der normalen operatoren. *Math. Ann.*, 102, 370–427.

J. von Neumann [1932]. Über einen Satz von Herrn M.H. Stone. *Ann. Math.*, (2)33, 567–573.

J. von Neumann [1936]. On a certain topology for rings of operators. *Ann. Math.*, 37, 111–115.

W.P. Novinger [1972]. Mean convergence in L^p spaces. *Proc. Amer. Math. Soc.*, 34, 627–628.

R. Pallu de la Barrière [1954]. Sur les algèbras d'opérateurs dans les espaces hilbertiens. *Bull Soc. Math. France*, 82, 1–52.

C.M. Pearcy [1978]. Some recent developments in operator theory. CBMS Series No. 36. Providence: American Mathematical Society.

A. Pelczynski [1960]. Projections in certain Banach spaces. *Studia Math.*, 19, 209–228.

R.R. Phelps [1965]. Extreme points in function algebras. *Duke Math. J.*, 32, 267–277.

R.S. Phillips [1940]. On linear transformations. *Trans. Amer. Math. Soc.*, 48, 516–541.

J. Pitcairn [1966]. A remark on linear spaces. *Amer. Math. Monthly*, 73, 875.

C.R. Putnam [1951]. On normal operators in Hilbert space. *Amer. J. Math.*, 73, 357–362.

C.R. Putnam [1968]. The spectra of operators having resolvents of first-order growth. *Trans. Amer. Math. Soc.*, **133**, 505–510.

J.D. Pyrce [1966]. Weak compactness in locally convex spaces. *Proc. Amer. Math. Soc.*, **17**, 148–155.

H. Radjavi and P. Rosenthal [1973]. *Invariant subspaces.* New York: Springer-Verlag.

T. Ransford [1984]. A short elementary proof of the Bishop–Stone–Weierstrass Theorem. *Math. Proc. Camb. Phil. Soc.*, **96**, 309–311.

C.J. Read [1984]. A solution to the invariant subspace problem. *Bull. London Math. Soc.*, **16**, 337–401.

C.J. Read [1986]. A short proof concerning the invariant subspace theorem. *J. London Math. Soc.* (2)**34**, 335–348.

R. Redheffer and P. Volkmann [1983]. Schür's generalization of Hilbert's inequality. *Proc. Amer. Math. Soc.*, **87**, 567–568.

M. Reed and B. Simon [1975]. *Methods of Modern Mathematical Physics II.* New York: Academic.

M. Reed and B. Simon [1979]. *Methods of Modern Mathematical Physics III. Scattering Theory.* New York: Academic.

C. Rickart [1960]. *General Theory of Banach Algebras.* Princeton: D. Van Nostrand.

J.R. Ringrose [1971]. *Compact Non-self-adjoint Operators.* New York: Van Nostrand-Reinhold.

A.P. Robertson and W. Robertson [1966]. *Topological Vector Spaces.* Cambridge University.

M. Rosenblum [1958]. On a theorem of Fuglede and Putnam. *J. London Math. Soc.*, **33**, 376–377.

P. Rosenthal [1968]. Completely reducible operators. *Proc. Amer. Math. Soc.*, **19**, 826–830.

W. Rudin [1962]. *Fourier Analysis on Groups.* New York: Interscience.

C. Ryll-Nardzewski [1967]. On fixed points of semigroups of endomorphisms of linear spaces. *Fifth Berkeley Sympos. Math. Statist. and Prob., Vol. II: Contrib. to Prob. Theory, Part I.* Berkeley: University of California, pp. 55–61.

S. Sakai [1956]. A characterization of W^*-algebras. *Pacific J. Math.*, **6**, 763–773.

S. Sakai [1971]. C^*-*algebras and* W^*-*algebras.* New York: Springer-Verlag.

D. Sarason [1966]. Invariant subspaces and unstarred operator algebras. *Pacific J. Math.*, **17**, 511–517.

D. Sarason [1972]. Weak-star density of polynomials. *J. Reine Angew. Math.*, **252**, 1–15.

D. Sarason [1974]. Invariant subspaces. *Topics in Operator Theory.* Math. Surveys, Vol. 13, 1–47. Providence: American Mathematical Society.

H.H. Schaeffer [1971]. *Topological Vector Spaces.* New York: Springer-Verlag.

R. Schatten [1960]. *Norm Ideals of Completely Continuous Operators.* Berlin: Springer-Verlag.

M. Schechter [1965]. Invariance of the essential spectrum. *Bull. Amer. Math. Soc.*, **71**, 365–367.

M. Schechter [1968]. Riesz operators and Fredholm perturbations. *Bull. Amer. Math. Soc.*, **74**, 1139–1144.

H.S. Shapiro and A.L. Shields [1961]. On some interpolation problems for analytic functions. *Amer. J. Math.*, **83**, 513–532.

B. Simon [1979]. *Trace Ideals and Their Applications. London Math. Soc. Lecture Notes, Vol. 35.* Cambridge University.

S. Simons [1967]. Krein's theorem without sequential convergence. *Math. Ann.*, **174**, 157–162.

A. Sobczyk [1941]. Projection of the space (m) on its subspace (c_0). *Bull. Amer. Math. Soc.*, **47**, 938–947.

P.G. Spain [1976]. A generalisation of a theorem of Grothendieck. *Quart. J. Math.*, (2)**27**, 475–479.

J.G. Stampfli [1974]. Compact perturbations, normal eigenvalues, and a problem of Salinas. *J. London Math. Soc.*, **9**, 165–175.

L.A. Steen [1973]. Highlights in the history of spectral theory. *Amer. Math. Monthly*, **80**, 359–381.

E.M. Stein and G. Weiss [1971]. *Introduction to Fourier Analysis on Euclidean Spaces.* Princeton University.

M.H. Stone [1932]. On one-parameter unitary groups in Hilbert space. *Ann. Math.*, (2)**33**, 643–648.

Z. Takeda [1951]. On a theorem of R. Pallu de la Barrière. *Proc. Japan Acad.*, **28**, 558–563.

R.C. Walker [1974]. *The Stone–Čech Compactification.* New York: Springer-Verlag.

J. Wermer [1952]. On invariant subspaces of normal operators. *Proc. Amer. Math. Soc.*, **3**, 276–277.

R.J. Whitley [1966]. Projecting m onto c_0. *Amer. Math. Monthly*, **73**, 285–286.

R.J. Whitley [1967]. An elementary proof of the Eberlein–Smulian Theorem. *Math. Ann.*, **172**, 116–118.

R.J. Whitley [1968]. The Krein–Smulian Theorem. *Proc. Amer. Math. Soc.*, **97**, 376–377.

J. Wichmann [1973]. Bounded approximate units and bounded approximate identities. *Proc. Amer. Math. Soc.*, **41**, 547–550.

A. Wilansky [1951]. The bounded additive operation on Banach space, *Proc. Amer. Math. Soc.*, **2**, 46.

F. Wolf [1959]. On the invariance of the essential spectrum under a change of boundary conditions of partial differential operators. *Indag. Math.*, **21**, 142–147.

K.W. Yang [1967]. A note on reflexive Banach spaces. *Proc. Amer. Math. Soc.*, **18**, 859–860.

W. Zelazko [1968]. A characterization of multiplicative linear functionals in complex Banach algebras. *Studia Math.*, **30**, 83–85.

List of Symbols

c_0 65

$l^p(I)$, l^p 66

$C^{(n)}[0,1]$ 66

$W_p^n[0,1]$ 66

c 67

$C_c(X)$ 67

spt f 67

$\mathscr{B}(\mathscr{X}, \mathscr{Y})$, $\mathscr{B}(\mathscr{X})$ 68

$\oplus_p \mathscr{X}_i, \oplus_0 \mathscr{X}_n$ 72

$\mathscr{X}^*$ 74

$L^\infty(M(X))$ 76

$A(K)$ 81

$\mathbb{C}_\infty$ 86

$\hat{\mu}$ 84

$\mathscr{M}^\perp$ 88

gra A 91

$C(X)$, $H(G)$ 101

$\sigma(\mathscr{X}, \mathscr{X}^*), \sigma(\mathscr{X}^*, \mathscr{X})$ 101

$[a, b]$ 101

$\mathrm{co}(A), \overline{\mathrm{co}}(A)$ 101

s 104

$C_c^{(\infty)}(\Omega), \mathscr{D}(\mathscr{K}), \mathscr{D}(\Omega)$ 116

$\mathscr{T}|\mathscr{X}$ 118

$\langle x, x^* \rangle, \langle x^*, x \rangle$ 124

wk, $\sigma(\mathscr{X}, \mathscr{X}^*)$ 124

wk*, $\sigma(\mathscr{X}^*, \mathscr{X})$ 125

cl A, wk-cl A 125

$A^\circ, {}^\circ B, {}^\circ A^\circ$ 126

$A^\perp, {}^\perp B$ 127

ball $\mathscr{X}$ 130

δ_x 137

βX 138

ext K 141

$\bar{f}$ 145

$h\mu$ 146

$\mathbb{R}_{\geqslant 0}$ 155

$f_x, {}_x f, f^\#$ 155, 224

$\mathscr{Y}'$ 166

A^* 167

$\mathscr{B}_0(\mathscr{X}, \mathscr{Y}), \mathscr{B}_0(\mathscr{X}), \mathscr{B}_{00}(\mathscr{X}, \mathscr{Y})$ 174

Lat T 178

$L^1(G)$ 190, 223

$\sigma(a), \sigma_l(a), \sigma_r(a)$ 195

$\rho(a), \rho_l(a), \rho_r(a)$ 195

$r(a)$ 197

$n(\gamma; a)$ 199

ins γ, out γ 200

ins Γ, out Γ 200

Hol(a) 201

$\sigma_{\mathscr{A}}(a)$ 205

$\| f \|_A$ 206

$K^{\wedge}$ 206

$\sigma_{\mathrm{ap}}(A)$ 208

$E(\Delta) = E(\Delta; A)$, $\mathscr{X}_\Delta$ 210

$E(\lambda)$, $\mathscr{X}_\lambda$ 210

$\Sigma, \Sigma(a)$ 219

rad $\mathscr{A}$ 220

$\tau^{\#}$ 221

$f * g$ 223, 337

$\mathbb{T}$ 223

$\hat{G}$ 227

$C^*(a)$ 236

$f(a)$ 238

Re $\mathscr{A}$ 240

$\mathscr{A}_+$ 240

a_+, a_- 241

$a \leqslant b$ 242

$|A|$ 242

$\mathscr{H}^{(n)}, A^{(n)}, \pi^{(n)}$ 248

$S_{\mathscr{A}}$ 252

$E_{g,h}$ 257

$\int \phi \, dE$ 258

$B(X, \Omega)$ 258

N_μ 265, 326

ess-ran(ϕ) 265

$\| \cdot \|_2$, $\mathscr{B}_2(\mathscr{H})$ 267

tr(A) 267

$\| \cdot \|_1$ $\mathscr{B}_1(\mathscr{H})$ 267

$g \otimes h$ 268

$[\mu_1] = [\mu_2]$ 269

$\mathscr{S}^{(n)}$ 275

$\mathscr{S}', \mathscr{S}''$ 276

$\mathscr{A}_\mu$ 277

$W^*(N)$ 285

$P^\infty(\mu)$ 291

$W(A)$ 291

$L^2(\mu; \mathscr{H})$, $L^\infty(\mu; \mathscr{B}(\mathscr{H}))$ 299

$\int \mathscr{H}(z) \, d\mu(z)$ 301

dom A 303

$B \subseteq A$ 304

Index

Graduate Texts in Mathematics

(*continued from page ii*)